Legume Genetics and Biology

Legume Genetics and Biology

From Mendel's Pea to Legume Genomics

Editors

Petr Smýkal
Eric J. Bishop von Wettberg
Kevin McPhee

MDPI • Basel • Beijing • Wuhan • Barcelona • Belgrade • Manchester • Tokyo • Cluj • Tianjin

Editors
Petr Smýkal
Palacky University
Czech Republic

Eric J. Bishop von Wettberg
University of Vermont
USA

Kevin McPhee
Montana State University
USA

Editorial Office
MDPI
St. Alban-Anlage 66
4052 Basel, Switzerland

This is a reprint of articles from the Special Issue published online in the open access journal *International Journal of Molecular Sciences* (ISSN 1422-0067) (available at: https://www.mdpi.com/journal/ijms/special_issues/pea_legume).

For citation purposes, cite each article independently as indicated on the article page online and as indicated below:

LastName, A.A.; LastName, B.B.; LastName, C.C. Article Title. *Journal Name* **Year**, *Article Number*, Page Range.

ISBN 978-3-03936-812-9 (Hbk)
ISBN 978-3-03936-813-6 (PDF)

Cover image courtesy of Petr Smýkal.

Contents

About the Editors

Petr Smýkal (Associate Professor). He received his Ph.D. in Plant Physiology and Molecular Biology at Charles University, Prague, CZ, in 1999. After 5 years of postdoctoral stays at ETH Zurich, Switzerland, and Albrecht Ludwig University of Freiburg, Germany, he then conducted research at Agritec Plant Research Ltd., Sumperk, CZ, for 10 years, working with plant genetic resources, characterizing and utilizing the genetic diversity of pea and flax germplasms. Since 2011, he has been at Department of Botany, Palacky University, in Olomouc, CZ. He is interested in study of legume seed dormancy (physical type of dormancy) and pod dehiscence, two key domestication traits, and uses a combination of anatomical, genetic, transcriptomic, and analytical chemistry tools. Seed dormancy is also studied as an adaptive trait in context of ecological genomics combining next-generation sequencing, seed testing, and geoinformatics using wild pea and Medicago truncatula models. Other favorites are crop wild relatives, particularly of legumes. Knowledge on Pisum, Cicer and Lens genus diversity is applied to broaden the respective crop genetic diversity.

Eric J. Bishop von Wettberg (Associate Professor) runs a research program focused on the consequences of genetic bottlenecks limiting genetic diversity and climate resilience in crop plants. He received his Ph.D. in Ecology from Brown University in 2007 and was a NIH National Research Service Award postdoc at the University of California at Davis from 2007 to 2009. He was a faculty member at Florida International University from 2010 to 2017. Broadly trained in genetics, ecology, and agroecology, he uses a combination of field, greenhouse, common garden, and laboratory approaches to improve crop plants in the face of ongoing climate change. Many crops, like the grain legume chickpea, have lost genetic variation as a result of human cultivation and selection. The lack of genetic variation reduces resilience of these crops to expected effects of climate change. His laboratory group are using a new collection of the wild relatives of chickpea to restore genetic variation to cultivated chickpea, and to understand the genetic basis of flowering time and drought tolerance. They are also using the same approach to improve winter peas as a forage and cover crop, to improve mung beans as a summer forage and sprout crop, and to increase the disease resistance of hops.

Kevin McPhee (Professor). Kevin McPhee is the Pulse Crop Breeder at Montana State University. He received his Bachelor of Science degree in Agronomy in 1991 from the University of Wyoming and his Ph.D. in Agronomy from the University of Idaho in 1995 with an emphasis on Plant Breeding and Genetics. He worked for the USDA Agricultural Research Service from 1995 to 2008, where his research focused on genetics and breeding of dry pea. From 2008 to 2016 Kevin held the position of Professor of Pulse Crop Breeding in the Department of Plant Sciences at North Dakota State University, where he conducted research on the genetics and breeding of pulse crops. In 2017, Kevin accepted the position of Professor of Pulse Crop Breeding at Montana State University. He established new pulse breeding programs at both NDSU and MSU focused on pea, lentil, and chickpea breeding for a range of research objectives.

International Journal of
Molecular Sciences

Editorial

Legume Genetics and Biology: From Mendel's Pea to Legume Genomics

Petr Smýkal [1,*], Eric J.B. von Wettberg [2] and Kevin McPhee [3]

[1] Department of Botany, Faculty of Sciences, Palacký University, 779 00 Olomouc, Czech Republic
[2] Department of Plant and Soil Sciences and Gund Institute for the Environment, University of Vermont, Burlington, VT 05405, USA; Eric.Bishop-Von-Wettberg@uvm.edu
[3] Plant Sciences and Plant, Pathology Department, Montana State University, Bozeman, MT 59717, USA; kevin.mcphee@montana.edu
* Correspondence: petr.smykal@upol.cz

Received: 29 April 2020; Accepted: 6 May 2020; Published: 8 May 2020

Abstract: Legumes have played an important part in cropping systems since the dawn of agriculture, both as human food and as animal feed. The legume family is arguably one of the most abundantly domesticated crop plant families. Their ability to symbiotically fix nitrogen and improve soil fertility has been rewarded since antiquity and makes them a key protein source. The pea was the original model organism used in Mendel's discovery of the laws of inheritance, making it the foundation of modern plant genetics. This Special Issue provides up-to-date information on legume biology, genetic advances, and the legacy of Mendel.

Keywords: genomics; legumes; nitrogen fixation; proteins

Introduction

Legumes have always been a part of everyday life, as human food and animal feed, being key protein sources. Legumes represent the second most important family of crop plants after Poaceae (grass family), accounting for approximately 27% of the world's crop production. While in cereals the major storage molecule is starch, which is deposited in the endosperm, in most of the grain legumes (pulses) the endosperm is transitory and consumed by the embryo during seed maturation. Legume seeds contain a high proportion of proteins (20–40%), and either lipids (soybean, peanut) or starch (or both) as a further carbon source [1]. The importance of legumes for agriculture as well as science has been recognized by the establishment of International Legume Society (ILS) in 2010 (https://www.legumesociety.org), followed by biannual conferences bringing together people working on broad aspects of legume biology. The last ILS conference was held in 2019 in Poland and this Special Issue has been made to reflect some of the presented work. The long-term strategy of ILS is linking together the different aspects of agricultural research on grain and forage legumes worldwide.

The Fabaceae is the third-largest family of flowering plants, with over 800 genera and 20,000 species. Currently, three major groups are recognized and regarded as subfamilies: the mimosoid legumes, Mimosoideae (sometimes regarded as the family Mimosaceae with four tribes and 3270 species); the papilionoid legumes, Papilionoideae (or the family Fabaceae/Papilionaceae with 28 tribes and 13,800 species); and the caesalpinioid legumes, Caesalpinoideae (or the family Caesalpiniaceae with four tribes and 2250 species) [2]. It is an extremely diverse family with a worldwide distribution, from arctic-alpine herbs to annual xerophytes and forest trees.

Legumes have played an important part in cropping systems since the dawn of agriculture. Records from the oldest civilizations of Egypt and eastern Asia demonstrate the ancient use of various beans, peas, vetches, soybeans, and alfalfa. One of the early Greek botanists, Theophrastus, in the third century before Christ, wrote of leguminous plants "reinvigorating" the soil and stated that beans were not a burdensome crop to the ground but even seemed to manure it. The Romans emphasized the use of leguminous plants for green manuring; they also introduced the systematic use of crop rotations, a practice that was forgotten for a time during the early Middle Ages and partly also in today´s agricultural practice.

Members of the Fabaceae were domesticated as grain legumes in parallel with cereal domestications [3–8]. There are 13 genera (in six legume tribes) that constitute major legume crops [1,2]. Among the first legumes to be domesticated were members of the galegoid tribe such as peas, faba beans, lentils, grass peas and chickpeas, which arose in the Fertile Crescent of Mesopotamian agriculture. These grain legumes (pulse legumes) accompanied cereal production and formed important dietary components of early civilizations in the Near East and the Mediterranean regions. Similar domestications of *Phaseolus* in the New World and *Glycine* in East Asia have had similar importance for human dietary diversity and security.

Cultivated legumes fulfill many human needs beyond being directly consumed by people. Many tree-sized species in the legume family are valuable for their hard, durable timber. Species from the genera *Aeschynomene*, *Arachis*, *Centrosema*, *Desmodium*, *Macroptilium*, and particularly *Stylosanthes* offer promise for improved tropical pasture systems. The barks of some species of acacias (*Acacia dealbata*, *A. decurrens*, and *A. pycnantha*) are sometimes used as sources of tannins, chemicals that are mostly used to manufacture leather from animal skins. Some important dyes are extracted from species in the legume family. One of the world's most important natural dyes is indigo, extracted from the foliage of the indigo (*Indigofera tinctoria*) of south Asia and to a lesser degree from American indigo (*I. suffruticosa*) of tropical South America. Derris or rotenone is a poisonous alkaloid extracted from *Derris elliptica* and *D. malaccensis* that has long been used by indigenous peoples of Southeast Asia as arrow and fish poisons. Rotenone is now used widely as a rodenticide to kill small mammals and as an insecticide to kill pest insects. Fenugreek (*Trigonella foenum graecum*), the seeds of which are used as a spice in curries. Legumes include also valuable fiber plants, such as the sunn-hemp of India (*Crotalaria juncea*) and Hemp sesbania (*Sesbania exaltata*) used by the Indians of the southwestern United States. Some legumes such as licorice (*Glycyrrhiza glabra*) and goatsrue (*Tephrosia virginiana*) have medicinal value; many others rank among ornamental plants (for example *Lathyrus odoratus*), and legumes are of great importance for honey production.

The pea (*Pisum sativum* L.) was the original model organism used in Mendel´s discovery (1866) of the laws of inheritance, making it the foundation of modern plant genetics. It had already been an object of experimental work before Mendel [9,10]. Despite their close phylogenetic relationships, crop legumes differ greatly in their genome size, base chromosome number, ploidy level, and reproductive biology. To establish a unified genetic system for legumes, two legume species in the Galegoid clade, *Medicago truncatula* and *Lotus japonicus*, from Trifolieae and Loteae tribes, respectively, were selected as model systems for studying legume genomics and biology [11,12]. Now, many legume crops have well-studied genetic systems. In a few cultivated legumes, comprehensive genetic analysis is limited due to the large size of their genomes. For soybeans, the most widely grown and economically important legume, a genome has been available since 2010 [13]. For the common bean (*Phaseolus vulgaris*), the most widely grown grain legume, a genome has been available since 2014 [14]. Many more legumes have been sequenced since. These genome sequences are now completed by a broad range of genomic resources, including tools for genome-wide association studies, diversity panels, and online databases [15]. These tools facilitated increasingly widespread efforts to implement molecular breeding in legumes. The existence of reference genomes is fundamental for the advancement of genetic mapping approaches using either classical biparental population or association mapping on wider panels. This has been shown in several papers in this issue [16,17] for soybean. Having genome-wide data on diversity on a sufficiently large and diverse set of accessions, along with accumulated phenotypic trait descriptions, provides the tools to conduct genome-wide association studies and genomic selection. This either might lead to the identification of candidate loci/genes governing studied traits or provide useful markers applicable for breeding [18,19].

The history of legume crop domestication is not only of theoretical interest to provide insight into evolution but also can be used in breeding of recently domesticated crops, as shown in lupine [20] and potentially applied to a broader range of crop wild relatives. Legumes are particular among the plant species in their ability to fix atmospheric nitrogen. Owing to their biology including symbiotic nitrogen fixation, legumes are vital components of sustainable agriculture. This has been acknowledged in all cropping systems. Although the fundamentals of bacteria and host plant symbiosis have been elicited, there are still numerous aspects to be studied, such as allelic variation of identified genes, as shown on red clover [21].

Since most of the legume crops are used as food or feed in form of mature, dry seeds, their nutritional composition is of great importance. The study of Sivasakthi et al. [22] shows an elegant application of basic knowledge of one of the genes underlying a classical Mendelian trait, green cotyledons, identified and applied in chickpea. Seed composition can be altered by water availability or other abiotic stresses, as shown in studies of lupine seeds [23]. Similarly, dissection of the molecular mechanisms of resistance to biotic and abiotic are of high relevance both in order to understand evolutionary mechanisms between pathogens/triggers and hosts as well as to facilitate the breeding process. Mutant lines are helpful in elucidation of gene function, as shown in soybeans [24]. Since pathogens display high variation potential and are able to quickly overcome single gene/allele resistance, it is important to identify the allelic variation of a given gene, as shown in powdery mildew resistance in peas [25]. Climate change is already impacting all crops including legumes. There is a great need to understand the mechanisms of stress avoidance/tolerance/resistance to minimalize this impact. The review of Kumar et al. [26] offers a view on breeding climate-resilient legume crops, which is vital particularly for tropical and subtropical countries already facing scarcity of water and soil resources. In current biology, there are commonly integrated various approaches in order to study complex biological pathways, such as that shown by the study of lupine flower development [27]. This work combines genomic, transcriptomic, and small RNA sequencing to understand the process of lupine flower ablation. Owing to progress in genomic methods such as next-generation sequencing, genetics and genomics is not limited to model species and is being applied to any species including crops with complex, polyploid genomes [28]. Evolutionary scenarios of speciation are a recurrent theme in biology, and especially in plants, there are often various pathways to speciation, including frequent hybridization and polyploidy. A central aspect of speciation is the establishment of gene-flow barriers. One of the ways to do this is the interaction between plastid and nuclear genomes leading to either viable to inviable progeny. In peas, the interaction between the chloroplast and nuclear-encoded genes results in either normal or albino/chlorotic plants. The study of Nováková et al. [29] shows the variation of respective genes in natural pea populations as well as identifying the influence of a domestication-imposed bottleneck.

Although Mendel's peas were the first "model" plant, legume biology has long lagged behind more successful models from the Brassicaceae family or economically important cereals. For Borlaug, grain legumes were the "slow runners" of the green revolution because of the limited extent to which they saw the genetic gains that have characterized breeding of cereals for the past century. However, owing to progress in genomic and phenotyping technologies together with recognition of their importance for ecology of natural or agronomical systems, they are gradually gaining ground. We look for seeing new work in legumes, including releases around the world of new legume varieties bred with genomic resources.

Author Contributions: E.J.B.v.W., K.M. and P.S. writing. All authors have read and agreed to the published version of the manuscript.

Funding: P.S. work is supported from Grant Agency of Czech Republic and Grant Agency of Palacký University, IGA-2020_003 project.

Conflicts of Interest: The authors declare no conflict of interest.

References

1. Smykal, P.; Coyne, C.J.; Ambrose, M.J.; Maxted, N.; Schaefer, H.; Blair, M.W.; Berger, J.; Greene, S.L.; Nelson, M.N.; Besharat, N.; et al. Legume Crops Phylogeny and Genetic Diversity for Science and Breeding. *Crit. Rev. Plant Sci.* **2015**, *34*, 43–104. [CrossRef]
2. Lewis, G.; Schrire, B.; Mackinder, B.; Lock, M. *Legumes of the World*; Royal Botanic Gardens: London, UK, 2005; ISBN 1900347806.
3. De Candolle, A. *Origin of Cultivated Plants*; American Association for the Advancement of Science: Appleton, WI, USA, 1890.
4. Vavilov, N.I. *The Origin, Variation, Immunity and Breeding of Cultivated Plants*; Starchester, K., Ed.; Chronica Botanica: Leyden, The Netherlands, 1951; Volume 13, pp. 1–364.
5. Smartt, J. *Grain Legumes: Evolution and Genetic Resources*; Cambridge University Press: Cambridge, UK, 1990.
6. Zohary, D.; Hopf, M.; Weiss, E. *Domestication of Plants in the Old World: The Origin and Spread of Domesticated Plants in Southwest Asia, Europe, and the Mediterranean Basin*, 4th ed.; Oxford University Press: Oxford, UK, 2012; ISBN 9780199549061.
7. Abbo, S.; van-Oss, R.P.; Gopher, A.; Saranga, Y.; Ofner, R.; Peleg, Z. Plant domestication versus crop evolution: A conceptual framework for cereals and grain legumes. *Trends Plant Sci.* **2014**, *19*, 351–360. [CrossRef] [PubMed]
8. Smýkal, P.; Nelson, M.N.; Berger, J.D.; Von Wettberg, E.J.B. The Impact of Genetic Changes during Crop Domestication. *Agronomy* **2018**, *8*, 119. [CrossRef]
9. Smykal, P. Pea (*Pisum sativum* L.) in Biology prior and after Mendel's Discovery. *Czech J. Genet. Plant Breed.* **2014**, *50*, 52–64. [CrossRef]
10. Smykal, P.; Varshney, R.K.; Singh, V.K.; Coyne, C.J.; Domoney, C.; Kejnovsky, E.; Warkentin, T. From Mendel's discovery on pea to today's plant genetics and breeding. *Appl. Genet.* **2016**, *129*, 2267–2280. [CrossRef] [PubMed]
11. Cook, D.R. *Medicago truncatula*—A model in the making! *Curr. Opin. Plant Biol.* **1999**, *2*, 301–304. [CrossRef]
12. Sato, S.; Nakamura, Y.; Kaneko, T.; Asamizu, E.; Kato, T.; Nakao, M.; Sasamoto, S.; Watanabe, A.; Ono, A.; Kawashima, K.; et al. Genome structure of the legume, *Lotus japonicus*. *DNA Res.* **2008**, *15*, 227–239. [CrossRef]
13. Schmutz, J.; Cannon, S.B.; Schlueter, J.; Ma, J.; Mitros, T.; Nelson, W.; Hyten, D.L.; Song, Q.; Thelen, J.J.; Cheng, J.; et al. Genome sequence of the palaeopolyploid soybean. *Nature* **2010**, *465*, 120. [CrossRef]
14. Schmutz, J.; McClean, P.E.; Mamidi, S.; Wu, G.A.; Cannon, S.B.; Grimwood, J.; Jenkins, J.; Shu, S.; Song, Q.; Chavarro, C.; et al. A reference genome for common bean and genome-wide analysis of dual domestications. *Nat. Genet.* **2014**, *46*, 707–713. [CrossRef]
15. Bauchet, G.J.; Bett, K.E.; Cameron, C.T.; Campbell, J.D.; Cannon, E.K.; Cannon, S.B.; Carlson, J.W.; Chan, A.; Cleary, A.; Close, T.J.; et al. The future of legume genetic data resources: Challenges, opportunities, and priorities. *Legume Sci.* **2019**, *1*, e16. [CrossRef]
16. Hina, A.; Cao, Y.; Song, S.; Li, S.; Sharmin, R.A.; Elattar, M.A.; Bhat, J.A.; Zhao, T. High-Resolution Mapping in Two RIL Populations Refines Major "QTL Hotspot" Regions for Seed Size and Shape in Soybean (*Glycine max* L.). *Int. J. Mol. Sci.* **2020**, *21*, 1040. [CrossRef] [PubMed]
17. Zhang, T.; Wu, T.; Wang, L.; Jiang, B.; Zhen, C.; Yuan, S.; Hou, W.; Wu, C.; Han, T.; Sun, S. A Combined Linkage and GWAS Analysis Identifies QTLs Linked to Soybean Seed Protein and Oil Content. *Int. J. Mol. Sci.* **2019**, *20*, 5915. [CrossRef] [PubMed]
18. Tafesse, E.G.; Gali, K.K.; Lachagari, V.B.R.; Bueckert, R.; Warkentin, T.D. Genome-Wide Association Mapping for Heat Stress Responsive Traits in Field Pea. *Int. J. Mol. Sci.* **2020**, *21*, 2043. [CrossRef] [PubMed]
19. Annicchiarico, P.; Nazzicari, N.; Laouar, M.; Thami-Alami, I.; Romani, M.; Pecetti, L. Development and Proof-of-Concept Application of Genome-Enabled Selection for Pea Grain Yield under Severe Terminal Drought. *Int. J. Mol. Sci.* **2020**, *21*, 2414. [CrossRef] [PubMed]
20. Plewiński, P.; Książkiewicz, M.; Rychel-Bielska, S.; Rudy, E.; Wolko, B. Candidate Domestication-Related Genes Revealed by Expression Quantitative Trait Loci Mapping of Narrow-Leafed Lupin (*Lupinus angustifolius* L.). *Int. J. Mol. Sci.* **2019**, *20*, 5670. [CrossRef] [PubMed]

21. Trněný, O.; Vlk, D.; Macková, E.; Matoušková, M.; Řepková, J.; Nedělník, J.; Hofbauer, J.; Vejražka, K.; Jakešová, H.; Jansa, J.; et al. Allelic Variants for Candidate Nitrogen Fixation Genes Revealed by Sequencing in Red Clover (*Trifolium pratense* L.). *Int. J. Mol. Sci.* **2019**, *20*, 5470. [CrossRef]

22. Sivasakthi, K.; Marques, E.; Kalungwana, N.; Carrasquilla-Garcia, N.; Chang, P.L.; Bergmann, E.M.; Bueno, E.; Cordeiro, M.; Sani, S.G.A.S.; Udupa, S.M.; et al. Functional Dissection of the Chickpea (*Cicer arietinum* L.) Stay-Green Phenotype Associated with Molecular Variation at an Ortholog of Mendel's I Gene for Cotyledon Color: Implications for Crop Production and Carotenoid Biofortification. *Int. J. Mol. Sci.* **2019**, *20*, 5562. [CrossRef]

23. Polit, J.T.; Ciereszko, I.; Dubis, A.T.; Leśniewska, J.; Basa, A.; Winnicki, K.; Żabka, A.; Audzei, M.; Sobiech, Ł.; Faligowska, A.; et al. Irrigation-Induced Changes in Chemical Composition and Quality of Seeds of Yellow Lupine (*Lupinus luteus* L.). *Int. J. Mol. Sci.* **2019**, *20*, 5521. [CrossRef]

24. Al Amin, G.M.; Kong, K.; Sharmin, R.A.; Kong, J.; Bhat, J.A.; Zhao, T. Characterization and Rapid Gene-Mapping of Leaf Lesion Mimic Phenotype of *spl-1* Mutant in Soybean (*Glycine max* (L.) Merr.). *Int. J. Mol. Sci.* **2019**, *20*, 2193. [CrossRef]

25. Sun, S.; Deng, D.; Duan, C.; Zong, X.; Xu, D.; He, Y.; Zhu, Z. Two Novel er1 Alleles Conferring Powdery Mildew (*Erysiphe pisi*) Resistance Identified in a Worldwide Collection of Pea (*Pisum sativum* L.) Germplasms. *Int. J. Mol. Sci.* **2019**, *20*, 5071. [CrossRef]

26. Kumar, J.; Choudhary, A.K.; Gupta, D.S.; Kumar, S. Towards Exploitation of Adaptive Traits for Climate-Resilient Smart Pulses. *Int. J. Mol. Sci.* **2019**, *20*, 2971. [CrossRef] [PubMed]

27. Glazińska, P.; Kulasek, M.; Glinkowski, W.; Wojciechowski, W.; Kosiński, J. Integrated Analysis of Small RNA, Transcriptome and Degradome Sequencing Provides New Insights into Floral Development and Abscission in Yellow Lupine (*Lupinus luteus* L.). *Int. J. Mol. Sci.* **2019**, *20*, 5122. [CrossRef] [PubMed]

28. Krishnamurthy, P.; Tsukamoto, C.; Ishimoto, M. Reconstruction of the Evolutionary Histories of UGT Gene Superfamily in Legumes Clarifies the Functional Divergence of Duplicates in Specialized Metabolism. *Int. J. Mol. Sci.* **2020**, *21*, 1855. [CrossRef] [PubMed]

29. Nováková, E.; Zablatzká, L.; Brus, J.; Nesrstová, V.; Hanáček, P.; Kalendar, R.; Cvrčková, F.; Majeský, Ľ.; Smýkal, P. Allelic Diversity of Acetyl Coenzyme A Carboxylase accD/bccp Genes Implicated in Nuclear-Cytoplasmic Conflict in the Wild and Domesticated Pea (*Pisum* sp.). *Int. J. Mol. Sci.* **2019**, *20*, 1773. [CrossRef] [PubMed]

International Journal of
Molecular Sciences

Review

Towards Exploitation of Adaptive Traits for Climate-Resilient Smart Pulses

Jitendra Kumar [1,*]**, Arbind K. Choudhary** [2,*]**, Debjyoti Sen Gupta** [1] **and Shiv Kumar** [3,*]

1   Indian Institute of Pulses Research, Kalyanpur, Kanpur 208 024, Uttar Pradesh, India; debgpb@gmail.com
2   ICAR Research Complex for Eastern Region, Patna 800 014, Bihar, India
3   Biodiversity and Integrated Gene Management Program, International Centre for Agricultural Research in the Dry Areas (ICARDA), P.O. Box 6299, Rabat-Institute, Rabat, Morocco
*   Correspondence: jitendra.kumar@icar.gov.in (J.K.); akicar1968@gmail.com (A.K.C.); sk.agrawal@cgiar.org (S.K.)

Received: 19 March 2019; Accepted: 28 May 2019; Published: 18 June 2019

Abstract: Pulses are the main source of protein and minerals in the vegetarian diet. These are primarily cultivated on marginal lands with few inputs in several resource-poor countries of the world, including several in South Asia. Their cultivation in resource-scarce conditions exposes them to various abiotic and biotic stresses, leading to significant yield losses. Furthermore, climate change due to global warming has increased their vulnerability to emerging new insect pests and abiotic stresses that can become even more serious in the coming years. The changing climate scenario has made it more challenging to breed and develop climate-resilient smart pulses. Although pulses are climate smart, as they simultaneously adapt to and mitigate the effects of climate change, their narrow genetic diversity has always been a major constraint to their improvement for adaptability. However, existing genetic diversity still provides opportunities to exploit novel attributes for developing climate-resilient cultivars. The mining and exploitation of adaptive traits imparting tolerance/resistance to climate-smart pulses can be accelerated further by using cutting-edge approaches of biotechnology such as transgenics, genome editing, and epigenetics. This review discusses various classical and molecular approaches and strategies to exploit adaptive traits for breeding climate-smart pulses.

Keywords: adaptive traits; gene/QTL; epigenetics; transgenics; genome editing; climate-smart pulses

---

1. Introduction

Pulses are cultivated worldwide as major or minor crops (Table 1) to provide for the nutrition and livelihood of millions of peoples. Pulses, being a rich source of protein (22–26%) and micronutrients (especially Fe and Zn), are a balanced food for vegetarians when complemented with cereals. Also, the green and dry plant parts of these crops are used as feed and fodder in many livestock production systems [1], and their cultivation has long helped to sustain cereal-based cropping systems through biological nitrogen fixation and carbon sequestration [2]. Most of these pulses originated in the Mediterranean region [3]. The reproductive phase of most such crop plants often occurs in the dry climate of the Mediterranean region during spring. This favors the evolution and survival of plants with "cleistogamous" flowers, as cleistogamy prevents desiccation of anthers and stigmas and encourages full seed set by autogamy [4]. Cleistogamy of pulses appears to be a relic of evolutionary antecedents. However, such cleistogamous flower buds do open for a small period, providing opportunities for occasional natural outcrossing, which occurs in almost all pulses (including various species of cultivated *Vigna*) to varying extents. This generates heterozygosity and brings about substantial heterogeneity in the population, resulting in the loss of newly developed cultivars if they go unnoticed. However, on the other hand, it makes them "resilient" to changing climate conditions, as heterozygosity in the population appears to confer resistance to environmental change [5]. Heterogeneity in plant populations accelerates opportunities for the selection of more stress-tolerant genotypes and thereby provides resilience to the crop as well as the ecosystem [6]. Crop plant resilience, therefore, appears to be brought about in nature by the shuffling and recombination of genes at many loci, leading to the creation of novel adaptive attributes which ultimately result in enhanced "adaptedness" for a few recombinants in the changed environmental condition.

Presently, the impact of global warming can be seen worldwide. For example, India has witnessed highly fluctuating weather conditions in the last decades [7]. It is evident that high temperatures have changed the rainfall pattern as well as distribution and have increased water scarcity. In the future, the shortage of water will increase drought-affected regions. Moreover, it will negatively impact those regions that have higher precipitation rates [8]. The impact of climate change on chemical and physical processes in soils and nutrient uptake from soils has previously been reviewed comprehensively [9]. In Myanmar, erratic rainfall due to climate change had a detrimental impact on pulse production efficiency [10]. Thus, aberrant weather conditions (global warming) are expected to pose serious threats to pulse productivity in the near future as rising temperatures will lead to production of poor biomass; reductions in days to flowering, rate of fertilization, and seed formation [11–15]; and intensifying vulnerability to disease and insect pests [1,16,17]. As per a Food and Agriculture Organization (FAO) report [18], climate change has put global food security more at risk; heightened the dangers of undernutrition in resource-poor regions of the world due to heat, drought, salinity, and waterlogging; and increased the threat of newly emerging diseases and insect pests. While assessing the impact of drought on crop yields, Kuwayama et al. [19] reported 0.1–1.2% yield reduction for corn and soybeans for each additional week of drought. According to Ambachew et al. [20], drought stress can cause 20–90% yield reduction in common bean, which in the worst scenario could go up to 100%. In other pulses, yield losses have been measured to the extent of 6–86% and 15–100% due to different abiotic and biotic stresses, respectively [21]. Although McKersie [8] has discussed a number of options for mitigating the effects of climate change on crop production, breeding for genotypic adaptation is one of the important strategies for dealing with future climate change [22]. It involves incorporating novel traits in crop varieties to enhance food productivity and stability. For breeding climate-resilient cultivars in pulses, it is imperative to bring about genetic improvements for adaptive traits [22,23]. Shunmugam et al. [24] reviewed the physiological traits that may facilitate breeding climate-resilient food legume crops for adaptation under abiotic stresses. The symbiont preference traits related to abiotic stresses have recently been studied in the model legume *Medicago truncatula* [25]. Cullis and Kunert [26] unlocked traits that impart drought tolerance by producing a range of secondary metabolites and proteinaceous inhibitors in response to environmental stresses in orphan legume crops. As climate change is the biggest threat to the production of both warm- and cool-season pulses in the coming years, the mining of adaptive traits in the germplasm to transfer them into newly bred cultivars is highly desirable. Information on this aspect of pulse crops is still scattered in the literature. In this review, we have therefore made an attempt to organize such dispersed information and discuss various strategies to exploit adaptive traits for breeding climate-resilient smart pulses.

2. Overview of Adaptive Traits in Pulses

Climate change can result in a wide range of abiotic stresses, such as drought, heat, cold, salinity, flood, and submergence, and biotic stresses, including increased attacks of pathogens and pests [27]. Therefore, breeding of adaptive traits is required for increasing the resilience of crops to current climate change conditions to help sustain productivity. Adaptive traits show their adaptive plasticity in changing environmental conditions and help crop plants survive and/or reproduce under biotic and abiotic stress conditions [26]. These adaptive traits can be agromorphological [28,29], physiological, and biochemical [24,26,30,31]. In pulses, breeders have attempted to improve many traits for the given target environment. Therefore, specific adaptive traits must be incorporated in the improved genotypes for each growing condition (Table 1).

The reproductive stage substantially influences seed yield in crop plants. It has been reported that drought stress during the pod-filling stage leads to pod abortion and thus reduces the number of seeds per plant, whereas terminal drought at the early podding stage resulted in an 85% decline in seed yield of chickpea [32]. Thus, pod-filling ability can be targeted as an agromorphological trait under moisture-deficient conditions for developing drought-resilient cultivars. In Mediterranean environmental conditions, leaf traits such as leaf area, leaf weight, and leaf growth rate have been identified for tolerance to drought stress [33]. Physiological traits, which are relatively stable across environments, provide greater breeding value [34]. A number of physiological traits including leaf parameters, seed set, pod abscisic acid concentrations, and root traits have been shown to impart tolerance to drought in chickpea [32,35]. The role of sucrose infusion has recently been identified in the salt tolerance of chickpea [36]. Prince et al. [37] performed an innovative analysis to decipher the mechanisms that underpin drought tolerance in legumes and established the role of root xylem plasticity in improving water-use efficiency in soybean plants subjected to water stress [37]. An extensive root system is a useful drought-avoiding physiological trait that helps to maintain the seed yield under drought conditions in pulses through enhanced extraction of soil water [38,39]. It is therefore desirable to exploit root and leaf traits while breeding for drought avoidance in pulses.

It has been established that proteins and metabolites are generated in different tissues of crop plants in response to environmental stresses [24]. Identification, quantitation, validation, and characterization for a wide range of proteins/metabolites from specific organ/tissue/cells under stress conditions can be useful biochemical traits for breeding climate-resilient crops. Protein differential expression analysis in response to various stresses at different growth stages has been studied in several pulse crops including chickpea, pea, green gram, and common bean [40]. Various morphophysiological traits imparting tolerance to such stresses have been identified in pulse crops for their wider adaptability considering the global trend of rising temperatures over the years [41–47]. One of the key physiological traits is photosynthetic activity, as high and low temperatures cause photodamage to photosystem-II (PS-II) [48,49]. In lentil, pollen and leaf traits could be helpful in identification of heat-tolerant genotypes [15]. Breeders have exploited early flowering traits in chickpea breeding programs, leading to the development of new chickpea varieties adapted to warmer, short-season environments that resulted in a chickpea revolution in southern India [50]. For lentil, the development of short-duration cultivars has increased the opportunity for the adaptation of lentil crops in rice-fallow areas owing to reductions in yield losses caused by forced maturity [51]. Adaptation towards freezing temperatures involves a number of structural and functional changes at the cellular level. During acclimation, organic compounds such as sugar, proline, and glycine betaine accumulate in plant cells and confer frost tolerance to surviving plants. One such organic compound, "glycine betaine", has been shown to mitigate cold stress damage in chickpea [52]. It is therefore obvious that the morphophysiological and biochemical parameters imparting adaptive value vary with the nature and kind of abiotic stress and pulse species, respectively. For breeding smart pulses for a specific situation, special adaptive features need to be exploited.

Table 1. Adaptive traits for different growing regions of important pulse crops.

Common/Scientific Name	Region	Adaptive Traits	Reference
Chickpea (*Cicer arietinum* L.)	Nontropical dry areas and semiarid tropics	Earliness; early vigor; spreading to erect growth habit; resistance to pod borer, AB, BGM, wilt, and root rot; tolerance to drought and heat; suitability for mechanical harvesting; herbicide tolerance	[53–60]
Lentil (*Lens culinaris* Medik.)	Nontropical dry areas and semiarid tropics	Earliness; early vigor; spreading to erect growth habit; resistance to wilt, root rot, *Stemphylium* blight, AB, rust, and black aphid; tolerance to drought and heat	[12,13,61,62]

Table 1. *Cont.*

Common/Scientific Name	Region	Adaptive Traits	Reference
Pea (*Pisum arvense* L.)	Cool, semiarid climates	Dwarfness, leaflessness, tendril, resistance to rust and powdery mildew, tolerance to terminal heat and drought, earliness	[63]
Mungbean (*Vigna radiata* Wilczek)	Arid and semiarid regions, wide adaptation, warm season	Short duration, MYMV and powdery mildew resistance, drought and heat tolerance, photo-thermo-insensitivity, preharvest sprouting	[64–66]
Blackgram (*Vigna mungo* (L.) Hepper)	Hot humid, semiarid regions	Short duration, MYMV and powdery mildew resistance, photo-thermo-insensitivity, tolerance to excess moisture stress	[64,67,68]
Pigeaonpea (*Cajanus cajan* (L.) Millsp.)	Semiarid and lower humidity tropic regions	Short-to-medium duration; short stature; resistance to PSB, wilt, SMD, pod borer, and pod fly	[69,70]
Grass pea (*Lathyrus sativus* L.)	Indian subcontinent and Mediterranean region	ODAP content, water-logging and drought tolerance	[63,71]
Common bean (*Phaseolus vulgaris* L.)	Most domesticated pulse for many tropical countries	Dwarfness; resistance to CBB; tolerance to cold, heat, and drought; earliness	[63,72–74]
Rice bean (*Vigna umbellata* (Thunb.) Ohwi and Ohashi)	Dry zones of the arid and semiarid regions	Tolerance to acid soils and drought, early maturity, high yield, determinate growth habit	[63,75]
Tepary bean (*Phaseolus acutifolius* A. Gray)	Dry season of tropical regions	Drought and CBB resistance, deep root system, tolerant to heat, high N_2 fixation, short growth period	[63,76,77]
Lima bean (*Phaseolus lunatus* L.)	Soils and climates of Piedmont of Georigia, Mexico, and Argentina	Plant types for marginal soil and limited water conditions, climbing types, bushy, compact types for intensive cultivation, large seed type, less cooking time	[63,78,79]
Runner bean (*Phaseolus coccineus* L.)	Cool climates of Italy and other parts	CBB resistance, high osmoregulation, heat tolerance and resistance to BCMV, dwarfness, early maturity	[63,80,81]
Adzuki bean (*Vigna angularis* Ohwi and Ohashi)	Subtropical and temperate climate zone	CBB resistance, drought tolerance	[63,82]
Hyacinth bean (*Lablab purpureus* (L.) Sweet)	Subhumid and semiarid conditions	Early maturity, drought tolerance, salinity tolerance	[63,83,84]
Horse gram (*Macrotyloma uniflorum* (Lamb.) Verds)	Low and erratic rainfall areas, better soils of the arid and semiarid regions	High tolerance towards acid soils, drought tolerance, green foliage till maturity, thermoinsensitivity, short maturity period, erect, nontendril plant type	[63,85,86]
Winged bean (*Psophocarpus tetragonolobus* (L.) D.C.)	Vietnam, parts of China	Erect type, determinate growth habit, high seed protein and oil content with high linoleic acid, photoperiodic responses	[63,87,88]
Cowpea (*Vigna unguiculata* (L.) Walp.)	Arid and semiarid regions, wide adaptation	Fast initial growth, early maturity, better source sink relations	[63,89,90]
Moth bean (Vigna aconitifolia (Jacq.) Marechal)	Arid tracts, low rainfall and warm climates	High photosynthates, tolerance to drought and heat, low fertility requirement, early and synchronous maturity, erect plant growth, tolerance to YMV	[63,91,92]

AB: *Aschochyta* blight, BGM: *Botrytis* greymold, BCMV: bean common mosaic virus, CBB: common bacterial blight, MYMV: mungbean yellow mosaic virus, ODAP: β-oxalyldiaminopropionic acid, PSB: *Phytophthora* stem blight, SMD: sterility mosaic disease, YMV: yellow mosaic virus.

3. Looking Back to Wild Species and Land Races for Adaptive Traits

The growing intensity of abiotic and biotic stresses calls for adoption of mitigation and adaptation strategies by incorporating resistance/tolerance to various stresses to increase resiliency and sustain the productivity of pulse crops in a changing climate scenario. These strategies will pave the way for efficiently meeting humankind's demand for a more plentiful and nutritious food supply [93–99]. Cultivated species of pulses have narrow genetic diversity to withstand current global warming challenges [100]. It is therefore necessary to look back to wild species and land races when searching for useful adaptive traits/genes. It is well documented that wild species have a reservoir of many useful genes [47,101] because they have evolved under natural selection to survive climatic extremes and can potentially provide further genetic gains [93,94,96]. Therefore, wild species need to be exploited in genetic improvement programs to alleviate the challenges of global warming and its related effects on pulses.

Wild relatives of crops have been used sparingly and typically in an ad hoc manner in many crop breeding programs [96,102,103]. In pulse crops, Sharma et al. [101] reported a number of wild accessions having high levels of resistance/tolerance to various stresses [101]. As wild relatives of chickpea and lentil are native to drought-prone areas, they possess useful traits for drought tolerance. According to Gorim et al. [104], an evaluation of wild relatives of lentil for root and shoot traits under water-deficit and fully watered conditions resulted in different patterns of root distribution into different soil horizons. The study revealed that wild lentil genotypes employed diverse strategies such as delayed flowering, reduced transpiration rates, reduced plant height, and deep root systems to either escape, evade, or tolerate drought conditions.

The use of wild species for targeted introgression of useful genes dates back to the work of Vavilov [105]. Thereafter, crop wild relatives have been used continually to transfer adaptive traits in a variety of crops including pulses. According to Maxted and Kell [96], more than 291 articles have come out on pulses regarding the identification and introgression of useful traits from 185 wild relative taxa to 29 crop species. Most of these studies have focused on disease and pest resistance (>50%), abiotic stress tolerance (10–15%), and yield traits (20%). Also, 74% of 104 molecular-assisted breeding studies (1995–2012) have dealt with introgression of traits from wild species that confer disease resistance, while the remaining studies covered abiotic stress tolerance, improved yield, and growth habit [103]. Thus far, resistance to many diseases and insect pests from wild relatives and unadapted germplasm has been successfully transferred into suitable genetic backgrounds [102,106–109]. Kumar et al. [109] recorded useful genetic variability for days to 50% flowering, secondary branches, number of pods/plant, biological yield/plant, grain yield/plant, and 100 seed weight in the indigenous gene pool of lentil. *Lens ervoides* (a wild species of cultivated lentil) has been exploited in Canada for transferring anthracnose resistance genes into cultivated backgrounds through embryo rescue techniques [110,111]. More recently, the crossing of cultivated species with *Lens tomentosus* accession "ILWL120" followed by ovule culture has resulted in the development of a number of prebreeding lines carrying diversity for flower color, seed coat, and cotyledon color [112]. In India, under an ICAR-ICARDA network project, many prebreeding lines (>500) developed by using crossable wild species of lentil have shown variability for yield-contributing traits. In chickpea and pigeonpea, wild relatives have been exploited for enhancing the adaptability of cultivated species against climate extremes under changing climate conditions [47,101]. Brumlop et al. [113] developed from a *C. arietinum* × *C. judaicum* cross the prebreeding line "IPC 71", which has an increased number of primary branches, pods per plant, and green seeds for further use in chickpea improvement programs. These reports and achievements substantiate the fact that wild species of pulses do carry potentially useful genes. Considering current climate variability and its manifold effects [101,114], such wild species need to be exploited for developing prebreeding lines of pulses for the new environment. Their utilization in pulse breeding programs may result in climate-resilient smart cultivars with a broad genetic base and the ability to sustain environmental extremes [101].

4. Conventional Breeding Approaches

Conventional breeding approaches have been used to tailor suitable plant types with the ability to adapt to different environmental niches/cropping systems. To this end, breeders focused mainly on highly heritable visually adaptive traits (agromorphological traits). The use of dwarfness and leaflessness (i.e., modification of leaflets into tendrils) traits in field pea resulted in the development of new plant types that allow penetration of sunlight to lower portions of the plant, provide natural mechanical support to preclude lodging, and prevent bird damage owing to a network of interlocked tendrils above the crop canopy [115]. Recently, Saxena et al. [116] recommended an ideal plant type of pigeonpea comprising rapid seedling growth; nondeterminate (NDT) growth habit; spreading or semispreading branches; a greater number of secondary and tertiary branches; long fruiting branches; more flower bunches; 5–6 pods/bunch; 4–6 seeds/pod; 12–14 g/100 seed weight; resistance to *Fusarium* wilt (FW), sterility mosaic (SM), and *Phytophthora* stem blight (PSB); deep root system/drought tolerance; and the ability to mitigate other abiotic stresses including waterlogging for pigeonpea–cereal intercropping systems. An early flowering exotic line "Precoz" (ILL 4605) of lentil has been utilized extensively to tailor plant architecture having vigorous growth, medium maturity, large seeds, and cold tolerance, particularly for Indo-Gangetic plains [109]. Earliness, which provides an escape mechanism from drought and terminal heat stresses, has been invariably used in almost all breeding programs to mitigate such stresses [1,117]. In chickpea, significant progress has been made in developing early maturing varieties that mature in 85–90 days in peninsular India [118]. Even extra short duration chickpea varieties, termed super-early types, have been reported in chickpea [119] and pigeonpea [120], and efforts in this direction are also underway in other pulse crops. Some super-early lines maturing within 100 days with a yield potential up to 1.5 t/ha have been reported by International Crops Research Institute for the Semi Arid Tropics (ICRISAT) in both determinate (DT) and NDT groups of pigeonpea [120]. For sustaining crop intensification under the rice-fallow system of eastern India, development of an early maturing variety (90–100 days) has been suggested in lentil, and efforts are being made to tailor genotypes having earliness and high biomass and harvest index [12]. Kashiwagi et al. [117] identified root traits to improve water uptake in chickpea under limited-moisture conditions. They used contrasting chickpea accessions vis-à-vis root biomass and rooting depth in a drought-avoidance breeding program to improve the root system of ensuing genotypes for cultivation in central and southern India.

In the coming years, the disease and pest scenario may be a serious problem due to climate change. Therefore, climate-smart pulses must carry resistance to diseases and insect pests. Germplasm screening under natural and artificial conditions to identify resistant sources for various diseases and insect pests has been a regular feature of resistance breeding programs for pulses [2,121]. Knowledge of the genetics of resistance traits and racial composition of pathogens has accelerated the development of cultivars having adaptability under epidemic conditions [2,122,123]. Recently, root rots (dry and black root rots) and collar rots have emerged as potentially damaging diseases in both chickpea and lentil. However, the literature pertaining to the racial description of the causal organisms (species of *Rhizoctonia*, *Fusarium*, and *Sclerotinia*) and resistant donors in both these crops is still scanty. These diseases need to be tackled through breeding in the days ahead. However, the complex nature of these diseases, the resistance mechanisms and traits (especially physiobiochemical traits), and the limited screening facilities are the major limitations to progress for pulses through conventional plant breeding. These limitations need to be overcome through phenomics-based breeding, which is currently employed for improving soybean [124].

5. Omics-Based Breeding Approaches for Adaptive Traits in Pulses

Different "omics" fields, namely, genomics, transcriptomics, epigenomics, proteomics, metabolomics, and phenomics, have emerged during the past years. These approaches have enhanced the precision and sped up the ongoing breeding programs of the major food crops such as wheat and rice [125]. Omics-based strategies can also be used to develop climate-smart pulses (Figure 1). These strategies have been categorized as current and emerging omics-based approaches and are discussed below.

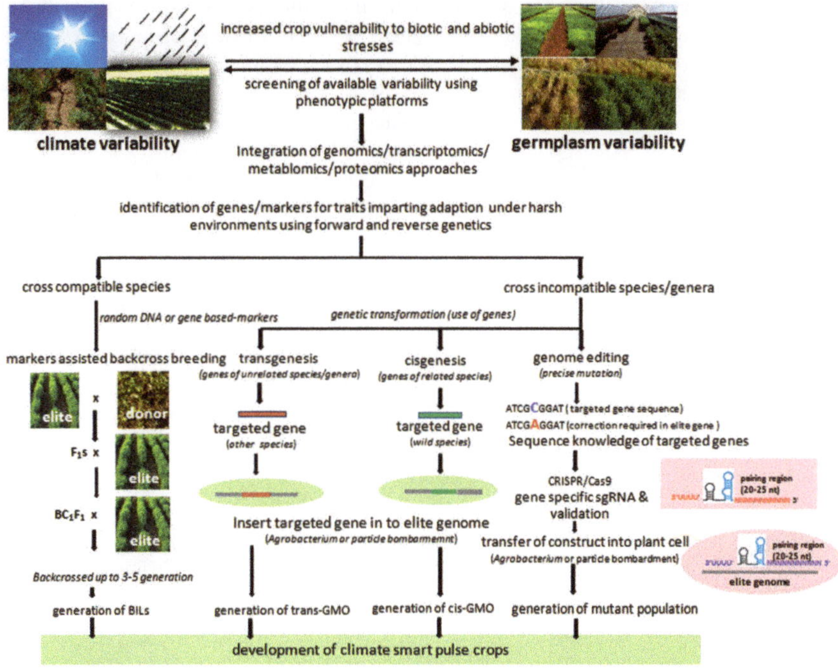

Figure 1. Omics-based approaches for the development of climate-smart pulse crops.

5.1. Current Genomics Approaches

During the last 25 years, substantial advances have been made in the genomic resources of pulse crops, leading to the development of various molecular markers and the availability of QTLs/genes that impart tolerance to various biotic and abiotic stresses. These genomic resources have been discussed earlier in detail [126,127]. Next-generation sequencing (NGS)-based genomics tools have enabled rapid and cost-effective identification of the functional and regulatory genes controlling abiotic stress resistance in many pulses [128]. These NGS tools have helped to develop SNP and INDEL markers [129] and expression atlases [130–132] and to understand the signaling pathways for tolerance to various environmental stresses in various legumes including pulses [128]. Genome sequences of major pulses, which are important genomic resources for translating the genomics into the field are now available in the public domain [133–136]. Genomics approaches have now become an integral part of the current conventional breeding program and can be used in different ways for shortening the period of genetic improvement and targeted manipulation of genomes for climate-smart pulses.

5.1.1. Molecular Markers Associated with Adaptive Traits in Pulses

Knowledge of genes controlling traits for wider adaptability is the prerequisite to develop climate-smart pulses. In the last three decades, efforts have been made to identify such genes/QTLs (Table 2) in chickpea, pigeonpea, and other pulse crops. In these crops, QTLs/genes for pods per plant (qPD4.1) and flowering (qFL4.1 and qFL5.1) in pigeonpea [137] and thermotolerance [138] in chickpea can help to construct new plant types suitable to changing environments [13]. Moreover, QTLs (*HQTL-1* and *HQTL-2*) have been identified for pollen viability in azuki bean [139]. In field pea, Javid et al. [140] validated markers associated with abiotic and biotic stresses for breeding programs. In this study, a molecular marker "PsMlo" showed an association with powdery mildew (PM) resistance and boron (B) tolerance, while several other markers were found associated with salinity tolerance across a diverse set of pea germplasm. The PsMlo1 marker predicted the PM and B phenotypic responses with high levels of accuracy (>80%) and thus showed its potential to facilitate improvement for PM resistance and B tolerance. More recently, Paul et al. [141] identified QTLs associated with heat tolerance in chickpea in a mapping population comprising recombinant inbred lines (RILs) evaluated under two heat-stress (late sown) and one nonstress (normal sown) environments. This resulted in the identification of 25 putative candidate genes responsible for heat stress in the two major genomic regions. The identified markers, which were linked to four major QTLs, can be utilized in breeding programs. For improving the adaptation of common bean to adverse environments, Diaz et al. [142] evaluated RILs under different abiotic stress conditions for a number of agrophysiological traits and identified molecular markers linked with QTLs for abiotic stress tolerance. In accessions of common bean from the northwestern Himalayas, Choudhary et al. [143] discovered the gene/QTLs for *Anthracnose*. These genes/QTLs need to be utilized in breeding programs for developing stress-tolerant cultivars of common bean.

Table 2. Genes/QTLs for adaptive traits identified in major and minor pulse crops.

Common Name	QTL/Gene	Trait	Method Used for Identification	Reference
Pea	*nod3*	Hyper nodulation mutation	Comparative genomics	[144]
	PsMlo	Powdery mildew resistance	Comparative genomics	[145,146]
	PsDREB2A	Drought response	Comparative genomics	[147]
Cowpea	Cowpea Co-like gene family	Photoperiod responsive	Sequencing along with comparative genomics	[148]
	Stg	Stem greenness after drought	QTL mapping	[149]
	Rdw	Dry weight recovery after drought	QTL mapping	[149]
	Mac 1–9	Resistance to *Macrophomina*	QTL mapping	[150]
	Major QTL	Cowpea leaf shape imparting drought tolerance	QTL mapping	[151]
	Dro-1, Dro-3, and *Dro7*	Stay-green	QTL mapping	[152]
	Hbs-1–Hbs-3	Heat-induced browning of seed coats	QTL mapping	[153]
	Thr-1–Thr-3	Foliar thrips	QTL mapping	[149]
	Major QTL	Aphid resistance	QTL mapping	[154]
	Major QTL	Resistance to root-knot nematodes	QTL mapping	[155]
	Fot31	*Fusarium* wilt		[151]
	Candidate genes	Resistance to root-knot nematodes	QTL mapping and transcriptome analysis	[156]
Pigeonpea	*Hsf* genes	Heat-response	Genome-wide analysis	[157]
	Dehydrin-like protein (*DLP*) gene and acid phosphatase class B family protein (*APB*) gene	Drought stress	Differentially expressed genes analysis	[158]
	Cyclophilin (*CcCYP*) gene	Multiple abiotic stress tolerance	cDNA expression analysis	[159]
	Pre-hevein-like protein PR-4 precursor (*PR-4*) and protease inhibitor/seed storage/LTP family protein (*Ltp*) genes	Defense against *Helicoverpa armigera*	Gene expression analysis using qPCR	[160]
Common bean	*Co-1–Co-10*	Resistance to anthracnose	Linkage mapping	[161]
	10 QTLs/genes	Resistance to anthracnose	Associations mapping	[143]
	Resistance gene analogs	Resistances to different pathogens	Associations mapping	[162]
Horse gram	9 genes	Response to drought stress	Transcriptome analysis	[163]
Adzuki bean	*VaAGL, VaPhyE,* and *VaAP2*	Flowering time and pod maturity	QTL mapping	[164]
Hyacinth bean	17 functionally relevant genes	Drought-stress response	Suppression subtraction hybridization (SSH) analysis	[83]
Chickpea	Aquaporins gene family	Biotic and abiotic stresses	Comprehensive genome-wide analysis	[165]
	CarERF116	Abiotic stress responsive	Genome-wide association analysis	[166]
	Major QTLs corresponding to flowering time genes (*efl-1, efl-3,* and *efl-4*)	Flowering time	QTL mapping	[167]
	CarLEA4	Plant developmental processes and abiotic stress responses	Gene expression analysis	[168]
	Differentially expressed genes	Drought stress response	Quantitative real-time PCR (qRT-PCR) analysis	[169]

Molecular marker technology helps plant breeders and gene bank curators to identify markers linked with morphological and physiological traits in the available germplasm that enhance crop adaptation under climate variability [170]. According to Kumar et al. [170], these linked markers can be used to develop "climate change ready" cultivars for cultivation. The application of molecular marker technology has resulted in the identification of markers linked to genes controlling several abiotic and biotic stresses [171–175] and other agronomic traits [53,174,176–180] in chickpea and pigeonpea. Also, the draft genome sequence of both Kabuli and Desi chickpeas and pigeonpea are available in the public domain [134,181,182]. These sequence data of chickpea and pigeonpea will assist in enhancing their productivity and lead to conserving food security in arid and semiarid environments. Many QTLs have been identified on several linkage groups (2, 3, 4, 6, and 8) for *Aschochyta* blight (AB) resistance [183], and marker-assisted backcrossing (MABC) has been used for conversion of targeted lines with respect to one or two traits without disturbing other native traits of the target variety in chickpea [184]. According to Varshney et al. [183], simultaneous genetic improvement for FW and AB resistance is possible through marker-assisted selection (MAS) in chickpea. To this end, they undertook two parallel MABC programs by targeting the *foc 1* locus and two QTL regions, namely, ABQTL-I and ABQTL-II to introgress resistance to FW and AB, respectively, in "C 214", an elite cultivar of chickpea. Phenotyping of lines developed through MAS led to the identification of some lines carrying both FW (race 1) and AB, resistance which would be tested further for yield and other agronomic traits under multilocation trials for possible release and cultivation. In an attempt to identify QTLs for root traits in chickpea, Serraj et al. [185] developed a RIL population from a cross between a long root genotype "ICC 4958" and a well-adapted, high yielding variety "Annigeri". This RIL population was used to map the genes/QTLs for root traits, leading to the identification of a "QTL hotspot" that explained a large part of the phenotypic variation for major drought tolerance traits, including the root traits. Kashiwagi et al. [117] used marker-assisted breeding to introgress this QTL hotspot into a leading Indian chickpea cultivar "JG 11". They demonstrated that introgression lines had shown a distinct yield advantage (>10%) over JG 11 in multilocation evaluations under terminal drought. These marker-based success stories of chickpea can also be replicated for improving stress tolerance in other pulse crops. Choudhary and co-workers [186,187] used root traits to screen pigeonpea genotypes against Al toxicity and established root exclusion as the possible mechanism for Al tolerance, whereas Daspute et al. [188] discovered Al-responsive citrate excretion as the biochemical basis of Al tolerance in pigeonpea. The information generated in pigeonpea may be utilized in other pulses for improvement of Al tolerance.

5.1.2. Gene(s) Related to Adaptive Traits

Since climate changes have a large influence on the creation of a number of biotic stresses, knowledge of genes that express themselves in different environmental conditions can help in breeding climate-resilient crops. Transcriptome analysis has helped to deliver functionally associated gene-based markers for breeding activities in lentil, field pea, and faba bean [189,190]. In several pulses (pea, lentil, chickpea, common bean, pigeonpea, and broad bean), transcriptomic data have been generated that can be used in gene-based marker discovery to assess genetic diversity, linkage mapping, and trait dissection [191]. In other crops, it has been widely used to identify the candidate genes that express themselves in specific environmental conditions (heat stress) or at particular plant growth stages [192]. Therefore, similar strategies can be employed for identification of such genes/traits imparting wider adaptation to pulse crops under global warming conditions. In chickpea, this approach has been used to identify candidate genes governing plant height and agromorphological traits [193,194], and in lentil, QTLs for B toxicity tolerance, flowering time, and seed characteristics [195,196]. Similar efforts have led to the identification of heat-responsive genes that are expressed in heat-sensitive and heat-tolerant genotypes during heat stress in chickpea [132,197]. In other legume crops, genes responsible for thermotolerance in soybean (*GmHsfA1*) and broad bean (*VfHsp17.9-CII*) have been cloned [198,199]. Naser and Shani [200] mentioned the importance of auxin-related genes that play an important role in plant growth, seed development, and abiotic stress response (drought and salinity tolerance). In pigeonpea, Pazhamala et al. [131] developed a compendium of 28,793 genes that express themselves during the reproductive stage in seed-forming tissues and identified a network of 28 flower-related genes. Similarly, Singh et al. [201] established a transcription factor database (i.e., PpTFDB) in pigeonpea that can be useful for functional genomic analysis in other legume crops [201]. Kudapa et al. [202] developed a comprehensive gene expression atlas by associating genome sequence with genes expressed across different plant developmental stages and organs covering the entire lifecycle of chickpea. They identified 15,947 unique numbers of differentially expressed genes and observed significant differences in gene expression patterns in the process of flowering, nodulation, and seed and root development. They could also identify candidate genes responsible for drought stress from the QTL hotspot region. These recent advances, including the development of gene expression atlases and signaling pathways involved in plants' responses to environmental stresses, will certainly facilitate the development of climate-smart pulses.

RNA-sequencing-based (NGS-based) transcriptome analysis is considered to be a superior approach to understand the gene function and molecular basis of many cellular responses in plants exposed to abiotic stresses. The gene expression analysis performed by Abdelrahman et al. [128] could identify a number of candidate genes for drought, salinity, cold, and heavy metal stress resistance in chickpea and other pulses. According to Singh et al. [203], transcriptome changes occur in response to seedling drought stress in lentil. They recognized the upregulation of genes involved in electron transport chains, oxidation-reduction processes, the TCA cycle, senescence and reduction of stomatal conductance, the downregulation of genes associated with gamma-aminobutyric acid synthesis, transcription binding and synthesis of cell wall proteins, and the negative regulation of abscisic acid responses in the drought-tolerant lentil genotype "PDL 2". Studies on the MLO gene family [145,204] have revealed that the *LcMLO1* and *LcMLO3* genes in lentil and the *PsMLO1* gene in pea are associated with PM resistance. In chickpea, Garg et al. [205] carried out a comparative transcriptome analysis of drought- and salinity-tolerant/sensitive genotypes at different developmental stages. They could identify genes encoding enzymes involved in the biosynthesis of sugar alcohols (inositol and trehalose), xyloglucan, and amino acids (proline and citrulline). The results of the transcriptome study represented a starting point to dissect the gene regulatory networks involved in drought and/or salinity stress in chickpea [205]. Transcriptome analysis performed in the nodules involving *Mesorhizobium ciceri* CP-31-(McCP-31)-chickpea and *M. mediterraneum* SWRI9-(MmSWRI9)-chickpea associations under P_i-deficient and -sufficient conditions could identify changes in the expression of genes in more-P_i-deficiency-sensitive MmSWRI9-induced nodules than in less-P_i-deficiency-sensitive McCP-31-induced nodules [206]. Recently, Mashaki et al. [169] studied transcriptome profiles in roots and shoots of two contrasting Iranian kabuli chickpea genotypes under water-limited conditions at the early flowering stage using an RNA-sequencing approach. They identified 4572 differentially expressed genes (DEGs) and grouped these DEGs into several subcategories depending upon the intensity of drought stress. Also, several transcription factors (TFs) controlling major metabolic pathways such as ABA, proline, and flavonoid biosynthesis have been identified, spotting DEGs in QTL hotspot regions (reported earlier) in chickpea. Thus, genes/TFs upregulated in the drought-tolerant genotype during drought stress in this study are potential candidates for enhancing tolerance to drought [169]. Moreover, an early flowering1 (Efl1) gene, which is an ortholog of the early flowering3 (ELF3) gene of *Arabidopsis* (*Arabidopsis thaliana*), has also been mapped and sequenced in chickpea [207]. It is therefore expected that integration of phenomics with transcriptomics, proteomics, and metabolomics will provide greater insight into the molecular changes occurring during the growth and development of various species of pulses under environmental stresses.

5.1.3. Transgenics for Increasing Adaptability of Pulses

Gram pod borer (*Helicoverpa armigera* Hubner) is the key insect pest of pigeonpea and chickpea, causing 17–35% yield losses [208]. No resistance sources to this insect pest are available in the cultivated germplasm and immediate wild progenitors of these two pulse crops. According to Choudhary et al. [121], wild relatives of pigeonpea, notably *Cajanus scaraeboides* and *C. platycarpus*, have morphologically adaptive features that impart resistance to pod borer. The resistance-imparting morphological traits in such wild species include density of nonglandular trichome *C* on pods (>5 times greater than that present on pods of cultivated accessions), width and waxiness of pod wall, and prominent pod constrictions. Attempts to develop pod-borer-resistant genotypes of pigeonpea and chickpea by conventional breeding methods have not been very successful due to crossable barriers and incompatibility with wild species. Moreover, the incomplete penetrance and variable expressivity of such wild genes in the cultivated background further complicates the outcome [121]. Therefore, a transgenic approach has been adopted to improve resistance to pod borer in both pigeonpea and chickpea. This has resulted in the development of transgenic lines of pigeonpea and chickpea carrying Bt genes, namely, *cry*1Ac, *cry*1Ab, and *cry*2Aa. The transgenic lines of chickpea with synthetic Bt genes either singly or in combination have exhibited a high level (98–100%) of mortality of *Helicoverpa* larvae [209,210]. Das et al. [211] reported that field trials of several such transgenic lines are underway at the Indian Institute of Pulses Research (IIPR), Kanpur. Efforts have also been made to utilize some of these effective lines in backcross breeding program for further improvements [211]. According to Singh et al. [212], transgenic plants expressing the Cry2Aa gene have been developed employing *Agrobacterium*-mediated in planta transformation approach in pigeonpea. Developed transgenic plants (T$_3$ lines) have demonstrated 80–100% mortality of the challenged larvae and improved the ability to prevent damage caused by the larvae. The selected transgenic plants accumulated Cry2Aa in the range of 25–80 μg/g [212]. Transgenic approach has also been used to tackle the problem of salt tolerance in chickpea and pigeonpea [213,214]. According to Bhatnagar-Mathur et al. [214], the osmoregulatory gene "*P5CSF129A*", encoding overproduction of proline transferred through genetic transformation, confers drought tolerance in chickpea. It is thus obvious that the ongoing efforts to develop effective transgenic lines for biotic and abiotic stresses will yield desired results very soon in both chickpea and pigeonpea.

5.2. Emerging Omics Approaches for Breeding of Adaptive Traits

Though knowledge of epigenomics, proteomics, metabolomics, and genome editing is still limited in pulses, these approaches have opened up new avenues for resolving the complexity of adaptive traits imparting tolerance to biotic and abiotic stresses. These diverse omics platforms have great potential for improving the current understanding of important traits, enabling us to develop new strategies for developing climate-smart pulses.

5.2.1. Integration of Proteomics and Metabolomics with Genomics for Enhancing Climate Resilience

Proteomics and metabolomics have emerged as cutting-edge areas of functional biology [40]. Integration of information obtained from proteomics and metabolomics with genomics data can enhance our understanding about plants' response to abiotic stresses [215]. Several studies have revealed a network of stress responsive genes/proteins/metabolites/transcription factors in various legumes, including soybean [215–217]. This can help to catalogue and prioritize the genes to exercise selection of superior traits for realizing genetic gains in crop breeding programs [218]. Recent advances in proteomics have included classification of proteins, comparison of protein profiles, post-translational modifications of proteins, identification of protein complexes and interacting networks, study of protein structure and functional groups, and their use in crop improvement [219].

In legumes, proteomic studies have unraveled the molecular mechanisms underlying tolerance to different biotic (AB, PM, FW, rust, mungbean yellow mosaic India virus, aphids, etc.) and abiotic (drought, waterlogging, salinity, cold, heat, mineral deficiency, heavy metal toxicity, and dark and UV–B irradiation) stresses [220,221]. After performing a proteomic analysis, Krishnan et al. [222] identified 373 proteins in pigeonpea seeds. They observed a large number of seed proteins showing significant homology for amino acid sequences with that of soybean seed proteins. They could recognize a large number of stress-related proteins which probably confer adaption to pigeonpea in drought-prone environments. More recently, Rathi et al. [223] identified and characterized proteins that enhance adaptation of grass pea under dehydration conditions. Based on their putative functions, they grouped these proteins into 22 functional categories, and 9.17% of these proteins showed their relation with dehydration-induced stress [223]. In faba bean, Li et al. [224] carried out leaf proteomic analysis under drought stress. They could classify quantified proteins mainly into five functional groups (regulatory proteins: 46.7%; energy metabolism: 23.3%; cell cytoskeleton: 6.7%; other functions: 20%; and unknown function: 3.3%). This study showed upregulation of chitinase, 50S ribosomal protein, Bet protein, and glutamate–glyoxylate amino-transferase under drought conditions, suggesting their important roles in drought tolerance [224]. According to Lin et al. [225], integration of transcriptomic and proteomic research has been fruitful for exploring bruchid-resistant genes in mungbean. This study will have a far-reaching impact on the control of bruchid (Callosobruchus spp.), which infests grains of almost all pulses in storage.

Various metabolic changes occur in plants when they are exposed to abiotic stresses [226]. Knowledge of metabolite profiles provides insight into the functional role of metabolites for traits imparting tolerance/resistance to abiotic and biotic stresses. In addition, integration of gene expression profiles with metabolite profiles helps to identify gene-to-metabolite associations/networks [40,227]. A few studies have been conducted to determine the metabolic profiles of legume crops, including pulses. Metabolic changes that take place during legume–rhizobial symbiosis have been studied under different conditions, including drought stress [228–233]. In common bean, Hernández et al. [229], after analyzing the nontargeted metabolite profile, identified changes in the roots and nodules of plants inoculated with *Rhizobium tropici* grown under P_i-deficient and -sufficient conditions. In this study, metabolic differences were observed between plants grown under these two contrasting conditions. Thirteen metabolites showed their role in those pathways that repressed or induced pathways in response to P_i deficiency. Nodules of P_i-deficient common bean plants showed a reduction in N-metabolism-related metabolites, which might contribute to a decrease in symbiotic nitrogen fixation (SNF) efficiency. In lentil, metabolite profiling of four different Mediterranean accessions performed by Muscolo et al. [234] showed that intermediates of the TCA cycle and glycolytic pathways decreased under drought and salinity conditions. Moreover, they recognized stress-specific metabolites such as threonate for NaCl and asparagine/ornithine and alanine/homoserine specifically to drought and salinity, respectively. In chickpea, Nasr Esfahani et al. [233] observed significant differences in C- and N-metabolism-related metabolites in the more-P_i-deficiency-susceptible *Mm*SWRI9-chickpea nodules and the less-P_i-deficiency-susceptible *Mc*CP-31-chickpea nodules under P_i deficiency. Moreover, they noted a remarkable increase in the level of organic acids in *Mc*CP-31-nodulated roots as compared with *Mm*SWRI9-nodulated roots under P_i deficiency. This study showed that a crosstalk among various signaling pathways involved in the regulation of *Mesorhizobium* chickpea exists for adaptation to P_i deficiency. Further in-depth knowledge at the genetic level can be useful for developing transgenic cultivars in leguminous crops to have adaptability under P_i deficiency by sustaining efficient SNF. In recent years, whole-genome sequences, genome-wide genetic variants, and cost-effective genotyping assays have emerged and provided an opportunity to utilize metabolomics information for the genetic enhancement of adaptive traits towards the ultimate aim of developing climate-resilient pulses [235].

5.2.2. Epigenomics for Improving Phenotypic Plasticity to Climate Change

Epigenetics denotes heritable changes in gene expression that can occur due to methylation of DNA or post-translational modification of histones involved in chromatin formation rather than changes in gene sequences [236]. Epigenetic variations can be reversible or transgenerational [237]. In reversible epigenetic variation, transcriptional memory may be responsive to cell fate decisions, developmental switches, or stress responses; otherwise, the gene expresses itself normally. Such epigenetic variations are not inherited by the next generation and, hence, are not useful for epigenetic breeding. Iwasaki [238] discussed the chromatin resetting mechanism related to the stability of epigenetic states under various stress conditions and revealed that silencing of reporter transgenes as well as endogenous loci occurs in response to various abiotic stresses such as high salinity, drought, heat, or UV radiation. However, such stress-induced transcriptional activation is mostly transient, and silencing is rapidly restored after resumption of optimal growth conditions. On the other hand, transgenerational epigenetic variation causes changes in gene expression that are stably transmitted to subsequent generations through mitosis or meiosis. Such epigenetic marks that create natural phenotypic variation play an important role in adaptations of plants in different environmental conditions [239–242]. Epialleles or epimutations, creating a mutant or an alternative phenotype, are generated through epigenetic changes [237,243]. According to Slotkin and Martienssen [244], these epialleles may result from changes in either genome or environmental conditions. Zhang and Hsieh [245] identified pure epialleles originating independently of any genetic variation in their model and other crop plant species. However, epialleles which are caused by genetic variations are difficult to detect without comprehensive genome structural analysis. Therefore, it is challenging to identify genomic loci that undergo epigenetic changes in response to environmental conditions [246]. According to Meyer [246], the use of such genomic loci in epigenetic breeding is a powerful strategy for developing climate-smart pulses under global warming conditions because such epialleles are able to improve the plant's ability to adapt to the inducing conditions in a heritable manner. Lele et al. [242] used amplified fragment length polymorphism (AFLP) and methylation-sensitive AFLP (MSAFLP) to identify the differences in genetic diversity caused by epigenetic or genetic variations and to study the role of epigenetic variation in the adaptation of *Vitex negundo* var. *heterophylla* (Chinese chaste tree) in different habitats. This study showed a relatively high level of genetic and epigenetic diversity but very low genetic and epigenetic differences between habitats within sites.

DNA methylation, a well-known form of epigenetic modification in plants, regulates genomic imprinting, expression of genes, and the process of disease development in plant species [247]. It influences transcription activity, morphological development, agronomic trait formation, the process of disease development, and environmental adaptation [247,248]. However, technological advancements have made it feasible to identify methylomes at a single-base resolution using BS-seq in soybean [248]. Methylome profiles studied among diverse accessions in key crop species including chickpea and soybean showed thousands of differentially methylated regions (DMRs) [248–250]. Genome-wide cytosine methylation analysis has been done on soybean for roots, stems, leaves, and cotyledons of developing seeds at single-base resolution. This study identified 2162 differentially methylated and hypomethylated regions, which provided significant insight into soybean gene expression [251]. Other studies have identified the role of DNA methylation in controlling cytoplasmic male sterility [252] and seed development in soybean [253]. Moreover, it has also played an important role in the polyploidy of soybean and common bean [254]. A whole-genome DNA methylation investigation performed by Shen et al. [248] identified 5412 DMRs which are useful in the domestication and improvement of soybean. The study also identified DMR-enriched genes belonging to carbohydrate metabolism [248]. In tetraploid cotton (a nonlegume crop), Song et al. [255] recognized 519 genes differing epigenetically between wild and cultivated species. Among these genes, a few methylated genes were responsible for traits such as flowering time and seed dormancy which helped in the domestication of cotton. This study also showed that DNA methylation changes the expression of the genes of wild species in response to environmental conditions or during the human selection. This study further revealed that the methylated gene helped cotton to adapt in natural tropical environments because DNA methylation of this gene does not encourage flowering under long-day conditions. Bhatia et al. [250] also identified DMR-associated genes involved in the development of the flower of chickpea. In addition to this, natural variation of epialleles has provided an opportunity for plant breeders to select and breed agronomically important traits [256,257]. As pulses have undergone domestication under varied agroclimatic conditions, a comprehensive investigation of methylated genes among accessions belonging to wild and cultivated species will help identify DMR-enriched genes that might affect their adaptedness to local climatic condition.

A breeding strategy was suggested by Raju et al. [258] to exploit epigenetic variations for increasing yield and stability in soybean. This strategy employed the MutS HOMOLOG1 (MSH1) system to induce epigenetic variation for agronomic traits. For epigenetic breeding, epi-lines were developed by crossing between wild type and *msh1*-acquired soybean memory lines, which showed a wide variation for multiple yield-related traits including pods per plant, seed weight, and maturity time in both greenhouse and field trials. Low extent of epitype-by-environment (e × E) interaction indicated higher yield stability. Furthermore, transcript profiling of the soybean epi-lines helped to identify genes involved in various metabolic pathways responsible for enhanced growth behavior across generations. This indicated the potentiality of MSH1-based epigenetic variation in plant breeding for enhanced yield and yield stability [258]. Thus, environmentally induced epigenetic variation can result in heritable phenotypic plasticity, which may play a major role in adaptation to environmental change [239,258,259]. Since pulses are grown in a wide range of environmental conditions and face many stresses throughout their lifecycle, breeding for epigenetic variations can be more useful for the ultimate aim of developing climate-smart pulse crops (Figure 2).

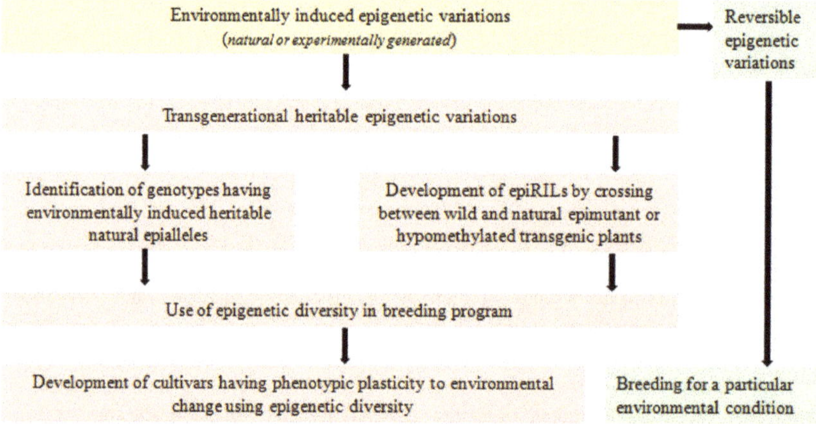

Figure 2. Epigenetic breeding for improving phenotypic plasticity to climate change in pulses.

5.2.3. Genome Editing Approaches for Adaptation

Genome editing has emerged as a new approach to bring about genetic changes at targeted regions of the genome and is being utilized as an alternative to classical plant breeding and the transgenic approach [260,261]. Genome editing includes insertion, removal, or replacement of a targeted gene. CRISPR/Cas9-based genome editing, used first in 2013 with *Arabidopsis* protoplasts and tobacco cells [262], is a highly advanced system and user-friendly tool for targeted gene manipulation in many plant species, including crop plants [263,264]. Genome editing could induce mutations in targeted genes with a frequency of 1.8–2.7% [265,266]. Among food crops, this approach was used for the first time in rice and wheat [267]. Initially, the rice *PDS* gene (*OsPDS*) was targeted with two sgRNAs (SP1 and SP2), which resulted in a 5% mutagenesis rate in protoplasts. Subsequently, three more rice genes (*OsBADH2*, *Os02g23823*, and *OsMPK2*) and one wheat gene (*TaMLO*) were targeted and mutated in protoplasts [267]. Recently, the CRISPR/Cas9 technology has been used successfully to edit five pyrabactin resistance 1-like (PYL) genes in rice. The mutants generated from the editing of *pyl1/4/6* exhibited the best growth and improved grain productivity up to 30% in natural paddy field conditions [266]. Such mutations (populations) can have better adaptive value in changing environmental conditions. The CRISPR/Cas9 system will likely be a promising alternative to conventional transgenic and breeding approaches that can deliver good results in this field. This system (CRISPR/Cas9) can also be useful for precise manipulation of genes governing adaptation of pulse crops to adverse environmental conditions.

Genetic transformation is now a routine activity in major pulses such as chickpea and pigeonpea, where transgenic plants have already been developed for insect resistance [209,210,212]. Therefore, gene families regulating the ABA pathway that play an important role under abiotic stress conditions identified earlier in legume crops may be targeted for gene editing in pulse crops to tackle the ill effects of the changing climate scenario [267–270]. The role of the MLO gene family has been identified in controlling powdery mildew resistance, and hence, this gene family may be used in pea and other pulses where powdery mildew is a serious problem [145]. Because the development of cultivars having resistance to pod borer in chickpea and pigeonpea is a major challenge due to the unavailability of resistance genetic resources in the gene pool, gene editing can target genes controlling susceptibility in the host plant of pulse crops as it is used to enhance resistance to viral diseases in plants [271]. Moreover, Wang et al. [272] applied genome editing to understand the basic mechanisms underpinning legume–rhizobia interactions. In several pulses, candidate genes imparting tolerance to abiotic and biotic stress as well as other agronomic traits have been identified [51,138,193,273–275]. The gene editing approach can be used to validate the function of these genes, as candidate genes controlling quantitative variations in nodulation have been validated using genome editing [276]. Identified mutant populations can also be useful as genetic resources for breeding improved cultivars and will help strengthen food security in the future.

6. Concluding Remarks

For food and nutritional security, it is essential to adopt mitigation and adaptation strategies for sustaining the production and productivity of pulses under changing climate conditions. However, pulse farmers, especially in South Asia and Africa, are poor in resources; hence, they have a limited capacity to adopt mitigation strategies. Consequently, we shall have to resolve the issues of climate change primarily through adaptation strategies. This calls for developing cultivars that can sustain food production in the future. During the past years, many adaptive traits have been targeted knowingly or unknowingly in plant breeding programs. However, breeding climate-resilient cultivars must address moving targets that differ across geographical locations [277,278]. This will help minimize the adverse impact of climate change on agriculture. In addition, we should lay more focus on the use of wild species and land races to enhance crop resilience through evolutionary breeding [279,280]. We should use modern science to bring back diversity in farmers' fields by developing an evolutionary population (EP) using a mixture of different genotypes of the same crop. As the genetic composition of an EP fluctuates year after year, genotypes having high adaptive value subsequently become predominant in stressful environments [280]. In common bean, such populations are currently grown [281], and farmers claim high yields under stressful conditions [282].

During the last three decades, considerable advances have been made in the genomics of pulse crops, and genome sequences of many pulse crops are now available in the public domain. This has resulted in the identification of genes/QTLs controlling various agromorphological traits. These advances have allowed breeders to incorporate multiple traits into an improved genetic background through genomics-assisted selection, thereby resulting in the development of stress-resilient pulse crops. For example, introgressed lines of soybean carrying the *Ncl* gene have the potential to regulate transport and accumulation of Na^+ and Cl^-. This has resulted in 3.6–5.5-fold greater yield advantages over conventional cultivars under salinity conditions. Such advances have made it possible to grow soybean in saline-affected areas [283]. Moreover, introgression of genes from tepary bean (*Phaseolus acutifolius* A. Gray) resulted in the development of elite common bean lines that are able to grow at 4 °C above the limit (18–19 °C) normally tolerated by this crop [284]. In chickpea, efforts to introgress the drought-tolerant QTL into the background of popular cultivars of Africa and Asia through marker-assisted selection have resulted in several chickpea introgression lines. In rainfed yield trials, these lines have shown at least a 10% yield advantage over the recurrent parent [285].

Genome editing technology based on CRISPR/Cas9 can be used to manipulate genes responsible for adaptation in adverse environmental conditions, and the resulting mutant populations can be screened under stressful conditions. Therefore, initiatives for developing climate-resilient varieties using genomics-based approaches merit special attention.

Environmentally induced epigenetic variation has been reported to play an important role in enhancing phenotypic plasticity to changing environments [286–291]. Though epigenetic variation has been studied and exploited in other crops, perhaps no reports are available on its use for the improvement of pulse crops. As pulses are grown across a wide range of environmental conditions, concerted efforts are required to study epigenetic variations in these crops. Such efforts may pave the way for climate-resilient smart pulses in the days ahead.

Funding: ICARDA: Rabat, Morocco provided the financial support for this work.

Acknowledgments: This work was undertaken as part of, and funded by the CGIAR Research Program on Grain Legumes and Dryland Cereals (GLDC) duly supported by CGIAR Fund Donors. Thanks are also due to the Head, Division of Crop Improvement, ICAR-Indian Institute of Pulses Research, Kanpur, for providing facilities and support.

Conflicts of Interest: The authors declare no conflict of interest. We further declare that the funders had no role in the design of the study; in the collection, analyses, or interpretation of data; in the writing of the manuscript, or in the decision to publish the results.

References

1. Ali, M.; Gupta, S. Carrying capacity of Indian agriculture: Pulse crops. *Curr. Sci.* **2012**, *25*, 874–881.
2. Choudhary, A.K.; Kumar, S.; Patil, B.S.; Bhat, J.S.; Sharma, M.; Kemal, S.; Ontagodi, T.P.; Datta, S.; Patil, P.; Chaturvedi, S.K.; et al. Narrowing yield gaps through genetic improvement for Fusarium wilt resistance in three pulse crops of the semi-arid tropics. *SABRAO J. Breed. Genet.* **2013**, *45*, 341–370.
3. Harlan, J.R. *Crops and Man*, 2nd ed.; American Society of Agronomy: Madison, WI, USA, 1992.
4. Allard, R.W. History of plant population genetics. *Annu. Rev. Genet.* **1999**, *33*, 1–27. [CrossRef] [PubMed]
5. Schierenbeck, K.A. Population-level genetic variation and climate change in a biodiversity hotspot. *Annal. Bot.* **2016**, *119*, 215–228. [CrossRef] [PubMed]
6. Bishop, J.; Potts, S.G.; Jones, H.E. Susceptibility of faba bean (*Vicia faba* L.) to heat stress during floral development and anthesis. *J. Agron. Crop Sci.* **2016**, *202*, 508–517. [CrossRef]
7. Van Oldenborgh, G.J.; Philip, S.; Kew, S.; van Weele, M.; Uhe, P.; Otto, F.; Singh, R.; Pai, I.; Cullen, H.; Achuta Rao, K. Extreme heat in India and anthropogenic climate change. *Nat. Hazards Earth Syst. Sci.* **2018**, *18*, 365. [CrossRef]
8. McKersie, B. Planning for food security in a changing climate. *J. Exp. Bot.* **2015**, *66*, 3435–3450. [CrossRef] [PubMed]
9. Pilbeam, D.J. Breeding crops for improved mineral nutrition under climate change conditions. *J. Exp. Bot.* **2015**, *66*, 3511–3521. [CrossRef]
10. Mar, S.; Nomura, H.; Takahashi, Y.; Ogata, K.; Yabe, M. Impact of erratic rainfall from climate change on pulse production efficiency in lower Myanmar. *Sustainability* **2018**, *10*, 402. [CrossRef]
11. Kumar, J.; Kant, R.; Kumar, S.; Basu, P.S.; Sarker, A.; Singh, N.P. Heat tolerance in lentil under field conditions. *Leg. Genom. Genet.* **2016**, *7*, 1–11. [CrossRef]
12. Kumar, J.; Gupta, S.; Gupta, P.; Dubey, S.; Tomar, R.S.; Kumar, S. Breeding strategies to improve lentil for diverse agro-ecological environments. *Indian J. Genet. Plant Breed.* **2016**, *76*, 530–549. [CrossRef]
13. Kumar, J.; Basu, P.S.; Gupta, S.; Dubey, S.; Gupta, D.S.; Singh, N.P. Physiological and molecular characterization for high temperature stress in *Lens culinaris*. *Funct. Plant Biol.* **2017**, *45*, 474–487. [CrossRef]
14. Nayyar, H.; Sehga, A.; Kumari, S.; Kumar, J.; Agrawal, S.K.; Singh, S.; Siddique, K.H. Effect of drought, heat and their interaction on the growth, yield and photosynthetic function of lentil (*Lens culinaris* Medikus) genotypes varying in heat and drought sensitivity. *Front. Plant Sci.* **2017**, *8*, 1776. [CrossRef]
15. Sita, K.; Sehgal, A.; Kumar, J.; Kumar, S.; Singh, S.; Siddique, K.H.; Nayyar, H. Identification of high-temperature tolerant lentil (*Lens culinaris* Medik) genotypes through leaf and pollen traits. *Front. Plant Sci.* **2017**, *8*, 744. [CrossRef] [PubMed]

16. Huang, Y.J.; Pirie, E.J.; Evans, N.; Delourme, R.; King, G.J.; Fitt, B.D. Quantitative resistance to symptomless growth of *Leptosphaeria maculans* (phoma stem canker) in *Brassica napus* (oilseed rape). *Plant Pathol.* **2009**, *58*, 314–323. [CrossRef]

17. Ramegowda, V.; Senthil, K.M. The interactive effects of simultaneous biotic and abiotic stresses on plants: Mechanistic understanding from drought and pathogen combination. *J. Plant Physiol.* **2015**, *176*, 47–54. [CrossRef] [PubMed]

18. FAO. *FAOSTAT*, Food and Agriculture Organization of the United Nations: Rome, Italy, 2016.

19. Kuwayama, Y.; Thompson, A.; Bernknopf, R.; Zaitchik, B.; Vail, P. Estimating the Impact of Drought on Agriculture Using the US Drought Monitor. *Am. J. Agric. Econ.* **2018**, *101*, 193–210. [CrossRef]

20. Ambachew, D.; Mekbib, F.; Asfaw, A.; Beebe, S.E.; Blair, M.W. Trait associations in common bean genotypes grown under drought stress and field infestation by BSM bean fly. *Crop J.* **2015**, *3*, 305–316. [CrossRef]

21. Rana, D.S.; Dass, A.; Rajanna, G.A.; Kaur, R. Biotic and abiotic stress management in pulses. *Indian J. Agron.* **2016**, *61*, 238–248.

22. Ramirez-Villegas, J.; Watson, J.; Challinor, A.J. Identifying traits for genotypic adaptation using crop models. *J. Exp. Bot.* **2015**, *66*, 3451–3462. [CrossRef]

23. Cutforth, H.W.; McGinn, S.M.; McPhee, K.E.; Miller, P.R. Adaptation of pulse crops to the changing climate of the Northern Great Plains. *Agron. J.* **2007**, *99*, 1684–1699. [CrossRef]

24. Shunmugam, A.; Kannan, U.; Jiang, Y.; Daba, K.; Gorim, L. Physiology based approaches for breeding of next-generation food legumes. *Plants* **2018**, *7*, 72. [CrossRef] [PubMed]

25. Batstone, R.T.; Dutton, E.M.; Wang, D.; Yang, M.; Frederickson, M.E. The evolution of symbiont preference traits in the model legume *Medicago truncatula*. *New Phytologist.* **2017**, *213*, 1850–1861. [CrossRef] [PubMed]

26. Cullis, C.; Kunert, K.J. Unlocking the potential of orphan legumes. *J. Exper. Bot.* **2017**, *68*, 1895–1903. [CrossRef]

27. Kole, C.; Muthamilarasan, M.; Henry, R.; Edwards, D.; Sharma, R.; Abberton, M.; Batley, J.; Bentley, A.; Blakeney, M.; Bryant, J.; et al. Application of genomics-assisted breeding for generation of climate resilient crops: Progress and prospects. *Front. Plant Sci.* **2015**, *6*, 563. [CrossRef] [PubMed]

28. Huang, S.; Gali, K.K.; Tar'an, B.; Warkentin, T.D.; Bueckert, R.A. Pea phenology: Crop potential in a warming environment. *Crop Sci.* **2017**, *57*, 1540–1551. [CrossRef]

29. Kumar, J.; Solanki, R.K. Evaluation of germplasm accessions for agro-morphological traits in lentil. *J. Food Leg.* **2014**, *27*, 275.

30. Kharzaei, H.; O'Sullivan, D.M.; Sillanpää, M.J.; Stoddard, F.L. Use of synteny to identify candidate genes underlying QTL controlling stomatal traits in faba bean (*Vicia faba* L.). *Theor. Appl. Genet.* **2014**, *127*, 2371–2385. [CrossRef]

31. Bargaz, A.; Zaman-Allah, M.; Farissi, M.; Lazali, M.; Drevon, J.J.; Maougal, R.T.; Georg, C. Physiological and molecular aspects of tolerance to environmental constraints in grain and forage legumes. *Int. J. Mol. Sci.* **2015**, *16*, 18976–19008. [CrossRef]

32. Pang, J.; Turner, N.C.; Khan, T.; Du, Y.L.; Xiong, J.L.; Colmer, T.D.; Devilla, R.; Stefanova, K.; Siddique, K.H.M. Response of chickpea (*Cicer arietinum* L.) to terminal drought: Leaf stomatal conductance, pod abscisic acid concentration, and seed set. *J. Exp. Bot.* **2017**, *68*, 1973–1985. [CrossRef]

33. Quan, W.; Liu, X.; Wang, H.; Chan, Z. Comparative physiological and transcriptional analyses of two contrasting drought tolerant alfalfa varieties. *Front. Plant Sci.* **2016**, *6*, 1256. [CrossRef]

34. Reynolds, M.; Langridge, P. Physiological breeding. *Curr. Opin. Plant Biol.* **2016**, *31*, 162–171. [CrossRef] [PubMed]

35. Chen, Y.; Ghanem, M.E.; Siddique, K.H.M. Characterising root trait variability in chickpea (Cicer arietinum L.) germplasm. *J. Exp. Bot.* **2017**, *68*, 1987–1999. [CrossRef] [PubMed]

36. Khan, H.A.; Siddique, K.H.M.; Colmer, T.D. Vegetative and reproductive growth of salt-stressed chickpea are carbon-limited: Sucrose infusion at the reproductive stage improves salt tolerance. *J. Exp. Bot.* **2017**, *68*, 2001–2011. [CrossRef] [PubMed]

37. Prince, S.J.; Murphy, M.; Mutava, R.N.; Durnell, L.A.; Valliyodan, B.; Shannon, J.G.; Nguyen, H.T. Root xylem plasticity to improve water use and yield in water-stressed soybean. *J. Exp. Bot.* **2017**, *68*, 2027–2036. [CrossRef] [PubMed]

38. Turner, N.C.; Wright, G.C.; Siddique, K.H.M. Adaptation of grain legumes (pulses) to water-limited environments. *Adv. Agron.* **2001**, *71*, 193–231. [CrossRef]

39. Kashiwagi, J.; Krishnamurthy, L.; Upadhyaya, H.D.; Krishna, H.; Chandra, S.; Vadez, V.; Serraj, R. Genetic variability of drought-avoidance root traits in the mini-core germplasm collection of chickpea (*Cicer arietinum* L.). *Euphytica* **2005**, *146*, 213–222. [CrossRef]

40. Ramalingam, A.; Kudapa, H.; Pazhamala, L.T.; Weckwerth, W.; Varshney, R.K. Proteomics and metabolomics: Two emerging areas for legume improvement. *Front. Plant Sci.* **2015**, *6*, 1116. [CrossRef]

41. Wheeler, T.R.; Craufurd, P.Q.; Ellis, R.H.; Porter, J.R.; Prasad, P.V. Temperature variability and the yield of annual crops. *Agric. Ecosyst. Environ.* **2000**, *82*, 159–167. [CrossRef]

42. Hatfield, J.L.; Boote, K.J.; Kimball, B.A.; Ziska, L.H.; Izaurralde, R.C.; Ort, D.; Thomson, A.M.; Wolfe, D. Climate impacts on agriculture: Implications for crop production. *Agron. J.* **2011**, *103*, 351–370. [CrossRef]

43. Bita, C.E.; Gerats, T. Plant tolerance to high temperature in a changing environment: Scientific fundamentals and production of heat stress-tolerant crops. *Front. Plant Sci.* **2013**, *4*, 273. [CrossRef] [PubMed]

44. Teixeira, E.I.; Fischer, G.; van Velthuizen, H.; Walter, C.; Ewert, F. Global hot-spots of heat stress on agricultural crops due to climate change. *Agric. For. Meteor.* **2013**, *170*, 206–215. [CrossRef]

45. Gaur, P.; Saminen, S.; Krishnamurthy, L.; Kumar, S.; Ghane, M.; Beebe, S.; Rao, I.; Chaturvedi, S.; Basu, P.; Nayyar, H.; et al. High temperature tolerance in grain legumes. *Leg. Perspect.* **2015**, *7*, 23–24.

46. Asseng, S.; Ewert, F.; Martre, P.; Rötter, R.P.; Lobell, D.B.; Cammarano, D.; Kimball, B.A.; Ottman, M.J.; Wall, G.W.; White, J.W.; et al. Rising temperatures reduce global wheat production. *Nat. Clim. Chang.* **2015**, *5*, 143–147. [CrossRef]

47. Choudhary, A.K.; Sultana, R.; Vales, M.I.; Saxena, K.B.; Kumar, R.R.; Ratnakumar, P. Integrated physiological and molecular approaches to improvement of abiotic stress tolerance in two pulse crops of the semi-arid tropics. *Crop J.* **2018**, *6*, 99–114. [CrossRef]

48. Mohanty, P.; Allakhverdiev, S.; Murata, N. Application of low temperatures during photoinhibition allows characterization of individual steps in photodamage and the repair of photosystem II. *Photogr. Res.* **2007**, *94*, 217–224. [CrossRef] [PubMed]

49. Murata, N.; Takahashi, S.; Nishiyama, Y.; Allakhverdiev, S.I. Photoinhibition of photosystem II under environmental stress. *Biochim. Biophys. Acta* **2007**, *1767*, 414–421. [CrossRef]

50. Berger, J.D.; Ali, M.; Basu, P.S.; Chaudhary, B.D.; Chaturvedi, S.K.; Deshmukh, P.S.; Dharmaraj, P.S.; Dwivedi, S.K.; Gangadhar, G.C.; Gaur, P.M.; et al. Genotype by environment studies demonstrate the critical role of phenology in adaptation of chickpea (*Cicer arietinum* L.) to high and low yielding environments of India. *Field Crops Res.* **2006**, *98*, 230–244. [CrossRef]

51. Kumar, J.; Gupta, S.; Biradar, R.S.; Gupta, P.; Dubey, S.; Singh, N.P. Association of functional markers with flowering time in lentil. *J. Appl. Genet.* **2018**, *59*, 9–21. [CrossRef]

52. Nayyar, H.; Chander, K.; Kumar, S.; Bains, T. Glycine betaine mitigates cold stress damage in Chickpea. *Agron. Sustain. Dev.* **2005**, *25*, 381–388. [CrossRef]

53. Cobos, M.J.; Winter, P.; Kharrat, M.; Cubero, J.I.; Gil, J.; Milian, T.; Rubio, J. Genetic analysis of agronomic traits in a wide cross of chickpea. *Field Crops Res.* **2009**, *111*, 130–136. [CrossRef]

54. Ahmad, F.; Gaur, P.; Croser, J. Chickpea (*Cicer arietinum* L.). In *Genetic Resources, Chromosome Engineering and Crop Improvement—Grain Legumes*; Singh, R., Jauhar, P., Eds.; CRC Press: Boca Raton, FL, USA, 2005; pp. 185–214.

55. Knights, E.J.; Southwell, R.J.; Schwinghamer, M.W.; Harden, S. Resistance to *Phytophthora medicaginis* Hansen and Maxwell in wild *Cicer* species and its use in breeding root rot resistant chickpea (*Cicer arietinum* L.). *Aust. J. Agric. Res.* **2008**, *59*, 383–387. [CrossRef]

56. Singh, R.; Sharma, P.; Varshney, R.K.; Sharma, S.K.; Singh, N.K. Chickpea improvement: Role of wild species and genetic markers. *Biotechnol. Genet. Eng. Rev.* **2008**, *25*, 267–314. [CrossRef] [PubMed]

57. Pande, S.; Galloway, J.J.; Gaur, P.M.; Siddique, K.H.M.; Tripathi, H.S.; Taylor, P.; MacLeod, M.W.J.; Basandrai, A.K.; Baker, A.; Joshi, S.; et al. Botrytis gray mold of chickpea: A review of biology, epidemiology, and disease management. *Aust. J. Agric. Res.* **2006**, *57*, 1137–1150. [CrossRef]

58. Whish, J.P.; Castor, P.; Carberry, P.S.; Peake, A.S. On-farm assessment of constraints to chickpea (*Cicer arietinum*) production in marginal areas of northern. *Aust. Exp. Agric.* **2007**, *43*, 505–520. [CrossRef]

59. Taran, B.; Warkentin, T.D.; Vandenberg, A.; Holm, F.A. Variation in chickpea germplasm for tolerance to imazethapyr and imazamox herbicides. *Can. J. Plant Sci.* **2010**, *90*, 139–142. [CrossRef]

60. Gaur, P.M.; Kumar, J.; Gowda, C.L.; Pande, S.; Siddique, K.H.; Khan, T.N.; Warkentin, T.D.; Chaturvedi, S.K.; Than, A.M.; Ketema, D. Breeding chickpea for early phenology: Perspectives, progress and prospects. In Proceedings of the Fourth International Food Legumes Research Conference, New Delhi, India, 18–22 October 2005.

61. Sarker, A.; Erskine, W. Lentil production in the traditional lentil world. In *Proceedings of Lentil Focus*; Brouwer, J.B., Ed.; Horham: Victoria, Australia, 2002; pp. 35–40.

62. Rubiales, D.; Fondevilla, S. Future prospects for ascochyta blight resistance breeding in cool season food legumes. *Front. Plant Sci.* **2012**, *3*, 27. [CrossRef] [PubMed]

63. Kumar, D.; Dixit, G.P. Genetic improvement of minor pulse crops-Retrospect and Prospects. In *Pulses in New Perspective*; Ali, M., Singh, B.B., Kumar, S., Dhar, V., Eds.; Indian Society of Pulses Research and Development: Kanpur, India, 2003; pp. 112–131.

64. Gupta, S.; Kumar, S. Urd bean breeding. In *Advances in Mung Bean and Urdbean*; Ali, M., Kumar, S., Eds.; Indian Institute of Pulses Research: Kanpur, India, 2006; pp. 149–168.

65. Reddy, K.S.; Dhanasekar, P.; Dhole, V.J. A review on powdery mildew disease resistance in mungbean. *J. Food Leg.* **2008**, *21*, 151–155.

66. Tickoo, J.L.; Lal, S.K.; Chandra, N.; Dikshit, H.K. Mung bean breeding. In *Advances in Mung Bean and Urd Bean*; Ali, M., Kumar, S., Eds.; Indian Institute of Pulses Research: Kanpur, India, 2006; pp. 110–148.

67. Sinha, R.P. Early maturity, dwarf mutant of urd bean [*V. mungo* (L.) Hepper]. *J. Nucl. Agric. Biol.* **1988**, *17*, 61–62.

68. Chadha, M.L.; Bains, T.S.; Sekhon, H.S.; Sain, S.K. Short duration mung bean for diversification of rice wheat systems. In *Milestones in Food Legumes Research*; Ali, M., Kumar, S., Eds.; Indian Institute of Pulses Research: Kanpur, India, 2009; pp. 151–177.

69. Upadhyaya, H.D.; Kashiwagi, J.; Varshney, R.K.; Gaur, P.M.; Saxena, K.B.; Krishnamurthy, L.; Gowda, C.L.; Pundir, R.P.; Chaturvedi, S.K.; Basu, P.S.; et al. Phenotyping chickpeas and pigeonpeas for adaptation to drought. *Front. Physiol.* **2012**, *3*, 179. [CrossRef] [PubMed]

70. Kassa, M.T.; Penmetsa, R.V.; Carrasquilla-Garcia, N.; Sarma, B.K.; Datta, S.; Upadhyaya, H.D.; Varshney, R.K.; von Wettberg, E.J.; Cook, D.R. Genetic patterns of domestication in pigeonpea (*Cajanus cajan* (L.) Millsp.) and wild *Cajanus* relatives. *PLoS ONE* **2012**, *7*, e39563. [CrossRef] [PubMed]

71. Gusmao, M. *Grass Pea (Lathyrus sativus cv. Ceora): Adaptation to Water Deficit and Benefit in Crop Rotation*; University of Western Australia: Perth, Australia, 2010.

72. Omae, H.; Kumar, A.; Shono, M. Adaptation to high temperature and water deficit in the common bean (*Phaseolus vulgaris* L.) during the reproductive period. *J. Bot.* **2012**, *2012*, 803413. [CrossRef]

73. Darkwa, K.; Ambachew, D.; Mohammed, H.; Asfaw, A.; Blair, M.W. Evaluation of common bean (*Phaseolus vulgaris* L.) genotypes for drought stress adaptation in Ethiopia. *Crop J.* **2016**, *4*, 367–376. [CrossRef]

74. Klaedtke, S.M.; Caproni, L.; Klauck, J.; De La Grandville, P.; Dutartre, M.; Stassart, P.M.; Chable, V.; Negri, V.; Raggi, L. Short-term local adaptation of historical common bean (*Phaseolus vulgaris* L.) varieties and implications for in situ management of bean diversity. *Int. J. Mol. Sci.* **2017**, *18*, 493. [CrossRef]

75. Isemura, T.; Kaga, A.; Tomooka, N.; Shimizu, T.; Vaughan, D.A. The genetics of domestication of rice bean, *Vigna umbellata*. *Ann. Bot.* **2010**, *106*, 927–944. [CrossRef] [PubMed]

76. Rao, I.; Beebe, S.; Polania, J.; Ricaurte, J.; Cajiao, C.; Garcia, R.; Rivera, M. Can tepary bean be a model for improvement of drought resistance in common bean? *Afr. Crop Sci. J.* **2013**, *21*, 265–281.

77. Hamama, A.A.; Bhardwaj, H.L. Tepary bean: A short duration summer crop in Virginia. In *Trends in New Crops and New Uses*; ASHS Press: Alexandria, VA, USA, 2002; pp. 429–431.

78. Carmo, M.D.; Gomes, R.L.; Lopes, Â.C.; Penha, J.S.; Gomes, S.O.; Assunção Filho, J.R. Genetic variability in subsamples of determinate growth lima bean. *Crop Breed. Appl. Biotech.* **2013**, *13*, 158–164. [CrossRef]

79. Ballhorn, D.J.; Kautz, S.; Heil, M.; Hegeman, A.D. Cyanogenesis of wild lima bean (*Phaseolus lunatus* L.) is an efficient direct defence in nature. *PLoS ONE* **2009**, *4*, e5450. [CrossRef] [PubMed]

80. Santalla, M.; Monteagudo, A.B.; González, A.M.; De Ron, A.M. Agronomical and quality traits of runner bean germplasm and implications for breeding. *Euphytica* **2004**, *135*, 205–215. [CrossRef]

81. Schwember, A.R.; Carrasco, B.; Gepts, P. Unraveling agronomic and genetic aspects of runner bean (*Phaseolus coccineus* L.). *Field Crops Res.* **2017**, *206*, 86–94. [CrossRef]

82. Isemura, T.; Kaga, A.; Konishi, S.; Ando, T.; Tomooka, N.; Han, O.K.; Vaughan, D.A. Genome dissection of traits related to domestication in azuki bean (*Vigna angularis*) and comparison with other warm-season legumes. *Ann. Bot.* **2007**, *100*, 1053–1071. [CrossRef] [PubMed]

83. Yao, L.M.; Wang, B.; Cheng, L.J.; Wu, T.L. Identification of key drought stress-related genes in the hyacinth bean. *PLoS ONE* **2013**, *8*, e58108. [CrossRef] [PubMed]

84. D'souza, M.R.; Devaraj, V.R. Role of calcium in increasing tolerance of hyacinth bean to salinity. *J. Appl. Biol. Biotech.* **2013**, *1*, 11–20. [CrossRef]

85. Gomashe, S.S.; Dikshit, N.; Chand, D.; Shingane, S.N. Assessment of genetic diversity using morpho-agronomical traits in horse gram. *Int. J. Curr. Microbiol. Appl. Sci.* **2018**, *7*, 2095–2103. [CrossRef]

86. Dikshit, N.; Katna, G.; Mohanty, C.S.; Das, A.B.; Sivaraj, N. Horse gram. In *Broadening the Genetic Base of Grain Legumes*; Springer: New Delhi, India, 2014; pp. 209–215.

87. Klu, G.Y. *Efforts to Accelerate Domestication of Winged Bean (Psophocarpus tetragonolobus (L.) DC.) by Means of Induced Mutations and Tissue Culture*; Wageningen University: Wageningen, The Netherlands, 1996.

88. Vatanparast, M.; Shetty, P.; Chopra, R.; Doyle, J.J.; Sathyanarayana, N.; Egan, A.N. Transcriptome sequencing and marker development in winged bean (*Psophocarpus tetragonolobus*; Leguminosae). *Sci. Rep.* **2016**, *6*, 29070. [CrossRef]

89. Hall, A. Phenotyping cowpeas for adaptation to drought. *Front. Physiol.* **2012**, *3*, 155. [CrossRef] [PubMed]

90. Kadam, N.N.; Xiao, G.; Melgar, R.J.; Bahuguna, R.N.; Quinones, C.; Tamilselvan, A.; Prasad, P.V.; Jagadish, K.S. Agronomic and physiological responses to high temperature, drought, and elevated CO_2 interactions in cereals. *Adv. Agron.* **2014**, *127*, 111–156. [CrossRef]

91. Gupta, N.; Shrivastava, N.; Singh, P.K.; Bhagyawant, S.S. Phytochemical evaluation of moth bean (*Vigna aconitifolia* L.) seeds and their divergence. *Bioch. Res. Int.* **2016**. [CrossRef] [PubMed]

92. Tiwari, B.; Kalim, S.; Bangar, P.; Kumari, R.; Kumar, S.; Gaikwad, A.; Bhat, K.V. Physiological, biochemical, and molecular responses of thermotolerance in moth bean (*Vigna aconitifolia* (Jacq.) Marechal). *Turk. J. Agric. For.* **2018**, *42*, 176–184. [CrossRef]

93. Tanksley, S.D.; Mccouch, S.R. Seed banks and molecular maps: Unlocking genetic potential from the wild. *Science* **1997**, *277*, 1063–1066. [CrossRef] [PubMed]

94. Pimentel, D.; Wilson, C.; McCullum, C.; Huang, R.; Dwen, P.; Flack, J.; Tran, Q.; Saltman, T.; Cliff, B. Economic and environmental benefits of biodiversity. *BioScience* **1997**, *47*, 747–757. [CrossRef]

95. Haussmann, B.I.G.; Parzies, H.K.; Presterl, T.; Susic, Z.; Miedaner, T. Plant genetic resources in crop improvement. *Plant Genet. Resour.* **2004**, *2*, 3–21. [CrossRef]

96. Maxted, N.; Kell, S.P. *Establishment of a Global Network for the in Situ Conservation of Crop Wild Relatives: Status and Needs*; FAO Commission on Genetic Resources for Food & Agriculture: Roman, Italy, 2009.

97. Tester, M.; Angridge, P.L. Breeding technologies to increase crop production in a changing world. *Science* **2010**, *327*, 818–822. [CrossRef] [PubMed]

98. Ford-Lloyd, B.V.; Schmidt, M.; Armstrong, S.J.; Barazani, O.; Engels, J.; Hadas, R.; Hammer, K.; Kell, S.P.; Kang, D.; Khoshbakht, K.; et al. Crop wild relatives—Undervalued, underutilized and under threat? *Biol Sci.* **2011**, *61*, 559–565. [CrossRef]

99. McCouch, S.; Baute, G.J.; Bradeen, J.; Bramel, P.; Bretting, P.K.; Buckler, E.; Burke, J.M.; Charest, D.; Cloutier, S.; Cole, G.; et al. Agriculture: Feeding the future. *Nature* **2013**, *499*, 23–24. [CrossRef]

100. Rana, J.C.; Gautam, N.K.; Gayacharan, M.S.; Yadav, R.; Tripathi, K.; Yadav, S.K.; Panwar, N.S.; Bhardwaj, R. Genetic resources of pulse crops in India: An overview. *Indian J. Genet. Plant Breed.* **2016**, *76*, 420–436. [CrossRef]

101. Sharma, S.; Upadhyaya, H.D.; Varshney, R.K.; Gowda, C.L. Pre-breeding for diversification of primary gene pool and genetic enhancement of grain legumes. *Front. Plant Sci.* **2013**, *4*, 309. [CrossRef]

102. Hajjar, R.; Hodgkin, T. The use of wild relatives in crop improvement: A survey of developments over the last 20 years. *Euphytica* **2007**, *156*, 1–3. [CrossRef]

103. Brumlop, S.; Reichenbecher, W.; Tappeser, B.; Finckh, M.R. What is the SMARTest way to breed plants and increase agrobiodiversity? *Euphytica* **2013**, *194*, 53–66. [CrossRef]

104. Gorim, L.Y.; Vandenberg, A. Evaluation of wild lentil species as genetic resources to improve drought tolerance in cultivated lentil. *Front. Plant Sci.* **2017**, *8*, 1129. [CrossRef]

105. Vavilov, N.I. Studies on the origin of cultivated plants (Russian). *Bull. Appl. Bot. Plant-Breed.* **1926**, *14*, 1–245.

106. Stalker, H.T. Utilization of wild species for crop improvement. *Adv. Agron.* **1980**, *33*, 111–147. [CrossRef]

107. Ladizinsky, G.; Pickersgill, B.; Yamamato, K. *World Crops: Cool Season Food Legumes*; Summerfield, R.J., Ed.; Kluwer Academic Publishers: Dordrecht, The Netherlands, 1988; pp. 967–978.

108. Singh, A.K.; Rana, R.S.; Mal, B.; Singh, B.; Agrawal, R.C. *Cultivated Plants and Their Wild Relatives in India—An Inventory*; Protection of Plant Varieties and Farmers' Rights Authority: New Delhi, India, 2013.

109. Kumar, J.; Srivastava, E.; Singh, M.; Kumar, S.; Nadarajan, N.; Sarker, A. Diversification of indigenous gene-pool by using exotic germplasm in lentil (*Lens culinaris* Medikus subsp. *culinaris*). *Physiol. Mol. Biol. Plants* **2014**, *20*, 125–132. [CrossRef] [PubMed]

110. Fiala, J.V.; Tullu, A.; Banniza, S.; Séguin-Swartz, G.; Vandenberg, A. Interspecies transfer of resistance to anthracnose in lentil (*Lens culinaris* Medic.). *Crop Sci.* **2009**, *49*, 825–830. [CrossRef]

111. Tullu, A.; Diederichsen, A.; Suvorova, G.; Vandenberg, A. Genetic and genomic resources of lentil: Status, use and prospects. *Plant Genet. Resour.* **2011**, *9*, 19–21. [CrossRef]

112. Suvorova, G. Hybridization of cultivated lentil *Lens culinaris* Medik. and wild species *Lens tomentosus* Ladizinsky. *Czech J. Genet. Plant Breed.* **2014**, *50*, 130–134. [CrossRef]

113. Chaturvedi, S.K.; Nadarajan, N. Genetic enhancement for grain yield in chickpea—Accomplishments and resetting research agenda. *Electron. J. Plant Breed.* **2010**, *1*, 611–615.

114. Pande, S.; Desai, S.; Sharma, M. Impacts of climate change on rainfed crop diseases: Current status and future research needs. In *National Symposium on Climate Change and Rainfed Agriculture*; CRIDA: Hyderabad, India, 2010; pp. 55–59.

115. Ali, M.; Kumar, S. Major Technological Advances in Pulses: Indian Scenario. In *Milestones in Food Legumes Research*; Ali, M., Kumar, S., Eds.; Indian Society of Pulses Research and Development: Kanpur, India, 2009; pp. 1–20.

116. Saxena, K.B.; Choudhary, A.K.; Saxena, R.K.; Varshney, R.K. Breeding pigeonpea cultivars for intercropping: Synthesis and strategies. *Breed. Sci.* **2018**, *68*, 159–167. [CrossRef]

117. Kashiwagi, J.; Krishnamurthy, L.; Purushothaman, R.; Upadhyaya, H.D.; Gaur, P.M.; Gowda, C.L.; Ito, O.; Varshney, R.K. Scope for improvement of yield under drought through the root traits in chickpea (*Cicer arietinum* L.). *Field Crops Res.* **2015**, *170*, 47–54. [CrossRef]

118. Kumar, J.; Sethi, S.C.; Johansen, C.; Kelly Rahman, M.M.; van Rheenen, H.A. Potential of short-duration chickpea varieties. *Indian J. Dryland Agric. Res. Dev.* **1996**, *11*, 28–32.

119. Kumar, J.; van Rheenen, H.A. A major gene for time of flowering in chickpea. *J. Hered.* **2000**, *91*, 67–68. [CrossRef] [PubMed]

120. Vales, M.I.; Srivastava, R.K.; Sultana, R.; Singh, S.; Singh, I.; Singh, G.; Patil, S.B.; Saxena, K.B. Breeding for earliness in pigeonpea: Development of new determinate and nondeterminate lines. *Crop Sci.* **2012**, *52*, 2507–2516. [CrossRef]

121. Choudhary, A.K.; Raje, R.S.; Datta, S.; Sultana, R.; Ontagodi, T. Conventional and molecular approaches towards genetic improvement in pigeonpea for insects resistance. *Am. J. Plant Sci.* **2013**, *4*, 372–385. [CrossRef]

122. Jiménez-Fernández, D.; Landa, B.B.; Kang, S.; Jiménez-Díaz, R.M.; Navas-Cortés, J.A. Quantitative and microscopic assessment of compatible and incompatible interactions between chickpea cultivars and *Fusarium oxysporum* f. sp. *ciceris* races. *PLoS ONE* **2013**, *16*, e61360. [CrossRef] [PubMed]

123. Nene, Y.L.; Reddy, M.V. Chickpea diseases and their control. In *The Chickpea*; Saxena, M.C., Singh, K.B., Eds.; CABI, Oxon: Watingford, UK, 1987; pp. 233–270.

124. Xavier, A.; Hall, B.; Hearst, A.A.; Cherkauer, K.A.; Rainey, K.M. Genetic architecture of phenomic-enabled canopy coverage in *Glycine max*. *Genetics* **2017**, *206*, 1081–1089. [CrossRef] [PubMed]

125. Parry, M.A.; Hawkesford, M.J. An integrated approach to crop genetic improvement. *J. Integr. Plant Biol.* **2012**, *54*, 250–259. [CrossRef]

126. Bohra, A.; Pandey, M.K.; Jha, U.C.; Singh, B.; Singh, I.P.; Datta, D.; Chaturvedi, S.K.; Nadarajan, N.; Varshney, R.K. Genomics-assisted breeding in four major pulse crops of developing countries: Present status and prospects. *Theor. Appl. Genet.* **2014**, *127*, 1263–1291. [CrossRef]

127. Bhat, J.A.; Shivaraj, S.M.; Ali, S.; Mir, Z.A.; Islam, A.; Deshmukh, R. Genomic resources and omics-assisted breeding approaches for pulse crop improvement. In *Pulse Improvement*; Springer: Cham, Switzerland, 2018; pp. 13–55.

128. Abdelrahman, M.; Jogaiah, S.; Burritt, D.J.; Tran, L.S.P. Legume genetic resources and transcriptome dynamics under abiotic stress conditions. *Plant Cell Environ.* **2018**, *41*, 1972–1983. [CrossRef]

129. Doddamani, D.; Khan, A.W.; Katta, M.A.; Agarwal, G.; Thudi, M.; Ruperao, P.; Edwards, D.; Varshney, R.K. CicArVarDB: SNP and InDel database for advancing genetics research and breeding applications in chickpea. *Database* **2015**, *2015*. [CrossRef]

130. O'Rourke, J.A.; Iniguez, L.P.; Fu, F.; Bucciarelli, B.; Miller, S.S.; Jackson, S.A.; McClean, P.E.; Li, J.; Dai, X.; Zhao, P.X.; et al. An RNA-Seq based gene expression atlas of the common bean. *BMC Genomics* **2014**, *15*, 866. [CrossRef]

131. Pazhamala, L.T.; Purohit, S.; Saxena, R.K.; Garg, V.; Krishnamurthy, L.; Verdier, J.; Varshney, R.K. Gene expression atlas of pigeonpea and its application to gain insights into genes associated with pollen fertility implicated in seed formation. *J. Exp. Bot.* **2017**, *68*, 2037–2054. [CrossRef] [PubMed]

132. Kudapa, H.; Garg, V.; Chitikineni, A.; Varshney, R.K. The RNA-Seq-based high resolution gene expression atlas of chickpea (*Cicer arietinum* L.) reveals dynamic spatio-temporal changes associated with growth and development. *Plant Cell Environ.* **2018**, *41*, 2209–2225. [CrossRef] [PubMed]

133. Soren, K.R.; Patil, P.G.; Das, A.; Bohra, A.; Datta, S.; Chaturvedi, S.K.; Nadarajan, N. *Advances in Pulses Genomic Research*; Indian Institute of Pulses Research: Kanpur, India, 2012.

134. Jain, M.; Misra, G.; Patel, R.K.; Priya, P.; Jhanwar, S.; Khan, A.W.; Shah, N.; Singh, V.K.; Garg, R.; Jeena, G.; et al. A draft genome sequence of the pulse crop chickpea (*Cicer arietinum* L.). *Plant J.* **2013**, *74*, 715–729. [CrossRef] [PubMed]

135. Kaila, T.; Chaduvla, P.K.; Saxena, S.; Bahadur, K.; Gahukar, S.J.; Chaudhury, A.; Sharma, T.R.; Singh, N.K.; Gaikwad, K. Chloroplast genome sequence of Pigeonpea (*Cajanus cajan* (L.) Millspaugh) and *Cajanus scarabaeoides* (L.) Thouars: Genome organization and comparison with other legumes. *Front Plant Sci.* **2016**, *7*, 1847. [CrossRef] [PubMed]

136. Varshney, R.K.; Pandey, M.K.; Bohra, A.; Singh, V.K.; Thudi, M.; Saxena, R.K. Toward the sequence-based breeding in legumes in the post-genome sequencing era. *Theor. Appl. Genet.* **2019**, *132*, 797–816. [CrossRef] [PubMed]

137. Kumawat, G.; Raje, R.S.; Bhutani, S.; Pal, J.K.; Mithra, A.S.; Gaikwad, K.; Sharma, T.R.; Singh, N.K. Molecular mapping of QTLs for plant type and earliness traits in pigeonpea (*Cajanus cajan* L. Millsp.). *BMC Genetics* **2012**, *13*, 84. [CrossRef] [PubMed]

138. Thudi, M.; Upadhyaya, H.D.; Rathore, A.; Gaur, P.M.; Krishnamurthy, L.; Roorkiwal, M.; Nayak, S.N.; Chaturvedi, S.K.; Basu, P.S.; Gangarao, N.V.; et al. Genetic dissection of drought and heat tolerance in chickpea through genome-wide and candidate gene-based association mapping approaches. *PLoS ONE* **2014**, *9*, e96758. [CrossRef]

139. Kaga, A.; Isemura, T.; Tomooka, N.; Vaughan, D.A. The domestication of the azuki bean (*Vigna angularis*). *Genetics* **2008**, *178*, 1013–1036. [CrossRef]

140. Javid, M.; Rosewarne, G.M.; Sudheesh, S.; Kant, P.; Leonforte, A.; Lombardi, M.; Kennedy, P.R.; Cogan, N.O.; Slater, A.T.; Kaur, S. Validation of molecular markers associated with boron tolerance, powdery mildew resistance and salinity tolerance in field peas. *Front. Plant Sci.* **2015**, *6*, 917. [CrossRef]

141. Paul, P.; Samineni, S.; Thudi, M.; Sajja, S.; Rathore, A.; Das, R.; Khan, A.; Chaturvedi, S.; Lavanya, G.; Varshney, R.; et al. Molecular mapping of QTLs for heat tolerance in chickpea. *Int. J. Mol. Sci.* **2018**, *19*, 2166. [CrossRef]

142. Diaz, L.M.; Ricaurte, J.; Tovar, E.; Cajiao, C.; Terán, H.; Grajales, M.; Polanía, J.; Rao, I.; Beebe, S.; Raatz, B. QTL analyses for tolerance to abiotic stresses in a common bean (*Phaseolus vulgaris* L.) population. *PLoS ONE* **2018**, *13*, e0202342. [CrossRef] [PubMed]

143. Choudhary, N.; Bawa, V.; Paliwal, R.; Singh, B.; Bhat, M.A.; Mir, J.I.; Gupta, M.; Sofi, P.A.; Thudi, M.; Varshney, R.K.; et al. Gene/QTL discovery for *Anthracnose* in common bean (*Phaseolus vulgaris* L.) from North-western Himalayas. *PLoS ONE* **2018**, *13*, e0191700. [CrossRef] [PubMed]

144. Bordat, A.; Savois, V.; Nicolas, M.; Salse, J.; Chauveau, A.; Bourgeois, M.; Potier, J.; Houtin, H.; Rond, C.; Murat, F.; et al. Translational genomics in legumes allowed placing in silico 5460 unigenes on the pea functional map and identified candidate genes in *Pisum sativum* L. *G3* **2011**, *2*, 93–103. [CrossRef] [PubMed]

145. Mohapatra, C.; Chand, R.; Singh, V.K.; Singh, A.K.; Kushwaha, C. Identification and characterisation of Mlo genes in pea (*Pisum sativum* L.) vis-à-vis validation of *Mlo* gene-specific markers. *Turk. J. Biol.* **2016**, *40*, 184–195. [CrossRef]

146. Sun, S.; He, Y.; Dai, C.; Duan, C.; Zhu, Z. Two major er1 alleles confer powdery mildew resistance in three pea cultivars bred in Yunnan Province, China. *Crop J.* **2016**, *4*, 353–359. [CrossRef]

147. Jovanovic, Z.; Stanisavljevic, N.; Mikic, A.; Radovic, S.; Maksimovic, V. The expression of drought responsive element binding protein ('DREB2A') related gene from pea ('*Pisum sativum*' L.) as affected by water stress. *Aust. J. Crop Sci.* **2013**, *7*, 1590–1596.

148. Timko, M.P.; Rushton, P.J.; Laudeman, T.W.; Bokowiec, M.T.; Chipumuro, E.; Cheung, F.; Town, C.D.; Chen, X. Sequencing and analysis of the gene-rich space of cowpea. *BMC Genomics* **2008**, *9*, 103. [CrossRef]

149. Muchero, W.; Ehlers, J.D.; Close, T.J.; Roberts, P.A. Mapping QTL for drought stress-induced premature senescence and maturity in cowpea [*Vigna unguiculata* (L.) Walp.]. *Theor. Appl. Genet.* **2009**, *118*, 849–863. [CrossRef]

150. Muchero, W.; Ehlers, J.D.; Roberts, P.A. QTL analysis for resistance to foliar damage caused by *Thrips tabaci* and *Frankliniella schultzei* (Thysanoptera: Thripidae) feeding in cowpea [*Vigna unguiculata* (L.) Walp.]. *Mol. Breed.* **2010**, *25*, 47–56. [CrossRef]

151. Pottorff, M.; Wanamaker, S.; Ma, Y.Q.; Ehlers, J.D.; Roberts, P.A.; Close, T.J. Genetic and physical mapping of candidate genes for resistance to *Fusarium oxysporum* f.sp. *tracheiphilum* race 3 in cowpea [*Vigna unguiculata* (L.) Walp]. *PLoS ONE* **2012**, *7*, e41600. [CrossRef]

152. Muchero, W.; Roberts, P.A.; Diop, N.N.; Drabo, I.; Cisse, N.; Close, T.J.; Muranaka, S.; Boukar, O.; Ehlers, J.D. Genetic architecture of delayed senescence, biomass, and grain yield under drought stress in cowpea. *PLoS ONE* **2013**, *8*, e70041. [CrossRef] [PubMed]

153. Pottorff, M.; Roberts, P.A.; Close, T.J.; Lonardi, S.; Wanamaker, S.; Ehlers, J.D. Identification of candidate genes and molecular markers for heat-induced brown discoloration of seed coats in cowpea [*Vigna unguiculata* (L.) Walp]. *BMC Genomics* **2014**, *15*, 328. [CrossRef] [PubMed]

154. Huynh, B.L.; Ehlers, J.D.; Ndeve, A.; Wanamaker, S.; Lucas, M.R.; Close, T.J.; Roberts, P.A. Genetic mapping and legume synteny of aphid resistance in African cowpea (*Vigna unguiculata* L. Walp.) grown in California. *Mol. Breed.* **2015**, *35*, 36. [CrossRef] [PubMed]

155. Huynh, B.L.; Matthews, W.C.; Ehlers, J.D.; Lucas, M.R.; Santos, J.R.; Ndeve, A.; Close, T.J.; Roberts, P.A. A major QTL corresponding to the Rk locus for resistance to root-knot nematodes in cowpea (*Vigna unguiculata* L. Walp.). *Theor. Appl. Genet.* **2016**, *129*, 87–95. [CrossRef] [PubMed]

156. Santos, J.R.; Ndeve, A.D.; Huynh, B.L.; Matthews, W.C.; Roberts, P.A. QTL mapping and transcriptome analysis of cowpea reveals candidate genes for root-knot nematode resistance. *PLoS ONE* **2018**, *13*, e0189185. [CrossRef] [PubMed]

157. Maibam, A.; Tyagi, A.; Satheesh, V.; Mahato, A.K.; Jain, N.; Raje, R.S.; Rao, A.R.; Gaikwad, K.; Singh, N.K. Genome-wide identification and characterization of heat shock factor genes from pigeonpea (*Cajanus cajan*). *Mol. Plant Breed.* **2015**, *6*, 1–11. [CrossRef]

158. Deeplanaik, N.; Kumaran, R.C.; Venkatarangaiah, K.; Shivashankar, S.K.; Doddamani, D.; Telkar, S. Expression of drought responsive genes in pigeonpea and in silico comparison with soybean cDNA library. *J. Crop Sci. Biotech.* **2013**, *16*, 243–251. [CrossRef]

159. Sekhar, K.; Priyanka, B.; Reddy, V.D.; Rao, K.V. Isolation and characterization of a pigeonpea cyclophilin (*CcCYP*) gene, and its over-expression in *Arabidopsis* confers multiple abiotic stress tolerance. *Plant Cell Environ.* **2010**, *33*, 1324–1338. [CrossRef]

160. Meitei, A.L.; Bhattacharjee, M.; Dhar, S.; Chowdhury, N.; Sharma, R.; Acharjee, S.; Sarmah, B.K. Activity of defense related enzymes and gene expression in pigeon pea (*Cajanus cajan*) due to feeding of *Helicoverpa armigera* larvae. *J. Plant Int.* **2018**, *13*, 231–238. [CrossRef]

161. Kelly, J.D.; Vallejo, V.A. A comprehensive review of the major genes conditioning resistance to anthracnose in common bean. *Hortic. Sci.* **2004**, *39*, 1196–1207. [CrossRef]

162. Miklas, P.N.; Kelly, J.D.; Beebe, S.E.; Blair, M.W. Common bean breeding for resistance against biotic and abiotic stresses: From classical to MAS breeding. *Euphytica* **2006**, *147*, 105–131. [CrossRef]

163. Bhardwaj, J.; Chauhan, R.; Swarnkar, M.K.; Chahota, R.K.; Singh, A.K.; Shankar, R.; Yadav, S.K. Comprehensive transcriptomic study on horse gram (*Macrotyloma uniflorum*): De novo assembly, functional characterization and comparative analysis in relation to drought stress. *BMC Genomics* **2013**, *14*, 647. [CrossRef] [PubMed]

164. Li, Y.; Yang, K.; Yang, W.; Chu, L.; Chen, C.; Zhao, B.; Li, Y.; Jian, J.; Yin, Z.; Wang, T.; et al. Identification of QTL and qualitative trait loci for agronomic traits using SNP markers in the adzuki bean. *Front. Plant Sci.* **2017**, *8*, 840. [CrossRef] [PubMed]

165. Deokar, A.A.; Tar'an, B. Genome-wide analysis of the aquaporin gene family in chickpea (*Cicer arietinum* L.). *Front. Plant Sci.* **2016**, *7*, 1802. [CrossRef] [PubMed]

166. Deokar, A.A.; Kondawar, V.; Kohli, D.; Aslam, M.; Jain, P.K.; Karuppayil, S.M.; Varshney, R.K.; Srinivasan, R. The *CarERF* genes in chickpea (*Cicer arietinum* L.) and the identification of *CarERF116* as abiotic stress responsive transcription factor. *Funct. Integr. Genom.* **2015**, *15*, 27–46. [CrossRef]

167. Mallikarjuna, B.P.; Samineni, S.; Thudi, M.; Sajja, S.B.; Khan, A.W.; Patil, A.; Viswanatha, K.P.; Varshney, R.K.; Gaur, P.M. Molecular mapping of flowering time major genes and QTLs in chickpea (*Cicer arietinum* L.). *Front. Plant Sci.* **2017**, *8*, 1140. [CrossRef] [PubMed]

168. Gu, H.; Jia, Y.; Wang, X.; Chen, Q.; Shi, S.; Ma, L.; Zhang, J.; Zhang, H.; Ma, H. Identification and characterization of a LEA family gene *CarLEA4* from chickpea (*Cicer arietinum* L.). *Mol. Biol. Rep.* **2012**, *39*, 3565–3572. [CrossRef]

169. Mashaki, K.M.; Garg, V.; Ghomi, A.A.; Kudapa, H.; Chitikineni, A.; Nezhad, K.Z.; Yamchi, A.; Soltanloo, H.; Varshney, R.K.; Thudi, M. RNA-Seq analysis revealed genes associated with drought stress response in kabuli chickpea (*Cicer arietinum* L.). *PLoS ONE* **2018**, *13*, e0199774. [CrossRef]

170. Kumar, J.; Choudhary, A.K.; Solanki, R.K.; Pratap, A. Towards marker-assisted selection in pulses: A review. *Plant Breed.* **2011**, *130*, 297–313. [CrossRef]

171. Anbessa, Y.; Taran, B.; Warkentin, T.D.; Tullu, A.; Vandenberg, A. Genetic analyses and conservation of QTL for *Ascochyta* blight resistance in chickpea. *Theor. Appl. Genet.* **2009**, *119*, 757–765. [CrossRef]

172. Kottapalli, P.; Gaur, P.M.; Katiyar, S.K.; Crouch, J.H.; Buhariwalla, H.K.; Pande, S.; Gali, K.K. Mapping and validation of QTLs for resistance to an Indian isolate of ascochyta blight pathogen in chickpea. *Euphytica* **2009**, *165*, 79–88. [CrossRef]

173. Anuradha, C.; Gaur, P.M.; Pande, S.; Kishore, K.; Ganesh, G.M.; Kumar, J.; Varshney, R.K. Mapping QTL for resistance to botrytis grey mould in chickpea. *Euphytica* **2011**, *182*, 1–9. [CrossRef]

174. Rehman, A.U.; Malhotra, R.S.; Bett, K.; Tar'an, B.; Bueckert, R.; Warkentin, T.D. Mapping QTL associated with traits affecting grain yield in chickpea (*Cicer arietinum* L) under terminal drought stress. *Crop Sci.* **2011**, *51*, 450–463. [CrossRef]

175. Vadez, V.; Soltani, A.; Krishnamurthy, L.; Sinclair, T.R. Modelling possible benefit of root related traits to enhance terminal drought adaption of chickpea. *Field Crops Res.* **2012**, *137*, 108–115. [CrossRef]

176. Bajaj, D.; Saxena, M.S.; Kujur, A.; Das, S.; Badoni, S.; Tripathi, S.; Upadhyaya, H.D.; Gowda, C.L.; Sharma, S.; Singh, S.; et al. Genome-wide conserved non-coding microsatellite (CNMS) marker-based integrative genetical genomics for quantitative dissection of seed weight in chickpea. *J. Exp. Bot.* **2014**, *66*, 1271–1290. [CrossRef] [PubMed]

177. Bajaj, D.; Upadhyaya, H.D.; Khan, Y.; Das, S.; Badoni, S.; Shree, T.; Kumar, V.; Tripathi, S.; Gowda, C.L.; Singh, S.; et al. A combinatorial approach of comprehensive QTL-based comparative genome mapping and transcript profiling identified a seed weight-regulating candidate gene in chickpea. *Sci. Rep.* **2015**, *5*, 9264. [CrossRef] [PubMed]

178. Das, S.; Upadhyaya, H.D.; Bajaj, D.; Kujur, A.; Badoni, S.; Laxmi, K.V.; Tripathi, S.; Gowda, C.L.L.; Sharma, S.; Singh, S.; et al. Deploying QTL-seq for rapid delineation of a potential candidate gene underlying major trait-associated QTL in chickpea. *DNA Res.* **2015**, *22*, 193–203. [CrossRef] [PubMed]

179. Kujur, A.; Bajaj, D.; Upadhyaya, H.D.; Das, S.; Ranjan, R.; Shree, T.; Saxena, M.S.; Badoni, S.; Kumar, V.; Tripathi, S.; et al. A genome-wide SNP scan accelerates trait-regulatory genomic loci identification in chickpea. *Sci. Rep.* **2015**, *5*, 11166. [CrossRef] [PubMed]

180. Kujur, A.; Bajaj, D.; Upadhyaya, H.D.; Das, S.; Ranjan, R.; Shree, T.; Saxena, M.S.; Badoni, S.; Kumar, V.; Tripathi, S.; et al. Employing genome-wide SNP discovery and genotyping strategy to extrapolate the natural allelic diversity and domestication patterns in chickpea. *Front. Plant Sci.* **2015**, *6*, 162. [CrossRef]

181. Varshney, R.K.; Chen, W.; Li, Y.; Bharti, A.K.; Saxena, R.K.; Schlueter, J.A.; Donoghue, M.T.; Azam, S.; Fan, G.; Whaley, A.M.; et al. Draft genome sequence of pigeonpea (*Cajanus cajan*), an orphan legume crop of resource-poor farmers. *Nat. Biotechnol.* **2012**, *30*, 83. [CrossRef]

182. Varshney, R.K.; Song, C.; Saxena, R.K.; Azam, S.; Yu, S.; Sharpe, A.G.; Cannon, S.; Baek, J.; Rosen, B.D.; Tar'an, B.; et al. Draft genome sequence of chickpea (*Cicer arietinum*) provides a resource for trait improvement. *Nat. Biotechnol.* **2013**, *31*, 240. [CrossRef]

183. Varshney, R.K.; Mohan, S.M.; Gaur, P.M.; Chamarthi, S.K.; Singh, V.K.; Srinivasan, S.; Swapna, N.; Sharma, M.; Singh, S.; Kaur, L.; et al. Marker-assisted backcrossing to introgress resistance to fusarium wilt race 1 and ascochyta blight in C214, an elite cultivar of chickpea. *Plant Genome* **2014**, *7*, 1–11. [CrossRef]

184. Varshney, R.K.; Nayak, S.N.; May, G.D.; Jackson, S.A. Next-generation sequencing technologies and their implications for crop genetics and breeding. *Trends Biotechnol.* **2009**, *27*, 522–530. [CrossRef] [PubMed]

185. Serraj, R.; Krishnamurthy, L.; Kashiwagi, J.; Kumar, J.; Chandra, S.; Crouch, J.H. Variation in root traits of chickpea (*Cicer arietinum* L) grown under terminal drought. *Field Crops Res.* **2004**, *88*, 115–127. [CrossRef]

186. Choudhary, A.K.; Singh, D.; Iquebal, M.A. Selection of pigeonpea genotypes for tolerance to aluminium toxicity. *Plant Breed.* **2011**, *130*, 492–495. [CrossRef]

187. Choudhary, A.K.; Singh, D.; Kumar, J. A comparative study of screening methods for tolerance to aluminum toxicity in pigeonpea. *Aust. J. Crop Sci.* **2011**, *5*, 1419–1426.

188. Daspute, A.A.; Yuriko, K.; Kumar, P.S.; Bashasab, F.; Yasufumi, K.; Mutsutomo, T.; Satoshi, I.; Choudhary, A.K.; Yamamoto, Y.Y.; Hiroyuki, K. Characterization of CcSTOP1; a C2H2 type transcription factor regulates *Al* tolerance gene in pigeonpea. *Planta* **2018**, *247*, 201–214. [CrossRef]

189. Kaur, S.; Cogan, N.O.I.; Pembleton, L.W.; Shinozuka, M.; Savin, K.W.; Materne, M.; Forster, J.W. Transcriptome sequencing of lentil based on second-generation technology permits large-scale unigene assembly and SSR marker discovery. *BMC Genomics* **2011**, *12*, 265. [CrossRef]

190. Kaur, S.; Pembleton, L.; Cogan, N.O.I.; Savin, K.W.; Leonforte, T.; Paull, J.; Materne, M.; Forster, J.W. Transcriptome sequencing of field pea and faba bean for discovery and validation of SSR genetic markers. *BMC Genomics* **2012**, *13*, 104. [CrossRef]

191. Sudheesh, S.; Verma, P.; Forster, J.W.; Cogan, N.O.; Kaur, S. Generation and characterisation of a reference transcriptome for lentil (*Lens culinaris* Medik.). *Int. J. Mol. Sci.* **2016**, *17*, 1887. [CrossRef]

192. Driedonks, N.; Rieu, I.; Vriezen, W.H. Breeding for plant heat tolerance at vegetative and reproductive stages. *Plant Reprod.* **2016**, *29*, 67–79. [CrossRef] [PubMed]

193. Kujur, A.; Upadhyaya, H.D.; Bajaj, D.; Gowda, C.L.; Sharma, S.; Tyagi, A.K.; Parida, S.K. Identification of candidate genes and natural allelic variants for QTLs governing plant height in chickpea. *Sci. Rep.* **2016**, *6*, 27968. [CrossRef] [PubMed]

194. Saxena, M.S.; Bajaj, D.; Das, S.; Kujur, A.; Kumar, V.; Singh, M.; Bansal, K.C.; Tyagi, A.K.; Parida, S.K. An integrated genomic approach for rapid delineation of candidate genes regulating agro-morphological traits in chickpea. *DNA Res.* **2014**, *21*, 695–710. [CrossRef] [PubMed]

195. Fedoruk, M.J.; Vandenberg, A.; Bett, K.E. Quantitative trait loci analysis of seed quality characteristics in lentil using single nucleotide polymorphism markers. *Plant Gen.* **2013**, *6*, 1–10. [CrossRef]

196. Kaur, S.; Cogan, N.O.I.; Stephens, A.; Noy, D.; Butsch, M.; Forster, J.W.; Materne, M. EST-SNP discovery and fine-resolution genetic mapping in lentil (*Lens culinaris* Medik.) enables candidate gene selection for boron tolerance. *Theor. Appl. Genet.* **2014**, *127*, 703–713. [CrossRef] [PubMed]

197. Agarwal, G.; Garg, V.; Kudapa, H.; Doddamani, D.; Pazhamala, L.T.; Khan, A.W.; Thudi, M.; Lee, S.H.; Varshney, R.K. Genome-wide dissection of AP2/ERF and HSP90 gene families in five legumes and expression profiles in chickpea and pigeonpea. *Plant Biotech. J.* **2016**, *14*, 1563–1577. [CrossRef] [PubMed]

198. Zhu, H.; Riely, B.K.; Burns, N.J.; Ane, J.M. Tracing non-legume orthologs of legume genes required for nodulation and arbuscular mycorrhizal symbioses. *Genetics* **2006**, *172*, 2491–2499. [CrossRef]

199. Kumar, R.; Lavania, D.; Negi, M.; Siddiqui, M.H.; Al-Whaibi, M.; Grover, A. Identification and characterization of a small heat shock protein 17.9-CII gene from faba bean (*Vicia faba* L.). *Acta Physiol. Plant.* **2015**, *37*, 190. [CrossRef]

200. Naser, V.; Shani, E. Auxin response under osmotic stress. *Plant Mol. Biol.* **2016**, *91*, 66–1672. [CrossRef]

201. Singh, A.; Sharma, A.K.; Singh, N.K.; Sharma, T.R. PpTFDB: A pigeonpea transcription factor database for exploring functional genomics in legumes. *PLoS ONE* **2017**, *12*, e0179736. [CrossRef]

202. Kudapa, H.; Azam, S.; Sharpe, A.G.; Taran, B.; Li, R.; Deonovic, B.; Cameron, C.; Farmer, A.D.; Cannon, S.B.; Varshney, R.K. Comprehensive transcriptome assembly of chickpea (*Cicer arietinum* L.) using Sanger and next generation sequencing platforms: Development and applications. *PLoS ONE* **2018**, *9*, e86039. [CrossRef] [PubMed]

203. Singh, D.; Singh, C.K.; Taunk, J.; Tomar, R.S.S.; Chaturvedi, A.K.; Gaikwad, K.; Pal, M. Transcriptome analysis of lentil (*Lens culinaris* Medikus) in response to seedling drought stress. *BMC Genomics* **2017**, *18*, 206. [CrossRef] [PubMed]

204. Polanco, C.; de Miera, L.E.; Bett, K.; de la Vega, M.P. A genome-wide identification and comparative analysis of the lentil MLO genes. *PLoS ONE* **2018**, *13*, e0194945. [CrossRef] [PubMed]

205. Garg, R.; Singh, V.K.; Rajkumar, M.S.; Kumar, V.; Jain, M. Global transcriptome and coexpression network analyses reveal cultivar-specific molecular signatures associated with seed development and seed size/weight determination in chickpea. *Plant J.* **2017**, *91*, 1088–1107. [CrossRef] [PubMed]

206. Nasr Esfahani, M.; Inoue, K.; Chu, H.D.; Nguyen, K.H.; Ha, C.V.; Watanabe, Y.; Burritt, D.J.; Herrera-Estrella, L.; Mochida, K.; Tran, L.P. Comparative transcriptome analysis of nodules of two Mesorhizobium-chickpea associations with differential symbiotic efficiency under phosphate deficiency. *Plant J.* **2017**, *91*, 911–926. [CrossRef] [PubMed]

207. Ridge, S.; Deokar, A.; Lee, R.; Daba, K.; Macknight, R.C.; Weller, J.L.; Tar'an, B. The chickpea early flowering 1 (*Efl1*) locus is an ortholog of *Arabidopsis ELF3*. *Plant Physiol.* **2017**, *175*, 802–815. [CrossRef]

208. Thakur, R.C.; Nema, K.K.; Singh, O.P. Present status of Helicoverpa armigera in pulses and strategies for its management in Madhya Pradesh. In *Helicoverpa Management: Current Status and Future Strategies. Proceedings of the First National Workshop, 30–31 August 1992*; Sachan, J.N., Ed.; Directorate of Pulses Research: Kanpur, India, 1992.

209. Acharjee, S.; Sarmah, B.K.; Kumar, P.A.; Olsen, K.; Mahon, R.; Moar, W.J.; Moore, A.; Higgins, T.J.V. Transgenic chickpeas (*Cicer arietinum* L.) expressing a sequence-modified *cry2Aa* gene. *Plant Sci.* **2010**, *178*, 333–339. [CrossRef]

210. Mehrotra, M.; Singh, A.K.; Sanyal, I.; Altosaar, I.; Amla, D.V. Pyramiding of modified *cry1Ab* and *cry1Ac* genes of *Bacillus thuringiensis* in transgenic chickpea (*Cicer arietinum* L.) for improved resistance to pod borer insect *Helicoverpa armigera*. *Euphytica* **2011**, *182*, 87–102. [CrossRef]

211. Das, A.; Datta, S.; Soren, K.R.; Patil, P.G.; Chaturvedi, S.K.; Nadarajan, N. *Gene Technology for Pulses Improvement*; IIPR: Kanpur, India, 2012.

212. Singh, S.; Kumar, N.R.; Maniraj, R.; Lakshmikanth, R.; Rao, K.Y.; Muralimohan, N.; Arulprakash, T.; Karthik, K.; Shashibhushan, N.B.; Vinutha, T.; et al. Expression of Cry2Aa, a Bacillus thuringiensis insecticidal protein in transgenic pigeon pea confers resistance to gram pod borer, *Helicoverpa armigera*. *Sci. Rep.* **2018**, *8*, 8820. [CrossRef]

213. Surekha, C.H.; Kumari, K.N.; Aruna, L.V.; Suneetha, G.; Arundhati, A.; Kishor, P.K. Expression of the *Vigna aconitifolia P5CSF129A* gene in transgenic pigeonpea enhances proline accumulation and salt tolerance. *Plant Cell Tissue Organ Cult.* **2014**, *116*, 27–36. [CrossRef]

214. Bhatnagar-Mathur, P.; Vadez, V.; Devi, M.J.; Lavanya, M.; Vani, G.; Sharma, K.K. Genetic engineering of chickpea (*Cicer arietinum* L.) with the *P5CSF129A* gene for osmoregulation with implications on drought tolerance. *Mol. Breed.* **2009**, *23*, 591–606. [CrossRef]

215. Behr, M.; Legay, S.; Hausman, J.F.; Guerriero, G. Analysis of cell wall-related genes in organs of Medicago sativa L. under different abiotic stresses. *Int. J. Mol. Sci.* **2015**, *16*, 16104–16124. [CrossRef] [PubMed]

216. Tran, L.S.P.; Mochida, K. Identification and prediction of abiotic stress responsive transcription factors involved in abiotic stress signaling in soybean. *Plant Signal. Behav.* **2010**, *5*, 255–257. [CrossRef] [PubMed]

217. Kosová, K.; Vítámvás, P.; Urban, M.O.; Prášil, I.T.; Renaut, J. Plant abiotic stress proteomics: The major factors determining alterations in cellular proteome. *Front. Plant Sci.* **2018**, *9*, 122. [CrossRef] [PubMed]

218. Zivy, M.; Wienkoop, S.; Renaut, J.; Pinheiro, C.; Goulas, E.; Carpentier, S. The quest for tolerant varieties: The importance of integrating "omics" techniques to phenotyping. *Front. Plant Sci.* **2015**, *6*, 448. [CrossRef] [PubMed]

219. Jorrín-Novo, J.V.; Pascual, J.; Sánchez-Lucas, R.; Romero-Rodríguez, M.C.; Rodríguez-Ortega, M.J.; Lenz, C.; Valledor, L. Fourteen years of plant proteomics reflected in proteomics: Moving from model species and 2DE–based approaches to orphan species and gel-free platforms. *Proteomics* **2015**, *15*, 1089–1112. [CrossRef] [PubMed]

220. Rathi, D.; Gayen, D.; Gayali, S.; Chakraborty, S.; Chakraborty, N. Legume proteomics: Progress, prospects, and challenges. *Proteomics* **2016**, *16*, 310–327. [CrossRef] [PubMed]

221. Larrainzar, E.; Wienkoop, S. A proteomic view on the role of legume symbiotic interactions. *Front. Plant Sci.* **2017**, *8*, 1267. [CrossRef]

222. Krishnan, H.B.; Natarajan, S.S.; Oehrle, N.W.; Garrett, W.M.; Darwish, O. Proteomic analysis of pigeonpea (*Cajanus cajan*) seeds reveals the accumulation of numerous stress-related proteins. *J. Agric. Food Chem.* **2017**, *65*, 4572–4581. [CrossRef]

223. Rathi, D.; Pareek, A.; Gayali, S.; Chakraborty, S.; Chakraborty, N. Variety-specific nutrient acquisition and dehydration-induced proteomic landscape of grasspea (*Lathyrus sativus* L.). *J. Proteomics* **2018**, *183*, 45–57. [CrossRef]

224. Li, P.; Zhang, Y.; Wu, X.; Liu, Y. Drought stress impact on leaf proteome variations of faba bean (*Vicia faba* L.) in the Qinghai–Tibet Plateau of China. *3 Biotech* **2018**, *8*, 110. [CrossRef] [PubMed]

225. Lin, W.J.; Ko, C.Y.; Liu, M.S.; Kuo, C.Y.; Wu, D.C.; Chen, C.Y.; Schafleitner, R.; Chen, L.F.; Lo, H.F. Transcriptomic and proteomic research to explore bruchid-resistant genes in mungbean isogenic lines. *J. Agric. Food Chem.* **2016**, *64*, 6648–6658. [CrossRef] [PubMed]

226. Rodziewicz, P.; Swarcewicz, B.; Chmielewska, K.; Wojakowska, A.; Stobiecki, M. Influence of abiotic stresses on plant proteome and metabolome changes. *Acta Phys. Plantarum* **2014**, *36*, 1–19. [CrossRef]

227. Abdelrahman, M.; Suzumura, N.; Mitoma, M.; Matsuo, S.; Ikeuchi, T.; Mori, M.; Murakami, K.; Ozaki, Y.; Matsumoto, M.; Uragami, A.; et al. Comparative de novo transcriptome profiles in *Asparagus officinalis* and *A. kiusianus* during the early stage of *Phomopsis asparagi* infection. *Sci. Rep.* **2017**, *7*, 2608. [CrossRef] [PubMed]

228. Pinheiro, C.; Passarinho, J.A.; Ricardo, C.P. Effect of drought and rewatering on the metabolism of *Lupinus albus* organs. *J. Plant Physiol.* **2004**, *161*, 1203–1210. [CrossRef] [PubMed]

229. Hernández, G.; Valdés-López, O.; Ramírez, M.; Goffard, N.; Weiller, G.; Aparicio-Fabre, R.; Fuentes, S.I.; Erban, A.; Kopka, J.; Udvardi, M.K.; et al. Global changes in the transcript and metabolic profiles during symbiotic nitrogen fixation in phosphorus-stressed common bean plants. *Plant Physiol.* **2009**, *151*, 1221–1238. [CrossRef] [PubMed]

230. Silvente, S.; Sobolev, A.P.; Lara, M. Metabolite adjustments in drought tolerant and sensitive soybean genotypes in response to water stress. *PLoS ONE* **2012**, *7*, e38554. [CrossRef]

231. Tawaraya, K.; Horie, R.; Saito, S.; Wagatsuma, T.; Saito, K.; Oikawa, A. Metabolite profiling of root exudates of common bean under phosphorus deficiency. *Metabolites* **2014**, *4*, 599–611. [CrossRef]

232. Tripathi, P.; Rabara, R.C.; Shulaev, V.; Shen, Q.J.; Rushton, P.J. Understanding water-stress responses in soybean using hydroponics system a systems biology perspective. *Front. Plant Sci.* **2015**, *6*, 1145. [CrossRef]

233. Nasr Esfahani, M.; Kusano, M.; Nguyen, K.H.; Watanabe, Y.; Ha, C.V.; Saito, K.; Sulieman, S.; Herrera-Estrella, L.; Tran, L.S. Adaptation of the symbiotic *Mesorhizobium*–chickpea relationship to phosphate deficiency relies on reprogramming of whole-plant metabolism. *Proc. Natl. Acad. Sci. USA* **2016**, *113*, E4610–E4619. [CrossRef]

234. Muscolo, A.; Junker, A.; Klukas, C.; Weigelt-Fischer, K.; Riewe, D.; Altmann, T. Phenotypic and metabolic responses to drought and salinity of four contrasting lentil accessions. *J. Exp. Bot.* **2015**, *66*, 5467–5480. [CrossRef]

235. Kumar, R.; Bohra, A.; Pandey, A.K.; Pandey, M.K.; Kumar, A. Metabolomics for plant improvement: Status and prospects. *Front. Plant Sci.* **2017**, *8*, 1302. [CrossRef] [PubMed]

236. Haig, D. The (dual) origin of epigenetics. *Cold Spring Harb. Symp. Quant. Biol.* **2004**, *69*, 67–70. [CrossRef] [PubMed]

237. Kakutani, T. Epi-alleles in plants: Inheritance of epigenetic information over generations. *Plant Cell Physiol.* **2002**, *43*, 1106–1111. [CrossRef] [PubMed]

238. Iwasaki, M. Chromatin resetting mechanisms preventing transgenerational inheritance of epigenetic states. *Front. Plant Sci.* **2015**, *6*, 380. [CrossRef] [PubMed]

239. Fujimoto, R.; Sasaki, T.; Ishikawa, R.; Osabe, K.; Kawanabe, T.; Dennis, E.S. Molecular mechanisms of epigenetic variation in plants. *Int. J. Mol. Sci.* **2012**, *13*, 9900–9922. [CrossRef] [PubMed]

240. Bossdorf, O.; Arcuri, D.; Richards, C.L.; Pigliucci, M. Experimental alteration of DNA methylation affects the phenotypic plasticity of ecologically relevant traits in *Arabidopsis* thaliana. *Evol. Ecol.* **2010**, *24*, 541–553. [CrossRef]

241. Kooke, R.; Johannes, F.; Wardenaar, R.; Becker, F.; Etcheverry, M.; Colot, V.; Vreugdenhil, D.; Keurentjes, J.J.B. Epigenetic basis of morphological variation and phenotypic plasticity in *Arabidopsis thaliana*. *Plant Cell* **2015**, *27*, 337–348. [CrossRef] [PubMed]

242. Lele, L.; Ning, D.; Cuiping, P.; Xiao, G.; Weihua, G. Genetic and epigenetic variations associated with adaptation to heterogeneous habitat conditions in a deciduous shrub. *Ecol. Evol.* **2018**, *8*, 2594–2606. [CrossRef]

243. Weigel, D.; Colot, V. Epialleles in plant evolution. *Genome Biol.* **2012**, *13*, 249. [CrossRef]

244. Slotkin, R.K.; Martienssen, R. Transposable elements and the epigenetic regulation of the genome. *Nat. Rev. Genet.* **2007**, *8*, 272. [CrossRef] [PubMed]

245. Zhang, C.; Hsieh, T.F. Heritable epigenetic variation and its potential applications for crop improvement. *Plant Breed. Biotechnol.* **2013**, *4*, 307–319. [CrossRef]

246. Meyer, P. Epigenetic variation and environmental change. *J. Exp. Bot.* **2015**, *66*, 3541–3548. [CrossRef] [PubMed]

247. Li, R.; Zhou, S.; Li, Y.; Shen, X.; Wang, Z.; Chen, B. Comparative methylome analysis reveals perturbation of host epigenome in chestnut blight fungus by a hypovirus. *Front. Microbiol.* **2018**, *9*, 1026. [CrossRef] [PubMed]

248. Shen, Y.; Zhang, J.; Liu, Y.; Liu, S.; Liu, Z.; Duan, Z.; Wang, Z.; Zhu, B.; Guo, Y.L.; Tian, Z. DNA methylation footprints during soybean domestication and improvement. *Genome Biol.* **2018**, *19*, 128. [CrossRef] [PubMed]

249. Piedra-Aguilera, Á.; Jiao, C.; Luna, A.P.; Villanueva, F.; Dabad, M.; Esteve-Codina, A.; Díaz-Pendón, J.A.; Fei, Z.; Bejarano, E.R.; Castillo, A.G. Integrated single-base resolution maps of transcriptome, sRNAome and methylome of tomato yellow leaf curl virus (TYLCV) in tomato. *Sci. Rep.* **2019**, *9*, 2863. [CrossRef] [PubMed]

250. Bhatia, H.; Khemka, N.; Jain, M.; Garg, R. Genome-wide bisulphite-sequencing reveals organ-specific methylation patterns in chickpea. *Sci. Rep.* **2018**, *8*, 9704. [CrossRef]

251. Song, Q.X.; Lu, X.; Li, Q.T.; Chen, H.; Hu, X.Y.; Ma, B.; Zhang, W.K.; Chen, S.Y.; Zhang, J.S. Genome-wide analysis of DNA methylation in soybean. *Mol. Plant* **2013**, *6*, 1961–1974. [CrossRef]

252. Li, Y.; Ding, X.; Wang, X.; He, T.; Zhang, H.; Yang, L.; Wang, T.; Chen, L.; Gai, J.; Yang, S. Genome-wide comparative analysis of DNA methylation between soybean cytoplasmic male-sterile line NJCMS5A and its maintainer NJCMS5B. *BMC Genomics* **2017**, *18*, 596. [CrossRef]

253. An, Y.Q.; Goettel, W.; Han, Q.; Bartels, A.; Liu, Z.; Xiao, W. Dynamic changes of genome-wide DNA methylation during soybean seed development. *Sci. Rep.* **2017**, *7*, 12263. [CrossRef]

254. Kim, K.D.; El Baidouri, M.; Abernathy, B.; Iwata-Otsubo, A.; Chavarro, C.; Gonzales, M.; Libault, M.; Grimwood, J.; Jackson, S.A. A comparative epigenomic analysis of polyploidy-derived genes in soybean and common bean. *Plant Physiol.* **2015**, *168*, 1433–1447. [CrossRef] [PubMed]

255. Song, Q.; Zhang, T.; Stelly, D.M.; Chen, Z.J. Epigenomic and functional analyses reveal roles of epialleles in the loss of photoperiod sensitivity during domestication of allotetraploid cottons. *Genome Biol.* **2017**, *18*, 99. [CrossRef] [PubMed]

256. Manning, K.; Tör, M.; Poole, M.; Hong, Y.; Thompson, A.J.; King, G.J.; Giovannoni, J.J.; Seymour, G.B. A naturally occurring epigenetic mutation in a gene encoding an SBP-box transcription factor inhibits tomato fruit ripening. *Nat. Genet.* **2006**, *38*, 948. [CrossRef] [PubMed]

257. Quadrana, L.; Almeida, J.; Asís, R.; Duffy, T.; Dominguez, P.G.; Bermúdez, L.; Conti, G.; Da Silva, J.V.C.; Peralta, I.E.; Colot, V.; et al. Natural occurring epialleles determine vitamin E accumulation in tomato fruits. *Nat. Commun.* **2014**, *5*, 4027. [CrossRef] [PubMed]

258. Raju, S.K.; Shao, M.R.; Sanchez, R.; Xu, Y.Z.; Sandhu, A.; Graef, G.; Mackenzie, S. An epigenetic breeding system in soybean for increased yield and stability. *Plant Biotechnol. J.* **2018**. [CrossRef] [PubMed]

259. Robertson, A.L.; Wolf, D.E. The role of epigenetics in plant adaptation. *Trends Evol. Biol.* **2012**, *4*, 4. [CrossRef]

260. Belhaj, K.; Chaparro-Garcia, A.; Kamoun, S.; Nekrasov, V. Plant genome editing made easy: Targeted mutagenesis in model and crop plants using the CRISPR/Cas system. *Plant Method.* **2013**, *9*, 39. [CrossRef]

261. Osakabe, Y.; Osakabe, K. Genome editing with engineered nucleases in plants. *Plant Cell Physiol.* **2014**, *56*, 389–400. [CrossRef]

262. Li, J.F.; Norville, J.E.; Aach, J.; McCormack, M.; Zhang, D.; Bush, J.; Church, G.M.; Sheen, J. Multiplex and homologous recombination-mediated genome editing in Arabidopsis and *Nicotiana benthamiana* using guide RNA and Cas9. *Nat. Biotech.* **2013**, *31*, 688–691. [CrossRef]

263. Osakabe, K.; Nishizawa-Yokoi, A.; Ohtsuki, N.; Osakabe, Y.; Toki, S. A mutated cytosine deaminase gene, codA (D314A), as an efficient negative selection marker for gene targeting in rice. *Plant Cell Physiol.* **2014**, *55*, 658–665. [CrossRef]

264. Ceasar, S.A.; Rajan, V.; Prykhozhij, S.V.; Berman, J.N.; Ignacimuthu, S. Insert, remove or replace: A highly advanced genome editing system using CRISPR/Cas9. *Biochim. Biophys. Acta* **2016**, *1863*, 2333–2344. [CrossRef]

265. Nekrasov, V.; Staskawicz, B.; Weigel, D.; Jones, J.D.; Kamoun, S. Targeted mutagenesis in the model plant *Nicotiana benthamiana* using Cas9 RNA-guided endonuclease. *Nat. Biotech.* **2013**, *31*, 691–693. [CrossRef] [PubMed]

266. Shan, Q.; Wang, Y.; Li, J.; Zhang, Y.; Chen, K.; Liang, Z.; Zhang, K.; Liu, J.; Xi, J.J.; Qiu, J.L.; et al. Targeted genome modification of crop plants using a CRISPR-Cas system. *Nat. Biotechnol.* **2013**, *31*, 686–688. [CrossRef] [PubMed]

267. Miao, C.; Xiao, L.; Hua, K.; Zou, C.; Zhao, Y.; Bressan, R.A.; Zhu, J.K. Mutations in a subfamily of abscisic acid receptor genes promote rice growth and productivity. *Proc. Natl. Acad. Sci. USA* **2018**, *115*, 6058–6063. [CrossRef] [PubMed]

268. Trenberth, K.E.; Fasullo, J.T.; Shepherd, T.G. Attribution of climate extreme events. *Nat. Clim. Chang.* **2015**, *5*, 725–730. [CrossRef]

269. Yang, Q.; Liu, K.; Niu, X.; Wang, Q.; Wan, Y.; Yang, F.; Li, G.; Wang, Y.; Wang, R. Genome-wide identification of PP2C genes and their expression profiling in response to drought and cold stresses in Medicago truncatula. *Sci. Rep.* **2018**, *8*, 12841. [CrossRef]

270. Luo, D.; Wu, Y.; Liu, J.; Zhou, Q.; Liu, W.; Wang, Y.; Yang, Q.; Wang, Z.; Liu, Z. Comparative transcriptomic and physiological analyses of *Medicago sativa* l. indicates that multiple regulatory networks are activated during continuous aba treatment. *Int. J. Mol. Sci.* **2019**, *20*, 47. [CrossRef] [PubMed]

271. Borrelli, V.M.; Brambilla, V.; Rogowsky, P.; Marocco, A.; Lanubile, A. The enhancement of plant disease resistance using CRISPR/Cas9 technology. *Front. Plant Sci.* **2018**, *9*, 1245. [CrossRef]

272. Wang, L.; Wang, L.; Zhou, Y.; Duanmu, D. Use of CRISPR/Cas9 for symbiotic nitrogen fixation research in legumes. *Progress Mol. Biol. Trans. Sci.* **2017**, *149*, 187–213.

273. Mir, R.R.; Kudapa, H.; Srikanth, S.; Saxena, R.K.; Sharma, A.; Azam, S.; Saxena, K.; Penmetsa, R.V.; Varshney, R.K. Candidate gene analysis for determinacy in pigeonpea (Cajanus spp.). *Theor. Appl. Genet.* **2014**, *127*, 2663–2678. [CrossRef]

274. Upadhyaya, H.D.; Bajaj, D.; Narnoliya, L.; Das, S.; Kumar, V.; Gowda, C.L.L.; Sharma, S.; Tyagi, A.K.; Parida, S.K. Genome-wide scans for delineation of candidate genes regulating seed-protein content in chickpea. *Front. Plant. Sci.* **2016**, *7*, 302. [CrossRef] [PubMed]

275. Alomari, D.Z.; Eggert, K.; Von Wirén, N.; Alqudah, A.M.; Polley, A.; Plieske, J.; Ganal, M.W.; Pillen, K.; Röder, M.S. Identifying candidate genes for enhancing grain Zn concentration in wheat. *Front. Plant Sci.* **2018**, *9*, 1313. [CrossRef] [PubMed]

276. Curtin, S.J.; Tiffin, P.; Guhlin, J.; Trujillo, D.I.; Burghardt, L.T.; Atkins, P.; Baltes, N.J.; Denny, R.; Voytas, D.F.; Stupar, R.M.; et al. Validating genome-wide association candidates controlling quantitative variation in nodulation. *Plant Physiol.* **2017**, *173*, 921–931. [CrossRef] [PubMed]

277. Ceccarelli, S. Drought. In *Plant Genetic Resources and Climate Change*; Jackson, M., Ed.; CAB International: Watingford, UK, 2014; pp. 221–235.

278. Suneson, C.A. An evolutionary plant breeding method. *Agron. J.* **1956**, *48*, 188–191. [CrossRef]

279. Döring, T.F.; Knapp, S.; Kovacs, G.; Murphy, K.; Wolfe, M.S. Evolutionary plant breeding in cereals—Into a new era. *Sustainability* **2011**, *3*, 1944–1971. [CrossRef]

280. Ceccarelli, S. GMO, organic agriculture and breeding for sustainability. *Sustainability* **2014**, *6*, 4273–4286. [CrossRef]

281. Ceccarelli, S. Increasing plant breeding efficiency through evolutionary-participatory programs. In *More Food: Road to Survival*; Pilu, R., Gavazzi, G., Eds.; Bentham Science Publishers: Charka, Sharjah, 2016; pp. 17–40.

282. Dwivedi, S.L.; van Bueren, E.T.; Ceccarelli, S.; Grando, S.; Upadhyaya, H.D.; Ortiz, R. Diversifying food systems in the pursuit of sustainable food production and healthy diets. *Trends Plant Sci.* **2017**, *22*, 842–856. [CrossRef]

283. Do, T.D.; Chen, H.; Hien, V.T.; Hamwieh, A.; Yamada, T.; Sato, T.; Yan, Y.; Cong, H.; Shono, M.; Suenaga, K.; et al. Nacl synchronously regulates Na+, K+, and Cl− in soybean and greatly increases the grain yield in saline field conditions. *Sci. Rep.* **2016**, *6*, 19147. [CrossRef]

284. Stokstad, E. Heat-beating beans resist climate change. *Science* **2015**, *347*. [CrossRef]

285. Thudi, M.; Gaur, P.M.; Krishnamurthy, L.; Mir, R.R.; Kudapa, H.; Fikre, A.; Kimurto, P.; Tripathi, S.; Soren, K.R.; Mulwa, R.; et al. Genomics-assisted breeding for drought tolerance in chickpea. *Funct. Plant Biol.* **2014**, *41*, 1178–1190. [CrossRef]

286. Tsaftaris, A.S.; Polidoros, A.N. DNA methylation and plant breeding. *Plant Breed. Rev.* **1999**, *18*, 87–176. [CrossRef]

287. Tsaftaris, A.S.; Polidoros, A.N.; Kapazoglou, A.; Tani, E.; Kovac̆evic, N.M. Epigenetics and plant breeding. *Plant Breed. Rev.* **2008**, *30*, 49. [CrossRef]

288. Jaligot, E.; Rival, A. Applying epigenetics in plant breeding: Balancing genome stability and phenotypic plasticity. In *Advances in Plant Breeding Strategies: Breeding, Biotechnology and Molecular Tools*; Springer: Cham, Switzerland, 2015; pp. 159–192.

289. Álvarez-Venegas, R.; De-la-Peña, C. Recent advances of epigenetics in crop biotechnology. *Front. Plant Sci.* **2016**, *7*, 413. [CrossRef] [PubMed]

290. Bilichak, A.; Kovalchuk, I. Transgenerational response to stress in plants and its application for breeding. *J. Exp. Bot.* **2016**, *67*, 2081–2092. [CrossRef] [PubMed]

291. Gallusci, P.; Dai, Z.; Génard, M.; Gauffretau, A.; Leblanc-Fournier, N.; Richard-Molard, C.; Vile, D.; Brunel-Muguet, S. Epigenetics for plant improvement: Current knowledge and modeling avenues. *Trends Plant Sci.* **2017**, *22*, 610–623. [CrossRef] [PubMed]

 International Journal of
Molecular Sciences

Article

Development and Proof-of-Concept Application of Genome-Enabled Selection for Pea Grain Yield under Severe Terminal Drought

Paolo Annicchiarico [1,*], **Nelson Nazzicari** [1], **Meriem Laouar** [2], **Imane Thami-Alami** [3], **Massimo Romani** [1] and **Luciano Pecetti** [1]

[1] Council for Agricultural Research and Economics (CREA), Research Centre for Animal Production and Aquaculture, viale Piacenza 29, 26900 Lodi, Italy; nelson.nazzicari@crea.gov.it (N.N.); mas.romani@libero.it (M.R.); luciano.pecetti@crea.gov.it (L.P.)

[2] Ecole Nationale Supérieure Agronomique (ENSA), Laboratoire d'Amélioration Intégrative des Productions Végétales (C2711100), Rue Hassen Badi, El Harrach, Alger DZ16200, Algeria; laouar_m@yahoo.fr

[3] Institut National de la Recherche Agronomique (INRA), Centre Régional de Rabat, Av. de la Victoire, Rabat BP 415, Morocco; thamialami_ma@yahoo.fr

* Correspondence: paolo.annicchiarico@crea.gov.it

Received: 13 February 2020; Accepted: 27 March 2020; Published: 31 March 2020

Abstract: Terminal drought is the main stress limiting pea (*Pisum sativum* L.) grain yield in Mediterranean environments. This study aimed to investigate genotype × environment (GE) interaction patterns, define a genomic selection (GS) model for yield under severe drought based on single nucleotide polymorphism (SNP) markers from genotyping-by-sequencing, and compare GS with phenotypic selection (PS) and marker-assisted selection (MAS). Some 288 lines belonging to three connected RIL populations were evaluated in a managed-stress (MS) environment of Northern Italy, Marchouch (Morocco), and Alger (Algeria). Intra-environment, cross-environment, and cross-population predictive ability were assessed by Ridge Regression best linear unbiased prediction (rrBLUP) and Bayesian Lasso models. GE interaction was particularly large across moderate-stress and severe-stress environments. In proof-of-concept experiments performed in a MS environment, GS models constructed from MS environment and Marchouch data applied to independent material separated top-performing lines from mid- and bottom-performing ones, and produced actual yield gains similar to PS. The latter result would imply somewhat greater GS efficiency when considering same selection costs, in partial agreement with predicted efficiency results. GS, which exploited drought escape and intrinsic drought tolerance, exhibited 18% greater selection efficiency than MAS (albeit with non-significant difference between selections) and moderate to high cross-population predictive ability. GS can be cost-efficient to raise yields under severe drought.

Keywords: drought tolerance; genotype × environment interaction; genetic gain; genomic selection; grain yield; inter-population predictive ability; marker-assisted selection; *Pisum sativum*

---

1. Introduction

The combined effect of population growth, change and instability of climate, reduced available irrigation water, land degradation, and inefficient and environment-unfriendly exogenous nitrogen inputs are threatening the global food security [1–3]. Greater cultivation of drought-tolerant, resilient legume crops would represent a key asset for facing these challenges, by increasing the sustainability of agriculture in terms of soil fertility, energy efficiency, greenhouse gas emissions, pollution, and crop diversity on the one hand and the efficiency and quality of food systems on the other [4–6]. This is especially true for countries of Europe and Northern Africa, where greater legume cultivation is required also to decrease their huge dependency on international markets for high-protein feedstuff [7,8]. Plant breeding has unanimously been indicated as the main avenue to decrease the economic gap with cereal crops that limits the cultivation of grain legumes in these countries [9,10]. Drought, which has been the main abiotic stress targeted by legume improvement programmes [11], has crucial importance for most of these countries, because drought-prone environments are expected to become common throughout Southern Europe and Northern Africa and to expand northward and eastward into central Europe as a consequence of climate change [12].

Field pea (*Pisum sativum* L.) has special interest for Southern Europe, where it displays higher grain yielding ability than other rain-fed cool-season grain legumes [13]. Further assets of this crop are moderately high rates of genetic yield gain [14,15], remarkable flexibility of utilization (as grain, hay, or silage) [16], and high energy value for animal nutrition [17]. Recent work highlighted high potential interest and farmers' appreciation for pea in North-African environments, too [16].

Genomic selection (GS) aims to predict breeding values for complex, polygenic traits by means of a statistical model constructed from phenotypic and genome-wide marker data of a germplasm sample representing a genetic base (training set), which, if sufficiently predictive, can then be applied for extensive genome-enabled selection within the target genetic base [18]. This selection strategy has represented a breakthrough for cattle production improvement [19]. The development of a high-throughput genotyping technique such as genotyping-by-sequencing (GBS) [20], by which large germplasm sets can be genotyped by thousands of single nucleotide polymorphism (SNP) markers at a lower cost than array-based techniques [21], has facilitated the application of GS in plant breeding, to select for overall crop performance or other complex traits (e.g., drought tolerance) rather than for specific traits linked to markers identified via comparative genomics or quantitative trait loci (QTL) discovery [22]. Pioneer studies for grain yield of legume crops were encouraging in this respect. The cross-environment predictive accuracy of top-performing GS models exceeded 0.45 for soybean breeding lines or landraces across sites of the USA [23,24], France [25], or China [26], and for white lupin landraces across Italian environments with contrasting climate or water availability [27]. It averaged 0.30 for three recombinant inbred line (RIL) populations of pea grown in climatically contrasting Italian environments [28], and 0.34 for various line populations of chickpea grown in Indian locations [29]. For grain yield of drought-prone pea germplasm, GS displayed an average predictive ability of 0.72 (estimated from intra-environment cross-validations) across three RIL populations grown in a managed-stress (MS) environment subjected to severe terminal drought [30], namely, the drought stress typical of Mediterranean-climate environments that implies increasing stress intensity during the reproductive stages of the crop cycle. Additionally, GS exhibited good ability to predict breeding values for other pea traits, such as phenology or individual seed weight [31]. On the whole, the available results suggested an advantage of GS over phenotypic selection in terms of predicted yield gains per unit time or unit cost, both for inbred and outbred legume crops [28,32]. However, information on actual yield progress derived from GS application would crucially contribute to verify the value of GS for legume yield improvement.

Higher yield of cool-season grain legumes under terminal drought may be achieved through different mechanisms that provide either drought escape or drought tolerance [11]. The report in [30] also included results of a genome-wide association study (GWAS), which revealed extensive co-localization of markers associated with high yield under stress and early flowering of pea. However, that study revealed also genetic variation for intrinsic drought tolerance (estimated as the yield deviation from the line value expected as a function of onset of flowering), along with putative QTL for this trait that could be exploited by marker-assisted selection (MAS). GS for intrinsic drought tolerance proved feasible too, although with lower predictive ability (averaging 0.27 across RIL populations) than GS for overall grain yield [30]. Intrinsic drought tolerance has greater practical interest than drought stress escape in inland regions of Southern Europe, where the exploitation of early flowering may be limited by greater susceptibility of autumn-sown early material to frost events [33].

The GS model for pea grain yield under severe terminal drought in [30], which was constructed from phenotyping data from one MS environment of Italy, would profit from refinement based on phenotyping data from drought-stressed agricultural environments. Yield data from MS environments can be valuable for phenotypic selection [34] and definition of GS models for drought-prone areas, because they are not subjected to the large genotype × year interaction caused by erratic rainfall that may feature in agricultural environments. However, a key prerequisite for their utilization is their ability to reproduce genotype yield responses as they occur in the target agricultural environments [35].

The main objectives of this study were (i) to improve the GS model for predicting pea grain yield under severe terminal drought that was reported in [30], by assessing the consistency of the phenotyping data used to build up that model with those recorded in two North-African agricultural environments, widening the amount of phenotyping data for GS model construction, and assessing cross-environment and cross-population (alias inter-population) predictive abilities; and (ii) to perform a proof-of-concept assessment of the value of the improved GS model and of MAS for intrinsic drought tolerance, on the basis of actual grain yields displayed under severe terminal drought by independent material that underwent GS, MAS, and phenotypic selection (PS). Additional objectives were (i) to assess the extent and pattern of genotype × environment (GE) interaction occurring across different drought-prone environments; and (ii) to verify the possible usefulness of a MS environment in Italy for yield-based PS targeted to North-African agricultural environments, as an indirect selection strategy that exploits the genetic correlation between a MS selection environment and the target agricultural sites.

2. Results

2.1. Multi-Environment Data Analysis of RIL Populations (Experiments 1, 2, and 3)

The site of Alger (Algeria) exhibited distinctly higher water availability over the crop cycle and higher crop mean yield than the Moroccan site of Marchouch (Table 1). The MS environment in Lodi (Italy), which aimed to generate severe terminal drought, was definitely more similar to Marchouch than Alger both for crop mean yield and water availability for the crop (Table 1). Yield values of top-yielding lines, i.e., those that could maximize the potential of each environment, confirmed that the MS environment and Marchouch were quite unfavorable ($\leq$0.91 t/ha; Table 1) compared to Alger (3.33 t/ha).

Table 1. Management, available water, air temperature in the last period of crop cycle and grain yield of pea experiments performed in a managed drought stress (MS) environment (Lodi, Italy) and two agricultural sites (Marchouch, Morocco; Alger, Algeria).

Exp.	Environment	Sowing Date [1]	Harvest Date [1]	Available Water (mm) [2]	Last Month's Mean Temperature (°C) [3]	Mean Yield (t/ha)	Yield of Top-Yielding Line (t/ha)
Exp. 1	MS Lodi	Feb. 25, 2015	Jun. 3, 2015	120	19.3	0.32	0.75
Exp. 2	Marchouch	Nov. 28, 2015	May 26, 2016	59	18.1	0.36	0.91
Exp. 3	Alger	Dec. 8, 2015	May 18, 2016	327	20.3	1.38	3.33
Exp. 4, 5, 6	MS Lodi	Apr. 12, 2017	Jun. 24, 2017	115	23.4	0.42[4]	-

[1] First sowing date and last harvest date, when spanning across various days. [2] Over the crop cycle; as irrigation under a rain-out shelter in Lodi, and rainfall in the other sites. [3] Average of mean daily temperature during the last month of crop cycle. [4] Mean of three experiments.

Among-line variation for grain yield and onset of flowering was observed within each of the three connected recombinant inbred line (RIL) populations (originated from paired crosses between the cultivars Attika, Isard, and Kaspa) in each environment ($p < 0.05$). The combined ANOVA for grain yield revealed highly significant ($p < 0.001$) genotype × environment (GE) interaction besides variation for genotype and environment main effects (Supplementary Table S1). The application of the Additive Main effects and Multiplicative Interaction (AMMI) model for partitioning the GE interaction variation showed that only the first GE interaction principal component (PC) axis was significant according to the F_R test (Supplementary Table S1). AMMI-modeled line yield responses as a function of the environment score on this PC axis, which are displayed in Figure 1 for the two top-performing lines in each environment or across environments, the parent cultivars and the control cultivar Spacial, indicated (i) the remarkable yield response of some lines compared to parent or control cultivars; (ii) the large extent of GE interaction of cross-over type (i.e., implying line rank changes) across environments, and (iii) the greater similarity of the MS environment with Marchouch than with Alger for GE interaction pattern. The latter result was confirmed by the genetic correlation for yield of the whole set of lines across pairs of sites, which was moderately positive between the MS environment and Marchouch (r_g = 0.50, p < 0.001), and non-significant (p > 0.05) between Alger and the MS environment (r_g = −0.02) or Marchouch (r_g = −0.14). Alger proved distinct from the other environments also because it displayed no phenotypic correlation of grain yield with onset of flowering (r_p = −0.09, p > 0.10), in contrast with the negative correlation (p < 0.001) observed for Marchouch (r_p = −0.45) and the MS environment (r_p = −0.81). All these results supported the redefinition of a GS model for grain yield under severe terminal drought based on pooled phenotypic data from the MS environment and Marchouch.

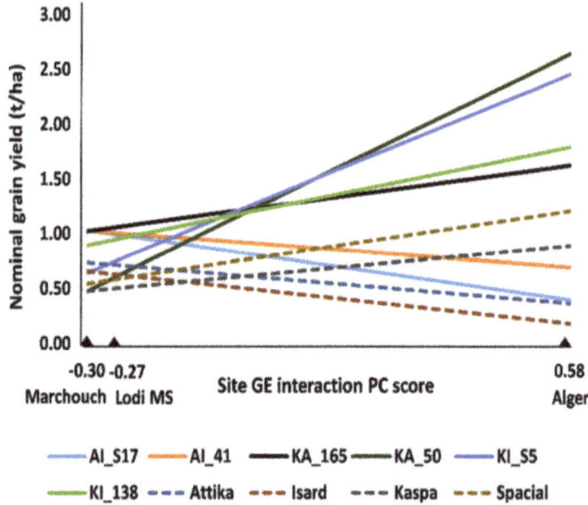

Figure 1. Additive Main effects and Multiplicative Interaction (AMMI)-modeled nominal grain yield of a set of top-performing pea lines out of 288 lines belonging to three connected recombinant inbred line (RIL) populations, including the two top-ranking lines in each site or over sites, three parent cultivars (Attika, Isard, and Kaspa) and one recent control cultivar (Spacial), grown in a managed drought stress (MS) environment of Lodi (Italy) and two agricultural environments of Marchouch (Morocco) and Alger (Algeria).

Indirect PS in the MS environment targeted to Marchouch was favored by higher broad-sense heritability under MS than in the agricultural site but was hindered by the only moderate genetic correlation between the selection and the target environment (which was r_g = 0.41 for values averaged across the individual RIL populations) (Table 2). As a result, indirect PS selection for Marchouch based on MS environment data was predicted to be 45% less efficient than direct PS in Marchouch (Table 2). The average r_g value close to zero was responsible for the extremely low predicted efficiency of indirect PS in the MS environment relative to direct PS for Alger (Table 2).

Table 2. Predicted efficiency (E_r) relative to direct phenotypic selection (PS) for pea grain yield in the target environment of (i) indirect PS in a managed drought stress (MS) environment (Lodi, Italy) for two agricultural sites (Marchouch, Morocco; Alger, Algeria); (ii) genomic selection (GS) using a model trained on line yield data from the target environment (A) or on data averaged across the MS environment and Marchouch (B), for three environments.

Target Environment	H_j^2	$H_{j'}^2$	$r_{g(j,j')}$	E_r, PS in MS [a]	r_{Ab} [b]		E_r, GS [c]	
					A	B	A	B
Marchouch	0.475	0.870	0.408	0.550	0.240	0.260	0.633	0.685
Alger	0.522	0.870	0.015	0.021	0.184	0.031	0.441	0.074
MS Lodi	0.870	–	–	–	0.741	0.713	1.066	1.023

H_j^2 and $H_{j'}^2$, broad sense heritability on a line mean basis for the target environment j and the selection environment j', respectively, for PS; $r_{g(j,j')}$, genetic correlation for line yields across j and j' environments; r_{Ab}, predictive ability of the top-performing of models constructed by Bayesian Lasso or Ridge Regression BLUP, considering models with five possible thresholds of genotype SNP missing data (10%, 20%, 30%, 40%, 50%) trained on joint data of three RIL populations (encompassing 288 lines overall). All values estimated for individual populations, reporting values averaged across populations. [a] Estimated as $(H_j\,H_{j'}\,r_{g(jj')})/H_j^2$. [b] Top-predicting models are reported in Table 3 for A; they are BL models with missing data thresholds of 50% for Marchouch, 10% for Alger and 20% for MS Lodi, for B. [c] Estimated as $(i_j''\,r_{Ab})/(i_j\,H_j^2)$, where i_j'' and i_j are standardized selection differentials used for GS and PS, respectively; $i_j'' = 2.197$ and $i_j = 1.755$, upon assumption of same overall costs for GS and PS and 2.8 lower cost per evaluated line of GS relative to PS.

Table 3. Predictive ability of the top-performing of models constructed by Bayesian Lasso or Ridge Regression BLUP for grain yield breeding value of pea lines belonging to three connected RIL populations in a managed drought stress (MS) environment (Lodi, Italy) and two agricultural sites (Marchouch, Morocco; Alger, Algeria), with model training on all RIL populations pooled in one data set or on the single populations.

Trait	Bayesian Lasso		Ridge Regression BLUP	
	All	Single	All	Single
Yield, MS Lodi	0.741	0.708	0.707	0.693
Yield, Marchouch	0.240	0.214	0.240	0.217
Yield, Alger	0.181	0.156	0.184	0.160
Mean yield, MS Lodi and Marchouch [1]	0.692	0.668	0.682	0.650

Averaged across results for three RIL populations encompassing 288 lines overall, considering models with five possible thresholds of genotype SNP missing data (10%, 20%, 30%, 40%, 50%). Fifty repetitions of 10-fold stratified cross-validations per analysis. [1] Using phenotypic data averaged across the two environments.

2.2. Predictive Ability of Genomic Selection Models (Experiments 1, 2, and 3)

As expected, the number of available polymorphic SNP markers issued from the genotyping-by-sequencing (GBS) analysis increased as a function of the threshold of allowed genotype SNP missing data, reaching the highest value for the threshold of 50% (Data repository S1). This SNP missing data threshold implied 4364 markers for the SNP calling criterion requiring at least six aligned reads per locus, and 7521 for the criterion requiring at least four reads. Only results for the six-read criterion are reported hereafter, because this criterion provided predictive ability values that were about equal or slightly higher than those provided by the four-read criterion in all analyses. Polymorphic markers for this criterion were 165 for the SNP missing data threshold of 10%, 647 for 20%, 1713 for 30%, and 3018 for 40%.

For all yield traits, Bayesian Lasso (BL) and Ridge Regression BLUP (rrBLUP) GS models tended to display a distinct increase of predictive ability passing from 10% to 20% of genotype SNP missing data, which could be attributed to the small marker number for the 10% threshold, along with modest or nil prediction improvement beyond the 20% threshold. This is shown in Supplementary Figure S1 for models trained on pooled data of the three RIL populations.

On average, GS model training on pooled data exhibited nearly 7% higher intra-environment predictive ability than model training on the individual populations, when comparing top-performing BL or rrBLUP models for the three test environments (Table 3).

Intra-environment prediction was maximized by BL for Lodi, whereas BL and rrBLUP achieved comparable predictive ability for Alger and Marchouch (Table 3). The best-predicting GS model for line mean yield across Lodi's MS environment and Marchouch (which was selected for GS in Exp. 4, 5, and 6) was BL trained on pooled population data with 20% SNP missing data threshold, whose predictive ability was 4% higher than that of the BL model with the same configuration but trained on individual populations (which was the alternative GS model used for Exp. 4).

In agreement with results of location similarity for GE interaction, the cross-environment predictive accuracy provided by best-predicting GS models was moderate for predicting Marchouch data from data of the MS environment or vice versa (range 0.35–0.46), and very low for predicting Alger data from MS environment data or vice versa (<0.06) or for predicting Alger data from Marchouch data or vice versa (<0.12) (Supplementary Table S2). The predictive ability of the top-performing GS model constructed from line mean yields across the MS environment and Marchouch was about 0.26 for Marchouch, 0.71 for the MS environment, and 0.03 for Alger (Table 2).

Cross-population predictive ability was investigated for the hypothesis of GS models trained on data of either one or two other connected RIL populations, considering by turns all possible combinations of training and validation populations. The assessment focused on line mean yield across Lodi's MS environment and Marchouch (whose data were exploited for GS proof-of-concept experiment work), and line yield in Alger (whose response pattern contrasted with that observed in the other two environments). Top-performing GS models for cross-population predictive ability were generally Bayesian Lasso with 20% to 40% genotype SNP missing rate. On average, the loss of prediction for the top-performing GS model passing from intra-population prediction (Table 3) to cross-population prediction (Table 4) was only 9% (0.630 vs. 0.692) for mean yield across Lodi's MS environment and Marchouch, and 18% (0.151 vs. 0.184) for yield in Alger, for models trained on joint data of two populations. The loss of prediction was distinctly greater, namely, 43% for mean yield across Lodi's MS environment and Marchouch, and 46% for yield in Alger, for models trained on one population.

Table 4. Cross-population predictive ability of the top-performing of models constructed by Bayesian Lasso or Ridge Regression BLUP for breeding value of pea lines belonging to three connected RIL populations, for grain yield in the agricultural site of Alger (Algeria) and mean grain yield across a managed drought stress (MS) environment (Lodi, Italy) and the site of Marchouch (Morocco). Average predictions for one RIL based on model training on data of one or two other connected RIL populations.

Trait	Training Populations	
	One	Two
Yield, Alger	0.099	0.151
Mean yield, MS Lodi, and Marchouch [1]	0.397	0.630

Averaged across results for three RIL populations encompassing 288 lines overall, considering models with five possible thresholds of genotype SNP missing data (10%, 20%, 30%, 40%, 50%). [1] Using phenotypic data averaged across the two environments.

2.3. Comparison of Genomic vs. Phenotypic Selection Based on Predicted Yield Gains (Experiments 1, 2, and 3)

Compared to PS in specific environments, the predicted efficiency of GS based on the best-performing site-specific GS model was about 37% lower for Marchouch, 56% lower for Alger, and 7% higher for the MS environment (Table 2). Interestingly, the gap in predicted efficiency of GS relative to PS for Marchouch was reduced by using the GS model that incorporated also data from the MS environment besides data from Marchouch (about 31% lower efficiency; Table 2). This model reduced very slightly (2%) the advantage of GS relative to direct PS for the MS environment, while showing very low relative efficiency for Alger (Table 2).

For Marchouch, indirect selection based on the top-performing GS model was predicted to be about 24% more efficient than indirect PS based on MS environment data (as indicated by relative efficiency of 0.685 vs. 0.550; Table 2). For Alger, GS based on the site-specific model was far more efficient than indirect PS based on MS environment data (relative efficiency of 0.441 vs. 0.021; Table 2), given the inability of the MS environment to reproduce the line yield responses for this site.

2.4. Comparison of Genomic vs. Phenotypic Selection Based on Actual Yield Gains (Experiment 4)

This experiment aimed to compare five groups of lines selected for grain yield under severe terminal drought according to different PS or GS criteria. ANOVA results for data excluding parent lines are reported in Supplementary Table S3. The five groups of lines differed at $p < 0.01$ for grain yield and onset of flowering and at $p < 0.05$ for aerial biomass, with no line group × RIL population interaction except for onset of flowering. The selected GS model constructed from line mean yields across the MS environment and Marchouch with model training on pooled RIL data produced lines with similar grain yielding ability but somewhat lower aerial biomass and earlier flowering ($p < 0.05$) compared to the GS model trained on the individual RIL populations (Table 5).

Table 5. Grain yield, aerial biomass, and onset of flowering under managed drought stress (MS) of pea line groups issued by genomic selection (GS) or phenotypic selection (PS) for grain yield under severe terminal drought or marker-assisted selection (MAS) for intrinsic drought tolerance.

| Line Group | Total no. of Lines | Yield (t/ha Dry Weight) | | Aerial Biomass (t/ha Dry Weight) | Onset of Flowering (dd from April 1) |
		Value	Difference to Parent Line Group		
Experiment 4 [1]					
PS in MS Lodi	9	0.749 **	0.495	3.264 **	26.2 **
GS, RIL population-specific model	9	0.655 **	0.401	3.299 **	27.4 **
PS across MS Lodi and Marchouch	9	0.653 **	0.399	3.216 *	27.4 **
GS, model trained on all populations	9	0.642 **	0.388	3.015	26.2 **
PS in Marchouch	9	0.540 **	0.286	3.094	28.5 **
Parent lines	3	0.254	-	2.819	30.8
LSD ($p < 0.05$)		0.104		0.195	0.5
Experiment 5 [2]					
GS, top-performing lines	6	0.353 *	0.128	2.786	31.2
GS, mid-performing lines	6	0.134	−0.091	2.747	35.5 **
GS, bottom-performing lines	6	0.121	−0.104	2.581	35.3 **
Parent lines	3	0.225	-	2.686	32.7
LSD ($p < 0.05$)		0.068		0.253	1.5
Experiment 6 [3]					
GS, top-performing lines	9	0.638 **	0.286	3.375 **	28.8
MAS, top-performing lines	9	0.595 **	0.243	3.031 **	28.3
GS/MAS mid-performing lines	6	0.462	0.110	3.059 **	28.6
GS, bottom-performing lines	9	0.290	−0.062	2.649	29.3
MAS, bottom-performing lines	9	0.208	−0.144	2.597	30
Parent lines	2	0.352	-	2.506	28.6
LSD ($p < 0.05$)		0.114		0.297	1.0

GS modelling based on data of independent lines evaluated in a MS experiment in Lodi and a field experiment in Marchouch (Exp. 1 and 2 in Table 1, respectively). LSD relates to line group mean comparison, excluding parent lines. Line group means followed by * and ** differ at $p < 0.05$ and $p < 0.01$, respectively, from the parent line mean according to Dunnett's test. [1] GS model trained on 205 lines from three RIL populations or on the single populations. GS and PS selection: three lines out of 30, for each of three RIL populations. GS and PS data averaged across populations. [2] GS model trained on 295 lines from three RIL populations. Each GS-based line group: two lines out of 30, for each of three connected crosses. GS data averaged across connected crosses. [3] GS model trained on 198 lines from two RIL populations. GS and MAS selection: applied to 24 lines previously selected for similar phenology out of 97 lines from another RIL population, selecting three lines for top- and bottom-performing groups, and two lines for the mid-performing group.

A meaningful comparison of GS vs. PS was obtained by comparing the average grain yield progress over the mean of parent lines of the two GS procedures (0.394 t/ha) with that of PS based on line mean yields across the MS environment and Marchouch (0.399 t/ha), which implied just 1% lower yield gain of GS relative to PS (Table 5). On average, the progress over parent lines of these selections was remarkable, implying over 2.5-fold higher grain yield, associated with a distinct shift towards earlier flowering and a trend towards higher aerial biomass (Table 5). On average, GS and PS produced material with comparable aerial biomass (3.157 t/ha for GS vs. 3.216 t/ha for PS; Table 5).

Specific PS for the MS environment maximized the grain yield gain over parent lines in the same test environment (Table 5). In comparison, specific PS for Marchouch exhibited 42% lower yield progress over parent lines (0.286 vs. 0.495 t/ha; Table 5).

2.5. Comparison of Material with Contrasting Genomic Predictions (Experiment 5)

The three line groups evaluated in Experiment 5, which were relative to putatively top-performing, mid-performing, and bottom-performing lines according to the GS model constructed from line mean yields across the MS environment and Marchouch, differed for grain yield and onset of flowering ($p < 0.01$) but not for aerial biomass, and displayed interaction with the Cross factor for grain yield and aerial biomass ($p < 0.05$; Supplementary Table S3). On average, the lines classed by GS in the top-performing group exhibited over 2.6-fold higher grain yield and four-day earlier onset of flowering than lines classed into the mid-performing or the bottom-performing group ($p < 0.01$), with no significant difference between mid- and bottom-performing groups (Table 5). In this experiment, the yield difference between selected line and parent line groups should not be interpreted in terms of GS gain over parent lines, because the genetic base that underwent GS here was intrinsically poorly adapted to severe drought because of earlier selection for cold tolerance (unlike the RIL populations of Experiments 4 or 6). However, GS (as represented by material classed in the top-performing group) allowed for a distinct grain yield progress over the mean of parent lines (57% higher yield; Table 5) under severe drought even in this late-flowering, cold-tolerant genetic base, also by means of a remarkable shift of the selected material towards earlier onset of flowering (Table 5).

2.6. Comparison of Genomic Selection vs. Marker-Assisted Selection for Intrinsic Drought Tolerance (Experiment 6)

In this experiment, MAS for intrinsic drought tolerance and the GS model constructed from line mean yields across the MS environment and Marchouch were applied to a set of lines featuring similar earliness of flowering. The five line groups, which were relative to putatively top-performing, mid-performing, or bottom-performing lines according to MAS or the GS, differed for grain and aerial biomass ($p < 0.01$) and onset of flowering ($p < 0.05$; Supplementary Table S3). Both GS and MAS groups of top-performing material exhibited distinctly higher grain yield than the other groups of lines and the set of parent lines ($p < 0.01$; Table 5). Compared to top-performing material from MAS, top-performing material from GS exhibited 11% higher aerial biomass ($p < 0.05$), and 7% higher grain yield (with 18% greater selection efficiency in terms of grain yield progress over the mean of parent lines, i.e., 0.286 t/ha vs. 0.243 t/ha), but the latter difference was not significant ($p > 0.05$; Table 5).

The success of GS and MAS selections was confirmed by the progressively lower grain yield across line groups that were classed as top-performing, mid-performing, and bottom-performing, respectively ($p < 0.05$; Table 5). The fact that also GS besides MAS capitalized on genetic variation for intrinsic tolerance to drought in this experiment was confirmed by the lack of shift towards earlier onset of flowering of its material classed as top-yielding (Table 5). Indeed, the very limited variation for flowering time available for exploitation by GS in this material could justify the smaller grain yield progress over parent lines achieved by GS in this experiment relative to Experiment 4 (0.286 vs. 0.388 t/ha for GS trained on pooled data of the three RIL populations; Table 5). The difference in aerial biomass between top-performing lines issued by GS and bottom-performing lines or parent lines was greater in Experiment 6 than in Experiments 4 or 5 (Table 5), indicating that selection for intrinsic drought tolerance had a special positive impact on plant vegetative growth.

3. Discussion

The outstanding GE interaction of cross-over type for pea grain yield across different drought-prone environments that was highlighted by AMMI analysis and genetic correlation results represents a challenge for phenotypic or genomic selection targeting these environments. The interaction was particularly high between Alger—which could be defined as a moderate-stress environment according to the yield value around 3.3 t/ha observed for top-performing material—and the other two environments—whose yield of top-performing material was below 1 t/ha—as indicated by genetic correlations close to zero and contrasting environment ordination on GE interaction PC 1. A limitation of this study was the lack of repetition in time of the experiments in the two agricultural locations, which did not allow to assess the extent of within-site GE interaction and mean yield variation and to verify the close relationship between environment similarity for GE interaction pattern and environment mean yield that was suggested by the results. However, wide GE interaction across environment mean yields in the range of 1–3 t/ha was repeatedly observed in cool-season cereals [36,37]. In pea, GE interaction for grain yield was reportedly modest for advanced breeding lines and elite cultivars across different drought-prone environments of Australia [38] but was large for different pea material across environments of Southern Europe [28,39–41] as well as within a different European region such as Poland [42].

Stress escape by early flowering was a key driver of specific adaptation to severely drought-prone environments in this study, as indicated by (i) its correlation with line grain yield in the MS environment and in Marchouch and its lack of correlation with yield in Alger, and (ii) the shift towards earlier flowering of material selected by GS for yield under severe terminal drought when tested in Experiments 4 and 5. However, the concurrent importance of intrinsic drought tolerance was highlighted by results of Experiment 6, in which the distinct yield progress under severe stress that was realized by material selected via GS or MAS could hardly capitalize on stress escape by earlier flowering. In pea, intrinsic drought tolerance was reportedly associated with traits such as osmotic adjustment, greater root spread, increased stomata diffusive resistance, and proline accumulation [43–45]. The first two traits are also known to enhance biomass production via greater effective use of water, unlike early flowering [46], which could justify the greater increase in aerial biomass of material selected for intrinsic drought tolerance relative to that selected also for drought escape.

The moderate genetic correlation and the similarity for GE interaction pattern of the two low-yielding sites (Marchouch and the MS environment) suggested that Mediterranean-climate environments with similar drought stress extent may represent a unique target region. However, indirect PS in one environment targeted to the other environment displayed distinctly lower efficiency than direct PS. In particular, indirect selection in the MS environment targeted to Marchouch exhibited 45% lower predicted efficiency than direct PS for Marchouch, whereas material issued by PS in Marchouch displayed 42% lower actual yield gain over parent lines in the MS environment of Experiment 4 than material issued by earlier PS in the MS environment. The sizeable GE interaction across these environments, which may be due to the large difference in temperature pattern between these geographically-distant sites and the important impact that such a difference may have on GE interactions for pea grain yield [40], sets a limit to the possibility of using a MS environment in Southern Europe to select for severely stressed environments of Northern Africa. However, PS performed in a MS environment that was geographically closer to its target environments may offer advantages relative to PS in agricultural environments, because of its lower experiment error that emerged in this study and the control over year-to-year rainfall variation that it offers.

The similar performance of the tested GS models, and the negligible or nil increase of predictive ability arising from imputing population structure information, agreed with earlier results for pea [30,31]. The lack of substantial rise of predictive ability beyond 20% genotype SNP missing data (implying

647 polymorphic markers) agreed as well with earlier findings for this material [30], suggesting that moderate marker numbers may be sufficient to approach prediction maximization for biparental RIL populations because of their narrower genetic variation and slow linkage disequilibrium decay relative to a broadly-based diversity panel. Actually, one such panel exhibited high GS prediction accuracy for pea seed weight and moderate accuracy for number of seeds per plant by using only 331 well-distributed markers [31].

GS models constructed from data of severely drought-prone environments such as Marchouch and the MS environment displayed nearly no value for a moderate drought-stress site such as Alger and vice versa, indicating that GS could hardly alleviate the difficulty to cope with the large GE interaction across stress levels. Breeders could use GS (or PS) to breed for (i) specific adaptation to severe-stress or moderate-stress environments, in the presence of high rainfall variation between sites and only moderate within-site rainfall variation in their target region (as it may be the case for geographically large target regions); or (ii) wide adaptation, by selecting for average value of the breeding values predicted by one GS model for severe-stress environments and another for moderate-stress environments (or by parallel PS selection across severe-stress and moderate-stress environments), in the opposite situation. Obviously, the latter option would imply much lower genetic progress in each environment type than the former.

Cross-population predictive ability has great practical interest for breeding programmes, as the transferability of GS models for predictions in other populations would decrease the cost of model development and would impact the strategies of GS implementation. In general, cross-population predictions tend to be poor across unrelated populations of inbred crops (e.g., [47]). We envisaged two scenarios for cross-population predictions, namely, model training on one or two connected RIL populations (which imply greater potential success relative to training on RIL populations that share no common parent with the target population). Particularly for the genome-enabled prediction of yield under severe drought (object of the proof-of-concept assessment), our results indicated high transferability of models trained on two RIL populations to the third connected population, which imply substantial potential savings in model training cost when exploiting connected RIL populations. Additionally, they encourage to verify whether substantial savings of model training costs may be achieved at a modest loss of cross-population predictive ability in other situations, for example, the GS exploitation of six biparental RIL populations that originated from four parents A, B, C, and D by means of intra-population predictions for two phenotyped populations, e.g., A × B and C × D, and by cross-population predictions for the other four populations (A × C, A × D, B × C, and B × D) based on the GS model constructed from joint data of the two phenotyped populations. The moderate GS model transferability across RIL populations sharing only one parent that was indicated by the 43% loss of predictive ability relative to intra-population prediction is close to the 37% loss that was reported for grain yield across Italian agricultural environments of the same populations [28] and ensured, anyway, a moderate prediction ability (0.397; Table 4).

The results of the proof-of-concept assessment of GS based on actual yield gains in the MS environment were encouraging for genome-enabled selection. Three experiments performed on independent material indicated consistently the remarkable yield progress over parent lines of material issued by GS. Two of them, designed to compare putative top-, mid-, and bottom-performing material according to genomic estimates of breeding values (Experiments 5 and 6), confirmed the ability of GS to identify top-performing lines. Finally, Experiment 4 indicated the comparable performance in the MS environment of PS and GS for mean yield across the MS environment and Marchouch. This experiment assumed same selection intensity for GS and PS (10% selected fraction for each RIL population), which would imply greater efficiency of GS over PS when considering the lower cost per test line and the shorter selection cycle (e.g., 0.5 years vs. one year or more) of GS relative to PS. This finding agreed largely with the somewhat greater efficiency of GS over PS according to predicted yield gains, whose assessment considered the different cost per test line of these selection approaches by assuming distinct selection intensity (while not accounting for the advantage of shorter selection by GS). In an earlier study on pea, GS outperformed PS in terms of predicted efficiency per unit time (for same selection cost) and correlation with line yield responses in independent environments, for grain yield selection across agricultural environments of Northern and Central Italy subjected to GE interaction mainly due to year-to-year variation for extent of low winter temperatures [28]. A preliminary comparison in terms of actual yield gains for another legume crop, i.e., alfalfa, was less encouraging for GS, which was successful for divergent selection of higher- and lower-yielding synthetic populations but produced distinctly lower genetic gain than PS [48]. Various comparisons of GS vs. PS for crop yield were reported for cereals. For wheat yield, Lozada et al. [49] reported 32% lower actual response to selection for wheat yield from GS relative to PS, whereas Michel et al. [50] found greater prediction accuracy for independent environments of GS relative to PS. For maize yield, Beyene et al. [51] reported the greater efficiency of GS over pedigree-based conventional PS when comparing actual yield gains from GS with ordinary gains reported for PS; Beyene et al. [52] found similar actual yield gains for GS and PS in a second study; and Môro et al. [53] observed 12% greater response to selection for GS relative to PS. Additionally, Sallam and Smith [54] reported similar actual yield gains of GS and PS for barley. These cereal studies would reveal additional merit for GS once accounting for its lower cost and shorter selection cycle. A few studies [48,49] provided evidence for the advantage over PS of genomic assisted selection, by which phenotypic yield data from preliminary trials are combined with genomic predictions.

This study could define and test a GS model for severe-stress environments, while data from at least another moderate-stress site besides Alger would be needed to define a GS model for this environment type. Although preliminary, our results are not encouraging for GS targeting Alger, on the basis of the modest predictive ability and over 50% lower predicted efficiency relative to PS of the GS model constructed from one-year data.

GS did not display a distinct and statistically significant superiority over MAS for grain yield related to intrinsic drought tolerance, although its estimated selection efficiency advantage was not quite negligible when expressed in terms of yield progress over the parent lines (+18%). However, GS produced material with significantly greater aerial biomass than MAS. Earlier comparisons of GS vs. MAS for production traits were reported for non-legume crops, where GS proved more efficient but with quite variable advantage. For example, the advantage of GS was in the range 18%–43% according to simulation results [55] and 14%–50% according to actual selection responses [56] for maize, while being over 2.5-fold according to wheat selection gains [49]. The only modest disadvantage of MAS relative to GS in this study suggests that the five genomic regions that were targeted by MAS (see Supplementary Table S4) may include important drought tolerance genes, whose discovery may be the target of further research.

In conclusion, both PS and GS for pea grain yield in the Mediterranean region are challenged by large GE interaction, whose size tends to increase as a function of the difference across environments for drought stress extent and environment yield potential. A GS model defined for severe-stress environments exhibited greater efficiency than PS when accounting for its shorter selection cycle and lower evaluation costs, as well as moderate to high transferability across connected RIL populations. Further research is warranted to compare GS vs. PS and to confirm model transferability across RIL populations on the ground of actual yield gains in severely drought-prone agricultural environments, as well as to compare wide-adaptation vs. specific adaptation strategies for GS or PS as a function of the target region of a breeding programme.

4. Materials and Methods

4.1. Multi-Environment Phenotyping and Data Analysis of RIL Populations (Experiments 1, 2, and 3)

Phenotyping data were generated for 288 semi-dwarf, semi-leafless lines belonging to three connected RIL populations originated by single-seed descent from paired crosses between Attika (a European cultivar described as a spring-type), Isard (a French winter-type cultivar), and Kaspa (an

Australian cultivar). These parent cultivars displayed fairly similar phenology and cycle duration along with high and stable grain yield and other positive agronomic characteristics across environments of Northern and Southern Italy [39,57]. The RIL populations are coded henceforth as 'A × I', 'K × A', and 'K × I' from the initials of their respective parents. The 288 lines represented a large subset of the 315 lines, 105 for each RIL population, that were phenotyped by Annicchiarico et al. [30]. In particular, this study included 96 lines for the A × I population, 92 for K × A, and 100 for K × I, for which enough seed was available for sowing in both North-African environments. In addition, the evaluation trials included the three parent cultivars, as well as the recent cultivar Spacial, which is characterized by excellent adaptation to Italian environments [41].

The lines were phenotyped for grain yield in three environments described as Exp. 1, 2, and 3 in Table 1. The first (Exp. 1) was a MS environment established in Lodi (Italy) as a large field-based phenotyping platform equipped with a rain-out shelter and a double rail irrigation boom. The management of Exp. 1, whose irrigation scheme mimicked the Mediterranean-climate rainfall pattern observed in the driest areas of Southern Italy, and its phenotyping results, were already reported in [30]. The second and third environments were the rain-fed agricultural sites of Marchouch (Exp. 2, Morocco, 33° 33′ N, 6° 41′ W) and Alger (Exp. 3, Algeria, 36° 45′ N, 3° 3′ E), respectively. Exp. 1 was sown in late winter of 2015 to avoid confounding effects of drought and cold stress, whereas Exp. 2 and 3 were autumn-sown (according to local practices) in 2015 in mild-winter environments that prevented the occurrence of cold stress. Exp. 1 involved smaller plots (0.8 × 0.2 m) and higher sowing density (100 germinating seeds/m^2) than Exp. 2 and Exp. 3 (plot size: 1.1 × 0.8 m; sowing density: 60 germinating seeds/m^2), owing to smaller room available in the MS environment. The experimental design was an alpha lattice with four replications for Exp. 1, and a randomized complete block (RCB) with three replications for Exp. 2 and 3. Dry grain yield was measured on a plot basis after estimating seed moisture by oven-drying seed samples at 90 °C for four days. Onset of flowering (as the number of days from April 1 to when 50% of plants in the plot had at least one open flower) was also recorded.

Grain yield and onset of flowering data of the RIL populations underwent a preliminary analysis of variance (ANOVA) that verified the occurrence of within-population variation for each experiment. Yield data of RIL material and the parent and control cultivars underwent a combined ANOVA including the factors genotype, environment, and block within environment. Experiment errors previously tested by Hartley's test proved to be not homogeneous ($p < 0.01$), implying some loss of sensitivity for the F tests of genotype main effects and GE interaction in the combined ANOVA [58] that had no practical importance because of the high statistical significance ($p < 0.001$) of these effects. GE interaction variation for yield was partitioned by Additive Main effects and Multiplicative Interaction (AMMI) analysis, expressing graphically the AMMI-modeled responses as nominal yields (which exclude the site main effect, irrelevant for entry ranking) as a function of the environment score on the first GE interaction principal component (PC 1) [59]. For sake of clarity, the graph included just a subset of top-performing genotypes. The significance of GE interaction principal component (PC) axes was tested by the F_R test [60]. The extent of GE interaction across pairs of environments was estimated by the genetic correlation (r_g) for yield responses of the whole set of lines as described in [61].

The interest of indirect PS for yield in the Italian MS environment for each of the two North-African environments relative to direct PS for yield in each agricultural environment was assessed by comparing predicted yield gains for each PS scenario. The predicted gain in environment j (represented by Alger or Marchouch) from one selection cycle of direct PS is [62]:

$$\Delta G_{Pj} = i_j \, H_j^2 \, \sigma_{p(j)} \tag{1}$$

where i_j is the standardized selection differential, H_j^2 is the broad-sense heritability on a line mean basis, and $\sigma_{p(j)}$ is the phenotypic standard deviation of the line mean values. The predicted yield gain in environment j from indirect selection in environment j' (represented by the MS environment) is [62,63]:

$$\Delta G_{Pjj'} = i_{j'} \, H_j \, H_{j'} \, r_{g(jj')} \, \sigma_{p(j)} \tag{2}$$

where $i_j{'}$ is the standardized selection differential in the environment j', H_j, and $H_{j'}$ are square root values of the broad-sense heritability on a line mean basis in the environments j and j', respectively, and $r_{g(jj')}$ is the genetic correlation for line yield responses across the two environments. We assumed $i_j = i_j{'}$ for both selection scenarios and used the ratio $(H_j\, H_{j'}\, r_{g(jj')})/H_j^2$ to estimate the predicted efficiency of indirect PS in the MS environment relative to direct PS in each target environment. This assessment ought to be considered as preliminary, as it could not account for GE interactions within each agricultural site arising from year-to-year climatic variation (which may be large, unlike those expected in a MS environment). Relevant r_g values estimated according to [61], H^2 values estimated by a restricted maximum likelihood method, and relative efficiency values, were assessed separately for each RIL population, reporting the values averaged across populations. H^2 values were also used to compute best linear unbiased prediction (BLUP) values according to [64], which were used for subsequent GS analyses. BLUP-based values of grain yield of the test material in the three cropping environments are reported in the Data repository S1 provided as supplementary material.

Statistical analyses of phenotyping data were carried out using SAS/STAT® software (SAS Institute Inc, Cary, NC, USA) [65] and, for AMMI analysis, CropStat software (International Rice Research Institute, Manila, The Philippines) [66].

4.2. Definition of GS and MAS Procedures

DNA was extracted from bulked stipules of four F_6 plants per genotype. Details of DNA isolation, GBS library construction, sequencing, genotype SNP calling, and SNP data filtering were reported in [30]. In brief, we adopted Elshire et al.'s [20] GBS protocol with modifications, using the *ApeKI* restriction enzyme and KAPA Taq polymerase. Raw reads (100 bp, single end read) were quality-filtered, de-multiplexed, and trimmed to 64 bp, grouping identical reads into one tag. We retained tags with 10 or more reads across all individuals for pairwise alignment aimed to find tag pairs that differed by 1 bp. The read distribution of the paired tags in each individual was used for SNP genotype calling, which, as in [30], was performed by each of two filtering criteria that removed markers with less than four or less than six aligned reads per locus, respectively. The latter, more conservative criterion aimed to minimize the risk of imperfect SNP calling arising from residual heterozygosity in the genotyped material. Markers that were monomorphic or with minor allele frequency <2.5% were removed. The data set was filtered for increasing levels of allowed genotype SNP missing values, excluding markers whose missing rate exceeded fixed thresholds of 10%, 20%, 30%, 40%, and 50%. SNP missing data were estimated using the K-Nearest neighbors imputation algorithm (K = 4) coupled with the simple matching coefficient distance function [67]. SNP marker data for the five thresholds of genotype missing data are provided in the Data repository S1.

We considered two GS models for yield prediction that stood out for predictive ability in a previous model comparison for pea grain yield limited to Lodi's MS environment [30], i.e., Bayesian Lasso (BL; [68]) and Ridge regression BLUP (rrBLUP; [69]). While rrBLUP assumes that the effects of all loci have a common variance, BL assumes relatively few markers with large effects, allowing different markers to have different effects and variances [70]. Bayesian models assign prior densities to markers effects, thereby inducing different types of shrinkage, obtaining the solution by sampling from the resulting posterior density [68].

For each of the two SNP calling criteria, we assessed GS models trained either on pooled data of the three RIL populations or the individual populations, with five possible genotype SNP missing data thresholds (10%, 20%, 30%, 40%, 50%). BL and rrBLUP models with different combinations of data training and SNP missing data thresholds were assessed for intra-environment predictive ability, as well as for ability to predict line mean yields across MS and Marchouch environments (whose application for actual GS of drought-tolerant lines was supported by GE interaction analysis results). We also assessed (i) the cross-environment predictive ability of GS models constructed in one environment to predict breeding values of independent lines in another environment; (ii) the cross-environment predictive accuracy r_{Ac} of the same GS models, by which the model predictive ability r_{Ab} is readjusted as a function of the square root of the broad-sense heritability on a line mean basis H_j of the predicted environment j by the formula $r_{Ac} = r_{Ab}/H_j$ [71]; and (iii) the ability of GS models constructed from line mean yields across MS and Marchouch environments to predict yield responses of independent lines in each agricultural site. Predictive ability values were measured as Pearson's correlation between observed and predicted phenotypes using cross-validations for the single RIL populations (to avoid bias associated with different population mean value) and then averaging results across populations. Finally, we explored the cross-population predictive ability of GS by assuming model training on data of one connected RIL population or on pooled data of two connected populations for all possible combinations of training and validation populations, for two traits of practical interest represented by yield in Alger and mean yield across MS and Marchouch environments. GS regression modelling, cross-validations, and predictive ability estimation were performed using the R package GROAN [72], adopting 50 repetitions of a 10-fold stratified cross-validation scheme for each analysis.

The MAS criterion for intrinsic drought tolerance was defined on the basis of the GWAS reported in [30] for 315 lines belonging to the same RIL populations, which found 10 linked SNP markers with association score ≥ 2.25. Further insight on the genomic position of these markers was obtained by aligning their sequence to the pea draft genome under construction by the International Pea Genome Sequencing Project coordinated by INRA and Tayeh et al.'s [73] consensus map [74]. MAS was based on seven aligned markers which belonged to five genomic regions (markers TP78343 and TP13485, on LG 5; TP94476, on LG 1; TP6268, on LG 3; TP63677 and TP51372, around 32-33 cM of LG 7 in Tayeh et al.'s [73] map; and TP6885, around 76-77 cM of LG 7 in Tayeh et al.'s [73] map; see Supplementary Table S4). Based on our late verification of the position of these markers on Kreplak et al.'s [75] pea reference genome using the alignment tool *bwa* [76], these markers aligned on five gene coding regions of chromosomes 2, 3, 5, and 7 (Supplementary Table S4) for which we report gene names and descriptions as available in the Pulsedb website (https://www.pulsedb.org/Analysis/989011). We envisaged MAS based on two criteria, i.e., the number of favorable alleles over the seven markers, and the number of favorable alleles over the five putative QTL belonging to the five genomic regions. Both criteria, however, identified the same sets of top- and bottom-ranking independent lines.

4.3. Comparison of Genomic vs. Phenotypic Selection Based on Predicted Yield Gains

An estimation of the predicted yield gain from one cycle of GS in environment j is [77]:

$$\Delta G_{Gj} = i_j'' \, r_{Ac} \, \sigma_{a(j)} \tag{3}$$

where i_j'' is the standardized selection differential used for GS, r_{Ac} is the GS model accuracy, and $\sigma_{a(j)}$ is the standard deviation of the line breeding values. Recalling that $r_{Ac} = r_{Ab}/H_j$ and $\sigma_{a(j)} = \sigma_{p(j)} H_j$, another expression of ΔG_{Gj} as a function of the GS model predictive ability r_{Ab} is:

$$\Delta G_{Gj} = i_j \, r_{Ab} \, \sigma_{p(j)} \tag{4}$$

which, when compared to the predicted gain from direct PS (ΔG_{Pj}) reported in Equation (1), indicates that the ratio of ($i_j'' \, r_{Ab}$) to ($i_j \, H_j^2$) could be used to estimate the predicted efficiency of GS relative to direct PS in the target environment. However, a fair comparison of PS vs. GS ought to be based on the same costs. PS based on one field experiment with three replications may imply about 2.5 [78] to 3.1 [32] greater cost per evaluated line than GBS-based GS, which, considering the average value of 2.8, implies 2.8 more test lines and 2.8 smaller selected fraction for GS relative to PS when assuming same evaluation costs. Hence, we hypothesized 10% selected fraction, i.e., $i_j = 1.755$ [60], for PS, and 3.6% selected fraction, i.e., $i_j'' = 2.197$, for GS, and used the ratio $(2.197 \, r_{Ab})/(1.755 \, H_j^2)$ to estimate the predicted efficiency of GS relative to PS. We envisaged GS for each environment based on best-performing environment-specific models or the best-performing model constructed from line mean yields across MS and Marchouch environments, computing r_{Ab} and H_j^2 values separately for each RIL population and reporting relative efficiency values averaged across populations. Particularly for agricultural sites, this comparison of selection strategies ought to be seen as preliminary, as it could not account for within-site GE interactions across cropping seasons neither for PS (where they would be accounted for in the denominator of H_j^2) nor for GS (where they would be accounted for by assessing r_{Ab} across different test years rather than through intra-environment cross-validations). Additionally, this comparison tended to underestimate the relative value of GS, as it did not account for the further advantage of shorter selection cycle duration offered by GS relative to PS.

4.4. Comparison of Genomic vs. Phenotypic Selection Based on Actual Yield Gains (Experiment 4)

In this study, GS was applied to independent lines using top-performing statistical models in the earlier assessment that were constructed from line mean yields across Lodi's MS environment and Marchouch. GS was applied using either the top-performing model trained on joint data of the three RIL populations, or the models trained on data of each separate population (RIL population-specific models). The set of 288 RILs was split into two subsets. The former set included 30 randomly chosen lines per RIL population, which acted as independent lines for GS and PS. The latter set included the remaining lines of the three populations, whose data were used for GS model definition. For each GS model, we selected for each RIL population the three lines out of 30 that were top-ranking for predicted yield. Likewise, we selected phenotypically for each RIL population the three lines out of 30 that were top-ranking for mean yield across MS and Marchouch environments. In addition, we performed environment-specific PS for Lodi's MS environment and for Marchouch, by selecting the three top-yielding lines out of 30 separately for each environment.

The experiment included the 45 lines issued by selecting three lines for each of three RIL populations for each of the five selection criteria (PS in the MS environment, in Marchouch, and across the two environments; GS based on population-specific and on joint-population data), and the RIL parent lines, which acted as a reference to assess yield gains. The material was evaluated for dry grain yield and aerial biomass (estimated on plant material oven-dried at 90 °C for four days) and onset of flowering in Exp. 4, which was performed in the same MS environment adopted for Exp. 1. Exp. 4 was designed as an RCB experiment with four replications, using same plot size and sowing density as Exp. 1. Compared to Exp. 1, Exp. 4 involved similar available water for the crop but over six-week later sowing (Table 1), which increased somewhat the extent of drought stress exerted on the selected material.

A first ANOVA that excluded parent line data aimed to compare the five selection criteria and to assess the interaction between selection criteria and RIL populations. It included the fixed factors line group (whose variants were defined by the selection criteria), RIL population, and line within group and RIL population, along with the random factor block. A second ANOVA that included parent lines as an additional line group and that held the factors line group, line within group and block aimed to compare each group of selected lines with the parent line group by Dunnett's multiple mean test.

4.5. Comparison of Material with Contrasting Genomic Predictions (Experiment 5)

The best-predicting GS model constructed from joint-population data of line mean yields across Lodi's MS experiment and Marchouch in the earlier assessment was applied to an independent set of 90 lines that included 30 lines from each of the same three crosses (A × I, K × A, and K × I). These new lines were issued from four generations of bulk selection under local autumn-sown field conditions in Lodi (F_2 to F_5 generation, starting with 800 F_2 seeds per cross), and were genotyped as F_6 plants by GBS as described earlier. Local selection for tolerance to low winter temperatures was expected to produce a shift towards later onset of flowering in this material relative to the mean phenology of the RIL populations issued from the same crosses. GS-based predictions were exploited to select the two top-ranking lines, two mid-ranking lines (ranks 15 and 16), and the two bottom-ranking lines out of 30 lines within each cross. The 18 selected lines and the three parent lines were evaluated for grain yield, aerial biomass, and onset of flowering in Lodi's MS environment by an RCB experiment with four replications (Exp. 5) whose management was identical to Exp. 4. The data analysis contemplated two

ANOVAs as described for Exp. 4, the only difference being the presence of a Cross factor instead of a RIL population factor in the first ANOVA (as the selection was not applied to RIL material in this case).

4.6. Comparison of Genomic Selection vs. Marker-Assisted Selection for Intrinsic Drought Tolerance (Experiment 6)

This study assessed both GS and MAS for intrinsic drought tolerance. The GS model was constructed using the top-predicting GS model in the earlier assessment that was trained on joint data of 192 lines belonging to the RIL populations K × A and K × I evaluated in Exp. 1, and was applied to lines of the RIL population A × I (which was selected because it displayed smaller variation in onset of flowering than the other populations in earlier work [30]). To limit the impact on grain yield of line variation for earliness, we applied GS and MAS to a subset of 24 lines out of the 96 lines available for the A × I population whose onset of flowering in Exp. 1 fell in the interval $m \pm s$, where m and s stand for mean and standard deviation values, respectively, of the phenology trait (which implied a range of 1.1 days among the 24 test lines). GS-based predictions for this subset of lines were exploited to select the three top-ranking lines, two mid-ranking lines, and three bottom-ranking lines. Likewise, MAS was used to predict three top-ranking lines that possessed all 14 favorable alleles for the seven target markers and the five relevant genomic regions, and three bottom-ranking lines that possessed no favorable alleles. The two GS-based mid-ranking lines acted as mid-ranking material also according to the MAS criterion, as they displayed six to eight favorable alleles overall. The 14 selected lines and the two parent lines were evaluated for grain yield, aerial biomass, and onset of flowering in Lodi's MS environment by an RCB experiment with four replications (Exp. 6) whose management was identical to Exp. 4 and 5.

A first ANOVA that excluded parent line data and included the fixed factors line group and line within group and the random factor block aimed to compare the five groups of lines. A second ANOVA including also parent line data compared each line group to the mean value of parent lines by Dunnett's multiple mean test.

Supplementary Materials: **Supplementary Materials** can be found at http://www.mdpi.com/1422-0067/21/7/2414/s1. **Table S1:** AMMI analysis for grain yield of 288 lines belonging to three connected RIL populations, three parent cultivars and one recent control cultivar grown in a managed drought stress environment of Lodi (Italy) and two agricultural environments of Marchouch (Morocco) and Alger (Algeria). **Table S2:** Cross-environment predictive ability (r_{Ab}) and predictive accuracy (r_{Ac}) of the top-performing of models constructed by Bayesian Lasso (BL) or Ridge Regression BLUP (rrBLUP) for grain yield breeding value of pea lines belonging to three connected RIL populations in a managed drought stress (MS) environment (Lodi, Italy) and two agricultural sites (Marchouch, Morocco; Alger, Algeria). **Table S3:** ANOVA F test results for grain yield, aerial biomass, and onset of flowering under managed drought stress (MS) of pea line groups belonging to different RIL populations or crosses. **Table S4:** Markers associated with intrinsic drought tolerance (as grain yield deviation from the value expected according to onset of flowering) and their location in five genomic areas of Tayeh et al.'s [73] consensus map and on Kreplak et al.'s [75] pea reference genome (for which we report also the name of the gene coding region). Donor genotype: A = Attika; I = Isard; K = Kaspa. See Supplementary Table 1 in Annicchiarico et al. [30] for SNP marker sequence. **Figure S1:** Predictive ability of two genomic selection models (Bayesian Lasso, BL, and Ridge Regression BLUP, rrBLUP) and five thresholds of genotype SNP missing data (10%, 20%, 30%, 40%, 50%), for grain yield of pea lines belonging to three connected RIL populations in a managed drought stress (MS) environment of Lodi (Italy) and the agricultural environments of Marchouch (Morocco) and Alger (Algeria) and for mean grain yield across the MS environment and Marchouch. Model training on data of three RIL populations (encompassing 288 lines overall), averaging validation results for individual populations based on 50 repetitions of 10-fold stratified cross-validations per individual analysis. **Data repository S1:** Phenotypic and genotypic data. Grain yield in three environments, and SNP marker data for five thresholds of genotype SNP missing data (10%, 20%, 30%, 40%, 50%), for 288 pea lines belonging to three connected RIL populations.

Author Contributions: Conceptualization and funding acquisition, P.A.; data generation, P.A., N.N., M.L., I.T.-A., M.R. and L.P.; data analysis, N.N. and P.A.; draft preparation, P.A., L.P. and N.N. All authors have read and agreed to the published version of the manuscript.

Funding: The multi-environment study and the development of GS and MAS procedures were funded by the FP7-ArimNet project 'Resilient, water- and energy-efficient forage and feed crops for Mediterranean agricultural systems (REFORMA)'. The assessment of selections issued by GS and MAS was funded by the FP7 project 'Legumes for the agriculture of tomorrow (LEGATO)' (grant agreement No. 613551).

Acknowledgments: We are grateful to E.C. Brummer, Y. Wei and B. Ferrari for their contribution to generation of GBS data, J. Kreplak and G. Aubert for verifying the position of key GBS-generated markers on the draft genome of pea, and P. Gaudenzi, M.R. Herma, A. Mebarkia, A. Passerini, S. Proietti, and M. Redha El-Amine, for technical assistance.

Conflicts of Interest: The authors declare no conflict of interest.

Abbreviations

rrBLUP	Ridge Regression BLUP (best linear unbiased prediction)
ANOVA	Analysis of variance
AMMI	Additive Main effects and Multiplicative Interaction
GBS	Genotyping-by-sequencing
MAS	Marker-assisted selection
RIL	Recombinant inbred line
SNP	Single nucleotide polymorphism
BL	Bayesian Lasso
GE	Genotype × environment
GS	Genomic selection
MS	Managed stress
PC	Principal component
PS	Phenotypic selection

References

1. Turral, H.; Burke, J.; Faurès, J.-M. *Climate Change, Water and Food Security*; FAO Water Reports No. 36; FAO: Rome, Italy, 2011.
2. Gomiero, T. Soil degradation, land scarcity and food security: Reviewing a complex challenge. *Sustainability* **2016**, *8*, 281. [CrossRef]
3. Lassaletta, L.; Billen, G.; Garnier, J.; Bouwman, L.; Velazquez, E.; Mueller, N.D. Nitrogen use in the global food system: Past trends and future trajectories of agronomic performance, pollution, trade, and dietary demand. *Env. Res. Lett.* **2016**, *11*, 095007. [CrossRef]
4. Cellier, P.; Schneider, A.; Thiébeau, P.; Vertès, F. Impacts environnementaux de l'introduction de légumineuses dans les systèmes de production. In *Les Légumineuses pour des Systèmes Agricoles et al. Imentaires Durables*; Schneider, A., Huyghe, C., Eds.; Editions Quae: Versailles, France, 2015; pp. 297–338.
5. Foyer, C.H.; Lam, H.-M.; Nguyen, H.T.; Siddique, K.H.M.; Varshney, R.K.; Colmer, T.D. Neglecting legumes has compromised human health and sustainable food production. *Nat. Plants* **2016**, *2*, 16112. [CrossRef] [PubMed]
6. Zander, P.; Amjath-Babu, T.S.; Preissel, S.; Reckling, M.; Bues, A.; Schläfke, N. Grain legume decline and potential recovery in European agriculture: A review. *Agron. Sustain. Dev.* **2016**, *36*, 26. [CrossRef]
7. FAO. *The State of Food and Agriculture*; Livestock in the balance; FAO: Rome, Italy, 2010.
8. Pilorgé, E.; Muel, F. What vegetable oils and proteins for 2030? Would the protein fraction be the future of oil and protein crops? *OCL* **2016**, *23*, D402. [CrossRef]
9. Schreuder, R.; De Visser, C. *EIP-AGRI Focus Group on protein crops: Final Report*; European Commission: Brussels, Belgium, 2014.
10. Magrini, M.-B.; Anton, M.; Choleza, C.; Corre-Hellou, G.; Duc, G.; Jeuffroy, M.-H. Why are grain-legumes rarely present in cropping systems despite their environmental and nutritional benefits? Analyzing lock-in in the French agrifood system. *Ecol. Econ.* **2016**, *126*, 152–162. [CrossRef]
11. Araújo, S.S.; Beebe, S.; Crespi, M.; Delbreil, B.; González, E.M.; Gruber, V. Abiotic stress responses in legumes: Strategies used to cope with environmental challenges. *Crit. Rev. Plant Sci.* **2015**, *34*, 237–280. [CrossRef]
12. Alessandri, A.; De Felice, M.; Zeng, N.; Mariotti, A.; Pan, Y.; Cherchi, A. Robust assessment of the expansion and retreat of Mediterranean climate in the 21st century. *Sci. Rep.* **2014**, *4*, 7211. [CrossRef]
13. Annicchiarico, P. Adaptation of cool-season grain legume species across climatically-contrasting environments of southern Europe. *Agron. J.* **2008**, *100*, 1647–1654. [CrossRef]

14. Warkentin, T.; Smykal, P.; Coyne, C.J.; Weeden, N.; Domoney, C.; Bing, D. Pea (*Pisum sativum* L.). In *Handbook of Plant Breeding: Grain Legumes*; De Ron, A., Ed.; Springer Science and Business Media: New York, NY, USA, 2015; pp. 37–83.

15. Annicchiarico, P. Feed legumes for truly sustainable crop-animal systems. *It. J. Agron.* **2017**, *12*, 880. [CrossRef]

16. Annicchiarico, P.; Thami-Alami, I.; Abbas, K.; Pecetti, L.; Melis, R.A.M.; Porqueddu, C. Performance of legume-based annual forage crops in three semi-arid Mediterranean environments. *Crop Pasture Sci.* **2017**, *68*, 932–941. [CrossRef]

17. Carrouée, B.; Crépon, K.; Peyronnet, C. Les protéagineux: Intérêt dans les systèmes de production fourragers français et européens. *Fourrages* **2003**, *174*, 163–182.

18. Meuwissen, T.H.E.; Hayes, B.J.; Goddard, M.E. Prediction of total genetic value using genome-wide dense marker maps. *Genetics* **2001**, *157*, 1819–1829. [CrossRef] [PubMed]

19. Wiggans, G.R.; Cole, J.B.; Hubbard, S.M.; Sonstegard, T.S. Genomic selection in dairy cattle: The USDA experience. *Ann. Rev. Anim. Biosci.* **2017**, *5*, 309–327. [CrossRef]

20. Elshire, R.J.; Glaubitz, J.C.; Sun, Q.; Poland, J.A.; Kawamoto, K.; Buckler, E.S. A robust, simple genotyping-by-sequencing (GBS) approach for high diversity species. *PLoS ONE* **2011**, *6*, 19379. [CrossRef]

21. Elbasyoni, I.S.; Lorenz, A.J.; Guttieri, M.; Frels, K.; Baenziger, P.S.; Poland, J.; Akhunov, E. A comparison between genotyping-by-sequencing and array-based scoring of SNPs for genomic prediction accuracy in winter wheat. *Plant Sci.* **2018**, *270*, 123–130. [CrossRef]

22. Varshney, R.K.; Kudapa, H.; Pazhamala, L.; Chitikineni, A.; Thudi, M.; Bohra, A. Translational genomics in agriculture: Some examples in grain legumes. *Crit. Rev. Plant Sci.* **2015**, *34*, 169–194. [CrossRef]

23. Jarquín, D.; Kocak, K.; Posadas, L.; Hyma, K.; Jedlicka, J.; Graef, G. Genotyping by sequencing for genomic prediction in a soybean breeding population. *BMC Genomics* **2014**, *15*, 740. [CrossRef]

24. Jarquín, D.; Specht, J.; Lorenz, A. Prospects of genomic prediction in the USDA soybean germplasm collection: Historical data creates robust models for enhancing selection of accessions. *G3 (Bethesda)* **2016**, *6*, 2329–2341. [CrossRef]

25. Duhnen, A.; Gras, A.; Teyssèdre, S.; Romestant, M.; Claustres, B.; Daydé, J. Genomic selection for yield and seed protein content in soybean: A study of breeding program data and assessment of prediction accuracy. *Crop Sci.* **2017**, *57*, 1325–1337. [CrossRef]

26. Ma, Y.; Reif, J.C.; Jiang, Y.; Wen, Z.; Wang, D.; Liu, Z. Potential of marker selection to increase prediction accuracy of genomic selection in soybean (*Glycine max* L.). *Mol. Breed.* **2016**, *36*, 113. [CrossRef] [PubMed]

27. Annicchiarico, P.; Nazzicari, N.; Ferrari, B.; Harzic, N.; Carroni, A.M.; Romani, M.; Pecetti, L. Genomic prediction of grain yield in contrasting environments for white lupin genetic resources. *Mol. Breed.* **2019**, *39*, 142. [CrossRef]

28. Annicchiarico, P.; Nazzicari, N.; Pecetti, L.; Romani, M.; Russi, L. Pea genomic selection for Italian environments. *BMC Genomics* **2019**, *20*, 603. [CrossRef] [PubMed]

29. Roorkiwal, M.; Rathore, A.; Das, R.R.; Singh, M.K.; Jain, A.; Srinivasan, S. Genome-enabled prediction models for yield related traits in chickpea. *Front. Plant Sci.* **2016**, *7*, 1666. [CrossRef]

30. Annicchiarico, P.; Nazzicari, N.; Pecetti, L.; Romani, M.; Ferrari, B.; Wei, Y. GBS-based genomic selection for pea grain yield under severe terminal drought. *The Plant Genome* **2017**, *10*. [CrossRef]

31. Burstin, J.; Salloignon, P.; Chabert-Martinello, M.; Magnin-Robert, J.-B.; Siol, M.; Jacquin, F. Genetic diversity and trait genomic prediction in a pea diversity panel. *BMC Genomics* **2015**, *16*, 105. [CrossRef]

32. Annicchiarico, P.; Nazzicari, N.; Wei, Y.; Pecetti, L.; Brummer, E.C. Genotyping-by-sequencing and its exploitation for forage and cool-season grain legume breeding. *Front. Plant Sci.* **2017**, *8*, 679. [CrossRef]

33. Annicchiarico, P.; Iannucci, A. Winter survival of pea, faba bean and white lupin cultivars across contrasting Italian locations and sowing times, and implications for selection. *J. Agric. Sci.* **2007**, *145*, 611–622. [CrossRef]

34. Cooper, M.; Stucker, R.E.; DeLacy, I.H.; Harch, B.D. Wheat breeding nurseries, target environments, and indirect selection for grain yield. *Crop Sci.* **1997**, *37*, 1168–1176. [CrossRef]

35. Annicchiarico, P.; Piano, E. Use of artificial environments to reproduce and exploit genotype × location interaction for lucerne in northern Italy. *Theor. Appl. Genet.* **2005**, *110*, 219–227. [CrossRef]

36. Ceccarelli, S. Wide adaptation: How wide? *Euphytica* **1989**, *40*, 197–205. [CrossRef]

37. Ceccarelli, S. Specific adaptation and breeding for marginal conditions. *Euphytica* **1994**, *77*, 205–219. [CrossRef]

38. Sadras, V.O.; Lake, L.; Leonforte, A.; McMurray, L.S.; Paull, J.G. Screening field pea for adaptation to water and heat stress: Associations between yield, crop growth rate and seed abortion. *Field Crops Res.* **2013**, *150*, 63–73. [CrossRef]

39. Annicchiarico, P.; Iannucci, A. Adaptation strategy, germplasm type and adaptive traits for field pea improvement in Italy based on variety responses across climatically contrasting environments. *Field Crops Res.* **2008**, *108*, 133–142. [CrossRef]

40. Iglesias-García, R.; Prats, E.; Flores, F.; Amri, M.; Mikić, A.; Rubiales, D. Assessment of field pea (*Pisum sativum* L.) grain yield, aerial biomass and flowering date stability in Mediterranean environments. *Crop Pasture Sci.* **2017**, *68*, 915–923. [CrossRef]

41. Pecetti, L.; Marcotrigiano, A.R.; Russi, L.; Romani, M.; Annicchiarico, P. Adaptation of field pea varieties to organic farming across different environments of Italy. *Crop Pasture Sci.* **2019**, *70*, 327–333. [CrossRef]

42. Bocianowski, J.; Księżak, J.; Nowosad, K. Genotype by environment interaction for seeds yield in pea (*Pisum sativum* L.) using additive main effects and multiplicative interaction model. *Euphytica* **2019**, *215*, 191. [CrossRef]

43. Rodríguez-Maribona, B.; Tenorio, J.L.; Conde, J.R.; Ayerbe, L. Correlation between yield and osmotic adjustment of peas (*Pisum sativum* L.) under drought stress. *Field Crops Res.* **1992**, *29*, 15–22. [CrossRef]

44. Grzesiak, S.; Iijima, M.; Kono, Y.; Yamauchi, A. Differences in drought tolerance between cultivars of field bean and field pea. A comparison of drought-resistant and drought-sensitive cultivars. *Acta Physiol. Plant.* **1997**, *19*, 349–357. [CrossRef]

45. Sánchez, F.J.; Manzanares, M.; de Andres, E.F.; Tenorio, J.L.; Ayerbe, L. Turgor maintenance, osmotic adjustment and soluble sugar and proline accumulation in 49 pea cultivars in response to water stress. *Field Crops Res.* **1998**, *59*, 225–235. [CrossRef]

46. Blum, A. Effective use of water (EUW) and not water-use efficiency (WUE) is the target of crop yield improvement under drought stress. *Field Crop Res.* **2009**, *112*, 119–123. [CrossRef]

47. Charmet, G.; Storlie, E.; Oury, F.X.; Laurent, V.; Beghin, D.; Chevarin, L. Genome-wide prediction of three important traits in bread wheat. *Mol. Breed.* **2014**, *34*, 1843–1852. [CrossRef] [PubMed]

48. Brummer, E.C.; Li, X.; Wei, Y.; Hanson, J.L.; Viands, D.R. The imperative of improving yield of perennial forage crops: Will genomic selection help? *Grassl. Sci. Eur.* **2019**, *24*, 370–372.

49. Lozada, D.N.; Mason, R.E.; Sarinelli, J.M.; Brown-Guedira, G. Accuracy of genomic selection for grain yield and agronomic traits in soft red winter wheat. *BMC Genet.* **2019**, *20*, 82. [CrossRef] [PubMed]

50. Michel, S.; Ametz, C.; Gungor, H.; Akgöl, B.; Epure, D.; Grausgruber, H. Genomic assisted selection for enhancing line breeding: Merging genomic and phenotypic selection in winter wheat breeding programs with preliminary yield trials. *Theor. Appl. Genet.* **2017**, *130*, 363–376. [CrossRef]

51. Beyene, Y.; Semagn, K.; Mugo, S.; Tarekegne, A.; Babu, R.; Meisel, B. Genetic gains in grain yield through genomic selection in eight bi-parental maize populations under drought stress. *Crop Sci.* **2015**, *55*, 154–163. [CrossRef]

52. Beyene, Y.; Gowda, M.; Olsen, M.; Robbins, K.R.; Pérez-Rodríguez, P.; Alvarado, G. Empirical comparison of tropical maize hybrids selected through genomic and phenotypic selections. *Front. Plant Sci.* **2019**, *10*, 1502. [CrossRef]

53. Môro, G.V.; Santos, M.F.; de Souza Jr, C.L. Comparison of genome-wide and phenotypic selection indices in maize. *Euphytica* **2019**, *215*, 76. [CrossRef]

54. Sallam, A.; Smith, K.P. Genomic selection performs similarly to phenotypic selection in barley. *Crop Sci.* **2016**, *56*, 2871–2881. [CrossRef]

55. Bernardo, R.; Yu, J. Prospects for genomewide selection for quantitative traits in maize. *Crop Sci.* **2007**, *47*, 1082–1090. [CrossRef]

56. Massman, J.M.; Jung, H.J.G.; Bernardo, R. Genomewide selection versus marker-assisted recurrent selection to improve grain yield and stover-quality traits for cellulosic ethanol in maize. *Crop Sci.* **2013**, *53*, 1–9. [CrossRef]

57. Annicchiarico, P. Scelta varietale in pisello e favino rispetto all'ambiente e all'utilizzo. *Inf. Agr.* **2005**, *61*, 47–52.

58. Cochran, W.G.; Cox, G.M. *Experimental Designs*, 2nd ed.; John Wiley & Sons: New York, NY, USA, 1957.

59. Gauch, H.G.; Piepho, H.-P.; Annicchiarico, P. Statistical analysis of yield trials by AMMI and GGE: Further considerations. *Crop Sci.* **2008**, *48*, 866–889. [CrossRef]

60. Piepho, H.-P. Robustness of statistical tests for multiplicative terms in the additive main effects and multiplicative interaction model for cultivar trials. *Theor. Appl. Genet.* **1995**, *90*, 438–443. [CrossRef]

61. Robertson, A. The sampling variance of the genetic correlation coefficient. *Biometrics* **1959**, *15*, 469–485. [CrossRef]

62. Falconer, D.S. *Introduction to Quantitative Genetics*, 3rd ed.; Longman: Harlow, UK, 1989.

63. Cooper, M.; DeLacy, I.H.; Basford, K.E. Relationships among analytical methods used to analyse genotypic adaptation in multi-environment trials. In *Plant Adaptation and Crop Improvement*; Cooper, M., Hammer, G.L., Eds.; CABI: Wallingford, UK, 1996; pp. 193–224.

64. DeLacy, I.H.; Basford, K.E.; Cooper, M.; Bull, I.K.; McLaren, C.G. Analysis of multi-environment trials – An historical perspective. In *Plant Adaptation and Crop Improvement*; Cooper, M., Hammer, G.L., Eds.; CABI: Wallingford, UK, 1996; pp. 39–124.

65. SAS Institute. *SAS/STAT® 9.3 User's Guide*; SAS Institute Inc.: Cary, NC, USA, 2011.

66. IRRI. *Cropstat Version 7.2*; International Rice Research Institute: Manila, The Philippines, 2009.

67. Schwender, H. Statistical Analysis of Genotype and Gene Expression Data. Available online: https://eldorado.tu-dortmund.de/handle/2003/23306 (accessed on 20 January 2020).

68. Park, T.; Casella, G. The Bayesian Lasso. *J. Am. Statist. Assoc.* **2008**, *103*, 681–686. [CrossRef]

69. Searle, S.R.; Casella, G.; McCulloch, C.E. *Variance Components*; John Wiley & Sons: New York, NY, USA, 2009.

70. Wang, X.; Xu, Y.; Hu, Z.; Xu, C. Genomic selection methods for crop improvement: Current status and prospects. *Crop J.* **2018**, *6*, 330–340. [CrossRef]

71. Lorenz, A.J.; Chao, S.; Asoro, F.G.; Heffner, E.L.; Hayashi, T.; Iwata, H.; Smith, K.P.; Sorrells, M.E.; Jannink, J.-L. Genomic selection in plant breeding: Knowledge and prospects. *Adv. Agron.* **2011**, *110*, 77–123. [CrossRef]

72. Nazzicari, N.; Biscarini, F. GROAN: Genomic Regression Workbench (Version 1.0.0). Available online: https://cran.r-project.org/package=GROAN (accessed on 20 January 2020).

73. Tayeh, N.; Aluome, C.; Falque, M.; Jacquin, F.; Klein, A.; Chauveau, A. Development of two major resources for pea genomics: The GenoPea 13.2K SNP Array and a high-density, high-resolution consensus genetic map. *Plant J.* **2015**, *84*, 1257–1273. [CrossRef]

74. Kreplak, J.; Aubert, G. (INRA, Dijon, France). Personal communication, 2019.

75. Kreplak, J.; Madoui, M.A.; Cápal, P.; Novák, P.; Labadie, K.; Aubert, G. A reference genome for pea provides insight into legume genome evolution. *Nature Genet.* **2019**, *51*, 1411–1422. [CrossRef]

76. Li, H.; Durbin, R. Fast and accurate short read alignment with Burrows-Wheeler Transform. *Bioinformatics* **2009**, *25*, 1754–1760. [CrossRef] [PubMed]

77. Heffner, E.L.; Lorenz, A.J.; Jannink, J.L.; Sorrells, M.E. Plant breeding with genomic selection: Gain per unit time and cost. *Crop Sci.* **2010**, *50*, 1681–1690. [CrossRef]

78. Bassi, F.M.; Bentley, A.R.; Charmet, G.; Ortiz, R.; Crossa, J. Breeding schemes for the implementation of genomic selection in wheat (*Triticum* spp.). *Plant Sci.* **2016**, *242*, 23–36. [CrossRef] [PubMed]

 International Journal of
Molecular Sciences

Article

Genomic Analysis of Vavilov's Historic Chickpea Landraces Reveals Footprints of Environmental and Human Selection

Alena Sokolkova [1], Sergey V. Bulyntsev [2], Peter L. Chang [3], Noelia Carrasquilla-Garcia [4], Anna A. Igolkina [1], Nina V. Noujdina [1,5], Eric von Wettberg [6], Margarita A. Vishnyakova [2], Douglas R. Cook [4,*], Sergey V. Nuzhdin [1,3,*] and Maria G. Samsonova [1,*]

[1] Department of Applied Mathematics, Peter the Great St. Petersburg Polytechnic University, 195251 St. Petersburg, Russia; alyonasok@yandex.ru (A.S.); igolkinaanna11@gmail.com (A.A.I.); nnoujdina@gmail.com (N.V.N.)

[2] Federal Research Centre All-Russian N.I. Vavilov Institute of Plant Genetic Resources (VIR), 190000 St. Petersburg, Russia; s_bulyntsev@mail.ru (S.V.B.); m.vishnyakova@vir.nw.ru (M.A.V.)

[3] Dornsife College of Letters Arts & Sciences, Program in Molecular and Computational Biology, University of Southern California, Los Angeles, CA 90089, USA; peterc@usc.edu

[4] Department of Plant Pathology, University of California Davis, Davis, CA 95616, USA; noecarras@ucdavis.edu

[5] Department of Geography, University of California Los Angeles, Los Angeles, CA 90095, USA

[6] Department of Plant and Soil Science, University of Vermont, Burlington, VT 05405, USA; Eric.Bishop-Von-Wettberg@uvm.edu

* Correspondence: drcook@ucdavis.edu (D.R.C.); snuzhdin@usc.edu (S.V.N.); m.g.samsonova@gmail.com (M.G.S.); Tel.: +1-530-754-6561 (D.R.C.); +7-812-2909645 (M.G.S.)

Received: 19 April 2020; Accepted: 28 May 2020; Published: 31 May 2020

Abstract: A defining challenge of the 21st century is meeting the nutritional demands of the growing human population, under a scenario of limited land and water resources and under the specter of climate change. The Vavilov seed bank contains numerous landraces collected nearly a hundred years ago, and thus may contain 'genetic gems' with the potential to enhance modern breeding efforts. Here, we analyze 407 landraces, sampled from major historic centers of chickpea cultivation and secondary diversification. Genome-Wide Association Studies (GWAS) conducted on both phenotypic traits and bioclimatic variables at landraces sampling sites as extended phenotypes resulted in 84 GWAS hits associated to various regions. The novel haploblock-based test identified haploblocks enriched for single nucleotide polymorphisms (SNPs) associated with phenotypes and bioclimatic variables. Subsequent bi-clustering of traits sharing enriched haploblocks underscored both non-random distribution of SNPs among several haploblocks and their association with multiple traits. We hypothesize that these clusters of pleiotropic SNPs represent co-adapted genetic complexes to a range of environmental conditions that chickpea experienced during domestication and subsequent geographic radiation. Linking genetic variation to phenotypic data and a wealth of historic information preserved in historic seed banks are the keys for genome-based and environment-informed breeding intensification.

Keywords: bioclimatic analysis; chickpea; GBS; GWAS; haploblock; SNP

---

1. Introduction

Landraces dominated agriculture for millennia, until the advent of intensive modern breeding in the mid 20th century, when reduced sets of elite cultivated varieties largely displaced the wider diversity of local genotypes [1]. Although the shift away from landraces was neither systematic nor synchronous, it is generally accepted that the subsequent convergence on a limited set of elite germplasm removed considerable useful variation [2]. In the early 20th century (1911–1940), N.I. Vavilov led a systematic effort to collect and preserve crop diversity, now maintained within the Vavilov Institute of Plant Genetic Resources (VIR) collection in St. Petersburg, Russia [3]. The geographic distribution and genetic diversity of most crops collected during this time frame are likely to reflect their historic patterns of cultivation established over the preceding millennia. Exploring these unique genetic resources provides an opportunity to revisit hypotheses about the radiation and secondary diversification of crops, not possible using later collections. Moreover, the expanded diversity of these early collections likely contains 'genetic gems' with the potential to enhance modern breeding efforts [4].

Here, we focus on biodiversity of *Cicer arietinum*, chickpea, which is among the world's most widely grown grain legumes and provides a vital source of dietary protein for ~15% of the world's population. Chickpea was first domesticated ~10 KYA, initially in southeastern Turkey, and then spread regionally throughout the Fertile Crescent. Although exact dates are unknown, archeological evidence suggests chickpea moved to India ~6000 years ago and to Ethiopia and North Africa ~3000 years ago [5]. Millennia of cultivation in these new areas, largely in isolation from each other, led to the establishment of new centers of secondary diversity, with accompanying differentiation of regionally specific landraces. Despite this generally accepted scenario, the relationships among the chickpea crops at these historic centers of cultivation are not fully resolved.

Chickpea domestication and breeding imposed a severe genetic bottleneck on the crop, with an estimated >95% of diversity lost between the crop wild progenitor and modern elite varieties [6]. Landraces represent an intermediate step to modern germplasm. An implicit, yet untested assumption is that chickpea landraces will have increased genetic diversity relative to modern elite germplasm. Moreover, we posit that geographic patterns of landrace diversity were shaped by post-domestication selection to adapt the crop to different agro-ecological environments and cultural preferences. Although Vavilov was unable to quantify the extent of diversity and differentiation, he and his contemporaries recognized the value of landraces as reserves of agriculturally-relevant traits, which motivated these early efforts in collection and conservation. Thus, chickpea landraces are expected to contain beneficial alleles, not segregating among modern elite varieties, which can be accessed and prioritized for crop improvement using genomics, phenotyping, and computational methods.

Here, we combine genomics, phenotyping, and computational biology to understand chickpea's agricultural variation one century ago, and from that analysis to infer the breadth and genetic bases of trait variation in the pre-modern era. Such knowledge can prioritize landrace haplotypes that contributed to diversification of chickpea as a crop, particularly haplotypes missing from modern breeding programs, thereby facilitating their use for crop improvement.

2. Results

2.1. Germplasm Resources and Phenotyping

To fully cover the biogeographic range of historic chickpea cultivation, we assembled 407 accessions collected between 1911 and 1940. Text descriptions of sampling locations, which were often local markets in small towns, were converted to geographic coordinates (Figure 1a). This set of accessions is enriched for genotypes under cultivation a minimum of one century ago in Turkey, India, Ethiopia, Uzbekistan, and Morocco, representing the major centers of post-domestication chickpea diversification and comprising 55% of the 407 analyzed accessions. Beyond the 147 Turkish and Ethiopian genotypes analyzed in an earlier study [4], we genotyped and/or phenotyped an additional 260 accessions spanning a total of 30 countries, with adjacent countries occasionally representing single extended historic agricultural systems (for examples, Ethiopia and Eritrea in eastern Africa, and several countries from the Fertile Crescent) (Table S1). The entire set of accessions was phenotyped under field conditions, genotyped, and used for further analysis.

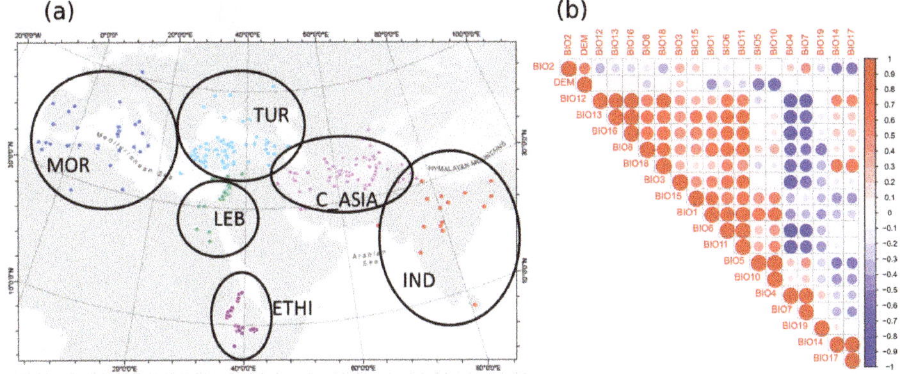

Figure 1. Sample distribution and correlation of bioclimatic variables. (**a**) Location of the chickpea samples around the world that were split into six geographically distinct groups. (**b**) The correlation between nineteen bioclimatic variables (bioclimatic variables and their abbreviations are presented in Table S2). Color intensity and the size of the asterisk are proportional to the correlation coefficients. ETHI, Ethiopia; IND, India; LEB, Lebanon; MOR, Morocco; TUR, Turkey; C_ASIA, Central Asia.

Correlation analyses of nineteen bioclimatic variables (bioclimatic variables and their abbreviations are presented in Table S2) from the range of chickpea collection sites revealed five groups of correlated variables (Figure 1b; Table S3). Three bioclimatic variables (BIO_2, BIO_{19}, DEM) were not strongly correlated to other variables. The first, third, and fifth groups (Table S3) correspond to temperature traits. The second and fourth groups (Table S3) consist of precipitation variables. While the first group (Table S3) consists of traits with moderate positive correlation (pairwise Spearman correlation coefficient, r > 0.4, Figure 1b), traits in the second group (Table S3) have stronger positive correlations (pairwise Spearman correlation coefficient, r > 0.7, Figure 1b), and traits in the remaining groups (Table S3) have the strongest positive correlations (pairwise Spearman correlation coefficient, r > 0.9, Figure 1b).

All 407 landraces accessions were phenotyped for thirty-six traits under field conditions in Kuban, Russia. The scored phenotypes and their abbreviations are presented in Table S4. Correlation analyses identified three groups of correlated traits (Figure 2). Phenotypic traits related to the color of plant organs and tissues were moderately correlated (pairwise Spearman correlation coefficient, r > 0.5, Figure 2) and form a single group. Quantitative traits characterizing the weights and sizes of whole plants and pods, as well as leaf size, also had moderate positive correlations (pairwise Spearman correlation coefficient, r > 0.4, Figure 2) and form two groups. Two phenological traits describing the duration of flowering and the duration of pod maturation had strong negative correlation (Spearman correlation coefficient, r = −0.76, Figure 2). Pod shape (PodSH) had moderate negative correlation with pod length (PDL) (Spearman correlation coefficient, r = −0.53, Figure 2) and pod width (PDW) (Spearman correlation coefficient, r = −0.55, Figure 2). Pod shape also had moderate negative correlation with thousand seeds weight (TSW) (Spearman correlation coefficient, r = −0.47, Figure 2). Phenotypic traits related to organ and tissue coloration had moderate negative correlation with traits describing the weights and sizes of plant and pods (pairwise Spearman correlation coefficient, r < −0.4, Figure 2).

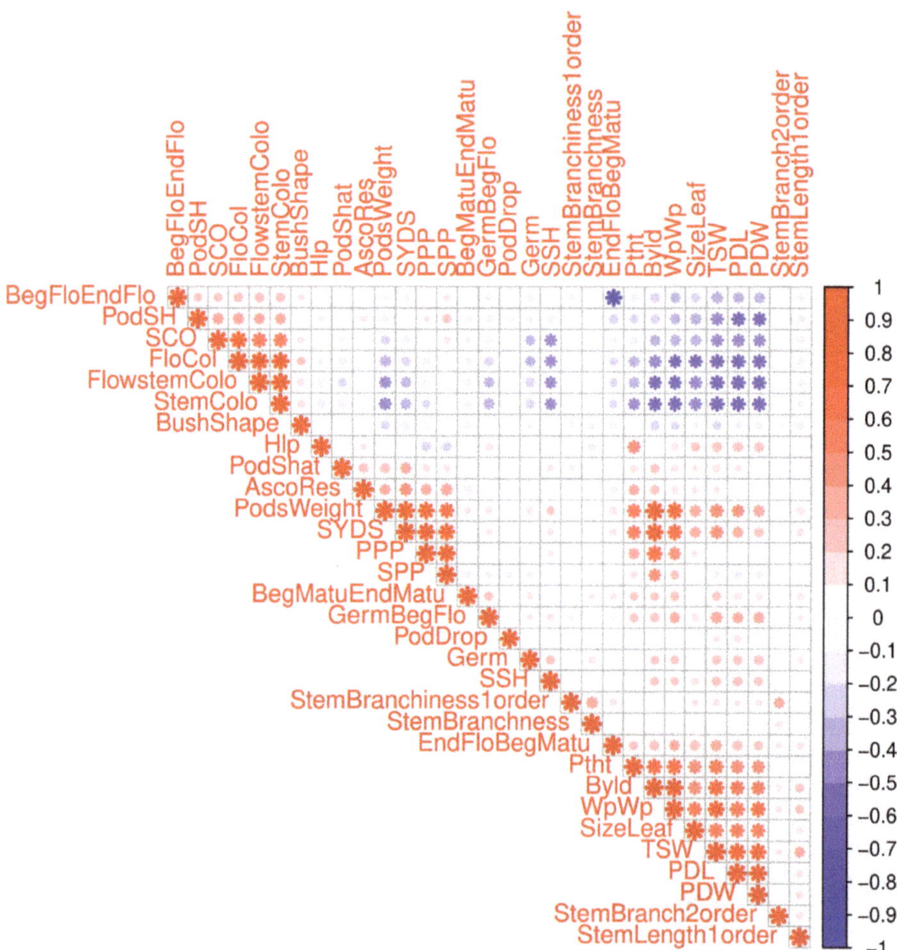

Figure 2. Correlation of thirty-one phenotypic traits. The scored phenotypes and their abbreviations are presented in Table S4. *Ascochyta*, the degree of damage (AsoDes) trait, was excluded from correlation analysis because it is the opposite value of *Ascochyta* resistance (AscoRes) trait. Moreover, we excluded overlapping time periods traits. Color intensity and the size of the asterisk are proportional to the correlation coefficients. PodSH, pod shape; SCO, seed color; SSP, number of seeds per plant; SSH, seed shape; TSW, thousand seeds weight; PDW, pod width; PDL, pod length.

2.2. Marker Polymorphism Analysis

Restriction site associated genotyping by sequencing (RAD-GBS) was used to survey polymorphism within the genomes of 407 accessions. SNPs were filtered to retain polymorphisms present in at least 90% of genotypes with a minor allele frequency of at least 3%. The resulting 2579 polymorphisms are distributed among all chromosomes, but with variable density that is especially elevated on chromosome 4 (Figure 3a). The elevated polymorphism content of chickpea chromosome 4 has been observed in previous studies (e.g., [4]). We hypothesized that selection and introgression via inadvertent hybridization between more and less advanced morphotypes might have resulted in agricultural improvement genes being aggregated to genomic 'agro islands', and in genotype-to-phenotype relationships resembling widespread pleiotropy.

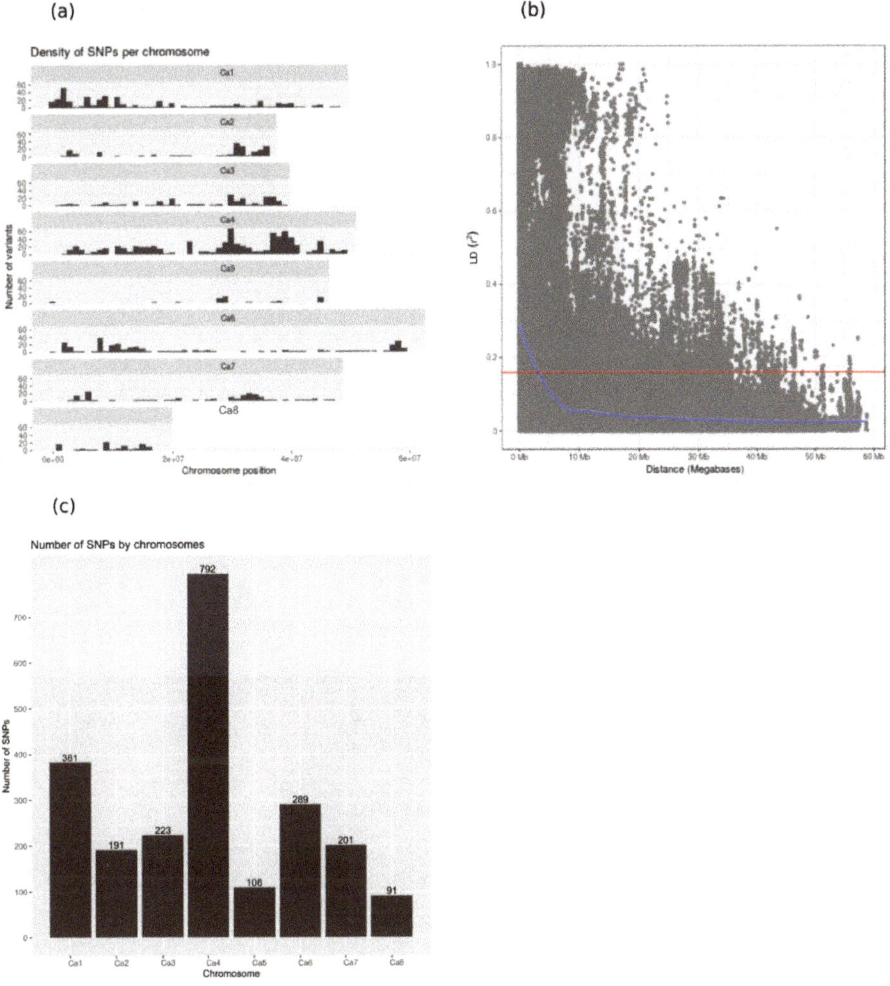

Figure 3. (**a**) Density of SNPs across the chickpea genome. Chromosome Ca6 is the longest chromosome in the chickpea genome (59.46 Mb) and chromosome Ca8 is the shortest (16.48 Mb). (**b**) Linkage disequilibrium (LD) (r^2) plots of the whole chickpea genome. The horizontal red line indicates the 95th percentile of the distribution of the unlinked r^2, which gives the critical value of r^2. (**c**) Distribution of SNPs along the eight chromosomes of the chickpea genome.

The sufficiency of this marker set for genetic tests depends in part on the scale of linkage disequilibrium (LD), because the relationship between physical distance and recombination frequency determines the precision of genetic association tests. LD is the non-random association between polymorphisms and can originate from demographic processes (e.g., shared ancestry and drift) or from selection (i.e., selective sweeps). In smaller populations of predominantly selfing organisms (including those that are the product of breeding), drift and selection typically have stronger effects than recombination, and thus LD extends to large genomic regions. Landraces are expected to exhibit especially extended LD. In line with these expectations, LD in chickpea landraces is very slow to decay (Figure 3b; Figure S1). Moreover, the marker density is uneven between chromosomes: from 91 SNPs on chromosome Ca8 to 792 SNPs on chromosome Ca4 (Figure 3c). Our sample size is comparable with other recent GWAS crop publications, hopefully resulting in adequate power.

2.3. Geographic Analyses

Patterns of population differentiation were analyzed using principle components (PCA) and visualized with unrooted trees. Figure 4 depicts the PCA plot for genetic data of the first versus second components and Figure S2 depicts a summary of variation and covariation attributed to the first five principle components. Interestingly, the accessions from the center of domestication, Turkey, are mainly divided into two clusters with light seeded Kabuli and Desi, which are smaller with dark seeds and purple flowers market classes intermixed with each cluster (Figure 4). The lack of distinctiveness between Desi and Kabuli adds further support to the same conclusion reached by Penmetsa et al. [7]. All groups containing Turkish accessions also contain minor representation from other regions, with the exception of a preponderance of landraces from North Africa in one of the Turkish groups. Notably, landraces from India and Ethiopia, which represent two of Vavilov's major sites of secondary diversification [8], are well resolved, though not exclusive of one another. Turkish accessions are absent from the group of Ethiopian landraces and constitute only a minor component of the Indian group, which is instead enriched in landraces from Central Asia. A portion of Central Asian accessions also occur in a distinct grouping dominated by the ancestral Desi form (Figure 4).

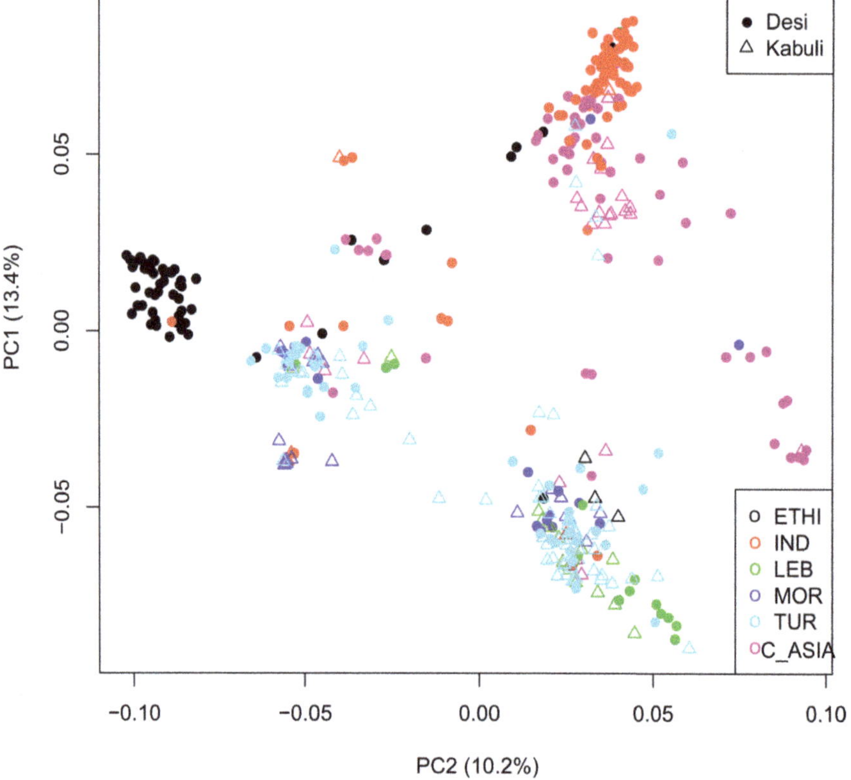

Figure 4. Scatter plots of the first two principal components of the principal component analysis (PCA) based on 2579 SNPs. Each dot represents an accession. Desi varieties are shown as asterisks and Kabuli as triangles.

These observations are consistent with the deduced pattern of molecular evolution. Maximum likelihood phylogenetic trees constructed with genome-wide SNP (Figure 5a) support inferences from the PCA analysis. Central Asian and Turkish accessions are broadly distributed throughout the tree, but notably absent from groups predominated by India and Ethiopia, consistent with more extensive diversity (Table S5) at the Turkish center of origin for the species, and with longstanding, but distinct secondary diversification in India, Central Asia, and Ethiopia. Chromosome 4 is known to have excess diversity relative to the rest of the genome [9,10], as indeed we observe here. Interestingly, certain of the relationships observed using genome-wide SNP are obscured in the tree constructed from chromosome 4 SNPs (Figure 5b). In particular, the previously coherent group of Ethiopian genotypes is divided more broadly within the tree and there is both greater subdivision within the Indian group and less distinction from the Central Asian landraces.

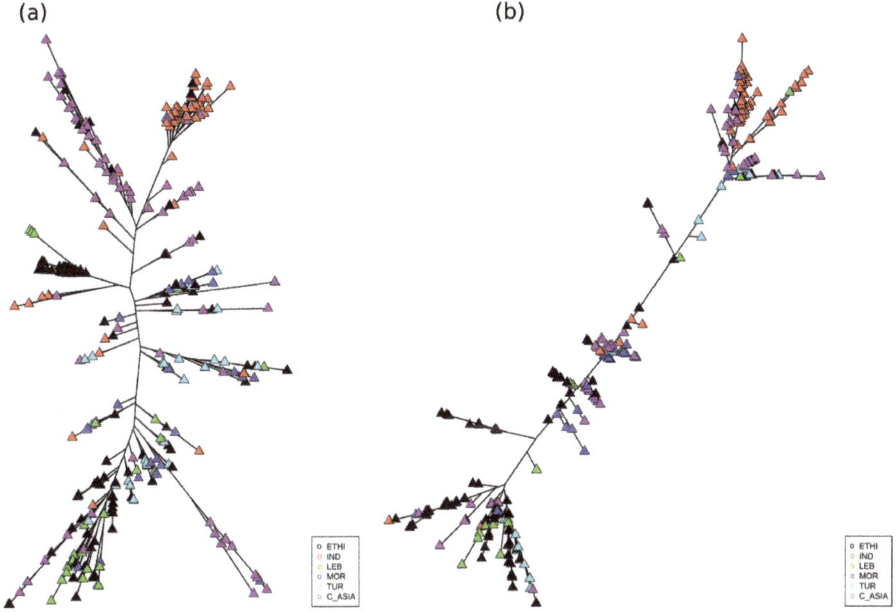

Figure 5. (**a**) Maximum likelihood phylogenetic tree showing relationships among accessions based on the whole genome SNPs and (**b**) on chromosome 4 SNPs.

2.4. Single Trait Associations

Genetic and phenotypic data were strongly concordant, as described in Table S6, which shows co-variances between genetic and phenotypic data.

To account for these effects, GWAS analysis was implemented with the first eight PCA axes scores used as covariates for all phenotypic and bioclimatic data (Figures S3–S18), revealing multiple significant associations among 70 SNPs with bioclimatic and phenotypic traits (Figures 6 and 7; Table S7). Twelve of 70 markers were found to have significant associations with two or more traits. SNP Ca2: 17161867 is associated with plant weight without pods (WpWp) as well as isothermality (BIO$_3$) and mean temperature of the warmest quarter (BIO$_{10}$) (see Table S2 and Table S4 for a full list of bioclimatic variables and phenotypes abbreviations). These genetic findings are supported by WpWp weakly negatively correlated with BIO$_3$ and BIO$_{10}$. SNP Ca3: 20549509 and SNP Ca6: 2908823 are associated with mean diurnal range (BIO$_2$) and BIO$_3$, which are themselves weakly positively correlated (Figure 1b). Three SNPs, two on the 8th chromosome (SNP Ca8: 9098790 and Ca8: 10314452) and one on the 4th chromosome (SNP Ca4: 30948593), are associated with two phenotypic variables:

biological yield (Byld) and plant weight without pods (WpWp), which are very strongly correlated and appear to derive from common genetic capacities (r = 0.92; Figure 2). Also on chromosome 4, Ca4: 33967674 is associated with the correlated group of phenotypes that includes plant weight traits (weight of seeds, pods, and the whole plant). SNP Ca6: 57117312 is associated with flower color (FloCol) and seed shape (SSH), which are themselves moderate negatively correlated (r = −0.45, Figure 2). SNP Ca7: 30930779 is associated with BIO_3, number of seeds per plant (SPP), and the group of phenotypes characterizing plant and organ weights. Three additional SNPs on chromosome 7 (SNP Ca7: 33337524, Ca7: 33340372, Ca7: 33457287) are associated with three bioclimatic variables, BIO_3, BIO_6, and BIO_{11}, which are part of a larger group of correlated variables (Figure 1b).

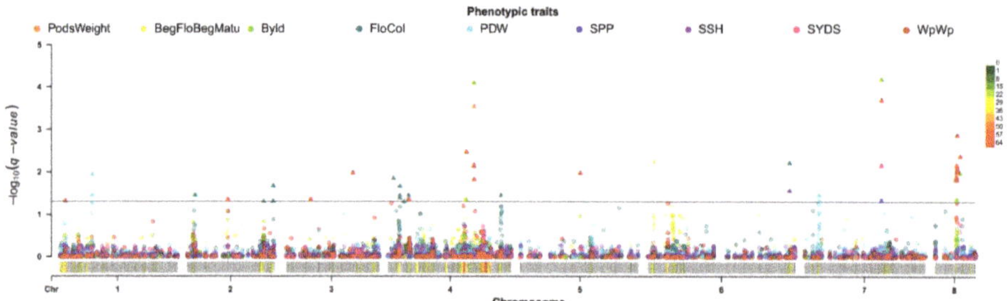

Figure 6. Summary of GWAS analyses with eight PCs as covariates for phenotype data (different colors correspond to different phenotype). SNPs with *q*-value < 0.05 are shown for each chromosome, marked as triangles. Chromosome density is attached on the bottom of the Manhattan plot.

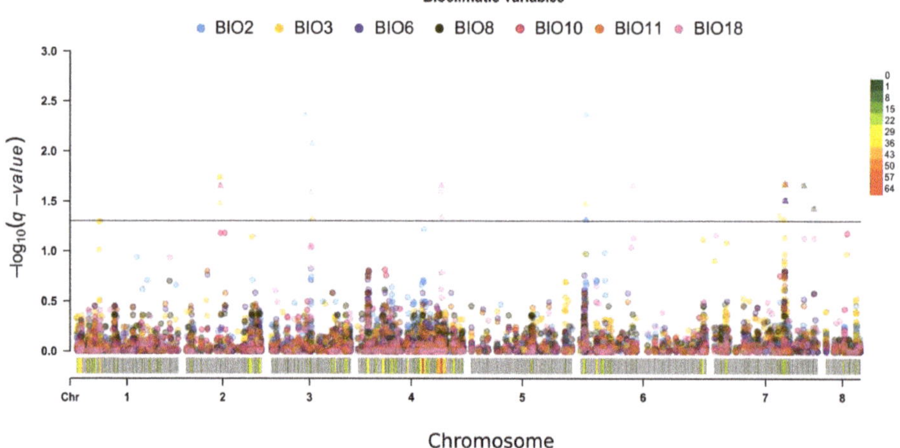

Figure 7. Summary of GWAS analyses with eight PCs as covariates for bioclimatic variables (different colors correspond to different bioclimatic variables). SNPs with *q*-value < 0.05 are shown for each chromosome, marked as triangles.

To incorporate geography explicitly into the analysis, we repeated the above GWAS, but with the addition of the first two axes of PCoA, which derive from the analysis of landrace geographic variation (Figures S19 and S20; Table S7). The results of these analyses were generally consistent with the results described above and are only introduced briefly here. An additional set of significant associations was found. Twelve SNPs are associated with pod length (PDL), nine on chromosome 6 and three on chromosome 7. Ten of these twelve SNPs exhibit significant linkage. Two SNPs on chromosome 7 are associated with secondary branching (StemBranch2order), but without strong linkage.

Because of extended LD, we cannot identify causal relationships between SNPs and phenotypes. Nevertheless, we explored the potential nature of the associated genes and found several important genes that have been reported in previous studies. For example, genes Ca_10410, Ca_10426, and Ca_10428 are present within haploblock Ca6:2541669 ... .Ca6:3024335, to which several SNPs associated with the beginning to flowering to the beginning to maturation phenotype and temperature related variables map (see Table S7). Ca_10410 (Ca6:2766285 ... .2768999) is involved in floral development and encodes flavin-binding kelch repeat F-box protein with high homology to circadian clock-associated FKF1 gene of soybean. Ca_10426 (Ca6:2881369 ... .2884463) encodes a XAP5 protein important for light regulation of the circadian clock that plays a global role in coordinating growth in response to the light environment. SNP Ca2: 17161867 associated with plant weight without pods (WpWp) and temperature related bioclimatic variables BIO_3 and BIO_{10}, as well as Ca2: 17161884 associated with the duration of flowering (BegFloEndFlo) and BIO_3 are all located within intron of gene Ca_16015. This gene encodes phosphoenolpyruvate carboxylase, enzyme involved in carbon fixation, and citric acid cycle biosynthesis flux [11]. The first intron of Ca_11533 gene encoding beta-D-xylosidase contains SNP Ca8: 9098790, which is associated with both WpWp and Byld. beta-D-Xylosidases are involved in the breakdown of xylan, a major component of plant cell-wall hemicelluloses [12]. SNP Ca1: 2218700, which is associated with WpWp, is located in the intergenic region upstream of gene Ca_00278 that encodes protein with polyphenol oxidase activity. In *Clematis terniflora* DC, decreasing activity of this enzyme elevates the plant photosynthesis by activating the glycolysis process, regulating Calvin cycle, and providing adenosine triphosphate (ATP) for energy metabolism. Besides, polyphenol oxidase is involved in the formation of brown melanin pigment in fruits and vegetables, plays a crucial role in the biosynthesis of secondary metabolites, and has a role in plant defense against biotic and abiotic stresses [13]. SNP Ca3: 10855323 associated with WpWp is located upstream of Ca_19358 gene encoding beta-*N*-acetylhexoamidase that catalyzes the hydrolysis of *N*-acetylglucosamine or *N*-acetylgalactosamine from the non-reducing terminal of oligosaccharides, glycoproteins, glycolipids, and other glycoconjugates. b-*N*-acetylhexosaminidase is highly active in dry or germinating seeds, where it participates in the degradation of reserve glycoproteins. Moreover, its activity is induced in the period of ripening in tomato and peaches [14]. The Ca_11539 (Ca8:9151680. ... 9159194) intron contains several SNPs associated with WpWp. This gene encodes an oligopeptidase degrading short peptides. SNP Ca4: 2145082 associated with flower color (FloCol) is located upstream of Ca_07836 gene, which is homologue of genes in *Pisum sativum* (protein A) and *Medicago truncatula* (bHLH-A), which are flower color associated genes [15].

2.5. Clustering of Phenotypes and Variables Sharing Enriched Haploblocks

The total number of the Haploview-inferred [16] haploblocks was 224, encompassing 1264 SNPs (mean per haploblock = 5.6) (Table S8). Filtering for more than six SNPs left 74 haploblocks (33% of total) as input to find haploblocks enriched for associated SNPs for each trait and variable using the fast gene set enrichment (FGSEA) method [17] (parameter for permutations = 100,000) (Table S9). Subsequent to bi-clustering of phenotypes and variables sharing enriched haploblocks, we defined several visually distinguished groups (Figure 8, Table S10). The first group contained two consecutive reproductive stages of plant development: the duration of flowering (BegFloEndFlo) and the duration from the end of flowering until the beginning of maturation (EndFloBegMatu). We hypothesize that the same genetic mechanisms influence the duration of both stages. The second group contains pod shattering (PodShat) and pod drop (PodDrop) traits as well as one-third of all bioclimatic factors, related to both temperature and precipitation, exclusive to a well correlated set from Figure 1b ($BIO_{6,8,11,12,13,16}$). Pod-related traits form a subgroup with three temperature-related bioclimatic factors: mean temperature (BIO_1), mean temperature of coldest month (BIO_6), and temperature annual range (BIO_7); this subgroup is similar in a set of enriched haploblocks with the group containing two additional heat-related bioclimatic factors, max temperature of warmest month (BIO_5), and mean temperature of warmest quarter (BIO_{10}). This grouping is consistent with a well-known relationship between high temperature and pod shattering/retention. A third group includes color-related traits, flower color (FloCol), peduncle color (FlowstemColo), seed color (SCO), and stem color (StemColo), which is expected, because genes in the phenylpropanoid pathway are implicated in the production of pigments in different plant organs. A fourth group aggregates *Ascochyta* blight resistance (AscoRes) and precipitation of the coldest quarter (BIO_{19}), which reflects a well understood relationship between *Ascochyta* incidence and rainfall during periods of reduced temperatures. Also of note is a group containing moisture stress-related covariates ($BIO_{14,17}$, precipitation of the driest month/quarter) and plant height (Ptht), which is expected to depend on moisture availability; interestingly, this group clusters with a group that contains phenotypic traits related to plant size (biological yield and pod size), which are traits related to the duration of vegetative growth and that are limited by moisture availability.

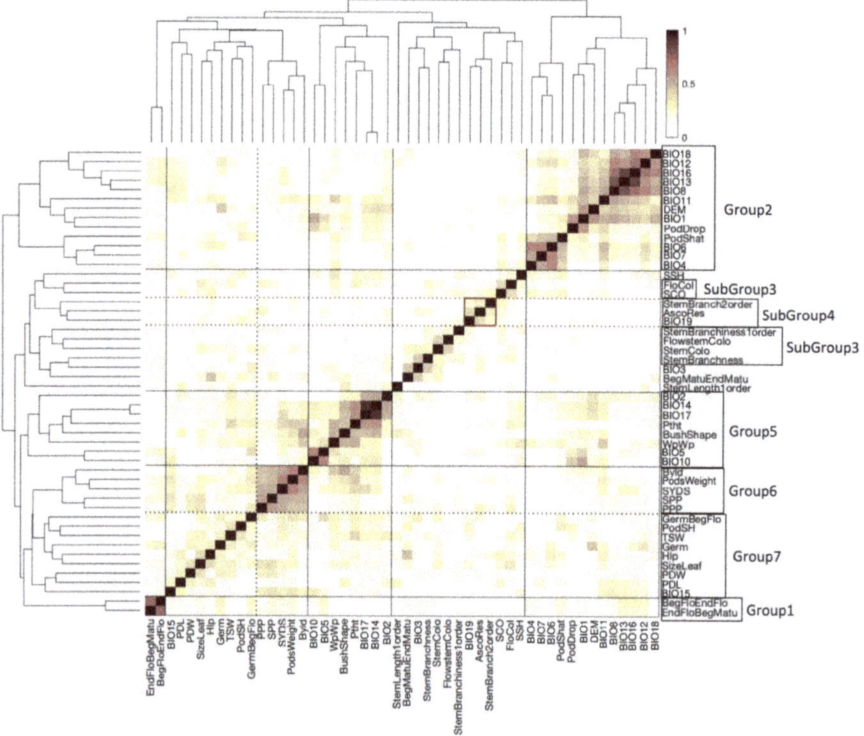

Figure 8. The degree of overlap in haploblocks enriched for SNPs associated with phenotypes and variables. Bi-clustering of similarity scores reveals several visually distinct groups of phenotypes. The haploblock similarity score is defined as a double sum of haploblocks simultaneously enriched for SNPs for both traits normalized to the amount of significantly enriched haploblocks for each trait. The degree of similarity is color coded.

3. Discussion

For many millennia, farmers and breeders have focused on selecting crops with desirable phenotypes [2]. With the successful domestication of numerous crops came the incremental loss of genetic and phenotypic variation. Genetic bottlenecks are especially common in selfing species such as grain legumes (e.g., [18]). Novel sources of variation for biotic and abiotic stress resistance are especially needed in chickpea, because the crop is often grown by resource-poor farmers, on marginal lands, and under low-input conditions. Broadening chickpea's genetic base should facilitate production of new varieties to address these needs, while also meeting changing consumer demands, new agricultural practices, and anticipated shifts in climatic conditions [6].

Chickpea landraces represent an expanded source of genetic and phenotypic variation that has not been systematically explored and has been used only in an ad hoc manner for modern breeding. The Vavilov Institute of Plant Genetic Resources is one of the world's primary libraries of lost genetic variation in food crops, capturing the genetic and functional diversity of regionally stratified agriculture typical of one century ago. It contains tens of thousands of legume accessions, including approximately one thousand chickpea accessions collected prior to intensive international breeding efforts [3]. The re-introduction of genetic material from the Vavilov Institute's collection into modern elite varieties could be a potent force for future agricultural improvement. To this end, we combine genomics, phenotyping, and computational biology to characterize the chickpea collection of Nikolay Vavilov and his colleagues, linking traits and environments to genes. Our results highlight the collection's currently latent potential of chickpea landraces, and underscore the value of this resource to meet the enormous challenges of 21st century agriculture. However, the identified candidate genes are needed in further validation and functional confirmation owing to such factors as one-year observation of phenotypes and long extend of LD in the germplasm.

Our observations contribute to an increasing understanding of genetic variation of quantitative and categorical traits in chickpea [19–21]. The present work adds a new dimension by incorporating a wider set of historical crop diversity, and by treating bioclimatic data at accession sampling sites as extended crop traits. In doing so, our GWAS hits highlight associations to genomic regions not discovered in prior GWAS and quantitative trait locus (QTL) analyses (Table S7). These hits map in the vicinity of genes involved in floral development, photosynthesis, cell wall or secondary metabolism, and carbohydrate biosynthesis, and some of them are close to already known QTLs. For example, SNP Ca4: 33967674, associated with yield, pod weight, plant weight without pods, and seed weight per plant, is located 752 kb downstream from known QTL (Table S11) governing pod number trait [22] and SNP Ca3: 28094292, associated with plant weight without pods, localizes 96 kb downstream of QTL (Table S11) containing cluster of FLOWERING LOCUS T (FT) genes and controlling phenology and growth habit [23]. SNP Ca4: 30948593 and SNP Ca8: 10314452, associated with yield, are located ~90 kb upstream from previously detected SNP (Table S11) and ~25 kb downstream from previously detected SNP, respectively (Table S11), also associated with yield [24]. SNP Ca6: 3024192, associated with beginning of flowering to the beginning of maturation phenotype, is located in the same haploblock Ca6_Block_3 (~87 kb upstream) as the previously detected SNP (Table S11), associated with days to 50% flowering [24]. Previously, we [25] published a study in which we were looking for associations between SNPs and bioclimatic covariates at collection sites. Two covariates, which include temperature characteristics, were jointly associated with one SNP on chromosome 8 (Ca8: 10314452). This SNP is associated with two phenotypic variables: biological yield (Byld) and plant weight without pods (WpWp) in the current study.

To rigorously test for associations, we implement a novel haploblock-based test that, we believe, will find much use in the crop genomics. The underlying statistics for the test are similar to the gene set enrichment analysis, where each haploblock represents a set of SNPs associated with a trait and all SNPs are ranked according to GWAS p-values. This analysis identified eleven haploblocks (Table S12) intersecting with previously reported GWAS hits. Haploblock Ca1_Block_18 and haploblock Ca4_Block_18 are enriched for SNPs associated with several phenotypes and bioclimatic variables, including thousand seeds weight phenotype. These haploblocks covers SNP on chromosome 1 and SNPs on chromosome 4, respectively, reported by Varshney et al. [24], associated with 100 seed weight (Table S12). Haploblock Ca3_Block_4, haploblock Ca4_Block_54 and haploblock Ca5_Block_4 are enriched for SNPs associated with several phenotypes and bioclimatic variables, including seeds weight per plant phenotype. These haploblocks overlay four SNPs on chromosome 3, three SNPs on chromosome 4, and eight SNPs on chromosome 5, respectively, reported by Varshney et al. [24], associated with yield per plant (Table S12). Haploblock Ca3_Block_7 is enriched for SNPs associated with the duration of vegetative growth, with seeds weight per plant, and with three bioclimatic variables (BIO_5, BIO_{13}, BIO_{16}). This haploblock covers two SNPs on chromosome 3, reported by Varshney et al. [24], associated with days to 50% flowering and with yield per plant, respectively (Table S12). Haploblock Ca3_Block_16 is enriched for SNPs associated with the duration of vegetative growth, as well as with plant height, plant weight without pods, and temperature-related bioclimatic variables BIO_3 and BIO_5. This haploblock intersects with a QTL for days to 50% flowering time (Table S12) reported from the GWAS analysis of Upadhyaya and colleagues [19]; Upadhyaya et al. nominated a particular candidate gene, SBP (SQUAMOSA promoter binding protein), though we advocate a more cautious approach that recognizes limitations of the study design and instead implicates haplotype intervals. Haploblock Ca4_Block_9 is enriched for SNPs associated with the duration of vegetative growth, with pod shattering, and with four bioclimatic variables (BIO_4, BIO_6, BIO_7, BIO_{12}). This haploblock covers SNP on chromosome 4 associated with days to 50% flowering (Table S12), reported by Varshney et al. [24]. Haploblock Ca7_Block_12 is enriched for SNPs associated with the duration of vegetative growth, with number of seeds per plant, with stem branchness, and with temperature-related bioclimatic variable BIO_3. This haploblock covers SNP on chromosome 7 associated with days to maturity (Table S12), reported by Varshney et al. [24]. The last haploblock, Ca8_Block_7, is enriched for traits related for branching and covers SNP on chromosome 8 reported by Bajaj et al. [20], associated with branch number (Table S12).

Previously, we [4] published a pilot study combining historic phenotypic data with reduced representation sequencing to establish a proof-of-principle for the results reported here. We employed a combination of genomics, computational biology, and phenotyping to characterize VIR's 147 chickpea accessions from Turkey and Ethiopia, representing chickpea's center of origin and a major location of secondary diversification, respectively. The majority of SNPs associated with multiple traits localized to a single chromosome 4 region. Here, we observe similar patterns with a larger sample of more diverse landraces and with a more comprehensive phenotypic and environmental dataset. We find multiple SNPs that are non-randomly distributed among several haploblocks, many of which are associated with multiple phenotypes (Table S9). The non-random clustering of phenotypes and variables (Figure 8) exactly arises as a result of such multi-trait associations. Although the grouping of traits and ancestral bioclimatic variables does not necessarily imply co-selection during domestication (e.g., [26]), these clusters may represent genetic complexes co-adapted to a range of environmental conditions that chickpea experienced during domestication and subsequent geographic radiation. Indeed, many of the trait–environment associations reflect well-known interactions between environmental factors and the crop's biology; for example, the relationships between *Ascochyta* blight occurrence and the duration of cool-wet periods, as well as the increased incidence of pod abortion and shattering under conditions of heat stress. Thus, by combining genomics with an explicit biogeographic framework encompassing climatic and phenotype covariates, we are able to suggest concordance between human selection, the crop's known biology, and environmental constraints.

4. Materials and Methods

4.1. Germplasm Resources and Phenotyping

We assembled a collection of VIR's chickpea germplasm originating from a range of countries including Ethiopia, Lebanon, Morocco, Turkey, India, and the broader Central Asia and Mediterranean regions (see Table S1). Phenotyping of the 407 chickpea genotype collection was conducted at the VIR Kuban experimental station with climatic conditions well suited for chickpea cultivation (see Text S1). During the vegetative period, thirty-six phenological, morphological, agronomical, and biological descriptors were measured. The scored phenotypes and their abbreviations are presented in Table S4.

4.2. Genotyping by Sequencing (GBS) and SNP Calling

The restriction site associated (RAD) GBS protocol from von Wettberg et al. [6] was used to generate reduced representation sequence data for 407 accessions (see Text S2). All Illumina data are available from the National Center for Biotechnology database under BioProject PRJNA388691. SNPs were called using the Genome Analysis Tool Kit (GATK) pipeline [27] and further filtered with VCFtools [28]. A total of 2579 SNPs accessions passed all filters, with 407 accessions remaining for further analysis.

4.3. Genetic Data Analyses

Principal component analysis (PCA) was conducted using the "SNPRelate" R library [29]. Custom scripts in Python [30] and R [31] were used to plot depth and distribution of SNPs on chromosomes.

Linkage disequilibrium (LD) was estimated using the squared correlation coefficient (r^2) between genotypes. VCFtools [28] was used to calculate intra-chromosomal and unlinked r^2 values. LD decay was assessed by plotting intra-chromosomal r^2 values against the physical distance (bp) between markers. The parametric 95th percentile of unlinked r^2 values distribution was taken as a critical value. The threshold beyond which the LD was accepted as real physical linkage was estimated to be $r^2 = 0.16$. The intersection of the smothering second degree local regression (LOESS) curve of intra-chromosomal r^2 values with this threshold was considered to be an estimate of the range of LD.

Relationships among genotypes were calculated and maximum likelihood phylogenetic trees were constructed using SNPhylo [32] based on filtered SNPs and drawn using R libraries "phytools" [33] and "ape" [34].

The nucleotide diversity (pi) was estimated from polymorphic sites and separately for each chromosome and geographical group using VCFtools [28]. By considering only polymorphic sites, we overestimate genomic diversity; however, these estimations can be used for between group comparisons. We applied the Mann–Whitney–Wilcoxon test [35] to make between group comparisons.

The Genome-wide complex trait analysis (GCTA) program [36] was used to estimate the proportion of variance in phenotypes explained by all genome-wide SNPs. First, phenotypic data were normalized. Then, the genetic relationships among individuals from genome-wide SNPs were calculated using GCTA-GRM (genetic relationship matrix) analysis. Finally, GCTA-GREML (genome-based restricted maximum likelihood) analysis was performed to estimate the proportion of variance in a phenotype explained by all GWAS SNPs (i.e., the SNP-based heritability).

4.4. Bioclimatic Analysis

Bioclimatic analysis was performed as described in Plekhanova et al. [4]; for details, see Text S3. Nineteen quantitative bioclimatic variables were used in the analysis (Table S2).

Shapiro–Wilk test for normality [37] was implemented to quantitative phenotypic traits and quantitative bioclimatic variables. Spearman correlation coefficients were calculated using the "rcorr" function from the "Hmisc" R library [38].

4.5. Mapping Approaches

GWAS analysis was performed using a single-locus linear mixed model, implemented in FaST-LMM toolset (factored spectrally transformed linear mixed models) [39]. Principal component analysis (PCA) of 2579 SNPs revealed that the first eight significant principal components (PCs) explained 48% of the variance of all markers. The LMM model was implemented with the first eight PCA axes scores used as covariates for all phenotypic and bioclimatic data. Principal coordinate analysis (PCoA), based on geographical distances between the accessions, was performed using the "pco" function from the "labdsv" library [40] in R, and revealed that the first two significant PCs explained 59% of the variance. We repeated the GWAS analysis including the first eight PCA axes scores and the first two PCoA axes scores as covariates for all traits. In both cases, we used genomic control parameter (λ_{GC}) and a false discovery rate (FDR) [41] of 0.05 to determine significant trait associated loci separately for each trait. Manhattan plots were performed using "CMplot" library [42] in R.

Annotation of significant associated markers was performed using the SNPEff program [43], as well as the legume information system (LIS) [44] and the LegumeIP [45] databases.

4.6. Biogeographic Analyses

In total, 407 accessions were split into six distinct groups reflecting geographic locations (Table S1): Ethiopia ("ETHI"), India ("IND"), Lebanon ("LEB"), Morocco ("MOR"), Turkey ("TUR"), and Central Asia ("C_ASIA"). The Mann–Whitney–Wilcoxon test [35] was used to identify differences among groups for each bioclimatic variable.

4.7. Haploblock Enrichment Analysis and Clustering of Enriched Haploblocks

To divide the genome into haplotype blocks (haploblocks) based on linkage disequilibrium, Haploview tools [16] were applied to the set of 2579 SNPs. Chromosomal regions with strong linkage were identified using default Haploview parameters (confidence interval for LD [0.7, 0.98]). Each haploblock was considered as the set of SNPs located within a given haploblock. We analysed haploblock enrichment for SNPs associated with trait or variable by applying the logic of gene-set enrichment analysis implemented in the FGSEA method [17], which takes as input data the list of all SNPs ranked by increasing GWAS *p*-values and the list of haploblocks. The method returns an enrichment score and FDR corrected *p*-value [41] for each haploblock. We performed FGSEA analysis for each trait (phenotype and bioclimatic variable), and haploblocks significantly enriched for associated SNPs were defined as those having positive enrichment scores and significantly low FDR corrected *p*-values (<0.05). The outcome of this analysis was that each phenotype or bioclimatic variable was characterized by a set of haploblocks significantly enriched with associated SNPs. To obtain groups of phenotypes and variables sharing sets of enriched haploblocks, we applied bi-clustering on the matrix of pairwise similarities between traits. To estimate the degree of overlap between haploblocks enriched for SNPs associated with different traits, we calculated the haploblock simalarity score as a sum of common haploblocks (i.e., haploblocks enriched for SNPs associated with both traits) divided by the sum of all haploblocks significantly enriched for SNPs associated with these two traits.

5. Conclusions

The Vavilov seed bank contains numerous landraces collected nearly one hundred years ago, and thus may contain 'genetic gems' with the potential to enhance modern breeding efforts. Here, we analyze 407 landraces, sampled from major historic centers of chickpea cultivation and secondary diversification. The collection was grown in the southern European part of Russia in 2016 with climatic conditions well suited for chickpea cultivation. GWAS conducted on both phenotypic traits and bioclimatic variables at landraces sampling sites as extended phenotypes resulted in 84 GWAS hits associated to various regions, most of which were not discovered in prior GWAS and QTL analyses. The novel haploblock-based test identified haploblocks enriched for SNPs associated with phenotypes and bioclimatic variables, of which eleven haploblocks intersect with previously reported GWAS hits on chromosomes Ca1, Ca3, Ca4, Ca5, Ca6, Ca7, and Ca8. Subsequent bi-clustering of traits sharing enriched haploblocks underscored both non-random distribution of SNPs among several haploblocks and their association with multiple traits. We suggest that these clusters of pleiotropic SNPs represent co-adapted genetic complexes to a range of environmental conditions that chickpea experienced during domestication and subsequent geographic radiation. We observed significant genomic diversity in Central Asia, which may have been a bridge for subsequent radiation in India and nearby areas. Linking genetic variation to phenotypic data and a wealth of historic information preserved in historic seed banks are the keys for genome-based and environment-informed breeding intensification.

Supplementary Materials: Supplementary materials can be found at http://www.mdpi.com/1422-0067/21/11/3952/s1.

Author Contributions: Formal analysis, A.S., P.L.C., A.A.I., N.V.N., E.v.W., S.V.N., and M.G.S.; Investigation, S.V.B., N.C.-G., M.A.V., and D.R.C.; Writing—original draft, A.S., E.v.W., D.R.C., S.V.N., and M.G.S. All authors have read and agreed to the published version of the manuscript.

Funding: RSF grant # 16-16-00007 to A.S., S.V.B., A.I., M.A.V., S.V.N. and M.G.S. supports dataset phenotyping, GWAS implementation and analysis, bioclimatic analysis, biogeographic analysis, and enrichment analysis of haploblocks. This work was also supported by a cooperative agreement from the United States Agency for International Development under the Feed the Future Program AID-OAA-A-14-00008 to D.R.C., E.v.W., and S.V.N.; by the Zumberge Foundation to S.V.N.; and by a grant from the U.S. National Science Foundation Plant Genome Program under Award IOS-1339346 to D.R.Cook and E.v.W. We also acknowledge support from the Government of Norway through the Global Crop Diversity Trust CWR14NOR23.3 07 to D.R.C. and E.v.W. E.v.W. is further supported by the USDA Hatch program through the Vermont State Agricultural Experimental Station.

Acknowledgments: We thank Victoria Scobeyeva for helping in DNA extraction.

Conflicts of Interest: The authors declare no conflict of interest.

References

1. Fairchild, D. *The World was My Garden: Travels of Plant Explorer*; LWW: New York, NY, USA, 1939; 495p.

2. Maxted, N.; Dulloo, M.E.; Ford-Lloyd, B.V. *Enhancing Crop Genepool Use: Capturing Wild Relative and Landrace Diversity for Crop Improvement*; CABI: Oxfordshire, UK, 2016; 469p.

3. Vishnyakova, M.A.; Burlyaeva, M.O.; Bulyntsev, S.V.; Seferova, I.V.; Plekhanova, E.S.; Nuzhdin, S.V. Chickpea landraces from centers of the crop origin: Diversity and differences. Sel'skokhozyaistvennaya biologiya. *Agric. Biol.* **2017**, *52*, 976–985.

4. Plekhanova, E.; Vishnyakova, M.A.; Bulyntsev, S.; Chang, P.L.; Carrasquilla-Garcia, N.; Negash, K.; Nuzhdin, S.V. Genomic and phenotypic analysis of Vavilov's historic landraces reveals the impact of environment and genomic islands of agronomic traits. *Sci. Rep.* **2017**, *7*, 4816. [CrossRef] [PubMed]

5. Redden, R.J.; Berger, J.D. History and origin of Chickpea. In *Chickpea Breeding & Management*; Yadav, S.S., Redden, R., Chen, W., Sharma, B., Eds.; CABI: Wallingford, UK, 2007; pp. 1–13.

6. Von Wettberg, E.J.; Chang, P.L.; Başdemir, F.; Carrasquila-Garcia, N.; Korbu, L.B.; Moenga, S.M.; Cordeiro, M.A. Ecology and community genomics of an important crop wild relative as a prelude to agricultural innovation. *Nat. Commun.* **2018**, *9*, 1–13. [CrossRef] [PubMed]

7. Varma Penmetsa, R.; Carrasquilla-Garcia, N.; Bergmann, E.M.; Vance, L.; Castro, B.; Kassa, M.T.; Coyne, C.J. Multiple post-domestication origins of kabuli chickpea through allelic variation in a diversification-associated transcription factor. *New Phytol.* **2016**, *211*, 1440–1451. [CrossRef] [PubMed]

8. Vavilov, N.I. The origin, variation, immunity and breeding of cultivated plants (Translated by S.K. Chestitee). *Chron. Botonica* **1951**, *13*, 1–366.

9. Kale, S.M.; Jaganathan, D.; Ruperao, P.; Chen, C.; Punna, R.; Kudapa, H.; Garg, V. Prioritization of candidate genes in 'QTL-hotspot' region for drought tolerance in chickpea (*Cicer arietinum* L.). *Sci. Rep.* **2015**, *5*, 15296. [CrossRef]

10. Thudi, M.; Khan, A.W.; Kumar, V.; Gaur, P.M.; Katta, K.; Garg, V.; Varshney, R.K. Whole genome re-sequencing reveals genome-wide variations among parental lines of 16 mapping populations in chickpea (*Cicer arietinum* L.). *BMC Plant Biol.* **2016**, *16*, 10. [CrossRef]

11. Chollet, R.; Vidal, J.; O'Leary, M.H. PHOSPHOENOLPYRUVATE CARBOXYLASE: A ubiquitous, highly regulated enzyme in plants. *Annu. Rev. Plant Physiol. Plant Mol. Biol.* **1996**, *47*, 273–298. [CrossRef]

12. Minic, Z.; Rihouey, C.; Do, C.T.; Lerouge, P.; Jouanin, L. Purification and characterization of enzymes exhibiting beta-D-xylosidase activities in stem tissues of Arabidopsis. *Plant Physiol.* **2004**, *135*, 867–878. [CrossRef]

13. Chen, X.; Yang, B.; Huang, W.; Wang, T.; Li, Y.; Zhong, Z.; Yang, L.; Li, S.; Tian, J. Comparative proteomic analysis reveals elevated capacity for photosynthesis in polyphenol oxidase expression-silenced *Clematis terniflora* DC. Leaves. *Int. J. Mol. Sci.* **2018**, *19*, 3897. [CrossRef]

14. Ryšlavá, H.; Valenta, R.; Hýsková, V.; Křížek, T.; Liberda, J.; Coufal, P. Purification and enzymatic characterization of tobacco leaf β-N-acetylhexosaminidase. *Biochimie* **2014**, *107 Pt B*, 263–269. [CrossRef]

15. Hellens, R.P.; Moreau, C.; Lin-Wang, K.; Schwinn, K.E.; Thomson, S.J.; Fiers, M.W.; Davies, K.M. Identification of mendel's white flower character. *PLoS ONE* **2010**, *5*, e13230. [CrossRef] [PubMed]

16. Barrett, J.C.; Fry, B.; Maller, J.; Daly, M.J. Haploview: Analysis and visualization of, L.D. and haplotype maps. *Bioinformatics* **2005**, *21*, 263–265. [CrossRef] [PubMed]

17. Sergushichev, A. An algorithm for fast preranked gene set enrichment analysis using cumulative statistic calculation. *BioRxiv* **2016**, 060012. [CrossRef]

18. Olsen, K.M.; Wendel, J.F. A bountiful harvest: Genomic insights into crop domestication phenotypes. *Annu. Rev. Plant Biol.* **2013**, *64*, 47–70. [CrossRef]

19. Upadhyaya, H.D.; Bajaj, D.; Das, S.; Saxena, M.S.; Badoni, S.; Kumar, V.; Parida, S.K. A genome-scale integrated approach aids in genetic dissection of complex flowering time trait in chickpea. *Plant Mol. Biol.* **2015**, *89*, 403–420. [CrossRef]

20. Bajaj, D.; Upadhyaya, H.D.; Das, S.; Kumar, V.; Gowda, C.L.L.; Sharma, S.; Parida, S.K. Identification of candidate genes for dissecting complex branchnumber trait in chickpea. *Plant Sci.* **2016**, *245*, 61–70. [CrossRef]

21. Kujur, A.; Upadhyaya, H.D.; Bajaj, D.; Gowda, C.L.L.; Sharma, S.; Tyagi, A.K.; Parida, S.K. Identification of candidate genes and natural allelic variants for QTLs governing plant height in chickpea. *Sci. Rep.* **2016**, *6*, 27968. [CrossRef]

22. Das, S.; Upadhyaya, H.D.; Srivastava, R.; Bajaj, D.; Gowda, C.L.; Sharma, S.; Singh, S.; Tyagi, A.K.; Parida, S.K. Genome-wide insertion-deletion (InDel) marker discovery and genotyping for genomics-assisted breeding applications in chickpea. *DNA Res.* **2015**, *22*, 377–386. [CrossRef]

23. Ortega, R.; Hecht, V.F.G.; Freeman, J.S.; Rubio, J.; Carrasquilla-Garcia, N.; Mir, R.R.; Penmetsa, R.V.; Cook, D.R.; Millan, T.; Weller, J.L. Altered Expression of an, F.T. Cluster underlies a major locus controlling domestication-related changes to chickpea phenology and growth habit. *Front. Plant Sci.* **2019**, *10*, 824. [CrossRef]

24. Varshney, R.K.; Thudi, M.; Roorkiwal, M.; He, W.; Upadhyaya, H.D.; Yang, W.; Doddamani, D. Resequencing of 429 chickpea accessions from 45 countries provides insights into genome diversity, domestication and agronomic traits. *Nat Genet.* **2019**, *51*, 857–864. [CrossRef] [PubMed]

25. Sokolkova, A.B.; Chang, P.L.; Carrasquila-Garcia, N.; Noujdina, N.V.; Cook, D.R.; Nuzhdin, S.V.; Samsonova, M.G. The signatures of ecological adaptation in the genomes of chickpea landraces. *Biophysics* **2020**, *65*, 237–240. [CrossRef]

26. Van-Oss, R.P.; Gopher, A.; Kerem, Z.; Peleg, Z.; Lev-Yadun, S.; Sherman, A.; Abbo, S. Independent selection for seed free tryptophan content and vernalization response in chickpea domestication. *Plant Breed.* **2018**, *137*, 290–300. [CrossRef]

27. McKenna, A.; Hanna, M.; Banks, E.; Sivachenko, A.; Cibulskis, K.; Kernytsky, A.; DePristo, M.A. The genome analysis toolkit: A MapReduce framework for analyzing next-generation DNA sequencing data. *Genome Res.* **2010**, *20*, 1297–1303. [CrossRef]

28. Danecek, P.; Auton, A.; Abecasis, G.; Albers, C.A.; Banks, E.; DePristo, M.A.; McVean, G. The variant call format and VCFtools. *Bioinformatics* **2011**, *27*, 2156–2158. [CrossRef]

29. Zheng, X.; Levine, D.; Shen, J.; Gogarten, S.; Laurie, C.; Weir, B. A High-performance computing toolset for relatedness and principal component analysis of SNP data. *Bioinformatics* **2012**, *28*, 3326–3328. [CrossRef]

30. Python Software Foundation. Python Language Reference, Version 2.7. Available online: http://www.python.org (accessed on 20 June 2018).

31. R Core Team. R: A Language and Environment for Statistical Computing. R Foundation for Statistical Computing, Vienna, Austria. 2018. Available online: https://www.R-project.org/ (accessed on 20 June 2018).

32. Lee, T.H.; Guo, H.; Wang, X.; Kim, C.; Paterson, A.H. SNPhylo: A pipeline to construct a phylogenetic tree from huge SNP data. *BMC Genom.* **2014**, *15*, 162. [CrossRef]

33. Revell, L.J. phytools: An, R. package for phylogenetic comparative biology (and other things). *Methods Ecol. Evol.* **2012**, *3*, 217–223. [CrossRef]

34. Paradis, E.; Schliep, K. Ape 5.0: An Environment for Modern Phylogenetics and Evolutionary Analyses in R. Bioinformatics. 2018. Available online: https://doi.org/10.1093/bioinformatics/bty633 (accessed on 15 June 2018).

35. Mann, H.B.; Whitney, D.R. On a test of whether one of two random variables is stochastically larger than the other. *Ann. Math. Stat.* **1947**, *18*, 50–60. [CrossRef]

36. Yang, J.; Lee, S.H.; Goddard, M.E.; Visscher, P.M. GCTA: A tool for genome-wide complex trait analysis. *Am. J. Hum. Genet.* **2011**, *88*, 76–82. [CrossRef]

37. Shapiro, S.S.; Wilk, M.B. An analysis of variance test for normality (complete samples). *Biometrika* **1965**, *52*, 591–611. [CrossRef]

38. Harrell, F.E., Jr. Hmisc: Harrell Miscellaneous. R Package Version 4.1-1. 2018. Available online: https://CRAN.R-project.org/package=Hmisc (accessed on 15 June 2018).

39. Lippert, C.; Listgarten, J.; Liu, Y.; Kadie, C.M.; Davidson, R.I.; Heckerman, D. FaST linear mixed models for genome-wide association studies. *Nat. Methods* **2011**, *8*, 833–835. [CrossRef] [PubMed]

40. Roberts, D.W. Labdsv: Ordination and Multivariate Analysis for Ecology. R Package Version 1.8-0. 2016. Available online: http://CRAN.R-project.org/package=labdsv (accessed on 15 June 2018).

41. Storey, J.D. The positive false discovery rate: A Bayesian interpretation and the q-Value. *Source Ann. Stat. Ann. Stat.* **2003**, *31*, 2013–2035. [CrossRef]

42. CMplot: Circle Manhattan Plot. Available online: https://github.com/YinLiLin/R-CMplot (accessed on 20 June 2018).

43. Cingolani, P.; Platts, A.; Wang, L.L.; Coon, M.; Nguyen, T.; Wang, L.; Ruden, D.M. A program for annotating and predicting the effects of single nucleotide polymorphisms, SnpEff: SNPs in the genome of Drosophila melanogaster strain w1118; iso-2; iso-3. *Fly Austin* **2012**, *6*, 80–92. [CrossRef]

44. Dash, S.; Campbell, J.D.; Cannon, E.K.; Cleary, A.M.; Huang, W.; Kalberer, S.R.; Weeks, N.T. Legume information system (LegumeInfo. org): A key component of a set of federated data resources for the legume family. *Nucl. Acids Res.* **2016**, *44*, D1181–D1188. [CrossRef]

45. Li, J.; Dai, X.; Liu, T.; Zhao, P.X. LegumeIP: An integrative database for comparative genomics and transcriptomics of model legumes. *Nucleic Acids Res.* **2012**, *40*, 1221–1229. [CrossRef]

International Journal of
Molecular Sciences

Article

Genome-Wide Association Mapping for Heat Stress Responsive Traits in Field Pea

Endale G. Tafesse [1], Krishna K. Gali [1], V.B. Reddy Lachagari [2], Rosalind Bueckert [1] and Thomas D. Warkentin [1,*]

[1] Department of Plant Sciences, College of Agriculture and Bio-resources, University of Saskatchewan, Saskatoon, SK S7N 5A8, Canada; endale.tafesse@usask.ca (E.G.T.); kishore.gali@usask.ca (K.K.G.); rosalind.bueckert@usask.ca (R.B.)
[2] AgriGenome Labs Pvt. Ltd., Hyderabad 500 078, India; vb.reddy@aggenome.com
* Correspondence: tom.warkentin@usask.ca; Tel.: +1-306-966-2371

Received: 14 February 2020; Accepted: 14 March 2020; Published: 17 March 2020

Abstract: Environmental stress hampers pea productivity. To understand the genetic basis of heat resistance, a genome-wide association study (GWAS) was conducted on six stress responsive traits of physiological and agronomic importance in pea, with an objective to identify the genetic loci associated with these traits. One hundred and thirty-five genetically diverse pea accessions from major pea growing areas of the world were phenotyped in field trials across five environments, under generally ambient (control) and heat stress conditions. Statistical analysis of phenotype indicated significant effects of genotype (G), environment (E), and G × E interaction for all traits. A total of 16,877 known high-quality SNPs were used for association analysis to determine marker-trait associations (MTA). We identified 32 MTAs that were consistent in at least three environments for association with the traits of stress resistance: six for chlorophyll concentration measured by a soil plant analysis development meter; two each for photochemical reflectance index and canopy temperature; seven for reproductive stem length; six for internode length; and nine for pod number. Forty-eight candidate genes were identified within 15 kb distance of these markers. The identified markers and candidate genes have potential for marker-assisted selection towards the development of heat resistant pea cultivars.

Keywords: pea; heat stress; genetic diversity; GWAS; genotyping-by-sequencing; marker-trait association; candidate-gene

---

1. Introduction

Pea (*Pisum sativum* L., $2n = 14$) is a major pulse crop widely grown in the temperate regions primarily for its nutritional values as a source of protein, slowly digestible starch, essential minerals, high fiber and low fat; and soil fertility benefits as it fixes atmospheric nitrogen [1–3]. However, as a cool season crop, pea is prone to heat and drought stress, with warm summers causing shortened life cycles, abortion of floral components and pods, and thus economic yield loss [4–6]. Due to global warming, the average surface temperature is predicted to increase by 3.7 °C by the end of this century, and thus heat stress is expected to be even more challenging in the future [7].

Genetic improvement of pea for heat and drought resistance is a promising approach to stabilize yield under environmental stresses. Pea germplasm has a wide range of diversity in morpho-anatomical, biochemical and physiological characteristics [8,9]. Among other things, such diversity has been explored to identify traits associated with heat response [10–12]. Pigments including chlorophylls, carotenoids, anthocyanins contribute to heat tolerance through heat dissipation and protection of vital plant components and processes [13,14]. Multi-environment studies on pea [10], and maize [15] revealed leaf color (greenness) as a trait linked to stress tolerance.

Chlorophyll represents pigment abundance and composition, and is used to drive photosynthesis, plant senescence, and yield potential [15,16]. Stay-green, a trait that delays plant senescence, is reported to be associated with improved yield under stress conditions [15]. Estimation of leaf chlorophyll concentration by the soil plant analysis development (SPAD) meter is reliable, and is strongly correlated with laboratory-based destructive methods [17].

Vegetative indices (VI), determined from different wavelengths of spectral reflectance, have been used as proxies to quantitatively and qualitatively assess traits linked with vegetation cover and plant vigor, pigment abundance and composition, and plant water status [18,19]. Thus, VIs indirectly indicate the overall physiological state of the plant under various environmental conditions. For example, photochemical reflectance index (PRI), derived from narrowband wavelengths, indicates photosynthetic efficiency and photosynthetic performance in stress [19]. Canopy temperature (CT) is a direct indicator of degree of stress in plants. If CT is greater than the air temperature, then the plants are under stress predominantly caused by heat and drought. Although the environment contributes to CT to a great extent, there exists significant variation in genotype response [12].

In pea and other crops, lodging is one of the plant factors that exacerbates heat stress by making the plant hold more heat in the canopy, and thereby leading to increased CT [12,20]. Heat and drought stress decreases reproductive stem and internode lengths [12], which are related to genes associated with gibberellin function [21,22]. Pod number, a major yield component in pea and other pulse crops, is an economic trait highly affected by heat stress [23,24]. Pod loss due to heat stress is mostly associated with pollen and stigma malfunction, and abortion of flowers, bud and pods [6,11].

Understanding of the genetic base of traits involved in pea stress response would assist breeders in developing heat resistant varieties. Genome-wide association study (GWAS) has been used as a tool for dissecting the genetic bases of various traits using the naturally occurring genetic diversity a species has accumulated over many generations [25,26]. Linkage disequilibrium (LD)-based association mapping provides high resolution, as it relies on the use of single nucleotide polymorphisms (SNP), and thus has the capacity to distinguish even between closely related individuals [27–30]. The advancement and inexpensive availability of high-throughput next generation sequencing (NGS) platforms enabled the use of SNPs for genetic diversity study and estimation of LD in pea and other crops [29,30]. Association mapping has been successfully used for identification of numerous genomic loci and underlying genes for complex traits in several crops including pea [25–35].

In pea, association and linkage mapping has been employed to uncover the genetic bases of several traits including agronomic and seed quality traits [30,35], disease resistance [32,36], seed mineral concentrations [37], seed lipid content [38], salinity tolerance [31], and frost tolerance [33]. Despite its importance, only limited studies have been carried out to identify genomic regions associated with pea stress tolerance [28]. Stress tolerance is complex and is controlled by many genes throughout the genome each with minor effects and each interacting with the environment [39]. The objectives of this study were to examine the G × E interaction in pigment and vegetative structures associated with stress response, to explore the genetic variation of stress tolerance present in a GWAS panel of 135 accessions, and to identify MTAs related with six stress responsive traits.

2. Results

2.1. Weather and Stress Condition of the Environments

The weather condition of the five environments during the pea growing season described by the average of daily maximum, minimum, 24 h mean temperatures, number of days when the daily maximum temperature was greater than 28 °C during the growing season, and total monthly precipitation is summarized in Table 1. In pea, significant yield loss due to heat stress is evident whenever the daily maximum air temperature exceeds 28 °C for several days during the growing season [5]. Impact of heat and drought is severe when it occurs during reproductive stages. Saskatoon 2015 was the most stressed environment as indicated by mean daily maximum air temperatures

> 27 °C, 18 days where air temperature was > 28 °C, and drier conditions during the reproductive stage. Similarly, 2017 Saskatoon was also under heat and drought stress during the reproductive stage with average air temperature ~26 °C, 16 days where air temperature was > 28 °C, and relatively low total precipitation. The remaining three environments were generally ambient and considered as control environments (Table 1).

Table 1. Seeding date, average maximum, minimum and 24 h daily mean temperatures, number of days when the daily maximum temperature was greater than 28 °C, and total monthly precipitation at different growth and development stages of pea at each environment.

Environment	Seeding Date	Growth and Development Stage	Number of Days Spent in the Growth and Development Stage	Daily Maximum Mean Temp. (°C)	Daily Minimum Mean Temp. (°C)	Daily 24 h Mean Temp. (°C)	Number of Days when Temp. was > 28 °C	Total Precipitation (mm)	Stress Situation
2015 Saskatoon	24-Apr		58	20.4b	5.3b	13.1b	7	23.1	Drought
2016 Rosthern	06-May	Germination to late vegetative stage	46	20.8ab	6.4ab	14.4ab	3	75.8	Control
2016 Saskatoon	26-Apr		50	21.5ab	6.1ab	14.6ab	8	63.7	Control
2017 Rosthern	21-May		44	22.3ab	7.4a	15.9a	5	62.1	Control
2017 Saskatoon	30-Apr		51	22.7a	6.2ab	14.7ab	9	58.5	Control
2015 Saskatoon			42	27.1a	14.0a	20.0a	18	41.3	Heat, drought
2016 Rosthern		Beginning of flowering to maturity	52	23.1d	13.4a	18.5b	4	126.2	Control
2016 Saskatoon			48	24.4cd	11.9b	18.2b	3	86.2	Control
2017 Rosthern			46	25.6bc	10.3c	18.3b	9	46.7	Drought
2017 Saskatoon			44	25.9ab	10.4c	18.6b	16	42.6	Heat, drought

Note: temp, temperature; mm, millimeter; Means of environmental variables that do not share a letter within a column under each growth stage are significantly different from each other. The data were analyzed by one-way ANOVA and followed with the Tukey–HSD test for the mean separations.

2.2. *Phenotypic Measurements, Analysis of Variance, and Marker Detection through Association Mapping*

Variance components of genotype (G), environment (E), and G × E interaction together with their significance on the six traits used in this study is presented in Table 2. For all traits analyzed, normality of residuals and homogeneity of variance were met.

Table 2. Variance components of environment, genotype, and their interaction and broad sense heritability (H^2) on SPAD, PRI, canopy temperature, reproductive stem length, internode length and pod number in 135 pea accessions.

Source	SPAD		PRI		Canopy Temperature		Reproductive Stem Length		Internode Length		Pod Number	
	Variance	% of Total	Variance	% of Total	Variance	% of Total	Variance	% of Total	Variance	% of Total	Variance	% of Total
Genotype (G)	19.88 ***	67.9	0.0000171 ***	4.8	0.095 ***	1.7	189.12 ***	63.4	1.69 ***	43.0	2.33 ***	36.6
Environment (E)	0.64 ***	2.2	0.000067 ***	18.7	4.70 ***	85.3	22.52 ***	7.6	0.19 **	4.8	0.79 ***	12.4
REP	0.05 **	0.2	0 ns	0.0	0.006 ns	0.1	8.72	2.9	0.11 **	2.7	0.00 ns	0.0
G × E	1.47 ***	5.0	0.00041 ***	11.4	0.00 ns	0.0	7.58 **	2.5	0 ns	0.0	0.07	1.1
Error	7.25	24.7	0.000233	65.1	0.71	12.9	145	23.6	1.94	49.5	3.18	49.9
Total	29.29		0.00036		5.51		298.19		3.93		6.36	
(H^2)	0.95		0.35		0.57		0.92		0.90		0.88	

Note: * Significant at the 0.05 level of probability; ** Significant at the 0.01 level of probability; *** Significant at the 0.001 level of probability; ns, not significant at the 0.05 level. SPAD, soil plant analysis development; PRI, photochemical reflectance index.

Descriptive statistics for minimum, maximum and mean values of phenotypic measurements on the traits of the GWAS panel across five environments are summarized in Table 3 and Figure 1.

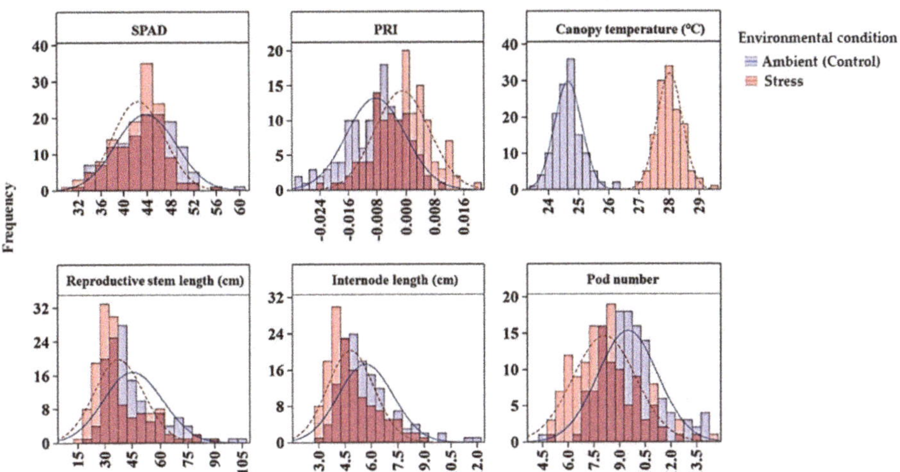

Figure 1. Distribution of average SPAD, PRI, canopy temperature, reproductive stem length, internode length and pod number of 135 GWAS accessions across ambient and stress environments. Note: The ambient (control) environments were 2016 Rosthern, 2016 Saskatoon and 2017 Rosthern; and the heat stress environments were 2015 and 2017 Saskatoon. PRI, photochemical reflectance index.

Table 3. Minimum, maximum and mean values of phenotypic traits of 135 pea accessions of the genome-wide association study panel.

Trait	Environment	Minimum	Maximum	Mean	Standard Deviation
SPAD	2015 Saskatoon	27.3	57.6	42.5	4.7
	2016 Rosthern	30.0	67.5	45.0	6.7
	2016 Saskatoon	31.0	61.1	43.7	4.8
	2017 Rosthern	32.5	56.8	42.9	5.0
	2017 Saskatoon	26.6	55.7	42.6	5.2
Photochemical reflectance index (PRI)	2015 Saskatoon	−0.039	0.028	0.000	0.012
	2016 Rosthern	−0.032	0.028	0.001	0.012
	2016 Saskatoon	−0.116	0.024	−0.019	0.024
	2017 Rosthern	−0.031	0.02	−0.006	0.01
	2017 Saskatoon	−0.037	0.026	−0.003	0.013
Canopy temperature (°C)	2015 Saskatoon	28.0	31.0	29.6	0.5
	2016 Rosthern	21.4	26.9	24.2	1.0
	2016 Saskatoon	22.3	28.4	24.6	1.2
	2017 Rosthern	23.5	26.9	25.1	0.6
	2017 Saskatoon	24.5	29.1	26.4	0.8
Reproductive stem length (cm)	2015 Saskatoon	13.2	90.7	37.9	15.0
	2016 Rosthern	16.0	117	48.9	19.7
	2016 Saskatoon	14.4	101	42.9	17.6
	2017 Rosthern	18.3	104	42.0	15.3
	2017 Saskatoon	14.6	99	36.0	15.1
Internode length (cm)	2015 Saskatoon	1.6	10.7	4.7	1.6
	2016 Rosthern	2.0	14.7	5.8	2.1
	2016 Saskatoon	1.9	14.7	5.1	2.0
	2017 Rosthern	2.4	14.9	6.0	2.0
	2017 Saskatoon	1.9	11.3	4.9	1.7
Pod number	2015 Saskatoon	3.0	13.0	7.8	1.8
	2016 Rosthern	3.5	18.5	9.8	2.8
	2016 Saskatoon	3.0	17.5	9.9	2.6
	2017 Rosthern	4.0	15.0	8.6	2.0
	2017 Saskatoon	4.5	18.5	8.3	2.4

Note: soil plant analysis development (SPAD), spectral reflectance and canopy temperature were taken four to six times in a season during reproductive stage on hot days at solar noon. A SPAD reading > 50 indicates a dark-green color and high chlorophyll concentration, a reading < 40 indicates a yellow-green color and low chlorophyll concentration. Reproductive stem length, internode length and pod number were measured on three plants per plot at physiological maturity. The overall weather classification of environments 2015 and 2017 at Saskatoon was heat stress, and the remaining three environments condition was ambient (control) for pea production. A SPAD value is an index of light transmittance at 650 nm and 940 nm. Similarly, PRI is an index derived from narrow-band (531 and 571 nm) spectral reflectance.

Chlorophyll concentration, measured by a SPAD meter, was affected by genotype, environment and their interaction; and the variance component analysis showed that maximum variation (67.9%) among the GWAS panel was due to the genotype effect, and the broad sense heritability was 0.95. Overall, genotype chlorophyll concentration ranged from 26.6 to 57.6 SPAD values under heat stress, and 30.0 to 67.5 under control conditions (Table 3). On average, the heat stressed environments had 3% less SPAD value than the ambient environments. Six markers (Chr5LG3_150942510, Chr5LG3_446272814, Chr5LG3_449362407, Chr5LG3_566189589, Chr5LG3_569788697, and Chr5LG3_572899434) were associated with SPAD in at least three out of the five environments, and on average each marker explained 7%–13% of the phenotypic variance (PV) measured as the difference in R-square of the model with the SNP and without the SNP. SNP markers Chr5LG3_566189589 and Chr5LG3_449362407 were associated with SPAD in 4 and 5 environments explaining 13% and 11% of the PVs, respectively (Table 4). PRI was also significantly affected by genotype, environment and by the G x E interaction. Variance components showed most of the variation in PRI was due to environmental factors, and the broad sense heritability was the least (0.35) compared with the other traits (Table 2). Two markers,

Chr6LG2_469101917, and Chr7LG7_263964018 were significantly associated with PRI at three out of the five environments. Each of the two markers explained 9% of PV (Table 4).

Table 4. Trait-linked SNP markers identified by association analysis of pea phenotypes associated with heat stress using the mixed linear model (MLM).

Trait	SNP Marker	Environment	p.value	R Square of Model with SNP	R Square of Marker [†]	Average R Square of Marker
SPAD	Chr5LG3_150942510	2016 Rosthern	3.77×10^{-4}	0.39	0.08	
		2016 Saskatoon	6.80×10^{-4}	0.45	0.06	
		2017 Saskatoon	2.15×10^{-4}	0.42	0.09	0.08
	Chr5LG3_446272814	2016 Saskatoon	1.89×10^{-4}	0.46	0.08	
		2017 Rosthern	2.46×10^{-4}	0.48	0.07	
		2017 Saskatoon	4.68×10^{-4}	0.41	0.08	0.08
	Chr5LG3_449362407	2015 Saskatoon	1.39×10^{-4}	0.42	0.09	
		2016 Rosthern	6.66×10^{-5}	0.41	0.1	
		2016 Saskatoon	3.27×10^{-5}	0.48	0.09	
		2017 Rosthern	1.24×10^{-6}	0.54	0.13	
		2017 Saskatoon	6.61×10^{-6}	0.46	0.13	0.11
	Chr5LG3_566189589	2015 Saskatoon	5.00×10^{-7}	0.56	0.15	
		2016 Rosthern	4.33×10^{-6}	0.45	0.14	
		2016 Saskatoon	1.23×10^{-5}	0.49	0.1	
		2017 Rosthern	9.83×10^{-6}	0.52	0.11	0.13
	Chr5LG3_569788697	2015 Saskatoon	1.22×10^{-4}	0.42	0.09	
		2016 Rosthern	5.03×10^{-4}	0.39	0.08	
		2016 Saskatoon	9.70×10^{-4}	0.45	0.06	
		2017 Rosthern	9.00×10^{-4}	0.47	0.06	0.07
	Chr5LG3_572899434	2015 Saskatoon	4.76×10^{-4}	0.41	0.08	
		2016 Rosthern	3.17×10^{-4}	0.39	0.08	
		2016 Saskatoon	5.09×10^{-4}	0.45	0.06	
		2017 Rosthern	2.98×10^{-4}	0.48	0.07	0.07
PRI	Chr6LG2_469101917	2016 Rosthern	8.99×10^{-4}	0.3	0.08	
		2017 Rosthern	8.85×10^{-5}	0.3	0.11	
		2017 Saskatoon	3.39×10^{-3}	0.16	0.07	0.09
	Chr7LG7_263964018	2016 Rosthern	8.99×10^{-4}	0.3	0.08	
		2017 Rosthern	8.85×10^{-5}	0.3	0.11	
		2017 Saskatoon	3.39×10^{-3}	0.16	0.07	0.09
Canopy temperature	Chr4LG4_415994869	2015 Saskatoon	1.16×10^{-3}	0.52	0.05	
		2016 Rosthern	1.08×10^{-3}	0.5	0.06	
		2016 Saskatoon	2.22×10^{-4}	0.44	0.08	0.06
	Chr5LG3_309595819	2015 Saskatoon	4.88×10^{-4}	0.53	0.06	
		2016 Rosthern	5.11×10^{-3}	0.48	0.04	
		2016 Saskatoon	4.39×10^{-4}	0.43	0.07	0.06
Reproductive stem length	Chr3LG5_18678117	2015 Saskatoon	2.18×10^{-4}	0.63	0.06	
		2016 Saskatoon	3.60×10^{-4}	0.62	0.05	
		2017 Rosthern	6.62×10^{-4}	0.7	0.04	
		2017 Saskatoon	8.42×10^{-5}	0.5	0.08	0.06
	Chr4LG4_29062302	2015 Saskatoon	5.85×10^{-4}	0.62	0.05	
		2016 Rosthern	2.58×10^{-3}	0.59	0.03	
		2016 Saskatoon	2.09×10^{-3}	0.61	0.04	
		2017 Rosthern	8.96×10^{-4}	0.7	0.03	
		2017 Saskatoon	3.11×10^{-3}	0.46	0.04	0.04
	Chr5LG3_566189271	2015 Saskatoon	1.72×10^{-4}	0.63	0.06	
		2016 Rosthern	3.71×10^{-4}	0.61	0.05	
		2016 Saskatoon	1.14×10^{-4}	0.63	0.06	
		2017 Rosthern	1.43×10^{-4}	0.71	0.04	0.05
	Chr5LG3_572669963	2015 Saskatoon	1.06×10^{-3}	0.62	0.05	
		2016 Saskatoon	1.03×10^{-4}	0.63	0.06	
		2017 Rosthern	2.53×10^{-4}	0.71	0.04	0.05
	Chr7LG7_20086906	2015 Saskatoon	6.08×10^{-4}	0.62	0.05	
		2016 Rosthern	4.27×10^{-3}	0.59	0.03	
		2016 Saskatoon	8.52×10^{-4}	0.61	0.04	
		2017 Rosthern	4.00×10^{-3}	0.69	0.03	0.04
	Chr7LG7_23295474	2015 Saskatoon	8.25×10^{-4}	0.62	0.05	
		2016 Saskatoon	4.84×10^{-4}	0.62	0.05	
		2017 Rosthern	3.82×10^{-4}	0.7	0.03	0.05
	Chr7LG7_96157380	2015 Saskatoon	2.72×10^{-4}	0.63	0.06	
		2016 Rosthern	2.15×10^{-3}	0.59	0.04	
		2016 Saskatoon	6.82×10^{-4}	0.62	0.05	
		2017 Rosthern	2.68×10^{-4}	0.71	0.04	0.05

Table 4. *Cont.*

Trait	SNP Marker	Environment	*p*.value	R Square of Model with SNP	R Square of Marker [†]	Average R Square of Marker
	Chr4LG4_62461234	2015 Saskatoon	8.58×10^{-3}	0.49	0.04	
		2016 Saskatoon	3.83×10^{-4}	0.48	0.07	
		2017 Saskatoon	3.18×10^{-4}	0.39	0.08	0.06
	Chr4LG4_63111072	2015 Saskatoon	3.86×10^{-4}	0.52	0.06	
		2017 Rosthern	3.54×10^{-3}	0.62	0.04	
		2017 Saskatoon	3.68×10^{-4}	0.39	0.08	0.06
	Chr4LG4_80759704	2016 Rosthern	3.50×10^{-3}	0.36	0.05	
		2016 Saskatoon	2.28×10^{-4}	0.49	0.03	
		2017 Rosthern	7.64×10^{-3}	0.62	0.08	0.06
Internode length	Chr5LG3_566189271	2015 Saskatoon	1.22×10^{-5}	0.55	0.09	
		2016 Rosthern	8.23×10^{-4}	0.38	0.07	
		2016 Saskatoon	4.72×10^{-5}	0.5	0.09	
		2017 Rosthern	2.29×10^{-3}	0.63	0.04	
		2017 Saskatoon	2.85×10^{-3}	0.36	0.05	0.07
	Chr6LG2_420562729	2015 Saskatoon	3.76×10^{-4}	0.52	0.06	
		2016 Saskatoon	3.87×10^{-3}	0.46	0.05	
		2017 Rosthern	8.96×10^{-4}	0.63	0.04	0.05
	Chr7LG7_197862543	2015 Saskatoon	4.69×10^{-4}	0.52	0.06	
		2016 Saskatoon	9.72×10^{-3}	0.45	0.05	
		2017 Saskatoon	1.39×10^{-3}	0.37	0.06	0.06
	Chr2LG1_4359822	2015 Saskatoon	8.14×10^{-4}	0.24	0.08	
		2016 Rosthern	1.75×10^{-3}	0.27	0.07	
		2016 Saskatoon	3.00×10^{-3}	0.16	0.08	0.08
	Chr2LG1_105547608	2015 Saskatoon	3.98×10^{-4}	0.25	0.09	
		2016 Saskatoon	3.01×10^{-3}	0.16	0.08	
		2017 Saskatoon	9.05×10^{-4}	0.22	0.09	0.09
	Chr2LG1_370541780	2015 Saskatoon	2.08×10^{-4}	0.26	0.1	
		2016 Saskatoon	7.58×10^{-4}	0.18	0.1	
		2017 Saskatoon	4.68×10^{-3}	0.19	0.06	0.09
	Chr2LG1_385949935	2015 Saskatoon	3.11×10^{-4}	0.26	0.1	
		2016 Saskatoon	8.17×10^{-5}	0.21	0.13	
		2017 Saskatoon	1.20×10^{-3}	0.18	0.05	0.10
	Chr2LG1_389336188	2015 Saskatoon	4.96×10^{-4}	0.25	0.09	
Pod number		2016 Saskatoon	2.71×10^{-3}	0.16	0.08	
		2017 Saskatoon	4.60×10^{-4}	0.23	0.1	0.09
	Chr2LG1_402022079	2015 Saskatoon	3.58×10^{-3}	0.22	0.06	
		2016 Rosthern	1.16×10^{-3}	0.27	0.07	
		2016 Saskatoon	5.15×10^{-4}	0.18	0.1	
		2016 Saskatoon	5.15×10^{-4}	0.18	0.1	0.08
	Chr3LG5_216337201	2015 Saskatoon	4.75×10^{-3}	0.22	0.07	
		2016 Rosthern	3.54×10^{-3}	0.26	0.06	
		2017 Saskatoon	3.49×10^{-4}	0.23	0.1	0.08
	Chr5LG3_530537682	2015 Saskatoon	3.32×10^{-3}	0.22	0.06	
		2016 Rosthern	3.80×10^{-3}	0.26	0.06	
		2016 Saskatoon	5.81×10^{-4}	0.18	0.1	0.07
	Sc04062_32372	2015 Saskatoon	4.27×10^{-4}	0.25	0.09	
		2016 Rosthern	8.51×10^{-3}	0.25	0.06	
		2016 Saskatoon	7.23×10^{-3}	0.14	0.06	
		2017 Saskatoon	1.70×10^{-5}	0.28	0.15	0.09

Note: All markers presented here were significant in at least three of five environments for a given trait. In each SNP designation, Chr and LG indicate chromosome and linkage group and the number after the _ is the base pair position. For non-chromosomal SNPs, Sc refers to scaffold followed by the scaffold number. Each locus is represented by one SNP marker of the LD block. [†]R-square value is presented as the difference of R-square explained by the model with and without SNP.

For canopy temperature (CT), the GWAS accessions significantly varied due to both genotype (G) and environment (E) effects, but not by the G x E interaction (Table 2). In general, under heat stress, the accessions' CT, measured four to six times in a season during reproductive stage on hot days at solar noon, ranged from 24.5 to 31.0 °C, whereas under ambient conditions, the CT ranged from 21.4 to 26.9 °C. This temperature difference indicated that CT is highly influenced by the environment effects with a relatively lower broad sense heritability of 0.57 (Table 2; Table 3; Figure 1). Two SNP markers (Chr4LG4_415994869 and Chr5LG3_309595819) were associated with CT in three of the five environments. The R-square value of the model with SNP ranged from 0.43 to 0.53, and each of the SNP markers explained 6% of PV.

Reproductive stem length was also affected by genotype and environment main effects and their interaction. The reproductive stem length under the stressed environments ranged from 13 to 99 cm, whereas under the control environments the range was from 14 to 117 cm, suggesting heat stress decreased the reproductive stem length. Analysis of variance components showed genotype and environment main effects respectively contributed to 63.4% and 7.6% of the variation in the GWAS panel. The broad sense heritability for reproductive stem length was 0.92. Seven SNP markers (Chr3LG5_18678117, Chr4LG4_29062302, Chr5LG3_566189271, Chr5LG3_572669963, Chr7LG7_20086906, Chr7LG7_23295474, and Chr7LG7_96157380) were associated with reproductive stem length in at least three of the five environments, and four of these SNPs were consistent in at least four of the five environments. SNP marker Chr4LG4_29062302 was found to be associated with the trait in all five environments with an average R-square of the model of 0.60. Overall, the R-square value of the model with SNP ranged up to 0.71 for reproductive stem length (Table 4).

Internode length was another trait significantly affected by genotype and environment main effects and their interaction. Under heat stress, the internode length ranged from 1.6 to 11.3 cm with a mean value of 11.0 cm, whereas under control conditions, the range was 1.9 to 14.9 cm with a mean value of 14.8 cm. Variance component analysis showed genotype and environment respectively contributed 43% and 4.8% of the variations to the GWAS panel. The broad sense heritability was 0.90. Six SNP markers (Chr4LG4_62461234, Chr4LG4_63111072, Chr4LG4_80759704, Chr5LG3_566189271, Chr6LG2_420562729, and Chr7LG7_197862543) were associated with internode length in at least three of the five environments. These markers were significantly associated with internode length in at least three of the five environments with the R-square value of the model with SNP ranged up to 0.63. SNP marker Chr5LG3_566189271 was identified in all five environments with an average R-square of 0.49.

Pod number was also significantly affected by genotype and environment main effects and their interaction. Variance component analysis showed genotype and environment, respectively, contributed 36.6% and 12.4% to the overall pod number variance in the GWAS panel. Compared with the three control environments, pod number under the heat stress environments decreased by 14.6%. The broad sense heritability in pod number was 0.88. Eight SNP markers (Chr2LG1_4359822, Chr2LG1_105547608, Chr2LG1_370541780, Chr2LG1_385949935, Chr2LG1_389336188, Chr2LG1_402022079, Chr3LG5_216337201, Chr5LG3_530537682, and Sc04062_32372) were associated with pod number in at least three of the five environments explaining 7% to 9% of PV, with an average R-square value of 21.9.

Manhattan plots showing the association of SNP markers with plant chlorophyll concentration and reproductive stem length in multiple trials, and the corresponding Q-Q plots are presented as examples from this research in Figures 2 and 3, respectively. The Q-Q plots represent the observed P values of each SNP marker against the expected P values. The Manhattan plots in Figure 2 showed the significant association of SNP markers on Chr 5 (LG3) with plant SPAD in each of the individual environments presented. The Manhattan plots in Figure 3 showed the significant association of SNP markers on multiple chromosomes with the reproductive stem length.

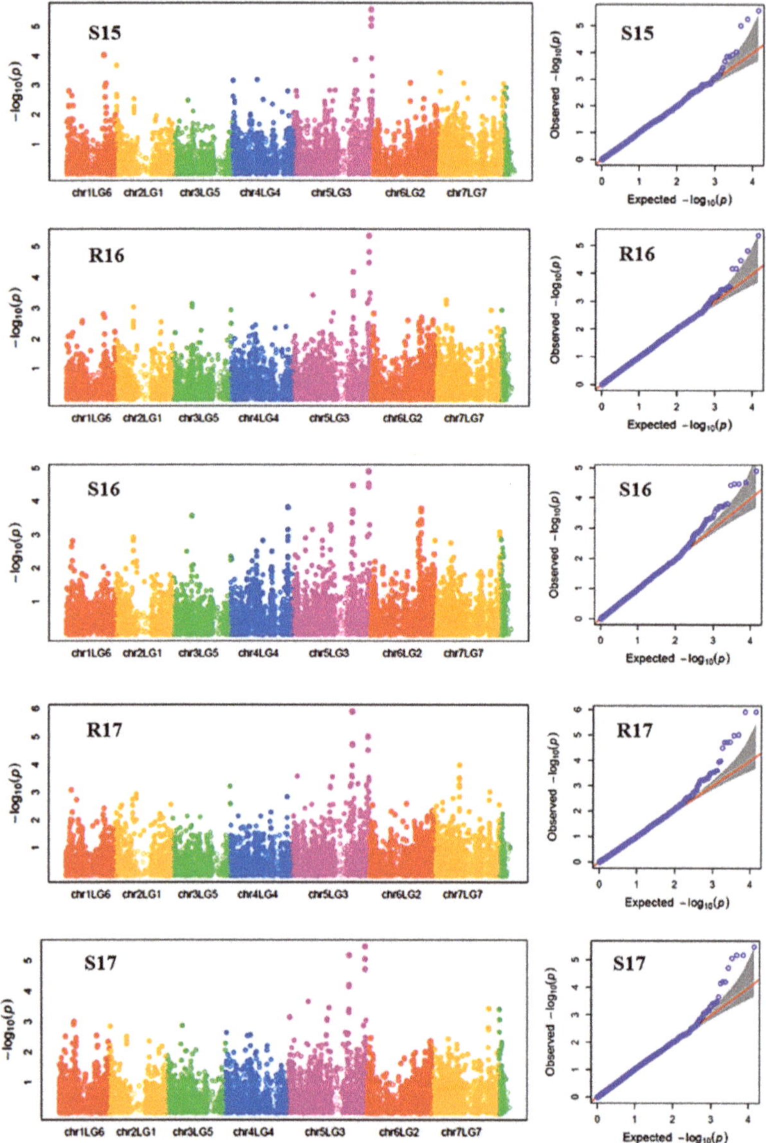

Figure 2. Manhattan plots and the corresponding Q-Q plots representing the identification of SNP markers associated with chlorophyll concentration measured by a SPAD meter. The Manhattan plots are based on association of 15,608 chromosomal and 1269 non-chromosomal SNPs with SPAD of 135 pea accessions in the multi-year, multi-environment trials. Note: S15, Saskatoon in 2015; R16, Rosthern in 2016; S16, Saskatoon in 2016; R17, Rosthern in 2017; and S17, Saskatoon in 2017.

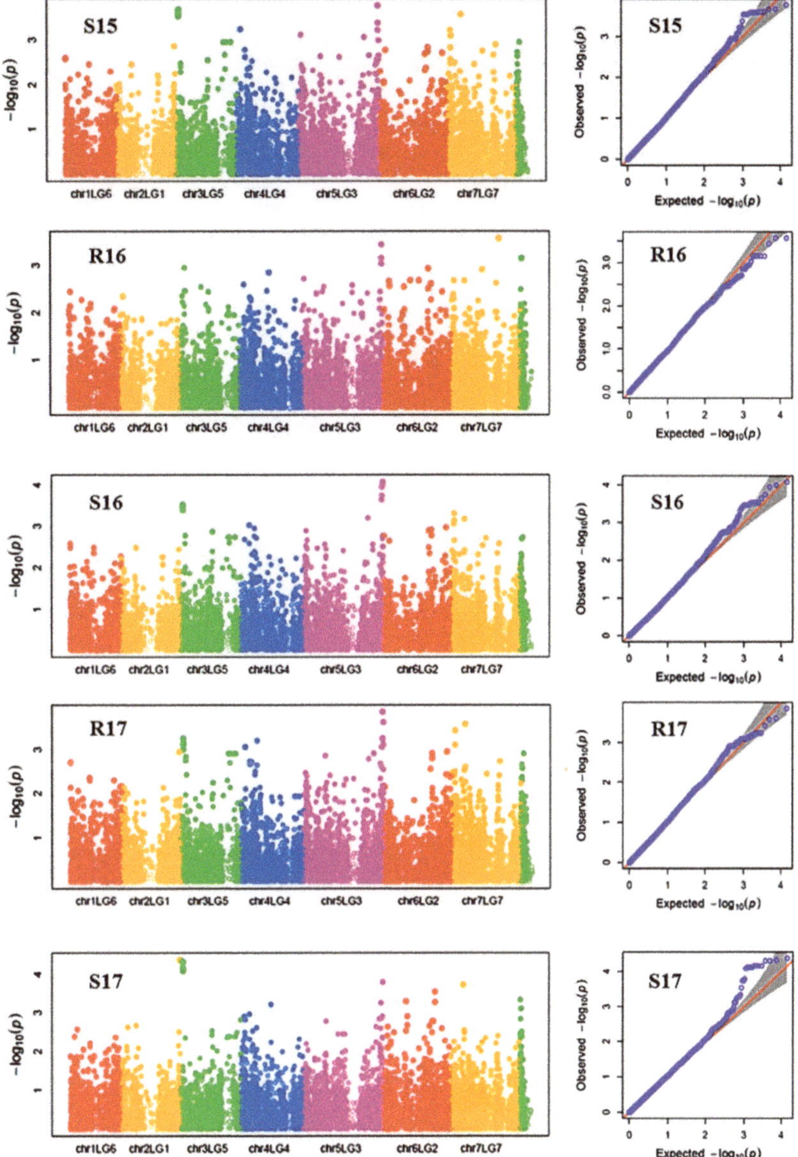

Figure 3. Manhattan plots and the corresponding Q-Q plots representing the identification of SNP markers associated with reproductive stem length. The Manhattan plots are based on association of 15608 chromosomal and 1269 non-chromosomal SNPs with reproductive stem length of 135 pea accessions in the multi-year, multi-environment trials. Note: R16, Rosthern in 2016; R17, Rosthern in 2017; S15, Saskatoon in 2015; S16, Saskatoon in 2016; and S17, Saskatoon in 2017.

Of all the MTAs that were observed in > 60% of the environments, the following markers had the greatest percent variation averaged over the selected environments for the respective traits: Chr5LG3_566189589 (13% PV) and Chr5LG3_449362407 (11% PV) for SPAD; Chr6LG2_469101917 and Chr7LG7_263964018 each with 9% PV for PRI; Chr4LG4_415994869 and

Chr5LG3_309595819 each with 6% PV for CT; Chr3LG5_18678117 (6% PV), Chr5LG3_572669963 (5% PV), and Chr7LG7_96157380 (5% PV) for reproductive stem length; Chr4LG4_63111072 (6% PV), Chr5LG3_566189271 (7% PV) and Chr4LG4_62461234 (6% PV) for internode length; and seven markers, Chr2LG1_105547608, Chr2LG1_370541780, Chr2LG1_385949935, Chr2LG1_389336188, Chr3LG5_216337201, Chr5LG3_530537682, and Sc04062_32372 each with 9% PV for pod number (Table 4).

Forty-eight unique genes were identified within a 15 kb region of the selected 32 SNP markers and are considered as candidate genes. The candidate genes identified for various traits included those encoding for transcription factor, translation initiation factor, heat shock protein, ribosomal protein, protein kinase, transmembrane protein, and others as listed in Table 5. Two genes, Psat5g299080 and Psat5g299040, which encode the proteins kinesin-related protein 4-like and PPR containing plant-like protein (putative tetratricopeptide-like helical domain-containing protein), were identified as potential candidate genes associated with internode length, reproductive stem length and chlorophyll content (SPAD).

Table 5. Candidate genes identified within 15 kb distance of the SNP markers identified for association with the traits of heat tolerance.

Trait[a]	SNP Marker	Gene_ID	Protein Names	Gene Names	Organism[b]	Gene Ontology IDs	Gene Ontology (GO)
SPAD	chr5LG3_446272814	Psat5g221440	Amidohydrolase ytcj-like protein (Fragment)	L195_g035501	Tp	GO:0016810	hydrolase activity, acting on carbon-nitrogen (but not peptide) bonds [GO:0016810]
	chr5LG3_449362407	Psat5g224400	cysteine-rich receptor-like protein kinase 25	LOC101505680	Ca	GO:0004672; GO:0005524; GO:0016021	integral component of membrane [GO:0016021]; ATP binding [GO:0005524]; protein kinase activity [GO:0004672]
		Psat5g224360	Pentatricopeptide repeat-containing protein at1g11290-like protein	L195_g006458	Tp	GO:0008270	zinc ion binding [GO:0008270]
		Psat5g224280	Pentatricopeptide repeat-containing protein at1g11290-like protein	L195_g022714	Tp	GO:0008270	zinc ion binding [GO:0008270]
	chr5LG3_566189589	Psat5g299080	Kinesin-related protein 4-like	L195_g011972	Tp		
		Psat5g299040	PPR containing plant-like protein (Putative tetratricopeptide-like helical domain-containing protein)	11431556 MTR_2g102210 MtrunA17_Chr2g0331911	Mt (Mtr)		
	chr5LG3_569788697	Psat5g301440	Embryo-specific 3 (Fragment)	L195_g051812	Tp		
		Psat5g301400	Nuclear pore protein	LOC101492584	Ca	GO:0005643; GO:0015031; GO:0016020; GO:0017056; GO:0051028	membrane [GO:0016020]; nuclear pore [GO:0005643]; structural constituent of nuclear pore [GO:0017056]; mRNA transport [GO:0051028]; protein transport [GO:0015031]
	chr5LG3_572899434	Psat5g303880	Putative sterile alpha motif/pointed domain-containing protein (SAM domain protein)	11433470 MTR_2g102140 MtrunA17_Chr2g0331871	Mt (Mtr)	GO:0045892	negative regulation of transcription, DNA-templated [GO:0045892]
		Psat5g303840	putative gamma-glutamylcyclotransferase At3g02910	LOC101506022	Ca	GO:0016740; GO:0061929	gamma-glutamylaminecyclotransferase activity [GO:0061929]; transferase activity [GO:0016740]
		Psat5g303800	protein NUCLEAR FUSION DEFECTIVE 4	LOC101504533	Ca	GO:0016021	integral component of membrane [GO:0016021]
		Psat5g303760	Uncharacterized protein	L195_g009520	Tp		

Table 5. *Cont.*

Trait[a]	SNP Marker	Gene_ID	Protein Names	Gene Names	Organism[b]	Gene Ontology IDs	Gene Ontology (GO)
PRI	chr6LG2_469101917	*Psat6g234040*	Putative GTP 3′,8-cyclase (EC 4.1.99.22)	MtrunA17_Chr1g0212051 Mt (Mtr)		GO:0006777	Mo-molybdopterin cofactor biosynthetic process [GO:0006777]
		Psat6g234000	Riboflavin biosynthesis protein ribF	L195_g000443	Tp	GO:0003919; GO:0009231	FMN adenylyltransferase activity [GO:0003919]; riboflavin biosynthetic process [GO:0009231]
	chr7LG7_263964018	*Psat7g148080*	TATA-binding-like protein	L195_g000140	Tp	GO:0005524	ATP binding [GO:0005524]
CT	chr4LG4_415994869	*Psat4g203800*	ethylene-responsive transcription factor-like protein At4g13040	LOC105851094	Ca	GO:0003677; GO:0003700; GO:0005634	nucleus [GO:0005634]; DNA binding [GO:0003677]; DNA-binding transcription factor activity [GO:0003700]
		Psat4g203760	NA	NA	NA	NA	NA
	chr5LG3_309595819	*Psat5g169800*	ABC transporter C family member 3-like isoform X1	LOC101491790	Ca	GO:0005524; GO:0016021; GO:0042626	integral component of membrane [GO:0016021]; ATP binding [GO:0005524]; ATPase activity, coupled to transmembrane movement of substances [GO:0042626]
		Psat5g169760	Retrovirus-related Pol polyprotein from transposon TNT 1-94	KK1_037587	Cc (Ci)	GO:0000943; GO:0003676; GO:0015074	retrotransposon nucleocapsid [GO:0000943]; nucleic acid binding [GO:0003676]; DNA integration [GO:0015074]
RSL	chr3LG5_18678117	*Psat3g006600*	uncharacterized protein LOC101515092	LOC101515092	Ca	GO:0016021	integral component of membrane [GO:0016021]
		Psat3g006560	L-allo-threonine aldolase-like protein (Putative aldehyde-lyase) (EC 4.1.2.-)	25499717 MTR_7g115690 MtrunA17_Chr7g0274621	Mt (Mtr)	GO:0006520; GO:0016829	lyase activity [GO:0016829]; cellular amino acid metabolic process [GO:0006520]
	chr4LG4_29062302	*Psat4g020520*	Alkaline-phosphatase-like protein (Putative Type I phosphodiesterase/nucleotide pyrophosphatase/phosphate transferase)	25494146 MTR_4g123557 MtrunA17_Chr4g0069621	Mt (Mtr)	GO:0006506; GO:0016021; GO:0051377	integral component of membrane [GO:0016021]; mannose-ethanolamine phosphotransferase activity [GO:0051377]; GPI anchor biosynthetic process [GO:0006506]
	chr5LG3_566189271	*Psat5g299080*	Kinesin-related protein 4-like	L195_g011972	Tp		

Table 5. *Cont.*

Trait[a]	SNP Marker	Gene_ID	Protein Names	Gene Names	Organism[b]	Gene Ontology IDs	Gene Ontology (GO)
		Psat5g299040	PPR containing plant-like protein (Putative tetratricopeptide-like helical domain-containing protein)	11431556 MTR_2g102210 MtrunA17_Chr2g0331911	Mt (Mtr)		
	chr5LG3_572669963	*Psat5g303680*	Putative sterile alpha motif/pointed domain-containing protein (SAM domain protein)	11430703 MTR_2g104230 MtrunA17_Chr2g0333351	Mt (Mtr)		
	chr7LG7_20086906	*Psat7g013080*	aldehyde dehydrogenase family 2 member C4-like	LOC101493969	Ca	GO:0016620	oxidoreductase activity, acting on the aldehyde or oxo group of donors, NAD or NADP as acceptor [GO:0016620]
		Psat7g013040	Cst complex subunit ctc1-like protein	L195_g004297	Tp	GO:0000723	telomere maintenance [GO:0000723]
	chr7LG7_23295474	*Psat7g015240*	Ribosomal L7Ae/L30e/S12e/Gadd45 family protein	L195_g030323	Tp		
		Psat7g015200	Tesmin/TSO1-like CXC domain protein	11408106 MTR_8g103320	Mt (Mtr)		
		Psat7g015160	NA	NA	NA	NA	NA
	chr7LG7_96157380	*Psat7g057080*	tRNA (Cytosine(34)-C(5))-methyltransferase-like protein	25501876 MTR_8g089980	Mt (Mtr)	GO:0003723; GO:0016428	RNA binding [GO:0003723]; tRNA (cytosine-5-)-methyltransferase activity [GO:0016428]
		Psat7g057040	tRNA (Cytosine(34)-C(5))-methyltransferase-like protein	25501876 MTR_8g089980	Mt (Mtr)	GO:0003723; GO:0016428	RNA binding [GO:0003723]; tRNA (cytosine-5-)-methyltransferase activity [GO:0016428]

Table 5. *Cont.*

Trait[a]	SNP Marker	Gene_ID	Protein Names	Gene Names	Organism[b]	Gene Ontology IDs	Gene Ontology (GO)
IL	chr4LG4_63111072	*Psat4g039600*	Eukaryotic translation initiation factor 3 subunit C (eIF3c) (Eukaryotic translation initiation factor 3 subunit 8) (eIF3 p110)	LOC101499912	Ca	GO:0001732; GO:0003743; GO:0005852; GO:0016282; GO:0031369; GO:0033290	eukaryotic 43S preinitiation complex [GO:0016282]; eukaryotic 48S preinitiation complex [GO:0033290]; eukaryotic translation initiation factor 3 complex [GO:0005852]; translation initiation factor activity [GO:0003743]; translation initiation factor binding [GO:0031369]; formation of cytoplasmic translation initiation complex [GO:0001732]
	chr4LG4_80759704	*Psat4g047680*	NA	NA	NA	NA	NA
		Psat4g047640	Ras GTPase-activating protein-binding protein 1-like	L195_g006539	Tp	GO:0003723	RNA binding [GO:0003723]
		Psat4g047600	Uncharacterized protein	L195_g056003	Tp	GO:0005739	mitochondrion [GO:0005739]
	chr5LG3_566189271	*Psat5g299080*	Kinesin-related protein 4-like	L195_g011972	Tp		
		Psat5g299040	PPR containing plant-like protein (Putative tetratricopeptide-like helical domain-containing protein)	11431556 MTR_2g102210 MtrunA17_Chr2g0331911	Mt (Mtr)		
	chr6LG2_420562729	*Psat6g211160*	Transmembrane amino acid transporter family protein	25485307 MTR_1g105980	Mt (Mtr)	GO:0016021	integral component of membrane [GO:0016021]
	chr7LG7_197862543	*Psat7g120120*	Uncharacterized protein	11443456 MTR_4g087360 MtrunA17_Chr4g0045601	Mt (Mtr)		
PN	chr2LG1_105547608	*Psat2g060680*	Uncharacterized protein	L195_g033306	Tp	GO:0003676; GO:0008270	nucleic acid binding [GO:0003676]; zinc ion binding [GO:0008270]
	chr2LG1_370541780	*Psat2g144160*	Pectin acetylesterase (EC 3.1.1.-)	LOC101497691	Ca	GO:0005576; GO:0005618; GO:0016021; GO:0016787; GO:0071555	cell wall [GO:0005618]; extracellular region [GO:0005576]; integral component of membrane [GO:0016021]; hydrolase activity [GO:0016787]; cell wall organization [GO:0071555]

Table 5. *Cont.*

Trait[a]	SNP Marker	Gene_ID	Protein Names	Gene Names	Organism[b]	Gene Ontology IDs	Gene Ontology (GO)
	chr2LG1_385949935	*Psat2g155280*	60S ribosomal protein l8-like	L195_g013966	Tp	GO:0003735; GO:0005840; GO:0006412	ribosome [GO:0005840]; structural constituent of ribosome [GO:0003735]; translation [GO:0006412]
	chr2LG1_389336188	*Psat2g157440*	Putative ATPase, AAA-type, core, AAA-type ATPase domain-containing protein (p-loop nucleoside triphosphate hydrolase superfamily protein)	11412855 MTR_5g020990 MtrunA17_Chr5g0404661	Mt (Mtr)	GO:0005524; GO:0016787	ATP binding [GO:0005524]; hydrolase activity [GO:0016787]
	chr2LG1_402022079	*Psat2g166600*	probable serine/threonine-protein kinase At1g01540 isoform X1	LOC101489894	Ca	GO:0004672; GO:0005524; GO:0016021	integral component of membrane [GO:0016021]; ATP binding [GO:0005524]; protein kinase activity [GO:0004672]
		Psat2g166560	PI-PLC X domain-containing protein At5g67130	LOC101489369	Ca	GO:0006629; GO:0008081	phosphoric diester hydrolase activity [GO:0008081]; lipid metabolic process [GO:0006629]
		Psat2g166520	Putative rapid ALkalinization Factor (RALF)	11409897 MTR_5g017160 MtrunA17_Chr5g0402121	Mt (Mtr)		
	chr2LG1_4359822	*Psat2g005000*	Nup133/Nup155-like nucleoporin	11434873 MTR_5g097260	Mt (Mtr)	GO:0005623; GO:0017056	cell [GO:0005623]; structural constituent of nuclear pore [GO:0017056]
		Psat2g004960	Cation-transporting ATPase plant (Putative calcium-transporting ATPase) (EC 3.6.3.8)	11434874 MTR_5g097270 MtrunA17_Chr5g0447521	Mt (Mtr)	GO:0000166; GO:0016021	integral component of membrane [GO:0016021]; nucleotide binding [GO:0000166]
	chr3LG5_216337201	*Psat3g111000*	Phosphomannomutase (EC 5.4.2.8)	11436930 MTR_7g076670	Mt (Mtr)	GO:0004615; GO:0005737; GO:0009298	cytoplasm [GO:0005737]; phosphomannomutase activity [GO:0004615]; GDP-mannose biosynthetic process [GO:0009298]
		Psat3g110960	bifunctional protein FolD 4, chloroplastic	LOC101496397	Ca	GO:0004488	methylenetetrahydrofolate dehydrogenase (NADP+) activity [GO:0004488]
	chr5LG3_530537682	*Psat5g270480*	Heat shock protein 70 (HSP70)-interacting protein, putative	25487616 MTR_2g090135	Mt (Mtr)		

Note: The pea genome sequence reported by Kreplak et al. [40] was used for identification of candidate genes. The reported gene annotation and nomenclature was followed. [a] SPAD, soil plant analysis development; PRI, photochemical reflectance index; CT, canopy temperature; RSL, reproductive stem length; IL, internode length; PN, pod number; [b] Tp, *Trifolium pratense* (Red clover); Ca, *Cicer arietinum* (Chickpea) (Garbanzo); Mt, *Medicago truncatula* (Barrel medic); Mtr, *Medicago tribuloides*; Cc, *Cajanus cajan* (Pigeon pea); Ci, *Cajanus indicus*.

2.3. Overall Association of Phenotypic Traits

Principal component analysis (PCA) based on the correlation of traits revealed the overall traits association and the genotype response across the five environments (Figure 4A,B). The first two PCs explained 61.9% of the total variability in the data. The loading plot illustrated traits association and how much each trait contributed to the PCs. The first PC was influenced mainly by SPAD, reproductive stem and internode lengths, whereas the second PC was influenced mainly by CT and pod number. SPAD positioned in an opposing direction (obtuse angle to straight line) to reproductive stem and internode lengths indicating a significant negative correlation between SPAD and the length measurements. Likewise, CT positioned in the opposite direction of pod number indicating their significant negative correlation. The hotter the canopy, the lower the pod number and thus seed yield (Figure 4A). Score plots illustrated genotype placement (response) across the environments (Figure 4B). The heat and or drought stressed environments (2015 Saskatoon and 2017 Saskatoon) positioned to the negative direction PC2 associating with high CT, whereas the control environments were associated greater pod number and SPAD value.

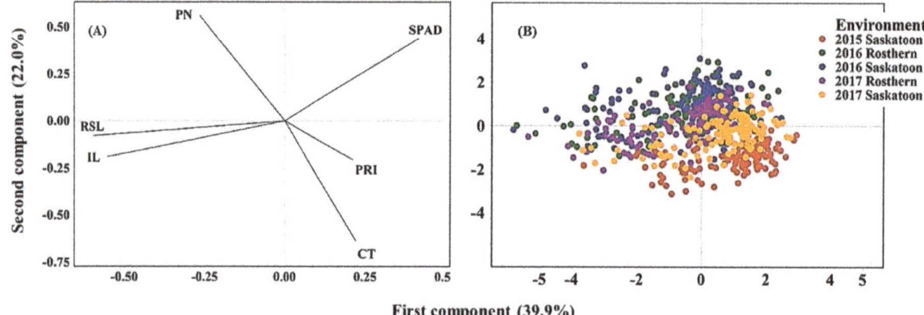

Figure 4. Loading (**A**) and Score (**B**) plots of principal component analysis illustrating the overall association of traits and genotype performance across environments. Note: PN, pod number; RSL, reproductive stem length; IL, internode length, CT, canopy temperature; PRI, photochemical reflectance index.

3. Discussion

As a cool season crop, pea is sensitive to heat stress which causes a significant yield loss. However, there exists substantial genetic variation among pea genotypes for heat tolerance [10,12,24,28]. A strategic assessment and use of available variation is essential for crop improvement through using allelic variation. With the availability of cost-effective, high-throughput SNP genotyping methods and genomic resources, GWAS has been an effective method for identifying genetic loci associated with traits of many crop species including legumes [29,30,36].

The present GWAS was undertaken to identify SNP markers associated with traits related with pea heat response using a panel of 135 genetically diverse pea accessions. The accessions were from breeding programs of major pea growing areas and, thus accounted genotypes with a wide range of heat sensitivity. Genotyping by sequencing (GBS) identified 16,877 good quality SNPs, of which 15,609 were distributed across seven chromosomes of pea and the remaining 1268 were non-chromosomal SNPs [30].

Linkage disequilibrium patterns of population structure and genetic relatedness information are important for association mapping to minimize the number of false positive associations [41], thus the LD of the 135 GWAS members was analyzed by chromosome, and the LD decay estimates of the 7 chromosomes ranged from 0.03 to 0.18 Mb [30]. Based on genetic relatedness the 16,877 SNPs in the GWAS panel were clustered into nine groups [30]. Similarly, Diapari et al. [37] clustered another 94 pea accessions into eight groups, and Siol et al. [42] grouped 917 *Pisum* accessions into 16 groups. The above groupings indicated the extent of genetic variability among pea accessions. The clustering did not necessarily correspond solely with the geographic origins of the individuals, but depended on additional factors of variability such as the objectives in different breeding programs [30].

In the present GWAS, we evaluated ten heat stress-responsive traits. The first six were: chlorophyll concentration by SPAD, PRI, CT, reproductive stem length, internode length, and pod number. The other four were: plant height, lodging, pod to node ratio, and water band index (WBI). From the latter four traits, five SNP markers on Chr 1 (LG6), Chr 2 (LG1), Chr 3 (LG5), Chr 5 (LG3) for lodging, and four SNP markers on Chr5 (LG3) for plant height were previously reported by Gali et al. [30], and no marker was detected to be significant in at least three of the five environments for pod to node ratio and WBI. As such, in the current paper we focused on the first six traits for phenotypic variation in the 135 pea accessions across five environments.

The five environments were grouped into ambient (three environments) and heat and or drought stress (two environments) conditions based on weather data and threshold temperature for heat stress in the field [5]. All traits had a wide range of phenotypic variation within each environment and stress level, which is essential for dissecting complex traits through association mapping. Overall, we identified 32 MTAs for six traits that have physiological and agronomic importance and are involved in pea heat response. A marker identified for a significant association with a given trait would be more reliable if the same marker is found in multiple environments [30]. Therefore, for the six traits we investigated, the SNP markers deemed significant were consistent in at least three environments, and these markers could potentially be used for marker-assisted selection of these traits in the effort of improving pea for heat tolerance.

In this study, the SPAD value was used to estimate chlorophyll concentration, a major component of chloroplasts, and can be used as a factor to determine crop adaptation to environmental stresses by retention of greenness [10,13,43]. Regression analysis on wheat reported that under heat stress, the SPAD value was associated with plasma and thylakoid membrane damage [44], which hinders light absorbing efficiency of photosystems (PSI and PSII), and hence reduced photosynthetic capacity ultimately leading to crop yield loss [11]. Understanding of the genetic bases that govern chlorophyll concentration may contribute to enhancing photosynthetic efficiency and thus minimize yield loss due to stressful environments.

We identified six MTAs that were related to SPAD value in repeated tests. All of the MTAs identified for SPAD were from Chr 5 (LG3). Bell et al. [45] reported that pea chlorophyll degradation under stress conditions is governed by the SGRL protein, a distinct class of the SGR gene which is induced by environmental stresses. The genomic location of SGRL was reported to be on LG3 which supports our result where all of the SPAD markers also reside on Chr 5 (LG3). The SGRL gene sequence (https://www.ncbi.nlm.nih.gov/nuccore/LN810021) location in the pea genome assembly spanned between the base pair positions Chr5LG3_151800929 and Chr5LG3_151804253, and is within close proximity (858 Kbp) of the SPAD associated marker ChrLG3_150942510. Using GWAS on soybean, Dhanapal et al. [29] identified 52 SNP markers associated with chlorophyll content.

Similarly, two loci were identified for association with PRI. One of these loci is on Chr 6 (LG2) and the second is on Chr 7 (LG7), and in both cases the markers were consistent in three environments. There are only a few reports that have used GWAS to identify markers associated with vegetation indices, namely, in soybean and wheat. In soybean, Herritt et al. [25] identified 31 SNPs linked with PRI, and on wheat, Gizaw et al. [34] reported the presence of markers associated with normalized chlorophyll-pigment ratio index (NCPI), and normalized difference vegetation index (NDVI). However, use of GWAS and vegetation indices has been lacking in cool season pulse crops. To the best of our knowledge, our report is the first to apply VIs in pea GWAS. The PRI is increasingly used as a predictor of crop photosynthetic efficiency which responds to environmental variables [19]. PRI is associated with photosynthetic protective mechanisms by dissipation of excess energy such as in the operation of xanthophyll cycle during stress. Violaxanthin de-epoxidase VDE is among the genes known to be involved in excess energy dissipation in the xanthophyll cycle [46].

Two MTAs, one each on Chr 5 (LG3) and Chr 4 (LG4), were detected for CT, a trait consistently used as an indicator of stress mainly of drought and heat stresses [12,14]. Generally, cooler canopy is associated with heat avoidance, and is an indicator of a healthy canopy with an optimal physiological state [12]. Again, to the best of our knowledge, no previous study exists on pea or other cool season legume crops that has reported genomic regions associated with CT. In a study using 24 pea cultivars across six environments, Tafesse [14] reported that leaf surface wax concentration is positively correlated with water band index, a proxy for leaf water retention, and contributes to a cooler canopy. WAX2 is among the genes controlling wax biosynthesis in Arabidopsis [47], and glossy13 is another gene with similar role reported in maize [48]. Lodging contributes to canopy heating in pea, and upright and semileafless cultivars with the *afila* gene have cooler CT [12,49]. Tar'an et al. [50] identified major loci associated with lodging resistance in pea on LG III, and one of the markers we identified for CT is also on LG III, suggesting genes controlling lodging also control CT.

We identified seven MTAs associated with reproductive stem length on chromosomes 3 (LG5), 4 (LG4), 5 (LG3), and 7 (LG7); and six MTAs associated with internode length on chromosomes 4 (LG4), 5 (LG3), 6 (LG2) and 7 (LG7). The markers associated with these two traits mostly were positioned on the same linkage groups, and a SNP marker Chr5LG3_566189271 was associated with both the traits. Using the current GWAS panel, Gali et al. [30] identified four MTAs associated with plant height that were on same linkage group as that of reproductive stem length and internode length. The SNP marker Chr5LG3_566189271 reported for plant height [34] was also associated with internode length in the current study. Both reproductive stem and internode lengths were significantly reduced by heat stress [12]. A cultivar's genetics affects internode length, and in pea the Le gene controls internode length [25], which directly affects reproductive stem length and plant height via its influence on gibberellic acid function on growth and determinacy/indeterminacy [25,51,52]. Using two pea recombinant inbred populations, Weeden [22] identified a major QTL on LG3 for a longer internode (Le), and a second QTL on LG4 for the recessive allele which caused plants to have shorter internodes.

We identified nine loci associated with pod number, of which six were on Chr 2 (LG1), one each on Chr 3 (LG5) and 5 (LG3), and one on a non-chromosomal scaffold. Plant pod number is the number of flower-bearing nodes multiplied by the average number of flowers per node. Previously, Jiang et al. [28] identified two unmapped QTLs for pod number using 92 diverse accessions. Also, Huang et al. [24] identified two QTLs for pod number based on a bi-parental mapping population on Chr 5 (LG3). The greater number of loci identified in this study was likely due to the use of a GWAS panel which represented a broad range of diversity in pod number ranging from 3 to 19 pods per plant, and where most of this variation is contributed by genetic factors. Benlloch [53] indicated that flower number per plant, which directly determines pod number, is controlled by two genes, Fn and Fna, and a single mutation of these genes increases flower number per plant. Pod number is a major yield component that has a strong correlation with seed yield, and is most affected by heat stress [12,23,24]. The reduction in pod number and yield was likely from heat stress-induced abortion of flower buds, flowers, and pods [4,23]. Pod set relies on pollen and stigma functioning optimally, both of which are very sensitive to heat stress [54].

In conclusion, in this GWAS we identified 32 MTAs and 48 candidate genes for traits associated with pea heat response. These results are expected to enhance the understanding of genetic loci controlling these traits. The identified candidate genes are involved in various biological functions and require further functional validation. The detected MTAs and candidate genes should be useful for marker-assisted selection for heat tolerant pea varieties.

4. Materials and Methods

4.1. Plant Materials

A panel of 135 diverse field pea accessions, as described by Gali et al. [30], were grown for two years (2016–2017) at two Rosthern (52°66′N, 106°33′W; Orthic Black Chernozem); and three years

(2015–2017) at Saskatoon (52°12'N, 106°63'W; Dark Brown Chernozem), Saskatchewan, in western Canada, for phenotypic evaluation. The combination of year-location forms five environments: 2015 Saskatoon; 2016 Rosthern; 2016 Saskatoon; 2017 Rosthern; and 2017 Saskatoon for phenotypic evaluation. Among the 135 accessions, 19 were from Australian pulse breeding programs, 77 were from eastern and western European countries, the Russian Federation and the UK, 15 were from the USA, 17 were from Canada (mostly from the Crop Development Centre, University of Saskatchewan), five were from Ethiopia, and two were from India. Thus, the accessions represented the major pea growing areas of the world. The accessions were commercial cultivars released over the past 50 years for local production, and were able to flower and mature under the five environments tested [30].

4.2. The Field Trials and Weather Conditions

The experimental design at each environment was a randomized complete block with two replications. Plot size was 1.37 m width × 3.66 m length, and the recommended seeding rate (100 seeds m^{-2}, targeting 80–85 plants m^{-2} on 0.25 m row spacing) was used. Weed control was achieved by management practices used in pea production in Saskatchewan as described by Tafesse et al. [12].

Weather data for 2015 Saskatoon starting from June 11 to the end of the growing season, 2016 Rosthern starting from June 21 to the end of the growing season, and 2016 Saskatoon starting from July 21 to the end of the season were collected from weather stations (Coastal Environmental Systems, Seattle, WA, USA) established at each site. Weather data of 2017 and the remaining 2015 and 2016 were obtained from Environment Canada database (https://climate.weather.gc.ca) recorded by the nearest stations to the trial sites. For Saskatoon, data from central Saskatoon station, and for Rosthern the mean of data from Saskatoon international airport and Prince Albert stations were used. The daily maximum air temperatures, amount of precipitation and number of days when air temperature exceeded 28 °C during the growing seasons were used to determine the degree of stress in each environment at different growth stages. The categorization of growth stages into vegetative (germination to end of vegetative growth) and reproductive (beginning of flowering to maturity) was conducted using the phenology data reported by Gali et al. [30]. Based on the weather data, 2015 and 2017 Saskatoon had heat and drought stress conditions and the remaining three environments were generally ambient and considered control environments (Table 1).

4.3. Phenotypic Measurements

Chlorophyll concentration was estimated non-destructively using a SPAD502Plus chlorophyll meter (Konica Minolta Sensing Americas Inc., USA). The SPAD value is a unitless index, calculated as the ratio of the intensity of light transmittance at red (650 nm) to infrared (940 nm) and gives a value that corresponds to the relative amount of chlorophyll present in the leaf. Hereafter, the chlorophyll concentration estimated by SPAD meter is referred to as 'SPAD'. The SPAD readings were taken four to six times each season, and for each measurement day the mean SPAD value was calculated by the instrument from three readings taken from three plants per plot on fully expanded stipules at the second or third node counting down from the apex of a main stem.

Similarly, spectral measurement was conducted repeatedly on leaf stipules using a portable spectroradiometer PSR-1100F (Spectral Evolution Inc, Lawrence, MA, USA). This device enabled hyperspectral readings with a range of 320-1,126 nm, and 1.6 nm sampling interval, and a total of 512 discrete narrow bands. PRI was calculated from the reflectance data according to Gamon et al. [19] as:

$$PRI = (R_{531} - R_{570})/(R_{531} + R_{570}) \tag{1}$$

where R is reflectance percentage and 531 and 570 are the wavelength bands in nm along the light spectrum. The PRI is used as a proxy for the xanthophyll cycle, a photosynthetic protective cycle that operates more during stress [19].

Canopy temperature (CT) was measured four to six times in each location in a season using a hand held infrared thermometer (Model 6110.4ZL, Everest Interscience Inc, Tucson, AZ, USA) as described by Tafesse et al. [12]. Measurements of SPAD, spectral reflectance, and CT were carried out repeatedly (four to six times in a season) during the reproductive stage, at solar noon on relatively hot days when air temperature is greater than 25 °C, and the mean value was used for analysis.

The other measurements taken at physiological maturity were: reproductive stem length (vine length from first flowering node to the tip of the main stem); internode length (determined as the ratio of reproductive stem length to reproductive node numbers); and pod number per plant (total pods counting all pods with at least one seed on the main stem). For these, each measured variable was the mean of three plants per plot sampled at random and lengths were measured in cm.

4.4. Phenotype Data Analysis

Before employing analysis of variance (ANOVA), homogeneity of variances and normality of residuals were tested using checked using Levene and Shapiro-Wilk tests, respectively [55,56]. Variance components of genotype, environment, the G × E interaction, block within environment, and the residual were analyzed using the generalized linear model (GLM) and by considering all factors as random effects. Broad sense heritability (H^2) was calculated as:

$$H^2 = \sigma g^2/(\sigma g^2 + \sigma ge^2/n + \sigma^2/nb) \tag{2}$$

where σg^2 is the genetic variance, and σge^2 is the variance of genotype and environment [57].

Over environments, combined ANOVA on SPAD, PRI, CT, reproductive stem length, internode length, and pod number was carried out using the Mixed procedure of SAS (Version 9.4, SAS Institute). Genotype, environment and G x E interaction were considered as fixed, and blocks as random factors. Principal component analysis (PCA) was performed with the multivariate function of Minitab (Version 19, Minitab LLC, USA) using means of traits to infer overall association among traits and genotype for the five environments. Based on significant eigenvalue (> 1), the first two principal components (PC) were selected for the minimum number of PCs to explain the greatest total variation in the data set.

4.5. Association Mapping

Genotyping of the 135 GWAS panel was performed by genotyping-by-sequencing (GBS, [58]), and 16,877 SNPs were reported based on a minimum read depth of five and minimum allele frequency of 0.05 [30]. The reported SNPs were used for association analysis using GAPIT (Genome Association and Prediction Integrated Tool—R package [30]) software. Association analysis for each trait was conducted using the mixed linear model (MLM). For MLM analysis, Q values were generated from structure analysis [59] and K (kinship coefficient matrix) values calculated by GAPIT and identity-by-state (IBS) methods were used. Principal co-ordinate values were used as co-variates. Although the result is not presented here, the model output of MLM was compared with the Super MLM model and the markers identified in both methods were mostly similar. The quantile-quantile (Q-Q) plots of each trait were drawn using the observed and expected $\log_{10}P$ values. Marker–trait associations were selected based on P value ($P \leq 0.001$) and repeated occurrence of the association in at least three of the five trials. The genes within 15 kb of the identified markers are reported as the candidate genes. The pea genome sequence reported by Kreplak et al. [40] was used for identification of candidate genes.

Author Contributions: Conceptualization, R.B. and T.D.W.; Data curation, E.G.T., K.K.G. and V.B.R.L.; Formal analysis, E.G.T.; Funding acquisition, R.B. and T.D.W.; Investigation, E.G.T. and K.K.G.; Methodology, E.G.T., K.K.G. and R.B.; Resources, R.B. and T.D.W.; Validation, K.K.G. and V.B.R.L.; Writing—original draft, E.G.T.; Writing—review & editing, K.K.G., V.B.R.L., R.B. and T.D.W. All authors have read and agreed to the published version of the manuscript

Funding: This research was funded by the Saskatchewan Agriculture Development Fund, Saskatchewan Pulse Growers, and the Western Grains Research Foundation.

Acknowledgments: The authors thank B. Louie and J. Denis for their technical assistance during the field measurements, and the Crop Development Centre, University of Saskatchewan Pulse Crop Breeding staff for plot maintenance.

Conflicts of Interest: The authors declare no conflict of interest. The funders had no role in the design of the study; in the collection, analyses, or interpretation of data; in the writing of the manuscript, or in the decision to publish the results.

Abbreviations

Chr	Chromosome
CT	Canopy temperature
GAPIT	Genome Association and Prediction Integrated Tool
GBS	Genotyping by sequencing
GLM	General linear model
GWAS	Genome wide association study
LD	Linkage disequilibrium
LG	Linkage group
MLM	Mixed linear model
MTA	Marker-trait association
PRI	Photochemical reflectance index
QTL	Quantitative trait loci
SNP	Single nucleotide polymorphism
SPAD	Soil plant analysis development

References

1. Cousin, R. Peas (*Pisum sativum* L.). *Field Crop. Res.* **1997**, *53*, 111–130. [CrossRef]
2. Dahl, W.J.; Foster, L.M.; Tyler, R.T. Review of the health benefits of peas (*Pisum sativum* L.). *Br. J. Nutr.* **2012**, *108*, 3–10. [CrossRef] [PubMed]
3. Smýkal, P.; Aubert, G.; Burstin, J.; Coyne, C.J.; Ellis, N.T.H.; Flavell, A.J.; Ford, R.; Hýbl, M.; Macas, J.; Neumann, P.; et al. Pea (*Pisum sativum* L.) in the Genomic Era. *Agronomy* **2012**, *2*, 74–115. [CrossRef]
4. Guilioni, L.; Wery, J.; Tardieu, F. Heat stress-induced abortion of buds and flowers in pea: Is sensitivity linked to organ age or to relations between reproductive organs? *Ann. Bot.* **1997**, *80*, 159–168. [CrossRef]
5. Bueckert, R.A.; Wagenhoffer, S.; Hnatowich, G.; Warkentin, T.D. Effect of heat and precipitation on pea yield and reproductive performance in the field. *Can. J. Plant. Sci.* **2015**, *95*, 629–639. [CrossRef]
6. Guilioni, L.; Wéry, J.; Lecoeur, J. High temperature and water deficit may reduce seed number in field pea purely by decreasinf plant growth rate. *Funct. Plant. Biol.* **2003**, *30*, 1151–1164. [CrossRef]
7. Core Writing Team; Pachauri, R.K.; Meyer, L.A. (Eds.) *IPCC, Climate change, contribution of working groups i, ii and iii to the fifth assessment report of the intergovernmental panel on climate change*; IPCC: Geneva, Switzerland, 2014; p. 151.
8. Leila, O.; Farida, A.; Hafida, R.B.; Aissa, A. Agro-morphological diversity within field pea (*Pisum sativum* L.) genotypes. *African J. Agric. Res.* **2016**, *11*, 4039–4047. [CrossRef]
9. Warkentin, T.D.; Smykal, P.; Coyne, C.J.; Weeden, N.; Domoney, C.; Bing, D.; Domoney, C.; Bing, D.; Leonforte, A.; Xuxiao, Z.; et al. Pea (Pisum. sativum L.). In *Grain legumes*; De Ron, A.M., Ed.; Springer: New York, NY, USA, 2015; pp. 37–83.
10. Sánchez, F.J.; Manzanares, M.; De Andrés, E.F.; Tenorio, J.L.; Ayerbe, L. Residual transpiration rate, epicuticular wax load and leaf colour of pea plants in drought conditions. influence on harvest index and canopy temperature. *Eur. J. Agron.* **2001**, *15*, 57–70. [CrossRef]
11. Wahid, A.; Gelani, S.; Ashraf, M.; Foolad, M.R. Heat tolerance in plants: An overview. *Environ. Exp. Bot.* **2007**, *61*, 199–223. [CrossRef]
12. Tafesse, E.G.; Warkentin, T.D.; Bueckert, R.A. Canopy architecture and leaf type as traits of heat resistance in pea. *Field Crop. Res.* **2019**, *241*, 107561. [CrossRef]
13. Havaux, M. Increased thermal deactivation of excited pigments in pea Leaves subjected to photoinhibitory treatments. *Plant. Physiol.* **1989**, *89*, 286–292. [CrossRef] [PubMed]

14. Tafesse, E.G. Heat stress resistance in pea (Pisum sativum L.) based on canopy and leaf traits. Doctoral dissertation, University of Saskatchewan, Saskatoon, SK, Canada, 2018.

15. Cerrudo, D.; Pérez, L.G.; Lugo, J.A.M.; Trachsel, S. Stay-green and associated vegetative indices to breed maize adapted to heat and combined heat-drought stresses. *Remote Sens.* **2017**, *9*, 235. [CrossRef]

16. Lichtenthaler, H.K. Chlorophylls and Carotenoids: Pigments of Photosynthetic Biomembranes. *Methods Enzymol.* **1987**, *148*, 350–382.

17. Wood, C.W.; Reeves, D.W.; Himelrick, D.G. Relationships between chlorophyll meter readings and leaf chlorophyll concentration, N status, and crop yield: A review. *Proc. Agron. Soc. New Zeal.* **1993**, *23*, 1–9.

18. Hatfield, J.L.; Gitelson, A.A.; Schepers, J.S.; Walthall, C.L. Application of spectral remote sensing for agronomic decisions. *Agron. J.* **2008**, *100*, 117. [CrossRef]

19. Gamon, J.A.; Serrano, L.; Surfus, J.S. The photochemical reflectance index: An optical indicator of photosynthetic radiation use efficiency across species, functional types, and nutrient levels. *Oecologia* **1997**, *112*, 492–501. [CrossRef]

20. Acreche, M.M.; Slafer, G.A. Lodging yield penalties as affected by breeding in Mediterranean wheats. *Field Crop. Res.* **2011**, *122*, 40–48. [CrossRef]

21. Lester, D.R.; Ross, J.J.; Davies, P.J.; Reid, J. 6 Mendel's Stem Length Gene (Le) Encodes a Gibberellin 3P-Hydroxylase; American Society of Plant Physiologists. *Plant. Cell.* **1997**, *9*, 1435–1443.

22. Weeden, N.F. Genetic changes accompanying the domestication of *Pisum sativum* L.: Is there a common genetic basis to the "domestication syndrome" for legumes? *Ann. Bot.* **2007**, *100*, 1017–1025. [CrossRef]

23. French, R.J. The contribution of pod numbers to field pea (*Pisum sativum* L.) yields in a short growing-season environment. *Aust. J. Agric. Res.* **1990**, *41*, 853–862. [CrossRef]

24. Huang, S.; Gali, K.K.; Tar'An, B.; Warkentin, T.D.; Bueckert, R.A. Pea phenology: Crop potential in a warming environment. *Crop. Sci.* **2017**, *57*, 1540–1551. [CrossRef]

25. Herritt, M.; Dhanapal, A.P.; Fritschi, F.B. Identification of genomic loci associated with the photochemical reflectance index by genome-wide association study in soybean. *Plant Genome* **2016**, *9*, 1–12. [CrossRef] [PubMed]

26. Korte, A.; Farlow, A. The advantages and limitations of trait analysis with GWAS: A review. *Plant. Methods* **2013**, *9*, 29. [CrossRef] [PubMed]

27. Brachi, B.; Morris, G.P.; Borevitz, J.O. Genome-wide association studies in plants: The missing heritability is in the field. *Genome Biol.* **2011**, *12*, 232. [CrossRef]

28. Jiang, Y.; Diapari, M.; Bueckert, R.A.; Tar'an, B.; Warkentin, T.D. Population structure and association mapping of traits related to reproductive development in field pea. *Euphytica* **2017**, *213*, 215. [CrossRef]

29. Dhanapal, A.P.; Ray, J.D.; Singh, S.K.; Hoyos-Villegas, V.; Smith, J.R.; Purcell, L.C.; Fritschi, F.B. Genome-wide association mapping of soybean chlorophyll traits based on canopy spectral reflectance and leaf extracts. *BMC Plant Biol.* **2016**, *16*, 174. [CrossRef]

30. Gali, K.K.; Sackville, A.; Tafesse, E.G.; Lachagari, V.B.R.; McPhee, K.; Hybl, M.; Mikić, A.; Smýkal, P.; McGee, R.; Burstin, J.; et al. Genome-Wide Association Mapping for Agronomic and Seed Quality Traits of Field Pea (*Pisum sativum* L.). *Front. Plant. Sci.* **2019**, *10*, 1538. [CrossRef]

31. Leonforte, A.; Sudheesh, S.; Cogan, N.O.; Salisbury, P.A.; Nicolas, M.E.; Materne, M.; Forster, J.W.; Kaur, S. SNP marker discovery, linkage map construction and identification of QTLs for enhanced salinity tolerance in field pea (*Pisum sativum* L.). *BMC Plant. Biol.* **2013**, *13*, 161. [CrossRef]

32. Sudheesh, S.; Lombardi, M.; Leonforte, A.; Cogan, N.O.; Materne, M.; Forster, J.W.; Kaur, S. Consensus genetic map construction for field pea (*Pisum sativum* L.), trait dissection of biotic and abiotic stress tolerance and development of a diagnostic marker for the er1 powdery mildew resistance gene. *PMBR.* **2015**, *33*, 1391–1403. [CrossRef]

33. Klein, A.; Houtin, H.; Rond, C.; Marget, P.; Jacquin, F.; Boucherot, K.; Huart, M.; Rivière, N.; Boutet, G.; Lejeune-Hénaut, I.; et al. QTL analysis of frost damage in pea suggests different mechanisms involved in frost tolerance. *Theor. Appl. Genet.* **2014**, *127*, 1319–1330. [CrossRef]

34. Gizaw, S.A.; Godoy, J.G.V.; Garland-Campbell, K.; Carter, A.H. Using spectral reflectance indices as proxy phenotypes for genome-wide association studies of yield and yield stability in pacific northwest winter wheat. *Crop. Sci.* **2018**, *58*, 1232–1241. [CrossRef]

35. Cheng, P.; Holdsworth, W.; Ma, Y.; Coyne, C.J.; Mazourek, M.; Grusak, M.A.; Fuchs, S.; McGee, R.J. Association mapping of agronomic and quality traits in USDA pea single-plant collection. *Mol. Breed.* **2015**, *35*, 75. [CrossRef]

36. Desgroux, A.; L'Anthoëne, V.; Roux-Duparque, M.; Rivière, J.P.; Aubert, G.; Tayeh, N.; Moussart, A.; Mangin, P.; Vetel, P.; Piriou, C.; et al. Genome-wide association mapping of partial resistance to *Aphanomyces euteiches* in pea. *BMC Genomics* **2016**, *17*, 124. [CrossRef] [PubMed]

37. Diapari, M.; Sindhu, A.; Warkentin, T.D.; Bett, K.; Tar'an, B. Population structure and marker-trait association studies of iron, zinc and selenium concentrations in seed of field pea (*Pisum sativum* L.). *Mol. Breed.* **2015**, *35*, 30. [CrossRef]

38. Ahmad, S.; Kaur, S.; Lamb-Palmer, N.D.; Lefsrud, M.; Singh, J. Genetic diversity and population structure of *Pisum sativum* accessions for marker-trait association of lipid content. *Crop. J.* **2015**, *3*, 238–245. [CrossRef]

39. Sita, K.; Sehgal, A.; HanumanthaRao, B.; Nair, R.M.; Prasad, P.V.; Kumar, S.; Gaur, P.M.; Farroq, M.; Siddique, K.H.M.; Varshney, R.K.; et al. Food Legumes and Rising Temperatures: Effects, adaptive functional mechanisms Specific to reproductive growth stage and strategies to improve heat tolerance. *Front. Plant. Sci.* **2017**, *8*, 1658. [CrossRef]

40. Kreplak, J.; Madoui, M.-A.; Capal, P.; Novak, P.; Labadie, K.; Aubert, G.; Bayer, P.E.; Gali, K.K.; Syme, R.A.; Main, D.; et al. A reference genome for pea provides insight into legume genome evolution. *Nat. Genetics* **2019**, *51*, 1411–1422. [CrossRef]

41. Flint-Garcia, S.A.; Thornsberry, J.M.; Buckler, E.S. Structure of Linkage Disequilibrium in Plants. *Annu. Rev. Plant. Biol.* **2003**, *54*, 357–374. [CrossRef]

42. Siol, M.; Jacquin, F.; Chabert-Martinello, M.; Smýkal, P.; Le Paslier, M.C.; Aubert, G.; Burstin, J. Patterns of genetic structure and linkage disequilibrium in a large collection of pea germplasm. *G3 Genes, Genomes, Genet.* **2017**, *7*, 2461–2471. [CrossRef]

43. Hasanuzzaman, M.; Nahar, K.; Alam, M.M.; Roychowdhury, R.; Fujita, M. Physiological, biochemical, and molecular mechanisms of heat stress tolerance in plants. *Int. J. Mol. Sci.* **2013**, *14*, 9643–9684. [CrossRef]

44. Talukder, S.K.; Babar, M.A.; Vijayalakshmi, K.; Poland, J.; Prasad, P.V.V.; Bowden, R.; Fritz, A. Mapping QTL for the traits associated with heat tolerance in wheat (*Triticum aestivum* L.). *BMC Genet.* **2014**, *15*, 1–13. [CrossRef] [PubMed]

45. Bell, A.; Moreau, C.; Chinoy, C.; Spanner, R.; Dalmais, M.; Le Signor, C.; Bendahmane, A.; Klenell, M.; Domoney, C. SGRL can regulate chlorophyll metabolism and contributes to normal plant growth and development in *Pisum sativum* L. *Plant. Mol. Biol.* **2015**, *89*, 539–558. [CrossRef] [PubMed]

46. Havaux, M.; Bonfils, J.P.; Lütz, C.; Niyogi, K.K. Photodamage of the photosynthetic apparatus and its dependence on the leaf developmental stage in the npq1 Arabidopsis mutant deficient in the xanthophyll cycle enzyme violaxanthin de-epoxidase. *Plant. Physiol.* **2000**, *124*, 273–284. [CrossRef] [PubMed]

47. Chen, X.B.; Goodwin, S.M.; Boroff, V.L.; Liu, X.L.; Jenks, M.A. Cloning and characterization of the WAX2 gene of Arabidopsis involved in cuticle membrane and WAX production. *Plant. Cell.* **2003**, *5*, 1170–1185. [CrossRef]

48. Li, L.; Li, D.L.; Liu, S.Z.; Ma, X.L.; Dietrich, C.R.; Hu, H.C.; Zhang, G.S.; Liu, Z.Y.; Zheng, J.; Wang, G.Y.; et al. The Maize glossy13 Gene, Cloned via BSR-Seq and Seq-Walking Encodes a Putative ABC Transporter Required for the Normal Accumulation of Epicuticular Waxes. *PLoS ONE* **2013**, *8*, e82333. [CrossRef]

49. Goldenberg, J.B. "*afila*" a new mutation in pea (*Pisum sativum* L.). *Biol. Genet.* **1965**, *1*, 27–31.

50. Tar'an, B.; Warkentin, T.; Somers, D.J.; Miranda, D.; Vandenberg, A.; Blade, S.; Woods, S.; Bing, D.; Xue, A.; DeKoeyer, D.; et al. Quantitative trait loci for lodging resistance, plant height and partial resistance to mycosphaerella blight in field pea (*Pisum sativum* L.). *Theor. Appl. Genet.* **2003**, *107*, 1482–1491. [CrossRef]

51. Reinecke, D.M.; Wickramarathna, A.D.; Ozga, J.A.; Kurepin, L.V.; Jin, A.L.; Good, A.G.; Pharis, R.P. Gibberellin 3-oxidase gene expression patterns influence gibberellin biosynthesis, growth, and development in pea. *Plant. Physiol.* **2013**, *163*, 929–945. [CrossRef]

52. Ingram, T.J.; Reid, J.B.; Murfet, I.C.; Gaskin, P.; Willis, C.L.; MacMillan, J. Internode length in *Pisum*. *Planta* **1984**, *160*, 455–463. [CrossRef]

53. Benlloch, R.; Berbel, A.; Ali, L.; Gohari, G.; Millán, T.; Madueño, F. Genetic control of inflorescence architecture in legumes. *Front. Plant. Sci.* **2015**, *6*, 1–14. [CrossRef]

54. Jiang, Y.; Lahlali, R.; Karunakaran, C.; Kumar, S.; Davis, A.R.; Bueckert, R.A. Seed set, pollen morphology and pollen surface composition response to heat stress in field pea. *Plant. Cell. Environ.* **2015**, *38*, 2387–2397. [CrossRef] [PubMed]
55. Levene, H. Robust tests for equality of variances. In *Contributions to probability and statistics*; Olkin, I., Ed.; Stanford Univ. Press: Palo Alto, CA, USA, 1960; pp. 278–292.
56. Shapiro, S.S.; Wilk, M.B. An analysis of variance test for normality (complete samples). *Biometrika* **1965**, *52*, 591–611. [CrossRef]
57. Wang, N.; Chen, B.; Xu, K.; Gao, G.; Li, F.; Qiao, J.; Yan, G.; Li, J.; Li, H.; Wu, X. Association mapping of flowering time QTLs and insight into their contributions to rapeseed growth habits. *Front. Plant. Sci.* **2016**, *7*, 1–11. [CrossRef] [PubMed]
58. Elshire, R.J.; Glaubitz, J.C.; Sun, Q.; Poland, J.A.; Kawamoto, K.; Buckler, E.S.; Mitchell, S.E. A robust, simple genotyping-by-sequencing (GBS) approach for high diversity species. *PLoS ONE* **2011**, *6*, 1–10. [CrossRef]
59. Lipka, A.E.; Tian, F.; Wang, Q.; Peiffer, J.; Li, M.; Bradbury, P.J.; Gore, M.A.; Buckler, E.S.; Zhang, Z. GAPIT: Genome association and prediction integrated tool. *Bioinformatics* **2012**, *28*, 2397–2399. [CrossRef]

International Journal of
Molecular Sciences

Article

Functional Dissection of the Chickpea (*Cicer arietinum L.*) Stay-Green Phenotype Associated with Molecular Variation at an Ortholog of Mendel's I Gene for Cotyledon Color: Implications for Crop Production and Carotenoid Biofortification

Kaliamoorthy Sivasakthi [1,†], Edward Marques [2,†], Ng'andwe Kalungwana [3], Noelia Carrasquilla-Garcia [4], Peter L. Chang [4], Emily M. Bergmann [4], Erika Bueno [2], Matilde Cordeiro [4], Syed Gul A.S. Sani [4], Sripada M. Udupa [5], Irshad A. Rather [6], Reyazul Rouf Mir [6], Vincent Vadez [1], George J. Vandemark [7], Pooran M. Gaur [1], Douglas R. Cook [4], Christine Boesch [3], Eric J.B. von Wettberg [2], Jana Kholova [1,*] and R. Varma Penmetsa [8,*]

[1] International Crops Research Institute for the Semi-Arid Tropics (ICRISAT), Patancheru 502 324, India; sakthibiotechbdu@gmail.com (K.S.); V.Vadez@cgiar.org (V.V.); P.Gaur@cgiar.org (P.M.G.)

[2] Department of Plant and Soil Science, University of Vermont, and Gund Institute for the Environment, Burlington, VT 05405, USA; Edward.Marques@uvm.edu (E.M.); Erika.Bueno@uvm.edu (E.B.); Eric.Bishop-Von-Wettberg@uvm.edu (E.J.B.v.W.)

[3] School of Food Science and Nutrition, University of Leeds, Leeds, LS2 9JT, UK; fsnak@leeds.ac.uk (N.K.); C.Bosch@leeds.ac.uk (C.B.)

[4] Department of Plant Pathology, University of California, Davis, CA 95616, USA; noecarras@ucdavis.edu (N.C.-G.); peterc@usc.edu (P.L.C.); embergmann@ucdavis.edu (E.M.B.); matilde.cordeiro@gmail.com (M.C.); gasani@ucdavis.edu (S.G.A.S.S.); drcook@ucdavis.edu (D.R.C.)

[5] International Center for Agricultural Research in the Dry Areas (ICARDA), P.O.Box 6299, Rue Hafiane Cherkaoui, 10112 Rabat, Morocco; S.Udupa@cgiar.org

[6] Division of Genetics & Plant Breeding, Sher-e-Kashmir University of Agricultural Sciences & Technology (SKUAST), Sopore 193 201, India; ratherirshad@gmail.com (I.A.R.); imrouf2006@gmail.com (R.R.M.)

[7] Grain Legume Genetics and Physiology Research, USDA-ARS, and, Washington State University, Pullman, WA 99164, USA; George.Vandemark@ars.usda.gov

[8] Department of Plant Sciences, University of California, Davis, CA 95616, USA

* Correspondence: j.kholova@cgar.org (J.K.); rvpenmetsa@ucdavis.edu (R.V.P.)

† These authors contributed equally to this work.

Received: 18 October 2019; Accepted: 1 November 2019; Published: 7 November 2019

Abstract: "Stay-green" crop phenotypes have been shown to impact drought tolerance and nutritional content of several crops. We aimed to genetically describe and functionally dissect the particular stay-green phenomenon found in chickpeas with a green cotyledon color of mature dry seed and investigate its potential use for improvement of chickpea environmental adaptations and nutritional value. We examined 40 stay-green accessions and a set of 29 BC2F4-5 stay-green introgression lines using a stay-green donor parent ICC 16340 and two Indian elite cultivars (KAK2, JGK1) as recurrent parents. Genetic studies of segregating populations indicated that the green cotyledon trait is controlled by a single recessive gene that is invariantly associated with the delayed degreening (extended chlorophyll retention). We found that the chickpea ortholog of Mendel's I locus of garden pea, encoding a SGR protein as very likely to underlie the persistently green cotyledon color phenotype of chickpea. Further sequence characterization of this chickpea ortholog CaStGR1 (CaStGR1, for carietinum stay-green gene 1) revealed the presence of five different molecular variants (alleles), each of which is likely a loss-of-function of the chickpea protein (CaStGR1) involved in chlorophyll catabolism. We tested the wild type and green cotyledon lines for components of adaptations to dry environments and traits linked to agronomic performance in different experimental systems and different levels of water availability. We found that the plant processes linked to disrupted CaStGR1 gene did not functionality affect transpiration efficiency or water usage. Photosynthetic pigments in grains, including provitaminogenic carotenoids important for human nutrition, were 2–3-fold higher in the stay-green type. Agronomic performance did not appear to be correlated with the presence/absence of the stay-green allele. We conclude that allelic variation in chickpea CaStGR1 does not compromise traits linked to environmental adaptation and agronomic performance, and is a promising genetic technology for biofortification of provitaminogenic carotenoids in chickpea.

Keywords: Mendel's I gene; cosmetic stay-green; biofortification; green cotyledon; carotenoids; pro-vitamin A; chickpea; *Cicer arietinum*

---

1. Introduction

The chickpea is an important source of nutrition and economic livelihood for developing countries [1]. In developing semiarid tropical (SAT) regions, chickpea is typically grown during the post-rainy season under rain-fed conditions [2]. As a result of this growing practice, fluctuations in crop yields largely reflect in-season water availability and crop adaptation to these conditions. Fluctuations in crop production threaten the nutritional and economic status of the already impoverished smallholder farming communities, which make up 80% of all Asian and African farmers [3]. One way to alleviate chickpea production fluctuations in SAT is through the introduction of cultivars with enhanced climate resilience and nutrient density. The utilization of functional stay-green phenotypes is a possible solution to enhance crops climate resilience due to its ability to conserve water and nutrients in drought conditions [4]. Functional stay-green technology is extensively studied and exploited by many crop improvement programs (mainly in cereals, sorghum: [5–10]; maize: [11–14]; wheat: [15–19]; rice: [20–23].

The biological basis (i.e., plant constitutive water and nutrient use dynamics) and benefits of the functional stay-green trait for the SAT agrisystems have been well documented [10,24–28]. In contrast, cosmetic-stay green which is linked to naturally occurring loss-of-function allelic variants [29] with dysfunctional chlorophyll degradation pathways, has rarely been studied in these conditions. This type of stay-green results in extended retention of chlorophyll in all plant organs (leaves, stems, grains) and delays age-related senescence as well as senescence caused by environmental factors (e.g., drought). The utility of cosmetic stay-green variants has been, thus far, limited to green color retention in ornamentals, vegetables, and turf-grasses [29]. However, green-seeded variants also occur in many legumes and pulses such as, chickpea, common bean, lima bean, lentil, cowpea, and pea. Seed greenness in pea has resulted into two major market categories, yellow and green pea, demonstrating the vast economic potential of this trait in other legumes and pulses.

The cosmetic stay-green trait might have much more practical implications than just visual appearance caused by extended chlorophyll retention [29–33]. For example, it is well known that chlorophyll biosynthesis and retention is co-regulated with carotenoids which facilitate scavenging of reactive oxygen species generated in the process of photon's capture by chlorophylls [34–36]. Therefore, we may expect that extended chlorophyll maintenance in any plant organ (including seeds) to be associated with extended maintenance of carotenoids (including β-carotene, i.e., provitamin A), which are of relevance to improving the human diet [37,38] as observed in chickpea [30,39]. On the other hand, the retention of chlorophyll and its associated pathways in cosmetic stay-green crops may impose drawbacks on crop agronomic performance, such as slow seedling establishment or arrested N-remobilization [29,40–47].

Therefore, in this study we aim to characterize the genetic, molecular and physiological basis of cosmetic stay-green trait in chickpea. We document allelic variation in the chickpea ortholog of the 'staygreen' protein that is invariantly associated with genotypes of the green cotyledon color.

We test the functional consequences of 'stay-green' on several key plant processes linked to water usage, transpiration efficiency, and other agronomic traits important for chickpea production in drought-prone regions of the semiarid tropical (SAT) agrisystems. Lastly, we examined stay-green's potential for natural biofortification of chickpea to alleviate the nutritional deficiencies commonly found in these systems.

2. Results

2.1. Delayed Degreening Phenotypes in Green Cotyledon Chickpea and Underlying Allelic Variation

2.1.1. Delayed Degreening Phenotypes in Green Cotyledon Chickpea

In the initial examination of two green-seeded accessions, PI 450,727 and W6 25975, we observed a delay in degreening of mature plant tissues after harvesting, including of leaves and pods. Subsequent senescence assays of fresh growing leaves (Figure 1) corroborated the initial observations of delayed degreening that were made on harvested whole plants.

Figure 1. Seed and leaf senescence phenotypes of normal and green chickpea. Dried mature seed of common chickpea with yellow cotyledons (**a**) and of the green cotyledon colored type (**b**). Differential degreening rates in detached leaves floated on water after 5 days in the absence of light from normal chickpea (**c**) and green cotyledon type (**d**), and from leaves wrapped in aluminum foil from yellow (**e**) and green chickpea (**f**). Asterisk in (**e**) and (**f**) mark leaves covered by foil for 5 days. Blue lines in each panel correspond to 1 cm.

To determine the extent of co-occurrence of delayed degreening of leaf tissues and the green cotyledon trait, we examined degreening in a broader set of green cotyledon chickpea. Using the detached leaf assay, examined degreening among 30 green-seeded chickpea germplasm available from the public gene banks, alongside four other germplasm lines with yellow cotyledon color (Table S1). In this experiment all 27 green cotyledon accessions (three other accessions did not germinate) exhibited delayed degreening, with detached leaves remaining remained visually green through day 7 of the detached leaf assay (Table S1). By contrast, each of the four yellow cotyledon accessions exhibited an apparently normal degreening phenotype, with progressive yellowing of leaves clearly evident by day 7 after the start of the experiment (Figure 1c,d). Furthermore, in a separate experiment, we examined degreening of leaves of this germplasm accessions using an "on-planta" assay, wherein leaves were wrapped in aluminum foil (to block out light and trigger degreening) and degreening assayed 5–10 days later (Figure 1e,f). Of 29 accessions assayed in this manner, all 26 green cotyledon lines exhibited persistent green leaves, while by contract, the three yellow cotyledon accessions exhibited yellow-colored leaves (Table S1). Together, the data from the two different assays for degreening invariantly correlated green cotyledon seed types with delayed degreening (senescence) of leaf tissues, and which contrasted with a more rapid (normal) senescence of leaves of the yellow cotyledon seed types. Moreover, this association held true in additional genotypes (breeding lines or cultivars) that were analyzed subsequently (Supplementary Table S1).

2.1.2. Identification of Chickpea Ortholog of the Staygreen (SGR) Protein

The delayed degreening observed to be associated with the green cotyledon colored chickpea was reminiscent of the 'stay green' phenotype. This suggested that the 'staygreen' gene as a potential candidate gene in chickpea, as this protein has been previously shown to underlie the green-cotyledon trait at Mendel's I locus in garden pea [48,49]. To identify chickpea sequence homologs of SGR protein, coding regions of SGR protein from pea and Medicago [33] were used in blast searches to identify chickpea transcript assemblies and genomic sequences from public databases. Alignment of messenger RNA sequences against the genomic sequence of chickpea indicated a gene structure comprised of four exons interspersed with three introns (Figure 2a and Supplementary Figure S1). Oligonucleotide primers were designed to encompass the entire coding region of the STG gene and used for PCR amplification from cDNA and genomic DNAs of yellow cotyledon chickpea. Amplified PCR products were Sanger sequenced and aligned against the transcript and genomic sequences of chickpea. The 100% correspondence of the sequence between the amplicons and those of the reference transcriptome and genomic sequences of chickpea confirmed the on-target amplification of the chickpea homolog. We designated this gene as CaStGR1 (for *Cicer arietinum Stay-Green* gene 1).

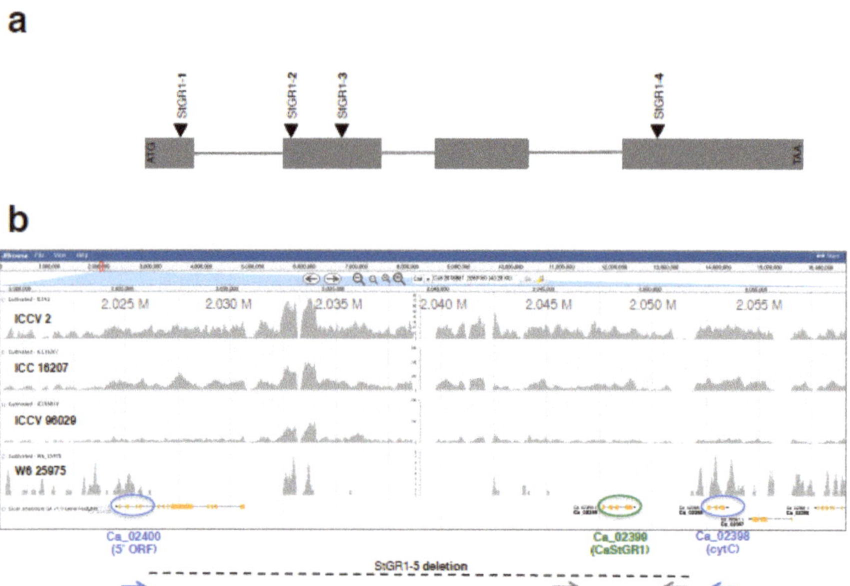

Figure 2. Gene structure and genomic context of type chickpea stay-green gene CaStGR1. (**a**) Schematic of the gene structure of CaStGR1 are shown in (**a**), with the four exons denoted by gray boxes and the three introns as thin lines. Locations of the four small deletion alleles CaStGR1 through CaStGR4 are denoted by triangles above the exons. (**b**) Whole genome Illumina short read skim sequencing read pileups of three normal yellow cotyledon colored chickpea genotypes (ICCV 2, ICC 16,207 and ICCV

96029) are aligned to the chickpea reference of 'CDC Frontier', alongside those from genotype W6 25,975 that harbors the large deletion allele CaStGR1-5. Predicted genes Ca-02399 (CaStGR1) and two flanking low copy genes Ca-02398 (cytC) and Ca-02400 (5' ORF) are marked by ovals. Location of oligonucleotides used in PCR amplification assays from the vicinity of CaStGR1 and falling within the large deletion are marked by gray arrows, and those from the deletion spanning amplification PCR are marked by blue arrows.

2.1.3. Association of CaStGR1 Sequence Variants with Green Cotyledon Chickpea Germplasm

Examination of the nucleotide sequence of the green-cotyledon line PI 450,727 indicated a single nucleotide (1-bp) deletion within the first exon of CaStGR1. This frameshifting mutation is predicted to result in missense changes (from amino acid residue 34) coupled with premature termination of translation (at amino acid residue 56) of 266 amino acid residues of a full-length, functional 'wild type' CaStGR1 protein.

To determine the prevalence of delayed degreening and of nucleotide variation in CaStGR1 more broadly among chickpea germplasm, we examined the rate of degreening in a set of 53 chickpea lines in total (Supplementary Table S1). This collection was predominantly germplasm from the US gene bank (34 accessions) that was supplemented with breeding lines (15 genotypes) and cultivars with green cotyledon color, with a smaller number of normal, tan/yellow cotyledon lines serving as controls (Supplementary Table S1).

A total of 33 genotypes of which 27 possessed green cotyledons, including genotypes PI 450,727 and W6 25,975 which were analyzed previously, along with six additional genotypes with yellow cotyledons, were assessed phenotypically in a leaf degreening assay. In this analysis, all of the 27 green cotyledon genotypes exhibited delayed degreening, whereas by contrast, all six of the yellow cotyledonary lines senesced rapidly with yellowing of detached leaves by day five of the experiment. Furthermore, the degreening phenotype of the 27 with green cotyledons were indistinguishable from that of the previously characterized genotypes PI 450,727 and W6 25,975 that were included alongside in this analysis. This invariant association between green cotyledon color and delay in degreening of detached leaves suggested that the additional 25 germplasm lines may harbor similar molecular variation previously observed in genotypes PI 450,727 and W6 25975.

PCR amplification with CaStGR1-specific oligos with genomic DNA as the template was conducted in 41 genotypes, of which 37 were green cotyledonary with the remaining four with yellow cotyledons. Amplification was consistently unsuccessful in 10 green cotyledon genotypes despite exhaustive PCR attempts, in a manner similar to that in the presumptive large-deletion in genotype W6 25,975 (Supplementary Table S1). Sanger sequencing of PCR amplicons revealed the presence of the 1-bp deletion previously identified in genotype PI 450,727 in an additional six genotypes (Supplementary Table S1 and Supplementary Figure S1). We designated this variant as CaStGR1-1 allele. Of the remaining 25 genotypes, the four genotypes with yellow cotyledons each had a nucleotide identical to that of 'wild type' staygreen gene (that we designated as allele CaStGR1), whereas the remaining 21 genotypes with green cotyledons contained either one of three nucleotide variants in the coding region of the CaStGR1 gene (Supplementary Table S1). Accession ICC 16,340 that was used as the source for breeding of green cotyledon chickpea at ICRISAT-India, along with four breeding lines (also from the ICRISAT-India chickpea breeding program) all shared a novel 8-bp deletion in exon 2 (Supplementary Table S1 and Figure S1) that we designated as allele CaStGR1-2. Ten other genotypes (9 germplasm accessions and the Canadian green-cotyledon cultivar "CDC Verano") shared another molecular lesion, consisting of a 1-bp deletion (Supplementary Table S1 and Supplementary Figure S1) that we designated as allele CaStGR1-3. Although this variant is also located within exon 2 of CaStGR1, it falls downstream in the coding sequence of the location of the 8-bp deletions observed among material from ICRISAT (allele CaStGR1-2; Supplementary Figure S1). The remaining six green cotyledon genotypes, that included three germplasm accessions and three breeding lines from the USDA-ARS breeding program in Pullman, Washington, USA, each harbored yet another molecular variant, in the form of a

1-bp deletion in exon 4 of CaStGR1 (Supplementary Table S1 and Supplementary Figure S1) which we designated as allele CaStGR1-4). Taken together, the PCR amplification and amplicon sequencing data identified five different molecular lesions in CaStGR1 (Figure 1a and Supplementary Figure S1) that occur exclusively among green cotyledon genotypes (Table 1 and Supplementary Table S1).

Table 1. Summary of nucleotide variants identified in CaStGR1 among chickpea germplasm. The color of cotyledons, designated allele names for the variants, the nature of molecular lesions found in each allele, and their frequencies among germplasm studied are listed.

Cotyledon Color	Allele	Nucleotide Variation in CaStGR1	Number of Genotypes
green	CaStGR1-1	1-bp "g" del in exon 1	7
green	CaStGR1-2	8-bp "ctaggttg" deletion in exon 2	5
green	CaStGR1-3	1-bp "c" deletion in exon 2	10
green	CaStGR1-4	1-bp "g" del in exon 4	6
green	CaStGR1-5	entire gene deleted	11
Yellow/Tan	CaStGR1 WT	"Wild Type"	6

2.1.4. Whole Genome Skim Sequencing Delimits the Extent of the Deletion in Allele CaStGR1-5

The absence of amplification in genotypes with the CaStGR1-5 allele with oligonucleotide primers located within the entire coding regions of CaStGR1 was suggestive of a larger sized deletion. To characterize the extend of this deletion we focused on genotype W6 25,975 that typifies this large-deletion allele. In an initial experiment, using the draft whole genome of chickpea genotype CDC-Frontier [50] as a guide, oligos sited in low copy sequences immediately adjacent (within few kbp) to CaStGR1 were designed and used in PCR amplification. Amplification products of the expected size (3-6 kbp in length) were consistently obtained from wild type ICCV 96,029 genotype and PI 450,727 harboring a 1-bp in exon 1 (allele CaStGR1-1). By contrast, no amplification products were obtained from W6 25975, indicating a deletion of larger and yet to be determined size.

To further characterize the extent of this deletion, a whole genome shotgun library was prepared using genomic DNA of the green cotyledon genotype W6 25,975 and sequenced with Illumina HiSeq platform. Sequences obtained were aligned against short read data from normal yellow cotyledon genotypes ICCV2, ICC 16,207 and ICCV 96029, and anchored to the draft whole genome sequence of chickpea genotype CDC-Frontier [50]. Analysis of the resulting pileup of short-read data localized the wild type CaStGR1 gene to between positions 2.047 and 2.049 Mbp on chickpea chromosome 8's pseudomolecule (Figure 2b). This multi-genotype sequence pileup data suggested a deletion of ~25 kbp in length, from ~2.026 Mbp within an adjacent predicted gene on one side, through CaStGR1 at ~2.047 Mbp, and into another predicted gene at ~2.052 Mbp on the other side of CaStGR1 (Figure 2b). Oligonucleotide primers were designed in the low copy predicted genes at ~2.026 Mbp and ~2.052 Mbp that flank CaStGR1, to encompass the ~25 kbp deduced deletion. PCR amplification with these deletion-spanning oligos yielded amplification products of the expected size (3–6 kbp) in genotype W6 25,975 but not in PI 450,727 (where the amplicon would be >25 kbp in size, beyond the capacity of PCR conditions used). The whole genome skim sequencing data together with the PCR results with the gap-spanning oligos corroborate that the CaStGR1-5 allele represents a large deletion of ~26 kbp in size that encompasses the entirety of the CaStGR1 gene (Figure 2b).

2.1.5. Genetic Cosegregation of Staygreen Sequence Variants with the Green-Cotyledon Trait

In two F2 populations that we examined, the green cotyledon trait segregates as a monogenic recessive trait. In the PI 450,727 x RS11 F2 population, of 47 F2s 35 were of yellow cotyledon color with the remaining 12 with green cotyledon color. In a second F2 population of 88 individuals derived from a cross between yellow cotyledon cultivar 'Royal' and the green cotyledon accession PI 359555, 63 F2s had yellow cotyledons and the remaining 25 F2s had green colors. These fit the 3:1 ratio that is expected for a monogenic recessive gene in the F2 generation (with chi-square values of 0.007 and 0.545; and *p*-values 0.933 and 0.460 for the PI 450,727 x RS11 and Royal x PI 359,555 F2 populations respectively).

The single nucleotide deletion identified in the green cotyledon accession PI 450,727 creates a Hpy-188I restriction enzyme recognition site, which allowed for the design of a CAPS (cleaved amplified polymorphic sequence) marker for the CaStGR1-1 variant allele. A F2 population of 47 individuals, derived from a cross between PI 450,727 (with green cotyledons) and accession RS11 (with normal yellow cotyledons), was phenotyped for cotyledon color and genotyped with the Hpy-188I CAPS marker. In this analysis, all 12 F2 individuals with green cotyledons were homozygous for the PI 450,727 allele, while the remaining 35 F2 individuals were either heterozygous or homozygous for the yellow cotyledon allele of RS11, as would be expected for a monogenic recessively inherited trait conditioning green cotyledon color.

We further examined cosegregation between cotyledon color and molecular variation in the CaStGR1 gene in additional F2 populations. A green cotyledonary breeding line with the CaStGR1-4 allele was crossed to the elite cultivars 'Nash' and 'Billybean' from which F2 populations were generated. Seeds of these F2s were scored for cotyledon color prior to sowing, and subsequently degreening of vegetative leaves assessed by the foil wrap assay. A KASP marker assay for the 1-bp deletion that occurs in this allele was developed and used to genotype these F2 individuals, and to examine the correlation with the seed cotyledon color and degreening phenotypes. In this analysis, all 52 individuals with green cotyledons and delayed degreening of leaves were homozygous for the 1-bp deletion allele. Of an additional 55 individuals with yellow cotyledons and rapid degreening of leaves, 24 individuals were homozygous for the wild type allele, with the remaining 31 individuals heterozygous for the two alleles. These observations are consistent with the expected monogenic recessive nature expected for the CaStGR1-4 allele. The loss-of-function of the protein in the 52 homozygotes for the deletion allele engendering phenotypes on seed color. By contrast, the presence of one or more of the wild type alleles in the other 55 individuals provides a functional protein, and the associated normal yellow cotyledon color and normal rate of degreening.

2.2. Characterization of Physiological Functions of Green Cotyledon Chickpea

The genetic and early phenotypic analysis indicated that green cotyledon chickpea is sharing a common suite of characteristics such as delayed degreening in leaf tissue, and which were in contrast to those observed in regular yellow cotyledon chickpeas. To determine the impacts of altered function of the chickpea stay-green gene in these green cotyledon lines, we undertook a set of studies to characterize the impacts on plant physiological functions and indicators of agronomic performance.

2.2.1. Plant Responsiveness to Soil and Atmospheric Drought (Experiment 1 and 2)

The main purpose of the response to soil and atmospheric drought experiments (experiment 1 and 2) was to characterize the crop capacity to restrict transpiration upon severing soil/atmospheric moisture deficit. The plant responsiveness to soil moisture deficit could be expressed as the soil moisture threshold (i.e., fraction of transpirable soil water; FTSW) when the plant transpiration significantly declines compared to transpiration of WW plants. Across the experiments, we documented a wide range of the genotypic responses to declining soil moisture. FTSW values of 0.43–0.64 were observed among germplasm (Figure 3a), which encompassed the narrower range of FTSW (0.54-0.58) observed in stay-green introgression lines (ILs) that originated from the Indian elite cultivars KAK2 and JKG1 (Figure 3b and Table 2). Within the germplasm lines, genotypes with functional StGR1-WT allele tended to limit their transpiration at a higher level of soil moisture (FTSW threshold higher than 0.5) although we couldn't statistically differentiate these lines from the other tested StGR1 allelic variants. In the series of experiments with introgression lines (ILs) based on Indian elite cultivars (KAK2 and JKG1), we found that FTSW thresholds of both cultivated recurrent parents (KAK2 and JKG1) was very narrow (0.54 ± 0.03) and significantly lower compared to the FTSW of the stay-green trait donor parent ICC 16,340 (0.58 ± 0.02) whereas there was no significant difference between ILs and the parental lines.

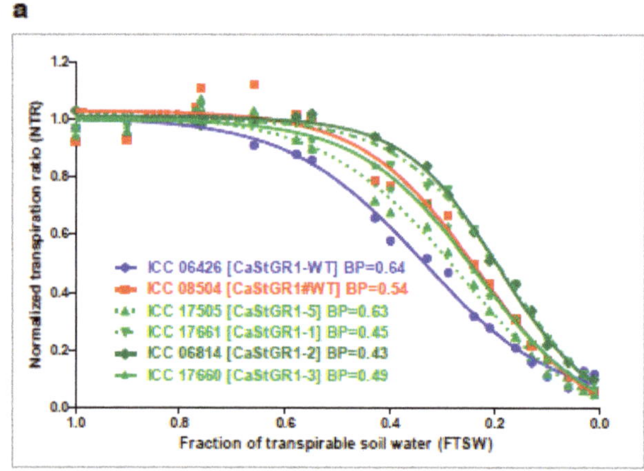

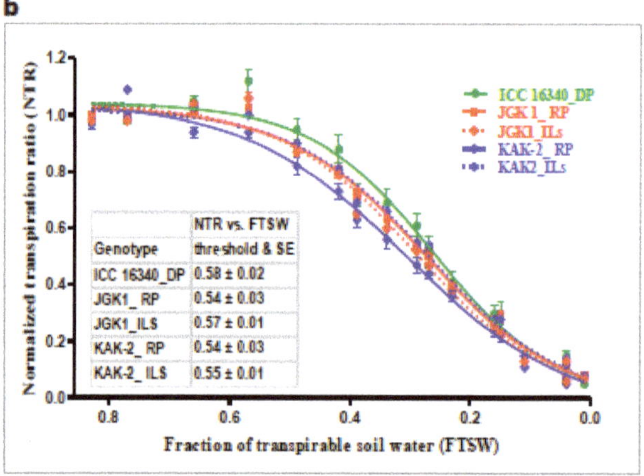

Figure 3. (**a**) Normalized transpiration ratio (NTR) versus fraction of transpirable soil water (FTSW) of chickpea genotypes differed in deletion of CaStGR1gene segments [ICC 08504-CaStGR-1-#Wild type (filled square with solid red line); ICC 06426-CaStGR-1-Wild type (filled round with solid blue line); ICC 17505-CaStGR-1-5 (filled upward triangle with dashed green line); ICC 17661-CaStGR-1-1 (filled down-word triangle with dashed green line); ICC 06814-CaStGR-1-2 (filled diamond with solid green line) and ICC 17660-CaStGR-1-3 (open round with solid green line)] exposed to progressive drying soil under glasshouse conditions. During detached leaf green assay, ICC 08504-CaStGR-1-#Wild type showed yellow colour in all leaflets fully. By contrast, ICC 06426-CaStGR-1-Wild type showed semi-green colour leaflets. Genotypes with CaStGR1-1 (ICC 17661), CaStGR1-2 (ICC 06814), CaStGR1-3 (ICC 17660), and CaStGR1-5 (ICC 17505) showed completely green colour in all the leaflet during detached leaf green assay. Values are transpiration data of five replicated plants for each genotype at each FTSW condition. The FTSW thresholds where transpiration initiated its decline were calculated with a plateau regression procedure from SAS. The regression lines of the relationships between NTR and FTSW were drawn by fitting NTR to FTSW data above and below the respective threshold for transpiration decline in each genotype with GraphPad Prism. The FTSW breakpoint (BP) are displayed in the figures. (**b**) Normalized transpiration ratio (NTR) versus fraction of transpirable soil water (FTSW) of stay green chickpea introgression lines (ILs) with different genetic background [stay green donor parent (DP) ICC 16,340 (square with solid green line); Recurrent parent (RP) JGK1 (square with solid red line); JGK1 background introgression lines JGK1-ILs (square with dashed red line); Recurrent

parent (RP) KAK2 (diamond with solid blue line); KAK2 background introgression lines KAK2-ILs (diamond with dashed red line)] exposed to progressive drying soil under glasshouse conditions. Values are transpiration data of five replicated plants for each genotype at each FTSW condition. The FTSW thresholds where transpiration initiated its decline were calculated with a plateau regression procedure from SAS. The regression lines of the relationships between NTR and FTSW were drawn by fitting NTR to FTSW data above and below the respective threshold for transpiration decline in tested genotype with GraphPad Prism. The FTSW breakpoint (BP) and their confidence intervals of regressions are displayed in the figures.

Table 2. Regression analysis of transpiration response to soil drying of green cotyledon trait donor genotype ICC 16340, recurrent yellow cotyledon elite cultivars KAK-2 and JGK1 and backcross introgression lines of the green cotyledon trait in these elite cultivar backgrounds.

Genotypes	NTR-FTSW Thresholds and Std. Error	Slope 1 and Std. Error	Slope 2 and Std. Error
ICC 16340_Stg-D-P	0.58 ± 0.02	1.92 ± 0.08	−0.59 ± 0.23
KAK2_R-P	0.54 ± 0.03	1.72 ± 0.05	0.20 ± 0.26
JGK 1_R-P	0.54 ± 0.03	1.88 ± 0.06	−0.12 ± 0.24
ICCX-060119-107 (KAK2)	0.54 ± 0.03	1.85 ± 0.08	−0.10 ± 0.22
ICCX-060119-113 (KAK2)	0.48 ± 0.03	1.98 ± 0.09	0.06 ± 0.17
ICCX-060119-116 (KAK2)	0.58 ± 0.03	1.64 ± 0.07	0.01 ± 0.19
ICCX-060119-123 (KAK2)	0.62 ± 0.05	1.69 ± 0.09	−0.32 ± 0.53
ICCX-060121-125 (JGK1)	0.51 ± 0.02	1.87 ± 0.06	−0.01 ± 0.17
ICCX-060121-128 (JGK1)	0.60 ± 0.03	1.63 ± 0.06	0.05 ± 0.28
ICCX-060121-129 (JGK1)	0.55 ± 0.03	1.72 ± 0.07	0.14 ± 0.18

Further, we tested the plant's capacity to regulate transpiration rate (TR [g of water transpired per cm-1 of canopy per h]) in conditions of a drying atmosphere (i.e., increasing vapor pressure deficit; VPD). Here, we documented wide range of variability in the tested material and across the range of conditions (outdoors typically ~0.5–3.0 kPa [Figure 4a,b] and in growth chambers 1.2 to 4.6 kPa [Figure 5a,b]). TR responses to VPD under natural atmospheric (outdoor) conditions and under controlled VPD (growth chamber) conditions showed a similar trend (Figure 4a,b and Figure 5a,b; Table 3). In germplasm, we found no consistent trend in material with ("wild type") stay-green allele or without (i.e., Loss-of-Function alleles CaStGR1-1 to CaStGR1-5) in the TR responsiveness to VPD (Figure 5a). Some StGR1 loss-of-function germplasm allelic variants were having TR higher while others lower than values observed for germplasm with a functional (wild type) stay-green gene. In experiment 2b and 2c's series encompassing the stay-green ILs, we found the stay-green donor ICC 16,340 had a higher TR and rapid TR increase upon increasing VPD compared to the recurrent elite parents and their stay-green derivatives in both outdoor and controlled (growth chamber) conditions (Figure 4a,b, Figure 5b; and Table 3). Interestingly, whereas ILs with the KAK2 background had TR and VPD values intermediate to those of the stay-green donor line ICC 16,340 and the elite cultivar KAK2 (Figure 4a), all the stay-green derivatives of JGK1 had even significantly lower TR across the VPD regimes compared to JGK1 elite parent (Figure 4b). Furthermore, in well watered (WW) conditions, there were no significant genotypic differences in the specific leaf weight (SLW) in germplasm, the JKG1-derived ILs had lower SLW compared to both of the parents (Supplementary Figure S2a,b).

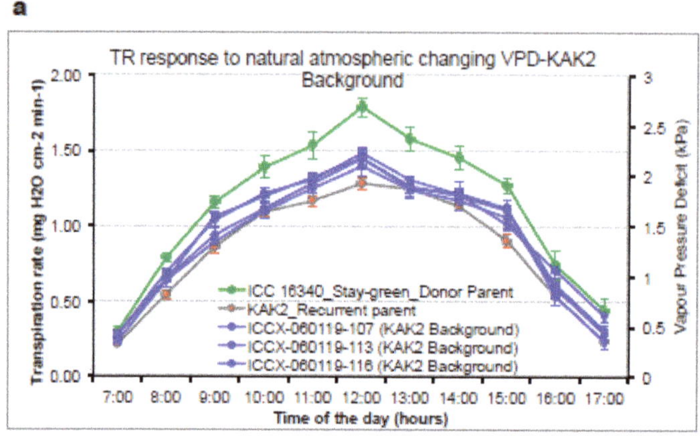

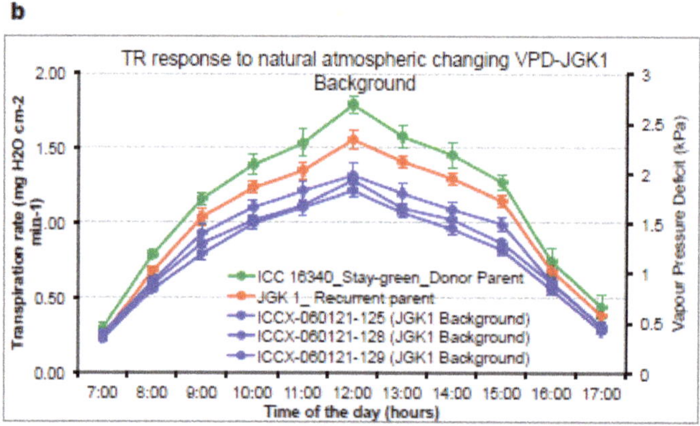

Figure 4. Transpiration rates (TR) of stay green chickpea introgression lines (ILs) with different genetic backgrounds of KAK2 elite cultivar (**a**), and 4JGK1 elite cultivar (**b**). Stay green donor parent (DP) ICC 16,340 (round with solid green line); Recurrent parent (RP) KAK2 (round with solid red line); KAK2 background introgression lines ICCX-060119-107, ICCX-060119-113, ICCX-060119-116 and ICCX-060119-123 (round with solid blue line); Recurrent parent (RP) JGK1 (round with solid red line); JGK1 background introgression lines ICCX-060121-125, ICCX-060121-128 and ICCX-060121-129 (round with solid blue line)] are response to natural changing in the atmospheric vapour pressure deficit (VPD) cycle. TRs were measured on well-watered plants grown in the glasshouse, which were temporarily transferred to outdoor conditions. There, plants were exposed to natural changing atmospheric VPD. TR and VPD data were used to draw a segmental or a single linear regression for all tested genotypes. Each data points represents the means (± SE) of eight replicates per genotype.

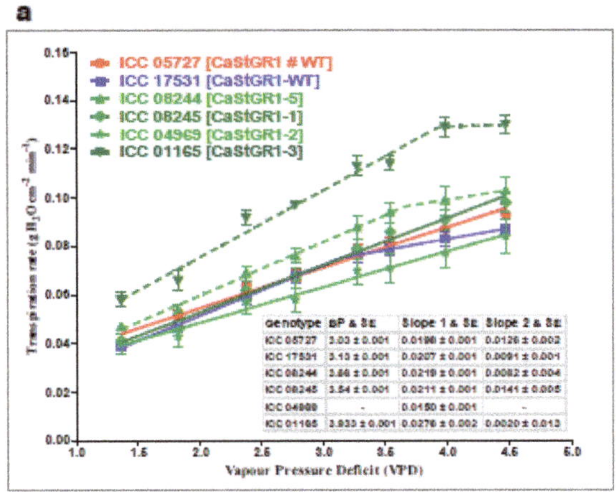

Genotype	BP & SE	Slope 1 & SE	Slope 2 & SE
ICC 05727	3.03 ± 0.001	0.0196 ± 0.001	0.0126 ± 0.002
ICC 17531	3.13 ± 0.001	0.0207 ± 0.001	0.0091 ± 0.001
ICC 08244	3.86 ± 0.001	0.0219 ± 0.001	0.0082 ± 0.004
ICC 08245	3.54 ± 0.001	0.0211 ± 0.001	0.0141 ± 0.005
ICC 04969	-	0.0150 ± 0.001	-
ICC 01165	3.833 ± 0.001	0.0276 ± 0.002	0.0020 ± 0.013

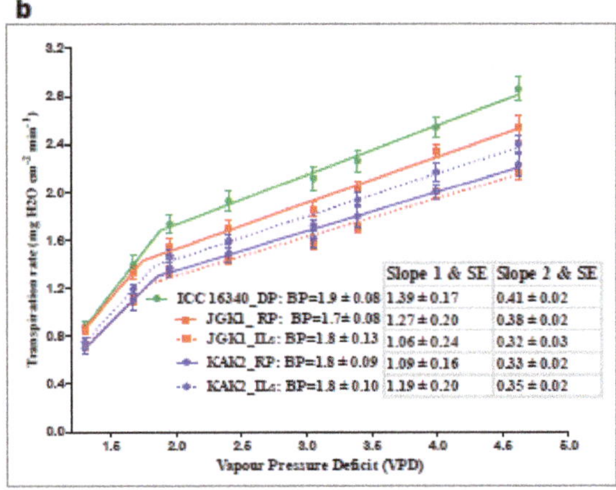

		Slope 1 & SE	Slope 2 & SE
ICC 16340_DP:	BP=1.9 ± 0.08	1.39 ± 0.17	0.41 ± 0.02
JGK1_RP:	BP=1.7± 0.08	1.27 ± 0.20	0.38 ± 0.02
JGK1_ILs:	BP=1.8 ± 0.13	1.06 ± 0.24	0.32 ± 0.03
KAK2_RP:	BP=1.8 ± 0.09	1.09 ± 0.16	0.33 ± 0.02
KAK2_ILs:	BP=1.8 ± 0.10	1.19 ± 0.20	0.35 ± 0.02

Figure 5. (**a**) Transpiration rates (TR) of six selected chickpea genotypes differed in deletion of CaStGR1gene segments [ICC 05727-CaStGR-1-#Wild type (round with solid red line); ICC 17531-CaStGR-1-Wild type (square with solid pink line); ICC 08244-CaStGR-1-5 (upward triangle with solid green line); ICC 08245-CaStGR-1-1 (diamond with solid blue line), ICC 04969-CaStGR-1-2 (star with solid orange line) and ICC 01165-CaStGR-1-3 (downward triangle with solid pink line)] in response to increasing VPD. During detached leaf green assay, ICC 05727-CaStGR-1-#Wild type showed yellow colour in all leaflets fully. By contrast, ICC 17531-CaStGR-1-Wild type showed semi-green colour leaflets. Genotypes with CaStGR1-1 (ICC 08245), CaStGR1-2 (ICC 04969), CaStGR1-3 (ICC 01165), and CaStGR11-5 (ICC 08244) showed completely green colour in all the leaflet during detached leaf green assay. TRs were measured on well-watered plants grown in the glasshouse, which were temporarily transferred to a growth chamber with control over temperature and relative humidity. There, plants were exposed to increasing VPD, set by modifying temperature and humidity. TR data are the mean of five replicate plants, computed hourly at each of the eight VPD levels. Data were used to draw a segmental or a single linear regression for all tested genotypes. Each data points represents the means (± SE) of five replicates per genotype. The slopes and breakpoint (BP) of regressions are displayed in the figures. (**b**) Transpiration rates (TR) of stay green chickpea introgression lines (ILs) with different genetic background [stay green donor parent (DP) ICC 16,340 (square with solid green line); Recurrent parent (RP) JGK1 (square with solid red line); JGK1 background introgression lines JGK1-ILs (square with dashed red line); Recurrent parent (RP) KAK2 (diamond with solid blue line); KAK2 background introgression lines KAK2-ILs (diamond with dashed red line)] are response to increasing VPD. TRs were

measured on well-watered plants grown in the glasshouse, which were temporarily transferred to a growth chamber with control over temperature and relative humidity. There, plants were exposed to increasing VPD, set by modifying temperature and humidity. Data were used to draw a segmental or a single linear regression for all tested genotypes. Each data points represents the means (± SE) of eight replicates per genotype. The slopes and breakpoint (BP) of regressions are displayed in the figures.

2.2.2. Variation in Plant Growth and Water-Use Related Traits in Lysimetric Facility (Experiment 3a, b)

In the lysimeter experiment under well-watered (WW) conditions with germplasm (Experiment 3a) and introgression lines (Experiment 3b), significant genotypic differences in the total amount of water required to reach maturity were observed (data no shown). However this was mostly conditioned by the different phenological development between germplasm and the ILs. The relationship between total water use and days to flowering was strongly correlated in germplasm ($R^2 = 0.63^*$; Supplementary Figure S3a) but only very weakly in the introgression lines ($R^2 = 0.10$ns; Supplementary Figure S3b). Under water stress (WS) differences in total amount of water extracted from lysimeters was independent of crop phenology but these did not coincide with the presence of particular CaStGR1 allele in any of the material used.

Under WW and WS, although differences were observed in total biomass accumulation and seed setting, these did not appear to be associated with the stay-green trait. However, the relative decline in total biomass accumulation due to water stress was very similar between all allelic variants with reduction in WS when compared to WW, of ~50% in germplasm and ~30% in IL material. In experiment 3b under WW treatments, the seed yield was largely explained by duration of phenological stages (Supplementary Figure S4). Interestingly in the same experiments under WS, the seed yield did not relate to crops phenology (Supplementary Figure S4) but related positively to seed number ($R^2 = 0.66^*$ in ILs; experiment 3a, $R^2 = 0.73$ in germplasm materials). In addition, TE [g biomass per kg of water transpired] and seed yield were strongly associated under WS conditions [$R^2 = 0.62^{***}$ in ILs (Figure 6a) and $R^2 = 0.37$ in germplasm], while there was a weak relationship between TE and seed yield under WW conditions (Figure 6b). Also, in experiment 3b under WS, there were several stay-green isolines in each elite genetic background, which had seed yield comparable or higher than the respective elite recurrents and stay-green donor (Supplementary Figure S5).

Table 3. Regression analysis of transpiration response to VPD in outdoor and growth chamber of green cotyledon trait donor genotype ICC 16340, recurrent yellow cotyledon elite cultivars KAK-2 and JGK1 and backcross introgression lines of the green cotyledon trait in these elite cultivar backgrounds.

Genotypes	TR Response to VPD at Outdoor		TR Response to VPD at Growth Chamber		
KAK-2 Background	**Mean TR & SE** **LSD (0.01) = 0.09**	**Slope at high VPD & SE** **LSD (0.01) = 0.59**	**Mean TR & SE** **LSD (0.001) = 0.35**	**Slope & SE** **LSD (0.01) = 0.04**	R^2
ICC 16340_Stay-green_Donor Parent	$1.31 \pm 0.04a$	$5.48 \pm 0.18a$	$2.02 \pm 0.06a$	$0.52 \pm 0.06a$	0.94
KAK2_Recurrent parent	$0.95 \pm 0.02b$	$4.18 \pm 0.08b$	$1.45 \pm 0.06b$	$0.41 \pm 0.04b$	0.94
ICCX-060119-107 (KAK2 Background)	$0.98 \pm 0.04b$	$4.16 \pm 0.18b$	$1.37 \pm 0.04b$	$0.43 \pm 0.04b$	0.95
ICCX-060119-113 (KAK2 Background)	$1.04 \pm 0.02b$	$4.23 \pm 0.06b$	$1.49 \pm 0.06b$	$0.46 \pm 0.04b$	0.95
ICCX-060119-116 (KAK2 Background)	$0.94 \pm 0.04b$	$4.19 \pm 0.20b$	$1.45 \pm 0.08b$	$0.47 \pm 0.05b$	0.94
ICCX-060119-123 (KAK2 Background)	$1.03 \pm 0.02b$	$4.16 \pm 0.16b$	$1.63 \pm 0.04b$	$0.40 \pm 0.05b$	0.91
JGK-1 Background	**Mean TR & SE** **LSD (0.001) = 0.12**	**Slope at high VPD & SE** **LSD (0.01) = 0.46**	**Mean TR & SE** **LSD (0.01) = 0.25**	**Slope & SE** **LSD (0.01) = 0.033**	R^2
ICC 16340_Stay-green_Donor Parent	$1.31 \pm 0.04a$	$5.48 \pm 0.18a$	$2.02 \pm 0.06a$	$0.52 \pm 0.06a$	0.94
JGK 1_Recurrent parent	$1.14 \pm 0.04b$	$4.77 \pm 0.16b$	$1.70 \pm 0.06b$	$0.45 \pm 0.04b$	0.95
ICCX-060121-125 (JGK1 Background)	$0.99 \pm 0.03bc$	$4.09 \pm 0.01c$	$1.53 \pm 0.06b$	$0.42 \pm 0.04b$	0.95
ICCX-060121-128 (JGK1 Background)	$0.90 \pm 0.03c$	$3.82 \pm 0.08c$	$1.52 \pm 0.08b$	$0.38 \pm 0.04b$	0.94
ICCX-060121-129 (JGK1 Background)	$0.90 \pm 0.01c$	$3.78 \pm 0.05c$	$1.52 \pm 0.05b$	$0.37 \pm 0.04b$	0.94

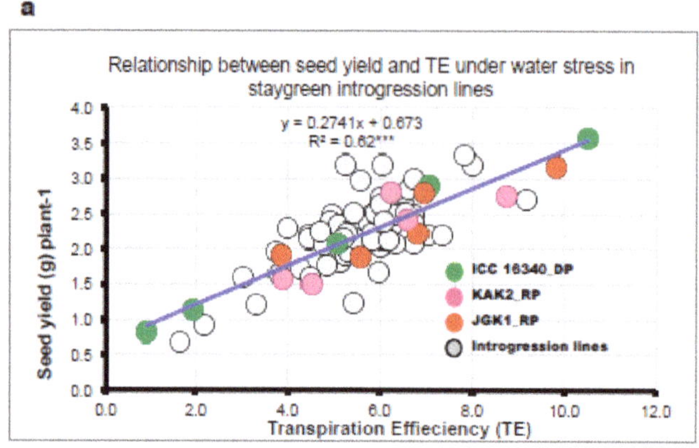

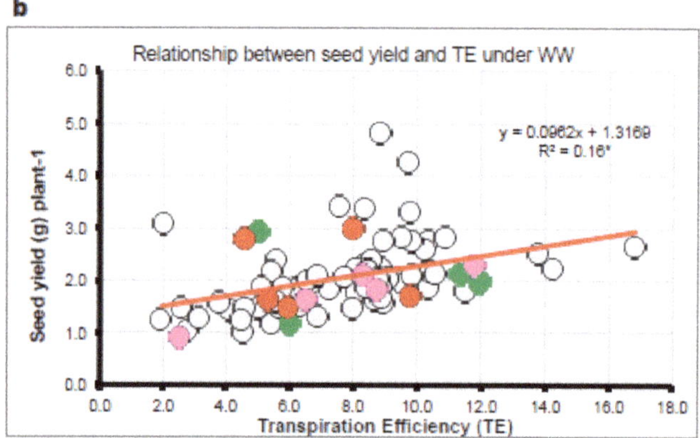

Figure 6. Relationships between seed yield and transpiration efficiency (TE) under (**a**) water stressed (WS) and (**b**) well watered conditions (WW) in stay green chickpea genotypes grown in the PVC cylinders (Lysimetric facility). The data used for these regression analyses are replicated data, obtained under WS and WW conditions. For each genotype, five replicates data points were used to draw the linear regressions. The stay green donor parent (ICC 16340) data are represented in green colour, recurrent parent (JGK1) data are represented in red colour, recurrent parent (KAK2) data are represented in pink colour and introgression lines (ILs) are represented in grey colour. The slopes and R^2 of regressions are displayed in the figures. R^2 values with * and *** (astric) symbols are significantly different at $p < 0.05$ and $p < 0.001$.

2.2.3. Evaluation of Canopy Growth Related Traits (Experiment 4)

The canopy growth parameters were examined only among stay-green ILs alongside the donor germplasm line ICC 16,340 and the recurrent elite cultivars JGK1 and KAK2. We found the donor parent ICC 16,340 had lower canopy growth rates than elite recurrent parents (JGK1, KAK2) with some of the ILs attaining higher growth rates compared to stay-green donor parent and recurrent parents (Figure 7a) and this reflected in the differences in canopy size and digital biomass estimates averaged across the time of observations (Figure 7b). The parental line JGK1 grew more slowly compared to the elite recurrent line KAK2 (Figure 7a). The stay-green derivative ILs in the KAK2 elite cultivar background had growth rates similar to those of the recurrent elite parent KAK2 (Figure 7a). Growth rates in stay-green ILs originated from the elite cultivar, JGK1 exceeded those observed in both parents, and at levels similar to those of in KAK2 stay-green ILs. This indicated that stay-green IL material had recovered its vigor (Figure 7a).

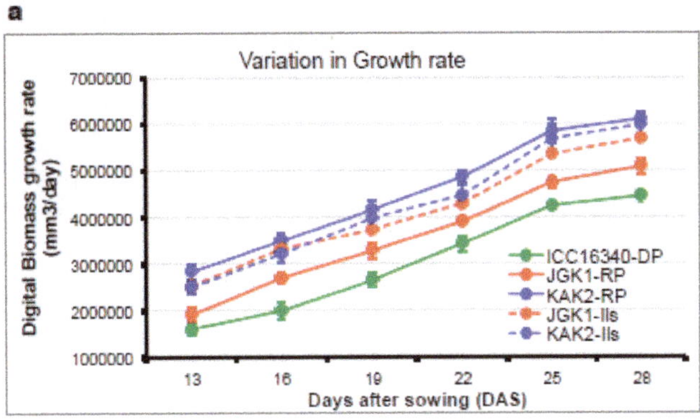

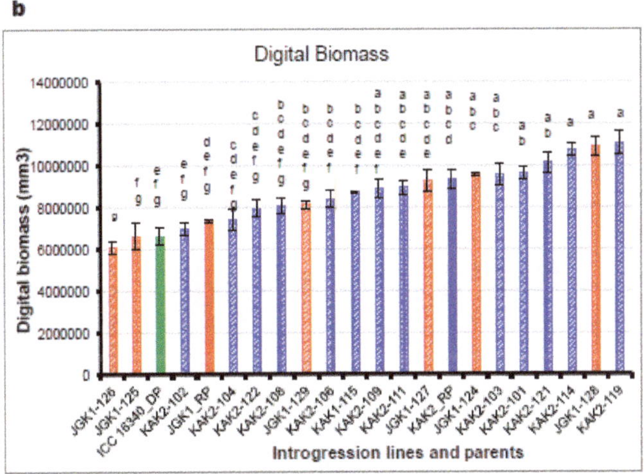

Figure 7. (**a**) Growth rate variation in digital biomass of stay green chickpea introgression lines (ILs) with different genetic background [stay green donor parent (DP) ICC 16,340 (round with solid green line); Recurrent parent (RP) JGK1 (round with solid red line); JGK1 background introgression lines JGK1-ILs (round with dashed red line); Recurrent parent (RP) KAK2 (round with solid blue line); KAK2 background introgression lines KAK2-ILs (round with dashed blue line)] are measured by LeasyScan phenotyping platform. Each data point represents the means (± SE) of four replicates per genotype. Data were used to draw a line graph for all tested genotypes. (**b**) Variation in digital biomass of stay green chickpea introgression lines (ILs) with different genetic background [stay green donor parent (DP) ICC 16,340 (bar filled with solid green colour); Recurrent parent (RP) JGK1 (bar filled with solid red colour); JGK1 background introgression lines JGK1-ILs (bar crossed lines with red colour); Recurrent parent (RP) KAK2 (bar filled with solid blue colour); KAK2 background introgression lines KAK2-ILs (bar crossed lines with blue colour)] are measured by LeasyScan phenotyping platform. Each data points represents the means (± SE) of four replicates per genotype. Data were used to draw a bar graph for all tested genotypes. Bars with different letters are significantly different ($p < 0.05$).

2.2.4. Evaluation in the Field Conditions (Experiment 5)

The IL plant material that was relatively more homogeneous for the main phenology-related characters was tested in the field alongside their recurrent parents (experiment 5; flowering 37–53 DAS, days to maturity 97-101). Some of the tested ILs attained similar or even higher grain yield under irrigated conditions (Figure 8a), which was partially positively driven by phenology differences [Relationship between accumulated biomass or seed yield and days to flowering; $R^2 = 0.51^*$ (Supplementary Figure S6a) and negatively related to harvest index [Relationship between accumulated biomass and harvest index (HI); $R^2 = 0.30^*$ (Supplementary Figure S6b)]. Water stress (WS) conditions reduced the grain yield cca 40–70%. Under WS conditions, the yield of stay-green ILs in relation with the days to flowering was much looser (Supplementary Figure S6c). We also observed the lack of correlation between the production traits (biomass and yield) and phenology parameters [Regression between accumulated biomass and days to flowering; $R^2 = 0.0001$ & regression between seed yield and days to flowering; $R^2 = 0.08$] while the relation between HI was maintained [Relationship between seed yield and harvest index (HI); $R^2 = 0.21$ (Supplementary Figure S6d)]. Interestingly, we found that the extent of yield reduction due to WS was similar between the parental lines and some of the stay-green introgression line progenies (Figure 8b), and was further positively related to plant capacity to grow in WW but negatively in WS (i.e., higher production potential, higher yield reduction due to WS while the "smaller" plants had suffered less yield reduction under WS).

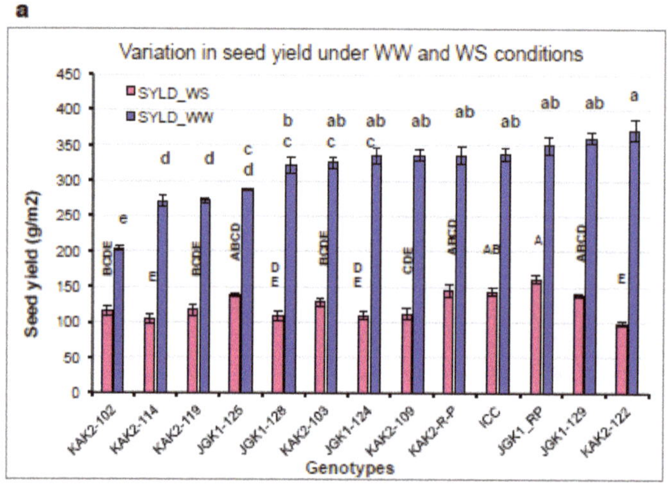

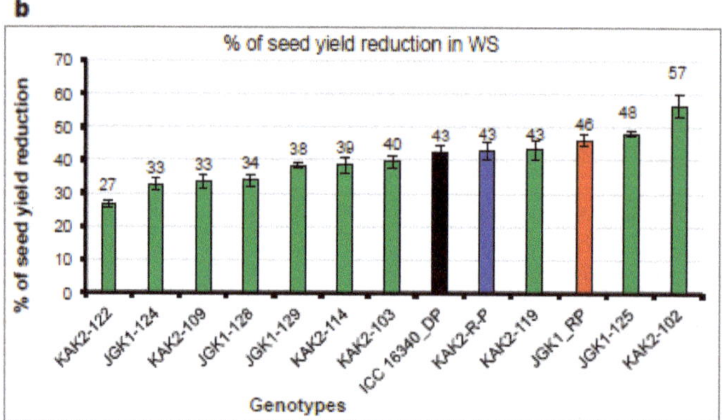

Figure 8. (**a**) Variation in seed yield under well water (bar filled with blue colour) and water stress (bar filled with pink colour) conditions. The data used for these bar graphs are mean data, obtained under well-watered (WW) and water stress (WS) conditions. For each genotype, three replicates data points were used to draw the bar graph. Bars with different capital letters (well-watered—WW) and small letters (water stressed—WS) alphabets are significantly different ($p < 0.05$) and same letters

represents non-significant. (**b**) Percentage of seed yield reduction under water stress (WS) conditions. The data used for these bar graph are mean data, obtained from well watered seed yield data were normalised against water-stressed seed yield data and then seed yield reduction values are presented in percentage. The data of stay green donor parent ICC 16,340 (bar filled with black colour); recurrent parent-JGK1 (bar filled with red colour); recurrent parent-KAK2 (bar filled with blue colour); stay-green introgression lines from both JGK1 and KAK2 genetic background–ILS (bar filled with green colour).

2.2.5. Leaf Pigments Content Under WW and WS Conditions (Experiments 1c, 3a,b,c)

Pigments in the Leaf Tissues and Grains.

Across all material tested, we found that plants grown outdoors (in lysimeters, experiment 3c) maintained much higher levels of photosynthetic pigments, especially carotenoids in leaves tissues, compared to plants cultivated in the glass-house (in lysimeters, experiment 3b) environments.

We found no differences between the levels of leaf pigments (i.e., chlorophyll a, chlorophyll b, total carotenoids) and their ratio (chlorophyll a/chlorophyll b ratio) in the materials carrying the CaStGR1-wt functioning allele and CaStGR1-1 to 5 malfunctioning allele (ILs and some germplasm) under WW. The methodology of stress imposition and the tissue sampling (the last fully developed leaf on the main stem) couldn't discriminate the stay-green material from wild-type under the WS conditions either. However, we found a higher chlorophyll_a and cholorophyll_b content in mature seeds of material carrying stay-green alleles compared to CaStGR1-wt in both germplasm (Supplementary Figure S7a,b) and stay-green ILs (Supplementary Figure S8a,b). Similarly, the grain total carotenoids content was ~10–30% higher in the stay-green loss-of-function variants (alleles CaStGR1-1 to 5) compared to wild type (CaStGR1-WT; Figure 9a) in germplasm and ILs. Furthermore, grain total caratenoid levels were not significantly affected by the conditions of cultivation (WW and WS) in the introgression lines (Figure 9b).

The detailed fractionation of carotenoids contents in ILs seeds revealed that there were ~3-fold higher levels of lutein and beta-carotene (provitamin A) in the seeds of green cotyledon introgression lines (ILs) compared to both of the yellow cotyledon colored elite cultivars (KAK2 and JGK1; Figure 10). By contrast, zeaxanthin content did not significantly vary between ILs with green cotyledons and the elite cultivars with yellow cotyledons (KAK2 and JGK1; Figure 10) used as recurrent parents in introgression line development.

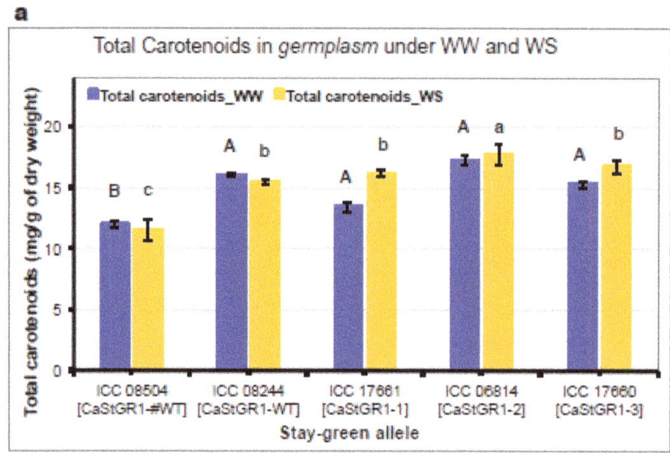

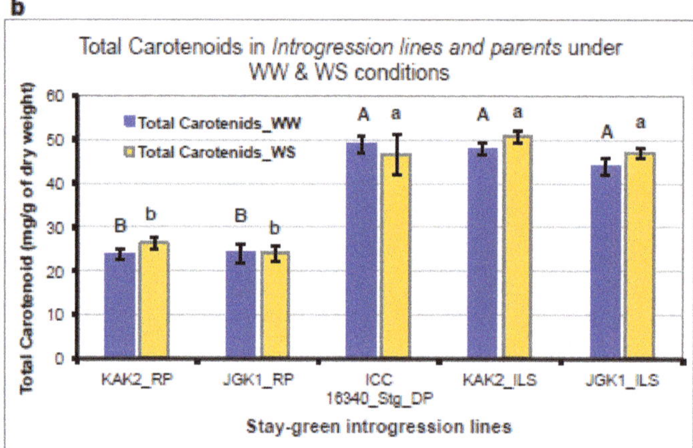

Figure 9. Variation in (**a**) seed total carotenoids content in germplasm [ICC 08,504 (CaStGR1-#WT), ICC 08,244 (CaStGR1-WT), ICC 17,661 (CaStGR1-1), ICC 06,814 (CaStGR1-2) and ICC 17,660 (CaStGR1-3)] and (**b**) stay green chickpea introgression lines (ILs) with different genetic background under well-watered (WW) and water-stressed (WS) conditions. During detached leaf green assay, ICC 08504-CaStGR-1-#Wild type showed yellow colour fully in all leaflets. By contrast, ICC 08244-CaStGR-1-Wild type showed semi-green colour leaflets. Genotypes with CaStGR1-1 (ICC 17661), CaStGR1-2 (ICC 06814) and CaStGR1-3 (ICC 17660) showed completely green colour in all the leaflet during detached leaf green assay In both graph (**a**) and (**b**), closed bars represents WW and open bars are represents WS. Each data points represents the means (± SE) of five replicates per genotype. Data were used to draw a bar graph for all tested genotypes. Bars with different capital letters (well water-WW) and small letters (water stressed-WS) alphabets are significantly different ($p < 0.05$) and same letters represents non-significant.

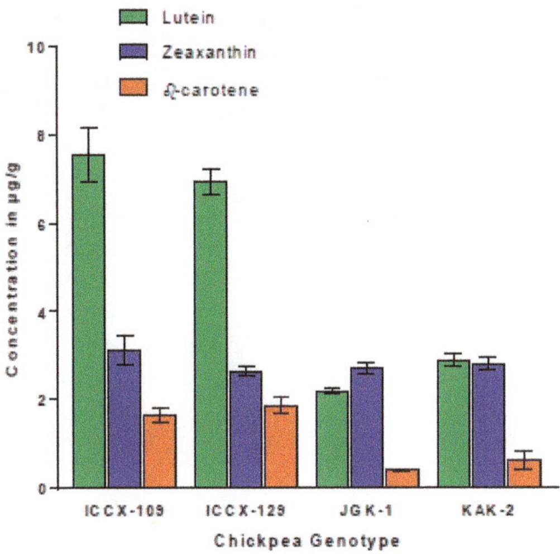

Figure 10. Variation in different carotenoids content (Lutein, Zeaxanthin and beta carotene) in seeds of stay green chickpea introgression lines (ILs) with different genetic background [ICCX-109 (KAK2 genetic background and ICCX-129 (JGK1 genetic background) and their recurrent parents (JGK1 and KAK2). The lutein pigment data are represents in light grey colour bars; Zeaxanthin pigment data are represents in black colour bars and beta carotene pigment data are represents in dark grey colour bars. Each data points represents the means (± SE) of three replicates per genotype.

3. Discussion

The two goals of the present study were to (1) understand the molecular and functional mechanisms underlying the delayed senescence in chickpea with the "cosmetic stay-green" trait [29,31] and, (2) to characterize the effects of the "cosmetic stay-green trait" on plant performance in semiarid agricultural systems. Since the majority of chickpea production occurs under water-limited rainfed conditions, (i.e., terminal drought), understanding responses to water limitations is critical to evaluating the potential of stay-green chickpea. Lastly, we also investigated the nutrient composition of stay-green chickpea, as a genetic biofortification technology to alleviate nutritional deficiencies for carotenoids in consumers.

3.1. Identification of 'Cosmetic Stay-Green' Allele in Chickpea

Recent developments in genome sequencing have provided deep sequence resources for several legumes, in terms of whole genome sequences and transcriptomes. These sequence data provide a valuable resource for both the comparative and evolutionary studies of genome structure and genes. Subsequent analysis of amplified chickpea sequences and their localization to the chickpea draft genome supported the identification of the cognate chickpea stay-green gene that exhibited a high degree of sequence similarity with the other legume stay-green orthologs, and localized to a syntenic position on chromosome 8 in the chickpea draft genome [50]. This genomic region corresponds to the large-effect QTL for carotenoid concentrations described among three F2 populations of chickpea [39], which contains the staygreen gene ortholog (LOC101509366; [39]). Our methods highlight the utility of draft or reference genomes for the more detailed study of individual genes from their initial identification to deduction of orthology from the evolutionary history.

In addition, we also conducted a whole genome skim sequencing, to delimit the extent of the deletion in allele StGR1-5. Initial and exhaustive PCR amplifications indicated this allele as probably encompassing the entire coding region of the chickpea ortholog, but whose boundaries were unknown. The use of whole genome skim sequencing of the genome for this allele allowed us to flank the large (several 10s of kbp; Figure 2b) deletion in a single experiment. This contrasts with earlier approaches such as primer amplicon 'walking' that given the large size of the deletion would not have yielded results or required the use of a large collection of oligos at varying distance surrounding the StGR1 gene.

The monogenic recessive nature of the green cotyledon trait is supported by observation of only yellow cotyledon phenotypes in the F1 individuals from crosses between yellow and green cotyledon chickpeas, and in cosegregation data in segregating progenies (described in results). Furthermore, the occurrence of green cotyledon phenotype in F1s obtained from crosses among alleles, and invariantly green cotyledon in their F2s supports our inference that the five molecular variants we identified and describe in this study comprise an allelic series in StGR1 gene.

The recessive behavior of the green cotyledon alleles of chickpea is consistent with a loss-of-function of the chickpea StGR1 gene in these phenotypic variants. This inference is corroborated by the likely impact of the deletions on the deduced amino acid sequence of the translated protein. The single nucleotide deletions in alleles StGR1-1 to allele StGR1-4 all occur within the coding regions, and consequently these deletions would result in a frame shift of the open reading frame (and premature truncation of the translated protein).

The identification of five different loss-of-function alleles in CaStGR1, and the absence of nesting (where more than one deletion allele occurs within a single genotype), implies that the green cotyledon trait arose independently at least four times in chickpea, and as naturally occurring variation among chickpea germplasm. The fifth gene-encompassing deletion allele StGR1-5 could represent a fifth independent origin of green cotyledon trait in chickpea. However, based on our data we cannot preclude the possibility that this allele may have arisen secondarily within the background of one of the other small 1-bp deletion alleles (StGR1-1 to StGR1-4). Additional analyses of the green cotyledon germplasm along with related germplasm might help to clarify this current ambiguity.

It is intriguing that green cotyledon breeding lines from the three different chickpea breeding programs (ICRISAT in India, USDA-ARS in USA, and the University of Saskatchewan in Canada) represent three different and distinct loss-of-function alleles of the stay-green gene as a source of the green cotyledon trait. This could be a reflection of the limited knowledge or availability of the sources of green cotyledon germplasm in these breeding programs. Alternatively, the use of the different alleles in each breeding program might reflect preferential use of distinct germplasm on the basis of other traits (e.g., for local adaptation, market type, disease reactions) present in the various germplasm sources. Indeed, our observation of varying phenology among green cotyledon germplasm could represent such additional phenotypic variation, along with seed size and color that also vary. In such a scenario, the distinct alleles for StGR1 gene are merely inadvertently co-selected for a desired common trait of green cotyledons from germplasm with additional characteristics.

Despite the recurrent selection at an orthologous StGR gene in multiple crop legumes for green cotyledon color, it is possible that additional genes exist that replicate this phenotype, or might modulate it. For example, in the more exhaustively studied Rice and *Arabidopsis* systems (e.g., [29,51–53]), genes other than the stay-green protein have also been implicated in the cotyledon color or persistence of chlorophyll machinery which would affect stay-green phenotypes. Furthermore, some aspects of the green cotyledon trait, and its manifestation at the level of whole seeds is also likely to depend on pigmentation in the overlying seed coat tissues. For example, in cowpea, distinct genes controlling green color in cotyledon and green color in seed coats have been described [42,54].

Our identification of the molecular nature of variation among green cotyledon chickpea should facilitate the use of molecular marker assisted selection (MAS) or backcrossing (MABC) for introgression of this trait in chickpea breeding. For example, in the current study we developed and tested a KASP marker for the StGR1-4 allele found in USDA-ARS breeding lines (Supplementary Tables S1 and S2). This assay is effective at monitoring the allele states (wt or 1-bp deletion) within exon 4 of the chickpea gene, and is being used for marker-assisted backcrossing in our program. Design and testing of similar KASP assays for the remaining single nucleotide deletions (alleles StGR1 -1, -2, -3) is being planned to facilitate similar use of MAS with these distinct allelic variants.

3.2. Green Cotyledon Trait as a Vavilovian Homologous Series of Variation

Green cotyledon market classes or types occur in several crop legumes, including garden pea [52], Medicago [33], chickpea [30,55], common bean [52], lima bean [52], and cowpea [54]. This recurrence suggests that the green cotyledon color trait arose from the repeated and independent selection from the white or yellow cotyledon forms that typify these crops and their wild relatives. The prevalence of repeated human selection for a common phenotype in multiple crops was suggested by the pioneering crop evolutionary botanist Nikolai Vavilov [56].

3.3. Stay-Green Alleles do not Affect the Plant Responsiveness to Soil and Atmospheric Drought

3.3.1. Plant Responsiveness to Soil and Atmospheric Drought

Any novel crop technology intended for practical utilization in complex agrisystems has to be appropriately tested to enhance the probability to be implemented and accepted. In many of the semiarid rain-fed agrisystems, one of the main limiting factors to crop productivity is soil moisture deficit [2,8,57–60]. To understand plant responses to decreased soil moisture, we have generated substantial evidence on plant functions that contribute to crop adaptations in these environments [61–65]. In the present study we evaluated whether stay-green phenotype in chickpea underlined by CaStGR1 gene might be functionally involved in any important environmental adaptations (i.e., responsiveness to soil and atmospheric drought). We found that in all tested material carrying the stay-green CaStGR1 gene (germplasm or cultivated crop types) we did not observe any association between allelic variation and plant responsiveness to soil/atmospheric drought which would have impacted crop production in dry environments. In the cultivated plant types, we found that CaStGR1-2 stay-green ILs inherited the level of environmental adaptations from the cultivated parent rather than from the donor of this stay-green allele (ICC 16340). In some particular cases, the level of adaptive features was even more pronounced than in the cultivated recurrent parent (JGK-1 and derived ILs; Figure 4b). We speculate that this "transgressive segregation" could have been, at least partially, driven by the higher capacity to grow and expand canopy of ILs originated from this cross (Figure 7b; see [65]).

3.3.2. Plant Water-Use Related Traits and Agronomic Performance

Crop functions linked to quantity and efficiency of water utilization (e.g., see above) determines its agronomic performance, especially in environments limited by the water availability [10,66,67]. As discussed above we showed that CaStGR1 allelic variation does not appear to affect the relatively simple plant functions which were previously documented to influence crop adaptations to dry environments [2,59,68]. However, since crop yield is a very complex trait, we have also tested the CaStGR1 allelic variants in the systems relevant for evaluation of crop agronomic characteristics (i.e., lysimteric system and field).

We found there were significant differences in grain and biomass yield in germplasm when tested under different irrigation regimes but none of the differences seemed to coincide with the presence of disrupted CaStGR1 allele (CaStGR1-1 to CaStGR1-5). These differences in germplasm production characteristics were mostly explained by the differences in phenological development. In the stay-green CaStGR-1-2 ILs derived on cultivated background, we found significant genotypic differences in the main production parameters with the recurrent parents attaining generally higher production (example on Figure 8a). Nevertheless, in each of these experiments there were at least few ILs in the genetic background of each of the two elite cultivars whose production was comparable to the elite recurrent parents under WW and WS treatment (which ILs were consistent). Interestingly, under WS treatment, yield of some ILs was similar to that of their respective recurrent parents despite the phenological development of these ILs was generally several days longer (~14 days). Further, we found that the relation between seed yield and flowering time was much looser than that of the germplasm (as in [69,70])-especially under WS where this correlation was hardly significant (e.g., Supplementary Figure S6). However, we found that the majority of variation in grain yield and yield components within this material was explained by TE, especially under WS (Figure 6a,b). We can speculate that higher TE in some of the tested ILs could have been the consequence of lower TR and increased transpiration responsiveness of some ILs to VPD (see above and Figure 4b). We can further speculate that the enhanced TE of some tested ILs could be a consequence of yet to be determined mechanisms induced by portions or interactions of genome remaining from the donor genotype since the recurrent background of IL material was not completely recovered at BC4-5:F2 (i.e., ~94–97% of recurrent background recovered).

Collectively these data indicate that across the range of tested conditions there is no significant trade-off between elevated carotenoid content and agronomic productivity. Yields were similar between lines with "wild type" CaStGR1 (with yellow cotyledons) and genotypes with loss-of-function alleles in the CaStGR1 gene (with green cotyledon and delayed degreening phenotypes).

3.3.3. Stay-Green Alleles Extend Retention of Chlorophyll and Provitaminogenic Carotenoids in Grains and Leaves

Several stay-green plant phenotypes have been described in different crops [52]. The common denominator of "stay-green" phenotype can be described as a plant's capacity to remain green (i.e., maintain chlorophylls) in particular circumstances (reviewed [29,71,72]). In general, we can consider two basic stay-green types; "cosmetic" and "functional". Cosmetic stay-green is underlined by any mechanism that avoids chlorophylls to degrade—therefore the plant tissues appear green even if desiccated. Functional stay-green is a consequence of plants ability to manage resources during the crop cycle (e.g., water and nitrogen; [8,25,27,28,73,74]).

We present evidence that the green-seeded chickpea material is of a "cosmetic" type and depended on the presence of disrupted CaStGR1 gene, an ortholog of Mendel's I locus of garden pea (see above), that affects the function of chlorophyll degrading enzyme [48] and resulted in retention of chlorophylls in dried plant tissues (grain and leaf). We were further interested in addressing whether the composition of chlorophylls *a* and *b* and the functionally related pigments (carotenoids) differed among plant tissues (grain and leaves) during a range of circumstances (irrigated and water stress).

Consistently, we found that the levels and the composition of pigments did not significantly differ between genotypes carrying disrupted CaStGR1 gene (allele 1–5) and wild-type under irrigated and even under water stress conditions (probably because for this estimation only the leaves from the top of the plants which still remained green even in wild-type were sampled). Nevertheless, we found that all stay-green genotypes, in general, maintained higher level of pigments in matured grains compared to wild-type in irrigated conditions (similarly in [30]). The pigments in the grain were not significantly affected by the conditions of cultivation (WW and WS) across the range of material tested and the grains produced by plants exhibiting stay-green phenotype had all 10–100% higher chlorophyll and total carotenoids contents compared to the respective wild-type checks (similarly in [30,75]). Further dissection indicated the stay-green ILs contained two to three fold higher levels of specific A-provitaminogenic carotenoids (beta-carotene) resembling or exceeding the levels achieved by "golden-rice" technology [39,76].

Additional studies are required to determine the extent to which these elevated levels of carotenoids translate into enhanced bioavailability of vitamin A for humans, factors influencing consumer acceptance of green cotyledon colored chickpeas as dry grains, and if green cotyledon chickpea may be associated with conditionally-reduced seed germination or seedling establishment as has been observed in some other crop legumes.

4. Materials and Methods

4.1. Plant Material: Chickpea Germplasm and Breeding Lines

Chickpea genotypes with the common yellow cotyledon color and those with the infrequently occurring green cotyledon color were obtained from gene banks (USDA GRIN in Pullman, Washington, and ICRISAT India) and from chickpea improvement programs (detailed in Supplementary Table S1). In the process of plant grow outs for seed multiplication, the gene bank accessions were visually screened for occurrence and confirmation of green cotyledon color in mature dry seeds. Furthermore, during such grow outs we examined degreening of leaves of this germplasm accessions using a detached leaf assay, wherein leaves were wrapped in aluminum foil (to block out light and trigger degreening) and the pigment loss/retention capacity ("degreening") assayed 5–10 days later. The same plants were tested for sequence polymorphism in the CaStGR1 gene (see below). In initial germplasm screen, eight lines with green cotyledon color representing four different allelic variants in the chickpea stay green candidate gene and two yellow cotyledon genotypes carrying the wild-type alleles were used in physiological studies (Supplementary Table S1). In these initial studies, as expected [30] elevated levels of total carotenoids among green cotyledon color lines relative to concurrently grown normal yellow chickpea lines was observed.

For the subsequent and more detailed analyses, breeding lines with contrasting yellow and green cotyledon color were used (Supplementary Table S1). These lines were derived from introgression of the green cotyledon trait from the germplasm accession ICC 16,340 into two Indian elite chickpea cultivars, JGK1 and KAK2 with yellow cotyledon colors. 25 BC4-5:F2 generation introgression lines and their parents were screened for phenology and agronomic traits. Based on homogeneous phenology (flowering time, duration of flowering) and agronomical traits (harvest index), genotypes were selected for further studies (Supplementary Table S3) details of genotypes used in different experiments).

4.2. Molecular Characterization of Candidate Gene and Genome:

The genotypes tested for variation in CaStGR1 allele are shown in S1 table. In these, the genomic DNA was extracted from the young leaflets using QIAGEN DNeasy Plant Kit following the manufacturer's recommended procedures, or from seed-derived cotyledon tissue (for the cultivar 'CDC Verano') using a phenol-chloroform based extraction protocol. PCR amplification for CaStGR1 were performed with ExTaq polymerase (Takara-Fisher) using oligonucleotide primers as detailed in Supplementary Table S2. PCR products were analyzed in 1% agarose gel electrophoresis. For Sanger sequencing, PCR amplicons were purified with ExoSAP kit (Affymetrix, Santa Clara, CA, USA) to remove any excess salts carried over from PCR reactions. Amplicons were Sanger sequenced using single primers at on-campus core sequencing facilities at the University of California and the University of Vermont. Chromatogram traces from amplicon sequencing were analyzed with the Sequencher (Gene Codes Corporation, Ann Arbor, MI USA) and Geneious 2019.1 software packages. Sanger sequence traces were curated manually to identify and verify the positions of variant nucleotides in sequencing data. Variants supported by at least two independently run sequencing reaction were recorded and used for enumerating allele distribution and frequencies.

Preparation of whole genomic libraries for Illumina sequencing and data analysis of Illumina short read data were as described previously [65]. Illumina reads were mapped to the *C. arietinum* 'CDC Frontier' reference genome assembly [50] using BWA MEM 0.7.9a-r786. Visualization of CaStGR1 and its flanking regions was done using an instance of GBrowse loaded with gene structural annotation available from the CDC Frontier reference.

For genotyping of the CaStGR1-1 allele as a CAPS marker, PCR products were digested with Hpy-188I restriction enzyme (New England Biolabs, USA) per manufacturer's recommended protocol. Digested PCR products were analyzed by gel electrophoresis in 1.35% agarose gels in 0.5× Tris Borate EDTA buffer stained with cybersafe reagent. Genotyping of the CaStGR1-4 allele in F2 population of wild type (yellow cotyledon) genotypes and green-cotyledon lines was conducted as a customized KASP assay (LGC Genomics, UK) using leaf tissue from greenhouse grown plants and oligos listed in Supplementary File S2.

4.3. Plant Growth Conditions for Physiological Assays (Experiments Listed in Table 3)

4.3.1. Experiments Conducted in Glass-House (Experiment 1, 2, 3a and b)

The glass-house environment was used to evaluate crop responsiveness to soil and atmospheric drought. In Experiments 1 and 2 (Supplementary Table S4), plants were grown in 8″ plastic pots filled with 5 kg of vertisol while for experiment 3a and b, plants were raised in PVC cylinders filled with 45 kg of vertisol. The experiments were set-up using completely randomized block design with treatments as separate blocks. The black soil (Vertisol) was collected from the ICRISAT farm and fertilized with DAP (di-ammonium phosphate) at the rate of 0.3 g per kg of soil in all experiments. Seeds were treated with fungicides (Thiram®; Sudhama Chemicals Pvt. Ltd. Gujarat, India) to avoid fungal contamination. Four seeds were sown in each pot, and a rhizobium inoculum (Strain No: IC 2002) was added to each pots to ensure adequate nodulation. Two weeks after sowing, plants were thinned to two plants per pot. Plants were maintained well-watered up to ~30 days after sowing. During the experiments duration, a data logger (Lascar Electronics Inc. Whiteparish, UK) was positioned within the plant canopy for the hourly recording of the air temperature and relative humidity (RH%) and these oscillated on average between 28–22 °C and 70–90% during the day–night cycle.

4.3.2. Experiments Conducted at LeasyField (Experiment 3c)

The Lysimetric facility located at International Crops Research Institute for the Semi-Arid Tropics (ICRISAT) Patancheru in India (17°30′N; 78°16′E; altitude 549 m). It offers an experimental setup to evaluate the basic crop agronomic features, monitor the crop capacity to convert water into biomass (g of dry mass per unit of water transpired) and to measure water use patterns during the cropping season. Plants were grown in lysimeters constructed from the PVC plumbing pipes with 20 cm diameter and 1.2 m length outdoors under a rain-out shelter (ROS) (Experiment 3c). The protocol for lysimeter soil preparation & filling, spacing arrangement, growing and weighing plants were followed according to [69,70] and [60,77]. Three seeds were sown in each cylinder and watered regularly and around 15 DAS thinned to one seedling per cylinder. The experiment was planned in a complete randomized block design. One block was assigned to a well-watered treatment (WW) and two blocks to water-stressed treatment (WS). The WS treatment was imposed by cessation of watering from 25 Days after sowing (DAS). WW plants were watered every week to maintain 80% field capacity until maturity. During the experiment's duration, the data logger was positioned within the plant canopy to record the day and night temperatures and relative humidity (RH%), which fluctuated under the natural day–night oscillations around average 31.7/15.5 °C and 40/85%.

4.3.3. Experiments Conducted at LeasyScan (Experiment 4)

LeasyScan is a high throughput phenotyping platform constructed to monitor crop canopy related parameters during the vegetative phase of development with high throughput and accuracy. Details of LeasyScan technology and set-up are elaborated in [61,62,64]. For experiment 4 the crop was raised in large trays (60 × 40cm, approximately 75 kg of vertisol; i.e., "miniplots") filled with vertisol using the recommended field management practices (20 kg·ha^{-1} of DAP and planting densities of 32 plants m^{-2}). The experimental design was an Alpha lattice with 4 replications to account for spatial variability. Plants were maintained under well water conditions throughout the experiment. Canopy size related parameters (i.e., 3D-Leaf area, digital biomass and leaf area index) were continuously measured from 15-40 DAS when the plants were harvested. During the crop grown period the daily temperature and humidity oscillated in between of 11/35.8 °C and 17.2/93.2% on average as per the records of the attached weather station (Model: WxPRO™; Campbell Scientific Ltd., Shepshed, UK).

4.3.4. Experiments Conducted in Field (Experiment 5)

The main crop agronomic features were measured in the field experiment that was planted in post-rainy 2017–18 season at ICRISAT field facilities. The field was solarized using a polyethene mulch during the preceding summer primarily to avoid the crop infection by *Fusarium oxysporum* f. sp, [78]. The basal dose of di-ammonium phosphate at the rate of 18kg N ha^{-1} and 20kg P ha^{-1} was applied before sowing. The field was prepared as broad bed and furrows with 1.2m wide beds flanked by 0.3m furrows. Within these beds, the plots of 4 rows of 4 m length were planted. Seeds were treated with Thiram® (Sudhama Chemicals Pvt. Ltd. Gujarat, India) to avoid fungal contamination during germination. The seeds were hand sown at a depth of 2–3 cm maintaining a row-to-row distance of 30 cm and a plant to plant distance of 10 cm (i.e., 33 plants m^{-2}). After sowing, furrow irrigation (60 mm) was given to ensure uniform seedling emergence. Subsequently, plants were grown under different irrigation regimes: water stress [WS; crop received only ~60 mm at the sowing], and well water [WW; crop received ~60mm at the sowing and additional ~20 mm irrigation every 20 days through perforated irrigation system]. The plots were kept weed-free by hand weeding and intensive protection was taken against pod borer (*Helicoverpa armigera*). The experiment was conducted in a randomized complete block design with three replications for each treatment (WW/WS).

4.4. Physiological Assays

4.4.1. Experiments to Test Plant Responsiveness to Soil Drought (Experiment 1b and 1c)

The main aim of "dry-down" experiments is to assess the capacity of genotypes to restrict the transpiration upon declining soil moisture, which could be a crucial adaptive trait for plants in particular water-limited environments. To test the transpiration restriction capacity of selected genotypes, these were organized in two experimental blocks; well-watered (WW) and water-stressed (WS) conditions. The day before the dry-down was initiated all pots were abundantly watered and the soil was allowed to drain overnight. The following day the soil surface of the pots were covered with plastic sheets, and then a uniform 2 cm layer of plastic beads to prevent soil evaporation. The pots were then weighed and this initial pot weight was considered as the soil-saturation level (field capacity) of the individual pots. Pot weight was recorded daily at the same time of day. Based on the daily weight loss, the well-watered plants were maintained at approximately 80% of the saturated weight (80% of the field capacity). For the WS treatment, the water available to the plant was gradually decreased by allowing a maximum daily water loss of 70g. The transpiration weight loss above 70g was compensated by adding an excess amount of transpired water to each pot. The experiment was terminated when transpiration of all WS plants was below 10% of their WW treated counterparts. After termination, the above-ground biomass of the plants was harvested, organs separated, and oven-dried at 60 °C for a minimum of 3 days. The traits assessed are detailed in Supplementary Table S4.

Additionally, during the dry-down experiments (in Experiment 1b and 1c), 30 mg leaf tissue (leaflets from the first fully developed leaf from the top of the main stem) from each replicate (i.e., in WW and WS) were collected twice WW and severe water stress (~0.25 NTR). Collected tissues were frozen by liquid nitrogen and conserved for later estimation of pigments (i.e., Chlorophylls and Carotenoids, see below). (http://gems.icrisat.org/allinstruments/controlled-imposition-of-water-stress/; methodology also used in e.g., [79–81])

4.4.2. Experiments to Test Plant Responsiveness to Atmospheric Drought (Experiment 2a,b,c)

While "dry-down" experiments (above, experiment 1b and c) were conducted to evaluate plant responsiveness to drying soil, complementary "transpiration responsiveness" experiments were designed to characterize the genotypic ability to limit transpiration upon drying atmosphere [increasing vapour pressure deficit (VPD)]. For this, the plants were evaluated during vegetative growth stage under well-watered conditions. Around 30-day-old plants grown in pots were watered to ~90% field capacity and soil evaporation minimized by applying the plastic sheets and beads similarly as in the regulated dry down experiment (above). Initially, the plant transpiration was evaluated outdoors during the cloud-less clear days in the natural circadian cycle or in the growth chambers (Conviron-PGW36 model, Controlled Environments Limited, Winnipeg Manitoba, Canada: see more details in http://www.conviron.com/sites/default/files/PGW36%20Data%20Sheet_1.pdf). In these experiments, temperature and humidity sensors were mounted at canopy level to record the actual conditions experienced by the crop canopy in 5 min intervals. In the outdoors conditions, plants were weighted in hourly intervals using 0.01 g precision scales (KERN 24100, Kern & Sohn GmbH, Balingen, Germany). Consequently, for the controlled environment testing, the same pots were placed into the growth chamber for one day to acclimate with the day/night temperature (°C) and relative humidity (RH%) of 32/26 °C and 60/80% respectively. Plants were then exposed to an increasing ladder of VPD ranging from 0.9 to 4.1 kPa by increasing temperature and decreasing RH% (80–30%) at hourly intervals for 8 h. Plant transpiration was also assessed hourly by swift weighing in between of the VPD transitioning regimes. At the end of the experiments, plants were harvested and leaf area (LA) was measured with a leaf area meter (LI-3100C area meter, LI-COR®Biosciences, Lincoln, NE, USA). Consequently, the plant transpiration rate was expressed as TR = T/LA [g of water transpired per unit of LA per hour] and regressed upon VPD during the particular time interval. In both germplasm and ILs (Experiment 2a and 2c), the specific leaf weight (SLW) was estimated as leaf dry weight (g)/leaf area (cm^{-2}).

(http://gems.icrisat.org/allinstruments/transpiration-response-to-increasing-vpd/; methodology also used in e.g., [61,62,79,80])

4.4.3. Experiments to Test Plant Baseline Agronomic Features and Water-Use Related Traits in Lysimetric Facility (Experiment 3a,b,c)

The unique lysimetric set-up allows estimating the plant water productivity while having access to relevant agronomic traits. The cylinders were covered with plastic sheets and beads similarly as in assay #1 and 2 and the water use monitoring started ~25 DAS. From this onwards, the cylinders were weighed weekly by lifting them with a block chained pulley using S-type load cell (Mettler-Toledo, CSE 100, Geneva, Switzerland) until crop maturity. The WW block of experimental plants was retained at 80% of field capacity. Under the WS treatment, the declining soil moisture was only monitored but not regulated, which contrasts with the regulated dry-down protocol used in the pot culture (see above #1). During the plant growth flowering dates were recorded for each plant. At the end of the experiment, plants were harvested, the crop residuals dried at 60 °C in an oven during minimum 72 h and the above ground biomass, grain and vegetative dry biomass were weighed (KERN 3600 g; 0.01 g precision balance, Kern & Sohn GmbH, Balingen, Germany). Plant transpiration was calculated from consecutive cylinder weight differences and water additions. Transpiration efficiency (TE; [gram of biomass per kilogram of water transpired; g/kg^{-1}]) and water use efficiency (WUE, [gram of seed weight per kilogram of water transpired; g/kg^{-1}]) was then calculated as the ratio of the total/grain dry biomass per unit of water transpired. Lastly, Harvest Index (HI) was calculated as the ratio of total dry grain biomass per the total dry weight of remaining above-ground biomass. (http://gems.icrisat.org/allinstruments/lysimetric-assessments/, methodology also used in e.g., [60,69,70,82,83]).

4.4.4. Experiments to Assess Plant Canopy at LeasyScan (Experiment 4)

The LeasyScan platform has been used to monitor traits indicating crop canopy traits related to "vigor". This is enabled by the optical system (PlantEye®; www.phenospex.com), which captures the dynamics of canopy growth during the crop vegetative growth-phase with high throughput and accuracy. We measured 3D-Leaf area (3D-L; canopy size reconstructed from 3D point-cloud distribution [mm^3]), projected leaf area (PL; canopy ground coverage [mm^2]) and plant height (PH; estimated from 3D point-cloud as height encompassing 95% of recorded points of given point-cloud) during 15-30 DAS (http://gems.icrisat.org/leasyscan/) methodology also used in e.g., [4,61,64]).

4.4.5. Agronomic Evaluation of ILs in Field Settings (Experiment 5)

Agronomic traits of selected stay-green introgression lines and their recurrent parents were evaluated using the precision field facility under optimal water input (WW) and under severe water shortage WS. Under both treatments, in each plot we monitored the phenology parameters (date to first flower, 50% flowering and 80% of the dried pods was recorded as maturity). At maturity, shoots were harvested plot wise and kept for drying at 60 °C for minimum of 3 days. Organs were separated, dry weights recorded and expressed in grams per meter square (g m^{-2}). 100 seed number was counted by seed counter (Data Count S60 seed Counter, Data technologies, Israel; http://www.datatechnologies.com/data_count_s_60_seed_counter.html), weighed and based on these the total seed number per square meter was calculated.

Harvest index was calculated:
$$HI = (Seed\ weight/total\ shoot\ biomass\ weight) \times 100\ [\%]. \tag{1}$$

4.5. Chlorophyll and Carotenoid Estimation in Leaves and Seeds (Measured in Experiment 1 and 3)

Photosynthetic pigment contents (chlorophyll a, chlorophyll b and total carotenoids) were assessed in the leaf tissues across various stages of plant exposure to declining soil content in lysimeters (un-regulated dry-down; Experiment 3) and in pot cultures (regulated dry-down; Experiment 1b and 1c). The grain pigments were assessed only in the experiments conducted at lysimetric experiments (Experiment 3a,b,c).

In Experiment 3c the leaf tissue samples were collected from each plant from the glasshouse lysimetric experiment. Chlorophyll a and b, as well as Carotenoids, were estimated from the samples using dimethyl sulfoxide (DMSO) method [84]. We standardized that around 18 mg of fresh leaf tissue/30mg of dry-seed powder extracted in a 5mL of DMSO resulted in suitable optical density (OD) between 0.3–0.9. The test-tubes with the exact weighted tissue and DMSO were placed in ~65 °C hot water bath and left for cca 3 h until the tissue became translucent ensuring all pigments were extracted into to the DMSO. The OD of extract was assessed spectrophotometrically (Shimadzu UV-2401 PC UV-Visible Spectrometer; Shimadzu Scientific Instruments) at 665.1 (Chlorophyll A), 649.1 (Chlorophyll B) and 480 (Total Carotenoids) and the contents were calculated as per [84].

The grain material from Experiment 3b was used to separate the main carotenoids using the High Performance Liquid Chromatography (HPLC) system. For this, the extraction of carotenoids was done according to the method of [85] with some modifications. Briefly, about 0.1 g of chickpea sample was weighed and placed in a screw-capped glass tube (~15 mL tube) and 1 mL ethanol containing 0.1% butylated hydroxytoluene (BHT) added to the solution. The mixture was saponified by adding 200 μL of 20% Potassium hydroxide (KOH) and mixed by vortexing. Extraction was completed by adding 1.5 mL hexane to the saponified solution, vortexed for 20 s and centrifuged at 2500 rpm for 5 min. Using a glass pipette, the upper hexane layer containing carotenoids was carefully removed and transferred to a new glass tube. Extraction was repeated 2 more times. The combined hexane extracts were then dried down under a stream of nitrogen gas. Purified β-apo-8′-carotenal was used (absorbance ~ 0.8; 100 μL) was used as an internal standard. The dried extract was reconstituted in 100 μL of 50:50 (*v/v*) methanol:dichloroethane and 10 μL of the sample injected into the HPLC system (duplicate injections per sample).

Chromatographic separation of carotenoids was carried out using the Ultra-Fast Prominence Liquid chromatography (Shimadzu, Kyoto, Japan) equipped with a SIL-20ac-xr Prominence auto-sampler, a DGU-20A5 Prominence degasser, a CTO-20AC column oven and an SPD-M20A Diode Array Detector (DAD). Separation of carotenoids was achieved at 25 °C on a C30 YMC carotenoid column (250 × 4.6 mm, i.d., 5 μm particle size, Waters, Ireland) on a gradient method with 95% Methanol as solvent A and 100% MTBE as solvent B. Identification of the carotenoids was based on the standards, their retention times and by comparing the absorption spectra with those in the literature. Quantification of the carotenoids were extrapolated from standard curves prepared from authentic standards after correcting for extraction efficiency based on the recovery of the internal standard. The processing of all chromatograms was done using Shimadzu LC Lab-Solutions software (also used in [26,84,85]).

4.6. Statistical Analysis

In the experiments 1b, 1c, 2a, 2c, 3a, 3b, 3c, 4 and 5, the differences between investigated genotypes were evaluated by simple/multiple-way ANOVA followed by the Tukey–Kramer test to evaluate the significance of genotypic differences (Statistical program package CoStat version 6.204 (Cohort Software, Monterey, CA, USA). The line graph (Experiment 2a, 4), bar graph (1b, 1c, 2a, 2c, 3a, 3b, 3c, 4 and 5) and simple linear regressions were fitted using Microsoft Excel 2013 (Microsoft Corp., Redmond, WA, USA). For treatment of temporal data from experiments 1b, 1c and experiments 2 a,c-i.e., transpiration response to atmospheric (Experiment 2a and 2c) and soil drought (Experiment 1b and 1c) we used methodologies described in [69,70,80,86,87]; specifically, a nonlinear regression analysis was done using GraphPad Prism version 6 (GraphPad Software Inc., San Diego, CA, USA), and Genstat 14.0 (VSN International Ltd., Hemel Hempstead, UK).

5. Conclusions

Chickpea production suffers greatly due to its cultivation predominantly as a rain-fed crop, particularly across developing countries. Significant progress has been made from crop agronomic practices and breeding to address the yield gap to ensure appropriate caloric intake of populations inhabiting these areas. Although caloric intake is slowly increasing, human nutrient deficiencies prevail in the same regions and remain largely unaddressed. Therefore, in this paper we tested the suitability of stay-green chickpea for cultivation in semiarid tropical regions, which as a genetic biofortification technology may help to reduce widespread vitamin-A deficiency while maintaining the levels of agronomic production. We tested a range of plant material with the stay-green character which was expressed as an extended maintenance of chlorophylls and carotenoids in dry seeds and leaves. We found this particular phenotype was controlled by variation in a single gene, CaStGR1, an ortholog of Mendel's I locus of garden pea, which occurred in 5 different allelic variants in the tested material. We also showed that across a range of environmental conditions the stay-green allelic variants were very likely neither influencing the mechanisms linked to drought stress adaptations nor negatively influencing important agronomic traits. Our evidence that the green-seeded CaStGR1 variants contain multiple-fold higher levels of the phytonutrients lutein, and provitamin A (beta-carotene) when compared to the more common yellow cotyledon chickpea indicate a higher nutritional value of the green cotyledon type. Further investigations of the bioavailability of vitamin A, multilocation trials for yield stability, and acceptability of the stay-green chickpea products in production regions by producers and consumers are warranted in order to establish the efficacy of genetic biofortification with stay-green chickpea for improving human nutrition and health.

Supplementary Materials: Supplementary materials can be found at http://www.mdpi.com/1422-0067/20/22/5562/s1.

Author Contributions: Conceptualization, V.V., C.B., J.K. and R.V.P.; Data curation, K.S., E.M., C.B., J.K. and R.V.P.; Formal analysis, K.S., E.M., C.B., J.K. and R.V.P.; Funding acquisition, S.M.U., V.V., G.J.V., D.R.C., E.J.B.v.W., J.K. and R.V.P.; Investigation, K.S., E.M., N.K., N.C.-G., P.L.C., E.M.B., E.B., M.C., S.G.A.S.S., S.M.U., I.A.R. and R.R.M.; Methodology, V.V., E.J.B.v.W., J.K. and R.V.P.; Project administration, V.V., J.K. and R.V.P.; Resources, G.J.V. and P.M.G.; Supervision, C.B., E.J.B.v.W., J.K. and R.V.P.; Visualization, K.S., P.L.C., and E.M.; Writing—original draft, K.S., E.M., E.J.B.v.W., J.K. and R.V.P.; Writing—review & editing, K.S., E.M., E.J.B.v.W., J.K. and R.V.P.

Funding: Supported by CGIAR Research Program on Grain Legumes and Dryland Cereals (CRP-GLDC) to J.K.; a cooperative agreement from the United States Agency for International Development under the Feed the Future Program AID-OAA-A-14-00008 to D.R.C., E.J.B.v.W., V.V. and R.V.P.; by a grant from the US National Science Foundation Plant Genome Program Award #IOS-1339346 to D.R.C., E.J.B.v.W., and R.V.P.; US National Institute of Food and Agriculture grant #2018-67013-27619 to J.K., E.J.B.v.W. and R.V.P.; and co-funded (award #OISE-16-62791-0, project #201603867) from the Civilian Research Development Foundation-Global to R.V.P. and (award # JRCAFS-23422) from the International Center for Biosaline Agriculture to S.U. Northeast SARE graduate research grant GNE-179 to Edward Marques. E.J.B.v.W. is further supported by the USDA Hatch program through the Vermont State Agricultural Experimental Station. Article Processing Charges (APC) paid by grant #2018-67013-27619 to J.K., E.J.B.v.W. and R.V.P.

Acknowledgments: We thank Antonia Palkovic, Paul Gepts, and members of the Gepts group (University of California, Davis, CA, USA) for their support of activities conducted at UC Davis, and Claire Coyne (USDA-ARS, Pullman, WA, USA) and Bunyamin Taran (University of Saskatchewan, SK, Canada) for the germplasm and breeding lines used in this study.

Conflicts of Interest: The authors declare no conflict of interest.

References

1. FAO. *Chickpea Value Chain Food Loss Analysis: Causes and Solutions*; FAO: Rome, Italy, 2017.
2. Hajjarpoor, A.; Vadez, V.; Soltani, A.; Gaur, P.; Whitbread, A.; Suresh Babu, D.; Kholová, J. Characterization of the main chickpea cropping systems in India using a yield gap analysis approach. *Field Crops Res.* **2018**, *223*, 93–104. [CrossRef]
3. FAOSTAT. *Food and Agricultural Commodities Production*; FAOSTAT: Rome, Italy, 2014.
4. Tharanya, M.; Kholova, J.; Sivasakthi, K.; Thirunalasundari, T.; Vadez, V. Pearl Millet. In *Water-Conservation Traits to Increase Crop Yields in Water-Deficit Environments: Case Studies*; Springer: Cham, Switherland, 2017; pp. 73–83.
5. Borrell, A.K.; Hammer, G.L. Nitrogen dynamics and the physiological basis of stay-green in Sorghum. *Crop Sci.* **2000**, *40*, 1295–1307. [CrossRef]
6. Burgess, M.G.; Rush, C.M.; Piccinni, G.; Schuster, G. Relationship between charcoal rot, the stay-green trait, and irrigation in grain sorghum. *Phytopathology* **2002**, *92*, S10.
7. Jordan, D.R.; Hunt, C.H.; Cruickshank, A.W.; Borrell, A.K.; Henzell, R.G. The relationship between the stay-green trait and grain yield in elite sorghum hybrids grown in a range of environments. *Crop Sci.* **2012**, *52*, 1153–1161. [CrossRef]
8. Kholová, J.; Tharanya, M.; Sivasakthi, K.; Srikanth, M.; Rekha, B.; Hammer, G.L.; McLean, G.; Deshpande, S.; Hash, C.T.; Craufurd, P.; et al. Modelling the effect of plant water use traits on yield and stay-green expression in sorghum. *Funct. Plant Biol.* **2014**, *1*, 1019–1034. [CrossRef]
9. McBee, G.G.; Waskom, R.M.; Miller, F.R.; Creelman, R.A. Effect of senescence and nonsenescence on carbohydrates in sorghum during late kernel maturity states. *Crop Sci.* **1983**, *23*, 372–376. [CrossRef]
10. Vadez, V.; Kholová, J.; Yadav, R.S.; Hash, C.T. Small temporal differences in water uptake among varieties of pearl millet (*Pennisetum glaucum* (L.) R Br) are critical for grain yield under terminal drought. *Plant Soil* **2013**, *371*, 447–462. [CrossRef]
11. Crafts-Brandner, S.J.; Below, F.E.; Wittenbach, V.A.; Harper, J.E.; Hageman, R.H. Differential Senescence of Maize Hybrids following Ear Removal. *Plant Physiol.* **1984**, *74*, 368–373. [CrossRef]
12. Gentinetta, E.; Ceppi, D.; Lepori, C.; Perico, G.; Motto, M.; Salamini, F. A major gene for delayed senescence in maize. Pattern of photosynthates accumulation and inheritance. *Plant Breed.* **1986**, *97*, 193–203. [CrossRef]
13. Rajcan, I.; Tollenaar, M. Source: Sink ratio and leaf senescence in maize. I. Dry matter accumulation and partitioning during grain filling. *Field Crops Res.* **1999**, *60*, 245–253. [CrossRef]
14. Zheng, H.J.; Wu, A.Z.; Zheng, C.; Dong, S.T. QTL mapping of maize (*Zea mays*) stay-green traits and their relationship to yield. *Plant Breed.* **2009**, *128*, 54–62. [CrossRef]
15. Adu, M.O.; Sparkes, D.L.; Parmar, A.; Yawson, D.O.; Science, B. 'Stay Green' in Wheat: Comparative Study of Modern Bread Wheat and Ancient Wheat Cultivars. *J. Agric. Biol. Sci.* **2011**, *6*, 16–24.
16. Bogard, M.; Jourdan, M.; Allard, V.; Martre, P.; Perretant, M.R.; Ravel, C.; Heumez, E.; Orford, S.; Snape, J.; Gaju, O.; et al. Anthesis date mainly explained correlations between post-anthesis leaf senescence, grain yield, and grain protein concentration in a winter wheat population segregating for flowering time QTLs. *J. Exp. Bot.* **2011**, *62*, 3621–3636. [CrossRef] [PubMed]
17. De Souza Luche, H.; da Silva, J.A.G., II; da Maia, L.C., III; de Oliveira, A.C., III. Stay-green: A potentiality in plant breeding. Stay-green: A potentiality in plant breeding Stay-green: Uma potencialidade no melhoramento genético de plantas. *J. Crop Prod.* **2015**, *45*, 1755–1760.
18. Christopher, J.T.; Manschadi, A.M.; Hammer, G.L.; Borrell, A.K. Developmental and physiological traits associated with high yield and stay-green phenotype in wheat. *Aust. J. Agric. Res.* **2008**, *59*, 354–364. [CrossRef]

19. Lopes, M.S.; Reynolds, M.P. Stay-green in spring wheat can be determined by spectral reflectance measurements (normalized difference vegetation index) independently from phenology. *J. Exp. Bot.* **2012**, *63*, 3789–3798. [CrossRef]

20. Fu, J.-D.; Yan, Y.-F.; Lee, B.-W. Physiological characteristics of a functional stay-green rice "SNU-SG1" during grain-filling period. *J. Crop Sci. Biotechnol.* **2009**, *12*, 47–52. [CrossRef]

21. Hoang, T.B.; Kobata, T. Stay-green in rice (*Oryza sativa* L.) of drought-prone areas in desiccated soils. *Plant Prod. Sci.* **2009**, *12*, 397–408. [CrossRef]

22. Mondal, W.A.; Dey, B.B.; Choudhuri, M.A. Proline accumulation as a reliable indicator of monocarpic senescence in rice cultivars. *Experientia* **1985**, *41*, 346–348. [CrossRef]

23. Wada, Y.; Wada, G. Varietal difference in leaf senescence during ripening period of advanced indica rice. *Jap. J. Crop Sci.* **1991**, *60*, 529–553. [CrossRef]

24. Borrell, A.K.; Mullet, J.E.; George-Jaeggli, B.; van Oosterom, E.J.; Hammer, G.L.; Klein, P.E.; Jordan, D.R. Drought adaptation of stay-green sorghum is associated with canopy development, leaf anatomy, root growth, and water uptake. *J. Exp. Bot.* **2014**, *65*, 6251–6263. [CrossRef]

25. Borrell, A.K.; van Oosterom, E.J.; Mullet, J.E.; George-Jaeggli, B.; Jordan, D.R.; Klein, P.E.; Hammer, G.L. Stay-green alleles individually enhance grain yield in sorghum under drought by modifying canopy development and water uptake patterns. *New Phytol.* **2014**, *203*, 817–830. [CrossRef] [PubMed]

26. Kholová, J.; McLean, G.; Vadez, V.; Craufurd, P.; Hammer, G.L. Drought stress characterization of post-rainy season (*rabi*) sorghum in India. *Field Crops Res.* **2013**, *141*, 38–46. [CrossRef]

27. Van Oosterom, E.J.; Borrell, A.K.; Chapman, S.C.; Broad, I.J.; Hammer, G.L. Functional dynamics of the nitrogen balance of sorghum. I. N demand of vegetative plant parts. *Field Crop Res.* **2010**, *115*, 19–28. [CrossRef]

28. Van Oosterom, E.J.; Chapman, S.C.; Borrell, A.K.; Broad, I.J.; Hammer, G.L. Functional dynamics of the nitrogen balance of sorghum. II. Grain filling period. *Field Crops Res.* **2010**, *115*, 29–38. [CrossRef]

29. Thomas, H.; Ougham, H. The stay-green trait. *J. Exp. Bot.* **2014**, *65*, 3889–3900. [CrossRef]

30. Ashokkumar, K.; Tar'an, B.; Diapari, M.; Arganosa, G.; Warkentin, T.D. Effect of Cultivar and Environment on Carotenoid Profile of Pea and Chickpea. *Crop Sci.* **2014**, *54*, 2225–2235. [CrossRef]

31. Hortensteiner, S. Stay-green regulates chlorophyll and chlorophyll-binding protein degradation during senescence. *Trends Plant Sci.* **2009**, *14*, 155–162. [CrossRef]

32. Segev, A.; Badani, H.; Kapulnik, Y.; Shomer, I.; Oren-Shamir, M.; Galili, S. Determination of Polyphenols, Flavonoids, and Antioxidant Capacity in Colored Chickpea (*Cicer arietinum* L.). *J. Food Sci.* **2010**, *75*, 115–119. [CrossRef]

33. Zhou, C.; Han, L.; Pislariu, C.; Nakashima, J.; Fu, C.; Jiang, Q.; Quan, L.; Blancaflor, E.B.; Tang, Y.; Bouton, J.H.; et al. From model to crop: Functional analysis of a STAY-GREEN gene in the model legume Medicago truncatula and effective use of the gene for alfalfa improvement. *Plant Physiol.* **2011**, *157*, 483–1496. [CrossRef]

34. Luo, Z.; Zhang, J.; Li, J.; Yang, C.; Wang, T.; Ouyang, B.; Li, H.; Giovannoni, J.; Ye, Z. A STAY-GREEN protein SlSGR1 regulates lycopene and β-carotene accumulation by interacting directly with SlPSY1 during ripening processes in tomato. *New Phytol.* **2013**, *198*, 442–452. [CrossRef] [PubMed]

35. Meier, S.; Tzfadia, O.; Vallabhaneni, R.; Gehring, C.; Wurtzel, E.T. A transcriptional analysis of carotenoid, chlorophyll and plastidial isoprenoid biosynthesis genes during development and osmotic stress responses in Arabidopsis thaliana. *BMC Syst. Biol.* **2011**, *5*, 77. [CrossRef] [PubMed]

36. Mur, L.A.J.; Aubry, S.; Mondhe, M.; Kingston-Smith, A.; Gallagher, J.; Timms-Taravella, E.; James, C.; Papp, I.; Hörtensteiner, S.; Thomas, H.; et al. Accumulation of chlorophyll catabolites photosensitizes the hypersensitive response elicited by Pseudomonas syringae in Arabidopsis. *New Phytol.* **2010**, *188*, 161–174. [CrossRef] [PubMed]

37. Ross, C. Vitamin A. In *Encyclopedia of Dietary Supplements*, 2nd ed.; Informa Healthcare: London, UK; New York, NY, USA, 2010; pp. 778–791.

38. Ross, A. Vitamin A and Carotenoids. In *Modern Nutrition in Health and Disease*, 10th ed.; Lippincott Williams & Wilkins: Baltimore, MD, USA, 2006; pp. 351–375.

39. Rezaei, M.K.; Deokar, A.; Arganosa, G.; Roorkiwal, M.; Pandey, S.K.; Warrkentin, T.D.; Varshney, R.K.; Tar'an, B. Mapping Quantitative Trait Loci for Carotenoid Concentration in Three F2 Populations of Chickpea. *Plant Genome* **2019**, *12*, 1–12.

40. Date, S.G. The Study of the Effect of Various Methods of Harvesting and Curing on the Color and Viability of Lima Bean Seed. Master's Thesis, Utah State University, Logan, UT, USA, 1962.

41. Davis, J.; Myers, J.R.; McClean, P.; Lee, R. Staygreen (sgr), a candidate gene for the Persistent Color phenotype in common bean. *Acta Hortic.* **2010**, *859*, 99–102. [CrossRef]

42. Fery, R.L.; Dukes, P.D. Genetic analysis of the green cotyledon trait in southern pea (*Vigna unguiculata* (L.) Walp.). *J. Am. Soc. Hortic. Sci.* **1994**, *119*, 1054–1056. [CrossRef]

43. Mae, T.; Ohira, K. Origin of the nitrogen a growing rice leaf and its relation to nitrogen nutrition. *Jpn. J. Soil Sci. Plant Nutr.* **1983**, *54*, 401–405.

44. Morita, K. Release of nitrogen from chloroplasts during senescence in rice (*Oryza sativa* L.). *Ann. Bot.* **1980**, *46*, 297–302. [CrossRef]

45. Sage, R.F.; Pearcy, R.W. The Nitrogen Use Efficiency of C_3 and C_4 Plants: I. Leaf Nitrogen, Growth, and Biomass Partitioning in *Chenopodium album* (L.) and *Amaranthus retroflexus* (L.). *Plant Physiol.* **1987**, *84*, 954–958. [CrossRef]

46. Tucker, C.L. Inheritance of white and green seed coat colors in lima beans. *Proc. Amer. Soc. Hortic. Sci.* **1965**, *87*, 286–287.

47. Tucker, C.L.; Sanches, R.L.; Harding, J. Effects of a gene for cotyledon color in lima beans, Phaseolus lunatus. *Crop Sci.* **1967**, *87*, 262–263. [CrossRef]

48. Armstead, I.; Donnison, I.; Aubry, S.; Harper, J.; Hortensteiner, S.; James, C.; Mani, J.; Moffet, M.; Ougham, H.; Roberts, L.; et al. Cross-species identification of Mendel's I locus. *Science* **2007**, *315*, 73. [CrossRef] [PubMed]

49. Sato, Y.; Morita, R.; Nishimura, M.; Yamaguchi, H.; Kusaba, M. Mendel's green cotyledon gene encodes a positive regulator of the chlorophyll-degradation pathway. *Proc. Natl. Acad. Sci. USA* **2007**, *104*, 14169–14174. [CrossRef] [PubMed]

50. Varshney, R.K.; Song, C.; Saxena, R.K.; Azam, S.; Yu, S.; Sharpe, A.G.; Cannon, S.; Baek, J.; Rosen, B.D.; Tar'an, B.; et al. Draft genome sequence of chickpea (*Cicer arietinum*) provides a resource for trait improvement. *Nat. Biotechnol.* **2013**, *31*, 240–246.

51. Jiang, H.; Li, M.; Liang, N.; Yan, H.; Wei, Y.; Xu, X.; Liu, J.; Xu, Z.; Chen, F.; Wu, G. Molecular cloning and function analysis of the stay green gene in rice. *Plant J.* **2007**, *52*, 197–209. [CrossRef]

52. Myers, J.R.; Aljadi, M.; Brewer, L. The Importance of Cosmetic Stay Green in Specialty Crops. *Plant Breed. Rev.* **2018**, *219*. [CrossRef]

53. Sakuraba, Y.; Schelbert, S.; Park, S.Y.; Han, S.H.; Lee, B.D.; Andrès, C.B.; Paek, N.C. STAY-GREEN and chlorophyll catabolic enzymes interact at light-harvesting complex II for chlorophyll detoxification during leaf senescence in Arabidopsis. *Plant Cell* **2012**, *24*, 507–518. [CrossRef]

54. Muchero, W.; Roberts, P.A.; Diop, N.N.; Drabo, I.; Cisse, N.; Close, T.J.; Muranaka, S.; Boukar, O.; Ehlers, J.D. Genetic architecture of delayed senescence, biomass, and grain yield under drought stress in cowpea. *PLoS ONE* **2013**, *8*, e70041. [CrossRef]

55. Singh, U.; Pundir, R.P.S. Amino acid composition and protein content of chickpea and its wild relatives. *Int. Chickpea Newsl.* **1991**, *25*, 19–20.

56. Smartt, J. Vavilov's law of homologous serious and de nova crop plant domestication. *Biol. J Linnean Society.* **1990**, *39*, 27–38. [CrossRef]

57. Kholova, J.; Zindy, P.; Malayee, S.; Baddam, R.; Murugesan, T.; Kaliamoorthy, S.; Hash, C.T.; Votrubová, O.; Soukup, A.; Kocová, M.; et al. Component traits of plant water use are modulated by vapour pressure deficit in pearl millet (*Pennisetum glaucum*(L.) R.Br.). *Funct. Plant Biol.* **2016**, *43*, 423–437. [CrossRef]

58. Vadez, V.; Halilou, O.; Hissene, H.M.; Sibiry-Traore, P.; Sinclair, T.R.; Soltani, A. Mapping Water Stress Incidence and Intensity, Optimal Plant Populations, and Cultivar Duration for African Groundnut Productivity Enhancement. *Front. Plant Sci.* **2017**, *8*, 432. [CrossRef] [PubMed]

59. Vadez, V.; Kholová, J.; Medina, S.; Kakkera, A.; Anderberg, H. Transpiration efficiency: New insight on an old story. *J. Exp. Bot.* **2014**, *65*, 6141–6153. [CrossRef] [PubMed]

60. Vadez, V.; Ratnakumar, P. High transpiration efficiency increases pod yield under intermittent drought in dry and hot atmospheric conditions but less so under wetter and cooler conditions in groundnut (*Arachis hypogaea* (L.)). *Field Crops Res.* **2016**, *193*, 16–23. [CrossRef]

61. Sivasakthi, K.; Zaman-Allah, M.; Tharanya, M.; Kholova, J.; Thirunalasundari, T.; Vadez, V. Chickpea. In *Water-Conservation Traits to Increase Crop Yields in Water-Deficit Environments: Case Studies*; Springer: Cham, Switherland, 2017; pp. 35–45.

62. Tharanya, M.; Kholová, J.; Sivasakthi, K.; Seghal, D.; Hash, C.T.; Raj, B.; Srivastava, R.K.; Baddam, R.; Thirunalasundari, T.; Yadav, R.; et al. Quantitative trait loci (QTLs) for water use and crop production traits co-locate with major QTL for tolerance to water deficit in a fine-mapping population of pearl millet (*Pennisetum glaucum* L. R.Br.). *Theor. Appl. Genet.* **2018**, *131*, 1509. [CrossRef]

63. Tharanya, M.; Sivasakthi, K.; Gloria, B.; Kholova, J.; Thirunalasundari, T.; Vadez, V. Pearl millet [*Pennisetum glaucum* (L.) R. Br.] contrasting for the transpiration response to vapour pressure deficit also differ in their dependence on the symplastic and apoplastic water transport pathways. *Funct. Plant Biol.* **2018**, *45*. [CrossRef]

64. Vadez, V.; Kholová, J.; Hummel, G.; Zhokhavets, U.; Gupta, S.K.; Hash, C.T. LeasyScan: A novel concept combining 3D imaging and lysimetry for high-throughput phenotyping of traits controlling plant water budget. *J. Exp. Bot.* **2015**, *66*, 5581–5593. [CrossRef]

65. Von Wettberg, E.J.; Chang, P.L.; Greenspan, A.; Carrasquila-Garcia, N.; Basdemir, F.; Moenga, S.; Bedada, G.; Dacosta-Calheiros, E.; Moriuchi, K.S.; Balcha, L.; et al. Ecology and community genomics of an important crop wild relative as a prelude to agricultural innovation. *Nat. Commun.* **2018**, *9*, 649. [CrossRef]

66. Vadez, V.; Deshpande, S.; Kholova, J.; Ramu, P.; Hash, C.T. Molecular breeding for stay-green: Progress and challenges in sorghum. In *Genomic Applications to Crop Breeding: Vol. 2. Improvement for Abiotic Stress, Quality and Yield Improvement*; John Wiley & Sons, Inc.: Hoboken, NJ, USA, 2013; pp. 125–141.

67. Vadez, V.; Kholová, J.; Zaman-Allah, M.; Belko, N. Water: The most important "molecular" component of water stress tolerance research. *Funct. Plant Biol.* **2013**, *40*, 1310–1322. [CrossRef]

68. Vadez, V.; Rao, J.S.; Bhatnagar-Mathur, P.; Sharma, K.K. DREB1A promotes root development in deep soil layers and increases water extraction under water stress in groundnut. *Plant Biol.* **2013**, *15*, 45–52. [CrossRef]

69. Zaman-Allah, M.; Jenkinson, D.M.; Vadez, V. A conservative pattern of water use, rather than deep or profuse rooting, is critical for the terminal drought tolerance of chickpea. *J. Exp. Bot.* **2011**, *62*, 4239–4252. [CrossRef] [PubMed]

70. Zaman-Allah, M.; Jenkinson, D.M.; Vadez, V. Chickpea genotypes contrasting for seed yield under terminal drought stress in the field differ for traits related to the control of water use. *Funct. Plant Biol.* **2011**, *38*, 270–281. [CrossRef]

71. Balazadeh, S. Stay-Green not always stay-green. *Mol. Plant.* **2014**, *7*, 1264–1266. [CrossRef] [PubMed]

72. Thomas, H.; Howarth, C.M. Five ways to stay green. *J. Exp. Bot.* **2000**, *51*, 329–337. [CrossRef]

73. Borrell, A.K. Stay-green alleles individually enhance grain yield in sorghum. *New Phytol.* **2014**, *203*, 817–830. [CrossRef]

74. Kamal, N.M.; Serag, Y.; Gorafi, A.; Tsujimoto, H.; Ghanim, A.M.A. Stay-Green QTLs Response in Adaptation to Post-Flowering Drought Depends on the Drought Severity. *BioMed Res. Int.* **2018**, *2018*, 1–15. [CrossRef]

75. Abbo, S.; Bonfil, D.J.; Berkovitch, Z.; Reifen, R. Towards enhancing lutein concentration in chickpea, cultivar and management effects. *Plant Breed.* **2010**, *129*, 407–411. [CrossRef]

76. Rezaei, M.K.; Deokar, A.; Tar'an, B. Identification and Expression Analysis of Candidate Genes Involved in Carotenoid Biosynthesis in Chickpea Seeds. *Front. Plant Sci.* **2016**, *7*, 1867. [CrossRef]

77. Vadez, V.; Rao, J.S.; Kholova, J.; Krishnamurthy, L.; Kashiwagi, J.; Ratnakumar, P.; Sharma, K.K.; Bhatnagar-Mathur, P.; Basu, P.S. Roots research for legume tolerance to drought: Quo vadis? *J. Food Legumes* **2008**, *21*, 77–85.

78. Sharma, S.B.; Sahrawat, K.L.; Burford, J.R.; Rupela, O.P.; Kumar Rao, J.V.D.K.; Sithanantham, S. Effects of Soil Solarization on Pigeonpea and Chickpea. Research Bulletin No. 11. *Int. Crops Res. Inst. Semi-Arid Trop.* **1988**, *11*, 1–23.

79. Belko, N.; Zaman-allah, M.; Diop, N.N.; Cisse, N.; Zombre, G.; Ehlers, J.D.; Vadez, V. Restriction of transpiration rate under high vapour pressure deficit and non-limiting water conditions is important for terminal drought tolerance in cowpea. *Plant Biol.* **2012**, *15*, 304–316. [CrossRef] [PubMed]

80. Kholová, J.; Hash, C.T.; Kakkera, A.; Kočová, M.; Vadez, V. Constitutive water conserving mechanisms are correlated with the terminal drought tolerance of pearl millet [*Pennisetum glaucum* (L.) R. Br.]. *J. Exp. Bot.* **2010**, *61*, 369–377. [CrossRef] [PubMed]

81. Vadez, V.; Sinclair, T.R. Leaf ureide degradation and N_2 fixation tolerance to water deficit in soybean. *J. Exp. Bot.* **2001**, *52*, 153–159. [PubMed]

82. Vadez, V.; Deshpande, S.P.; Kholová, J.; Hammer, G.L.; Borrell, A.K.; Talwar, H.S.; Hash, C.T. Stay-green quantitative trait loci's effects on water extraction, transpiration efficiency and seed yield depend on recipient parent background. *Funct. Plant Biol.* **2011**, *38*, 553–566. [CrossRef]

83. Vadez, V.; Krishnamurthy, L.; Hash, C.T.; Upadhyaya, H.D.; Borrell, A.K. Yield, transpiration efficiency, and water-use variations and their interrelationships in the sorghum reference collection. *Crop Pasture Sci.* **2011**, *62*, 645–655. [CrossRef]

84. Wellburn, A.R. The Spectral Determination of Chlorophylls *a* and *b*, as well as Total Carotenoids, Using Various Solvents with Spectrophotometers of Different Resolution. *J. Plant Physiol.* **1994**, *144*, 3307–3313. [CrossRef]

85. Kurilich, A.C.; Juvik, J.A. Quantification of carotenoid and tocopherol antioxidants in Zea mays. *J. Agric. Food Chem.* **1999**, *47*, 1948–1955. [CrossRef]

86. Kholová, J.; Hash, C.T.; Kumar, P.L.; Yadav, S.R.; Kočová, M.; Vadez, V. Terminal drought-tolerant pearl millet [*Pennisetum glaucum* (L.) R. Br.] have high leaf ABA and limit transpiration at high vapor pressure deficit. *J. Exp. Bot.* **2010**, *61*, 1431–1440. [CrossRef]

87. Sivasakthi, K.; Thudi, M.; Tharanya, M.; Kale, S.M.; Kholová, J.; Halime, M.H.; Jaganathan, D.; Baddam, R.; Thirunalasundari, T.; Gaur, P.M.; et al. Plant vigour QTLs co-map with an earlier reported QTL hotspot for drought tolerance while water saving QTLs map in other regions of the chickpea genome. *BMC Plant Biol.* **2018**, *18*, 29. [CrossRef]

International Journal of
Molecular Sciences

Article

Reconstruction of the Evolutionary Histories of UGT Gene Superfamily in Legumes Clarifies the Functional Divergence of Duplicates in Specialized Metabolism

Panneerselvam Krishnamurthy [1,*,†], Chigen Tsukamoto [2] and Masao Ishimoto [1]

[1]   Institute of Crop Science, NARO, 2-1-2 Kannondai, Tsukuba 305-8518, Japan; ishimoto@affrc.go.jp
[2]   Faculty of Agriculture, Iwate University, Morioka 020-8550, Japan; chigen@iwate-u.ac.jp
*   Correspondence: pselva7@gmail.com; Tel.: +91-431-2618125
†   Present address: Crop Improvement Division, ICAR–National Research Centre for Banana, Tiruchirappalli 620-012, India.

Received: 21 January 2020; Accepted: 5 March 2020; Published: 8 March 2020

Abstract: Plant uridine 5′-diphosphate glycosyltransferases (UGTs) influence the physiochemical properties of several classes of specialized metabolites including triterpenoids via glycosylation. To uncover the evolutionary past of UGTs of soyasaponins (a group of beneficial triterpene glycosides widespread among Leguminosae), the UGT gene superfamily in *Medicago truncatula, Glycine max, Phaseolus vulgaris, Lotus japonicus,* and *Trifolium pratense* genomes were systematically mined. A total of 834 nonredundant UGTs were identified and categorized into 98 putative orthologous loci (POLs) using tree-based and graph-based methods. Major key findings in this study were of, (i) 17 POLs represent potential catalysts for triterpene glycosylation in legumes, (ii) UGTs responsible for the addition of second (*UGT73P2*: galactosyltransferase and *UGT73P10*: arabinosyltransferase) and third (*UGT91H4*: rhamnosyltransferase and *UGT91H9*: glucosyltransferase) sugars of the C-3 sugar chain of soyasaponins were resulted from duplication events occurred before and after the hologalegina–millettoid split, respectively, and followed neofunctionalization in species-/ lineage-specific manner, and (iii) UGTs responsible for the C-22-*O* glycosylation of group A (arabinosyltransferase) and DDMP saponins (DDMPtransferase) and the second sugar of C-22 sugar chain of group A saponins (*UGT73F2*: glucosyltransferase) may all share a common ancestor. Our findings showed a way to trace the evolutionary history of UGTs involved in specialized metabolism.

Keywords: family 1 glycosyltransferases; legumes; putative ortholog loci; soyasaponins; specialized metabolites; triterpenoids

---

1. Introduction

Glycosyltransferases (GTs) (EC 2.4.x.y) are ubiquitous enzymes of a superfamily that generally mediate the transfer of carbohydrate moieties from nucleotide-activated donor molecules to a broad range of saccharide or non-saccharide acceptor molecules and form glycosidic linkages via two distinct catalytic mechanisms-defined inversion or retention [1,2]. They are present in all phyla and influence the physio-chemical properties of acceptor molecules through which entail in diverse pivotal cellular processes [3]. Though GTs are extremely divergent in terms of sequence similarity, most of its members exhibit well-conserved secondary and tertiary structures and adopt either the characterized GT-A or GT-B fold [1,4–6].

The carbohydrate-active enzyme (CAZy) database classifies the GTs from diverse species based on their amino acid sequence conservation [7]. As of March 2020, a total of 110 numbered GT families have been identified and the number will likely increase in the future (http://www.cazy. org/GlycosylTransferases.html). Of these, GTs utilizing uridine 5′-diphosphate (UDP)-conjugated carbohydrates as the sugar donors are referred as family 1 GTs (alias UGTs). They are generally cytosolic in nature, widespread in the plant kingdom, and constitute the largest GT family [8,9]. Plant UGTs are assigned between families 71–100, 701–1000 and 7001–10000 in the current classification system (https://prime.vetmed.wsu.edu/resources/udp-glucuronsyltransferase-homepage). They are inverting GTs exhibiting GT-B fold and consist of a characteristic 44-amino acid consensus sequence, designated as the plant secondary product glycosyltransferase (PSPG) box, at the C terminus [10,11]. The highly divergent N-terminal and the well-conserved C-terminal PSPG box are acknowledged to be engaged in the determination of sugar acceptor and sugar donor, respectively [11]. Plant UGTs glycosylate multitude of acceptor molecules including phytohormones and diverse specialized metabolites by which influence the acceptor molecules stability, solubility, storage, transport, compartmentalization, and bioactivity [8,10,12,13]. They also have important functions in detoxification of xenobiotics and facilitate plant protection [8,14,15].

Plants naturally synthesize a tremendous number of triterpenoids through specialized metabolism that often exists as glycosidic conjugates (i.e., saponins) and have potential functions in different sectors of day-to-day life applications [16–18]. Like that of steroids, the committed biosynthesis pathway of triterpenoids stems from the mevalonate pathway-derived precursor 2,3-oxidosqualene [19]. Several triterpene scaffolds generate from 2,3-oxidosqualene by one of many oxidosqualene cyclase (OSC) enzymes, but the OSC namely β-amyrin synthase yields the most common scaffold β-amyrin [20,21]. The members of cytochrome P450 monooxygenase (CYP450) and UGT families decorate the pentacyclic C_{30} skeleton of β-amyrin by oxygenation and glycosylation, respectively, at various active sites depending on the genetic background of the given genera/species. Though the vast diversity of triterpenoids is broadly achieved by OSCs, CYP450s, and UGTs, the diversification created by UGTs is exponential and by far the most. For example, in soybean (*Glycine max*), the combinatorial activity of three different CYP450s produces only two soyasapogenols (namely A and B) from β-amyrin whereas the epistatic activity of eight different UGTs on soyasapogenol A (SA) and B (SB) could generate >50 triterpene glycosides [22]. Triterpene-related UGTs not only enhance the diversification of triterpenoids and its pharmacological values, but are also involved in plant defense against take-all-diseases [23] and herbivores [24].

Soyasaponins are oleanane-type pentacyclic triterpene glycosides implicated in diverse pharmaceutical benefits [25], several characters of root growth [26] and in undesirable taste properties of soybean-based food products [27]. They are widespread among the species of Leguminosae including the model legumes barrel medic (*Medicago truncatula*) and birdsfoot trefoil (*Lotus japonicus*), but abundant principally in the seeds of *G. max*. At least nine different UGTs have been assumed to be involved in the biosynthesis of soyasaponins. Of these, seven UGTs have been characterized to date (*UGT73P2* and *UGT91H4* [28], *UGT73F2* and its allelic variant *UGT73F4* [29], *UGT73P10* [30], *UGT91H9* [31], and *UGT73B4* [32]), excluding the UGTs responsible for the C-3-O- and C-22-O-glycosylation of SA/SB and SA respectively. Though the biochemical and genetical characteristics of soyasaponin UGTs are studied well, how they evolved upon large-scale [whole-genome duplication (WGD) alias polyploidization)] or small-scale (e.g., segmental/ tandem) duplication events remains to be studied. Also, the corresponding homologs of soyasaponin UGTs in the model legumes *M. truncatula/L. japonicus* are yet to be discovered.

Leguminosae (alias Fabaceae) is the third largest flowering plant family consists of more than 750 genera and 19,500 species [33]. Leguminosae plants biosynthesize a vast diversity of specialized metabolites as glycosidic conjugates in taxa-specific manner [34]. Both the model legumes *M. truncatula* and *L. japonicus*, as well as the economically important oil seed legume crop *G. max*, all belong to a legume subfamily Papilionoidea which experienced two WGD events—one at ~59 [papilionoid-specific WGD (PWGD)] and the other at ~13 [glycine-specific WGD (GWGD)] million years ago (MYA) [35]. To explore the effect of WGD events on soyasaponin UGTs, a systematic genome-wide survey of UGT gene superfamily was conducted using the latest genome versions of *M. truncatula* (MtUGTs), *G. max*

(GmUGTs), *L. japonicus* (LjUGTs), common bean (*Phaseolus vulgaris*; PvUGTs), and red clover (*Trifolium pratense*; TpUGTs). All the identified UGTs were assigned to putative ortholog loci (POLs) for the first time, which disclosed the mode of expansion of UGTs, gene gain/loss and intron addition/deletion events in *M. truncatula* and *G. max*. POL assignments underscore the evolutionary origin of soyasaponin UGTs and functional divergence of their homologs. In addition, it showed a way for future studies to easily pick up candidate ortholog UGTs across legumes to unravel their functions and extends our understanding in the evolution of UGT gene family.

2. Results

2.1. Genome-Wide Identification of UGT Gene Family in Five Papilionoid Legumes

With the help of PSPG sequence and several other criteria (see Materials and Methods Section 4.1), a total of 243, 208, 168, 94, and 121 authentic UGTs were identified for *M. truncatula*, *G. max*, *P. vulgaris*, *L. japonicus*, and *T. pratense* respectively in this study (Table 1). These numbers shall be treated as the least because many sequences (36 for *M. truncatula*, 34 for *G. max*, 5 for *P. vulgaris*, 64 for *L. japonicus* and 50 for *T. pratense*) in all five species were excluded based on one or more criteria (Tables S1–S5). Following the guidelines of UGT nomenclature committee, no UGTs were named in this study because we believe that the final designation of their nomenclature should be made after their functional characterization by in vitro and/or in vivo techniques.

Table 1. Number of plant UGTs in different phylogenetic groups.

No.	Plant Species Name *	No. UGTs in Different Phylogenetic Groups																		Total UGTs	Ref.
		A	B	C	D	E	F	G	H	I	J	K	L	M	N	O	P	Q	R		
1	Mimulus guttatus	10	2	3	11	14	–	12	1	4	2	9	17	2	1	9	3	–	–	100	[36]
2	Camellia sinensis	15	5	2	20	23	2	13	2	2	2	1	27	3	–	6	6	–	3	132	[37]
3	Vitis vinifera	23	3	4	8	46	5	15	7	14	4	2	31	5	1	2	11	–	–	181	[36]
4	Linum usitatissimum	16	5	6	21	22	1	19	6	9	4	5	19	3	1	–	–	–	–	137	[38]
5	Populus trichocarpa	12	2	6	14	49	–	42	5	5	6	2	23	6	1	3	2	–	–	178	[36]
6	Cucumis sativus	10	1	2	12	13	–	11	5	–	2	1	17	2	1	3	5	–	–	85	[36]
7	Arabidopsis thaliana	14	3	3	13	22	3	6	19	1	2	2	17	1	1	–	–	–	–	107	[36]
8	Brassica rapa	12	4	4	24	31	1	9	18	1	3	3	26	2	2	–	–	–	–	140	[39]
9	Brassica napus	17	10	10	36	48	2	14	35	2	3	6	61	4	3	–	–	–	–	251	[39]
10	Brassica oleraca	15	7	4	23	32	–	8	23	1	2	3	32	2	2	–	–	–	–	154	[39]
11	Cajanus Cajan	2	2	1	36	33	–	9	–	5	–	–	12	2	–	6	12	–	–	120	[40]
12	Glycine max	25	3	1	43	36	1	15	3	18	3	2	19	4	1	5	3	–	–	182	[36]
	Glycine max	5	1	2	38	46	6	16	2	4	–	–	18	5	–	6	–	–	–	149	[41]
	Glycine max	21	3	–	46	52	8	16	3	17	7	–	19	5	1	6	4	–	–	208	This study
13	Phaseolus vulgaris	19	3	2	33	33	5	18	3	15	3	–	17	4	1	6	5	–	1	168	This study
14	Lotus japonicus	9	3	–	25	22	2	9	1	2	1	0	10	1	1	6	1	–	1	94	This study
15	Medicago truncatula	28	4	–	55	55	2	39	3	5	9	–	33	2	1	3	3	–	1	243	This study
16	Trifolium pratense	11	3	–	29	39	1	13	3	1	2	–	12	1	–	2	3	–	1	121	This study
17	Malus domestica	33	4	7	13	55	6	40	14	11	12	6	16	13	1	5	5	–	–	241	[36]
18	Prunus mume	16	2	3	17	23	3	18	10	4	?	8	17	3	?	–	–	–	–	130	[42]
19	Prunus persica	10	2	4	19	29	4	34	9	5	7	7	18	14	1	1	4	–	–	168	[43]
20	Oryza sativa	14	9	8	26	38	–	20	7	9	3	1	23	5	2	6	9	–	–	180	[36]
21	Triticum aestivum	22	3	2	17	37	2	4	5	7	5	–	19	3	1	3	13	36	–	179	[44]
22	Sorghum bicolor	10	4	6	24	50	–	17	12	8	3	1	26	6	3	8	2	–	–	180	[36]
23	Zea mays	8	3	5	18	34	2	12	9	9	3	1	23	3	4	5	1	7	–	147	[45]

*—Species ordered in phylogenetic relevance; '–'—UGTs not detected/absent in the respective species; '?'—Unknown in the corresponding paper.

The UGT family of *G. max* [36,41,46] and *L. japonicus* [47] has been described previously. We did not go for a detailed comparison with the results of Yin et al. [46,47] because of discrepancies in the screening criteria of those studies (e.g., they considered all proteins having PSPG motif as UGTs irrespective of the protein length) with that of the current study. Though Caputi et al. [36] and Rehman et al. [41] utilized the first genome version of *G. max*, the former identified 183 UGTs while the latter identified 149 UGTs. Since it remains unclear how Rehman et al. [41] underestimated the number of GmUGTs, we compared our results with that of Caputi et al. [36]. Out of 183, 160 sequences were also identified in this study; seven sequences were redundant, 16 were absent, and 48 were new in the second *G. max* genome assembly (Wm82.a2.v1). This suggests that the number of UGTs identified in this study may vary in the future genome assemblies of the corresponding species.

2.2. Phylogenetic Relationship of the UGTs in Five Papilionoid Legumes

Plant UGTs from diverse species could form at least 18 distinct groups (designated A–R) in unrooted phylogenetic analyses (Table 1). Earlier studies identified 14 (A–N) of the 18 UGT groups using Arabidopsis genome [36]. Perhaps, whole-genome examination from other higher plants identified four new UGT groups named O–R: groups O and P observed in many higher plants including rice [36] while the existence of groups Q (only in maize [45] and wheat [44]) and R (only in tea [37]) are restricted. In this study, the five-legume species found to retain 14–16 phylogenetic groups (A–R, except K and Q) (Figure 1; Table 1). Notably, (i) all the five legumes lacked groups K and Q, (ii) group C members only found in *P. vulgaris*, and (iii) groups N and R absent respectively in *T. pratense* and *G. max*. Interestingly, search in other legumes identified group K members only in Arachis species, group C members in pigeon pea (*Cajanus cajan*) and Vigna species while no legume species carried group Q. This suggests many legumes lost groups K and C during their course of evolution, and the presence of group Q could be specific to monocots. The number of individuals within each group among the five legumes has varied (Table 1). Nevertheless, the highest number of UGTs was observed in groups E and D followed by groups L, G, and A. This coincides with Caputi et al. [36] that those five groups in each species have expanded more than any other groups during the evolution of higher plants. Among the five legumes, *M. truncatula* had relatively many members in groups G and L while *G. max* and *P. vulgaris* had that in group I, suggesting that those groups may have expanded evolutionarily in species- and lineage-specific manner, respectively.

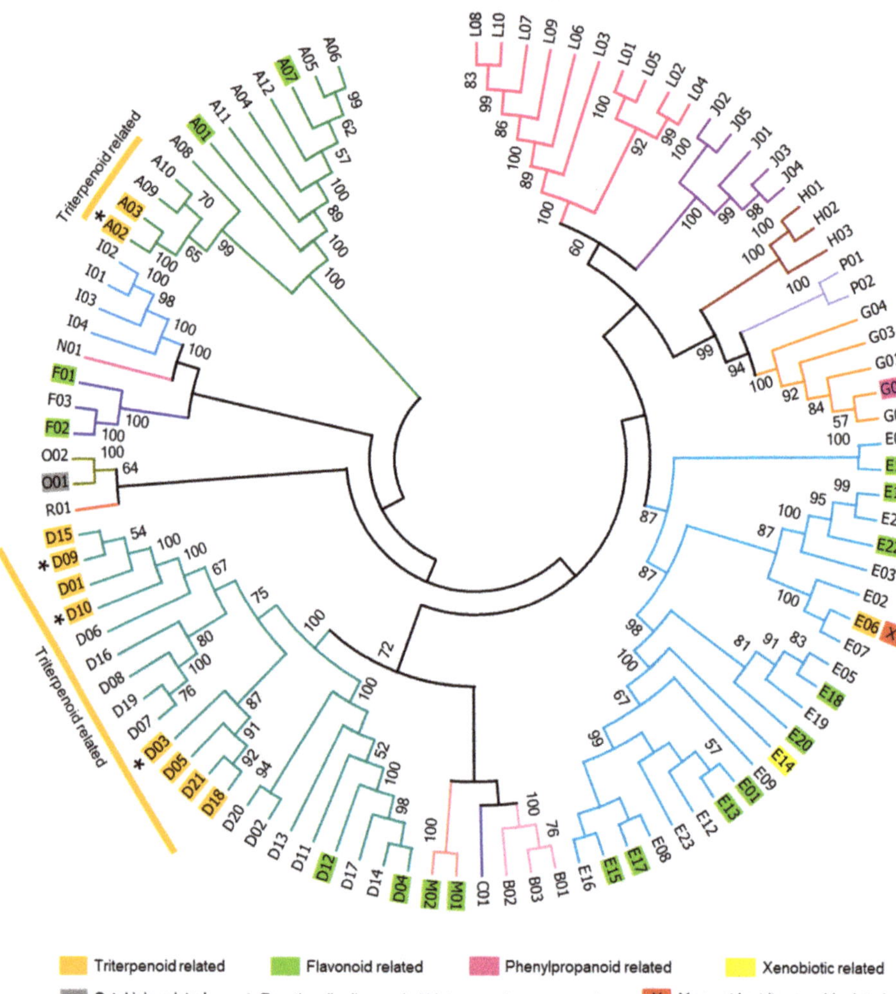

Figure 1. Evolutionary relationship of putative ortholog loci (POL) of five legume UGTs. A total of 196 full-length amino acid sequences covering all phylogenetic groups (A–R, excluding K and Q) and POLs (*n* = 98) were selected from *Glycine max* (number of UGTs = 86), *Medicago truncatula* (76), *Phaseolus vulgaris* (9), *Lotus japonicus* (8), *Cajanus cajan* (3), *Trifolium pratense* (3), *Vigna radiata* (3), *Cicer arietinum* (2), *Arachis duranensis* (4), *Lupinus angustifolius* (1), and *Trifolium subterraneum* (1). Each POL included two sequences, each from different species (See Text S1). Subtrees (i.e., UGT pairs) were compressed with corresponding POL numbers to understand the POL relationship. POLs highlighted in orange, green, purple, yellow, and gray backgrounds denote that at least one UGT from that POL has been characterized for the glycosylation of triterpenoids, flavonoids, phenylpropanoids, xenobiotics, and cytokinin's respectively. The first letter of each POL represents their phylogenetic groups.

2.3. Putative Ortholog Loci Assignments for UGTs of Papilionoid Legumes

Although *M. truncatula*, *P. vulgaris*, *L. japonicus*, and *T. pratense* have undergone similar WGD events, the retention of a high number of UGTs in *M. truncatula* (Table 1) suggest that MtUGT family may have expanded through multiple species-specific small-scale duplication events during its evolution course. Concurrently, despite the recent GWGD event, *G. max* retained relatively less UGTs than *M. truncatula*. This implies *G. max* may have lost several UGTs during its evolution. To validate these presumptions and to trace the evolutionary histories of UGTs in legumes, assigning putative ortholog loci (POL) is essential. Because legumes experienced different duplication events, the gene number may vary among the species, but the gene loci number would be evolutionarily more stable. Several platforms such as POG [48] and PLAZA [49] were developed to trace the cluster of ortholog groups among species using different criteria including the genome/gene synteny search between species. Additionally, the Phytozome gene family [50] and context viewer in Legume Information System (LIS) database [51] were helpful to get basic insight into orthologs but not feasible when the gene family has too many duplicates. These platforms, databases, and tools were certainly helpful but not sufficient to confidently assign POL for all the identified legume UGTs due to several species-specific duplication events. After several trial and error attempts to subdue the shortcomings in POL assignments, we observed that the multi-species phylogenetic clustering and the full-length amino acid percent identity were together effective in assigning POL for legume UGTs.

Using the proposed scenario (see Materials and Methods Section 4.3), 98 POLs were estimated by combining *M. truncatula*, *G. max*, *P. vulgaris*, *L. japonicus*, and *T. pratense* UGTs (Table S6; Figures S1–S11). Of these, 35 POLs had members in all five species, 25 POLs had members in either of four species, 16 POLs had members in either of three species, and 12 POLs had members in either of two species (Figure S12). Albeit using five different yet closely related species, ten POLs (one each for *M. truncatula* and *L. japonicus*, two for *T. pratense*, and three each for *G. max* and *P. vulgaris*) lacked corresponding orthologs within the five species (Figure S12). They were assigned to POL based on the UGTs of other legume species such as pigeon pea and mung bean (*Vigna radiata*) (Table S6). This suggests that analyzing UGT family of other papilionoid and non-papilionoid species may reveal new POLs. Furthermore, members in some POLs (e.g., E20 and L04) shared relatively less amino acid identity with their co-members [they were included in the same POL due to the absence of true orthologs in other legumes (Table S6)], and members in some POLs (e.g., D03 and G02) formed large clusters with complex relationship while in some POLs (e.g., D02, D06 and I03) they formed short clusters. These imply that some of the current POLs can be divided into more POLs or combined into other existing POLs in future and therefore the number of POLs identified in this study should be treated as the least.

Of the 98 POLs, the highest number of POL was found for the major groups E (*n* = 23) and D (*n* = 21), as the number of UGTs present in these groups was high. Groups A and L sustained respectively 12 and 10 POLs while all the remaining groups sustained 1–5 POLs (Table 2 and Table S6). Noteworthy is that albeit the number of UGTs in groups G and J had huge difference, both groups consisted of only 5 POLs each. Further observation clearly showed that the retention, expansion, or lose of POL in each phylogenetic group was merely species-specific followed by lineage-specific. For example, (i) groups G and L in *M. truncatula* had only 5 and 9 POLs but contained 39 and 33 UGTs, respectively, reflecting the species-specific expansion; because, such expansion was not observed for *G. max*, *P. vulgaris*, *L. japonicus* and *T. pratense*; and (ii) group I in *G. max* and *P. vulgaris* retained 15–17 UGTs in 4 POLs whereas *M. truncatula*, *L. japonicus* and *T. pratense* retained 1–2 POLs with 1–5 UGTs suggesting that the expansion of group I was specific to *G. max*/*P. vulgaris* lineage.

Table 2. Distribution of UGT POLs among the legumes.

Phylogenetic Groups	Distribution of UGT POLs among the Phylogenetic Groups [a]					
	M. truncatula	*G. max*	*P. vulgaris*	*L. japonicus*	*T. pratense*	Total
A	10 (28)	10 (21)	09 (19)	08 (09)	07 (11)	12
B	03 (04)	03 (03)	02 (03)	02 (03)	03 (03)	03
C	–	–	01 (02)	–	–	01
D	14 (55)	19 (46)	16 (33)	15 (25)	13 (29)	21
E	18 (55)	21 (52)	16 (33)	12 (22)	14 (39)	23
F	01 (02)	03 (08)	03 (05)	02 (02)	01 (01)	03
G	05 (39)	04 (16)	03 (18)	03 (09)	04 (13)	05
H	03 (03)	02 (03)	03 (03)	01 (01)	03 (03)	03
I	02 (05)	04 (17)	04 (15)	02 (02)	01 (01)	04
J	04 (09)	05 (07)	02 (03)	01 (01)	02 (02)	05
K	–	–	–	–	–	–
L	09 (33)	08 (19)	09 (17)	07 (10)	07 (12)	10
M	02 (02)	02 (05)	02 (04)	01 (01)	01 (01)	02
N	01 (01)	01 (01)	01 (01)	01 (01)	-	01
O	02 (03)	02 (06)	02 (06)	02 (06)	02 (02)	02
P	01 (03)	02 (04)	02 (05)	01 (01)	01 (03)	02
Q	–	–	–	–	–	–
R	01 (01)	–	01 (01)	01 (01)	01 (01)	01
Total	76 (243)	86 (208)	76 (168)	59 (94)	60 (121)	98

[a] Numbers within the brackets denotes the number of UGTs corresponded to the loci.

2.4. Expansion of UGTs in M. truncatula and G. max

Because of the sequencing coverage, completeness, and higher resolution, we only focused MtUGTs and GmUGTs from here for all further analyses with fewer exceptions. POL assignments revealed an interesting criterion: the 243 UGTs of *M. truncatula* traced back to 76 POLs whereas that of 208 GmUGTs were traced back to 86 POLs (Table 2; Table S6). This emphasizes the fact that UGT family in *M. truncatula* expanded more but lost some POLs during its evolution. Perhaps, our findings show that the 76 POLs in *M. truncatula* were dispersed as 33 single-copy, 14 double-copy, and 29 multi-copy POLs (Table S6). In *G. max*, 40 were single-copy, 21 were double-copy, and 25 were multi-copy POLs. Notably, 19 POLs were single-copy in both species. In *M. truncatula*, four multi-copy POLs namely G02, D03, D06, and L01 had 30, 17, 15, and 11 members respectively (Table S6). These four POLs represent 30% of UGTs in total number of MtUGTs (73 in 243) whereas that represent only 8.7% in *G. max* (18 in 208). We thus attributed these four POLs as the predominant source for the higher number of UGTs in *M. truncatula*. No UGT members had been found for 22 POLs in *M. truncatula* and 12 POLs in *G. max*. Of these, 16 POLs had no members in *M. truncatula* but had in *G. max* whereas 6 POLs had no members in *G. max* but had in *M. truncatula*; 6 POLs lacked members from both species. This shows that the retention or loss of POLs in *M. truncatula* and *G. max* was species- or lineage-specific.

2.5. Analysis of Intron Gain/Loss Events in M. truncatula and G. max

Introns present in the coding sequences were considered for this study. The majority of UGTs in *M. truncatula* (n = 140; 57.6%) and *G. max* (114; 54.8%) had no introns. Among the intron containing UGTs, 87 out of 103 (84.5%) in *M. truncatula* and 76 out of 94 in *G. max* (80.9%) had one intron. Nine UGTs contained 2, five contained 3, and two contained 5 introns in *M. truncatula* (Table S1) while *G. max* had two introns in 11, three in 5, and four in 2 UGTs (Table S2).

Intron gain or loss events were inferred by the comparison of members present in the given POL across five legumes (Table S7). The 98 POLs of UGTs were first classified into three types: no-intron POLs (n = 42), one-intron POLs (n = 26) and mixed-intron POLs (n = 30). Based on our criteria (see Materials and Methods Section 4.4), 11 one-intron UGTs from *M. truncatula*, and 9 one-intron UGTs from *G. max* were found as intron-gained genes. This implies that 12.6% (11 in 87) of one-intron MtUGTs and 11.8% (9 in 76) of one-intron GmUGTs gained introns evolutionarily. Though experimental validation is required, this finding suggest that no-intron UGTs can become one-intron UGTs evolutionarily. In addition, 16 MtUGTs and 18 GmUGTs were also identified as intron-gained genes which consisted of 2–5 introns. Our findings reveal that, six (G, H, I, J, N, and P) and two (O and R) phylogenetic groups could be designated as one-intron and no-intron containing groups, respectively.

2.6. Chromosomal Locations and Gene Duplication Analyses in M. truncatula and G. max

UGTs distributed throughout all the chromosomes (Ch) of *M. truncatula* (Figure S13; Table S1) and *G. max* (Figure S14; Table S2). The UGTs density per chromosome was highly uneven in both species. In *M. truncatula*, Ch6 (n = 42) had the highest number of UGTs followed by Ch5/Ch8 (n = 40) and Ch7 (n = 37). Ch1 (n = 14) and Ch3 (n = 15) had the least number of UGTs, while Ch2 and Ch4 had 20 and 25 UGTs, respectively. In *G. max*, Ch8 had the maximum number of UGTs (n = 21) followed by Ch3 (n = 18) and Ch2 (n = 17), whereas the least number of UGTs found in Ch4, Ch5, Ch17, and Ch20 which had 3–4 UGTs. All other chromosomes had 6–16 UGTs. Scaffolds represent 10 and one UGTs in *M. truncatula* and *G. max* respectively.

All the double-copy and multi-copy POLs in *M. truncatula* and *G. max* were selected for gene duplication analysis (Table S8). The members in double-copy POLs shared 60–94% amino acid identity at full-length protein level whereas that of multi-copy POLs shared 54–99% within each POL in both species. The variation in sequence conservation among UGTs in the given POL suggest that the duplicated copies diverged rapidly after the duplication. In *M. truncatula*, 61 sequences were identified as segmental duplicates and 149 sequences were identified as tandem duplicates. This shows that the UGT family in *M. truncatula* has been expanded majorly through tandem duplication (61.3%; 149 in 243) and subtly through segmental duplication (25.1%; 61 in 243) events. A similar trend was observed for the UGT family expansion in *G. max*, in which tandem duplication contributed 51.4% (107 in 208) whereas segmental duplication contributed 29.3% (61 in 208). Of the 29 multi-copy POLs in *M. truncatula*, eight POLs (A06, A10, E10, I02, J02, L07, L08, and P01) involved only in tandem duplication events; four POLs (A08, D02, G05, and L04) involved only in segmental duplication events; and the remaining 17 POLs experienced both events (Table S8). Among the 25 multi-copy POLs in *G. max*, nine POLs (A01, D03, D04, D06, D07, E10, E13, E19, and I02) involved only in tandem duplication, one POL (E17) involved only in segmental duplication and the remaining 15 POLs involved in both events. By using chromosomal positions and the gene order, it appears that the members in 17 multi-copy POLs (which experienced both events) in *M. truncatula* were first scattered on different *M. truncatula* chromosomes via segmental duplication and then concentrated through tandem duplication (e.g., D06). Whereas, it appears that most members (if not all) of 15 multi-copy POLs (which experienced both events) in *G. max* were first underwent tandem duplication and then translocated into other chromosomes by segmental duplication or by GWGD (e.g., D03).

2.7. Duplication History and Functional Divergence of Triterpene Related UGT POLs in M. truncatula and G. max

Albeit the genomes of *M. truncatula* and *G. max* retained hundreds of putative UGT sequences, only a handful of them have been studied for their functions (10 in *M. truncatula* and 27 in *G. max*) to date. In the case of triterpene glycosylation, only three MtUGTs and eight GmUGTs were characterized. These 11 UGTs were clustered and evolutionarily close to 11 POLs (A02, A03, D01, D03, D05, D09, D10, D15, D18, D21, and E06) (Figure 1).

The two members [*Glyma.08G181000: UGT91H4* and *Glyma.10g104700: UGT91H9* (Figure 2A)] in A02 of *G. max* catalyze the addition of rhamnose or glucose, respectively, at the terminal position of C-3 sugar chain of SA and SB in vitro and in vivo [28,31]. The members of A02 from 14 legumes (Table S9) formed two sister clades (i.e., A02-I and A02-II) with high bootstrap support in phylogenetic analysis (Figure 2A). The A02-I locus corresponding the homologs of *UGT91H4* was located in syntenic blocks across all legumes and had one or two homologs in all the analyzed legumes except *M. truncatula*, which had two synteny and six non-synteny homologs (Figure 2B). Whereas, the A02-II locus corresponding *UGT91H9* homologs had single homologs only in millettoid species (e.g., *G. max*, *P. vulgaris,* and Vigna species) (Table S9) and found in syntenic blocks only between *P. vulgaris* and cow pea (*Vigna unguiculate*). The A02-I and A02-II members shared high amino acid identity (>70%) and showed a segmental duplication relationship in *G. max* and adzuki bean (*Vigna angularis*) and a tandem duplication relationship in *P. vulgaris* and cowpea (Figure 2B). Divergence time analysis estimates the duplication (whether it was tandem or segmental) may have occurred at ~44–47 MYA (Table S10). The eight UGTs in A02 of *M. truncatula* shared 71–90% amino acid identity and appear to be resulted from tandem as well as segmental duplication events (Table S8). None of these eight UGTs were characterized to date. The expression of a tandem duplicate *Medtr2g008220* and *Medtr2g008225* (*UGT91H6*) [both shared 90% amino acid identity; duplication time estimated as ~10.1 MYA (Table S10)] was highly correlated with triterpene biosynthetic genes [52]. Also, they shared 77% and 72% amino acid identity respectively to *UGT91H4* and found together with it in syntenic blocks (Figure 2B) suggesting that one of these two or both genes might have similar functions to that of *UGT91H4*.

A03 was a single-copy POL across all legumes and none of its members have been functionally characterized. Noteworthy, missense mutations in the PSPG box or its proximal regions of *Glyma.15g051400* (a A03 member of *G. max*) did not affect the soyasaponin profile [53]. The A03 member in *M. truncatula* (*Medtr2g008226*: *UGT91H5*) showed high co-expression values with triterpene biosynthetic genes [52] and tightly linked with a tandem duplicate of A02-I members (i.e., *Medtr2g008220* and *Medtr2g008225*), implying that A03 members may glycosylate triterpenes. Microsynteny analysis revealed that A03 locus tandemly linked to A02-I locus in diverse species including the early diverged legumes blue lupin (*Lupinus angustifolius*) and cultivated peanut (*Arachis hypogea*) (Figure 2B). Divergence time analysis estimates the tandem duplication event may have occurred ~42–84 MYA (Table S10).

D03 retained 17 members in *M. truncatula* and seven members in *G. max* (Table S9). One of these members from *M. truncatula* (*Medtr2g035020*: *UGT73F3*) was shown to glucosylate the C-28 position of hederagenin in vitro and in vivo [52]. Concurrently, a D03 member from *G. max* (*Glyma.07G254600*: *UGT73F2*) and its allelic variant *UGT73F4* were characterized to attach glucose or xylose respectively at the terminal position of C-22 sugar chain of SA in in vitro and in vivo [29]. These suggest that the catalytic functions of D03 members had been diverged and neofunctionalized during their course of evolution in species-specific manner. D03 existed as a multi-copy POL in hologalegina (eg. *M. truncatula, T. pratense* and *L. japonicus*) and millettoid species (eg. *G. max* and *P. vulgaris*) but a single-copy POL in the early diverged legumes (Table S9). Many of D03 members of *M. truncatula* and *T. pratense* were non-synteny homologs (Table S9) and showed a complex phylogenetic relationship (Figure S3). Even the syntenic D03 homologs from 14 legumes did not resolve well phylogenetically; however, they were divided into D03-I, D03-II, and D03-III clades based on the amino acid percent identity of D03 members (Figure 3A). The 17 MtUGTs in D03 shared 61–84% amino acid percent identity and may have resulted from more than one segmental and tandem duplication events (Table S8). *UGT73F3* shared 81–84% amino acid identity with its neighboring UGTs *Medtr2g034990* and *Medtr2g035040*, suggesting that these three UGTs may have resulted from a tandem duplication event occurred at ~11–16 MYA (Table S10). *UGT73F2* was tandemly located with three UGTs (*Glyma.07G254700*, *Glyma.07G254800*, and *Glyma.07G254900*) and all these showed high amino acid identity with another tandem duplicates located at the 17th chromosome (*Glyma.17G019400*, *Glyma.17G019500*, and *Glyma.17G019600*) suggesting that one of these genes first underwent tandem duplication and then copied into another chromosome by segmental duplication or GWGD. This notion is supported well by the gene-collinearity between Ch07 and Ch17 (Figure 3B). Divergence time analysis in *M. truncatula, G. max, P. vulgaris,* and chickpea (*Cicer arietinum*) estimates that the tandem duplication may have occurred ~40–104 MYA (Table S10). Non-sense mutations in *Glyma.07G254700*, *Glyma.07G254900*, *Glyma.17G019400*; *Glyma.17G019500* and *Glyma.17G019600* does not affect the saponin composition in mature soybean seeds implying that these genes might be not involved in soyasaponin biosynthesis [53]. The *Glyma.07G254800* was assumed as a pseudogene because the gene was not amplified using different primer sets [53].

A. Phylogenetic relationship of A02 and A03 homologues in 14 legumes

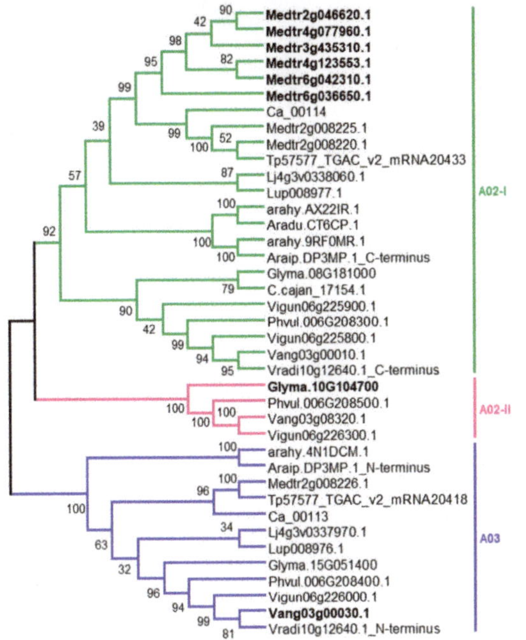

B. Microsynteny relationship of A02 and A03 loci among legumes

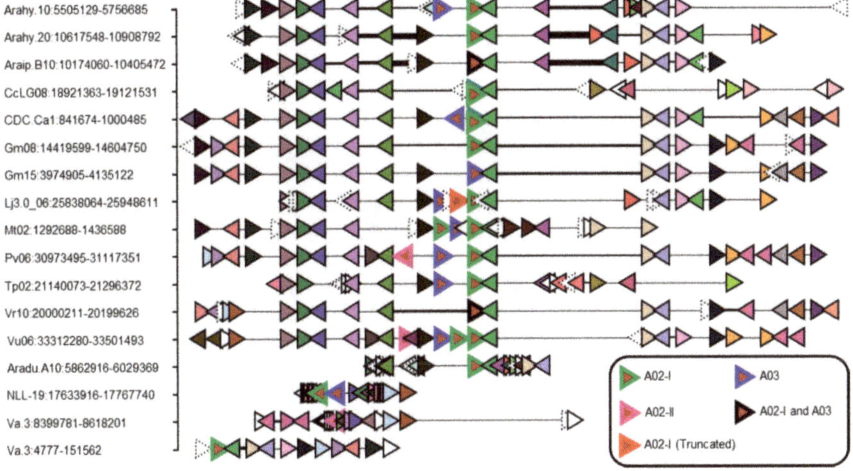

Figure 2. Evolutionary history of putative ortholog loci (POL) A02 and A03 in legumes. (**A**). Phylogenetic relationship of A02 and A03 homologs in 14 legumes. Bolded genes are non-synteny homologs with any of the 14 legumes. The full-length sequences of *Araip.DP3MP.1* and *Vradi10g12640.1* could be a sequencing error; their N-terminus and C-terminus shared high amino acid percent identity with *UGT91H5* (A02) and *UGT91H4* (A03) members respectively. See Table S9 for species and gene ID's information. (**B**). Microsynteny relationship of A02 and A03 loci across legumes. Microsyntenic genome segments are retrieved and centered using *Phvul.006G208300*. Orthologous/paralogous gene pairs are indicated through the use of a common color. Uncolored and cracked genes are singletons and orphans respectively in this genomic region. Species and genomic positions are mentioned in

the left side of each segment. From top to bottom, Arahy—*Arachis hypogea*, Araip—*Arachis ipaensis*, CcLG—*Cajanus cajan*, CDC.Ca—*Cicer arietinum*, Gm—*Glycine max*, Lj—*Lotus japonicus*, Mt—*Medicago truncatula*, Pv—*Phaseolus vulgaris*, Tp—*Trifolium pratense*, Vr—*Vigna radiata*, Vu—*Vigna unguiculate*, Aradu—*Arachis duranensis*, NLL—*Lupinus angustifolius*, and Va—*Vigna angularis*.

A. Phylogenetic relationship of syntenic D03 homologs in legumes

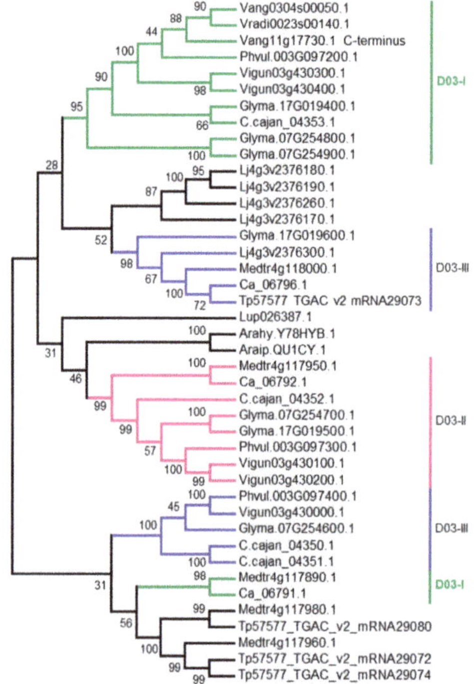

B. Microsynteny relationship of D03 loci among legumes

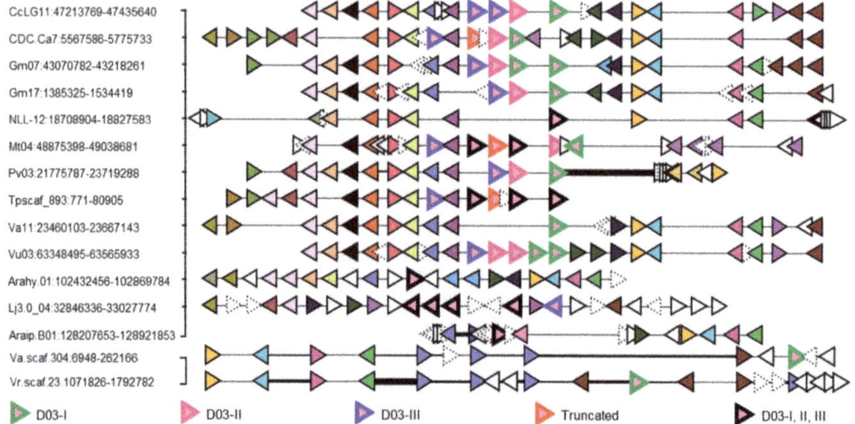

Figure 3. Evolutionary history of putative ortholog locus (POL) D03 in legumes. (**A**). Phylogenetic relationship of syntenic D03 homologs in 14 legumes. The full-length sequence of Vang11g17730.1 (1283 amino acids length) could be a sequencing error; only its C-terminus (471 amino acids) shared high amino acid percent identity with D03 members. Based on amino acid percent identity, D03 members

are sub-grouped into D03-I, D03-II, and D03-III. Unresolved members are not sub-grouped. See Table S9 for species and gene id's information. (**B**). Microsynteny relationship of D03 locus across legumes. Microsyntenic genome segments are retrieved and centered using Phvul.003G097300 and Araip.QU1CY in panel 1 and using Vang0304s00050 in panel 2. Orthologous/paralogous gene pairs are indicated through use of a common color. Uncolored and distorted genes are singletons and orphans respectively in this genomic region. Species and genomic positions are mentioned in the left side of each segment. From top to bottom, CcLG—*Cajanus cajan*, CDC.Ca—*Cicer arietinum*, Gm—*Glycine max*, NLL—*Lupinus angustifolius*, Mt—*Medicago truncatula*, Pv—*Phaseolus vulgaris*, Tp—*Trifolium pratense*, Va—*Vigna angularis*, Vu—*Vigna unguiculate*, Arahy—*Arachis hypogea*, Lj—*Lotus japonicus*, Araip—*Arachis ipaensis* and Vr—*Vigna radiata*.

D05 was a single-copy POL (Table S6). Its member (*Medtr4g031800*: *UGT73K1*) in *M. truncatula* reported to glycosylate hederagenin, SB, and soyasapogenol E in vitro [54]. Recently, the member of D05 (*Glyma.16G033700*) in *G. max* was reported to attach DDMP moieties at the C-22 hydroxyl position of SB in vivo [32]. In another independent study, we have identified that the D05 members are DDMP transferases and have homologs in diverse legume species including the early diverged ones [53]. These suggest that *UGT73K1* could be a candidate gene for DDMP transferase in *M. truncatula*. Noteworthy, *Glyma.16G033700* has been wrongly named as *UGT73B4* in Sundaramoorthy et al. [32].

D09 (*Glyma.11G053400*: *UGT73P2*) and D10 (*Glyma.01g046300*: *UGT73P10*) members of *G. max* catalyze the addition of galactose and arabinose sugars respectively at the second position of the C-3 sugar chain of SB in vitro and/or in vivo [28,30]. The members of D01, D09, D10, and D15 from 14 legumes shared considerable amino acid identity and formed sister clades in phylogenetic analysis with high bootstrap support (Figures 1 and 4). This implies that these four POLs may share a common ancestor and that the members of D01 and D15 may have a potential for triterpene glycosylation like that of D09/D10. Supporting this assumption, (i) a D01 member from *M. truncatula* (*Medtr8g044140*: *UGT73P1*) had been proposed to be involved in triterpene glycosylation because of its elevated co-expression with that of other triterpene biosynthetic genes upon methyl jasmonic acid treatment in *M. truncatula* root cell suspension cultures [54], and (ii) D15 members were tandemly linked to D09 in blue lupin and Arachis species (Figure 5). The presence/absence of D01, D09, D10, and D15 homologs in 14 legumes (Table S9) denote D09 was evolutionarily old and conserved, D15 was evolutionarily old but lost in many legumes and D01/D10 may have originated (via segmental duplication) after the PWGD but before the split of hologalegina and millettoid species (i.e., <59–48 MYA). However, assuming D01/D10/D15 were stemmed from D09, the divergence time analysis estimated that they were duplicated from D09 at 38–100, 54–81, and 73–101 MYA, respectively.

G. max retained single copy each for D09, D10, and D15 (Table S6). *M. truncatula* retained one copy for D09, three copies for D10, and none for D15. The D09 of *M. truncatula* (*Medtr5g016660*) shared 75.6% amino acid identity to *UGT73P2* and co-expressed highly with soyasaponin biosynthesis genes (Table S11) implying that *Medtr5g016660* might have similar functions to that of *UGT73P2*. The three D10 sequences in *M. truncatula* shared 74–75% amino acid identity and appear to be raised from tandem (*Medtr5g039900* and *Medtr5g040030*) and segmental duplication (*Medtr6g035295*) events occurred at ~15–18 MYA (Table S10). None of these genes were studied previously. However, based on the syntenic relationship and amino acid identity, we assume *Medtr5g039900* and *Medtr5g040030* were the most probable candidates to carry out the similar functions of *UGT73P10*. Both *G. max* and *M. truncatula* retained two copies in D01 that may have segmentally duplicated from one another at ~58 and ~11 MYA respectively (Table S10). Nonsense mutations in D01 (i.e., *Glyma.10G280400* and *Glyma.15G221300*) and D15 (*Glyma.01G188800*) members of *G. max* does not affect the soyasaponin composition [53] implying that these genes might be not involved in soyasaponin biosynthesis. The in vivo activity of UGT73P2 was never characterized before. We thus identified missense mutations causing various amino acid changes in PSPG box or its proximal region of *UGT73P2* but none of them affected the soyasaponin composition [22]. We reckon that the in vivo characterization of *UGT73P2* is essential to validate its function.

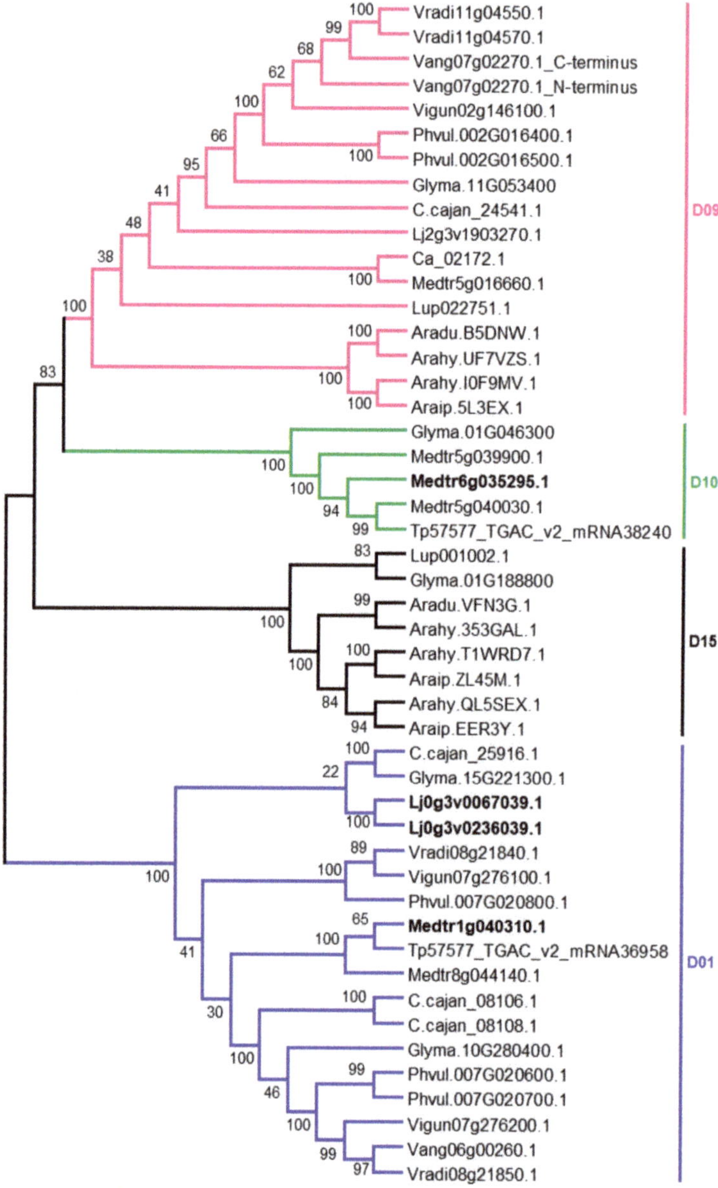

Figure 4. Phylogenetic relationship of putative ortholog loci (POL) D01, D09, and D10 homologs in 14 legumes. The full-length sequence of Vang07g02270.1 (901 amino acids length) could be a sequencing error; its C-terminus (409 amino acids) and N-terminus (492 amino acids) both shared high amino acid percent identity with D09 members. Bolded genes are non-synteny homologs with any of the 14 legumes. See Table S9 for species and gene ID's information.

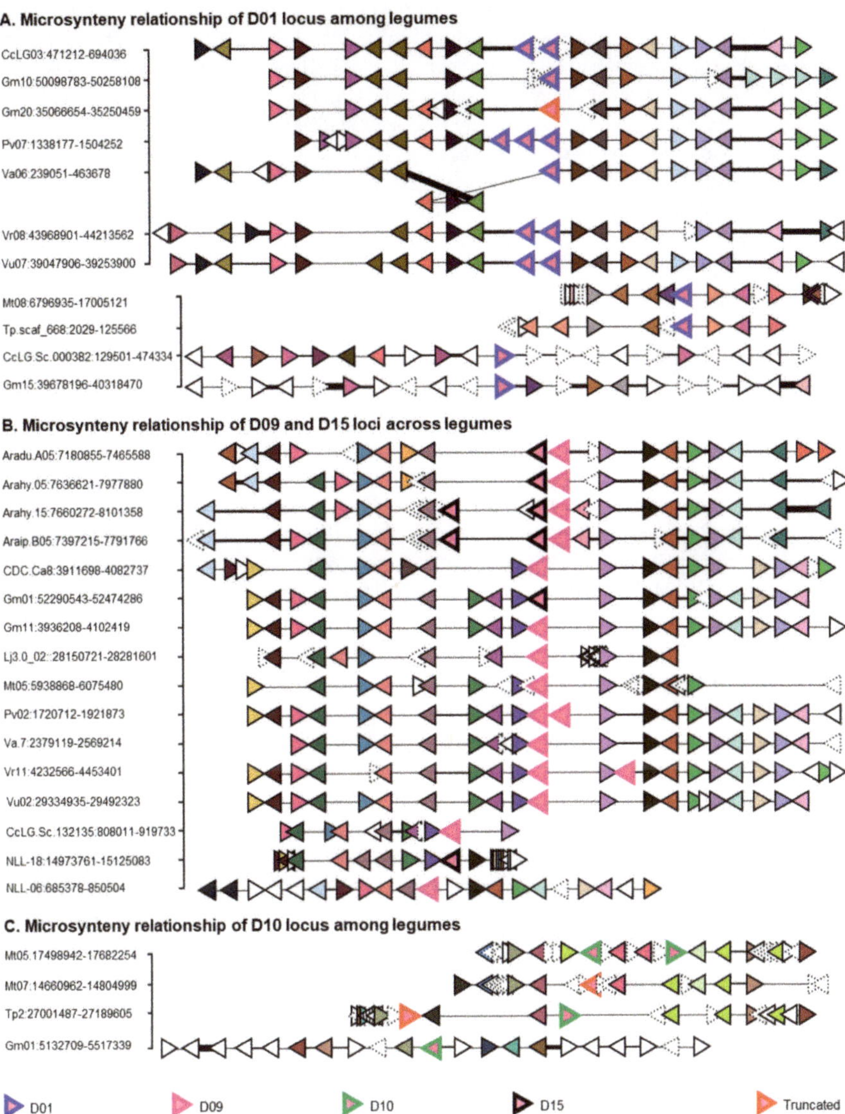

Figure 5. Microsynteny relationship of putative ortholog loci (POL) D01, D09, D10, and D15 in 14 legumes. (**A**). Microsyntenic genome segments having D01 locus are retrieved and centered using *Phvul.007G020600* (first panel) and *Medtr8g044140* (second panel). (**B**). Microsyntenic genome segments having D09 and D15 loci are retrieved and centered using Phvul.002G016400. (**C**). Microsyntenic genome segments having D10 locus are retrieved and centered using *tripr.gene37002* (Tp57577_TGAC_v2_mRNA38240). A–C, Orthologous/paralogous gene pairs are indicated through use of a common color. Uncolored and distorted genes are singletons and orphans respectively in these genome regions. Species and genomic positions are mentioned in the left side of each segment. Aradu—*Arachis duranensis*, Arahy—*Arachis hypogea*, Araip—*Arachis ipaensis*, CDC.Ca—*Cicer arietinum*, CcLG—*Cajanus cajan*, Gm—*Glycine max*, Lj—*Lotus japonicus*, NLL—*Lupinus angustifolius*, Mt—*Medicago truncatula*, Pv—*Phaseolus vulgaris*, Tp—*Trifolium pratense*, Va—*Vigna angularis*, Vr—*Vigna radiata*, and Vu—*Vigna unguiculate*. See Table S9 for species and gene ID's information.

D18 of *G. max* included two members (*Glyma.08G348500* and *Glyma.08G348600*). One of these members (*Glyma.08G348500*) was co-expressed highly with *UGT73F2* and characterized to attach arabinose at the C-22-O position of SB-glycoside in vitro [23]. The resulting product was subsequently utilized by *UGT73F2* which attaches glucose as the second sugar to the C22-O-arabinose of SB in in vitro. Intriguingly, the products generated neither by *Glyma.08G348500* nor by *Glyma.08G348500+UGT73F2* are never identified in vivo in *G. max*. In this context, we reckon that *Glyma.08G348500* may not carry out its in vitro function in in vivo and it may arabinosylate the C-22-O position of SA or SA-glycosides in vivo in *G. max*. *Glyma.08G348500* and *Glyma.08G348600* were tandem duplicates sharing 63% amino acid identity (Table S8) and may have originated from one another at ~49 MYA (Table S10). D18 had one member each in *P. vulgaris*, Vigna species and in *T. subterraneum* but none in many other legumes (e.g., *M. truncatula*, *L. japonicus* and chickpea) (Table S6) implying that D18 may have undergone deletion process in those species.

D21 had homologs only in *T. pratense* and *T. subterraneum*. The relevance of D21 members in triterpene glycosylation yet to be discovered.

E06 of *M. truncatula* retained two members (*Medtr5g070040* and *Medtr5g070090*). One of these members (*Medtr5g070090: UGT71G1*) was proposed to be specific to medicagenic acid based on the integrated transcript and metabolite profiling in methyl jasmonic acid-treated *M. truncatula* root cell suspension cultures [54]. Since E06 was an only POL from group E as triterpene related and could glycosylate flavones and isoflavones with higher efficiency than triterpenes in in vitro [54], the in vivo function of *UGT71G1* or its homologs is necessitated to unarguably consider E06 as a triterpene related POL. *Medtr5g070040* and *Medtr5g070090* were tandem duplicates sharing 77% amino acid identity (diverged at ~14 MYA). In case of *G. max*, E06 retained three members resulted from tandem duplication (*Glyma.02G225800* and *Glyma.02G226000*) and segmental duplication/ GWGD (*Glyma.14G192800*) events.

3. Discussion

Hundreds of putative UGT sequences have been uncovered from the whole-genome sequence of several plant species including legumes (Table 1). All these studies have reported the species-specific expansion of UGT members in each phylogenetic group based on the identified UGT gene numbers within the phylogenetic groups among species. Albeit not without merits, these studies are never being sufficient enough to completely uncover the expansion, gene gain/ loss, and intron addition/deletion histories of UGTs. Identifying the putative ortholog groups across species and putative paralog groups within species is a valid and promising approach to estimate the mode of gene family expansion to determine gene functional differentiation to trace gene gain/loss events across species and to transfer functional information of well-studied genes from one species to non-studied species [55–57]. Previously, 24 ortholog groups were proposed using the UGTs from primitive and higher plant species and provided an overview like that of the phylogenetic group analysis [9]. However, assigning POLs to trace back all the identified UGTs from diverse species into a common ancestor and establishing their one-to-one, one-to-many, and many-to-many relationship across species are challenging and often jeopardized by the presence of multiple non-similar duplication events among the species. As a primary step, we thus herein assigned POLs for legume UGTs based on the tree-based (i.e., multi-species phylogenetic relationship) and graph-based (i.e., amino acid identity percentage) strategies to unravel the expansion pattern of UGT family and to decipher how triterpene UGTs evolved over the period in legumes.

3.1. Expansionary and Evolutionary Dynamics of the UGT Gene Family in M. truncatula and G. max: Insights from POL Assignments

Since legumes experienced different WGD events, it is mandatory to trace back each UGT genes from different legume genomes to a common ancestor that would help to deepen our knowledge on the evolutionary histories of UGTs in legumes. Hence, the 834 UGTs of all five legumes were traced back to 98 POLs. Since we were uncertain about the quality and completeness of genome assembly in *P. vulgaris*, *L. japonicus*, and *T. pratense*, we utilized POL assignment to determine the evolution and expansion patterns of UGT gene family in *M. truncatula* and *G. max*. Also, we wanted to clarify how *M. truncatula* retained most number of UGTs (n = 243) than its close relative *T. pratense*, both of which diverged around 23 MYA [58] and how *G. max* retained lesser UGTs (n = 208) than *M. truncatula* despite the recent glycine-specific WGD event. POL assignments clearly showed that, despite the high number of UGTs, *M. truncatula* lost 21 POLs during its course of evolution whereas *G. max* lost only 12. This suggests the higher/ lower number of UGTs in one species not necessarily correspond to that of the increased/decreased POLs. A simple comparison of POL numbers in other three legumes (*P. vulgaris*, *L. japonicus* and *T. pratense*) suggests that losing UGT POLs may be insubstantial but most legumes (if not all) would experience POL loss events either in species- or lineage-specific manner.

To pinpoint the expansion and mode of expansion, we subcategorized UGT POLs based on the gene copy number into single-, double-, and multi-copy POLs. *M. truncatula* and *G. max* retained respectively 33 and 40 single-copy POLs, 14 and 21 double-copy POLs, and 29 and 25 multi-copy POLS. Close observation of POLs revealed four key factors: (i) groups D and E contained the most number of UGTs in both species like other plant species (Table 1); but they still retained 5–10 single-copy POLs in those groups, (ii) expansion of group G was *M. truncatula*-specific but POL assignment showed that only G02 underwent rampant expansion while G01 and G04 carried each single UGT, (iii) expansion of group I was specific to *G. max/P. vulgaris* lineage but only I02 and I03 POLs expanded more whereas I01 and I04 POLs carried each single UGT, and (iv) eight single-copy POLs in *G. max* were expanded more in *M. truncatula* with 44 UGTs (~five-fold increase) while only two single copy POLs in *M. truncatula* were contained 3–4 UGTs in *G. max*. These results emphasize the fact that the increase in UGT number in *M. truncatula* was mostly achieved by POL-specific expansion and such POLs may be regarded as duplication susceptible. Despite the expansion, some POLs are tended to be duplication resistance (i.e., the single-copy POLs) in both species and such POLs may carry out important functions in plant growth, development, and protection. Nevertheless, current study predicts 86.4% [210 (149 tandem and 61 segmental) were duplicates in 243] of MtUGTs and 81.7% [170 (107 tandem and 63 segmental) were duplicates in 208] of GmUGTs were duplicates (Table S8). These sequences showed different degree of sequence conservation suggesting that they may have undergone rapid Ka and Ks nucleotide substitutions after the duplication event and retained for sub-or neo-functionalization.

More than 60% of UGTs in *M. truncatula* and 50% of UGTs in *G. max* were identified as tandem duplicates and formed cluster on different chromosomes suggesting that unequal crossover accelerated and contributed more in the expansion of UGT gene family in those species. Notably, the POLs G02 and D06 could respectively formed a cluster with 20 and 13 UGTs tandemly on chromosomes 6 and 8 in *M. truncatula*. Whereas, in *G. max*, the largest tandem cluster was formed with only six members (E10 and E13). Many tandem duplicates of GmUGTs found in synteny blocks of two corresponding chromosomes (e.g., A01, D03, and I02) suggesting that the segmental duplication or GWGD event also provided considerable contribution in the expansion of UGT gene family in *G. max*.

Intron addition/deletion events are a part of gene evolution. Previous studies (e.g., Li et al. [43]) mostly examined the intron addition/deletion histories of UGTs based on the mapping of introns positions. These studies provide information about the conserved intron position and approximate intron addition/deletion events. However, this information does not clarify intron deletion in no-intron UGTs or intron addition in one-intron UGTs. In this context, in this study, the UGT POLs were subcategorized into three types namely no-intron, one-intron, and multi-intron POLs (Table S7). This analysis showed that ~12% of no-intron UGTs in *M. truncatula* and *G. max* underwent intron addition and became one-intron UGTs during evolution. Although we could not detect any intron deletion events, POL assignments defined six phylogenetic groups (G, H, I, J, N, and P) as one-intron UGTs and two groups (O and R) as no-intron UGTs in legumes. Further results from diverse plant species are necessitated to determine whether this phenomenon is universal among higher plants. Nevertheless, this information will undoubtedly assist future studies to trace the intron addition/deletion history of UGTs in diverse species.

3.2. Evolutionary Insights into the Sugar Chain Biosynthesis of Soyasaponins

The contribution of gene duplication followed by neofunctionalization (i.e., positive selection) is evident in the diversification of several groups of specialized metabolites [59] including triterpenoids [60]. However, no solid examples are currently available to emphasize the importance of gene duplication and neofunctionalization in the UGTs-oriented diversification of triterpenoids. Hence, in this study, we tried to establish the history and consequence of duplication on soyasaponin UGTs in triterpene glycosylation (Figure 6).

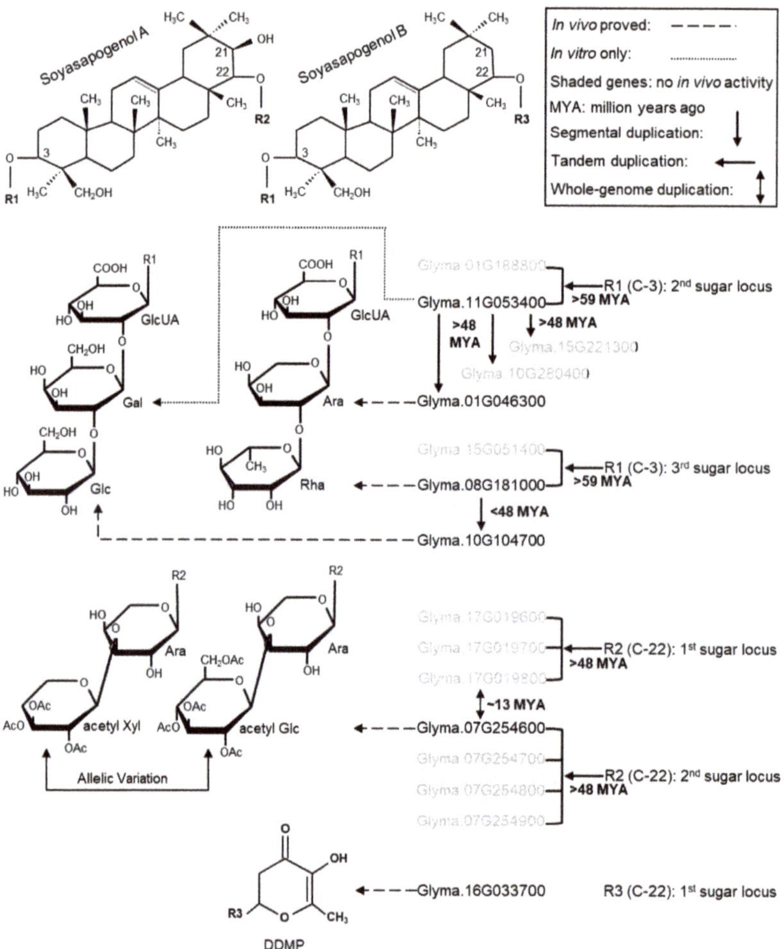

Figure 6. Duplication history and evolutionary fates of soyasaponin-related UGT loci. From top to bottom, structures of soyasapogenols A and B, major C-3 sugar chains, C-22 sugar chains and DDMP sugars are shown. Soyasaponin-related UGTs with their duplication history are shown on the right side of sugar chains. Divergence period of each duplication events is inferred based on the presence of corresponding homologs in other legumes (see Discussion Sections 3.2.1 and 3.2.2). Shaded genes are

not involved in soyasaponin biosynthesis in vivo [53]. UGTs are connected to responsible sugars by modified arrows: the enzymatic activity of dashed arrow UGTs is proved by in vivo experiments while the activity of round-dot arrow UGT is proved only by in vitro experiments. *Glyma.16G033700* may correspond a single-copy POL (D05) and its closest neighbor is the multi-copy POL D03. No UGTs have been characterized for the C-3-*O*-glycosylation of SA/SB and C-22-*O*-glycosylation of SA to-date.

To reconstruct the evolutionary history of soyasaponin UGTs, it is mandatory to consider not only the gene duplication events but also the prevalence of soyasaponins in *G. max* and other legumes. Soyasaponins comprising bidesmosidic SA-glycosides (also known as group A saponins) and monodesmosidic SB-glycosides (also known as DDMP saponins) are predominantly accumulated in the seeds of *G. max*. Although many soyasaponin components identified in *G. max*, majority of them accumulated much lesser quantity in vivo and only the Aa/Ab and βg components correspond the maximum proportion of total group A and DDMP saponins respectively [22]. DDMP saponins and its derivatives group B or group E saponins are widespread among legumes while group A saponins are restricted to the subgenus *Soja* that includes the *G. max* and its wild relative *Glycine soja* (Gs).

3.2.1. Evolution of the C-3 Sugar Chain of Soyasaponins

Six genuine C-3 sugar chains (four are tri-saccharide and two are di-saccharide) comprising of five sugars (glucuronic acid as first; galactose or arabinose as second; and rhamnose or glucose as third) are identified from soyasaponins (Figure 6). DDMP saponins having galactose (catalyzed by *UGT73P2*) as second sugar in the C-3 sugar chain (i.e., either of αg, βg and γg) are identified in many papilionoid legumes including the early diverged ones [e.g., cladrastis (*Styphnolobium japonicum*), genistoid (e.g., blue lupin) and dalbergioid (peanut) species] while those having arabinose (catalyzed by *UGT73P10*) at the same position (i.e., either of αa, βa, and γa) are exclusively reported in millettoid species (e.g., Phaseoleae and Desmodium species) except for *Amorpha fruticosa* (a dalbergioid species) [61]. Also, the homologs of *UGT73P2* (POL D09) were identified in syntenic blocks across legumes whereas *UGT73P10* homologs (POL D10) were identified in hologalegina and millettoid species (Figure 5; Table S9). These suggest that galactose being present at second position in the C-3 sugar chain catalyzed by *UGT73P2* is evolutionarily old and conserved. Since the specificity of *UGT73P10* towards soyasaponins was relatively lesser than *UGT73P2* (because all major soyasaponin components (i.e., Aa/Ab and βg) had only galactose as second sugar) and the loss of *UGT73P10* homologs was prevalent in many species (Table S9), we suspect *UGT73P10* must have stemmed out from *UGT73P2* and underwent gene deletion in many species but retained for neofunctionalization in some species especially in the millettoid lineage. Divergence time analysis estimates that the segmental duplication of *UGT73P2* may have occurred at ~54–81 MYA (i.e., before the PWGD) (Table S10). This coincides with the presence of βa in *A. fruticose* [61]. However, the identification of either of αa, βa, and γa, and the true homologs of *UGT73P10* in several early diverged legume species will clarify whether the duplication event occurred before the papilionoid speciation. Until then, based on the presence or absence of *UGT73P10* homologs in 14 legumes, we tentatively assume *UGT73P10* may have duplicated from *UGT73P2* at >48 MYA (i.e., before the hologalegina-millettoid split) (Figure 6).

Like the galactose of C-3 sugar chain, the widespread occurrence of βg indicates rhamnose (catalyzed by *UGT91H4*) being present at the third position was evolutionarily old and conserved, whereas the restricted occurrence of αg and αa in legumes [61] indicates that glucose (catalyzed by *UGT91H9*) at the same position was evolutionarily recent. Notably, the homologs of *UGT91H4* (POL A02–clade I) were identified in syntenic blocks across legumes whereas *UGT91H9* homologs (POL A02–clade II) were identified only in millettoid species (Figure 2B; Table S9). These UGTs showed a segmental or tandem duplication relationship in the millettoid species and the duplication event may have occurred ~44–47 MYA (i.e., after the hologalegina–millettoid split). To support this notion, soyasaponins having glucose at the third position were never identified in legumes other than the millettoid species [61]. Though group A and DDMP saponins have glucose as third sugar in their C-3 sugar chain, only group A saponins (i.e., Aa and Ab) accumulated in high concentration in vivo while none of the DDMP saponins with glucose at the same position accumulated predominantly. Considering these facts, we assume *UGT91H9* must have stemmed out from *UGT91H4* and followed neofunctionalization with high specificity towards group A saponins (Figure 6).

3.2.2. Evolution of the C-22 Sugar Chain of Soyasaponins

In addition to the C-21 hydroxyl position, group A and DDMP saponins are mainly differenced at the C-22-*O* position of their aglycones where the former has arabinose while the latter has DDMP (Figure 6). Distribution of soyasaponins among legumes [61] suggests DDMP moiety at the C-22-*O* position of soyasapogenols (catalyzed by *UGT73K's* [32,53]; POL D05) is evolutionarily old and conserved. The members of D05 shared considerable amino acid identity to D03 members and clustered neighborly (Figure 1) suggesting that both were evolutionarily and phylogenetically related. POL D03 of *G. max* retained a set of tandem duplicated genes in two different chromosomes (i.e., Ch07 and Ch17). One of the genes from the sets was characterized for the addition of second (xylose/glucose catalyzed by *UGT73F4/UGT73F6*; from Ch07) sugar of the C-22 sugar chain of group A saponins (Figure 6). We believe the identification of a gene responsible for the C-22-*O*-arabinosylation will shed more lights on the biosynthetic origin of group A saponins. Of note, though SA identified in other than glycine species (e.g., *M. truncatula* [62] and lupine [63]), group A saponins were only identified in glycine species. Notably, lupine accumulates SA with general C-3 sugar chain (i.e., Rha-Gal-GlcUA-) but had only one sugar at the C-22-*O* position and that too xylose not arabinose [63]. These suggest the gene of C-22-*O*-arabinosylation may have evolved by species-specific functional divergence.

3.3. Triterpene Related UGT POLs and Their Functional Divergence

UGTs modulate the functionalities of different triterpene aglycones by glycosylating them at various active sites depending on the genetic background of given genera/species. Intriguingly, the discovery of UGTs for specific triterpenes in legumes is scarce. For example, the legume model plant *M. truncatula* accumulates at least ten different triterpene aglycones including medicagenic acid, hederagenin, and soyasapogenols, attached with various hexose sugars at various active sites [62]; yet, only three UGTs have been characterized for triterpenes (Table S1). To accelerate/ease the search of UGTs of beneficial triterpenoids in legumes, we herein utilized our POL assignments to estimate the candidate UGTs for legumes triterpenes.

To narrow down triterpene-related UGTs in legumes, we first collected all the studied UGTs of *M. truncatula, G. max, L. japonicus, P. vulgaris,* and *T. pratense* from published literature and mapped the information in the POL tree. UGTs of 15 in *M. truncatula*, 34 in *G. max*, 7 in *L. japonicus* and 1 in *P. vulgaris* have been studied so far; of these, 3 MtUGTs, 8 GmUGTs and 1 LjUGT are characterized for triterpene glycosylation (Tables S1–S5). These 12 UGTs were evolutionarily related to 11 POLs (Figure 1). Earlier studies show that a part of group A, D and E members of legume UGTs are capable to glycosylate triterpenes (Figure 1). Since none of group E members of legumes characterized for triterpene glycosylation in vivo and *UGT71G1* (only this member was attributed as triterpene related and belonged to POL E06) glycosylated flavones and isoflavones with higher efficiency than triterpenes in in vitro [54], we believe that group E members shall not be specific for triterpenes. We also underline that researchers shall not conclude the functions of a given UGT solely based on the in vitro experiments, without its or its homologs in vivo functional analysis. This notion could be supported by several examples. To describe few, (i) *UGT73K1* glucosylated hederagenin, SB and soyasapogenol E in vitro [54] but its homologs attached DDMP moieties to SB in vivo [32,48], and (ii) *UGT73F2* glycosylated isoflavones in vitro [64] but it was later reported to be specific for soyasaponins using in vivo and in vitro experiments [29].

Based on the phylogenetic clustering of characterized UGTs, we propose here that at least four POLs from group A (A02, A03, A09, and A10) and 13 POLs from group D (D01, D03,

D05–D10, D15, D16, D18, D19, and D21) could glycosylate diverse triterpene scaffolds in different legume species. During our study to discover soyasaponin UGTs in *G. max*, we found that the members of A03 (*Glyma.15G051400*), D01 (*Glyma.10G280400* and *Glyma.15G221300*), D03 (*Glyma.07G254700–Glyma.07G254900* and *Glyma.17G019400–Glyma.17G019600*), D08 (*Glyma.02G104600*), and D15 (*Glyma.01G188800*) are not involved in soyasaponin biosynthesis in vivo and one of them glycosylated either of soyasaponin aglycones or glycosides in vitro [53]. In this context, we presume that the homologs of these genes in other species may glycosylate diverse triterpenes in vitro and/or in vivo. Supporting this notion, a homolog of D08 in *Glycyrrhiza uralensis* (*GuUGAT*; GenBank ID: ANJ03631.1–79.3% amino acid identity to *Glyma.02G104600*) [65] glycosylated C-3-*O* position of glycyrrhetinic acid in vitro and a homolog of D03 namely *UGT73F17* (GenBank ID: AXS75258.1–69.4% amino acid identity to *Glyma.17G019500*) from G. *uralensis* glycosylated C-30 of glycyrrhizic acid [66]. Though the in vivo functions of *GuUGAT* and *UGT73F17* remain to be studied, these data imply D03 and D08 members may underwent species-specific functional divergence.

4. Materials and Methods

4.1. Identification of Putative UGTs in Five Legumes

Proteins containing the PF00201 (UDP-glucuronosyl/glucosyltransferase) domain were retrieved for *M. truncatula, G. max, P. vulgaris, L. japonicus,* and *T. pratense* from the LIS database (http:// legumeinfo.org/) [51]. Concurrently, a stand-alone blast-p search was performed with the PSPG sequence of known UGTs (*UGT73F3* for *M. truncatula, L. japonicus, T. pratense* and *UGT73F2* for *G. max* and *P. vulgaris*) against the available respective proteome data in Phytozome v12.1 database (https: //phytozome.jgi.doe.gov/pz/portal.html#) [50] and Miyakogusa database (http://www.kazusa.or.jp/lotus; for *L. japonicus*) [67]. These databases were searched with default settings except the function '# the number of alignments to show' in Phytozome that was set at 300. All the retrieved primary sequences (i.e., spliced transcripts ignored) were manually checked; and, proteins that had incomplete PSPG compare to their orthologs/paralogs and proteins whose first amino acid is not methionine were excluded from the study. Additionally, short-length proteins (i.e., proteins having less than 350 amino acids), and too lengthy proteins (i.e., proteins having more than 600 amino acids) were excluded. If UGTs of same species share 100% amino acid identity at full-length protein level, one of UGTs from the given identical pair was excluded further. Criteria of each excluded sequence from this study were described in Tables S1–S5.

4.2. Phylogenetic Analysis

To determine the evolutionary relationship and the presence/absence of UGT's phylogenetic groups, the selected amino acid sequences of five legume species were aligned together or separately with 14 AtUGTs (groups A-N), three ZmUGTs (groups O-Q), and one CsUGT (group R) by MUSCLE and used to construct neighbor-joining (NJ)-oriented unrooted phylogenetic trees. All multiple sequence alignments and phylogenetic trees generation were performed by MEGA6 program [68]. Trees were constructed under Poisson model, uniform rates and pairwise deletion options with 1000 bootstrap replicates which values were expressed as percentages in each node.

4.3. Assignment of POL for Legume UGTs

To assign UGTs POL, multi-species phylogenetic trees were constructed separately for each phylogenetic group in MEGA6 (Figures S1–S11). Trees were generated as mentioned in the previous section. A POL was assigned based on phylogenetic clustering and the amino acid percent identity of full-length proteins, by applying two conditions that sequence conservation shall be relatively high among UGTs in the given POL across species and the UGTs must cluster together in the phylogenetic analyses. Sequence identities were inferred from Clustal Omega (http://www.ebi.ac.uk/Tools/msa/ clustalo/) [69]. POLs were named within each group in chronological order according to MtUGTs present in the given POL. If no MtUGTs are present in the given POL, the next species UGTs positions were utilized for the POL naming.

4.4. Estimation of Intron Addition or Deletion Events

To estimate intron addition or deletion events, first, intron information of all the identified UGTs was mapped in the POL assignments data (Table S7). Second, POLs were classified into three types namely (i) no-intron POLs, (ii) one-intron POLs, and (iii) mixed-intron POLs, based on the intron numbers present in the given POL across species. If most (if not all) members from three or more species in the given POL sustain no or one intron, that given POL was defined as no-intron POL or one-intron POL, respectively. Mixed-intron POLs sustain members with or without introns, which could not let us to make any concrete decision. Third, intron addition or deletion events were examined for no-intron and one-intron POLs: if the members do not follow the designation of a given POL, they were considered as intron-gained or intron-lost members. Additionally, if any members from any intron POL types had more than one intron, they were considered as intron gained members.

4.5. Chromosomal Mapping, Gene Duplication, and Divergence Time Analyses

The physical locations of UGTs were plotted on chromosomes by Map Chart 2.2 software [70] using the chromosomal coordinates of MtUGTs and GmUGTs that were respectively inferred from their most recent genome versions. UGT members in each POL within species could be considered as paralogs and across species could be considered as orthologs. The term homologs represent both paralogs and orthologs within and across species. Duplicated copies separated by four or fewer other gens were attributed as tandem duplicates while other copies were attributed as segmental duplicates.

The amino acid sequences of each duplicated pair or each duplicated group were aligned separately by MUSCLE in MEGAX [71] using neighbor-joining method, with the first and/or final 10 amino acids in each alignment were checked manually and modified if necessary. These alignments were then used to guide the alignment of corresponding coding sequences in RevTrans 2.0b server [72]. The resulting coding sequence alignments were utilized for the calculation of synonymous (dS) and nonsynonymous (dN) nucleotide substitution rates per site using yn00 tool implemented in PAML package [73]. The obtained dS values of Nei-Gojobori method were used in the formula $T = dS/(2 \times \lambda) \times 10^{-6}$ to estimate the divergence time (T) of duplicated pairs. Assuming the PWGD occurred at ~58 MYA, the λ (rate of dS nucleotide substitutions per site per year) was 1.08×10^{-8}, 5.85×10^{-9}, 8.46×10^{-9}, 6.05×10^{-9} and 8.12×10^{-9} for *M. truncatula* [74], *G. max* [75], *P. vulgaris* [75], *C. arietinum* [76] and *A. duranensis/A. ipaensis* [77] species, respectively.

4.6. Microsynteny Analyses

Microsynteny relationship of triterpenoid-related UGTs among legumes was inferred from the online tool Genome Context Viewer (https://legumeinfo.org/lis_context_viewer/instructions) [78]. Gene IDs of *G. max*, *M. truncatula* or *P. vulgaris* belonged to triterpenoid-related POLs were subjected and the tool was run with default settings. The resulting output files were aligned using MS office.

5. Conclusions

Based on the multi-species phylogenetic relationship and amino acid identity percentage, POLs were successfully assigned to each UGTs identified in this study. The loss/retention of POLs, addition/deletion of introns and the multiplication of UGTs in a given POL were merely species-specific followed by lineage-specific. Notably, a rampant duplication in four POLs accounted for 30% of total UGTs in *M. truncatula* while that never happened for other legumes. In *M. truncatula* and *G. max*, 43–47% of POLs retained single copies and the remaining of them retained two or multiple copies accounting 80–85% of the total number of UGTs. Tandem duplication majorly contributed to the expansion of UGT family in *M. truncatula* (61.3%) and *G. max* (51.4%). Besides the expansion, both species lost many UGTs and different POLs in species-specific manner during their course of evolution. UGTs reported to diversify the C-3 sugar chain of soyasaponins were all resulted from two independent duplication events while the UGTs reported for the C-22-*O* glycosylation of soyasaponins were evolutionarily close. The members from 13 group D and 4 group A POLs could be triterpene related. In sum, our study paved a way to decipher evolutionary dynamics of UGTs, emphasized the contribution of duplication and neofunctionalization of UGTs in triterpene glycoside diversification and will assist in precise selection of candidate UGTs for various specialized metabolites across legumes.

Supplementary Materials: Supplementary materials can be found at http://www.mdpi.com/1422-0067/21/5/1855/s1.

Author Contributions: P.K. conceived, designed and performed all the works. C.T. and M.I. supervised the experiments and provided critical comments. All authors have read and agreed to the published version of the manuscript.

Funding: This work was supported by the Japan Society for the Promotion of Science (JSPS) KAKENHI (Grants-in-Aid for JSPS Research Fellows [15F P15390]) to P.K. and M.I.

Conflicts of Interest: The authors declare no conflict of interest.

References

1. Coutinho, P.M.; Deleury, E.; Davies, G.J.; Henrissat, B. An evolving hierarchical family classification for glycosyltransferases. *J. Mol. Biol.* **2003**, *328*, 307–317. [CrossRef]

2. Schuman, B.; Alfaro, J.A.; Evans, S.V. Glycosyltransferase structure and function. In *Topics in Current Chemistry*; Peters, T., Ed.; Springer: Berlin/Heidelberg, Germany, 2007; Volume 272, pp. 211–257.

3. Wagner, G.K.; Pesnot, T. Glycosyltransferases and their assays. *ChemBioChem* **2010**, *11*, 1939–1949. [CrossRef] [PubMed]

4. Lairson, L.L.; Henrissat, B.; Davies, G.J.; Withers, S.G. Glycosyltransferases: Structures, functions, and mechanisms. *Annu. Rev. Biochem.* **2008**, *77*, 521–555. [CrossRef] [PubMed]

5. Liang, D.M.; Liu, J.H.; Wu, H.; Wang, B.B.; Zhu, H.J.; Qiao, J.J. Glycosyltransferases: Mechanisms and applications in natural product development. *Chem. Soc. Rev.* **2015**, *44*, 8350–8374. [CrossRef] [PubMed]

6. Osmani, S.A.; Bak, S.; Møller, B.L. Substrate specificity of plant UDP-dependent glycosyltransferases predicted from crystal structures and homology modeling. *Phytochemistry* **2009**, *70*, 325–347. [CrossRef]

7. Lombard, V.; Golaconda Ramulu, H.; Drula, E.; Coutinho, P.M.; Henrissat, B. The carbohydrate-active enzymes database (CAZy) in 2013. *Nucleic Acids Res.* **2013**, *42*, D490–D495.

8. Bowles, D.; Lim, E.K.; Poppenberger, B.; Vaistij, F.E. Glycosyltransferases of lipophilic small molecules. *Annu. Rev. Plant Biol.* **2006**, *57*, 567–597. [CrossRef]

9. Yonekura-Sakakibara, K.; Hanada, K. An evolutionary view of functional diversity in family 1 glycosyltransferases. *Plant J.* **2011**, *66*, 182–193. [CrossRef]

10. Gachon, C.M.; Langlois-Meurinne, M.; Saindrenan, P. Plant secondary metabolism glycosyltransferases: The emerging functional analysis. *Trends Plant Sci.* **2005**, *10*, 542–549. [CrossRef]

11. Wang, X. Structure, mechanism and engineering of plant natural product glycosyltransferases. *FEBS Lett.* **2009**, *583*, 3303–3309. [CrossRef]

12. Bowles, D.; Isayenkova, J.; Lim, E.K.; Poppenberger, B. Glycosyltransferases: Managers of small molecules. *Curr. Opin. Plant Biol.* **2005**, *8*, 254–263. [CrossRef] [PubMed]

13. Härtl, K.; MacGraphery, K.; Rüdiger, J.; Schwab, W. Tailoring natural products with glycosyltransferases. In *Biotechnology of Natural Products*, 1st ed.; Schwab, W., Lange, B.M., Wüst, M., Eds.; Springer International Publishing AG: Heidelberg, Germany, 2018; pp. 219–263.

14. Brazier-Hicks, M.; Offen, W.A.; Gershater, M.C.; Revett, T.J.; Lim, E.K.; Bowles, D.J.; Davies, G.J.; Edwards, R. Characterization and engineering of the bifunctional *N*- and *O*-glucosyltransferase involved in xenobiotic metabolism in plants. *Proc. Natl. Acad. Sci. USA* **2007**, *104*, 20238–20243. [CrossRef] [PubMed]

15. Li, X.; Michlmayr, H.; Schweiger, W.; Malachova, A.; Shin, S.; Huang, Y.; Dong, Y.; Wiesenberger, G.; McCormick, S.; Lemmens, M.; et al. A barley UDP-glucosyltransferase inactivates nivalenol and provides Fusarium Head Blight resistance in transgenic wheat. *J. Exp. Bot.* **2017**, *68*, 2187–2197. [CrossRef] [PubMed]

16. Goossens, A.; Osbourn, A.; Michoux, F.; Bak, S. Triterpene messages from the EU-FP7 project TriForC. *Trends Plant Sci.* **2018**, *23*, 273–276. [CrossRef] [PubMed]

17. Osbourn, A.; Goss, R.J.M.; Field, R.A. The saponins—Polar isoprenoids with important and diverse activities. *Nat. Prod. Rep.* **2011**, *28*, 1261–1268. [CrossRef]

18. Seki, H.; Tamura, K.; Muranaka, T. Plant-derived isoprenoid sweetners: Recent progress in biosynthetic gene discovery and perspectives on microbial production. *Biosci. Biotech. Biochem.* **2018**, *82*, 927–934. [CrossRef]

19. Thimmappa, R.; Geisler, K.; Louveau, T.; O'Maille, P.; Osbourn, A. Triterpene biosynthesis in plants. *Annu. Rev. Plant Biol.* **2014**, *65*, 225–257. [CrossRef]

20. Salmon, M.; Thimmappa, R.B.; Minto, R.E.; Melton, R.E.; Hughes, R.K.; O'Maille, P.E.; Hemmings, A.M.; Osbourn, A. A conserved amino acid residue critical for product and substrate specificity in plant triterpene synthases. *Proc. Natl. Acad. Sci. USA* **2016**, *113*, E4407–E4414. [CrossRef]

21. Xue, Z.; Duan, L.; Liu, D.; Guo, J.; Ge, S.; Dicks, J.; O'Maille, P.; Osbourn, A.; Qi, X. Divergent evolution of oxidosqualene cyclases in plants. *New Phytol.* **2012**, *193*, 1022–1038. [CrossRef]

22. Krishnamurthy, P.; Fujisawa, Y.; Takahashi, Y.; Abe, H.; Yamane, K.; Mukaiyama, K.; Son, H.R.; Hiraga, S.; Kaga, A.; Anai, T.; et al. High throughput screening and characterization of a high-density soybean mutant library elucidate the biosynthesis pathway of triterpenoid saponins. *Plant Cell Physiol.* **2019**, *60*, 1082–1097. [CrossRef]

23. Louveau, T.; Orme, A.; Pfalzgraf, H.; Stephenson, M.J.; Melton, R.; Saalbach, G.; Hemmings, A.M.; Leveau, A.; Rejzek, M.; Vickerstaff, R.J.; et al. Analysis of two new arabinosyltransferases belonging to the carbohydrate-active enzyme (CAZY) glycosyl transferase family 1 provides insights into disease resistance and sugar donor specificity. *Plant Cell* **2018**, *30*, 3038–3057. [CrossRef]

24. Augustin, J.M.; Drok, S.; Shinoda, T.; Sanmiya, K.; Nielsen, J.K.; Khakimov, B.; Olsen, C.E.; Hansen, E.H.; Kuzina, V.; Ekstrøm, C.T.; et al. UDP-glycosyltransferases from the UGT73C subfamily in *Barbarea vulgaris* catalyse sapogenin 3-O-glucosylation in saponin-mediated insect resistance. *Plant Physiol.* **2012**, *160*, 1881–1895. [CrossRef] [PubMed]

25. Guang, C.; Chen, J.; Sang, S.; Cheng, S. Biological functionality of soyasaponins and soyasapogenols. *J. Agri. Food Chem.* **2014**, *62*, 8247–8255. [CrossRef] [PubMed]

26. Rahman, A.; Tsurumi, S. The unique auxin influx modulator chromosaponin I: A physiological overview. *Plant Tissue Cult.* **2002**, *12*, 181–194.

27. Yano, R.; Takagi, K.; Takada, Y.; Mukaiyama, K.; Tsukamoto, C.; Sayama, T.; Kaga, A.; Anai, T.; Sawai, S.; Ohyama, K.; et al. Metabolic switching of astringent and beneficial triterpenoid saponins in soybean is achieved by a loss-of-function mutation in cytochrome P450 72A69. *Plant, J.* **2017**, *89*, 527–539. [CrossRef] [PubMed]

28. Shibuya, M.; Nishimura, K.; Yasuyama, N.; Ebizuka, Y. Identification and characterization of glycosyltransferases involved in the biosynthesis of soyasaponin I in *Glycine max*. *FEBS Lett.* **2010**, *584*, 2258–2264. [CrossRef] [PubMed]

29. Sayama, T.; Ono, E.; Takagi, K.; Takada, Y.; Horikawa, M.; Nakamoto, Y.; Hirose, A.; Sasama, H.; Ohashi, M.; Hasegawa, H.; et al. The *Sg-1* glycosyltransferase locus regulates structural diversity of triterpenoid saponins of soybean. *Plant Cell* **2012**, *24*, 2123–2138. [CrossRef] [PubMed]

30. Takagi, K.; Yano, R.; Tochigi, S.; Fujisawa, Y.; Tsuchinaga, H.; Takahashi, Y.; Takada, Y.; Kaga, A.; Anai, T.; Tsukamoto, C.; et al. Genetic and functional characterization of Sg-4 glycosyltransferase involved in the formation of sugar chain structure at the C-3 position of soybean saponins. *Phytochemistry* **2018**, *156*, 96–105. [CrossRef] [PubMed]

31. Yano, R.; Takagi, K.; Tochigi, S.; Fujisawa, Y.; Nomura, Y.; Tsuchinaga, H.; Takahashi, Y.; Takada, Y.; Kaga, A.; Anai, T.; et al. Isolation and characterization of the soybean Sg-3 gene that is involved in genetic variation in sugar chain composition at the C-3 position in soyasaponins. *Plant Cell Physiol.* **2018**, *59*, 792–805. [CrossRef]

32. Sundaramoorthy, J.; Par, G.T.; Komagamine, K.; Tsukamoto, C.; Chang, J.H.; Lee, J.D.; Kim, J.H.; Seo, H.S.; Song, J.T. Biosynthesis of DDMP saponins in soybean is regulated by a distinct UDP-glycosyltransferase. *New Phytol.* **2019**, *222*, 261–274. [CrossRef]

33. Christenhusz, M.J.M.; Byng, J.W. The number of known plant species in the world and its annual increase. *Phytotaxa* **2016**, *261*, 201–217. [CrossRef]

34. Wink, M. Evolution of secondary metabolites in legumes (Fabaceae). *S. Afr. J. Bot.* **2013**, *89*, 164–175. [CrossRef]

35. Wang, J.; Sun, P.; Li, Y.; Liu, Y.; Yu, J.; Ma, X.; Sun, S.; Yang, N.; Xia, R.; Lei, T.; et al. Hierarchically aligning 10 legume genomes establishes family-level genomics platform. *Plant Physiol.* **2017**, *174*, 284–300. [CrossRef] [PubMed]

36. Caputi, L.; Malnoy, M.; Goremykin, V.; Nikiforova, S.; Martens, S. A genome-wide phylogenetic reconstruction of family 1 UDP-glycosyltransferases revealed the expansion of the family during the adaptation of plants to life on land. *Plant J.* **2012**, *69*, 1030–1042. [CrossRef] [PubMed]

37. Cui, L.; Yao, S.; Dai, X.; Yin, Q.; Liu, Y.; Jiang, X.; Wu, Y.; Qian, Y.; Pang, Q.; Gao, L.; et al. Identification of UDP-glycosyltransferases involved in the biosynthesis of astringent taste compounds in tea (*Camellia sinensis*). *J. Exp. Bot.* **2016**, *67*, 2285–2297. [CrossRef] [PubMed]

38. Barvkar, V.T.; Pardeshi, V.C.; Kale, S.M.; Kadoo, N.Y.; Gupta, V.S. Phylogenomic analysis of UDP glycosyltransferase 1 multigene family in *Linum usitatissimum* identified genes with varied expression patterns. *BMC Genomics* **2012**, *13*, 175. [CrossRef]

39. Rehman, H.M.; Nawaz, M.A.; Shah, Z.H.; Ludwig-Muller, J.; Chung, G.; Ahmad, M.Q.; Yang, S.H.; Lee, S.I. Comparative genomic and transcriptomic analyses of Family-1 UDP glycosyltransferase in three Brassica species and Arabidopsis indicates stress-responsive regulation. *Sci. Rep.* **2018**, *8*, 1875. [CrossRef]

40. Song, Z.; Niu, L.; Yang, Q.; Dong, B.; Wang, L.; Dong, M.; Fan, X.; Jian, Y.; Meng, D.; Fu, Y. Genome-wide identification and characterization of UGT family in pigeonpea (*Cajanus cajan*) and expression analysis in abiotic stress. *Trees* **2019**, *33*, 987–1002. [CrossRef]

41. Rehman, H.M.; Nawaz, M.A.; Bao, L.; Shah, Z.H.; Lee, J.M.; Ahmad, M.Q.; Chung, G.; Yang, S.H. Genome-wide analysis of family-1 UDP-glycosyltransferases in soybean confirms their abundance and varied expression during seed development. *J. Plant Physiol.* **2016**, *206*, 87–97. [CrossRef]

42. Zhang, Z.; Zhuo, X.; Yan, Z.; Zhang, Q. Comparative genomic and transcriptomic analyses of family-1 UDP glycosyltransferase in *Prunus mume*. *Int. J. Mol. Sci.* **2018**, *19*, 3382. [CrossRef]

43. Wu, B.; Gao, L.; Gao, J.; Xu, Y.; Liu, H.; Cao, X.; Zhang, B.; Chen, K. Genome-wide identification, expression patterns, and functional analysis of UDP glycosyltransferase family in peach (*Prunus persica* L. Batsch). *Front. Plant Sci.* **2017**, *8*, 389. [CrossRef] [PubMed]

44. He, Y.; Ahmad, D.; Zhang, X.; Zhang, Y.; Wu, L.; Jiang, P.; Ma, H. Genome-wide analysis of family-1 UDP glycosyltransferases (UGT) and identification of UGT genes for FHB resistance in wheat (*Triticum aestivum* L.). *BMC Plant Biol.* **2018**, *18*, 67. [CrossRef] [PubMed]

45. Li, Y.; Li, P.; Wang, Y.; Dong, R.; Yu, H.; Hou, B. Genome-wide identification and phylogenetic analysis of Family-1 UDP glycosyltransferases in maize (*Zea mays*). *Planta* **2014**, *239*, 1265–1279. [CrossRef] [PubMed]

46. Yin, Q.; Shen, G.; Di, S.; Fan, C.; Chang, Z.; Pang, Y. Genome-wide identification and functional characterization of UDP-glucosyltransferase genes involved in flavonoid biosynthesis in *Glycine max*. *Plant Cell Physiol.* **2017**, *58*, 1558–1572. [CrossRef] [PubMed]

47. Yin, Q.; Shen, G.; Chang, Z.; Tang, Y.; Gao, H.; Pang, Y. Involvement of three putative glucosyltransferases from the UGT72 family in flavonol glucoside/rhamnoside biosynthesis in *Lotus japonicus* seeds. *J. Exp. Bot.* **2017**, *68*, 597–612. [PubMed]

48. Tomcal, M.; Stiffler, N.; Barkan, A. POGs2: A web portal to facilitate cross-species inferences about protein architecture and function in plants. *PLoS ONE* **2013**, *8*, e82569. [CrossRef]

49. Van Bel, M.; Diels, T.; Vancaester, E.; Kreft, L.; Botzki, A.; Van de Peer, Y.; Coppens, F.; Vandepoele, K. PLAZA 4.0: An integrative resource for functional, evolutionary and comparative plant genomics. *Nucleic Acids Res.* **2017**, *46*, D1190–D1196. [CrossRef]

50. Goodstein, D.M.; Shu, S.; Howson, R.; Neupane, R.; Hayes, R.D.; Fazo, J.; Mitros, T.; Dirks, W.; Hellsten, U.; Putnam, N.; et al. Phytozome: A comparative platform for green plant genomics. *Nucleic Acids Res.* **2012**, *40*, D1178–D1186. [CrossRef]

51. Dash, S.; Campbell, J.D.; Cannon, E.K.; Cleary, A.M.; Huang, W.; Kalberer, S.R.; Karingula, V.; Rice, A.G.; Singh, J.; Umale, P.E.; et al. Legume information system (LegumeInfo.org): A key component of a set of federated data resources for the legume family. *Nucleic Acids Res.* **2016**, *44*, D1181–D1188. [CrossRef]

52. Naoumkina, M.A.; Modolo, L.V.; Huhman, D.V.; Urbanczyk-Wochniak, E.; Tang, Y.; Sumner, L.W.; Dixon, R.A. Genomic and coexpression analyses predict multiple genes involved in triterpene saponin biosynthesis in Medicago truncatula. *Plant Cell* **2010**, *22*, 850–866. [CrossRef]

53. Ishimoto M's Research Group. Identification of UGTs involved in soyasaponin glycosylation. Unpublished Work.

54. Achnine, L.; Huhman, D.V.; Farag, M.A.; Sumner, L.W.; Blount, J.W.; Dixon, R.A. Genomics-based selection and functional characterization of triterpene glycosyltransferases from the model legume Medicago truncatula. *Plant J.* **2005**, *41*, 875–887. [CrossRef] [PubMed]

55. Proost, S.; Van Bel, M.; Sterck, L.; Billiau, K.; Van Parys, T.; Van de Peer, Y.; Vandepoele, K. PLAZA: A comparative genomics resource to study gene and genome evolution in plants. *Plant Cell* **2009**, *21*, 3718–3731. [CrossRef] [PubMed]

56. Trachana, K.; Larsson, T.A.; Powell, S.; Chen, W.H.; Doerks, T.; Muller, J.; Bork, P. Orthology prediction methods: A quality assessment using curated protein families. *Bioessays* **2011**, *33*, 769–780. [CrossRef] [PubMed]

57. Walker, N.S.; Stiffler, N.; Barkan, A. POGs/PlantRBPs: A resource for comparative genomics in plants. *Nucl. Acid Res.* **2007**, *55*, D852–D856. [CrossRef]

58. De Vega, J.J.; Ayling, S.; Hegarty, M.; Kudrna, D.; Goicoechea, J.L.; Ergon, A.; Rognli, O.A.; Jones, C.; Swain, M.; Geurts, R.; et al. Red clover (*Trifolium pratense* L.) draft genome provides a platform for trait improvement. *Sci. Rep.* **2015**, *5*, 17394. [CrossRef]

59. Ober, D. Seeing double: Gene duplication and diversification in plant secondary metabolism. *Trends. Plant Sci.* **2005**, *10*, 444–449. [CrossRef]

60. Hamberger, B.; Bak, S. Plant P450s as versatile drivers for evolution of species-specific chemical diversity. *Philos. Trans. R. Soc. B* **2013**, *368*, 20120426. [CrossRef]

61. Okubo, K.; Yoshiki, Y. Oxygen-radical-scavenging activity of DDMP-conjugated saponins and physiological role in leguminous plant. In *Saponins Used in Food and Agriculture*; Waller, G.R., Yamasaki, K., Eds.; Plenum Press: New York, NY, USA, 1996; Volume 405, pp. 141–154.

62. Pollier, J.; Morreel, K.; Geelen, D.; Goossens, A. Metabolite profiling of triterpene saponins in Medicago truncatula hairy roots by liquid chromatography fourier transform ion cyclotron resonance mass spectrometry. *J. Nat. Prod.* **2011**, *74*, 1462–1476. [CrossRef]

63. Kinjo, J.; Kishida, F.; Watanabe, K.; Hashimoto, F.; Nohara, T. Five new triterpene glycosides from Russell lupine. *Chem. Parm. Bull.* **1994**, *42*, 1874–1878. [CrossRef]

64. Dhaubhadel, S.; Farhangkhoee, M.; Chapman, R. Identification and characterization of isoflavonoid specific glycosyltransferase and malonyltransferase from soybean seeds. *J. Exp. Bot.* **2004**, *59*, 981–994. [CrossRef]

65. Xu, G.; Cai, W.; Gao, W.; Liu, C. A novel glucuronosyltransferase has an unprecedented ability to catalyse continuous two-step glucuronosylation of glycyrrhetinic acid to yield glycyrrhizin. *New Phytol.* **2016**, *212*, 123–135. [CrossRef] [PubMed]

66. He, J.; Chen, K.; Hu, Z.M.; Li, K.; Song, W.; Yu, L.Y.; Leung, C.H.; Ma, D.L.; Qiao, X.; Ye, M. UGT73F17, a new glycosyltransferase from *Glycyrrhiza uralensis*, catalyzes the regiospecific glycosylation of pentacyclic triterpenoids. *Chem. Commun.* **2018**, *54*, 8594–8597. [CrossRef] [PubMed]

67. Sato, S.; Nakamura, Y.; Kaneko, T.; Asamizu, E.; Kato, T.; Nakao, M.; Sasamoto, S.; Watanabe, A.; Ono, A.; Kawashima, K.; et al. Genome structure of the legume, *Lotus japonicus*. *DNA Res.* **2008**, *15*, 227–239. [CrossRef] [PubMed]

68. Tamura, K.; Stecher, G.; Peterson, D.; Filipski, A.; Kumar, S. MEGA6: Molecular evolutionary genetics analysis version 6.0. *Mol. Biol. Evol.* **2013**, *30*, 2725–2729. [CrossRef] [PubMed]

69. Sievers, F.; Higgins, D.G. Clustal omega. *Curr. Protoc. Bioinform.* **2014**, *48*, 3.13.1–3.13.16. [CrossRef] [PubMed]

70. Voorrips, R.E. MapChart: Software for the graphical presentation of linkage maps and QTLs. *J. Hered.* **2002**, *93*, 77–78. [CrossRef]

71. Kumar, S.; Stecher, G.; Li, M.; Knyaz, C.; Tamura, K. MEGA X: Molecular evolutionary genetics analysis across computing platforms. *Mol. Biol. Evol.* **2018**, *35*, 1547–1549. [CrossRef]

72. Wemersson, R.; Pedersen, A.G. RevTrans-Constructing alignments of coding DNA from aligned amino acid sequences. *Nucleic Acids Res.* **2003**, *31*, 3537–3539. [CrossRef]

73. Yang, Z. PAML 4: Phylogenetic analysis by maximum likelihood. *Mol. Biol. Evol.* **2007**, *24*, 1586–1591. [CrossRef]

74. Young, N.D.; Debelle, F.; Oldroyd, G.E.D.; Geurts, R.; Cannon, S.B.; Udvardi, M.K.; Benedito, V.A.; Mayer, K.F.X.; Gouzy, J.; Schoof, H.; et al. The Medicago genome provides insight into the evolution of rhizobial symbioses. *Nature* **2011**, *480*, 520–524. [CrossRef]

75. Schmutz, J.; McClean, P.E.; Mamidi, S.; Wu, G.A.; Cannon, S.B.; Grimwood, J.; Jenkins, J.; Shu, S.; Song, Q.; Chavarro, C.; et al. A reference genome for common bean and genome-wide analysis of dual domestications. *Nat. Genet.* **2014**, *46*, 707–713. [CrossRef] [PubMed]
76. Jain, M.; Misra, G.; Patel, R.K.; Priya, P.; Jhanwar, S.; Khan, A.W.; Shah, N.; Singh, V.K.; Garg, R.; Jeena, G.; et al. A draft genome sequence of the pulse crop chickpea (*Cicer arietinum* L.). *Plant J.* **2013**, *74*, 715–729. [CrossRef] [PubMed]
77. Bertioli, D.J.; Cannon, S.B.; Froenicke, L.; Huang, G.; Farmer, A.D.; Cannon, E.K.S.; Liu, X.; Gao, D.; Clevenger, J.; Dash, S.; et al. The genome sequences of *Arachis duranensis* and *Arachis ipaensis*, the diploid ancestors of cultivated peanut. *Nat. Genet.* **2016**, *48*, 438–446. [CrossRef] [PubMed]
78. Cleary, A.; Farmer, A. Genome Context Viewer: Visual exploration of multiple annotated genomes using microsynteny. *Bioinformatics* **2018**, *34*, 1562–1564. [CrossRef]

International Journal of
Molecular Sciences

Article

High-Resolution Mapping in Two RIL Populations Refines Major "QTL Hotspot" Regions for Seed Size and Shape in Soybean (*Glycine max* L.)

Aiman Hina [1], Yongce Cao [2], Shiyu Song [1], Shuguang Li [1], Ripa Akter Sharmin [1], Mahmoud A. Elattar [1], Javaid Akhter Bhat [1,*] and Tuanjie Zhao [1,*]

1   Ministry of Agriculture (MOA) Key Laboratory of Biology and Genetic Improvement of Soybean (General), State Key Laboratory for Crop Genetics and Germplasm Enhancement, Soybean Research Institute, National Center for Soybean Improvement, Nanjing Agricultural University, Nanjing 210095, China; aimanhina@yahoo.com (A.H.); songshiyu0706@126.com (S.S.); dawn0524@126.com (S.L.); ripa.sharmin@gmail.com (R.A.S.); mahmoud891987@gmail.com (M.A.E.)

2   Shaanxi Key Laboratory of Chinese Jujube; College of Life Science, Yan'an University, Yan'an 716000, China; caoyongce@yau.edu.cn

*   Correspondence: javid.akhter69@gmail.com (J.A.B.); tjzhao@njau.edu.cn (T.Z.); Tel.: +86-198-2504-6530 (J.A.B.); +86-25-8439-9531 (T.Z.)

Received: 27 December 2019; Accepted: 1 February 2020; Published: 4 February 2020

Abstract: Seed size and shape are important traits determining yield and quality in soybean. However, the genetic mechanism and genes underlying these traits remain largely unexplored. In this regard, this study used two related recombinant inbred line (RIL) populations (ZY and K3N) evaluated in multiple environments to identify main and epistatic-effect quantitative trait loci (QTLs) for six seed size and shape traits in soybean. A total of 88 and 48 QTLs were detected through composite interval mapping (CIM) and mixed-model-based composite interval mapping (MCIM), respectively, and 15 QTLs were common among both methods; two of them were major ($R^2 > 10\%$) and novel QTLs (viz., $qSW\text{-}1\text{-}1_{ZN}$ and $qSLT\text{-}20\text{-}1_{K3N}$). Additionally, 51 and 27 QTLs were identified for the first time through CIM and MCIM methods, respectively. Colocalization of QTLs occurred in four major QTL hotspots/clusters, viz., "QTL Hotspot A", "QTL Hotspot B", "QTL Hotspot C", and "QTL Hotspot D" located on Chr06, Chr10, Chr13, and Chr20, respectively. Based on gene annotation, gene ontology (GO) enrichment, and RNA-Seq analysis, 23 genes within four "QTL Hotspots" were predicted as possible candidates, regulating soybean seed size and shape. Network analyses demonstrated that 15 QTLs showed significant additive x environment (AE) effects, and 16 pairs of QTLs showing epistatic effects were also detected. However, except three epistatic QTLs, viz., $qSL\text{-}13\text{-}3_{ZY}$, $qSL\text{-}13\text{-}4_{ZY}$, and $qSW\text{-}13\text{-}4_{ZY}$, all the remaining QTLs depicted no main effects. Hence, the present study is a detailed and comprehensive investigation uncovering the genetic basis of seed size and shape in soybeans. The use of a high-density map identified new genomic regions providing valuable information and could be the primary target for further fine mapping, candidate gene identification, and marker-assisted breeding (MAB).

Keywords: Soybean; seed shape; seed size; QTL mapping; high-density genetic map; QTL hotspot; epistatic interactions; candidate genes

---

1. Introduction

Soybean (*Glycine max* L.) is one of the most economically important crops, being a rich source of both edible oil and protein, and can fix atmospheric nitrogen through a symbiotic association with microorganisms in the soil, and are used as a model plant for legume research [1]. However, over the past five decades, a continuous decline in soybean production in China has been recorded [2]. Besides, annually, China imports more than 80% of soybeans and its products to meet its domestic demands; hence, there is an immediate need to increase the domestic production of soybean to make the country self-sufficient [2]. Yield-related traits are the key target of plant breeders to improve soybean yield/production. In this regard, traits related to seed size and shape are the crucial parameters determining seed-weight and yield in soybean [3,4]. In soybean, seed size traits such as length (SL); width (SW) and thickness (ST); and seed shape traits, viz., length-to-width (SLW), length-to-thickness (SLT), and width-to-thickness (SWT) ratios determine seed appearance, quality, and yield in soybeans [5]. Seed size is also a vital fitness trait in flowering plants and plays a crucial role in adaptation to a particular environment [6]. However, seed size and shape are complex quantitative traits governed by polygenes and highly influenced by the environment (E) and genotype × environment (G × E) interactions [7,8]. Specific soy-based food products made from soybean are also determined mainly by seed size and shape [9,10]. For example, for the production of fermented soybeans (natto) and sprouts, small-seeded cultivars are suitable, while for soymilk, green soybeans (edamame), boiled soybeans (nimame), and soybean curd (tofu), large-seeded varieties are used [11–13]. Additionally, these traits influence the germination ability and seedling vigor, and that, in turn, plays an essential role in determining the competitive strength of the seedlings for light, nutrient resources, and stress tolerance [14–16].

Quantitative trait loci (QTL) analysis has proved as a powerful technique to elucidate complex trait architecture. Over the past two decades, recent advances in marker technology and statistical methods have allowed the identification of many QTLs related to seed size and shape traits. The USDA Soybean Genome Database (SoyBase, http://www.soybase.org) presently document more than 400 QTLs for seed size and shape, and the majority of them are not confirmed (http://www.soybase.org). The previous studies used mostly low-resolution and low-density molecular markers such as simple sequence repeats (SSRs) that often result in larger confidence intervals and make the use of these QTLs less effective in crop improvement [3,5,17,18]. For example, Mian et al. [19] reported 16 QTLs for seed size and shape on 12 different chromosomes of soybean. Hoeck et al. [20] identified 27 QTLs associated with seed size distributed on 16 soybean chromosomes, and Li et al. [21] detected three QTLs for SL on Chr07, Chr13, and Chr16. Lü et al. [18] identified 19 main-effect QTLs (M-QTLs) and three epistatic-effect QTLs (E-QTLs) for SL on eight chromosomes. Xie et al. [22] finely mapped QTLs for soybean seed size traits on Chr06 in the recombinant inbred line (RIL) population derived from a cross between Lishuizhongzihuang and Nannong493-1. Likewise, Che et al. [17] identified 16 QTLs for seed shape, distributed on seven linkage groups in soybeans by using the RIL population. Hu et al. [7] mapped 10 QTLs for seed shape on six chromosomes in soybeans. However, only a few yield-related stable QTLs have been identified in different genetic backgrounds and environments [23]. Hence, it is vital to identify and validate QTLs in multiple backgrounds and environments for their potential use in marker-assisted breeding (MAB). Lastly, the earlier studies mostly focused on the identification of main-effect QTLs for seed size/shape in soybean; however, minimal efforts have been made to understand complex genetic interaction effects, such as epistasis and environment effects [24–26].

The inheritance of quantitative traits varies from simple to complex; however, the phenotypic variation of most quantitative traits is complex, governed by many factors [27]. In addition to main-effect QTLs, phenotypic variation (PV) of complex traits is also governed by QTL by QTL (epistatic) and QTL by environment (QTL × E) interactions, which contribute significantly to complex trait variations [28]. By considering these QTL interactions in the QTL mapping model of complex traits will lead to increased precision of QTL mapping [29]. Therefore, these factors cannot be considered only as the main obstacles to dissect the genetic architecture of complex traits, but they also affect the accuracy of breeding value estimation, and thus, hinder the efficiency of breeding programs. Hence, it is imperative to consider these factors while dissecting the genetic basis of complex traits and their uses in improving plant performance. In recent years, epistatic and QTL × E interaction effects are under consideration in several crop species, including soybeans, for QTL mapping [30]. Therefore, extensive efforts are required to study such QTL interaction effects for their effective exploitation in soybean breeding.

Development of high-density genetic maps, and their use in the detection of QTLs/genes, have allowed a detailed and broader understanding of the genetic basis underlying complex quantitative traits. Furthermore, the analysis of genes has partitioned the related traits into individual Mendelian factors [31]. Nevertheless, limited reports are targeting the mapping of QTLs related to seed size and shape based on the high-density map in different genetic backgrounds. Besides, to mine candidate genes for seed size and shape in soybeans, negligible efforts were made. By keeping the above in view, the present study has used a high-density linkage map of two RIL populations, viz., ZY and K3N, evaluated in multiple environments to identify main and epistatic-effect QTLs, as well as their interactions with the environment, to mine candidate genes for seed size and shape in soybeans. These results will be helpful in MAB for developing soybean varieties with improved yield and quality, as well as to clone underlying genes for seed size and shape in soybean.

2. Results

2.1. Evaluation of Phenotypic Variation for RIL Populations

Mean, range (minimum and maximum value), standard deviation, skewness, kurtosis, heritability (h^2), and coefficient of variation (CV%) associated with six soybean seed shape and size traits of two RIL populations (ZY and K3N), along with their parents, evaluated across three different environments, viz., 2012FY, 2012JP, and 2017JP, are presented in Table S1.

The difference in average phenotypic values between the contrasting parents of both RIL populations for all six traits was evident and consistent across all three individual environments (Table S1). The trait value of several RILs exceeded their parents for all studied traits in both directions, suggesting transgressive segregation in both RIL populations (Figure 1). All six traits related to seed size and shape showed different levels of distribution in both RIL populations (ZY and K3N), with mostly skewness and kurtosis <1, and the majority have CV >3%, which is typical for quantitative traits, indicating the suitability of these populations for QTL mapping (Figure 1 and Table S1a,b).

Combined ANOVA results revealed that variations among the RILs of both populations were highly significant ($p < 0.0001$ or $p < 0.05$) for all six traits (Table S2a,b). The environmental differences and G × E interaction effects were also highly significant for all the studied traits, except SLW, SLT, and SWT in the case of the K3N population (Table S2a,b). Heritability in a broad sense (h^2) for both RIL populations in individual, as well as combined, environments was above 60%, indicating high heritability for all studied traits (Table S2a,b). The correlation coefficient (r^2) among the six traits related to seed size and shape for both RIL populations are presented in Table S3. Correlation analysis has shown a significant positive correlation between any two seed shape size traits, and a significant negative correlation exists between seed shape and seed size traits (Table S3).

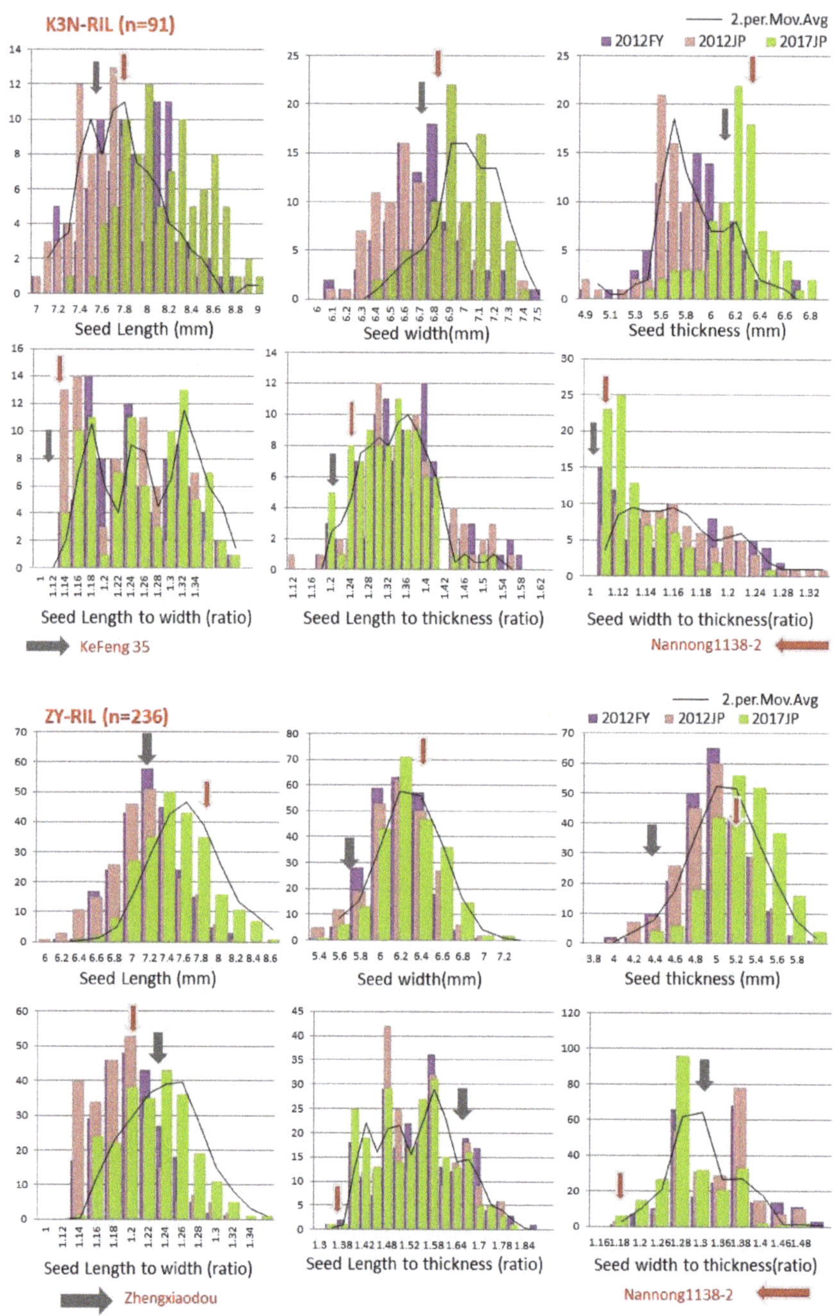

Figure 1. Frequency distribution of seed length (SL), seed width (SW), seed thickness (ST), seed length-to-width (SLW), seed length-to-thickness (SLT), and seed width-to-thickness (SWT) in ZY and K3N recombinant inbred line (RIL) populations across three different environments (2012FY, 2012JP, and 2017JP). Trend lines show the moving average. Arrows represent mean value of corresponding parent. Horizontal and vertical axis represent trait value and number of genotypes, respectively.

2.2. QTL Mapping of Seed Size by CIM

The high-density genetic maps of ZY and K3N populations were used to perform a linkage analysis for the identification of QTLs associated with SL, SW, and ST in soybeans. In total, we identified 50 main-effect QTLs associated with three seed size traits, viz., SL, SW, and ST, explaining the phenotypic variation (PV/R^2) of 4.46–22.64%, mapped on 18 soybean chromosomes in both ZY and K3N populations across three environments, viz., 2012FY, 2012JP, and 2017JP (Table 1 and Figure 2). For seed length (SL), 14 main-effect QTLs were detected on ten different chromosomes (Table 1). Among them, $qSL\text{-}9\text{-}1_{ZY,\ K3N}$ was stable and had significant QTL with an average $R^2 = 10.01\%$ and are consistently found in two individual environments (2012FY and 2017JP), as well as in both RIL populations (ZY and K3N) (Table 1). Additionally, $qSL\text{-}13\text{-}1_{ZY}$, expressing a PV of 8.26%, was detected in two different environments (2017JP and 2012JP) in the ZY population (Table 1). Moreover, one minor stable QTL, $qSL\text{-}4\text{-}1_{ZY}$, expressing an average PV of 6%, was consistently identified in all three studied environments, viz., 2012FY, 2012JP, and 2017JP (Table 1). Four major QTLs, viz., $qSL\text{-}11\text{-}1_{K3N}$, $qSL\text{-}17\text{-}1_{K3N}$, $qSL\text{-}18\text{-}1_{K3N}$, and $qSL\text{-}20\text{-}1_{K3N}$, with $R^2 > 10\%$, were environmental-sensitive and identified in only one environment in the K3N population (Table 1). The remaining seven minor QTLs ($R^2 < 10\%$), viz., $qSL\text{-}6\text{-}1_{ZY}$, $qSL\text{-}6\text{-}2_{ZY}$, $qSL\text{-}6\text{-}3_{ZY}$, $qSL\text{-}9\text{-}2_{ZY}$, $qSL\text{-}13\text{-}2_{ZY}$, $qSL\text{-}14\text{-}1_{ZY}$, and $qSL\text{-}15\text{-}1_{ZY}$, were also identified in a single environment in the ZY population (Table 1).

Table 1. Main-effect quantitative trait loci (M-QTLs) identified for three seed-size traits (seed length (SL), seed width (SW), and seed thickness (ST)) in ZY and K3N recombinant inbred line (RIL) populations across multiple environments.

Trait	QTL [a]	Chr (LG) [b]	Pos (cM) [c]	LOD [d]	Add [e]	R^2(%) [f]	Confidence Interval (cM) [g]	Physical Range(bp) [h]	Env [i]	Ref [j]
SL	*qSL-4-1*$_{ZY}$	4 (C1)	61.81	4.92	0.1	7.21	60.4–82.2	42,941,550–44,864,597	2012FY	[3]
			72.01	3.47	0.09	4.49			2012JP	
			79.41	4.52	0.1	6.3			2017JP	
	qSL-6-1$_{ZY}$	6 (C2)	23.61	4.12	0.1	5.43	18.7–25.9	5,404,972–7,692,663	2012JP	[3]
	qSL-6-2$_{ZY}$	6 (C2)	59.51	3.78	0.08	5.18	58.9–66.6	17,259,711–38,704,696	2012FY	[32]
	qSL-6-3$_{ZY}$	6 (C2)	66.21	6.22	0.12	8.35	65.7–66.7	38,704,696–41,044,201	2012JP	[32]
	qSL-9-1$_{ZY,K3N}$	9(K)	27.21	4.95	0.1	7.03	24.6–32.3	5,252,918–5,818,109	2017JP	THIS STUDY
			31.11	4.02	0.15	13	30–36.4		2012FY	THIS STUDY
	qSL-9-2$_{ZY}$	9 (K)	82.81	3.51	−0.08	4.82	79.2–86	38,148,965–40,891,870	2012FY	THIS STUDY
	qSL-11-1$_{K3N}$	11 (B1)	83.81	4.16	0.14	12.57	77.5–85.5	10,660,406–15,527,096	2012JP	THIS STUDY
	qSL-13-1$_{ZY}$	13 (F)	48.81	3.5	0.09	5.07	47–49.8	20,463,309–22,44,2989	2017JP	[32]
			48.81	8.14	0.14	11.46	48–49.8		2012JP	
	qSL-13-2$_{ZY}$	13 (F)	124.41	6.24	0.11	8.89	123.4–124.8	42,740,832–43,643,315	2012FY	[3]
	qSL-14-1$_{ZY}$	14 (B2)	184.31	4.94	−0.11	7.2	181.8–185.3	19,020,008–26,651,167	2017JP	THIS STUDY
	qSL-15-1$_{ZY}$	15 (E)	26.31	3.44	0.09	5.05	18.8–37.5	4,951,107–9,734,486	2017JP	THIS STUDY
	qSL-17-1$_{K3N}$	17 (D2)	101.41	3.8	0.14	11.95	99.2–103.3	38,148,257–39,028,119	2012FY	THIS STUDY
	qSL-18-1$_{K3N}$	18 (G)	84.31	3.69	−0.13	11.84	83.3–88.8	15,974,989–35,229,774	2017JP	THIS STUDY
	qSL-20-1$_{K3N}$	20 (I)	61.21	7.19	−0.18	22.64	55.9–67.7	36,184,890–38,300,982	2012JP	THIS STUDY

Table 1. *Cont.*

Trait	QTL [a]	Chr (LG) [b]	Pos (cM) [c]	LOD [d]	Add [e]	R^2(%) [f]	Confidence Interval (cM) [g]	Physical Range(bp) [h]	Env [i]	Ref [j]
SW	$qSW\text{-}1\text{-}1_{ZY}$	1 (D1a)	95.31	6.35	0.09	8.63	89.9–99.5	49,641,073–51,122,075	2012JP	THIS STUDY
	$qSW\text{-}2\text{-}1_{K3N}$	2 (D1b)	97.11	5.24	−0.09	17.32	96–102	42,094,237–43,533,158	2017JP	THIS STUDY
	$qSW\text{-}4\text{-}1_{ZY}$	4 (C1)	61.81	4.04	0.06	5.85	60.2–65.1	42,941,550–47,127,389	2012FY	[3]
	$qSW\text{-}5\text{-}1_{K3N}$	5(A1)	56.31	3.64	0.08	11.26	53.4–61	34,233,479–36,140,865	2017JP	THIS STUDY
	$qSW\text{-}6\text{-}1_{ZY}$	6 (C2)	16.31	10.53	0.12	15.35	15.6–16.6	5,651,662–5,975,443	2012JP	[3,32]
	$qSW\text{-}6\text{-}2_{ZY}$	6 (C2)	23.31	10.76	0.11	14.45	20.9–24.7	6,147,315–7,6,92,663	2012JP	[32]
	$qSW\text{-}8\text{-}1_{K3N}$	8 (A2)	25.61	4.17	−0.1	12.61	20.5–27.8	6,386,731–8,823,572	2012JP	THIS STUDY
	$qSW\text{-}9\text{-}1_{K3N}$	9 (k)	29.31	5.22	0.15	17.24	29.2–37	32,901,15–58,181,09	2012FY	THIS STUDY
	$qSW\text{-}9\text{-}2_{ZY}$	9 (k)	46.61	3.77	0.07	5.39	44.6–52	21,069,019–30,126,684	2012JP	THIS STUDY
	$qSW\text{-}10\text{-}1_{K3N}$	10 (O)	55.21	3.83	−0.1	12.85	52–59.2	32,040,762–38,080,781	2012FY	THIS STUDY
	$qSW\text{-}13\text{-}1_{ZY}$	13 (F)	48.41	7.07	0.09	9.59	48–49.1	20,443,593–22,442,989	2012JP	[33]
			51.31	3.57	0.07	5.32	49.8–52.6		2017JP	
	$qSW\text{-}13\text{-}2_{ZY}$	13 (F)	124.31	3.57	0.06	5.19	123.5–124.8	42,740,832–43,643,315	2012FY	THIS STUDY
	$qSW\text{-}17\text{-}1_{ZY}$	17 (D2)	2.01	6.22	0.08	10.92	0–3.3	33,39,67–2,389,816	2012FY	THIS STUDY
	$qSW\text{-}17\text{-}2_{ZY}$	17 (D2)	9.81	5.76	0.08	10.33	5.1–12.1	20,877,60–34,333,86	2012FY	THIS STUDY

Table 1. *Cont.*

Trait	QTL [a]	Chr (LG) [b]	Pos (cM) [c]	LOD [d]	Add [e]	R^2(%) [f]	Confidence Interval (cM) [g]	Physical Range(bp) [h]	Env [i]	Ref [j]
ST	qST-1-1$_{ZY}$	1 (D1a)	86.61	4.61	0.09	6.83	82.4–89.6	48,271,814–49,736,597	2012JP	THIS STUDY
	qST-1-2$_{ZY}$	1 (D1a)	92.81	4.58	0.09	6.17	89.6–98.3	49,7363,57–50,776,854	2012JP	THIS STUDY
	qST-2-1$_{K3N}$	2 (D1b)	97.11	5.29	−0.11	15.59	93.6–97.8	41,894,158–42,544,803	2017JP	[33]
	qST-3-1$_{K3N}$	3 (N)	21.31	3.93	0.1	10.98	14.6–23.4	24,562,76–59,471,80	2017JP	THIS STUDY
	qST-4-1$_{ZY}$	4 (C1)	62.81	3.45	0.07	4.46	58.2–65.1	42,894,734–47,127,389	2012JP	THIS STUDY
	qST-5-1$_{K3N}$	5(A1)	93.41	4.92	0.13	15.8	91.1–94	38,801,307–39,045,621	2012JP	THIS STUDY
	qST-6-1$_{ZY}$	6 (C2)	16.31	6.16	0.11	9.68	15.1–16.6	5,651,662–5,975,443	2012JP	[3]
	qST-6-2$_{ZY}$	6 (C2)	23.61	8.14	0.12	11.53	21.9–26	6,164,792–7,843,,389	2012JP	[3]
			23.61	4.46	0.08	11.12	21.5–26.2		2012FY	
	qST-6-3 $_{K3N}$	6 (C2)	129.81	4.17	−0.11	12.49	128.3–132.3	49,654,656–50,477,277	2017JP	[3]
	qST-8-1$_{K3N}$	8 (A2)	13.41	3.44	−0.1	10	7.5–15.5	3,060,492–5,128,185	2012JP	[3]
	qST-11-1$_{ZY}$	11 (B1)	23.81	5.44	0.09	8.04	23–31.2	10,235,376–15,990,255	2017JP	THIS STUDY
	qST-12-1 $_{K3N}$	12 (H)	84.01	5.28	0.14	16.67	80.4–86	34,404,607–35,936,212	2012FY	THIS STUDY
	qST-12-2$_{K3N}$	12 (H)	89.51	6.26	0.14	9	88.5–92.3	35,660,845–36,343,427	2012FY	THIS STUDY
	qST-12-3$_{K3N}$	12 (H)	96.51	4.32	0.12	13.69	93.1–110.1	36,343,428–38,545,317	2012FY	THIS STUDY
	qST-13-1$_{ZY}$	13 (F)	19.21	3.61	0.08	6.39	10.1–33	7,974,412–1,484,336	2012FY	THIS STUDY
	qST-13-2$_{ZY}$	13 (F)	48.41	4.23	0.09	10.28	46.7–9.7	71,87,17–22,442,989	2012JP	[3]
	qST-13-3$_{ZY}$	13 (F)	51.31	5.17	0.09	7.87	50.1–52.6	22,197,750–23,410,888	2017JP	[33]
			53.61	3.57	0.08	4.86	52.9–55.6		2012JP	

Table 1. *Cont.*

Trait	QTL [a]	Chr (LG) [b]	Pos (cM) [c]	LOD [d]	Add [e]	R^2(%) [f]	Confidence Interval (cM) [g]	Physical Range(bp) [h]	Env [i]	Ref [j]
	qST-16-1$_{K3N}$	16 (J)	77.01	4.8	0.12	14.24	66–78.2	31,905,448–35,735,751	2012JP	[33]
	qST-17-1$_{ZY}$	17 (D2)	3.01	5.29	0.09	7.63	0.7–3.3	27,02,76–2,389,816	2012FY	THIS STUDY
	qST-17-2$_{ZY}$	17 (D2)	8.81	5.24	0.1	8.41	3.3–17.4	2,389,537–5,085,098	2012FY	[33]
	qST-18-1$_{K3N}$	18 (G)	72.11	5.39	0.13	14.43	73–82	55,571,932–57,042,462	2012FY	THIS STUDY
	qST-18-2$_{ZY}$	18 (G)	78.01	4.78	0.09	6.74	73.5–82	11,268,490–46,240,347	2012FY	THIS STUDY

a: QTLs detected in different environments at the same, adjacent, or overlapping marker intervals were considered the same QTL; b: chromosome; c: position of the QTL; d: the log of odds (LOD) value at the peak likelihood of the QTL; e: indicates additive; f: phenotypic variance (%) expressed by the QTL; g: 1-LOD support confidence intervals (confidence interval length); h: physical position of QTL; i: environment; and j: references from www.soybase.org.

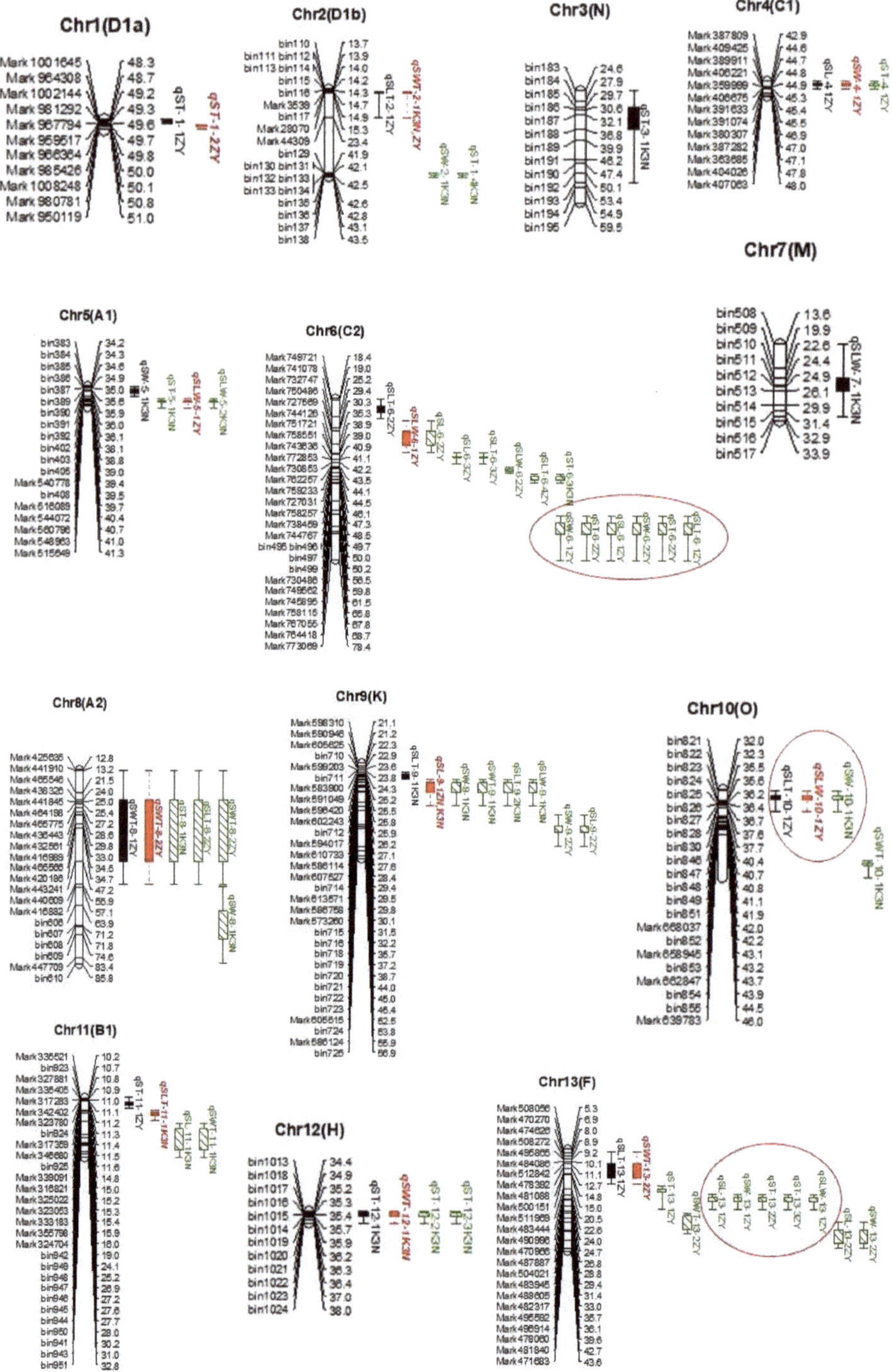

Figure 2. *Cont.*

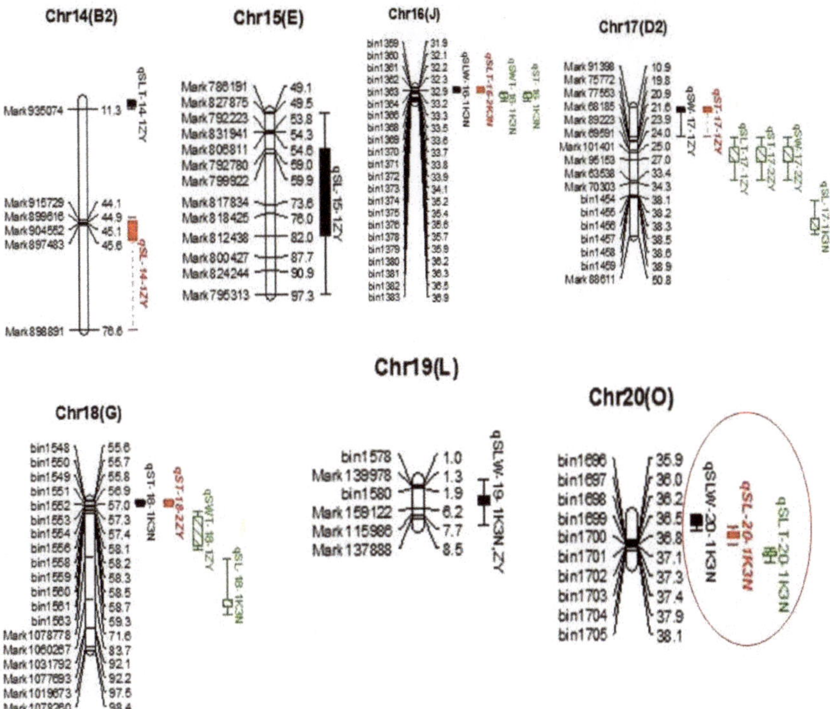

Figure 2. Location of quantitative trait loci (QTLs) on the genetic linkage map of the ZY and K3N RIL populations. Distances among markers are indicated using the physical location to the right of the linkage groups; names of markers are shown on the left. Only those SNP/SLAF markers are shown that were in and around the QTL regions. The red circles indicate the four QTL hotspots/clusters. Colored bars represent different QTLs.

In both ZY and K3N populations, a total of 14 main-effect QTLs associated with SW were identified, distributed on ten different chromosomes/LG (Table 1). Among them, *qSW-13-1*$_{ZY}$ was detected in two individual environments, viz., 2012JP and 2017JP, in ZY population and expressed an average of 7.45% of PV (Table 1). However, nine major QTLs, viz., *qSW-2-1*$_{K3N}$, *qSW-5-1*$_{K3N}$, *qSW-6-1*$_{ZY}$, *qSW-6-2*$_{ZY}$, *qSW-8-1*$_{K3N}$, *qSW-9-1*$_{K3N}$, *qSW-10-1*$_{K3N}$, *qSW-17-1*$_{ZY}$, and *qSW-17-2*$_{ZY}$, with $R^2 > 10\%$, were identified only in one environment and expressed PV that varies from 10.33–17.32% in both RIL populations (Table 1). Four minor QTLs, viz., *qSW-1-1*$_{ZY}$, *qSW-4-1*$_{ZY}$, *qSW-9-2*$_{ZY}$, and *qSW-13-2*$_{ZY}$, were also detected as environment-sensitive and expressing a PV of 5.19–8.63% (Table 1).

For ST, we identified 22 main-effect QTLs in both RIL populations across three environments, distributed on 13 LG (Table 1). One stable major (*qST-6-2*$_{ZY}$) and minor (*qST-13-3*$_{ZY}$) QTLs were consistently detected in two individual environments in the ZY population with an average R^2 of 11.32% and 6.36%, respectively (Table 1). Moreover, ten major QTLs: *qST-2-1*$_{K3N}$, *qST-3-1*$_{K3N}$, *qST-5-1*$_{K3N}$, *qST-6-3*$_{K3N}$, *qST-8-1*$_{K3N}$, *qST-12-1*$_{K3N}$, *qST-12-3*$_{K3N}$, *qST-13-2*$_{ZY}$, *qST-16-1*$_{K3N}$, and *qST-18-1*$_{K3N}$ were identified in only one individual environment in the K3N population, with PV ranging from 10.00–16.67% (Table 1). Besides, ten minor QTLs, viz., *qST-1-1*$_{ZY}$, *qST-1-2*$_{ZY}$, *qST-4-1*$_{ZY}$, *qST-6-1*$_{ZY}$, *qST-11-1*$_{ZY}$, *qST-12-2*$_{K3N}$, *qST-13-1*$_{ZY}$, *qST-17-1*$_{ZY}$, *qST-17-2*$_{ZY}$, and *qST-18-2*$_{ZY}$, expressing PV in the range of 4.46–9.68%, were environment-sensitive (Table 1).

Among 50 QTLs identified for all three seed size traits, 31 QTLs were novel identified for the first time, and the remaining 19 QTLs are reported earlier in the same physical genomic interval (Table 1). Moreover, 25 out of 50 QTLs were major, with $R^2 > 10\%$, and the remaining 25 were minor QTLs, with $R^2 < 10\%$. However, we detected several major QTLs in the K3N population (18), compared to the ZY. Notably, the most prominent QTL with the highest logarithm of odds (LOD) score (10.76) in a 23.31cM region was located on Chr06, named $qSW\text{-}6\text{-}2_{ZY}$, expressing 14.45% of PV (Table 1). The majority of QTLs showed a positive additive effect with favorable alleles from parent Zhengxiaodou, except ten QTLs ($qSL\text{-}9\text{-}2_{ZY}$, $qSL\text{-}14\text{-}1_{ZY}$, $qSL\text{-}18\text{-}1_{K3N}$, $qSL\text{-}20\text{-}1_{K3N}$, $qSW\text{-}2\text{-}1_{K3N}$, $qSW\text{-}8\text{-}1_{K3N}$, $qSW\text{-}10\text{-}1_{K3N}$, $qST\text{-}2\text{-}1_{K3N}$, $qST\text{-}6\text{-}3_{K3N}$, and $qST\text{-}8\text{-}1_{K3N}$) that displayed negative additive effects with beneficial alleles from Nannong1138-2 (Table 1).

2.3. QTL Mapping of Seed Shape by CIM

In total, we identified 38 QTLs associated with three seed shape traits, viz., SLW, SLT, and SWT on 15 different chromosomes in both RIL populations (ZY and K3N) across all three individual environments (Table 2 and Figure 2). A single QTL expressed a PV that varies from 3.44% ($qSLT\text{-}16\text{-}1_{K3N}$) to 26.84% ($qSLW\text{-}20\text{-}1_{K3N}$) (Table 2). For SLW, we identified 11 QTLs located on nine different chromosomes (Table 2). A major and stable QTL, $qSLW\text{-}6\text{-}1_{ZY}$, was detected consistently on Chr06 in all three individual environments (2012FY, 2012JP, and 2017JP) in the ZY population and expressed a PV of 16.03% (Table 2). Besides, another major stable QTL, $qSLW\text{-}20\text{-}1_{K3N}$, was identified on Chr20 in two individual environments (2012 JP and 2017JP), expressing an average PV of 19.24% in the K3N population (Table 2). The $qSLW\text{-}19\text{-}1_{K3N,ZY}$ was identified in both RIL populations, as well as two individual environments (2012FY and 2017JP), with an average PV of 9.17% (Table 2). The remaining eight QTLs were environment-sensitive (identified in only one individual environment); out of them, three QTLs, viz., $qSLW\text{-}7\text{-}1_{K3N}$, $qSLW\text{-}9\text{-}1_{K3N}$, and $qSLW\text{-}16\text{-}1_{K3N}$, were major, with $R^2 > 10\%$ (Table 2).

In the case of SLT, we identified a total of 16 QTLs distributed on 11 different chromosomes in both RIL populations across three individual environments (Table 2). Among them, $qSLT\text{-}10\text{-}1_{ZY}$ and $qSLT\text{-}20\text{-}1_{K3N}$ were significant and stable QTLs having $R^2 > 10\%$, as well as detected in three and two individual environments, respectively (Table 2). Additionally, four significant QTLs, viz., $qSLT\text{-}9\text{-}1_{K3N}$, $qSLT\text{-}9\text{-}2_{K3N}$, $qSLT\text{-}11\text{-}1_{K3N}$, and $qSLT\text{-}13\text{-}1_{ZY}$, expressing a PV of 10.29–12.62%, were detected only in one individual environment (Table 2). The remaining ten QTLs were minor, having $R^2 < 10\%$ detected in only one individual environment (Table 2).

For SWT, a total of 11 QTLs on nine different chromosomes were mapped in both RIL populations (Table 2). Among these QTLs, $qSWT\text{-}2\text{-}1_{K3N,ZY}$ and $qSWT\text{-}8\text{-}1_{ZY}$ were the stable QTLs identified in three and two individual environments, respectively; additionally, $qSWT\text{-}2\text{-}1_{K3N,ZY}$ was identified in both RIL populations. Besides, four out of 11 QTLs, viz., $qSWT\text{-}9\text{-}1_{K3N}$, $qSWT\text{-}10\text{-}1_{K3N}$, $qSWT\text{-}11\text{-}1_{K3N}$, and $qSWT\text{-}16\text{-}1_{K3N}$, were major ($R^2 > 10\%$) but were environment-sensitive, detected only in K3N-RIL populations (Table 2). The remaining five minor QTLs, viz., $qSWT\text{-}8\text{-}2_{ZY}$, $qSWT\text{-}12\text{-}1_{K3N}$, $qSWT\text{-}13\text{-}1_{ZY}$, $qSWT\text{-}13\text{-}2_{ZY}$, and $qSWT\text{-}18\text{-}1_{ZY}$, were detected in one individual environment with $R^2 > 10\%$ (Table 2).

Overall, 38 QTLs were associated with three different seed shape traits in both the K3N and ZY populations; out of them, 20 QTLs have been reported for the first time, while earlier studies have already reported the remaining 18 QTLs (Table 2). Moreover, 17 out of 38 QTLs were major, with $R^2 > 10\%$, and four of them, viz., $qSLW\text{-}6\text{-}1_{ZY}$, $qSLW\text{-}20\text{-}1_{K3N}$, $qSLT\text{-}10\text{-}1_{ZY}$, and $qSLT\text{-}20\text{-}1_{K3N}$, were detected stably in more than one individual environment. The most prominent major and stable QTL was $qSLW\text{-}20\text{-}1_{K3N}$ (novel QTL), with the highest LOD value of 9.01 in an individual environment, identified at 53.61 cM position on Chr20 and expressing a PV of 26.84% (Table 2). The 16 QTLs have positive additive effects with beneficial alleles inherited from KeFeng35, whereas the remaining 22 QTLs possess negative additive effects with favorable alleles derived from Nannong1138-2 (Table 2).

Table 2. M-QTLs identified for three seed-shape traits (seed length-to-width (SLW), seed length-to-thickness (SLT), and seed width-to-thickness (SWT)) in ZY and K3N RIL populations across multiple environments.

Trait	QTL [a]	Chr (LG) [b]	Pos (cM) [c]	LOD [d]	Add [e]	R^2(%) [f]	Confidence Interval (cM) [g]	Physical Range (bp) [h]	Env [i]	Ref [j]
SLW	$qSLW$-5-1$_{ZY}$	5(A1)	60.01	3.55	−0.01	4.77	56.6–62.9	39,366,066–41,2966,26	2012FY	THIS STUDY
	$qSLW$-5-2$_{K3N}$	5(A1)	92.81	6.73	−0.02	9.15	89.9–93.4	38,337,588–39,465,963	2017JP	THIS STUDY
	$qSLW$-6-1$_{ZY}$	6 (C2)	63.61	6.59	0.02	18.68	50.1–65.6	13,274,690–38,704,696	2012FY	[3,32]
			63.61	8.03	0.02	12.17	62.6–65.9		2017JP	
			66.21	5.94	0.02	17.24	57.6–66.7		2012JP	
	$qSLW$-6-2$_{ZY}$	6 (C2)	77.01	5.66	0.01	8.15	76.4–77.8	46,087,483–46,232,257	2017JP	[3]
	$qSLW$-7-1$_{K3N}$	7 (M)	10.01	4.8	0.02	12.78	7.5–20.1	1,361,954–3,819,224	2017JP	THIS STUDY
	$qSLW$-9-1$_{K3N}$	9 (K)	98.41	3.97	0.02	12.97	88.2–102.2	38,138,667–41,052,048	2012FY	THIS STUDY
	$qSLW$-10-1$_{ZY}$	10 (O)	53.91	4.64	−0.01	6.6	49–61.7	41,983,494–45,988,221	2017JP	THIS STUDY
	$qSLW$-13-1$_{ZY}$	13 (F)	68.61	4.04	0.01	6.44	62.2–70.8	23,963,991–26,852,039	2012JP	[33]
	$qSLW$-16-1$_{K3N}$	16 (J)	68.21	6.39	−0.02	16.92	66.5–73.9	31,905,448–33,541,661	2012JP	[33]
	$qSLW$-19-1$_{K3N,ZY}$	19(L)	0.01	3.89	0.02	12.25	0–10.8	1–1,939,363	2012FY	THIS STUDY
			5.81	3.55	−0.02	6.09	0–9.6		2017JP	
	$qSLW$-20-1$_{K3N}$	20 (I)	53.61	9.01	−0.03	26.84	52.5–55.1	35,924,513–38,138,435	2012JP	THIS STUDY
			60.21	4.64	−0.02	11.64	53.6–64.8		2017JP	
SLT	$qSLT$-2-1$_{ZY}$	2 (D1b)	63.91	4.04	−0.03	6.45	61.8–65.7	14,715,990–15,293,225	2012FY	[33]
	$qSLT$-6-1$_{ZY}$	6 (C2)	29.01	3.7	−0.02	5.27	25.6–30.3	6,779,201–8,789,201	2012JP	[3]
	$qSLT$-6-2$_{ZY}$	6 (C2)	62.31	3.92	0.02	5.71	62–65.6	18,806,329–29,376,980	2012JP	[3]
	$qSLT$-6-3$_{ZY}$	6 (C2)	70.31	4.71	0.02	6.79	68.6–70.6	39,478,712–42,3014,72	2012JP	[3]
	$qSLT$-6-4$_{ZY}$	6 (C2)	82.61	4.98	0.03	7.82	79.3–83.7	47,288,454–48,097,950	2012JP	[3]
	$qSLT$-8-1$_{ZY}$	8 (A2)	3.51	5.3	0.03	8.21	0.5–10.3	1,281,677–4,722,531	2017JP	THIS STUDY
	$qSLT$-8-2$_{ZY}$	8 (A2)	16.11	3.83	0.02	5.56	13.2–20	4,722,281–8,343,142	2012FY	[3]
	$qSLT$-9-1$_{K3N}$	9 (K)	24.31	4.8	−0.03	11.44	22.6–28.6	2,378,279–3,574,689	2012JP	THIS STUDY

Table 2. *Cont.*

Trait	QTL [a]	Chr (LG) [b]	Pos (cM) [c]	LOD [d]	Add [e]	R^2(%) [f]	Confidence Interval (cM) [g]	Physical Range (bp) [h]	Env [i]	Ref [j]
	$qSLT$-9-2$_{K3N}$	9 (K)	84.31	4.05	0.03	12.62	81.9–88.4	36,947,988–40,302,752	2012FY	THIS STUDY
	$qSLT$-10-1$_{ZY}$	10 (O)	53.91	4.55	−0.03	17	48.9–59.4	41,983,494–45,988,221	2012FY	THIS STUDY
			53.91	4.61	−0.02	16.8	49.5–57.8		2012JP	
			53.91	4.96	−0.03	7.4	47.5–60.1		2017JP	
	$qSLT$-11-1$_{K3N}$	11 (B1)	82.61	3.77	0.03	11.7	77.6–84.5	10,660,406–15,086,914	2012FY	THIS STUDY
	$qSLT$-13-1$_{ZY}$	13 (F)	16.21	5.44	−0.03	10.29	6.5–23.3	8,857,191–5,270,536	2012FY	[33]
	$qSLT$-14-1$_{ZY}$	14 (B2)	57.21	5.29	−0.03	7.31	55.7–69	76,63,93–45,068,56	2012JP	THIS STUDY
	$qSLT$-16-1$_{K3N}$	16 (J)	69.61	3.25	−0.05	3.44	68.7–84.1	32,060,131–37,397,385	2012JP	[33]
	$qSLT$-17-1$_{ZY}$	17 (D2)	5.11	4.58	−0.03	6.61	4.4–16.4	2,498,772–5,085,098	2012FY	[33]
	$qSLT$-20-1$_{K3N}$	20 (I)	64.21	3.82	−0.02	10.22	59–65.5	35,673,231–38,972,972	2017JP	
			66.11	7.28	−0.04	18.54	59.3–71.9		2012JP	THIS STUDY
SWT	$qSWT$-2-1$_{K3N, ZY}$	2 (D1b)	63.91	3.89	−0.02	6.4	56.4–65.7	14,715,990–23,379,924	2012FY	[33]
			67.81	3.92	−0.01	5.9	67–70.3		2017JP	
			67.81	5.74	−0.02	8.55	65.7–70.3		2012JP	
	$qSWT$-8-1$_{ZY}$	8 (A2)	2.11	3.46	0.01	5.2	0–13.4	1,315,065–8,343,142	2017JP	THIS SYUDY
			2.51	4.73	0.01	10.88	1.2–9.3		2012JP	
	$qSWT$-8-2$_{ZY}$	8 (A2)	16.61	3.48	0.02	5.32	13.2–20	4,722,281–8,343,142	2012FY	THIS STUDY
	$qSWT$-9-1$_{K3N}$	9 (K)	71.51	5.44	−0.01	16.29	65.8–72.2	32,505,690–36,079,751	2017JP	THIS STUDY
	$qSWT$-10-1$_{K3N}$	10 (O)	87.71	4.11	0.01	12.47	80.7–99.2	40,440,079–44,537,290	2017JP	THIS STUDY
	$qSWT$-11-1$_{K3N}$	11 (B1)	103.61	3.78	−0.01	10.89	101.5–105.3	18,952,782–33,288,718	2017JP	THIS STUDY
	$qSWT$-12-1$_{K3N}$	12 (H)	87.31	5	−0.02	4.55	85.8–92.4	34,926,974–36,343,427	2012FY	THIS STUDY
	$qSWT$-13-1$_{ZY}$	13 (F)	16.21	4.93	−0.02	9.84	6.4–23.4	8,857,191–5,270,536	2012FY	[33]
	$qSWT$-13-2$_{ZY}$	13 (F)	29.81	4.26	−0.02	8.07	17.8–37.2	6,777,564–1,267,746	2017JP	[33]
	$qSWT$-16-1$_{K3N}$	16 (J)	74.81	4.66	−0.02	13.59	72.7–78.6	33,458,104–35,735,751	2012JP	[33]
	$qSWT$-18-1$_{ZY}$	18 (G)	78.71	3.74	−0.02	5.86	72.4–83.7	56,974,254–46,749,768	2012FY	THIS STUDY

a: QTLs detected in different environments at the same, adjacent, or overlapping marker intervals were considered the same QTL; b: chromosome; c: position of the QTL; d: the log of odds (LOD) value at the peak likelihood of the QTL; e: indicates additive; f: phenotypic variance (%) expressed by the QTL; g: 1-LOD support confidence intervals (confidence interval length); h: physical position of QTL; i: environment; and j: references from www.soybase.org.

2.4. MCIM Mapping and Comparison of CIM and MCIM Methods

To further validate the QTLs detected by CIM, we performed another method of mixed-model-based composite interval mapping (MCIM) to dissect the additive effect QTLs and QTL x E interactions. By using the MCIM method, we identified a total of 48 additive effect QTLs distributed on 15 chromosomes for all six traits related to seed size and shape in both the RIL populations and all three environments, which expressed 1.69 to 29.35% of the PV (Table 3). Moreover, the additive effect of different QTLs was either negative or positive; for example, 30 and 18 QTLs have positive and negative additive effects, respectively. Hence, indicating that both parents contribute beneficial alleles for seed size and shape traits in ZY and K3N populations (Table 3). Out of 48 QTLs, 10 QTLs were significant, with $R^2 > 10\%$, whereas the remaining 38 QTLs were minor, with $R^2 < 10\%$ (Table 3).

Among these 48 QTLs, 15 QTLs showed significant additive by environment interaction (AE) effects (Table 3). However, four QTLs viz., *qSL-13-4$_{ZY}$*, *qSW-13-3$_{ZY}$*, *qST-13-4$_{ZY}$*, and *qST-10-1$_{K3N}$* revealed AE effect at all environments, while seven and four QTLs showed AE effect in two and one specific environments, respectively (Table 3). The AE effect of these 15 QTLs associated with seed size and shape traits could express the PV that varies from 0.01 to 4.15%. The remaining 33 QTLs identified through the MCIM approach do not possess any AE effect; hence, they are environmentally stable QTLs (Table 3).

Lastly, we performed a comparative analysis of QTLs detected by CIM and MCIM approaches. A total of 88 and 48 QTLs were identified by CIM and MCIM, respectively. Among these QTLs, 15 QTLs were common and are detected by both methods in the same physical genomic interval, indicating the reliability and stability of these QTLs. Besides, by comparing the physical genomic regions of QTLs identified in both populations (ZY and K3N) and mapping methods (CIM and MCIM), two QTLs, viz., *qSW-1-1$_{ZY}$* and *qSLT-20-1$_{K3N}$*, were detected in common, with $R^2 > 10\%$, identified for the first time. Hence, these QTLs were considered as the most stable and novel QTLs that could be utilized potentially for gene cloning and MAB of soybean seed size and shape traits.

Table 3. Additive and additive x environment interaction effects of QTLs associated with seed shape traits in two RIL populations.

RIL	Trait	QTL	Chr	Pos (cM)	Physical Range (bp)	Flanking Marker	Additive Effect		AE Effect				Ref
							A	PVE (%)	AE1	AE2	AE3	PVE (%)	
ZY	SL	qSL-4-2_{ZY}	4	39.22	19283057–19438579	Mark386837–Mark359625	0.18	22.47	NS	NS	NS	0	[3]
		qSL-6-1_{ZY}	6	65.93	18831802–19024597	Mark750099–Mark741078	0.09	6.04	NS	NS	NS	1.69	[32]
		qSL-9-2_{ZY}	9	82.03	39583668–39417035	Mark596882–Mark570229	0.08	3.89	NS	NS	NS	0	THIS STUDY
		qSL-13-3_{ZY}	13	78.75	31540635–31390215	Mark473668–Mark478524	0.08	4.65	−0.03 **	0.03 **	NS	0.82	[3]
		qSL-13-4_{ZY}	13	4.001	13278146–14133381	Mark471284–Mark486987	0.06	2.62	0.02 **	−0.01 *	−0.01 *	0.45	[3]
		qSL-14-2_{ZY}	14	12.14	47107927–47737451	Mark941949–Mark890565	−0.08	3.97	0.01 **	NS	−0.01 *	0.11	THIS STUDY
		qSL-15-1_{ZY}	15	30.54	7595034–8201285	Mark817834–Mark818425	0.07	2.89	NS	NS	NS	0.04	[3]
		qSL-17-2_{ZY}	17	54.6	13439080–13701008	Mark96769–Mark84717	0.06	2.61	NSN	NS	NS	0	[34]
	SW	qSW-1-2_{ZY}	1	67.22	17489394–20846299	Mark974988–Mark977669	0.03	1.69	NS	0.03 **	−0.04 **	2.78	THIS STUDY
		qSW-1-1_{ZY}	1	100.2	51095466–51296512	Mark1014325–Mark988734	0.14	29.35	NS	NS	NS	0	THIS STUDY
		qSW-4-1_{ZY}	4	64.09	48139232–48129734	Mark370122–Mark383650	0.08	9.23	NS	NS	NS	0.1	[3]
		qSW-4-3_{ZY}	4	13.9	3681724–5115633	Mark411964–Mark375360	0.13	25.78	NS	NS	−0.01 *	0.25	[3]
		qSW-6-1_{ZY}	6	23.63	6820998–6873235	Mark743934–Mark764418	0.07	6.39	NS	0.05 **	−0.05 **	4.15	[32]
		qSW-9-3_{ZY}	9	58.19	30576094–33868843	Mark584128–Mark577210	0.04	2.86	NS	NS	NS	0	THIS STUDY
		qSW-13-3_{ZY}	13	80.04	32094249–32215414	Mark487690–Mark477270	0.09	12.55	−0.01 *	0.03 **	−0.02 **	1.67	THIS STUDY
		qSW-17-3_{ZY}	17	26.12	5914087–5923952	Mark80080–Mark102449	0.07	6.28	0.03 **	NS	−0.02 **	1.4	[3]
		qSW-20-1_{ZY}	20	82.14	41039353–41758780	Mark244793–Mark230802	0.04	2.48	NS	NS	−0.01 *	0.16	THIS STUDY
	ST	qST-1-3_{ZY}	1	66.15	24564849–25434623	Mark988529–Mark966630	0.06	3.58	NS	NS	NS	0	THIS STUDY
		qST-4-2_{ZY}	4	44.13	30867521–32458924	Mark404804–Mark410274	0.06	3.43	0.01 **	NS	−0.02 **	0.47	[3]
		qST-6-4_{ZY}	6	28.98	8734791–8864415	Mark778253–Mark768262	0.09	7.87	NS	NS	NS	0	THIS STUDY
		qST-13-4_{ZY}	13	74.5	29909876–29933390	Mark492165–Mark480651	0.08	6.47	−0.04 **	0.02 **	0.01 **	1.31	[33]
		qST-17-3_{ZY}	17	26.12	5914087–5923952	Mark80080–Mark102449	0.08	5.27	NS	NS	NS	0.03	[3]
		qST-18-2_{ZY}	18	73.25	11452216–56974484	Mark107824–Mark1041506	0.14	17.36	NS	NS	NS	0	THIS STUDY
		qST-20-1_{ZY}	20	46.1	32964872–34278811	Mark222598–Mark260922	0.09	6.68	NS	NS	NS	0	[33]

Table 3. *Cont.*

RIL	Trait	QTL	Chr	Pos (cM)	Physical Range (bp)	Flanking Marker	Additive Effect		AE Effect				Ref
							A	PVE (%)	AE1	AE2	AE3	PVE (%)	
	SLW	$qSLW$-6-3$_{ZY}$	6	68.61	39116570–39478973	Mark735467–Mark752696	0.02	16.59	NS	NS	NS	0.02	[33]
		$qSLW$-9-2$_{ZY}$	9	26.46	5253206–5593340	Mark586124–Mark605515	0.02	6.23	NS	NS	NS	0.01	THIS STUDY
		$qSLW$-10-1$_{ZY}$	10	34.22	40118012–41534124	Mark647482–Mark674118	−0.01	3.59	NS	NS	NS	0.22	THIS STUDY
		$qSLW$-14-1$_{ZY}$	14	2.62	46512262–46896487	Mark910347–Mark906347	−0.03	3.59	NS	NS	−0.01	0.22	THIS STUDY
	SLT	$qSLT$-2-2$_{ZY}$	2	54.96	13892893–14031565	Mark45679–Mark24395	−0.04	14.3	NS	NS	NS	0	THIS STUDY
		$qSLT$-6-5$_{ZY}$	6	72.26	43230405–43114475	Mark747633–Mark770074	0	7.18	NS	NS	NS	0	[34]
		$qSLT$-8-3$_{ZY}$	8	107.06	44543723–45326890	Mark457232–Mark423205	0.03	8.95	NS	NS	NS	0	THIS STUDY
		$qSLT$-10-1$_{ZY}$	10	34.22	40118012–41534124	Mark647482–Mark674118	−0.03	8.65	NS	NS	NS	0	THIS STUDY
		$qSLT$-10-2$_{ZY}$	10	3.31	4454862–4625724	Mark637694–Mark659817	0.03	9.56	NS	NS	NS	0	[34] -
		$qSLT$-14-2$_{ZY}$	14	118.74	27709972–27836278	Mark905511–Mark937850	−0.02	4.16	NS	NS	NS	0	THIS STUDY
		$qSLT$-17-1$_{ZY}$	17	12.06	3433165–1091815	Mark91398–Mark70303	−0.02	3.77	NS	NS	NS	0	[33]
		$qSLT$-18-1$_{ZY}$	18	72.71	11452216–56974484	Mark107824–Mark1041506	−0.04	3.77	NS	NS	NS	0	THIS STUDY
	SWT	$qSWT$-2-2$_{ZY}$	2	45.8	10992717–11277233	Mark4308–Mark33823	−0.04	3.72	NS	NS	NS	0	THIS STUDY
		$qSWT$-8-3$_{ZY}$	8	111.68	45326630–45114110	Mark466182–Mark457232	0.01	4.91	NS	NS	NS	0	THIS STUDY
		$qSWT$-13-3$_{ZY}$	13	118.22	39676002–42053780	Mark492087–Mark510247	−0.01	4.03	NS	NS	NS	0.36	[34]
		$qSWT$-14-1$_{ZY}$	14	132.78	43567951–44159326	Mark927308–Mark890844	−0.01	3.64	NS	NS	NS	0.01	THIS STUDY
		$qSWT$-18-1$_{ZY}$	18	73.25	56974254–11452436	Mark107824–Mark1041506	−0.03	18.77	NS	NS	NS	0.79	THIS STUDY
K3N	SL	qSL-17-3$_{K3N}$	17	118.01	40207655–41906774	bin1466–bin1467	0.08	5.71	NS	NS	NS	0	THIS STUDY
		qSL-20-1$_{K3N}$	20	55.91	36184890–36777026	bin1698–bin1699	−0.09	6.72	NS	−0.01 *	0.03 **	1.02	THIS STUDY
	ST	qST-10-1$_{K3N}$	10	59.22	36682803–37647030	bin827–bin829	−0.08	7.05	−0.03 **	0.02 **	0.02 **	2.06	THIS STUDY
	SLW	$qSLW$-5-2$_{K3N}$	5	94.46	38132148–38801307	bin402–bin403	−0.01	7.06	NS	NS	NS	0.01	[34]
		$qSLW$-16-1$_{K3N}$	16	69.6	32318950–33186025	bin1362–bin1363	−0.01	2.2	NS	NS	NS	0.02	[34]
	SLT	$qSLT$-16-1$_{K3N}$	16	75.68	33853674–35244129	bin1371–bin1372	−0.04	19.15	NS	−0.01 *	NS	1.49	THIS STUDY
		$qSLT$-20-1K3N	20	68.27	37878839–38300982	bin1704–bin1705	−0.03	11.77	NS	NS	NS	0.05	THIS STUDY

Chr.: chromosome. * $p < 0.05$ and ** $p < 0.01$. PVE indicates phenotypic variation expressed by additive effects. AE1: 2012FY, AE2: 2012JP, and AE3: 2017JP.

2.5. Epistatic Interaction Effects

A total of 16 pairs of epistatic QTLs were detected for seed size and shape in both RIL populations (Table 4). Out of these 16 pairs, four epistatic QTL pairs, viz., qSL-2-1$_{K3N}$ and qSL-2-2$_{K3N}$, qST-9-1$_{K3N}$ and qST-12-4$_{K3N}$, $qSLT$-2-3$_{K3N}$ and $qSLT$-7-1$_{K3N}$, and $qSWT$-6-1$_{K3N}$ and $qSWT$-8-4$_{K3N}$, possess both significant AA and AAE interaction effects with PV of 1.71–9.70% and 1.68–12.03% expressed, respectively (Table 4). However, the remaining 12 QTLs pairs had only significant AA effects and did not possess any significant AAE interaction effects (Table 4). Hence, the above findings indicate that environment and epistatic interaction effects have considerable influence on the regulation of phenotypic expressions of seed size and shape traits in soybeans. Though, except for three QTLs, viz., qSL-13-3$_{ZY}$, qSL-13-4$_{ZY}$, and qSW-13-4$_{ZY}$, all the remaining additive-effect QTLs did not show any epistatic effects.

Table 4. Estimated epistatic effects (AA) and environmental (AAE) interactions of QTLs for seed shape and size traits across all environments.

RIL	QTL_i	Chr_i	Pos_i	Marker Interval_i	QTL_j	Chr_j	Pos_j	Marker Interval_j	(AA) Effect			(AAE) Effect			
									AA	PVE (%)	AE1	AE2	AE3	PVE (%)
ZY	$qSL\text{-}13\text{-}3_{ZY}$	13	78.75	Mark473668–Mark478524	$qSL\text{-}13\text{-}4_{ZY}$	13	4.001	Mark471284–Mark486987	−0.05	1.84	NS	NS	NS	-
	$qST\text{-}1\text{-}4_{ZY}$	1	67.22	Mark962092–Mark962281	$qST\text{-}13\text{-}5_{ZY}$	13	118.5	Mark492087–Mark510247	−0.07	4.16	NS	NS	NS	-
	$qST\text{-}1\text{-}5_{ZY}$	1	88.72	Mark1006029–Mark981292	$qST\text{-}13\text{-}4_{ZY}$	13	32.59	Mark489890–Mark487004	−0.07	4.11	NS	NS	NS	0.02
	$qST\text{-}6\text{-}5_{ZY}$	6	26.19	Mark738100–Mark750615	$qST\text{-}9\text{-}1_{ZY}$	9	98.04	Mark591384–Mark570193	−0.07	4.96	NS	NS	NS	0.06
	$qSLT\text{-}6\text{-}6_{ZY}$	6	11.85	Mark729845–Mark775156	$qSLT\text{-}10\text{-}3_{ZY}$	10	14.29	Mark628845–Mark623926	0.02	3.98	NS	NS	NS	-
	$qSLT\text{-}15\text{-}1_{ZY}$	15	26.29	Mark799922–Mark817834	$qSLT\text{-}19\text{-}1_{ZY}$	19	64.16	Mark114395–Mark141336	0.03	9.59	NS	NS	NS	-
	$qSWT\text{-}1\text{-}1_{ZY}$	1	67.22	Mark990284–Mark986367	$qSWT\text{-}13\text{-}4_{ZY}$	13	83.91	Mark484073–Mark489301	0.02	6.7	NS	NS	NS	0.25
	$qSWT\text{-}13\text{-}5_{ZY}$	13	53.11	Mark505121–Mark508857	$qSWT\text{-}20\text{-}1_{ZY}$	20	83.56	Mark250253–Mark257473	0.02	6.2	NS	NS	NS	0.01
K3N	$qSL\text{-}2\text{-}1_{K3N}$	2	40.91	bin87–bin88	$qSL\text{-}2\text{-}2_{K3N}$	2	95.94	bin130–bin131	−0.11	9.7	−0.05 **	0.01 **	NS	3.02
	$qST\text{-}9\text{-}1_{K3N}$	9	71.51	bin748–bin749	$qST\text{-}12\text{-}4_{K3N}$	12	0.6	bin962–bin963	0.04	1.71	NS	0.03 **	NS	2.41
	$qSLW\text{-}4\text{-}1_{K3N}$	4	50.29	bin285–bin286	$qSLW\text{-}15\text{-}1_{K3N}$	15	148.55	bin1298–bin1299	0.03	4.34	NS	NS	NS	-
	$qSLW\text{-}7\text{-}2_{K3N}$	7	37.3	bin526–bin527	$qSLW\text{-}12\text{-}1_{K3N}$	12	55.09	bin990–bin991	−0.01	6.33	NS	NS	NS	0.08
	$qSLT\text{-}2\text{-}3_{K3N}$	2	4.2	bin164–bin165	$qSLT\text{-}7\text{-}1_{K3N}$	7	94.74	bin568–bin569	−0.01	3.21	NS	0.01 **	NS	1.68
	$qSWT\text{-}6\text{-}1_{K3N}$	6	97.8	bin470–bin471	$qSWT\text{-}8\text{-}4_{K3N}$	8	109.71	bin672–bin673	−0.04	7.56	−0.01 *	NS	0.01**	12.03
	$qSWT\text{-}11\text{-}2_{K3N}$	11	97.12	bin934–bin935	$qSWT\text{-}17\text{-}1_{K3N}$	17	5.47	bin1386–bin1387	0.01	7.1	NS	NS	NS	1.11
	$qSWT\text{-}11\text{-}3_{K3N}$	11	106.76	bin952–bin953	$qSWT\text{-}14\text{-}2_{K3N}$	14	45.84	bin1170–bin1171	0.01	6.48	NS	NS	NS	0.74

Chr_i and Chr_j indicate the two sites involved in epistatic interactions and Pos indicates genetic position for each of the sites. * $p < 0.05$ and ** $p < 0.01$. PVE indicates phenotypic variation expressed by epistatic effects. AE1: 2012FY, AE2: 2012JP, and AE3: 2017JP.

2.6. Colocalization of QTLs in QTL cluster/Hotspot

A QTL cluster/hotspot is defined as a densely populated QTL region of the chromosome that contains multiple QTLs associated with various traits. In this study, we observed colocalization of QTLs on four QTL Clusters/hotspots located on different chromosomes, viz., Chr6, Chr10, Chr13, and Chr20, and were named Cluster-06/QTL Hotspot A, Cluster-10/QTL Hotspot B, Cluster-13/QTL Hotspot C, and Cluster-20/QTL Hotspot D, respectively (Table 5). The highest concentration of QTLs for seed size and shape traits was identified in "QTL Hotspot A" of Chr06, spanning the physical interval of 2.19Mb (Figure 3). This QTL hotspot harbors six QTLs (three major and three minor), viz., $qSW\text{-}6\text{-}1_{ZY}$, $qST\text{-}6\text{-}1_{ZY}$, $qSL\text{-}6\text{-}1_{ZY}$, $qSW\text{-}6\text{-}2_{ZY}$, $qST\text{-}6\text{-}2_{ZY}$, and $qSLT\text{-}6\text{-}1_{ZY}$, associated to seed size and shape traits, expressing a PV of 5.43–15.35% (Table 5). Another set of QTL-rich regions possessing five QTLs (two major and three minor), viz., $qSL\text{-}13\text{-}1_{ZY}$, $qSW\text{-}13\text{-}1_{ZY}$, $qST\text{-}13\text{-}2_{ZY}$, $qST\text{-}13\text{-}3_{ZY}$, and $qSLW\text{-}13\text{-}1_{ZY}$ was "QTL Hotspot C", with a length of 6.3 Mb (Table 5 and Figure 3). However, both "QTL Hotspot B" and "QTL Hotspot D" contain three QTLs each associated with studied traits and spanning the physical interval of 4.0Mb and 2.3Mb expressed PV of 6.60–17.03% and 10.22–26.84%, respectively (Table 5). Furthermore, all these four "QTL cluster/hotspots" comprise many significant QTLs identified in more than one individual environment. QTLs within "QTL Hotspot B" were identified in both ZY and K3N populations (Table 5 and Figure 3). Hence, these four major "QTL hotspots" are the stable genomic regions governing the inheritance of seed shape and size in soybeans.

Table 5. Four QTL hotspots/clusters detected in ZY and K3N RIL populations across multiple environments.

QTL Cluster Name	Chr_Bin Range	QTL Name	Physical Range (bp)	LOD	Additive Effect	R² (%)
		$qSW\text{-}6\text{-}1_{ZY}$		10.53	0.12	15.35
		$qST\text{-}6\text{-}1_{ZY}$		6.16	0.11	9.68
Cluster-06/QTL Hotspot A	Chr06_Mark730486-Mark767055(ZY)	$qSL\text{-}6\text{-}1_{ZY}$	5651662–7843389	4.12	0.1	5.43
		$qSW\text{-}6\text{-}2_{ZY}$		10.76	0.11	14.45
		$qST\text{-}6\text{-}2_{ZY}$		8.14	0.12	11.53
				4.46	0.08	11.12
		$qSLT\text{-}6\text{-}1_{ZY}$		3.70	−0.02	5.27
				4.55	−0.03	17.03
Cluster-10/QTL Hotspot B	Chr10_ Mark668037-Mark662847(ZY) Chr10_bin821-bin828(K3N)	$qSLT\text{-}10\text{-}1_{ZY}$	41983494–45988221	4.61	−0.02	16.83
				4.96	−0.03	7.40
		$qSLW\text{-}10\text{-}1_{ZY}$		4.64	−0.01	6.60
		$qSW\text{-}10\text{-}1_{K3N}$		3.83	−0.1	12.85
		$qSL\text{-}13\text{-}1_{ZY}$		3.5	0.09	5.07
				8.14	0.14	11.46
Cluster-13/QTL Hotspot C	Chr13_ Mark477148-Mark495958(ZY)	$qSW\text{-}13\text{-}1_{ZY}$	20463309–26852039	7.07	0.09	9.59
				3.57	0.07	5.32
		$qST\text{-}13\text{-}2_{ZY}$		4.23	0.09	10.28
		$qST\text{-}13\text{-}3_{ZY}$		5.17	0.09	7.87
				3.57	0.08	4.86
		$qSLW\text{-}13\text{-}3_{ZY}$		4.04	0.01	6.44
		$qSLW\text{-}20\text{-}1_{K3N}$		9.01	−0.03	26.84
Cluster-20/QTL Hotspot D	Chr20_bin1697-bin1704(K3N)		35957343–38300982	4.64	−0.02	11.64
		$qSL\text{-}20\text{-}1_{K3N}$		7.19	−0.18	22.64
		$qSLT\text{-}20\text{-}1_{K3N}$		3.82	−0.02	10.22
				7.28	−0.04	18.54

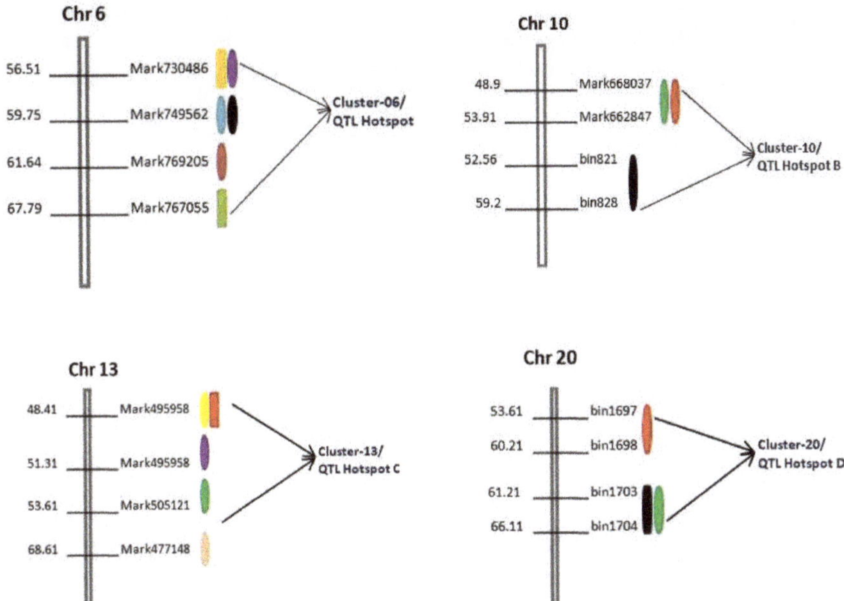

Figure 3. Diagram showing the physical location of four QTL clusters/hotspot regions (cluster-06, cluster-10, cluster-13, and cluster-20) on four different chromosomes viz., Chr6, Chr10, Chr13, and Chr20 identified in two RIL populations across multiple environments Different colors indicate different QTLs within same region.

2.7. Candidate Gene Mining within Major "QTL Hotspots"

The whole-genome sequence and gene annotations availability makes it possible to identify possible candidate genes within major genomic regions. In the present study, all the model genes along with their gene annotations were downloaded from Phytozome and Soybase. In total, we identified 2406 gene models within the physical genomic interval of all four major "QTL hotspots" (Table S4). An online web-based toolkit agriGO V2.0 was used for a gene ontology (GO) enrichment analysis to visualize the biological process, molecular function, and cellular component main categories (Figure 4). Among all the genes present within the four "QTL hotspots", only the 831, 193, 192, and 118 genes from "QTL Hotspot A", "QTL Hotspot B", "QTL Hotspot C", and "QTL Hotspot D", respectively, had GO annotations available (Figure 4). In all the four major "QTL hotspots", a higher percentage of genes were associated within the terms cellular process, metabolic process, cell part, cell, catalytic activity, and binding (Figure 4), suggesting a vital role of these terms in regulating seed size and shape in soybeans.

Based on the gene annotations, available literature, and GO enrichment analysis, we predicted 26, 19, 35, and 18 candidate genes from "QTL Hotspot A", "QTL Hotspot B", "QTL Hotspot C", and "QTL Hotspot D," respectively (Table S6). These genes function directly or indirectly in regulating seed development, as well as seed shape and size, such as mitotic cell division, storage of proteins and lipids, transport, metabolic process, signal transduction of plant hormones, degradation of the ubiquitin-proteasome pathway, and fatty acid beta-oxidation (Table S6). To further refine the above-predicted candidate genes list, we retrieved RNA-Seq data of these candidate genes from Soybase (www.soybase.org) [35].

Based on RNA-seq analysis, 23 genes out of above 88 predicted candidate genes showed significantly higher gene expression/fold-change in the seed development stages, root nodules, leaf, and pod shell. These genes include nine *(Glyma06g02390, Glyma06g08290, Glyma06g04810, Glyma06g03700, Glyma06g02790, Glyma06g06160, Glyma06g07200, Glyma06g09650, and Glyma06g10700)*; two *(Glyma10g35360 and Glyma10g36440)*; six *(Glyma13g17750, Glyma13g17980, Glyma13g21770, Glyma13g18730, Glyma13g21700, and Glyma13g22790)*; and six *(Glyma20g28550, Glyma20g28460, Glyma20g28640, Glyma20g27300, Glyma20g29750, and Glyma20g30100)* genes from "QTL Hotspot A", "QTL Hotspot B", "QTL Hotspot C", and "QTL Hotspot D", respectively (Figure 5 and Table 6). Hence, these 23 genes might be the possible candidate genes regulating seed size and shape in soybean. However, they need further functional validation to check their actual roles in governing seed size and development.

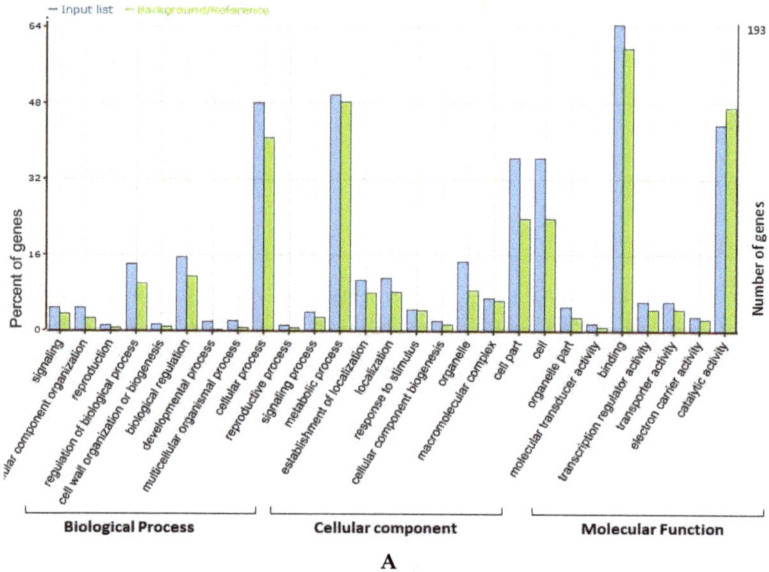

Figure 4. *Cont.*

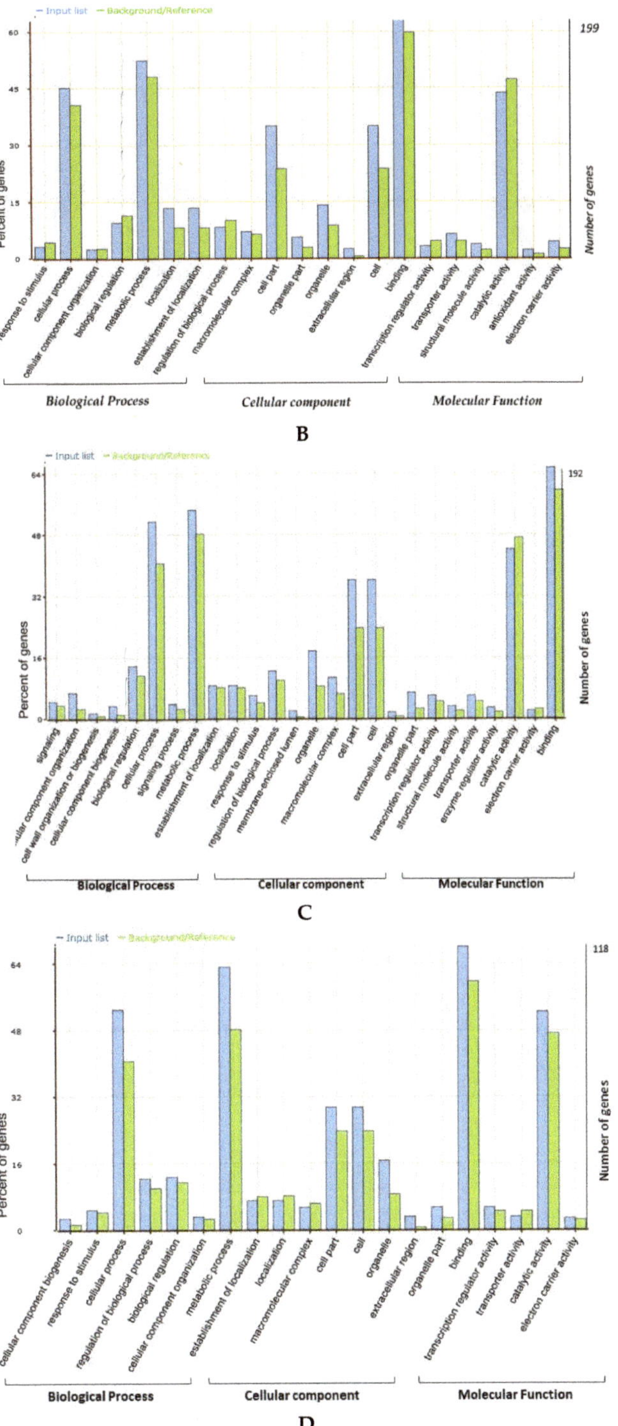

Figure 4. agriGO annotation information. (**A**) Cluster-06 (QTL Hotspot A), (**B**) Cluster-10 (QTL Hotspot B), (**C**) Cluster-13 (QTL Hotspot C), and (**D**) Cluster-20 (QTL Hotspot D).

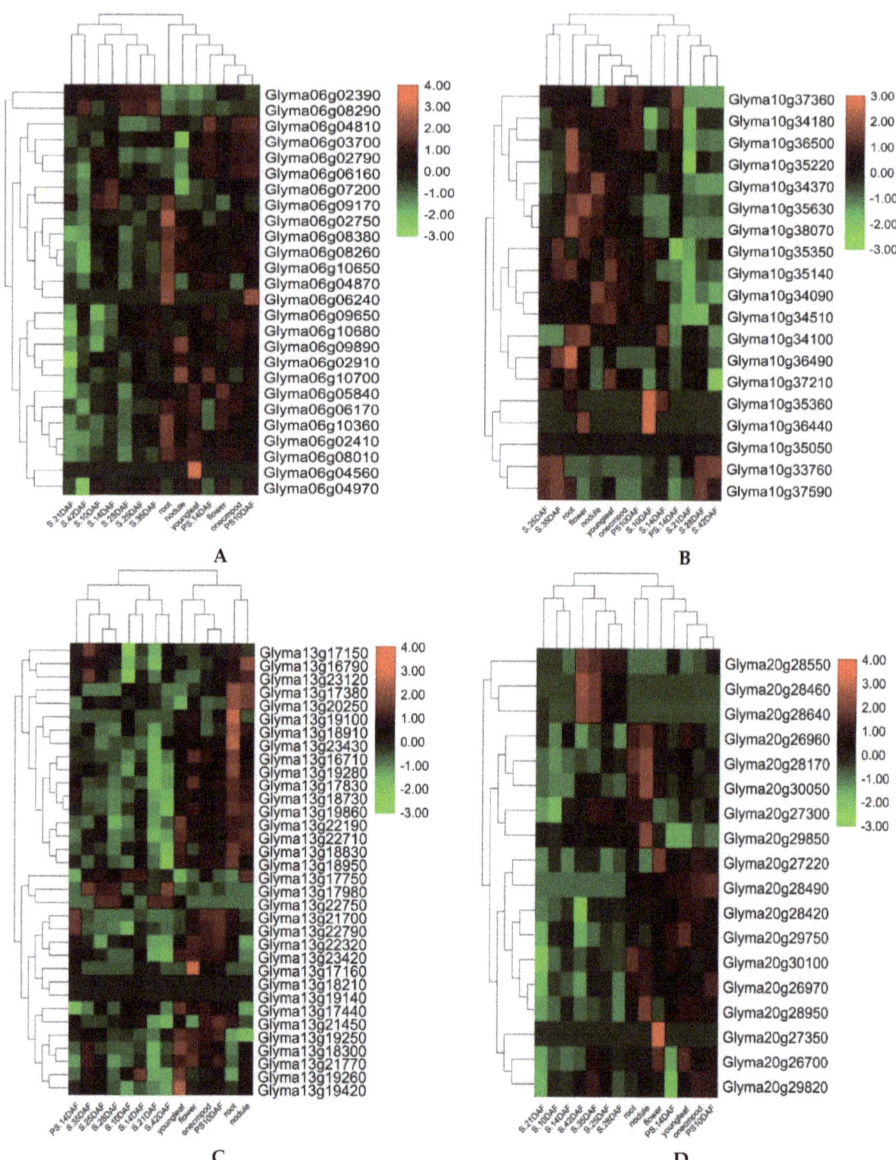

Figure 5. Heat map exhibiting the expression profiles of 23 candidate genes among the different soybean tissues and development stages from four QTL hotspots/clusters. (**A**) Cluster-06 (QTL Hotspot A), (**B**) Cluster-10 (QTL Hotspot B), (**C**) Cluster-13 (QTL Hotspot C), and (**D**) Cluster-20 (QTL Hotspot D). Heat map was generated using the RNA-sequencing data downloaded from online dataset SoyBase. Youngleaf—young leaf, Onecmpod—1 cm of pod, PS—pod shell, DAF—days after flowering, and S—seed.

Table 6. Predictive gene annotation information.

Cluster/QTL Hotspot	Mapped IDs	Gene Functional Annotation
Cluster-06/QTL Hotspot A	*Glyma06g02390*	RING/U-box superfamily protein/protein ubiquitination
	Glyma06g08290	Lipid storage
	Glyma06g04810	Seed coat development extracellular region
	Glyma06g03700	Seed development ovule development
	Glyma06g02790	Response to auxin stimulus
	Glyma06g06160	Ubiquitin-protein ligase activity
	Glyma06g07200	Response to ethylene stimulus ubiquitin-protein ligase activity
	Glyma06g09650	microtubule nucleation, response to auxin stimulus, cytokinin mediated signaling pathway
	Glyma06g10700	Response to brassinosteroid stimulus sterol biosynthetic process
Cluster-10/QTL Hotspot B	*Glyma10g35360*	Response to auxin stimulus cellular component
	Glyma10g36440	Auxin biosynthesis
Cluster-13/QTL Hotspot C	*Glyma13g17750*	Response to auxin stimulus protein dimerization activity
	Glyma13g17980	Embryo development
	Glyma13g21770	Endosperm development embryo development
	Glyma13g18730	Ubiquitin-dependent protein catabolic process
	Glyma13g21700	Response to ethylene stimulus-response to auxin stimulus
	Glyma13g22790	Protein kinase activity
Cluster-20/QTL Hotspot D	*Glyma20g28550*	Seed maturation protein
	Glyma20g28460	Lipid storage
	Glyma20g28640	Lipid storage
	Glyma20g27300	Lipid metabolic process seed maturation cell growth
	Glyma20g29750	Multidimensional cell growth polysaccharide biosynthetic process regulation of hormone levels
	Glyma20g30100	Embryo development seed development protein phosphorylation

3. Discussion

Seed shape and size is an economically important trait determining the yield and quality in soybeans. Hence, developing soybean cultivars with improved seed shapes and sizes is considered as a critical objective of soybean breeding programs. However, to develop improved cultivars, it is a prerequisite to have a detailed understanding of genetic architecture, as well as a mechanism underlying the trait of interest. Both seed shape and size are complex quantitative traits, governed by multiple genes and are highly environmentally sensitive. Although, over the past decades, many QTLs related to soybean seed shape/size have been reported but not stable and confirmed, due to small-sized mapping populations and low-density genetic maps, and, hence, not implied for breeding improved seed shapes and sizes in soybean. Thus, the present study aimed to utilize high-density intraspecific linkage maps of ZY and K3N populations, evaluated in three different environments, to identify the stable significant main-effect QTLs, "QTL Hotspots", and epistatic-effect QTLs, as well as their interactions with the environment; additionally, find possible candidate genes for soybean seed sizes and shape traits. In this study, ANOVA results revealed a significant difference among the RILs of both ZY and K3N populations for all six traits ($P < 0.01$, Table S2). Similar to previous studies, our study also reported that all six traits related to seed size/shape were significantly affected by G, E, and G × E [7,36]. Frequency distribution of all six traits (SL, SW, ST, SLW, SLT, and SWT) showed the characteristics of continuous variations, and all these traits have transgressive segregation in both directions, which indicates that both parents contributed favorable alleles for these traits (Figure 1). These findings are in agreement with the prior findings, which also stated continuous distribution and transgressive segregation for seed size/shape traits among RILs of soybeans in multiple environments [3,17,22]. In our study, the estimated heritability of all six traits was high (>60%) in both RIL populations across all three environments (Table S1), which was consistent with previous studies [7]. The high heritability suggests that if the trial repeated in the same growing/environment conditions, there would be a high possibility of achieving the same kind of phenotypic results. A highly significant correlation (either positive or negative) between any two seed shapes or seed size traits and between seed size and shape traits is in accordance, as previously reported by Xu et al. [5].

QTL mapping is a practical approach and has been frequently employed for the detection of QTLs/genes underlying the quantitative traits in crop plants. However, the efficiency and accuracy of QTL mappings are influenced, mainly by parental diversity and marker density [26]. The quality of genetic maps has a significant impact on the accuracy of QTL detection, and, consequently, increasing marker density can intensify the resolution of QTL mapping [37]. Hence, it is a prerequisite to utilize high-density linkage maps for improving the efficiency and accuracy of linkage mapping and MAS. In this study, high-density genetic maps of ZY and K3N populations were used, consisting of 3255 SLAF and 1733 bin markers, respectively. The markers in both linkage maps, viz., ZY and K3N, were integrated to all 20 LGs and covered the total length of 2144.85 and 2362.44 cM, respectively, with an average distance between adjacent markers of 0.66 cM and 1.36 cM, respectively.

The use of high-density bin-maps assisting in QTL identification with tightly linked markers provided a good foundation for analyzing quantitative traits. However, to reduce environmental errors, RILs were planted in three environments (consisting of different locations and years), and each of the environments was statistically different. Jansen et al. [38] described that the QTL position and effects could be accurately evaluated if the phenotypic data collected in various environments were different from a statistical perspective. Although, markers associated with the QTLs regulating the seed sizes and shapes in soybeans have been mapped on all linkage groups (Soybase, www.soybase.org). However, for cross-validation and improving the accuracy of QTL mapping results, we used two different methods for QTL mapping, viz., CIM and MCIM. A total of 88 and 48 QTLs were detected by CIM and MCIM methods, respectively, associated with all six traits related to seed size and shape (Tables 1–3). Among these QTLs, 15 common QTLs were verified through both CIM and MCIM, indicating that these QTLs were stable and utilized effectively as potential candidate regions for enhancing seed sizes and shapes in soybeans. The QTL results of our study revealed better matches with the SoyBase database (www.soybase.org; Tables 1 and 2); however, 51 (CIM) and 27 (MCIM) QTLs were identified for the very first time (Tables 1–3). These novel QTLs collectively expressed more than 90% of PV for seed size and shape, suggesting their potential value for the development of improved soybean cultivars. Among these novel QTLs, qSL-9-$1_{ZY,K3N}$, qST-6-2_{ZY}, $qSLW$-6-1_{ZY}, $qSLW$-20-1_{K3N}, $qSLT$-10-1_{ZY}, and $qSLT$-20-1_{K3N} were reported as stable and major QTLs, identified in more than one individual environment, with $R^2 > 10\%$. Besides, by comparing the physical genomic regions of QTLs identified in both populations (ZY and K3N) and mapping methods (CIM and MCIM), two major and novel QTLs, viz., qSW-1-3_{ZY} and $qSLT$-20-3_{K3N}, were characterized commonly in both mapping methods. These above seven unique and stable QTLs significantly represent potential loci for the improvement of seed sizes and shapes in soybeans. Hence, identification of many new and unique QTLs in the present study suggests distinct genetic architecture in the population derived from the diverse Chinese cultivated soybean genotypes and the need to use more germplasm for revealing the complex genetic basis of soybeans. The favorable alleles for seed size and shape traits were contributed by both parents of two RIL populations, viz., ZY and K3N. Therefore, it is critical to note that not only the higher phenotype parent contributes beneficial alleles but also the contribution of favorable alleles by lower phenotype parents cannot be disregarded; similar results are also described earlier [30].

The stability of the QTL is essential for use in a breeding program. In addition to novel stable QTLs identified for both seed size and shape traits, this study also identified 37 and 21 QTLs through the CIM and MCIM methods, which have been previously colocalized in the same physical interval by earlier studies (see references in Tables 1–3). Out of these colocalized QTLs, 12 and 3 QTLs detected by the CIM and MCIM methods were major ($R^2 > 10\%$). Therefore, our results showed the reliability of QTL mapping. Furthermore, these QTLs can be utilized as principal targets to identify the candidate genes and MAS in future studies.

It has been demonstrated that epistatic and QTL by environment interaction effects are the two crucial genetic factors that make an enormous contribution to the phenotypic variation observed in complex traits, and the knowledge of those interaction effects is vital for understanding the genetic mechanism of complex traits [39,40]. Previous studies revealed that the seed sizes and shapes of soybeans is significantly affected by the environment [36]. Moreover, knowledge of specific QTL by environment interactions can guide the search of varieties adapted to particular environments. The QTLs with more significant additive effects are often considered more stable [41,42]. For example, the $qSW\text{-}1\text{-}1_{ZY}$ and $qST\text{-}18\text{-}2_{ZY}$ (additive effect: 0.14) identified in both CIM and MCIM methods; though, $qSLT\text{-}6\text{-}5_{ZY}$ (additive effect: 0.001) was detected only in the MCIM method (ZY only) (Table 3). The genetic architecture of seed size and shape also includes epistatic interactions between QTLs [11,43]. Hence, ignoring intergenic interactions will lead to the overestimation of individual QTL effects, and the underestimation of genetic variance [44], consequently, could result in a substantial drop in the genetic response to MAS, particularly at late generations [45]. In the present study, 16 pairs of digenetic epistatic QTLs pairs were identified for seed size and shape in both populations and expressed phenotypic variations that varied from 1.71 to 9.70% (Table 4). Except for $qSL\text{-}13\text{-}3_{ZY}$, $qSL\text{-}13\text{-}4_{ZY}$, and $qST\text{-}13\text{-}4_{ZY}$, all the remaining epistatic QTLs do not possess additive effects alone, suggesting that these loci might serve as modifying genes that interact with other genes to affect the phenotypes of seed sizes and shapes (Table 4). All 16 pairs have significant AA, but only four QTL pairs, viz., $qSL\text{-}2\text{-}1_{K3N}$ and $qSL\text{-}2\text{-}2_{K3N}$, $qST\text{-}9\text{-}1_{K3N}$ and $qST\text{-}12\text{-}4_{K3N}$, $qSLT\text{-}2\text{-}3_{K3N}$ and $qSLT\text{-}7\text{-}1_{K3N}$, and $qSWT\text{-}6\text{-}1_{K3N}$ and $qSWT\text{-}8\text{-}4_{K3N}$, hold significant AAE interaction effects. However, the total AAE phenotypic variations expressed by these four epistatic pairs was 19.14%. These results show that epistatic and environmental interactions are fundamental for understanding the genetic basis of seed sizes and shapes in soybeans, demonstrating that these effects should be considered in a QTL mapping program and could increase the accuracy of phenotypic value predictions in MAS.

Colocalization of QTLs on chromosomes for different traits related to seed size and shape were also observed in this study. This colocalization of QTLs linked to related traits on chromosomes was reported earlier in soybeans and referred to as "QTL cluster/hotspots" [46]. In this study, we scrutinized a few genomic regions containing QTL clusters and found four QTL clusters/hotspots on four different chromosomes, viz., Chr06, Chr10, Chr13, and Chr20 (Figure 3 and Table 5). The QTLs within each cluster/hotspot are associated with three or more traits related to seed sizes and shapes in soybeans. The highest number of six and five QTLs were observed in "QTL Hotspot A" and "QTL Hotspot C", respectively, harboring QTLs for more than three traits related to seed size and shape (Figure 3 and Table 5). The other two hotspots, viz., "QTL Hotspot B" and "QTL Hotspot D", contain three QTLs, each for three different traits related to seed size and shape (Table 5). These QTLs clusters/hotspots have not reported and added to the growing knowledge of the genetic control of these traits. The phenomenon of the QTL clustering might represent a linkage of genes/QTLs or result from the pleiotropic effects of a single QTL in the same genomic region [47]. This colocalization of QTLs for different seed size and shape traits was following the fact that they were highly significantly correlated with each other (Table 1). These "QTL hotspot" regions showed that the QTLs linkage/pleiotropy could facilitate the enhancement of seed size and shape. Previously, some of the QTLs for other traits have also been identified in the same region of "QTL Hotspot A" on chromosome 06, which are related to seed oil and protein content [48,49] and days to flowering [50]. In the case of "QTL Hotspot B", QTLs related to seed weight and seed yield [51], length of the reproductive stage [33], days to flowering, and maturity [33] were reported in the same physical interval.

Similarly, earlier studies have also reported QTLs for seed weight [7] and seed volume [33] in the "QTL Hotspot C" region on Chr13. In "QTL Hotspot D", QTLs related to seed maturity [33] and seed oil content [52] have formerly reported. Seed oil and protein content in soybeans have reported a significant correlation with seed size and shape [53], as seed oil and protein content represents a major component of soybean seeds, representing 38–42% and 18–22%, respectively; hence, these traits are directly related to seed sizes and shapes in soybeans [13]. Both seed size and shape are important yield component traits [54] and it has been reported that days to flowering and maturity is directly correlated to yield in soybeans [55,56], signifying the potential probability of common genic factors for these traits and also showing the necessity to promote further study for these regions. These QTL clusters have provided some valuable information to define genome regions associated with different traits. Based on the comprehensive analysis of clusters in this study, breeding programs targeting an increase of seed sizes and shapes with high yield and superior quality can focus on hotspot clustering and select QTLs around the region. Finally, the existence of QTL clusters/hotspots has provided proof that genes related to some crop traits are more densely concentrated in certain genomic regions of crop genomes than others [33,51].

Identification of candidate genes underlying the QTL region is of great interest for practical plant breeding. Earlier studies based on QTL mapping of seed size and shape did not practice mining for candidate genes [22,54], and, to date, only a few seed size/shape-related genes have been isolated from the soybean. For example, the *Ln* gene has a large effect on the number of seeds per pod and seed size/shape [57], and, recently, the *PP2C-1* (protein phosphatase type-2 C) allele from wild soybean accession ZYD7 were found to contribute toward the increase in seed size/shape [58]. Based on the gene annotations, available literature, and GO enrichment analysis, the present study identified the possible candidate genes regulating the seed sizes and shapes in soybeans that underlies the four categorized "QTL hotspots". Gene ontology (GO) analysis revealed that most of the genes underlying the above four "QTL hotspots" belong to the terms cellular process, metabolic process, cell part, cell, catalytic activity, and binding, and these elements are reported as being vital in seed development [59–61]. A total of 2406 gene models were mined within the physical interval of the four "QTL hotspots." Out of them, 88 were considered as possible candidate genes, based on the GO enrichment analysis, gene function, and available literature. These 88 predicted candidate genes have functions that are directly or indirectly involved in seed development, influencing the shape and size of seeds, such as lipid storage, transport and metabolic processes, signal transduction of plant hormones, degradation of the ubiquitin-proteasome pathway, fatty acid beta-oxidation, the brassinosteroid-mediated signaling pathway, and the auxin biosynthetic process (Table S5). From the available gene expression data (RNA-seq), 23 of the 88 predicted candidate genes expressed significantly higher gene expression, particularly in seed development stages, root nodules, leaf, and pod shell (Figure 5 and Table S5). Out of these 23 genes, five genes, viz., *Glyma06g04810*, *Glyma06g03700*, *Glyma13g17980*, *Glyma13g21770*, and *Glyma20g30100* have functions that are related to seed development, ovule development, endosperm, and embryo development, which have been reported to directly contribute to seed sizes and shapes in crop plants, including soybeans [62,63]. Likewise, *Glyma06g02390*, *Glyma06g06160*, *Glyma06g07200*, and *Glyma13g18730* encode RING/U-box superfamily proteins/protein ubiquitination. The ubiquitin pathway has recently been known to play an essential part in seed size determination [60]. Several factors involved in ubiquitin-related activities have been revealed to determine seed sizes in *Arabidopsis* and rice [60]. Genes, viz., *Glyma06g08290*, *Glyma20g28460*, *Glyma20g28640*, *Glyma20g29750*, *Glyma20g28550*, *Glyma13g22790*, *Glyma20g29750*, and *Glyma20g27300*, function in lipid storage, seed maturation, and cell growth, which have formerly been reported to determine seed size and shape in oilseeds, including soybeans [64]. For example, overexpression of *GmMYB73* promotes lipid accumulation in soybean seeds, which leads to increased seed sizes in soybeans [65]. Genes, viz., *Glyma06g09650*, *Glyma10g35360*, *Glyma10g36440*, *Glyma13g17750*, and *Glyma13g21700*, are involved in auxin biosynthesis, responses to auxin stimulus, and responses to ethylene stimulus. The auxin regulates seed weights and sizes in *Arabidopsis* [22,66]. *Glyma06g10700* functions to regulate the brassinosteroid stimulus, which positively governs seed size [62]. Hence, based on the gene function, GO, and RNA-Seq analysis, the above 23 genes were considered as the most potentially possible candidate genes for regulating the seed sizes and shapes in soybeans. However, it requires further validation and verification to confirm their actual roles in seed sizes/shapes in soybeans, as well as their future uses for the improvement of seed quality traits. Some of these genes were already included in our ongoing project for functional validation to ascertain their effects on the seed sizes and shapes. Hence, the precise identification of QTLs in a specific physical interval through the use of a high-density map would make it easy to identify candidate genes.

4. Materials and Methods

4.1. Plant Material and Experimental Conditions

In the present study, two related RIL populations, viz., ZY and K3N, consisting of 236 and 91 lines, respectively, were used for elucidating the genetic basis of seed shapes and sizes in soybeans. The ZY and K3N populations were derived through a single seed descent (SSD) method by crossing a common higher seed size parent Nannong1138-2 (N) with two cultivated soybean varieties, viz., Zhengxiaodou (Z) and KeFeng35 (K3), having smaller seed sizes [67]. All the plant material was received from Soybean Germplasm Gene Bank, located at the National Centre for Soybean Improvement (Ministry of Agriculture), Nanjing Agricultural University, Nanjing, China. The $F_{6:9}$–$F_{6:11}$ generations of both RIL populations were planted in three different environments, viz., Jiangpu Experimental Station, Nanjing, Jiangsu Province (Latitude 33°03′ N; Longitude 63°118′ E) in 2012 and 2017 (2012JP and 2017JP) and Fengyang Experimental Station, Chuzhou, Anhui Province (Latitude 32°87′ N; Longitude 117°56′ E) in 2012 (2012FY). Both RIL populations, along with their parents, were planted in a single-line plot of 1 m in length and 0.5 m in width in a randomized complete block design with three replications. In each environment, standard cultural and agronomic practices were trailed, as previously described [68,69].

4.2. Phenotypic Evaluation and Statistical Analysis

For the phenotypic assessment of seed size and shape, we collected seeds from the randomly selected ten plants harvested from the middle of each block across three different environments (2012JP, 2012FY, and 2017JP) in both RIL populations. The seed size traits include seed length (SL), seed width (SW), and seed thickness (ST), whereas seed shape was assessed using three different ratios, viz., seed length/seed width (SLW), seed length/seed thickness (SLT), and seed width/seed thickness (SWT). The SL, SW, and ST were measured in millimeters (mm) using the vernier caliper instrument, according to Kaushik et al. [39]. However, SLW, SLT, and SWT were estimated from the individual values of the SL, SW, and ST, respectively, by following Omokhafe and Alika [41].

Descriptive statistics, such as mean, range (maximum and minimum values), coefficient of variation (CV%), skewness, and kurtosis for above seed size and shape traits in both RIL populations, including their parents, were calculated using the SPSS17.0 software (http://www.spss.com) [42]. For each environment, an analysis of variance (ANOVA) was carried out using a generalized linear model (GLM) program of SAS PROC (SAS Institute Inc. v. 9.02, 2010, Cary, NC, USA). The ANOVA for the combined environment (CE) was also performed in SAS software using mixed PROC with random factors: lines, environments, replication within environments, and the line-by-environment interaction. Pearson correlation coefficient (r) among traits was calculated from the average data using PROC

CORR in combined environments. The broad-sense heritability (h^2) in RIL populations was estimated using the following equation:

$$h^2 = \sigma_G^2 / \left(\sigma_G^2 + \sigma_{GE}^2 / n + \sigma_e^2 / \mathrm{nr} \right) \tag{1}$$

where σ_G^2 is the genotypic variance, σ_{GE}^2 is the variance of the genotype-by-environment interaction, σ_e^2 is the error variance, n is the number of environments, and r is the number of replications within an environment [44].

4.3. SNP Genotyping and Bin Map Construction

Genetic map construction began with the extraction of DNA from the young and fresh leaves of both RIL populations, along with their parents, by following the protocol of Zhang et al. [45]. DNA library construction, high-throughput sequencing (RAD-Seq), high-quality SNP acquisition, and SLAF/bin marker integration for ZY and K3N populations, respectively, were performed as described by Huang et al. [70] and Cao et al. [30]. These SLAF and bin markers were employed to develop the linkage maps of the ZY and K3N populations, respectively, using JoinMap 4.0 [71]. High-density genetic maps of the ZY and K3N populations contained 3255 SLAF and 1733 bin markers, respectively. The total length of the ZY and K3N maps were 2144.85 and 2362.44 cM, with an average distance between the adjacent markers as 0.659 and 1.36 cM, respectively (Table S7). The average length of each linkage group was 162.75 and 86.65 cM for ZY and K3N linkage maps, with the mean marker density of each linkage group as 107.24 and 118.122, respectively (Table S7).

4.4. QTL Mapping for Seed Size and Shape

For QTL analysis, we used the WinQTLCart 2.5 software [47] and QTLNetwork 2.2 [72]. The model of composite interval mapping (CIM) was used to identify the main-effect QTLs (M-QTLs) with a 10 cM window at a walking speed of 1cM for the WinQTLCart 2.5 software. The LOD threshold was premeditated using 1000 permutations for an experimental-wise error rate of $P = 0.05$ to determine whether the QTL was significantly associated with the traits [73]. The model of mixed linear composite interval mapping (MCIM) was applied to identify significant additive effect QTLs, epistatic QTLs (AA), genotype-by-environment interaction effects (additive by the environment (AE) and AA by the environment (AAE)) in the QTLNetwork 2.2 [74]. The physical location of M-QTLs on each chromosome were drawn by using MapChart 2.1 software [75].

QTLs were named by following standard nomenclature [76], with minor modifications. For example, for the QTL denoted as $qSW\text{-}1\text{-}1_{ZY}$, q indicates QTL, SW stands for the trait (seed width), -1 show the chromosome on which the QTL detected, -1 also indicates the order of QTL identified on the chromosome for each trait, and$_{ZY}$ represents the ZY-RIL population in which QTL was detected.

4.5. Mining of Candidate Genes for Major QTLs

QTLs identified in two or more than two environments with $R^2 > 10\%$ were considered as significant and stable QTLs [77]. By utilizing the online resource databases of Phytozome (http://phytozome.jgi.doe.gov) and SoyBase (http://www.soybase.org), we downloaded all the genomic data within the physical interval position of the major "QTL hotspots", and candidate genes were predicted based on the gene annotations (http://www.soybase.org and https://phytozome.jgi.doe.gov), as well as previously published literature. Gene ontology (GO) information was derived from SoyBase through online resources: GeneMania (http://genemania.org/); Gramene (http://archive.gramene.org/db/ontology); the Kyoto Encyclopedia of Genes and Genomes website (KEGG, www.kegg.jp); and the National Centre for Biotechnology Information (NCBI: https://www.ncbi.nlm.nih.gov). These were used to screen the predicted candidate genes further. Gene ontology (GO) enrichment analysis was conducted for all the genes within the four major "QTL hotspots", viz., "QTL Hotspot A", "QTL Hotspot B", "QTL Hotspot C", and "QTL Hotspot D", using agriGO V2.0 (http://systemsbiology.cau.edu.cn/agri-GOv2/) [78]. The freely available RNA-Seq dataset at the SoyBase website was obtained to analyze the expression of predicted candidate genes in different soybean tissues, as well as the development stages. A heat map for the visualization of fold-change in the expression patterns of these predicted candidate genes was constructed by using TBtools_JRE1.6 software [79].

5. Conclusions

In conclusion, the present study is a detailed investigation for elucidating the genetic architecture of seed sizes and shapes in soybean. In aggregate, 88 and 48 QTLs were detected through CIM and MCIM, respectively, including 15 common QTLs, with two major ($R^2 > 10\%$) and novel QTLs, viz., $qSW\text{-}1\text{-}1_{ZY}$ and $qSLT\text{-}20\text{-}1_{K3N}$. Besides, 51 and 27 QTLs, identified through CIM and MCIM, respectively, were reported for the first time. All identified QTLs were clustered into four major "QTL cluster/hotspots" and represent the major and stable genomic regions governing the inheritance of soybean seed sizes and shapes. Hence, these "QTL hotspot" regions could be of significant consideration for future soybean breeding. Our study predicted 23 genes as the possible candidates, regulating seed sizes and shapes within the genomic region of four "QTL hotspots"; however, they need further functional validation to clarify their actual roles in seed development. Moreover, our results showed that 15 QTLs exhibited significant AE effects, and 16 pairs of QTLs possessed an epistatic effect. However, except for three QTLs, viz., $qSL\text{-}13\text{-}3_{ZY}$, $qSL\text{-}13\text{-}4_{ZY}$, and $qSW\text{-}13\text{-}4_{ZY}$, all the remaining epistatic QTLs showed no main effects. Hence, the hotspot regions and novel significant stable QTLs identified in the present study will be the main focus of soybean breeders for fine mapping, gene cloning, and the MAB of soybean varieties with improved seed quality and yield.

Supplementary Materials: The following are available online at http://www.mdpi.com/1422-0067/21/3/1040/s1, Table S1: Descriptive statistics, broad-sense heritability (h2) of Seed Shape & size traits evaluated in two recombinant inbred lines (RILs) ZY & K3N. The RILs and the parent were grown in different field environments of years and locations. * &** represent significance at 5% and 1%, respectively; Table S2: Combined analysis of variance (ANOVA) for Seed shape & size trait (SL,SW,ST,SLW,SLT,SWT) in ZY & K3N RIL Populations across three different environments (2012FY, 2012JP and 2017JP); Table S3: Correlation analysis among six Seed size/shape traits (SL,SW,ST,SLW,SLT,SWT). * &** represent significance at 5% and 1%, respectively; Table S4: Model genes within cluster-06, cluster-10, cluster-13 and cluster-20 regions in both RIL Populations (ZY & K3N); Table S5: Candidate genes within cluster-06, cluster-10, cluster-13 and cluster-20 in ZY and K3N-RIL Populations and their relative expression data retrived from RNA-seq data available in soybase; Table S6: Predicted candidate genes within cluster-06, cluster-10, cluster-13 and cluster-20 regions in both RI Populations (ZY & K3N) based on known functional annotation; Table S7: Distribution of SLAF and bin markers mapped on soybean chromosomes/linkage groups in ZY & K3N Populations, respectively.

Author Contributions: T.Z. and J.A.B. conceived and designed the project. A.H. and S.S. performed the experiments. A.H., Y.C., S.S., S.L., R.A.S., and M.A.E. analyzed the data. A.H. drafted the manuscript. T.Z. and, J.A.B. revised the paper. All authors have read and agreed to the published version of the manuscript.

Funding: This research was funded by the National Key Research and Development Program (2018YFD0201006), the National Natural Science Foundation of China (Grant Nos. 31871646 and 31571691), the MOE Program for Changjiang Scholars and Innovative Research Team in University (PCSIRT_17R55), the Fundamental Research Funds for the Central Universities (KYT201801), and the Jiangsu Collaborative Innovation Center for Modern Crop Production (JCIC-MCP) Program.

Conflicts of Interest: The authors declare no conflicts of interest. The funders had no role in the design of the study; in the collection, analyses, or interpretation of data; in the writing of the manuscript, or in the decision to publish the results.

Abbreviations

SL	Seed Length
SW	Seed Width
ST	Seed Thickness
SLW	Seed Length-to-Width
SLT	Seed Length-to-Thickness
SWT	Seed Width-to-Thickness
G x E	Genotype x Environment

QTL	Quantitative Trait Loci
USDA	United States Department of Agriculture
SSRs	Simple Sequence Repeats
M-QTL	Main-Effect QTL
E-QTL	Epistatic-Effect QTL
RIL	Recombinant Inbred Line
PV	Phenotypic Variation
SSD	Single Seed Decent
RCBD	Randomized Complete Block Design
Mm	millimeters
CV	Coefficient of Variation
ANOVA	Analysis of Variance
GLM	Generalized Linear Model
CE	Combined Environment
MSG	Multiplexed Shotgun Genotyping
SNP	Single Nucleotide Polymorphisms
CIM	Composite Interval Mapping
LOD	Logarithm of the Odds

References

1. Wang, D.; Bales-Arcelo, C.; Zhang, Z.; Gu, C.; DiFonzo, C.D.; ZHANG, G.; Yang, Z.; Liu, M.; Mensah, C. Sources of Aphid Resistance in Soybean Plants. Google Patents. 2019. Available online: https://peshkin.mech.northwestern.edu/patents/10108288. (accessed on 1 October 2019).

2. Liu, D.; Yan, Y.; Fujita, Y.; Xu, D. Identification and validation of QTLs for 100-seed weight using chromosome segment substitution lines in soybean. *Breed. Sci.* **2018**, *68*, 442–448. [CrossRef] [PubMed]

3. Salas, P.; Oyarzo-Llaipen, J.; Wang, D.; Chase, K.; Mansur, L. Genetic mapping of seed shape in three populations of recombinant inbred lines of soybean (Glycine max L. Merr.). *Theor. Appl. Genet.* **2006**, *113*, 1459–1466. [CrossRef] [PubMed]

4. Yan, S.; Zou, G.; Li, S.; Wang, H.; Liu, H.; Zhai, G.; Guo, P.; Song, H.; Yan, C.; Tao, Y. Seed size is determined by the combinations of the genes controlling different seed characteristics in rice. *Theor. Appl. Genet.* **2011**, *123*, 1173. [CrossRef] [PubMed]

5. Xu, Y.; Li, H.-N.; Li, G.-J.; Wang, X.; Cheng, L.-G.; Zhang, Y.-M. Mapping quantitative trait loci for seed size traits in soybean (Glycine max L. Merr.). *Theor. Appl. Genet.* **2011**, *122*, 581–594. [CrossRef] [PubMed]

6. Tao, Y.; Mace, E.S.; Tai, S.; Cruickshank, A.; Campbell, B.C.; Zhao, X.; Van Oosterom, E.J.; Godwin, I.D.; Botella, J.R.; Jordan, D.R. Whole-genome analysis of candidate genes associated with seed size and weight in sorghum bicolor reveals signatures of artificial selection and insights into parallel domestication in cereal crops. *Front. Plant. Sci.* **2017**, *8*, 1237. [CrossRef]

7. Hu, Z.; Zhang, H.; Kan, G.; Ma, D.; Zhang, D.; Shi, G.; Hong, D.; Zhang, G.; Yu, D. Determination of the genetic architecture of seed size and shape via linkage and association analysis in soybean (Glycine max L. Merr.). *Genetica* **2013**, *141*, 247–254. [CrossRef]

8. Sax, K. The association of size differences with seed-coat pattern and pigmentation in Phaseolus vulgaris. *Genetics* **1923**, *8*, 552.

9. Cui, Z.; James, A.; Miyazaki, S.; Wilson, R.F.; Carter, T. Breeding specialty soybeans for traditional and new soyfoods. In *Soybeans as Functional Foods and Ingredients*; Liu, K., Ed.; AOSC Press: Champaign, IL, USA, 2004; pp. 274–332.

10. Gandhi, A. Quality of soybean and its food products. *Int. Food Res. J.* **2009**, *16*, 11–19.

11. Liang, H.; Xu, L.; Yu, Y.; Yang, H.; Dong, W.; Zhang, H. Identification of QTLs with main, epistatic and QTL by environment interaction effects for seed shape and hundred-seed weight in soybean across multiple years. *J. Genet.* **2016**, *95*, 475–477. [CrossRef]

12. Teng, W.; Feng, L.; Li, W.; Wu, D.; Zhao, X.; Han, Y.; Li, W. Dissection of the genetic architecture for soybean seed weight across multiple environments. *Crop. Pasture Sci.* **2017**, *68*, 358–365. [CrossRef]

13. Wu, D.; Zhan, Y.; Sun, Q.; Xu, L.; Lian, M.; Zhao, X.; Han, Y.; Li, W. Identification of quantitative trait loci underlying soybean (Glycine max [L.] Merr.) seed weight including main, epistatic and QTL× environment effects in different regions of Northeast China. *Plant. Breed.* **2018**, *137*, 194–202. [CrossRef]

14. Coomes, D.A.; Grubb, P.J. Colonization, tolerance, competition and seed-size variation within functional groups. *Trends Ecol. Evol.* **2003**, *18*, 283–291. [CrossRef]

15. Gómez, J.M. Bigger is not always better: Conflicting selective pressures on seed size in Quercus ilex. *Evolution* **2004**, *58*, 71–80. [CrossRef] [PubMed]

16. Haig, D. Kin conflict in seed development: An interdependent but fractious collective. *Annu. Rev. Cell Dev. Biol.* **2013**, *29*, 189–211. [CrossRef]

17. Che, J.; Ding, J.; Liu, C.; Xin, D.; Jiang, H.; Hu, G.; Chen, Q. Quantative trait loci of seed traits for soybean in multiple environments. *Genet. Mol. Res.* **2014**, *13*, 4000–4012. [CrossRef]

18. Lü, H.-Y.; Liu, X.-F.; Wei, S.-P.; Zhang, Y.-M. Epistatic association mapping in homozygous crop cultivars. *PLoS ONE* **2011**, *6*, e17773.

19. Mian, M.; Bailey, M.; Tamulonis, J.; Shipe, E.; Carter, T.; Parrott, W.; Ashley, D.; Hussey, R.; Boerma, H. Molecular markers associated with seed weight in two soybean populations. *Theor. Appl. Genet.* **1996**, *93*, 1011–1016. [CrossRef]

20. Hoeck, J.A.; Fehr, W.R.; Shoemaker, R.C.; Welke, G.A.; Johnson, S.L.; Cianzio, S.R. Molecular marker analysis of seed size in soybean. *Crop. Sci.* **2003**, *43*, 68–74. [CrossRef]

21. Li, C.; Jiang, H.; Zhang, W.; Qiu, P.; Liu, C.; Li, W.; Gao, Y.; Chen, Q.; Hu, G. QTL analysis of seed and pod traits in soybean. *Mol. Plant. Breed.* **2008**, *6*, 1091–1100.

22. Xie, F.-T.; Niu, Y.; Zhang, J.; Bu, S.-H.; Zhang, H.-Z.; Geng, Q.-C.; Feng, J.-Y.; Zhang, Y.-M. Fine mapping of quantitative trait loci for seed size traits in soybean. *Mol. Breed.* **2014**, *34*, 2165–2178. [CrossRef]

23. Tanksley, S.D.; Miller, J.; Paterson, A.; Bernatzky, R. Molecular mapping of plant chromosomes. In *Chromosome Structure and Function*; Springer: Boston, MA, USA, 1988; pp. 157–173.

24. Li, W.-H.; Wei, L.; Li, L.; You, M.-S.; Liu, G.-T.; Li, B.-Y. QTL mapping for wheat flour color with additive, epistatic, and QTL× environmental interaction effects. *Agric. Sci. China* **2011**, *10*, 651–660. [CrossRef]

25. Panthee, D.R.; Marois, J.J.; Wright, D.L.; Narváez, D.; Yuan, J.S.; Stewart, C.N. Differential expression of genes in soybean in response to the causal agent of Asian soybean rust (Phakopsora pachyrhizi Sydow) is soybean growth stage-specific. *Theor. Appl. Genet.* **2009**, *118*, 359. [CrossRef] [PubMed]

26. Zhang, Y.; Li, W.; Lin, Y.; Zhang, L.; Wang, C.; Xu, R. Construction of a high-density genetic map and mapping of QTLs for soybean (Glycine max) agronomic and seed quality traits by specific length amplified fragment sequencing. *BMC Genom.* **2018**, *19*, 641. [CrossRef] [PubMed]

27. Xu, Y.; Crouch, J.H. Marker-assisted selection in plant breeding: From publications to practice. *Crop. Sci.* **2008**, *48*, 391–407. [CrossRef]

28. Yang, B.Z.; Zhao, H.; Kranzler, H.R.; Gelernter, J. Practical population group assignment with selected informative markers: Characteristics and properties of Bayesian clustering via STRUCTURE. *Genet. Epidemiol.* **2005**, *28*, 302–312. [CrossRef]

29. Wang, R.-L.; Stec, A.; Hey, J.; Lukens, L.; Doebley, J. The limits of selection during maize domestication. *Nature* **1999**, *398*, 236. [CrossRef]

30. Cao, Y.; Li, S.; Chen, G.; Wang, Y.; Bhat, J.; Karikari, B.; Kong, J.; Junyi, G.; Zhao, T.-J. Deciphering the Genetic Architecture of Plant Height in Soybean Using Two RIL Populations Sharing a Common M8206 Parent. *Plants* **2019**, *8*, 373. [CrossRef]

31. Li, H.; Liu, H.; Han, Y.; Wu, X.; Teng, W.; Liu, G.; Li, W. Identification of QTL underlying vitamin E contents in soybean seed among multiple environments. *Theor. Appl. Genet.* **2010**, *120*, 1405–1413. [CrossRef]

32. Moongkanna, J.; Nakasathien, S.; Novitzky, W.; Kwanyuen, P.; Sinchaisri, P.; Srinives, P. SSR markers linking to seed traits and total oil content in soybean. *Thai J. Agric. Sci.* **2011**, *44*, 233–241.

33. Fang, C.; Ma, Y.; Wu, S.; Liu, Z.; Wang, Z.; Yang, R.; Hu, G.; Zhou, Z.; Yu, H.; Zhang, M. Genome-wide association studies dissect the genetic networks underlying agronomical traits in soybean. *Genome Biol.* **2017**, *18*, 161. [CrossRef] [PubMed]

34. Jun, T.H.; Freewalt, K.; Michel, A.P.; Mian, R. Identification of novel QTL for leaf traits in soybean. *Plant. Breed.* **2014**, *133*, 61–66. [CrossRef]

35. Severin, A.J.; Woody, J.L.; Bolon, Y.-T.; Joseph, B.; Diers, B.W.; Farmer, A.D.; Muehlbauer, G.J.; Nelson, R.T.; Grant, D.; Specht, J.E. RNA-Seq Atlas of Glycine max: A guide to the soybean transcriptome. *BMC Plant. Biol.* **2010**, *10*, 160. [CrossRef] [PubMed]

36. Niu, Y.; Xu, Y.; Liu, X.-F.; Yang, S.-X.; Wei, S.-P.; Xie, F.-T.; Zhang, Y.-M. Association mapping for seed size and shape traits in soybean cultivars. *Mol. Breed.* **2013**, *31*, 785–794. [CrossRef]

37. Fasoula, V.A.; Harris, D.K.; Boerma, H.R. Validation and designation of quantitative trait loci for seed protein, seed oil, and seed weight from two soybean populations. *Crop. Sci.* **2004**, *44*, 1218–1225. [CrossRef]

38. Jansen, R.; Van Ooijen, J.; Stam, P.; Lister, C.; Dean, C. Genotype-by-environment interaction in genetic mapping of multiple quantitative trait loci. *Theor. Appl. Genet.* **1995**, *91*, 33–37. [CrossRef]

39. Kaushik, N.; Kumar, K.; Kumar, S.; Kaushik, N.; Roy, S. Genetic variability and divergence studies in seed traits and oil content of Jatropha (*Jatropha curcas* L.) accessions. *Biomass Bioenergy* **2007**, *31*, 497–502. [CrossRef]

40. Tanksley, S.D. Mapping polygenes. *Annu. Rev. Genet.* **1993**, *27*, 205–233. [CrossRef]

41. Omokhafe, K.; Alika, J. Clonal variation and correlation of seed characters in Hevea brasiliensis Muell. Arg. *Ind. Crop. Prod.* **2004**, *19*, 175–184. [CrossRef]

42. Palanga, K.K.; Jamshed, M.; Rashid, M.; Gong, J.; Li, J.; Iqbal, M.S.; Liu, A.; Shang, H.; Shi, Y.; Chen, T. Quantitative trait locus mapping for Verticillium wilt resistance in an upland cotton recombinant inbred line using SNP-based high density genetic map. *Front. Plant. Sci.* **2017**, *8*, 382. [CrossRef]

43. Chen, W.; Yao, Q.; Patil, G.B.; Agarwal, G.; Deshmukh, R.K.; Lin, L.; Wang, B.; Wang, Y.; Prince, S.J.; Song, L. Identification and comparative analysis of differential gene expression in soybean leaf tissue under drought and flooding stress revealed by RNA-Seq. *Front. Plant. Sci.* **2016**, *7*, 1044. [CrossRef]

44. Nyquist, W.E.; Baker, R. Estimation of heritability and prediction of selection response in plant populations. *Crit. Rev. Plant. Sci.* **1991**, *10*, 235–322. [CrossRef]

45. Zhang, W.-K.; Wang, Y.-J.; Luo, G.-Z.; Zhang, J.-S.; He, C.-Y.; Wu, X.-L.; Gai, J.-Y.; Chen, S.-Y. QTL mapping of ten agronomic traits on the soybean (Glycine max L. Merr.) genetic map and their association with EST markers. *Theor. Appl. Genet.* **2004**, *108*, 1131–1139. [CrossRef] [PubMed]

46. Zhang, X.; Hina, A.; Song, S.; Kong, J.; Bhat, J.A.; Zhao, T. Whole-genome mapping identified novel "QTL hotspots regions" for seed storability in soybean (Glycine max L.). *BMC Genom.* **2019**, *20*, 499. [CrossRef] [PubMed]

47. Wang, S. Windows QTL Cartographer 2.5. Raleigh,NC: Department of Statistics, North Carolina State University. 2007. Available online: https://brcwebportal.cos.ncsu.edu/qtlcart/WQTLCart.htm. (accessed on 1 June 2019).

48. Bandillo, N.; Jarquin, D.; Song, Q.; Nelson, R.; Cregan, P.; Specht, J.; Lorenz, A. A population structure and genome-wide association analysis on the USDA soybean germplasm collection. *Plant. Genome* **2015**, *8*. [CrossRef]

49. Priolli, R.H.G.; Campos, J.; Stabellini, N.; Pinheiro, J.B.; Vello, N.A. Association mapping of oil content and fatty acid components in soybean. *Euphytica* **2015**, *203*, 83–96. [CrossRef]

50. Zhang, J.; Song, Q.; Cregan, P.B.; Nelson, R.L.; Wang, X.; Wu, J.; Jiang, G.-L. Genome-wide association study for flowering time, maturity dates and plant height in early maturing soybean (Glycine max) germplasm. *BMC Genom.* **2015**, *16*, 217. [CrossRef]

51. Zhou, Z.; Jiang, Y.; Wang, Z.; Gou, Z.; Lyu, J.; Li, W.; Yu, Y.; Shu, L.; Zhao, Y.; Ma, Y. Resequencing 302 wild and cultivated accessions identifies genes related to domestication and improvement in soybean. *Nat. Biotechnol.* **2015**, *33*, 408. [CrossRef]

52. Leamy, L.J.; Zhang, H.; Li, C.; Chen, C.Y.; Song, B.-H. A genome-wide association study of seed composition traits in wild soybean (Glycine soja). *BMC Genom.* **2017**, *18*, 18. [CrossRef]

53. Hacisalihoglu, G.; Settles, A.M. Quantification of seed ionome variation in 90 diverse soybean (Glycine max) lines. *J. Plant. Nutr.* **2017**, *40*, 2808–2817. [CrossRef]

54. Zhao, X.; Li, W.; Zhao, X.; Wang, J.; Liu, Z.; Han, Y.; Li, W. Genome-wide association mapping and candidate gene analysis for seed shape in soybean (Glycine max). *Crop. Pasture Sci.* **2019**, *70*, 684–693. [CrossRef]

55. Cober, E.R.; Morrison, M.J. Regulation of seed yield and agronomic characters by photoperiod sensitivity and growth habit genes in soybean. *Theor. Appl. Genet.* **2010**, *120*, 1005–1012. [CrossRef] [PubMed]

56. Copley, T.R.; Duceppe, M.-O.; O'Donoughue, L.S. Identification of novel loci associated with maturity and yield traits in early maturity soybean plant introduction lines. *BMC Genom.* **2018**, *19*, 167. [CrossRef] [PubMed]

57. Jeong, N.; Suh, S.J.; Kim, M.-H.; Lee, S.; Moon, J.-K.; Kim, H.S.; Jeong, S.-C. Ln is a key regulator of leaflet shape and number of seeds per pod in soybean. *Plant. Cell* **2012**, *24*, 4807–4818. [CrossRef] [PubMed]

58. Lu, X.; Xiong, Q.; Cheng, T.; Li, Q.-T.; Liu, X.-L.; Bi, Y.-D.; Li, W.; Zhang, W.-K.; Ma, B.; Lai, Y.-C. A PP2C-1 allele underlying a quantitative trait locus enhances soybean 100-seed weight. *Mol. Plant.* **2017**, *10*, 670–684. [CrossRef]

59. Fan, C.; Xing, Y.; Mao, H.; Lu, T.; Han, B.; Xu, C.; Li, X.; Zhang, Q. GS3, a major QTL for grain length and weight and minor QTL for grain width and thickness in rice, encodes a putative transmembrane protein. *Theor. Appl. Genet.* **2006**, *112*, 1164–1171. [CrossRef]

60. Li, N.; Li, Y. Ubiquitin-mediated control of seed size in plants. *Front. Plant. Sci.* **2014**, *5*, 332. [CrossRef]

61. Mao, H.; Sun, S.; Yao, J.; Wang, C.; Yu, S.; Xu, C.; Li, X.; Zhang, Q. Linking differential domain functions of the GS3 protein to natural variation of grain size in rice. *Proc. Natl. Acad. Sci. USA* **2010**, *107*, 19579–19584. [CrossRef]

62. Jiang, W.-B.; Huang, H.-Y.; Hu, Y.-W.; Zhu, S.-W.; Wang, Z.-Y.; Lin, W.-H. Brassinosteroid regulates seed size and shape in Arabidopsis. *Plant. Physiol.* **2013**, *162*, 1965–1977. [CrossRef]

63. Meng, Y.; Chen, F.; Shuai, H.; Luo, X.; Ding, J.; Tang, S.; Xu, S.; Liu, J.; Liu, W.; Du, J. Karrikins delay soybean seed germination by mediating abscisic acid and gibberellin biogenesis under shaded conditions. *Sci. Rep.* **2016**, *6*, 22073. [CrossRef]

64. Siloto, R.M.; Findlay, K.; Lopez-Villalobos, A.; Yeung, E.C.; Nykiforuk, C.L.; Moloney, M.M. The accumulation of oleosins determines the size of seed oilbodies in Arabidopsis. *Plant. Cell* **2006**, *18*, 1961–1974. [CrossRef] [PubMed]

65. Liu, Y.-F.; Li, Q.-T.; Lu, X.; Song, Q.-X.; Lam, S.-M.; Zhang, W.-K.; Ma, B.; Lin, Q.; Man, W.-Q.; Du, W.-G. Soybean GmMYB73 promotes lipid accumulation in transgenic plants. *BMC Plant. Biol.* **2014**, *14*, 73. [CrossRef] [PubMed]

66. Schruff, M.C.; Spielman, M.; Tiwari, S.; Adams, S.; Fenby, N.; Scott, R.J. The AUXIN RESPONSE FACTOR 2 gene of Arabidopsis links auxin signalling, cell division, and the size of seeds and other organs. *Development* **2006**, *133*, 251–261. [CrossRef] [PubMed]

67. Arnaud-Santana, E.; Coyne, D.; Eskridge, K.M.; Vidaver, A. Inheritance; low correlations of leaf, pod, and seed reactions to common blight disease in common beans; and implications for selection. *J. Am. Soc. Hortic. Sci.* **1994**, *119*, 116–121. [CrossRef]

68. Lihua, C.Y.D. The principle of high-yielding soybean and its culture technique. *Acta Agron. Sin.* **1982**, *1*. Available online: http://zwxb.chinacrops.org (accessed on 1 June 2019).

69. Liu, X.; Jin, J.; Wang, G.; Herbert, S. Soybean yield physiology and development of high-yielding practices in Northeast China. *Field Crop. Res.* **2008**, *105*, 157–171. [CrossRef]

70. Huang, X.; Feng, Q.; Qian, Q.; Zhao, Q.; Wang, L.; Wang, A.; Guan, J.; Fan, D.; Weng, Q.; Huang, T. High-throughput genotyping by whole-genome resequencing. *Genome Res.* **2009**, *19*, 1068–1076. [CrossRef]

71. Van Ooijen, J. JoinMap®4, Software for the calculation of genetic linkage maps in experimental populations. *Kyazma Bvwageningen* **2006**, *33*. Available online: https://www.kyazma.nl/index.php/JoinMap/ (accessed on 1 March 2019).

72. Yang, J.; Hu, C.; Hu, H.; Yu, R.; Xia, Z.; Ye, X.; Zhu, J. QTLNetwork: Mapping and visualizing genetic architecture of complex traits in experimental populations. *Bioinformatics* **2008**, *24*, 721–723. [CrossRef]

73. Churchill, G.A.; Doerge, R.W. Empirical threshold values for quantitative trait mapping. *Genetics* **1994**, *138*, 963–971.

74. Xu, H.; Zhu, J. Statistical approaches in QTL mapping and molecular breeding for complex traits. *Chin. Sci. Bull.* **2012**, *57*, 2637–2644. [CrossRef]

75. Voorrips, R. MapChart: Software for the graphical presentation of linkage maps and QTLs. *J. Hered.* **2002**, *93*, 77–78. [CrossRef] [PubMed]

76. McCouch, S.R.; Chen, X.; Panaud, O.; Temnykh, S.; Xu, Y.; Cho, Y.G.; Huang, N.; Ishii, T.; Blair, M. Microsatellite marker development, mapping and applications in rice genetics and breeding. *Plant Mol. Biol.* **1997**, *35*, 89–99.

77. Qi, Z.; Xiaoying, Z.; Huidong, Q.; Dawei, X.; Xue, H.; Hongwei, J.; Zhengong, Y.; Zhanguo, Z.; Jinzhu, Z.; Rongsheng, Z. Identification and validation of major QTLs and epistatic interactions for seed oil content in soybeans under multiple environments based on a high-density map. *Euphytica* **2017**, *213*, 162.

78. Tian, T.; Liu, Y.; Yan, H.; You, Q.; Yi, X.; Du, Z.; Xu, W.; Su, Z. agriGO v2. 0: A GO analysis toolkit for the agricultural community, 2017 update. *Nucleic Acids Res.* **2017**, *45*, W122–W129. [CrossRef] [PubMed]
79. Chen, C.; Xia, R.; Chen, H.; He, Y. TBtools, a Toolkit for Biologists integrating various biological data handling tools with a user-friendly interface. *BioRxiv* **2018**, 289660.

International Journal of
Molecular Sciences

Article

A Combined Linkage and GWAS Analysis Identifies QTLs Linked to Soybean Seed Protein and Oil Content

Tengfei Zhang [1,†], Tingting Wu [1,†], Liwei Wang [1,†], Bingjun Jiang [1], Caixin Zhen [1], Shan Yuan [1], Wensheng Hou [1,2], Cunxiang Wu [1], Tianfu Han [1] and Shi Sun [1,*]

[1] Ministry of Agriculture and Rural Affairs Key Laboratory of Soybean Biology, Institute of Crop Sciences, Chinese Academy of Agricultural Sciences, Beijing 100081, China; wutingting@caas.cn (T.W.); lwwmaize@163.com (L.W.); 18331121822@163.com (C.Z.); yuanshan@caas.cn (S.Y.); houwensheng@caas.cn (W.H.); wucunxiang@caas.cn (C.W.); hantianfu@caas.cn (T.H.)

[2] National Center for Transgenic Research in Plants, Institute of Crop Sciences, Chinese Academy of Agricultural Sciences, Beijing 100081, China

*   Correspondence: sunshi@caas.cn; Tel.: +86-10-8210-8589

†   These authors contributed equally to this work.

Received: 26 September 2019; Accepted: 21 November 2019; Published: 25 November 2019

Abstract: Soybean is an excellent source of vegetable protein and edible oil. Understanding the genetic basis of protein and oil content will improve the breeding programs for soybean. Linkage analysis and genome-wide association study (GWAS) tools were combined to detect quantitative trait loci (QTL) that are associated with protein and oil content in soybean. Three hundred and eight recombinant inbred lines (RILs) containing 3454 single nucleotide polymorphism (SNP) markers and 200 soybean accessions, including 94,462 SNPs and indels, were applied to identify QTL intervals and significant SNP loci. Intervals on chromosomes 1, 15, and 20 were correlated with both traits, and QTL *qPro15-1*, *qPro20-1*, and *qOil5-1* reproducibly correlated with large phenotypic variations. SNP loci on chromosome 20 that overlapped with *qPro20-1* were reproducibly connected to both traits by GWAS ($p < 10^{-4}$). Twenty-five candidate genes with putative roles in protein and/or oil metabolisms within two regions (*qPro15-1*, *qPro20-1*) were identified, and eight of these genes showed differential expressions in parent lines during late reproductive growth stages, consistent with a role in controlling protein and oil content. The new well-defined QTL should significantly improve molecular breeding programs, and the identified candidate genes may help elucidate the mechanisms of protein and oil biosynthesis.

Keywords: soybean; protein content; oil content; quantitative trait loci (QTL); linkage analysis; genome-wide association study (GWAS); candidate genes

---

1. Introduction

With an average composition of approximately 40% protein and 20% oil, soybean (*Glycine max* (L.) Merr.) is the most important source of vegetable protein and edible oil, accounting for 68% of total global protein consumption [1] and more than half of global oilseed production [2]. Breeders have the goal of producing soybean varieties with high-protein and oil content, traits that are quantitatively controlled by multiple genes that have small effects and are significantly influenced by the environment [3–5]. A strong negative correlation between protein and oil content has been verified in previous studies [6,7], suggesting that some quantitative trait loci (QTL) may inversely affect protein and oil content. Identifying and studying QTL associated with protein or oil content is important for directing molecular breeding, and identifying genes and gene functions that affect protein and oil content.

To find genetic markers that are near genes controlling traits of interest, linkage analysis can be performed using biparental segregating populations [8]. Since Diers et al. [3] first used linkage analysis to discover a major QTL connected to soybean protein and oil content on chromosome (Chr.) 20, studies have been conducted to detect QTL near various types of markers, including amplified fragment length polymorphism (AFLP) markers, restriction fragment length polymorphism (RFLP) markers, and simple sequence repeat (SSR) markers in biparental populations [7,9–13]. The Soybase website has listed 255 and 322 QTL linked to protein and oil content, respectively, involving every chromosome in the biparental population (http://soybase.org/, 8 July 2019). However, the limited overlap of protein/oil-interrelated markers and sparse density of molecular markers used in previous reports have inhibited the identification of candidate genes within the wide QTL intervals and limited the increase in protein or oil content resulting from marker-assisted selection (MAS) [14]. Requirements for the construction of secondary mapping populations and the use of map-based cloning have slowed down application to breeding. Using recently developed high-density single nucleotide polymorphism (SNP) markers based on high-throughput sequencing, Seo et al. [2] identified 23 protein and oil QTL within small regions that covered 14 linkage groups using 1570 SNP markers, including *qHPO20*, a QTL significant for both protein and oil content that overlapped a previously reported QTL [13,15–17]. Wang et al. [18] constructed two high-density genetic maps that contained 4000 more SNP markers, examined loci related to soybean evolutionary traits, and predicted candidate genes that related to these traits. Patil et al. used a high-resolution bin map (3343 SNP markers) to detect 18 QTL connected to soybean seed protein, oil, and sucrose content QTL that were then confirmed by a genome-wide association study (GWAS) [5].

GWAS, based on linkage disequilibrium (LD), is a prevailing strategy to find genetic variations that affect complex traits by using genome-wide markers combined with phenotypes [8,19]. Hansen et al. [20] first successfully applied GWAS to plant genetics, tightly linking the *B* gene to the annual growth habit of sea beet using genome-wide AFLP markers. In recent years, GWAS has been applied to analyze complex quantitative traits in soybean such as protein, oil, fatty acid, and amino acid content and salinity tolerance in the different wild, landrace, and elite soybean lines, yielding putative candidate genes based on bioinformatic analysis in order to identify their action mechanisms [21–23]. Lee et al. [24] gathered 621 soybean accessions from maturity group I–IV and 34014 SNP markers to identify QTL for protein, oil, and amino acid content. They also detected some QTL on Chr. 5, 10, 15, and 20 that coincided with previous results [3,25–27].

Compared with linkage analysis, association analysis does not require the construction of a mapping population and can analyze multiple alleles from the same locus simultaneously [19]. Due to abundant recombination accumulated during the long-term evolution of natural populations, the results are of higher resolution that can even be located within individual genes [8,28]. However, population structure and genetic relationships may lead to false positive results in association analyses [29]; hence, it is best to combine linkage analysis and association analysis for the most accurate QTL results. Combined analysis methods have successfully mapped loci to associated traits in rice [30,31] and maize [32,33], but it is rarely employed to study soybean protein and oil content. In this study, we combine linkage analysis and GWAS methods to detect QTL and identify candidate genes that are linked to protein and oil content.

2. Results

2.1. Phenotypic Variation of Protein and Oil Contents in Two Panels

Three hundred and eight recombinant inbred lines (RILs) and 203 soybean accessions were used in this study. The seed protein and oil content of two panels grown over three years in three different locations are summarized in Table 1. The protein content in the two parent lines for the RILs, Zigongdongdou (ZGDD) and Heihe27 (HH27), averaged over different locations, was 44.55% and 40.81%, respectively, and the average oil content was 18.69% and 20.07%, respectively. Differences in protein and oil content between the parent lines were significant in Sanya in 2016, 2017, 2018 (16SY, 17SY, 18SY), and Xinxiang in 2018 (18XX) ($p < 0.01$), but not in Xiangtan in 2017 (17XT). Data for the parent lines grown in Xinxiang in 2017 (17XX) were not available. The mean content for RILs was between the parents, with transgressive segregation expanding the range. The skewness and kurtosis indicate that the data conforms to a normal distribution that is apparent in the histograms in Figure 1, suggesting that both protein and oil content are controlled by multiple genes that can be analyzed by linkage analysis. The protein and oil content in the association panel also follow a normal distribution that is conducive to GWAS and have a wide phenotypic variation for traits as indicated by the variance, range, and coefficient of variance (CV) observed (Table 1).

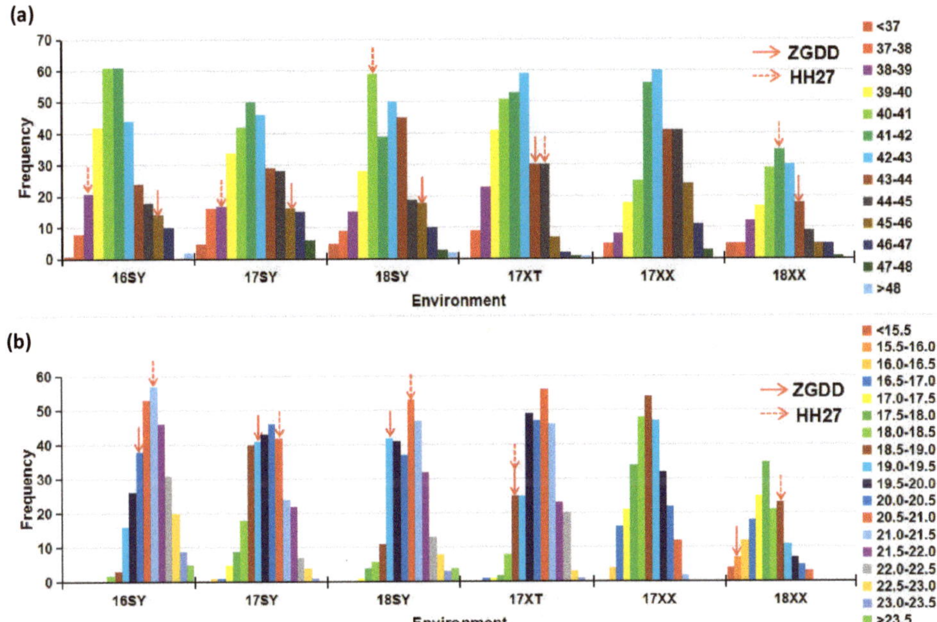

Figure 1. Histogram of recombinant inbred lines (RILs)' protein content (**a**) and oil content (**b**) in six environments. 16SY, 17SY, 18SY, 17XT, 17XX, 18XX represent different environments of Sanya, Xiangtan, Xinxiang in 2016, 2017, and 2018. Zigongdongdou (ZGDD) and Heihe27 (HH27) are the parents of the RILs. Bars in different colors represent different content of protein/oil.

Variance analysis (ANOVA) (Table 1) revealed that significant differences ($p < 0.01$) were found in genotype, environment, and genotype × environment interactions for the two traits. Broad-sense heritability (H^2) of both traits was high (0.83~0.90), demonstrating that genetic factors play a vital role in the accumulation of protein and oil in these lines.

Table 1. Descriptive statistics and variance analysis for protein and oil content of two panels in multiple environments.

| Population | Trait | Environment [a] | Parents | | Means (%) | Variance | Range (%) | CV [c] (%) | Skewness | Kurtosis | F Value of Variance Analysis | | | H^2 [d] |
			HH27 [b] (%)	ZGDD [b] (%)							Genotype (G)	Environment (E)	G*E	
RILs [g]	Protein	16SY	38.55	45.33	41.54 ± 0.12	4.52	36.29~48.64	5.12	0.49	0.11	14.33 ***,f	47.63 ***	2.75 ***	0.83
		17SY	38.94	45.17	41.89 ± 0.14	6.33	35.90~48.00	6.00	0.15	−0.46				
		18SY	40.15	45.01	41.93 ± 0.14	5.65	33.48~48.50	5.67	0.07	0.20				
		17XT	44.65	43.68	41.58 ± 0.11	4.03	37.35~48.38	4.83	0.16	−0.17				
		17XX	NA [e]	NA	42.62 ± 0.12	4.14	37.53~47.29	4.77	−0.10	−0.34				
		18XX	41.74	43.57	41.48 ± 0.17	5.14	33.19~47.23	5.46	−0.18	0.72				
	Oil	16SY	21.40	20.21	21.13 ± 0.06	1.17	18.13~24.38	5.11	0.08	−0.08	18.27 ***	1533.15 ***	2.57 ***	0.87
		17SY	20.53	19.13	19.93 ± 0.07	1.47	16.12~23.17	6.09	0.00	−0.18				
		18SY	20.73	19.43	20.54 ± 0.07	1.37	17.49~24.26	5.69	0.16	0.04				
		17XT	18.83	18.91	20.38 ± 0.06	1.18	16.76~23.14	5.32	−0.20	−0.14				
		17XX	NA	NA	18.71 ± 0.06	1.12	16.19~21.44	5.65	−0.04	−0.49				
		18XX	18.86	15.77	17.81 ± 0.09	1.39	15.13~20.90	6.62	0.17	−0.09				
Accessions	Protein	18SY	-	-	42.02 ± 0.23	12.42	33.40~51.33	8.39	0.19	−0.45	20.28 ***	3.94 **	3.15 ***	0.86
		17XT	-	-	42.21 ± 0.17	7.08	35.51~49.22	6.30	−0.01	−0.33				
		17XX	-	-	42.10 ± 0.20	8.83	36.10~48.83	7.06	0.22	−0.66				
		18XX	-	-	42.46 ± 0.21	8.58	30.68~49.84	6.90	−0.51	1.01				
	Oil	18SY	-	-	20.73 ± 0.11	2.84	15.65~23.94	8.12	−0.54	−0.33	27.54 ***	535.37 ***	2.95 ***	0.90
		17XT	-	-	21.14 ± 0.09	2.04	17.78~25.35	6.75	0.10	−0.25				
		17XX	-	-	20.10 ± 0.10	2.31	16.14~23.53	7.56	−0.28	−0.48				
		18XX	-	-	19.20 ± 0.12	2.63	15.29~22.93	8.44	−0.03	−0.54				

[a] 16SY, 17SY, 18SY, 17XT, 17XX, 18XX—different environments of Sanya, Xiangtan, Xinxiang in 2016, 2017, and 2018. [b] HH27—Heihe27; ZGDD—Zigongdongdou. [c] CV—coefficient of variation. [d] H^2—broad-sense heritability. [e] NA—not available. [f] ** $p < 0.01$; *** $p < 0.001$. [g] RILs—recombinant inbred lines.

2.2. Genetic Map and QTL Analysis of Protein and Oil Contents

Seven thousand one hundred and twenty-three SNP markers were filtered to construct a genetic map. Markers with severe segregation distortion ($x^2 > 100$) were removed through Joinmap 4.1. The final map included 3454 SNP markers covering 20 linkage groups (LGs) that spanned 2208.16 cM of the genome with an average distance of 0.64 cM between adjacent markers. There was an average of 173 SNP markers in each LG, ranging from 70 (on Chr. 11) to 260 (on Chr. 3) [34].

Using the genetic map, we identified QTL that were co-detected by two algorithms: inclusive composite interval mapping (ICIM) and a mixed model based on composite interval mapping (MCIM) and/or consistently detected in multiple environments, and combined QTL that exist in two adjacent intervals as the same QTL. This resulted in the identification of seven protein content QTL and eight oil content QTL that were located on 11 chromosomes (Table 2 and Figure 2). The limit of detection (LOD) value (the threshold for ICIM) of these QTL ranged from 2.90 to 35.35 while the F values (the threshold for MCIM) were from 4.80 to 26.20, and these QTL explained 1.56% to 23.98% of the phenotypic variation. The QTL with positive values for the additive effect indicates that the ZGDD parent contributes to the allele that is conducive to the trait. The QTL on Chr. 1, 15, and 20 are linked to both protein and oil content. Among those QTL, *qPro15-1/qOil15-1* contributed to a high phenotypic variation explanation (PVE) (13.40%~17.81%), was localized to a narrow physical region (from 2691560 bp to 3476238 bp), and was indicted by both algorithms and three different environments. Therefore, this QTL interval was further examined to identify candidate genes. A second, oil-related QTL *qOil5-1*, was also detected by two algorithms, apparent in every environment, and contributed a large PVE ranging from 7.04% to 23.98%. *qPro15-1*, *qPro20-1*, and *qOil5-1* were significant QTL intervals identified in at least three environments, having high LOD/F value and contributing more than 7% PVE (Table 2 and Supplementary Figure S1).

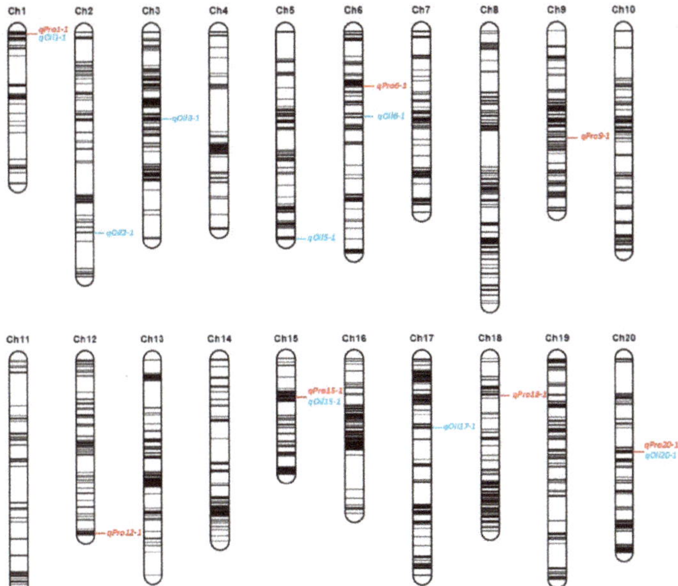

Figure 2. Location of quantitative trait loci (QTL) related to protein and oil contents. QTL in red color were protein while in blue were oil.

Table 2. Co-detected QTL identified by linkage analysis in two-algorithm and/or multiple growth environments.

Trait	QTL Name	Chr. (LG) [a]	Method [b]	Location (cM)	Marker Interval (cM)	Physical Region (bp)	LOD/F Value [c]	PVE (%) [d]	Additive Effect [e]	Environment [f]	Reference [g]
Protein	qPro1-1	1 (D1a)	ICIM	1	0~1.5	1488983~1566969	2.90~6.49	2.73~5.52	−0.41~−0.49	1, 3	Seed protein 3-4
			MCIM	7.5	6.5~8.4	2605140~2852655	5.00	-	−0.24	-	
	qPro6-1	6 (C2)	ICIM	33	32.5~33.5	5836780~5931027	4.93	4.17	0.42	1	cqSeed protein-005, Seed protein 30-5
			MCIM	32.1	31.7~32.2	5609477~5632020	5.10	-	0.26	-	
	qPro9-1	9 (K)	ICIM	62~68	59.5 ~70.5	38117239~41020511	3.63~8.96	3.43~8.40	0.38~0.41	1, 2, 4	Seed protein 33-3, Seed protein 34-6
			MCIM	61.7	60.7~62.7	38117239~39894925	10.40	-	0.42	-	
	qPro12-1	12 (H)	ICIM	105~107	103.5~107	38776571~39867556	3.64~3.78	3.05~3.58	0.36~0.46	1, 3	Seed protein 6-1
	qPro15-1	15 (E)	ICIM	23~26	22.5~26.5	2691560~3476238	9.00~19.05	13.40~17.81	0.79~0.89	3, 4, 5	Seed protein 30-3
			MCIM	26.2	26.1~26.3	3311604~3350307	26.20	-	0.52	-	
	qPro18-1	18 (G)	ICIM	22	21.5~22.5	5577815~5618246	6.06	5.80	−0.59	3	Seed protein 47-6
			MCIM	22.3	22.0~23.3	5618246~5979842	4.80	-	−0.24	-	
	qPro20-1	20 (I)	ICIM	54~61	48.5~62.5	34734798~37115770	6.14~8.62	7.24~9.39	0.56~0.75	1, 2, 3	Seed protein 26-5, Seed protein 34-11
			MCIM	58.7	57.7~59.7	36089907~37115770	15.30	-	0.34	-	
Oil	qOil1-1	1 (D1a)	ICIM	1~10	0~14.5	1488983~3316074	2.85~2.99	1.56~1.67	0.14~0.17	1, 3	Seed oil 23-2
	qOil2-1	2 (D1b)	ICIM	121	116.5~126.5	43783867~45442501	2.56	3.64	0.19	5	cqSeed oil-014, Seed oil 39-6
			MCIM	112.8	111.8~113.1	42545649~43226016	5.40	-	0.14	-	
	qOil3-1	3 (N)	ICIM	52~54	50.5 ~55.5	33430615~34447425	2.62~2.72	1.86~2.78	0.15~0.21	2, 4	Seed oil 43-30
	qOil5-1	5 (A1)	ICIM	117~126	116.5~126	40003403~41813079	3.89~35.35	7.04~23.98	−0.27~−0.63	1, 2, 3, 4, 5, 6	Seed oil 39-1, Seed oil 35-2, Seed oil 13-1
			MCIM	125.9	124.9~126.4	40566361~41813079	25.10	-	−0.40	-	
	qOil6-1	6 (C2)	ICIM	50~52	44.5~52.5	8313637~9652882	3.24~4.09	4.16~6.68	−0.26~−0.34	2, 6	cqSeed oil-016
	qOil15-1	15 (E)	ICIM	26	25.5~26.5	2691560~3240013	19.25	15.97	−0.44	4	cqSeed oil-007, Seed oil 2-3
			MCIM	26.2	26.1~26.3	3311604~3350307	28.40	-	−0.28	-	
	qOil17-1	17 (D2)	ICIM	41	39.5~41.5	7100839~8674575	3.19	1.72	0.18	3	Seed oil 23-3
			MCIM	45.1	44.1~46.1	7453724~9120650	5.30	-	0.13	-	
	qOil20-1	20 (I)	ICIM	56~62	51.5~62.5	34734798~37115770	3.93~5.20	2.30~2.87	−0.20	1, 3	Seed oil 27-4, Seed oil 24-6

[a] Chr. (LG), chromosome (linkage group). [b] inclusive composite interval mapping (ICIM) and a mixed model based on composite interval mapping (MCIM) were used. [c] limit of detection (LOD) value was the threshold by ICIM, and F-value was the threshold by MCIM, respectively, with the critical threshold value LOD = 2.5 and F = 4.7, respectively. [d] PVE, explanation of phenotypic variation. [e] Positive value means the ZGDD allele contributed to the trait. [f] 1, 2, 3, 4, 5, 6 represented 16SY, 17SY, 18SY, 17XT, 17XX, 18XX, respectively. [g] Reported quantitative trait loci (QTL) in Soybase databse (https://www.soybase.org/) that overlapped our QTL here.

2.3. Genome-Wide Association Study (GWAS) Results

Two hundred and three soybean accessions consisting of a diverse range of protein and oil content were genotyped, yielding 3,977,183 SNPs and 491,910 indels. After filtering for missing rates $\leq$ 10%, minor allele frequencies $\geq$ 5% and LD pruning, 94,462 SNPs and indels were available for GWAS.

Principal component (PC) analysis was conducted with 94,462 SNPs and indels and three outlier cultivars (Hai 94, Wuhuasiyuehuang, and Suidaohuang) were identified and removed from the association panel. The first three PCs dominate the population structure (Figure 3a), they divide the population into two main groups which exhibit a geographic distribution pattern (Figure 3b). The first subgroup primarily consisted of cultivars from the northeast region of China (NER) and the USA, while the second subgroup mainly included cultivars from the Huang-Huai region of China (HHR) and the south region of China (SR). However, a few accessions from NER and USA (e.g., liaodou15 and Hood) were sorted into the second subgroup and a few accessions from HHR and SR (e.g., qihuang10 and taiwan75) were placed in the first subgroup (Figure 3b,c), perhaps due to their parents' origin area. Population structure analysis indicated K = 2 was the modeling choice (Figure 3c and Supplementary Figure S2), and the result was confirmed by the PC analysis. The heat map of the population shows their kinship that can distribute into two subpopulations (Figure 3d). The physical distance of total LD decay, where r^2 dropped below 0.1, was approximately 132 kb (Figure 3e), and we also detected the LD decay of accessions from NER, USA, HHR, and SR with the relative LD decay distances of 180, 190, 171, and 161kb, respectively.

To minimize false positives due to population structure, we performed a general linear model (GLM) and a mixed linear model (MLM) and found that MLM effectively reduced false positive SNPs. A threshold of $-\log (P)$ = 4 was determined as the criteria for detecting significant signals of protein and oil content. Further, we conducted GWAS on two sub-population panels (the NER-USA and HHR-SR sub-populations). A total of 19, 12, and 36 SNP loci distributed on 17 chromosomes were detected in the NER-USA and HHR-SR sub-populations and total population, respectively (Supplementary Table S1 and Supplementary Figure S3). The p-values of all significant SNP loci were from 9.37×10^{-7} to 9.90×10^{-5}. One SNP on Chr. 5, 9, 13, 16, 18, and three SNPs on Chr. 20 (41133383, 35512580, and 34990940) were associated with both traits in one environment, and these significant SNP loci on Chr. 20 were associated with both protein and oil content in 17XX and 18XX.

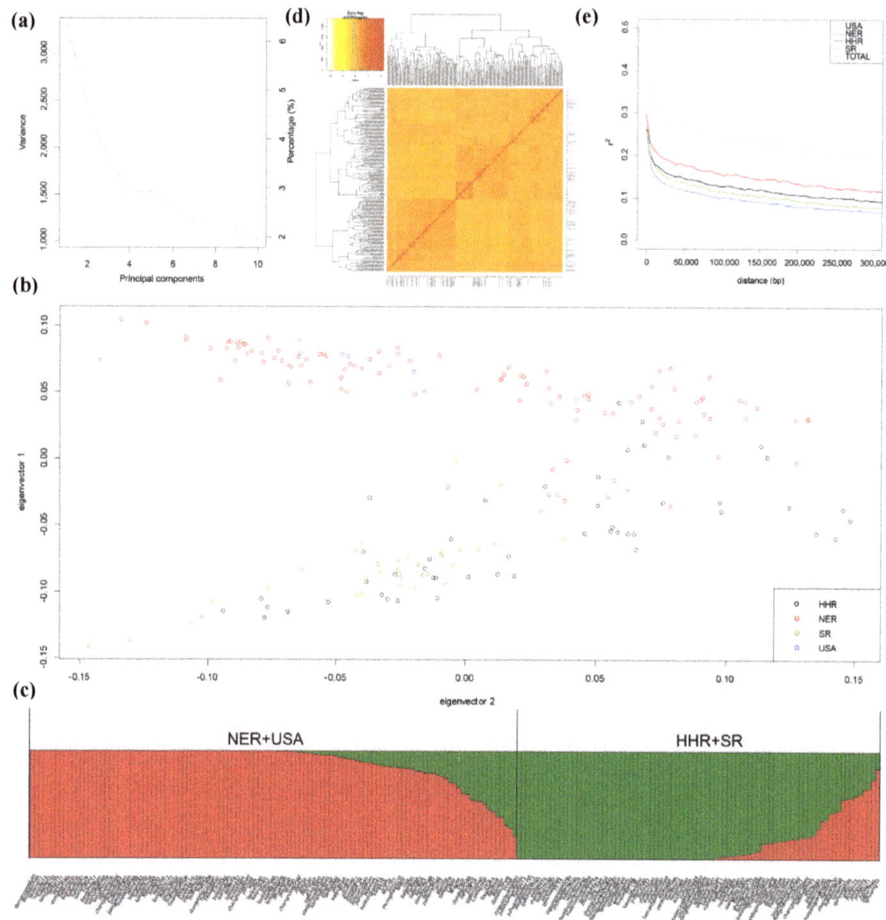

Figure 3. The principal component (PC) analysis (**a**,**b**), population structure analysis (**c**), heat map of the kinship matrix of the 203 soybean accessions (**d**), and linkage disequilibrium (LD) decay (**e**) of the association panel.

2.4. Co-Detected Results by Linkage Analysis and GWAS

We combined the results of linkage analysis and GWAS to identify SNP regions that were co-detected by both analyses (Table 3).

Table 3. Co-detected SNP loci regions by linkage analysis and GWAS.

Chr. [a]	Trait	Method [b]	Environment [c]	Markers Interval (cM)/SNP Number [d]	SNP Loci Region/Location (bp)	LOD/F Value [e]	PVE (%) [f]	Additive Effect [g]
2	Oil	ICIM	5	116.5~126.5	43783867~45442501	2.56	3.64	0.19
		GWAS	4	1	45017225	-	-	-
6	Protein	ICIM	1	32.5~33.5	5836780~5931027	4.93	4.17	0.42
		MCIM	-	31.7~32.2	5609477~5632020	5.10	-	0.26
	Oil	GWAS	3, 5	2	5713084~5992538	-	-	-
9	Protein	ICIM	1, 2, 4	59.5~70.5	38117239~41020511	3.63~8.96	3.43~8.40	0.38~0.41
		MCIM	-	60.7~62.7	38117239~39894925	10.40	-	0.42
	Oil	GWAS	6	1	40301013	-	-	-
20	Protein	ICIM	1, 2, 3	48.5~62.5	34734798~37115770	6.14~8.62	7.24~9.39	0.56~0.75
		MCIM	-	57.7~59.7	36089907~37115770	15.30	-	0.34
		GWAS	5, 6	5	34990940~35578946	-	-	-
	Oil	ICIM	1, 3	51.5~62.5	34734798~37115770	3.93~5.20	2.30~2.87	−0.20
		GWAS	5	4	34801441~35512580	-	-	-

[a] Chr. chromosome. [b] Inclusive composite interval mapping (ICIM) and a mixed model based on composite interval mapping (MCIM) were two algorithms in linkage analysis. [c] 1, 2, 3, 4, 5, 6 represented 16SY, 17SY, 18SY, 17XT, 17XX, 18XX, respectively. [d] Markers interval is the QTL interval in linkage analysis, SNP number is the significant SNP loci number in the SNP loci region. [e] LOD value is the threshold by ICIM and F-value is the threshold by MCIM, respectively, with the critical threshold value LOD = 2.5 and F = 4.7, respectively. [f] PVE, explanation of phenotypic variation. [g] Positive value means ZGDD allele contributed to the trait.

Four significant SNP loci regions distributed on Chr. 2, 6, 9 and 20 were co-detected, and all the SNP loci detected by GWAS were distributed in the QTL intervals obtained by linkage analysis. The co-detected SNP regions on Chr. 2 and Chr. 6 had a weak PVE (<5%), but likely included genes that exerted modest effects. The region on Chr. 20 was linked to both protein and oil content, with a higher PVE for protein content (7.24~9.39%).

2.5. Candidate Genes and Expression Levels

We next searched for candidate genes in a co-detected SNP region from Chr. 20 and an extra strongly indicated QTL interval from Chr. 15. Based on the LD decay distance of total population and four regions, we extended the regions about 200 kb that from 34.60 to 35.40 Mb including the SNP loci on Chr. 20. The physical region of *qPro15-1/qOil15-1* on Chr. 15 was from 2.60 to 3.50 Mb. We focused on genes that were indicated by annotation information to be involved in protein or oil metabolism as candidate genes. Nine and 16 genes were selected on Chr. 15 and Chr. 20, respectively, that were predicted to have one of these four categories of function: structural components, metabolic enzymes, material transporters, and regulators of gene expression (Table 4).

Quantitative real-time PCR (qRT-PCR) was applied to measure the relative expression of the 25 candidate genes, identifying eight genes that had significant differences in the expression levels between the two parent lines at late reproductive growth stages in pods. *Glyma.15g033200*, *Glyma.15g034100*, *Glyma.20g105300*, *Glyma.20g106900*, and *Glyma.20g107600* shared a pattern of low expression levels with no significant difference between the two parents up to 25 days after the R3 period, then the expression level of these genes in ZGDD increased sharply compared to HH27 (Figure 4a). As shown in Figure 4b, the expression of *Glyma.15g034600* and *Glyma.20g103200* was higher in HH27, but generally low in both parents until 25 days after the R3 period. Then, 30 days after the R3 period, gene expression switched to be significantly higher in ZGDD only to reverse again by the 35th day to be highly expressed in HH27 relative to ZGDD. *Glyma.15g040100* has its own distinct pattern. For the first 10 days after the R3 period, both parents had high expression levels, then the expression decreased with expression in HH27 being significantly higher than in ZGDD (Figure 4c). The detailed expression patterns for the other five genes are shown in Supplementary Figure S4.

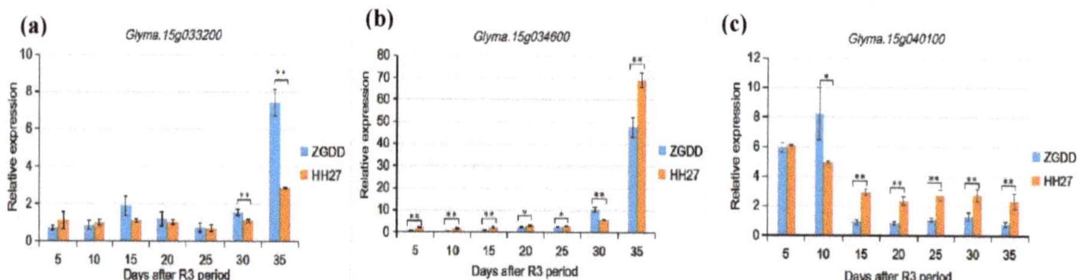

Figure 4. Relative expression patterns of candidate genes. *Glyma.15g033200*, *Glyma.15g034100*, *Glyma.20g105300*, *Glyma.20g106900*, and *Glyma.20g107600* express as (**a**), *Glyma.15g034600* and *Glyma.20g103200* as (**b**), *Glyma.15g040100* as (**c**). * $p < 0.05$, ** $p < 0.01$.

Table 4. Candidate genes that may control protein/oil content within the SNP regions on Chr. 15 and 20.

Trait	Gene	Start (bp)	Stop (bp)	Annotation
Oil	Glyma.15g034100	2722009	2727957	acyltransferase activity, diacylglycerol and triacylglycerol biosynthesis
	Glyma.15g034400	2740960	2746344	aldehyde dehydrogenase family, aldehyde dehydrogenase [NAD (P)+] activity
	Glyma.15g034600	2765299	2770528	drug transmembrane transport, associated with the transport of citric acid and malic acid
	Glyma.15g042500	3339156	3341447	fatty acid, lipid biosynthetic process, transferase activity, 3-oxoacyl-[acyl-carrier-protein] synthase activity
	Glyma.20g107600	35025241	35029762	Arabidopsis phospholipase-like protein, regulation of gene expression
	Glyma.20g107800	35038552	35042864	hydroxypyruvate reductase, glycerate dehydrogenase, glyoxylate reductase, NADP activity
	Glyma.20g108800	35116048	35118928	mitochondrial pyruvate transmembrane transport
	Glyma.20g109900	35222837	35228540	lipid metabolic process, steroid biosynthetic process, mevalonate pathway
	Glyma.20g110000	35229423	35231861	acetyltransferase activity
	Glyma.20g111000	35315630	35319063	fatty acid desaturase, lipid metabolic process
Protein	Glyma.15g033200	2656030	2657795	structural constituent of ribosome, 28S ribosomal protein
	Glyma.15g039000	3068347	3075209	60S ribosomal protein
	Glyma.15g040100	3164697	3168839	ACT domain-containing protein, metabolic process like protein synthesis and degradation.
	Glyma.15g041500	3255042	3256599	ribosomal large subunit assembly, 60S ribosomal protein L23
	Glyma.15g042300	3307111	3308840	structural constituent of ribosome, 60S ribosomal protein L35
	Glyma.20g103200	34605252	34609867	tryptophan biosynthetic process, anthranilate synthase activity
	Glyma.20g105300	34757381	34771672	ACT-like protein, serine/threonine kinase family protein
	Glyma.20g106200	34862155	34865242	amino acid transmembrane transport
	Glyma.20g106900	34962043	34967985	translation initiation factor 3 (IF-3) family protein
	Glyma.20g109600	35200934	35205885	ubiquitin-dependent protein catabolic process, proteasome complex, proteolysis activity
	Glyma.20g110100	35232344	35233758	nutrient reservoir activity, cupins superfamily protein, storage protein
	Glyma.20g110400	35261156	35268971	ACT domain-containing protein, metabolic process like protein synthesis and degradation.
	Glyma.20g111900	35396205	35400722	cationic amino acid transporter, amino acid transmembrane transporter activity
Protein/Oil	Glyma.20g106800	34935548	34940516	protein dephosphorylation, phosphatase activity, pyruvate dehydrogenase
	Glyma.20g110200	35235204	35239070	lipoate biosynthetic, radical SAM superfamily protein, transferase activity

ACT—Aspartate kinase, Chorismate mutase and TyrA (prephenate dehydrogenase), SAM—S-adenosyl methionine, NADP—nicotinamide adenine dinucleotide phosphate, IF—initiation factor.

3. Discussion

3.1. The Accuracy of QTL Analysis and GWAS is Improved by Using Phenotypic Data from Different Locations and Employing Ample SNP Markers

The two soybean populations we studied were planted in three different geographical locations. Previous studies have shown that soybean varieties originating from higher latitudes possess lower protein content and higher oil content and that the protein content of the same soybean variety is negatively correlated with latitude, altitude, day length, and the oil content, while being positively correlated with temperature and moisture [7,35,36]. Song et al. [37] also found that crude protein content was positively correlated with accumulated temperature ≥ 15 °C and mean daily temperature. In our study, the protein and oil content of ZGDD/HH27 were different when grown in three different locations, especially in XT (Table 1); the comprehensive climate factors in XT must have led to the significant change in protein and oil content of HH27, producing a wide phenotypic variation in the RILs used for QTL analysis. Growth in multiple environments also expanded the phenotypic variation of the association panel (Table 1). The use of plants grown over multiple years at different locations helped us reduce environmental factors to identify genes that consistently affect these traits in our QTL and GWAS analyses.

Increasing the number of markers also improves the accuracy of QTL analysis and GWAS. The application of AFLP, RFLP, and SSR markers in QTL localization of soybean protein and oil content has previously been limited, resulting in imprecise QTL region identification [29,38]. In this study, 3454 SNP markers obtained by simplified genome resequencing were used to increase the resolution of the genetic map (0.64 cM of average distance between adjacent markers) and reduce the physical interval of the QTL (average distance was about 1.5 Mb). For GWAS, LD decay distance determines the minimum saturation marker density and using more markers produces a higher probability of detecting functional sites [19]. Nearly one hundred thousand filtered SNP and indel markers were used in this study, improving the precision of GWAS to study the complex traits.

3.2. Refined QTL Intervals and SNP Loci for Protein and Oil Content Were Identified

Although multiple protein and oil content QTL have been previously discovered, few have been effectively used in breeding due to their small phenotypic effects and poor reproducibility, so identifying QTL with consistent, large effects is desirable [7,39]. In our linkage analysis study, QTL that were detected by both ICIM and MCIM algorithms and/or stably detected in multiple growth environments were identified. All of the QTL intervals overlapped or were close to regions reported by previous studies (Table 2). Here, we followed up on the QTL located on Chr. 15 and 20 that were significantly and consistently correlated to both protein and oil content. Diers et al. [3] located an oil-related QTL near the RFLP marker Pb on Chr. 15, which was close to the physical region of *qPro15-1/qOil15-1* identified in this study and a similar region was further associated with soybean protein or oil content in other QTL studies [2,12,26]. Some SNP loci correlated to fatty acid and amino acid phenotypes were also included in this region [15,22], but candidate genes were not discovered. A second QTL, *qPro20-1/qOil20-1* identified in our study, also overlapped with a previously reported QTL region. Reinprecht et al. [40] detected a protein and oil QTL adjacent to the marker Satt270 which included a protein content QTL identified by Lu et al. [41] and an oil content QTL found by Qi et al. [16]. Using high-density SNP markers obtained by genome resequencing, Patil et al. [5] also detected a protein content QTL in the physical region from 33.8 to 37.4 Mb on Chr. 20 in two environments. Our refined QTL will make MAS breeding more accurate and efficient. As to an oil QTL on Chr. 5, GWAS studies have shown that some SNP loci in this region regulated oil content and hence they searched for the candidate genes [29,42]. Zhang et al. [22] associated a SNP locus at Chr. 5: 41883826 bp with oil content, and identified a candidate gene *Glyma.05g245000* that annotated as 3-Oxo-5-asteroid-4-dehydrogenase. Lee et al. [24] discovered five significant oil-related SNP loci positioned within 41.75~41.89 Mb on Chr. 5, and they listed some previous QTL for protein and oil content that were in the same region identified by our linkage analysis results. This supports the reliability of our results and suggests that there must be some oil regulating genes in this region.

In the GWAS results, the SNP loci on Chr. 20 interested us because they were embedded in the QTL intervals that had high PVE. Priolli et al. [43] detected a SNP locus that associated with fatty acid components near marker Satt270 on Chr. 20 and that is located in the SNP loci region identified in this study. However, since other GWAS studies have identified a different region of Chr. 20 from 29 to 34 Mb [5,15,24,29,42], this study has likely discovered a novel region to excavate for candidate genes.

3.3. The Candidate Genes Differentially Expressed at Late Reproductive Growth Stage between Both Parents Will Be Further Analyzed

Comparing the results identified as QTL intervals and SNP loci, we decided the co-detected SNP loci region 34.60~35.40 Mb extended by approximately 200 kb on Chr. 20, and the *qPro15-1/qOil15-1* interval 2.60~3.50 Mb on Chr. 5, an extra region, were the best novel regions for candidate genes, rather novel QTL intervals, that might control protein and/or oil metabolism. The synthesis and catabolism of protein and oil are complex biochemical processes [44,45] and we looked for the genes that might play roles in protein and/or oil metabolism based on annotated information.

To test the 25 genes predicted to influence the accumulation of protein and oil content, we looked for differential relative expression levels during R3 to R8 growth stages in the two parents ZGDD and HH27, resulting in the identification of eight candidate genes for further study. Previous studies have shown that the accumulation of protein and oil content is most concentrated during the late reproductive growth stage and they are negatively correlated at this stage [46,47]. The energy needed to produce oil in seed mainly comes from saccharides and protein, and some varieties of protein can be degraded into acetyl-CoA, which is the raw material of oil [46,48]. Protein and oil accumulate in the developing seeds of pods, but the surrounding pods can transport matter into the seeds, so we extracted RNA from the whole pods. Here, we discovered that eight genes had significantly different expression levels at a late reproductive growth stage between the two parents, ZGDD derived from low latitude of China that was grown in short-day conditions (12 h light/12 h dark) and HH27 derived from high latitude of China that was grown in long-day conditions (16 h light/8 h dark). Among these eight genes, five genes of *Glyma.15g033200* (structural constituent of ribosome), *Glyma.15g034100* (acyltransferase activity), *Glyma.20g105300* (serine/threonine kinase family protein), *Glyma.20g106900* (translation initiation factor 3 (IF-3) family protein), and *Glyma.20g107600* (phospholipase-like protein) had similar expression patterns (Figure 4a), the significantly up-regulated expression of these genes in ZGDD at the late reproductive growth stage might be the reason for high protein content of ZGDD. The significantly up-regulated expression of two genes *Glyma.15g034600* (drug transmembrane transport, transport of citric acid, and malic acid) and *Glyma.20g103200* (tryptophan, anthranilate synthase) in HH27 at the late reproductive growth stage (Figure 4b) might contribute to the high oil content of HH27. *Glyma.15g040100* (ACT domain-containing protein, metabolic process like protein synthesis and degradation), expressed stably but significantly higher in HH27 (Figure 4c), could also have created the higher oil content in HH27. These results provide preliminary evidence for the possible roles of these genes played in the accumulation of protein and/or oil content. However, environmental conditions have a great influence on protein and oil content. Since the parents were grown with two different photoperiod treatments, a possible role for photoperiod will be addressed in follow up experiments to determine whether gene expression and accumulation of protein and oil content are related to photoperiod or the variety itself. Different protein and oil content between cultivars from diverse regions may be correlated to variations in photoperiods.

4. Materials and Methods

4.1. Plant Materials and Field Trials

The plant material included a linkage panel and an association panel. The linkage panel consisted of RILs from 308 $F_{2:7}$ lines derived from a cross between HH27 (protein content is 39%, oil content is 21%) and ZGDD (protein content is 45%, oil content is 19%). RILs and their parents were grown in six environments: Sanya (SY, 18°23′N, 109°11′E), Hainan province in 2016, 2017, and 2018; Xiangtan (XT, 27°47′N, 112°55′E), Hunan province in 2017; and Xinxiang (XX, 35°18′N, 113°55′E), Henan province in 2017 and 2018. The association panel, composed of 203 soybean accessions that included 182 accessions from China (94 from NER, 50 from HHR, 38 from SR) and 21 accessions from the USA (Supplementary Table S2), was planted in 18SY, 17XT, 17XX, and 18XX. Both panels were grown in a randomized complete block design with two replications. The arrangement was 1.5 m long rows with 0.5 m row spacing and 0.1 m of distance between individuals.

4.2. Phenotypic Data and Analysis

Fourier transform-near infrared reflectance (FT-NIR) spectrometry (Bruker, Karlsruhe, German) was applied to scan the near infrared absorption spectra of the dry seeds. Under the Quant 2 method of OPUS v. 4.2 software (Bruker, Karlsruhe, German), the samples' protein and oil content data were calculated using the dry basis model [49]. Each RIL and soybean accession from each replication of each environment was detected three times using about 150~200 dry seeds per detection, with the average used in statistical analysis. The histogram of phenotypic data was constructed using EXCEL (Microsoft, Redmond, WA, USA). Statistical analysis of phenotypic data and ANOVA was conducted using SAS v. 9.4 (SAS Institute, Cary, NC, USA), with type III analysis being employed. The H^2 of protein and oil contents was calculated using the following equation [50]:

$$H^2 = \sigma_G^2 / (\sigma_G^2 + \sigma_{G*E}^2/e + \sigma_e^2/re$$

in which σ_G^2 is genetic variance, σ_{G*E}^2 is genotype × environment variance, σ_e^2 is error variance, r is the number of replications, and e is the number of environments.

4.3. Genotyping and Linkage Analysis

For RILs, 2b-RAD technology [51] was applied to do simplified genome resequencing. Qualified libraries were paired-end sequenced on the Illumina Hiseq Xten platform to obtain high-quality SNP markers widely distributed throughout the genome. Using Joinmap v. 4.1 [52] to construct the genetic map, markers beyond the LOD threshold 5.0 were scattered into 20 LGs [34]. A regression algorithm and the Kosambi function were used to calculate the map distances (in cM) between adjacent markers.

QTL were predicted using two software packages based on different algorithms. IciMapping v. 4.1 [53] used the ICIM algorithm, with the following default parameters: mapping method was ICIM-ADD, step was 1 cM, PIN was 0.001, and LOD threshold was manual input of 2.5. QTLNetwork v. 2.1 [54] employed the MCIM algorithm, with the permutation performed 1000 times, the F-threshold set to 4.7, and testing window and walk speed set to 10 and 1 cM, respectively.

4.4. Genotyping and GWAS

The 203 soybean cultivars were sequenced with the high-throughput next-generation sequencing platform of Hi-Seq 2000 with an average sequencing depth of 10-fold and genotyped using the Genome Analysis Toolkit (GATK) pipeline [55]. After being trimmed by TRIMMOMATIC (parameter: illuminaclip: adaptor. seq: 2:30:10 trailing: 3 sliding window: 4:10 MINLEN: 20), clean reads were mapped to the soybean reference genome (v. Wm82.a2.v1) by BWA [56] with default parameters, SNPs/InDels were called by GATK (-stand_call_conf set to 30.0, -stand_emit_conf set to 10.0, and -glm set to BOTH). The variations were then recalibrated by a Gaussian mixture model, and outliers were discarded. Variants were further filtered by BCFtools (v. 1.2, QUAL $\geq$ 50.0, DP $\geq$ 5.0, QD $\geq$ 5.0, MQ $\geq$ 30, MAF $\geq$ 0.03, Coverage $\geq$ 90%). InDels longer than 6 bp were discarded. More than four million SNPs and indels were obtained. SNPs and indels were filtered with missing rates $\leq$ 10% and minor allele frequencies $\geq$ 5% using PLINK [57]. The sequencing data of 125 accessions used in this study have been deposited into the NCBI database under Short Read Archive (SRA) accession number SRP062560, and the sequencing data of the rest 78 accessions used only in this study have been deposited into SRA database in NCBI under accession number PRJNA589345. Linkage disequilibrium value was calculated using the LD composite method in SNPRelate software and highly-linked SNPs were pruned, and LD plots were modified via locally weighted scatterplot smoothing (LOWESS) using R software and testing smoothing parameters fixed to 0.01 [58]. PCA was conducted by SNPRelate, and 3 cultivars of population bias were removed based on the PCA result. The software fastSTRUCTURE was used to analyze the population structure (K = 2, 3, 4, 5) [59] and it was verified based on the PCA result. Association signals of seed protein and oil were identified based on 94,462 SNPs and indels from 200 samples with MLM by the first three PCs and kinship in GAPIT [60]. The LD analysis was calculated by using the squared allele frequency correlation (r^2) in PopLDdecay [61]. The critical threshold was set as $p < 10^{-4}$ to declare the significant SNP loci in GWAS.

4.5. Identification and Verification of Candidate Genes

Based on the LD decay distance, 200 kb upstream and downstream of regions near the significant SNP loci on Chr. 20 were explored to identify genes whose functional annotation related to the metabolism of protein and/or oil in the soybean reference genome Williams 82 (http://www.soybase.org/). The functional annotation was from TAIR (www.arabidopsis.org/), GO (http://geneontology.org/), PFAM (http://pfam.xfam.org/), PANTHER(http://www.pantherdb.org/) databases and KOG (clusters of orthologous groups for eukaryotic complete genomes) annotation. Similarly, the QTL interval on Chr. 15 were also scanned to identify candidate genes. qRT-PCR was applied to identify the relative expression of candidate genes in the pods of two parents: ZGDD planted in a short-day greenhouse (12 h light/12 h dark) and HH27 planted in a long-day greenhouse (16 h light/8 h dark) to simulate their suitable light conditions in order to get protein and oil content close to that in the originate region (the parent ZGDD originates from low latitude area of China (Zigong, Sichuan province, 29°20′N, 104°46′E), it is sensitive to photoperiods and can only blossom and mature in a short-day condition; the parent HH27 derives from high latitude area of China (Heihe, Heilongjiang province, 50°14′N, 127°31′E), it is insensitive to photoperiods and can blossom and mature in both long-day and short-day conditions.). Pods were picked from the middle nodes of the main stem every five days from the R3 through the R8 stage, with three replicates for each plant [62]. The entire pods were used for the isolation of total RNA using TransZol Up (Transegen Biotech, Beijing, China). First-strand cDNA was synthesized from 1 μg of the total RNA using a FastQuant RT Kit (Tiangen Biotech, Beijing, China). For qRT-PCR, 10 μL reaction volume was applied using KAPA SYBR® FAST qPCR Kits (KAPA Biosystems, Wilmington, MA, USA) with the following components: 1 μL of 1:5 diluted cDNA, 0.2 μL of each primer (10 μM), 5 μL of 2 × SYBR FAST qPCR Master Mix, 0.2 μL of 50 × ROX Low Reference Dye, and water to a final volume of 10 μL. QuantStudio 7 Flex (Applied Biosystems, Waltham, MA, USA) was used to run the qRT-PCR with following conditions: hold stage was 95 °C for 3 min; PCR stage was 40 cycles of 95 °C for 5 s and 60 °C for 30 s; melt curve stage was 95 °C for 15 s, 60 °C for 1 min and 95 °C for 15 s. All PCR reactions were run in triplicate. Data were analyzed using the $2^{-\Delta\Delta Ct}$ method with the mRNA level of the GmActin (*Glyma.18g290800*) gene used as the internal control. The primers used are shown in Supplementary Table S3.

5. Conclusions

In summary, using linkage analysis and GWAS, we detected 15 reproducible and significant QTL intervals and 67 significant SNP loci that affect the protein and/or oil content of soybeans. We searched the co-detected SNP region on Chr. 20 and an extra QTL interval on Chr. 15 to identify 25 candidate genes that may regulate the accumulation of soybean protein and oil. Among them, eight genes had differential expression patterns in the parent lines (ZGDD and HH27) at late reproductive growth stages. Further experiments with these gene candidates should lead to a better understanding of the molecular mechanisms of protein and oil biosynthesis in soybean. The refined QTL intervals and SNP loci in our study could also improve molecular breeding based on these markers.

Supplementary Materials: Supplementary materials can be found at http://www.mdpi.com/1422-0067/20/23/5915/s1.

Author Contributions: Conceptualization, S.S.; formal analysis, T.Z. and T.W.; data curation, T.Z., T.W., L.W. and B.J.; methodology, S.S., T.Z., T.W. and L.W.; investigation, T.Z., and C.Z.; resources, S.S., S.Y. and C.W.; supervision, S.S.; validation, T.Z. and T.W.; funding acquisition, T.H., and W.H.; writing—original draft, T.Z. and T.W.; writing—review and editing, S.S., T.W., L.W., B.J. and T.H.

Funding: This research was funded by Major Science and Technology Projects of China (2016ZX08004-003), Ministry of Science and Technology of China (2017YFD0101400), and the CAAS (Chinese Academy of Agriculture Sciences) Agricultural Science and Technology Innovation Project.

Acknowledgments: We thank Lijuan Qiu and. Zhangxiong Liu for supplying some germplasms. Thanks research assistants Jinlu Tao, Haiyan Zeng, Yanfeng Zhou and Xuegang Sun for their field management.

Conflicts of Interest: The authors declare no conflict of interest. The funders had no role in the design of the study; in the collection, analyses, or interpretation of data; in the writing of the manuscript, or in the decision to publish the results.

References

1. Zhang, Y.H.; Liu, M.F.; He, J.B.; Wang, Y.F.; Xing, G.N.; Li, Y.; Yang, S.P.; Zhao, T.J.; Gai, J.Y. Marker-assisted breeding for transgressive seed protein content in soybean [*Glycine max* (L.) Merr.]. *Theor. Appl. Genet.* **2015**, *128*, 1061–1072. [CrossRef] [PubMed]

2. Seo, J.H.; Kim, K.S.; Ko, J.M.; Choi, M.S.; Kang, B.K.; Kwon, S.W.; Jun, T.H. Quantitative trait locus analysis for soybean (*Glycine max*) seed protein and oil concentrations using selected breeding populations. *Plant Breed.* **2019**, *138*, 95–104. [CrossRef]

3. Diers, B.W.; Keim, P.; Fehr, W.R.; Shoemaker, R.C. RFLP analysis of soybean seed protein and oil content. *Theor. Appl. Genet.* **1992**, *83*, 608–612. [CrossRef] [PubMed]

4. Chaudhary, J.; Patil, G.B.; Sonah, H.; Deshmukh, R.K.; Vuong, T.D.; Valliyodan, B.; Nguyen, H.T. Expanding Omics Resources for Improvement of Soybean Seed Composition Traits. *Front. Plant Sci.* **2015**, *6*, 504. [CrossRef]

5. Patil, G.; Vuong, T.D.; Kale, S.; Valliyodan, B.; Deshmukh, R.; Zhu, C.; Wu, X.; Bai, Y.; Yungbluth, D.; Lu, F.; et al. Dissecting genomic hotspots underlying seed protein, oil, and sucrose content in an interspecific mapping population of soybean using high-density linkage mapping. *Plant Biotechnol. J.* **2018**, *16*, 1939–1953. [CrossRef]

6. Hartwig, E.E.; Hinson, K. Association Between Chemical Composition of Seed and Seed Yield of Soybeans1. *Crop Sci.* **1972**, *12*, 829. [CrossRef]

7. Patil, G.; Mian, R.; Vuong, T.; Pantalone, V.; Song, Q.; Chen, P.; Shannon, G.J.; Carter, T.C.; Nguyen, H.T. Molecular mapping and genomics of soybean seed protein: A review and perspective for the future. *Theor. Appl. Genet.* **2017**, *130*, 1975–1991. [CrossRef]

8. Sonah, H.; O'Donoughue, L.; Cober, E.; Rajcan, I.; Belzile, F. Identification of loci governing eight agronomic traits using a GBS-GWAS approach and validation by QTL mapping in soya bean. *Plant Biotechnol. J.* **2015**, *13*, 211–221. [CrossRef]

9. Ning, H.L.; Bai, X.L.; Li, W.B.; Xue, H.; Zhuang, X.; Li, W.X.; Liu, C.Y. Mapping QTL Protein and Oil Contents Using Population from Four-way Re-combinant Inbred Lines for Soybean (*Glycine max* L. Merr.). *Acta Agron. Sin.* **2016**, *42*, 1620. [CrossRef]

10. Lin, Y.H.; Zhang, L.J.; Wei, L.I.; Zhang, L.F.; Ran, X.U. QTLs mapping related to protein content of soybeans. *Soyb. Sci.* **2010**, *29*, 207–209.

11. Liang, H.Z.; Wang, S.; Yu, Y.; Lian, Y.; Wang, T.F.; Wei, Y.; Gong, P.T.; Liu, X.Y.; Fang, X.J. QTL mapping of isoflavone, oil and protein content in soybean. *Sci. Agric. Sin.* **2009**, *42*, 2652–2660.

12. Pathan, S.M.; Vuong, T.; Clark, K.; Lee, J.D.; Shannon, J.G.; Roberts, C.A.; Ellersieck, M.R.; Burton, J.W.; Cregan, P.B.; Hyten, D.L.; et al. Genetic Mapping and Confirmation of Quantitative Trait Loci for Seed Protein and Oil Contents and Seed Weight in Soybean. *Crop Sci.* **2013**, *53*, 765. [CrossRef]

13. Panthee, D.R.; Pantalone, V.R.; West, D.R.; Saxton, A.M.; Sams, C.E. Quantitative Trait Loci for Seed Protein and Oil Concentration, and Seed Size in Soybean. *Crop Sci.* **2005**, *45*, 2015. [CrossRef]

14. Wang, X.; Jiang, G.L.; Green, M.; Scott, R.A.; Song, Q.; Hyten, D.L.; Cregan, P.B. Identification and validation of quantitative trait loci for seed yield, oil and protein contents in two recombinant inbred line populations of soybean. *Mol. Genet. Genom.* **2014**, *289*, 935–949. [CrossRef]

15. Hwang, E.Y.; Song, Q.; Jia, G.; Specht, J.E.; Hyten, D.L.; Costa, J.; Cregan, P.B. A genome-wide association study of seed protein and oil content in soybean. *BMC Genom.* **2014**, *15*, 1. [CrossRef]

16. Qi, Z.M.; Wu, Q.; Han, X.; Sun, Y.N.; Du, X.Y.; Liu, C.Y.; Jiang, H.W.; Hu, G.H.; Chen, Q.S. Soybean oil content QTL mapping and integrating with meta-analysis method for mining genes. *Euphytica* **2011**, *179*, 499–514. [CrossRef]

17. Warrington, C.V.; Abdel-Haleem, H.; Hyten, D.L.; Cregan, P.B.; Orf, J.H.; Killam, A.S.; Bajjalieh, N.; Li, Z.; Boerma, H.R. QTL for seed protein and amino acids in the Benning x Danbaekkong soybean population. *Theor. Appl. Genet.* **2015**, *128*, 839–850. [CrossRef]

18. Wang, W.; Liu, M.; Wang, Y.; Li, X.; Cheng, S.; Shu, L.; Yu, Z.; Kong, J.; Zhao, T.; Gai, J. Characterizing Two Inter-specific Bin Maps for the Exploration of the QTLs/Genes that Confer Three Soybean Evolutionary Traits. *Front. Plant Sci.* **2016**, *7*, 242. [CrossRef]

19. Yang, X.H.; Yan, J.B.; Zheng, Y.P.; Yu, J.M.; Li, J.S. Reviews of association analysis for quantitative traits in plants. *Acta Agron. Sin.* **2007**, *33*, 523–530.

20. Hansen, M.; Kraft, T.; Ganestam, S.; Säll, T.; Nilsson, N.O. Linkage disequilibrium mapping of the bolting gene in sea beet using AFLP markers. *Genet. Res.* **2001**, *77*, 61–66. [CrossRef]

21. Han, S.F. Genome-Wide Association Studies for Fatty Acid Component Traits in Soybean. Master's Thesis, Nanjing Agricultural University, Nanjing, China, 2013.

22. Zhang, J.; Wang, X.; Lu, Y.; Bhusal, S.J.; Song, Q.; Cregan, P.B.; Yen, Y.; Brown, M.; Jiang, G.L. Genome-wide Scan for Seed Composition Provides Insights into Soybean Quality Improvement and the Impacts of Domestication and Breeding. *Mol. Plant* **2018**, *11*, 460–472. [CrossRef] [PubMed]

23. Valliyodan, B.; Qiu, D.; Patil, G.; Zeng, P.; Huang, J.; Dai, L.; Chen, C.; Li, Y.; Joshi, T.; Song, L.; et al. Landscape of genomic diversity and trait discovery in soybean. *Sci. Rep.* **2016**, *6*, 23598. [CrossRef] [PubMed]

24. Lee, S.; Van, K.; Sung, M.; Nelson, R.; LaMantia, J.; McHale, L.K.; Mian, M.A.R. Genome-wide association study of seed protein, oil and amino acid contents in soybean from maturity groups I to IV. *Theor. Appl. Genet.* **2019**, *132*, 1639–1659. [CrossRef] [PubMed]

25. Kim, M.; Schultz, S.; Nelson, R.L.; Diers, B.W. Identification and fne mapping of a soybean seed protein QTL from PI 407788A on chromosome 15. *Crop Sci.* **2016**, *56*, 219–225. [CrossRef]

26. Tajuddin, T. Analysis of quantitative trait loci for protein content in soybean seeds using recombinant inbred lines. *J. Agron. Indones.* **2005**, *33*, 19–24. [CrossRef]

27. Van, K.; McHale, L.K. Meta-Analyses of QTLs Associated with Protein and Oil Contents and Compositions in Soybean [*Glycine max* (L.) Merr.] Seed. *Int. J. Mol. Sci.* **2017**, *18*, 1180. [CrossRef]

28. Xu, X.; Zeng, L.; Tao, Y.; Vuong, T.; Wan, J.; Boerma, R.; Noe, J.; Li, Z.; Finnerty, S.; Pathan, S.M.; et al. Pinpointing genes underlying the quantitative trait loci for root-knot nematode resistance in palaeopolyploid soybean by whole genome resequencing. *Proc. Natl. Acad. Sci. USA* **2013**, *110*, 13469–13474. [CrossRef]

29. Cao, Y.; Li, S.; Wang, Z.; Chang, F.; Kong, J.; Gai, J.; Zhao, T. Identification of Major Quantitative Trait Loci for Seed Oil Content in Soybeans by Combining Linkage and Genome-Wide Association Mapping. *Front. Plant Sci.* **2017**, *8*, 8. [CrossRef]

30. Sajise, A.G.C.; Gregorio, G.B.; Kretzschmar, T.; Ismail, A.M.; Wissuwa, M.; Lee, J.S. Genetic dissection for zinc deficiency tolerance in rice using bi-parental mapping and association analysis. *Theor. Appl. Genet.* **2017**, *130*, 1903–1914.

31. Lou, Q.; Chen, L.; Mei, H.; Wei, H.; Feng, F.; Wang, P.; Xia, H.; Li, T.; Luo, L. Quantitative trait locus mapping of deep rooting by linkage and association analysis in rice. *J. Exp. Bot.* **2015**, *66*, 4749–4757. [CrossRef]

32. Wang, H.; Xu, S.; Fan, Y.; Liu, N.; Zhan, W.; Liu, H.; Xiao, Y.; Li, K.; Pan, Q.; Li, W.; et al. Beyond pathways: Genetic dissection of tocopherol content in maize kernels by combining linkage and association analyses. *Plant Biotechnol. J.* **2018**, *16*, 1464–1475. [CrossRef] [PubMed]

33. Deng, M.; Li, D.; Luo, J.; Xiao, Y.; Liu, H.; Pan, Q.; Zhang, X.; Jin, M.; Zhao, M.; Yan, J. The genetic architecture of amino acids dissection by association and linkage analysis in maize. *Plant Biotechnol. J.* **2017**, *15*, 1250–1263. [CrossRef] [PubMed]

34. Wang, L.W.; Sun, S.; Wu, T.T.; Liu, L.P.; Sun, X.G.; Cai, Y.P.; Li, J.C.; Xu, X.; Yuan, S.; Chen, L.; et al. Natural variation in GmPRR37 affects photoperiodic flowering and contributes to regional adaptation of soybean. *Plant J.* **2019**. under review.

35. Hu, M.X.; Wan, C.W. The effect of different ecogeographic enviroment on the seed quality of soybeans in China. *Soybean Sci.* **1990**, *9*, 39–49.

36. Zu, S.H. Tne agroclimatic analysis on tne oil content of soybean and its geographical distribution in Heilongjiang province. *Soybean Sci.* **1983**, *2*, 266–276.

37. Song, W.; Yang, R.; Wu, T.; Wu, C.; Sun, S.; Zhang, S.; Jiang, B.; Tian, S.; Liu, X.; Han, T. Analyzing the Effects of Climate Factors on Soybean Protein, Oil Contents, and Composition by Extensive and High-Density Sampling in China. *J. Agric. Food Chem.* **2016**, *64*, 4121–4130. [CrossRef]

38. Kale, S.M.; Jaganathan, D.; Ruperao, P.; Chen, C.; Punna, R.; Kudapa, H.; Thudi, M.; Roorkiwal, M.; Katta, M.A.; Doddamani, D.; et al. Prioritization of candidate genes in "QTL-hotspot" region for drought tolerance in chickpea (*Cicer arietinum L.*). *Sci. Rep.* **2015**, *5*, 15296. [CrossRef]

39. Wang, J.; Chen, P.; Wang, D.; Shannon, G.; Zeng, A.; Orazaly, M.; Wu, C. Identification and mapping of stable QTL for protein content in soybean seeds. *Mol. Breed.* **2015**, *35*, 92. [CrossRef]

40. Reinprecht, Y.; Poysa, V.W.; Yu, K.; Rajcan, I.; Ablett, G.R.; Pauls, K.P. Seed and agronomic QTL in low linolenic acid, lipoxygenase-free soybean (*Glycine max* (L.) Merrill) germplasm. *Genome* **2006**, *49*, 1510–1527. [CrossRef]

41. Lu, W.; Wen, Z.; Li, H.; Yuan, D.; Li, J.; Zhang, H.; Huang, Z.; Cui, S.; Du, W. Identification of the quantitative trait loci (QTL) underlying water soluble protein content in soybean. *Theor. Appl. Genet.* **2013**, *126*, 425–433. [CrossRef]

42. Bandillo, N.; Jarquin, D.; Song, Q.; Nelson, R.; Cregan, P.; Specht, J.; Lorenz, A. A Population Structure and Genome-Wide Association Analysis on the USDA Soybean Germplasm Collection. *Plant Genome* **2015**, *8*, 1–13. [CrossRef]

43. Priolli, R.H.G.; Campos, J.B.; Stabellini, N.S.; Pinheiro, J.B.; Vello, N.A. Association mapping of oil content and fatty acid components in soybean. *Euphytica* **2015**, *203*, 83–96. [CrossRef]

44. Yu, Z.R.; Huang, X.T. *Modern Biochemistry*; Chemical Industry Press: Beijing, China, 2001; pp. 238–246.

45. Zhu, Y.; Li, Y.; Zheng, X. *Modern Molecular Bology*, 3rd ed.; Higher Education Press: Beijing, China, 2010; pp. 126–153.

46. Yu, F.Y.; Xin, X.J.; Zhang, D.J.; Zhou, S.Q.; Qiu, H.M. Dynamic accumulation of dry matter, oil and protein in soybean seed. *Res. Agric. Mod.* **2009**, *30*, 637–640.

47. Qiu, L.; Wang, J.; Meng, Q. A preliminary study on accumulation characteristics of protein and fat in developing soybean seeds. *Sci. Agric. Sin.* **1990**, *23*, 28–32.

48. Bruening, W.P.; Egli, D.B. Accumulation of Nitrogen and Dry Matter by Soybean Seeds with Genetic Differences in Protein Concentration. *Crop Sci.* **2007**, *47*, 359–366.

49. Zhang, J.W.; Han, F.X.; Sun, J.M.; Han, G.Z.; Yu, S.X.; Yu, F.K.; Yan, S.R.; Yang, H. Genetic variation of protein and fat content in soybean mini core collections. *J. Plant Genet. Resour.* **2014**, *15*, 405–410.

50. Nyquist, W.E.; Baker, R. Estimation of heritability and prediction of selection response in plant populations. *Crit. Rev. Plant Sci.* **1991**, *10*, 235–322. [CrossRef]

51. Wang, S.; Meyer, E.; McKay, J.K.; Matz, M.V. 2b-RAD: A simple and flexible method for genome-wide genotyping. *Nat. Methods* **2012**, *9*, 808–810. [CrossRef]

52. Stam, P. Construction of integrated genetic linkage maps by means of a new computer package: JoinMap. *Plant J.* **1993**, *3*, 739–744. [CrossRef]

53. Wang, J.K. Inclusive Composite Interval Mapping of Quantitative Trait Genes. *Acta Agron. Sin.* **2009**, *35*, 239–245. [CrossRef]

54. Zhu, J. Mixed model approaches of mapping genes for complex quantitative traits. *J. Zhejiang Univ. Sci. B* **1999**, *33*, 327–335.

55. McKenna, A.; Hanna, M.; Banks, E.; Sivachenko, A.; Cibulskis, K.; Kernytsky, A.; Garimella, K.; Altshuler, D.; Gabriel, S.; Daly, M.; et al. The Genome Analysis Toolkit: A MapReduce framework for analyzing next-generation DNA sequencing data. *Genome Res.* **2010**, *20*, 1297–1303. [CrossRef] [PubMed]

56. Li, H.; Durbin, R. Fast and accurate short read alignment with Burrows-Wheeler transform. *Bioinformatics* **2009**, *25*, 1754–1760. [CrossRef] [PubMed]
57. Purcell, S.; Neale, B.; Todd-Brown, K.; Thomas, L.; Ferreira, M.A.R.; Bender, D.; Maller, J.; Sklar, P.; De Bakker, P.I.W.; Daly, M.J.; et al. PLINK: A Tool Set for Whole-Genome Association and Population-Based Linkage Analyses. *Am. J. Hum. Genet.* **2007**, *81*, 559–575. [CrossRef]
58. Cleveland, W.S. LOWESS: A Program for Smoothing Scatterplots by Robust Locally Weighted Regression. *Am. Stat.* **1981**, *35*, 54. [CrossRef]
59. Raj, A.; Stephens, M.; Pritchard, J.K. fastSTRUCTURE: Variational inference of population structure in large SNP data sets. *Genetics* **2014**, *197*, 573–589. [CrossRef]
60. Lipka, A.E.; Tian, F.; Wang, Q.; Peiffer, J.; Li, M.; Bradbury, P.J.; Gore, M.A.; Buckler, E.; Zhang, Z. GAPIT: Genome association and prediction integrated tool. *Bioinformatics* **2012**, *28*, 2397–2399. [CrossRef]
61. Zhang, C.; Dong, S.S.; Xu, J.Y.; He, W.M.; Yang, T.L. PopLDdecay: A fast and effective tool for linkage disequilibrium decay analysis based on variant call format files. *Bioinformatics* **2019**, *35*, 1786–1788. [CrossRef]
62. Zhao, X.; Jiang, H.; Feng, L.; Qu, Y.; Teng, W.; Qiu, L.; Zheng, H.; Han, Y.; Li, W. Genome-wide association and transcriptional studies reveal novel genes for unsaturated fatty acid synthesis in a panel of soybean accessions. *BMC Genom.* **2019**, *20*, 68. [CrossRef]

International Journal of
Molecular Sciences

Article

Candidate Domestication-Related Genes Revealed by Expression Quantitative Trait Loci Mapping of Narrow-Leafed Lupin (*Lupinus angustifolius* L.)

Piotr Plewiński [†], Michał Książkiewicz [*,†], Sandra Rychel-Bielska, Elżbieta Rudy and Bogdan Wolko

Department of Genomics, Institute of Plant Genetics, Polish Academy of Sciences, 60-479 Poznan, Poland; pple@igr.poznan.pl (P.P.); sryc@igr.poznan.pl (S.R.-B.); Elarudy@op.pl (E.R.); bwol@igr.poznan.pl (B.W.)
* Correspondence: mksi@igr.poznan.pl; Tel.: +48-616-550-268
† These authors contributed equally to this work.

Received: 21 October 2019; Accepted: 9 November 2019; Published: 12 November 2019

Abstract: The last century has witnessed rapid domestication of the narrow-leafed lupin (*Lupinus angustifolius* L.) as a grain legume crop, exploiting discovered alleles conferring low-alkaloid content (*iucundus*), vernalization independence (*Ku* and *Julius*), and reduced pod shattering (*lentus* and *tardus*). In this study, a *L. angustifolius* mapping population was subjected to massive analysis of cDNA ends (MACE). The MACE yielded 4185 single nucleotide polymorphism (SNP) markers for linkage map improvement and 30,595 transcriptomic profiles for expression quantitative trait loci (eQTL) mapping. The eQTL highlighted a high number of *cis-* and *trans*-regulated alkaloid biosynthesis genes with gene expression orchestrated by a regulatory agent localized at *iucundus* locus, supporting the concept that *ETHYLENE RESPONSIVE TRANSCRIPTION FACTOR RAP2-7* may control low-alkaloid phenotype. The analysis of *Ku* shed light on the vernalization response via *FLOWERING LOCUS T* and *FD* regulon in *L. angustifolius*, providing transcriptomic evidence for the contribution of several genes acting in C-repeat binding factor (*CBF*) cold responsiveness and in UDP-glycosyltransferases pathways. Research on *lentus* selected a *DUF1218* domain protein as a candidate gene controlling the orientation of the sclerified endocarp and a homolog of *DETOXIFICATION14* for purplish hue of young pods. An *ABCG* transporter was identified as a hypothetical contributor to sclerenchyma fortification underlying *tardus* phenotype.

Keywords: vernalization responsiveness; alkaloid content; pod shattering; gene expression; quantitative trait loci

1. Introduction

The narrow-leafed lupin, *Lupinus angustifolius* L., is a grain legume crop, appreciated as an organic fertilizer that improves soil structure and productivity, as well as providing a source of protein for human and animals. This species has witnessed rapid domestication during the last century. Several important agronomic traits have been identified and transferred into improved germplasm [1]. These traits include, among others, vernalization independence (overlapping loci *Ku* and *Julius*), low-alkaloid content (*iucundus*), reduced pod shattering (*tardus* and *lentus*), soft seededness (*mollis*), white flower color (*leucospermus*) and anthracnose resistance (*Lanr1*).

Vernalization responsiveness is the natural adaptation to climatic conditions, based on the requirement of a prolonged low temperature period during germination to induce flowering [2,3]. Natural dominant mutation in the so-called *Ku* or *Julius* loci diminished the need of vernalization and enabled temperature-independent sowing of *L. angustifolius* [4].

A high level of quinolizidine alkaloids is a typical feature of primitive populations in many lupin species, as these chemical compounds protect plants from pests and fungi [5], however, alkaloids are major antinutritional factors and provide bitter taste [6,7]. Three unlinked low-alkaloid recessive alleles were identified in *L. angustifolius*, and one of them, *iucundus*, was extensively implemented in breeding [8,9]. Some germplasm resources having less than 0.01% of grain alkaloid have been developed [10,11].

Shattering of dry pods is natural process of seed dispersal, however, it is a very undesired trait in modern agriculture because it dramatically decreases harvested yield. Two unlinked recessive alleles contribute to reduced pod shattering in *L. angustifolius*, namely *tardus*, affecting sclerenchyma strips of the dorsal and ventral pod seams, and *lentus*, modifying the orientation of the sclerified endocarp of the pod [12,13].

L. angustifolius is natively adapted to the Mediterranean climate which has hot dry summers, because of one of its survival strategies which is impermeability of seed coat to water. Hard-seeded germplasm has a long dormancy period and irregular germination. Recessive soft-seediness allele *mollis* confers water permeability and efficient seed germination [14]. It is the most difficult domestication *L. angustifolius* allele for breeding because the desired phenotype is maternally determined [15].

The agronomic potential of *L. angustifolius* has been reduced by high susceptibility to anthracnose, caused by the pathogenic fungus, *Colletotrichum lupini* (Bondar) Nirenberg, Feiler and Hagedorn [16]. The resistance to anthracnose in *L. angustifolius* was revealed to be controlled by several single dominant genes that were discovered in different germplasm resources, namely, *Lanr1* in cultivar Tanjil, *AnMan* in cv. Mandelup, and *LanrBo* in the breeding line Bo7212 [17–19].

Several genes contribute to *L. angustifolius* seed and flower color. The most widely exploited is the recessive allele *leucospermus*, affecting anthocyanin synthesis and resulting in bright seeds and white flowers [1].

To generate numerous molecular markers for agronomic trait selection in narrow-leafed lupin breeding programs, microsatellite-anchored fragment length polymorphisms (MFLP) fingerprinting has been exploited [20,21]. Trait-associated markers have been developed for *iucundus* (marker iucLi) [22], *Ku* (KuHM1) [23], *mollis* (MoLi) [15], *lentus* (LeM1, LeM2 and LeLi) [13,24], and *tardus* (TaLi, TaM1 and TaM2) [25,26]. Narrow-leafed lupin genomic studies have been greatly facilitated by the incremental development of linkage map carrying sequence-defined markers [27–30], construction of nuclear genome bacterial artificial chromosome (BAC) libraries [31,32], and assembly of the draft genome sequence [30,33,34].

Recently, a new method of transcriptome-based genotyping-by-sequencing, called massive analysis of cDNA ends (MACE), has been developed [35]. The MACE provides markers anchored in 3′-ends of transcribed sequences, and therefore is directly matching active RNA content of the genome. First implementations highlighted the relevance of the MACE for sequence polymorphism detection, gene expression quantification, transcript-based marker development, and candidate gene identification [36–41]. In this study, the MACE protocol was used for development of polymorphic gene-based markers and for quantification of gene expression in mapping population of *L. angustifolius*. These new data were exploited for construction of a linkage map and for determination of expression quantitative trait loci (eQTLs) related to selected domestication traits.

2. Results and Discussion

2.1. Development of New Polymorphic Markers

The MACE protocol was applied for 89 RILs and for parental lines of *L. angustifolius* mapping population (83A:476 × P27255), yielding 11,864 markers. A total of 9304 markers were localized within gene sequences whereas 2560 markers were found in loci lacking any annotation. There were 4185 MACE markers retained after application of total missing data threshold (counting heterozygotes and no data scores), followed by inference of consensus segregation for genes represented by several single nucleotide polymorphism, SNPs. There were 3532 genes represented by single markers, four genes were represented by pairs of markers with heterogeneous segregation patterns, and 645 markers were localized in unannotated loci. The annotation of markers is provided in the Table S1.

The MACE is a method providing sequences anchored in the 3′-ends of mRNA and can be used to develop sequence-defined markers, as well as to quantify gene expression [35,36]. In this study both applications of the MACE protocol were exploited, providing molecular markers and gene expression scores related to the same RNA isolates. The MACE marker set was supplemented with 10 newly developed BAC-end derived PCR markers, namely five dCAPS (019A15_3, 026O16_3, 034M08_5, 043N19_3, and 103O20_3), four CAPS (061O23_3, 085K20_5, 085L14_3, and 128I22_5), and one allele-specific PCR marker (085L14_5). The BAC-end based marker allele sequences were deposited in NCBI Genbank under accession numbers (MN518055-MN518073). Information on primer pair sequences, PCR primer annealing temperature, PCR product lengths, enzyme used for polymorphism detection, and restriction product lengths for both alleles is provided in Table S2. For the past 15 years, the use of BAC-derived PCR markers has been a method of choice in studies involving *L. angustifolius* genome physical and linkage mapping. Because the *L. angustifolius* karyotype carries numerous small and very uniform chromosomes, the BAC-derived markers have been frequently used as chromosome-specific landmarks to validate physical linkage of particular genome regions, as well as to facilitate assignment of particular chromosomes to linkage groups [31,42–47]. The BAC-derived markers have also been exploited for fine mapping of a region carrying a candidate gene for vernalization independence *Ku* locus, as well as for comparative mapping of genes from isoflavonoid and fatty acid synthesis pathways [48–51]. In this study, BAC-derived markers were developed to localize on the linkage map some clones identified during our previous studies and confirmed to carry repetitive elements.

2.2. Construction of a Linkage Map

Markers used for linkage map development included 4,185 MACE and 10 BAC-end PCR markers developed in this study as well as previously published data including seven trait loci (*Ku, tardus, lentus, mollis, leucospermus, iucundus* and *Lanr1*) [21], eight trait related markers (TaM1, TaM2, LeM1, LeM2, KuHM1, AntjM2, MoA, MoLi) [13,15,24,25,52,53], and 109 BAC-derived markers anchoring particular linkage groups to chromosomes [43–47,49–51,54]. The segregation data for markers used for linkage mapping and the calculated χ^2 *p*-values of distortion from expected 1:1 segregation are provided in Table S3. There were 4309 markers localized on the genetic map, which constituted 20 linkage groups, carrying from 144 to 304 markers (215 on average) and 16 markers remained unmapped (Table 1).

The segregation pattern of 59.7% of the markers was redundant, and therefore the map contains 1735 loci, namely from 60 to 120 loci per linkage group. The lengths of linkage groups vary from 78.5 to 156.38 cM, reaching 2163.63 cM in total. The results of linkage mapping are provided in Table S4. A high percentage of markers matching particular linkage groups with corresponding pseudochromosomes, reaching from 96.6% (chromosome NLL-16) to 100% (chromosomes NLL-14 and NLL-20), highlighted the collinearity between published *L. angustifolius* draft genome sequence [34] and this version of linkage map (Figure 1a). Major issues were found for chromosomes NLL-16 (block of eight adjacent markers representing ~300 kbp localized in the linkage group NLL-01), NLL-15 (block of six adjacent markers covering ~200 kbp mapped in the linkage group NLL-17), and NLL-12 (five markers mapped in the linkage group NLL-14). As many as 209 unassembled scaffolds were localized on the genetic map, namely from six to 22 scaffold per linkage group (Figure 1b). A comparison of the number of markers assigned to particular chromosomes, scaffolds, and linkage groups is provided in Table S5.

Table 1. Characteristics of the MACE-based *L. angustifolius* linkage map.

Linkage Group	Number of Markers	Number of Loci	Number of Genes	Length (cM)	Number of Scaffolds
NLL-01	289	120	251	156.38	11
NLL-02	207	73	185	132.78	21
NLL-03	204	77	169	115.02	22
NLL-04	193	95	157	126.93	8
NLL-05	144	70	130	101.29	10
NLL-06	304	119	267	144.66	10
NLL-07	191	85	155	104.25	10
NLL-08	231	101	194	119.29	8
NLL-09	192	93	169	140.98	10
NLL-10	154	75	121	85.64	6
NLL-11	269	107	216	119.57	8
NLL-12	271	88	220	85.86	15
NLL-13	230	81	200	78.5	11
NLL-14	160	76	121	87.5	10
NLL-15	248	87	207	86.65	10
NLL-16	240	96	204	99.42	9
NLL-17	227	70	176	91.57	6
NLL-18	196	77	151	91.34	9
NLL-19	144	60	122	92.08	10
NLL-20	199	85	169	103.92	9
Total	4309	1735	3590	2163.63	209

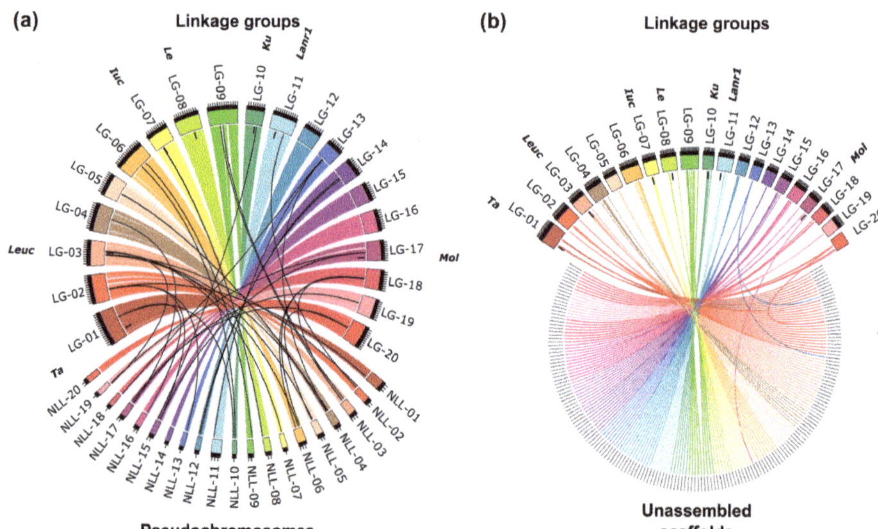

Figure 1. Collinearity links matching narrow-leafed lupin linkage groups (LG-01–LG-20) and: (**a**) pseudochromosomes (NLL-01–NLL-20) and (**b**) unassembled scaffolds. Ribbons symbolize homologous links identified by DNA sequence similarity. Chromosomes and linkage groups are drawn to scale indicated by ticks (10 Mbp and 10 cM). Postions of the following major domestication loci are indicated: *Tardus* (*Ta*), *leucospermus* (*Leuc*), *iucundus* (*Iuc*), *lentus* (*Le*), *Ku*, *Lanr1*, and *mollis* (*Mol*).

Mapping data from the most recent *L. angustifolius* linkage maps [30,34] were not incorporated to our map due to a limited number of RIL lines common for all three studies (about 70), as well as due to observed inconsistency in segregation patterns between physically linked markers originating from different studies, indicating diverse genetic origin of some RILs having the same numbers assigned, putatively resulting from seed admixture or cross-pollination during seed multiplication. Seeds of the mapping population were shared between the Department of Agriculture and Food Western Australia and the Institute of Plant Genetics, Polish Academy of Sciences, in the year 2003, and maintained independently thereafter. As lupin breeding was recently licensed to the private sector in Australia it may be currently impossible to access original set of RILs developed for this mapping population. Similar issues with possible cross-pollination during mapping population development have also been reported for 43 RILs from the recently published linkage map of yellow lupin, *L. luteus* [55], as well as for one RIL in white lupin, *L. albus* [56]. Nevertheless, the total number of RILs used in the most recent *L. angustifolius* genome mapping study, namely 87 lines with only 78 lines overlapping with previous mapping studies, was too low to provide the high resolution required for significant improvement of genome assembly, and resulted in high marker redundancy, reaching 89.9% [30].

2.3. Gene Expression Profiling, Gene Ontology Enrichment, and Expression Quantitative Trait Loci Mapping

The MACE analysis provided normalized gene expression levels for all RILs analyzed. Namely, 30,595 genes revealed nonzero expression for at least 1 RIL, 25,024 genes for at least 30% of RILs, 23,557 genes for at least 50% of RILs, and 15,686 genes for all RILs. The normalized gene expression values for mapping population and parental lines (83A:476 and P27255) are provided in Table S6. The gene expression patterns in the mapping population were associated with domestication trait segregation (wild alleles used as positive values). Genes with a statistically significant association (FDR p-value threshold of 0.01) were identified for all domestication traits analyzed, namely 98 genes for *iucundus*, 50 for *Ku*, 35 for *leucospermus*, 29 for *lentus*, 17 for *tardus*, 11 for *Lanr1* and five for *mollis*. The values of the *t*-Student test association between domestication trait segregation and gene expression patterns, including FDR correction and statistical significance analysis, are provided in Table S7. The gene ontology (GO) enrichment analysis of genes with expression pattern associated with *iucundus* trait segregation highlighted lysine biosynthesis and lysine metabolism, as well as cofactor binding and coenzyme binding, as the most overrepresented processes and functions, respectively (Table S8). This was an expected outcome as quinolizidine alkaloids are derived from lysine via a series of chemical reactions [57]. GO analysis for *tardus*-associated genes revealed iron-sulfur cluster assembly, metallo-sulfur cluster assembly, and cofactor biosynthesis process enrichments. No statistically significant GO enrichments were identified for genes associated with *Ku*, *leucospermus*, *lentus*, *mollis* and *Lanr1* traits.

Composite interval mapping revealed the presence of numerous eQTL peaks close to domestication trait loci. Within a genetic linkage distance of 2 cM from a particular domestication trait locus, from one (*mollis*) to 61 (*iucundus*) genes had eQTL peaks localized (Table 2). The LOD values for eQTL permutation test are provided in Table S9, whereas data on eQTL localization on the linkage map are provided in Table S10. Some potential candidate genes were identified for all analyzed loci, except anthracnose resistance locus *Lanr1*. As plants were not inoculated to allow long-range gene profiling during their development, including the generative phase, genes related to anthracnose were putatively not activated in the experiment. Anthracnose resistance will be addressed in another study. Here, genes identified for *iucundus*, *Ku*, *lentus*, *tardus*, *mollis* and *leucospermus* are discussed.

Table 2. Expression quantitative trait loci localized near (≤2 cM) major domestication trait loci.

Domestication Trait	Genes with eQTL [1] Peak	Mean eQTL Peak LOD Value	Maximum eQTL Peak LOD Value	Genes with *cis* Genomic Positions	Genes with *trans* Genomic Positions	Genes in Unassigned Scaffolds
iucundus	61	14.77	37.70	11	45	5
Ku	25	11.20	40.97	11	11	3
leucospermus	9	13.18	35.34	5	3	1
lentus	6	16.36	42.97	5	1	0
tardus	2	25.05	31.82	2	0	0
Lanr1	4	10.20	13.48	4	0	0
mollis	1	9.59	9.59	0	1	0

[1] eQTL, expression quantitative trait locus.

2.4. Genes Identified for Low-Alkaloid Iucundus Locus

The high number of genes revealed for *iucundus* locus might be related to the complexity of the alkaloid biosynthesis process and the number of genes involved. Taking into consideration the position of gene coding sequences in the genome, only a relatively small fraction of eQTL genes revealed for *iucundus* (18%) was *cis*-regulated, whereas the vast majority (74%) was *trans*-regulated. Furthermore, from 34 genes for *iucundus* that had association values between their expression patterns and trait segregation above 0.5 or below −0.5, as many as 30 revealed a positive association with wild, high alkaloid phenotype. Moreover, as many as 16 genes highly associated with *iucundus* revealed to have their major eQTL locus explaining more than 50% of their observed expression variance localized directly at *iucundus,* or very close to it (Table 3). Many of these genes are hypothesized to be involved in alkaloid biosynthesis process. Such an observation strongly supports a hypothesis that *iucundus* locus in *L. angustifolius* encodes a single regulatory agent controlling this complex secondary metabolic pathway and differentiating between high and low quinolizidine alkaloid biosynthesis profiles.

Table 3. Genes showing the highest gene expression association and eQTL peak co-localization (≤2 cM) with low-alkaloid *iucundus* locus.

Protein	Group	Peak cM	Peak LOD	PVE [1] %	Association Based on *t*-Student Test	Protein Annotation
OIW21347.1	NLL-07	40.2	37.7	84.2	−0.82	voltage-gated potassium channel subunit beta 1, KAB1
OIV96299.1	NLL-07	40.1	37.3	73.4	0.81	lysine/ornithine decarboxylase, LDC
OIV96574.1	NLL-07	40.1	35.4	60.5	0.84	purine permease 1, PUP1
OIW10551.1	NLL-07	39.6	29.4	67.2	0.80	MLP-like protein 423, MLP423
OIV89004.1	NLL-07	39.6	29.2	56.5	0.78	ethylene-responsive transcription factor, RAP2-7
OIW02927.1	NLL-07	39.6	27.3	69.9	0.69	amino acid permease, AAP
OIW07732.1	NLL-07	41.3	26.3	81.9	−0.73	uncharacterized protein
OIW13431.1	NLL-07	40.2	25.9	59.3	0.78	dihydroflavonol 4-reductase, DFR
OIV89008.1	NLL-07	40.2	25.8	66.2	−0.77	Fe superoxide dismutase 2, FSD2
OIV90042.1	NLL-07	39.6	24.6	59.6	0.72	aspartate kinase 1, AK1
OIV95196.1	NLL-07	39.6	24.3	57.2	0.77	HXXXD-type ACYL-TRANSFERASE, LaAT
OIW13432.1	NLL-07	40.1	24.0	50.2	0.77	dihydroflavonol 4-reductase, DFR
OIW03412.1	NLL-07	40.1	23.5	50.7	0.75	homeobox protein knotted-1-like, KNAT1
OIW15620.1	NLL-07	40.2	22.7	56.6	0.67	uncharacterized protein
OIW20548.1	NLL-07	40.1	21.7	50.3	0.72	diaminopimelate decarboxylase 1, DAPDC1
OIW02909.1	NLL-07	40.1	20.4	53.3	0.72	cytochrome P450, CYP71B23
OIV89669.1	NLL-07	39.6	20.4	48.0	0.70	cinnamoyl-CoA reductase 1, CCR1
OIV96948.1	NLL-07	40.1	19.5	38.5	0.73	copper amine oxidase, MHK10.21
OIW20507.1	NLL-07	40.1	18.3	40.1	0.70	anthocyanidin reductase, BAN
OIW20661.1	NLL-07	41.3	18.0	49.4	0.70	4-hydroxy-tetrahydrodipicolinate synthase, DHDPS
OIV97872.1	NLL-07	39.6	17.2	43.5	0.72	purine permease 1, PUP1
OIV93156.1	NLL-07	40.8	14.9	45.1	0.64	LL-diaminopimelate aminotransferase, DapL
OIV97100.1	NLL-07	39.6	13.9	44.4	0.64	MYB transcription factor 34, MYB34
OIW07643.1	NLL-07	39.0	13.7	36.3	0.64	4-hydroxy-tetrahydrodipicolinate synthase, DHDPS
OIV89772.1	NLL-07	39.6	13.3	40.7	0.51	pentatricopeptide repeat-containing protein
OIW10549.1	NLL-07	40.2	12.7	38.4	0.57	MLP-like protein 423, MLP423
OIV96820.1	NLL-07	40.2	11.8	30.5	0.57	glutamate synthase 1, GLT1
OIW21355.1	NLL-07	39.6	11.8	32.3	0.56	carboxylesterase 1, CXE1
OIW10098.1	NLL-07	40.1	11.8	28.3	0.53	aspartate-semialdehyde dehydrogenase, ASDH
OIW04462.1	NLL-07	39.6	11.7	30.8	0.57	VQ motif-containing protein
OIW10550.1	NLL-07	40.1	10.2	14.9	0.74	MLP-like protein 423, MLP423
OIW20088.1	NLL-07	39.6	9.6	27.4	0.53	short-chain dehydrogenase reductase, SDR
OIV89148.1	NLL-07	41.3	8.2	22.4	0.51	MLP-like protein 31, MLP31
OIW02362.1	NLL-07	39.6	7.9	21.4	−0.55	DMR6-like OXYGENASE 2, 2OG

[1] PVE, proportion of explained variance.

Among the genes with expression positively associated with high alkaloid phenotypes in the RIL population, the highest LOD values of eQTLs were revealed for Lup009726 (OIV96299.1, LOD 37.3), Lup028431 (OIV96574.1, LOD 35.4), Lup015923 (OIW10551.1, LOD 29.4), and Lup007628 (OIV89004.1, LOD 29.2) (Figure 2a). The Lup009726 product revealed 99.5% sequence identity to lysine/ornithine decarboxylase LDC (BAK32797.1) protein which catalyzes the first step of quinolizidine alkaloid biosynthesis [57]. Expression of the *LDC* gene has been confirmed to be associated with alkaloid content in *L. angustifolius* by several independent studies [58–60]. Lup028431 encodes purine permease transporter 1 (PUP1), PUP proteins which are generally involved in alkaloid biosynthesis and transport. Nicotine uptake permease from *Nicotiana tabacum* (*NtPUP1*), for example, affects nicotine metabolism, as well as regulates the *ETHYLENE RESPONSE FACTOR 189*, a key transcription factor in nicotine biosynthesis pathway [61,62]. Indeed, Lup028431 was selected in another study as potential *L. angustifolius* quinolizidine alkaloid biosynthetic gene because it revealed similar expression pattern to the *LDC* gene [63]. Lup015923 has been annotated as *MAJOR LATEX PROTEIN 423* (*MLP423*, AT1G24020), which is hypothesized to be involved in stress responsive activation of biosynthetic pathway of coumestrol, a coumestan isoflavone in soybean [64]. Lup015923 was recently highlighted as one of candidate quinolizidine alkaloid biosynthesis genes in *L. angustifolius* due to highly elevated expression in bitter P27255 accession [58]. Lup007628 (OIV89004.1) is the *ETHYLENE RESPONSIVE TRANSCRIPTION FACTOR RAP2-7*, a candidate locus for *iucundus*, evidenced by a gene expression study involving transcriptome sequencing of four accessions and quantitative RT-PCR profiling of 14 accessions differing in alkaloid content, as well as by molecular marker development and linkage mapping [59,60]. Interestingly, closely located to *RAP2.7* at *iucundus* locus, another *cis*-regulated component, *Fe SUPEROXIDE DISMUTASE 2* (Lup007632, OIV89008.1), revealed similarly high LOD and explained eQTL variance values, but opposite direction of association. Other *iucundus*-associated genes have included Lup005321 (OIW13431.1) and Lup005322 (OIW13432.1) encoding homologs of *DIHYDROFLAVONOL 4-REDUCTASE* which is one of the key genes from anthocyanin biosynthesis pathway [65]. The set of genes with highly significant eQTLs localized in the *iucundus* region also includes a Lup021586 (OIV95196.1) gene encoding *HXXXD*-type *ACYL-TRANSFERASE* (*LaAT*, AB581532.1). The expression profile of *LaAT* has been highly associated with alkaloid content in *L. angustifolius* [58–60,66]. Moreover, one of the homologs of this gene, LAGI01_35805, has been recently designated as a candidate gene underlying low-alkaloid *pauper* locus in *L. albus*, as evidenced by linkage mapping and validation survey in a set of 127 bitter and 23 sweet accessions [56].

In addition to Lup007628 and Lup007632, nine other *cis*-regulated genes revealed eQTLs localized at *iucundus* locus, including three hypothetical components of alkaloid biosynthesis pathways, Lup007706 (OIW07664.1), Lup017658 (OIW07643.1), and Lup032669 (OIW21355.1). Lup007706 encodes a representative of a *S*-adenosyl-L-methionine-dependent methyltransferases superfamily protein. N-methylation of quinolizidine alkaloids was confirmed to occur in crude protein extracts from *Laburnum anagyroides* carrying *S*-adenosyl-L-methionine: cytisine N-methyltransferase [67]. Moreover, a homolog of *S*-adenosyl-L-methionine-dependent N methyltransferase catalyzes a nitrogen methylation involved in vindoline alkaloid biosynthesis in Madagascar periwinkle (*Catharanthus roseus*) [68]. Lup017658 encodes a *4-HYDROXY-TETRAHYDRODIPICOLINATE SYNTHASE* gene which is generally involved in biosynthesis of L-lysine, a precursor of quinolizidine alkaloids. This gene revealed considerably elevated expression in bitter accessions of *L. angustifolius*, indicating its hypothetical involvement in alkaloid biosynthesis pathway [59]. Lup032669 encodes the *CARBOXYLESTERASE 1* gene. *CARBOXYLESTERASE 1* was evidenced to be involved in one of the final three steps of noscapine alkaloid biosynthesis [69]. To summarize, this study highlighted a relatively high number of alkaloid biosynthesis genes with gene expression orchestrated by a regulatory agent(s) localized at *iucundus* locus. This study provided novel evidences supporting the concept that *RAP2.7* may control low-alkaloid *iucundus* phenotype, however, further evidence would require *cis*-trans tests. Nonetheless, such studies are hampered by very low transformation efficiency in narrow-leafed lupin [70].

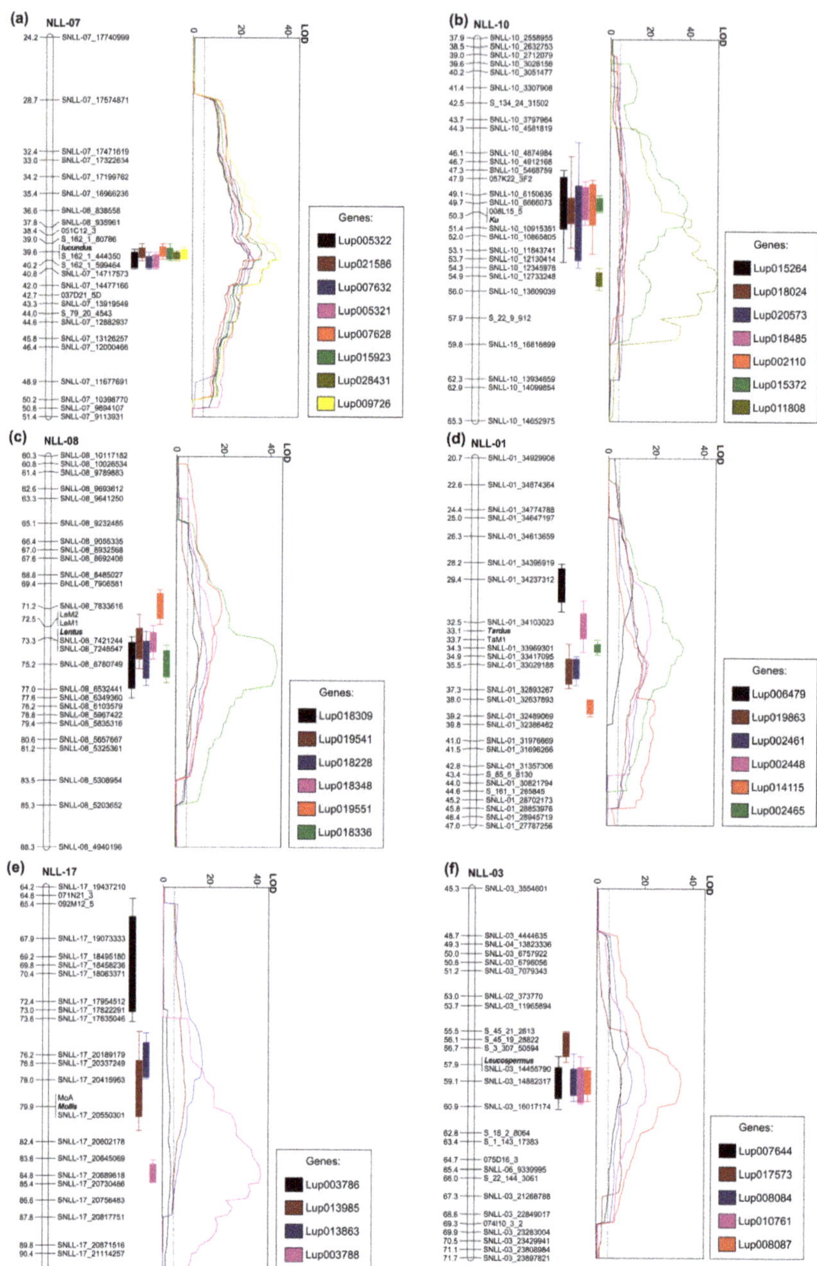

Figure 2. Major expression quantitative trait loci (eQTLs) revealed for narrow-leafed lupin domestication trait loci: (**a**) main alkaloid content *iucundus* locus, (**b**) vernalization responsiveness *Ku* locus, (**c**) pod shattering *lentus* locus, (**d**) pod shattering *tardus* locus, (**e**) soft seededness *mollis* locus, and (**f**) white flower color *leucospermus* locus. Linear plots show LOD values (threshold 4.8), whereas vertical bar graphs visualize eQTL ranges (outer, LOD$_{max}$-2 and inner, LOD$_{max}$-1) on corresponding linkage group fragments. Linkage groups are drawn to scale indicated by ticks and labels.

2.5. Genes Revealed for Vernalization Independence Ku Locus

The P27255 parent is late flowering and requires vernalization for flowering induction, whereas the 83A:476 parent is early flowering and vernalization independent. In this study, seeds were subjected to vernalization procedure to ensure transition from vegetative to generative phase in all analyzed RILs. Such an approach could result in diminishing of some differences in expression profiles of vernalization-responsive genes between early and late flowering RILs. However, despite this partial pre-sowing vernalization, relatively a large number of genes revealed to have their eQTL peaks closely localized to *Ku* locus (Table 4, Figure 2b). Contrary to *iucundus*, the same ratio of *cis-* and *trans*-regulation for major eQTL loci was observed (44%). Genes showing the highest gene expression association and eQTL peak co-localization (≤2 cM) with *Ku* included Lup011808 (OIW03171.1, LOD 41.0) and Lup015372 (OIW20567.1, LOD 38.9). Lup015372 encodes hypothetical uncharacterized protein, whereas Lup011808 a homolog of *A. thaliana CALCIUM/CALMODULIN-REGULATED RECEPTOR-LIKE KINASE 1, CRLK1. CRLK1* confers cold responsiveness in plants via C-repeat binding factors (*CBF*) pathway [71]. Overexpression of the *CBF* in *Arabidopsis* delays flowering by promoting the expression of *FLOWERING LOCUS C* (*FLC*), indicating a link between cold signaling and flowering time regulation [72]. One of the downstream genes in this pathway is *INDUCER OF CBF EXPRESSION 1* (*ICE1*) which integrates cold signals into *FLC*-mediated flowering pathway [73]. Another cross-talk between cold response and flowering initiation pathways, involving a *SUPPRESSOR OF OVEREXPRESSION OF CONSTANS 1* (*SOC1*) gene was also identified [74]. A recent study confirmed that *CBF* pathway affects flowering time but does not affect vernalization response in *Arabidopsis* [75]. Moreover, a calcium and calmodulin-binding protein kinase (*NtCBK1*) from *N. tabacum* functions as a negative regulator of flowering; high levels of *NtCBK1* in the shoot apical meristem extended the vegetative phase of growth [76]. Indeed, *CRLK1*, in this study, was revealed to be positively associated with late flowering phenotype. These observations bring attention to the hypothetical involvement of calcium and calmodulin link with *FLC* pathway in flowering time regulation in *L. angustifolius*. Legume genomes do not contain *FLC* homologs but other genes from this pathway, including activators and repressors of *FLC*, are present [77].

Other *Ku* eQTLs include Lup011781 (OIW03144.1), Lup011739 (OIV96743.1), Lup011836 (OIW03199.1), and Lup002110 (OIW20134.1) sequences. Lup011781 has been identified as a homolog of *MHM17-10* (AT5G56980) gene. It is pathogen-associated molecular pattern-induced gene with unknown function, putatively participating in jasmonic acid pathway [78]. Lup011739 encodes a homolog of *GALACTURONOSYLTRANSFERASE-LIKE 10*, which is involved in cell wall organization and its expression is regulated by *FLAVIN-BINDING KELCH REPEAT, F-BOX 1* (*FKF1*), blue light receptor, and well-known photoperiodic flowering time regulator [79]. Lup011836 encodes general transcription factor IIH subunit 2 (*GTF2H2*), performing basic functions in transcription and nucleotide excision repair of damaged DNA. Lup002110 has been annotated as a representative of the UDP-glycosyltransferases protein family. One of the *A. thaliana* UDP-glycosyltransferases, *UGT87A2*, was revealed to be involved in the regulation of flowering in vernalization and gibberellin pathways via the flowering repressor *FLC* [80]. UGT87A2 vs. OIW20134.1 protein alignment revealed 96% coverage, 31% identical sites, and 46% positive sites.

As plants in this study were subjected to a moderate vernalization procedure, differences in gene expression resulting from variation in vernalization responsiveness should be reduced as compared with those expected for nonvernalized plants. This reduction might be highlighted by a relatively low LOD value (as compared with other eQTLs from this region) of the major *L. angustifolius* gene underlying vernalization responsiveness, a homolog of *A. thaliana FLOWERING LOCUS T* (*FT*), *LanFTc1* (Lup015264, OIW03334.1, LOD 7.0) [49,81]. As expected, *LanFTc1* revealed negative association with late flowering phenotype. Similar LOD values were also revealed for Lup018024 (OIV92673.1), Lup020573 (OIW03269.1), and Lup018485 (OIW19675.1) genes. Lup018024 encodes a homolog of bZIP transcription factor, *FD*, which mediates signals from the *FT* gene at the shoot apex and promotes plant flowering in general [82]. However, in this study, *FD* expression revealed a positive association with late flowering phenotype. In *Arabidopsis*, the *FT-FD* complex induces the transcription of several floral-promoting genes, such as *SOC1* and *FRUITFULL (FUL)*, which accelerate flowering, as well as *APETALA1 (AP1)* and *LEAFY (LFY)*, which control floral meristem identity [83]. Indeed, Lup018485 revealed similarity to *AGAMOUS-LIKE 8 (AGL8)/ FUL* and *AP1* MADS-box transcription factors and was found to be negatively associated with late flowering phenotype. Lup020573 was annotated as MYB60 transcription factor. MYB transcription factors perform various regulatory functions in plants in responses to biotic and abiotic stresses, development, differentiation, metabolism, defense, etc. [84].

Table 4. Genes showing the highest gene expression association and eQTL peak LOD co-localization (≤2 cM) with vernalization independence *Ku* locus.

Protein	Group	Peak cM	Peak LOD	PVE [1] %	Association Based on *t*-Student Test	Protein Annotation
OIW03171.1	NLL-10	50.8	41.0	92.1	0.85	calcium/calmodulin-regulated receptor-like kinase 1, CRLK1
OIW20567.1	NLL-10	49.7	38.9	65.1	−0.89	uncharacterized protein
OIV96743.1	NLL-10	51.4	18.4	53.5	−0.71	galacturonosyltransferase 10-like, GAUT10
OIW03144.1	NLL-10	51.3	16.7	45.2	0.62	MHM17-10, AT5G56980
OIW03199.1	NLL-10	48.4	16.4	50.5	0.69	general transcription factor IIH subunit 2, GTF2H2
OIW03193.1	NLL-10	50.3	10.6	31.5	−0.53	pentatricopeptide repeat-containing protein
OIW20134.1	NLL-10	49.7	9.7	28.9	0.52	UDP-Glycosyltransferase 85A2, UGT85A2
OIV89838.1	NLL-10	51.4	9.6	30.6	0.51	reticulon family protein
OIW19675.1	NLL-10	50.3	8.9	22.6	−0.56	MADS-box transcription factor AGAMOUS-LIKE 8, AGL8
OIW03269.1	NLL-10	52.0	8.5	24.6	0.51	MYB transcription factor 60, MYB60
OIV92673.1	NLL-10	50.3	7.4	22.2	0.53	protein FD-like, FD
OIW03334.1	NLL-10	49.1	7.0	26.7	−0.51	flowering locus protein T, LanFTc1

[1] PVE, proportion of explained variance.

To summarize, analysis of *Ku* eQTLs shed light on vernalization pathway in *L. angustifolius*, providing transcriptomic evidence for the contribution of several genes acting upstream of *FLC* in *CBF* and *UDP*-glycosyltransferases pathways. The study also revealed transcriptomic contribution of conserved mechanism of *FT-FD* regulon on transition from vegetative to generative development phase in *L. angustifolius*.

2.6. Genes Profiled for Lentus and Tardus Pod Shattering Loci

The recessive *lentus (le)* allele changes the orientation of the sclerified endocarp in the pod, substantially reducing torsional forces after drying [12]. In this study, the following three genes highly associated with *lentus* were identified to have major eQTL peaks localized in the proximity of this locus: Lup018336 (OIW06948.1, LOD 43.0), Lup018348 (OIW06960.1, LOD 17.4), and Lup018228 (OIW06846.1, LOD 13.0) (Table 5, Figure 2c). All these genes originated from the same region at chromosome NLL-08, carrying *lentus*. Lup018336 encoded a homolog of *A. thaliana* fiber protein carrying *DUF1218* domain. The genome of *A. thaliana* contained 15 members of *DUF1218* genes. Members of the *DEAL* subfamily of the *DUF1218* confer bilateral symmetry of *Arabidopsis* leaves by controlling proper coordination of cell proliferation between different domains of the leaf lamina margin [85]. Another group of *DUF1218* genes has been related to secondary cell wall biosynthesis and includes *AtUNKA* (At4g27435), *MODIFYING WALL LIGNIN-1*, and *MODIFYING WALL LIGNIN-2* (At1g31720/MWL-1 and At4g19370/MWL-2) [86–88]. Lup018348 encodes a homolog of *DETOXIFICATION14*, a member of the multidrug and toxic compound extrusion (*MATE* efflux) family [89]. MATE transporters perform various functions including phytohormone transport, secondary metabolite transport, xenobiotic detoxification, aluminium tolerance, disease resistance, tip growth processes, and senescence [90]. Some MATE proteins have been involved in the transport of anthocyanins or proanthocyanidins to vacuoles and in the flavonoid metabolism pathways [91,92]. Anthocyanins are accumulated in cell vacuoles and are responsible for diverse pigmentation from orange to red, purple, and blue [93]. Interestingly, Lup018336 revealed positive gene expression association with pod shattering phenotype, whereas Lup018348 was positively associated with nonshattering pods. These results are in line with the general observation that *le* allele affects a pod pigmentation, resulting in a purplish hue of young pods and a bright yellowish-brown color on the internal surface of mature pods. Lup018348 may be responsible for this pigmentation, whereas Lup018336 for pod shattering in *L. angustifolius*.

Table 5. Genes showing the highest gene expression association and eQTL peak LOD co-localization (≤2 cM) with anthracnose resistance *Lanr1*, white flower color *leucospermus*, soft seededness *mollis*, and pod shattering *tardus* and *lentus* loci.

Trait	Protein	Group	Peak cM	Peak LOD	PVE [1] %	Association Based on *t*-Student Test	Protein Annotation
Lanr1	OIW02433.1	NLL-11	41.7	13.5	38.0	0.57	adenine nucleotide alpha hydrolases-like
Lanr1	OIW02411.1	NLL-11	41.6	9.0	29.0	−0.51	galactosyltransferase family protein
le	OIW06948.1	NLL-08	75.2	43.0	75.8	0.76	fiber protein Fb34, DUF1218
le	OIW06960.1	NLL-08	73.3	17.4	45.8	−0.66	MATE efflux family protein DETOXIFICATION14, DTX14
le	OIW06846.1	NLL-08	75.2	13.0	38.6	0.56	CDP-diacylglycerol–glycerol-3-phosphate 3-phosphatidyltransferase 2, PGPS2
leuc	OIW21684.1	NLL-03	59.1	35.3	75.1	−0.86	ubiquitin-60S ribosomal protein L40A, RPL40A
leuc	OIW15321.1	NLL-03	59.6	20.8	53.4	0.76	nascent polypeptide-associated complex subunit alpha 2, NACA2
leuc	OIW15287.1	NLL-03	59.1	14.7	40.4	0.59	F-box/WD repeat-containing protein
leuc	OIV97389.1	NLL-03	56.1	13.1	36.6	−0.54	protein NRT1/ PTR FAMILY 3.1-like, NPF3.1
leuc	OIV89020.1	NLL-03	59.1	10.4	28.6	0.57	GATA type zinc finger transcription factor, WLIM2a
mol	OIW15058.1	NLL-17	79.0	9.6	34.6	−0.50	FAM32A, 7-dehydrocholesterol reductase, DWARF5
ta	OIW17837.1	NLL-01	34.3	31.8	70.9	−0.79	BolA-like family protein 2, BolA2
ta	OIW17820.1	NLL-01	33.1	18.3	44.6	0.66	G-family ATP-binding ABC transporter 5, ABCG5

[1] PVE, proportion of explained variance.

The recessive *tardus* (*ta*) allele affects the sclerenchyma strips of the dorsal and ventral pod seams, greatly increasing the fusion of two pod halves and moderately hampering their separation when drying [12]. Two genes revealed high association and eQTL peak co-localization with *tardus*, namely Lup002465 (OIW17837.1) and Lup002448 (OIW17820.1) (Table 5, Figure 2d). Lup002465 encodes BolA-like family protein with unknown function. Lup002448 is a G family ATP-binding ABC transporter. Such a transporter in rice (*RCN1*) is required for hypodermal suberization of roots [94]. Similarly, *ABCG1* confers suberin formation in potato tuber periderm [95]. Some ABCG transporters are involved in sclerenchyma fiber development via monolignol transport in lignin biosynthesis pathway [96,97]. Moreover, one of ABCG transporters has been revealed to be involved in the silicon-induced formation of Casparian bands in the exodermis of rice [98]. ABCG transporters also perform other diverse functions, including abiotic and biotic stress responses, however, these examples provide non-negligible support to select Lup002448 as a candidate gene involved in *tardus* trait.

2.7. Gene Related to the Soft Seededness Mollis Allele

Recessive allele *mollis* provides water permeable testa at maturity [14,99]. Seed dormancy in legumes is related to the deposition of phenolics and, hypothetically, development of suberin-impregnated layers of palisade cells as observed in pea and soybean [100,101]. Only one highly associated gene was revealed by eQTL analysis, Lup013985 (OIW15058.1), annotated as a protein FAM32A/7-dehydrocholesterol reductase (homolog of *A. thaliana DWARF5* gene) (Table 5, Figure 2e). Because the sequence homology of these genes between *L. angustifolius* and *Arabidopsis* is quite low, it is difficult to elucidate a particular function by comparative analysis. It can be concluded that it is putatively a gene involved in plant sterol metabolism. Plant sterols are essential structural components that influence biophysical properties of membranes such as permeability and fluidity [102]. Mutation in one of enzymes contributing to steryl glycoside biosynthesis pathway, UDP-Glc:sterol glycosyltransferase, alters embryonic development, seed suberin accumulation, and cutin formation in the seed coat, resulting in abnormal permeability [103]. Recently, it has been evidenced that a maternally deposited endosperm cuticle underlies this seed coat permeability in *A. thaliana* [104]. *Mollis* is also maternally determined and as such is considered to be the most difficult *L. angustifolius* domestication gene for selection by phenotype observation. Lup013985 cannot be considered to be a candidate gene conferring *mollis* allele, because it is located in different chromosome than *mollis* locus, however, it might be considered to be a hypothetical *trans*-regulated component eventually contributing to *mollis* phenotype.

2.8. Genes with eQTL Loci Matching White Flower Color Leucospermus Allele

Recessive *leucospermus* allele confers white flower and bright seed pigmentation in *L. angustifolius*. A similar trait in pea was conferred by a basic-helix-loop-helix (bHLH) transcription factor [105] but eQTL analysis did not highlight any bHLH transcription factor with LOD peak close to *leucospermus* locus. Two genes with expression positively associated with recessive allele revealed eQTL peaks close to *leucospermus*, namely Lup008087 (OIW21684.1) and (Lup017573) OIV97389.1 (Table 5, Figure 2f). Lup008087 encodes ubiquitin-60S ribosomal protein L40 (RPL40A) isoform and is localized in Scaffold_168_4 mapped in this study in linkage group NLL-03 close to *leucospermus*, however, a particular biological function of *RPL40* gene is unknown. Lup017573, from the chromosome NLL-15, revealed similarity to the NRT1 and PTR family proteins. Three eQTLs revealed a negative association with recessive allele, including Lup008084 (OIW15287.1) annotated as F-BOX/WD repeat-containing protein. Interestingly, a single mutation in an F-BOX domain-containing protein, OsFBX310, confers brown hull phenotype in rice resulting from a high content of total flavonoids and anthocyanins [106], however, putatively due to the large evolutionary distance between monocots and dicots, sequence alignment reveled very limited similarity between the OsFBX310 and OIW15287.1 protein sequences.

2.9. Applicability of MACE for Gene-Based Studies

This study is the first report on exploitation of the MACE for *L. angustifolius* genome and transcriptome analysis. As a method for gene expression analysis, the MACE was first used in chronic kidney disease survey [35] and in de novo transcriptome analysis of *Calliphora vicina* pupae [37]. The MACE was also exploited for stem rust transcriptomic response in perennial ryegrass (*Lolium perenne*), highlighting a candidate *LpPg1* resistance gene and yielding numerous SNPs which were further transformed into PCR-based molecular markers [36]. The MACE protocol was also applied in pea (*Pisum sativum*) providing single nucleotide variants subsequently converted into CAPS markers [38]. Furthermore, MACE-based studies in pea resulted in the identification of a new mutant allele of the key nodulation gene *Sym33* [107]. The MACE was also used for transcriptomic profiling of *Phaseolus vulgaris* seeds and *Solanum lycopersicum* pollen [39,40]. The MACE was also exploited for GWAS, tagging several candidate genes for salt stress tolerance in *Triticum aestivum* [41].

Several previous *L. angustifolius* genotyping approaches were based on diversity arrays technology (DArT) profiling. DArT studies have highlighted low genetic diversity in narrow-leafed lupin breeding material as compared with primitive and wild germplasm [108]. This domestication bottleneck resulted from narrow genetic variability of exploited resources and significantly limited adaptation range in this crop [109,110]. The DArT-seq has also been exploited for genome-wide association studies (GWAS) targeting several narrow-leafed lupin phenology and yield traits, but it did not provide any candidate gene with significant associations between a marker and a quantitative trait [111,112].

In this study, the MACE was revealed to be an advantageous technique for marker development and gene expression profiling. The eQTL mapping highlighted numerous genes involved in the vernalization response and alkaloid biosynthesis, providing a valuable contribution for further advancement of knowledge on the complexity of molecular networks controlling these two biological processes. Taking into consideration the recent improvements in deciphering the molecular basis underlying early flowering and low-alkaloid phenotypes, as well as addressing results reported here, *L. angustifolius* can serve as a reference model for such studies across the whole genus. Moreover, information about candidate genes identified in *L. angustifolius* can be translated to other legume species as these processes are generally conserved.

2.10. Recommendations for Improving Narrow-Leafed Lupin As a Crop

During the process of *L. angustifolius* domestication several agronomic traits were identified and transferred into improved germplasm by classical selection approaches. Current breeding materials and cultivars usually carry desired alleles of all major domestication traits in homozygous state (*Ku, iucundus, lentus, tardus, mollis, leucospermus* and *Lanr1*). However, domestication process was highly focused on these traits and resulted in approximately threefold reduction in genome-wide diversity across domesticated accessions as compared with their wild relatives [112]. Further improvement of this species as a crop will require harnessing of primitive germplasm and subsequent reselection of domesticated alleles in the progenies. One of the most challenging issue is related to the influence of global warming on temperature and rainfall patterns in all major areas where lupins are currently cultivated. This issue could be partially resolved by SNP-based selection of wild accessions of narrow-leafed lupin with well-established local adaptation to warm and dry climate of the eastern Mediterranean basin [111]. Novel opportunities for reducing the time required for transition between phenological phases could also be uncovered by exploitation of natural variability in genes from vernalization and cold pathways highlighted in this study, particularly *LanFTc1, CRLK1, FD, UGT85A2, GAUT10,* and *MYB60*. Moreover, a common issue related to dry and warm weather patterns, which are expected to occur more frequently due to changing climate, is pod dehiscence. Identified candidate genes for *lentus* (a homolog carrying DUF1218 domain) and *tardus* (an *ABCG5* transporter) await further genotypic and phenotypic exploration in wide genetic background because mapping population represents only a small fraction of diversity existing in *L. angustifolius* germplasm.

3. Materials and Methods

3.1. Plant Material

The reference 83A:476 × P27255 recombinant inbred line (RIL) population (n = 89, F$_8$) of *L. angustifolius* [27] delivered by the Department of Agriculture and Food Western Australia was used in the study. This population was developed from a cross between a domesticated Australian breeding line (83A:476) and a wild accession from Morocco (P27255). The line P27255 is late flowering and vernalization-responsive (recessive allele *ku*), pod shattering (dominant alleles *Tardus* and *Lentus*), hard seeded (*Mollis*), blue flower and dark seed color *Leucospermus*), high alkaloid (*Iucundus*) and anthracnose susceptible (*lanr1*). 83A:476 has an opposite allele combination (*Ku, tardus, lentus, mollis, leucospermus, iucundus, Lanr1*). Both parental lines are homozygous in relation to these alleles.

3.2. Controlled Environment Experiment

Seeds of mapping population and parental lines (83A:476 × P27255) were vernalized for 16 days at 4 °C in darkness on Petri dishes with moist filter paper. Filter paper (Chemland, Stargard, Poland) was changed every four days to maintain phytosanitary conditions. Following vernalization, plants were transferred to pots (2 plants per 11 cm × 11 cm pot, about 8 cm between plants) and grown in controlled conditions (photoperiod 16 h, temperature +25 °C day and +18 °C night) at the Wielkopolska Center of Advanced Technologies in Poznań, Poland. Tissue was sampled from young leaves two times a day, 4 h after beginning of photoperiod and 1 h before the end of photoperiod on the 28th, 36th, and 44th day from sowing. Five biological replicates were collected.

3.3. Massive Analysis of cDNA Ends

Frozen plant tissue (50 mg, −80 °C) was homogenized using TissueLyser II (Qiagen, Hilden, Germany) and two stainless steel beads (ø 5 mm) placed in a 2 mL tube (Eppendorf, Hamburg, Germany). RNA isolation was performed using SV Total RNA Isolation System (Promega, Madison, WI, USA) according to the protocol. The concentration of RNA was measured using NanoDrop 2000 (ThermoFisher Scientific, Waltham, MA, USA) and A260/A280 ratio. RNA quality was visualized by 1% agarose gel electrophoresis (1X TAE) of denatured samples. RNA concentration was equalized to 400 ng/μL in nuclease-free water. Samples from particular line (representing 5 terms and 5 biological replicates) were bulked together in equal aliquots. 10 μL of mixture (4 μg of RNA) was subjected to the MACE protocol. The MACE profiling and SNP calling was outsourced (GenXPro, Frankfurt, Germany). The MACE reads were aligned to the *L. angustifolius* genome assembly [34] (http://www.lupinexpress.org). Normalization procedure was as follows [113]: The average raw count of each gene within a library was divided by the geometric mean of all counts in all samples and the median of the quotients was calculated per library. Each raw count was then divided by the library-specific median value.

3.4. Molecular Markers and Linkage Mapping

The 83A:476-like scores were assigned as "a", the P27255-like scores as "b", and the heterozygotes as "h". If several cosegregating MACE markers in particular gene were identified, the marker with the lowest percentage of missing data was chosen to infer consensus segregation representing each cluster. To provide a mapping file, heterozygote scores were removed. Accepted missing data threshold was 11%. Chi-square (χ^2) values for Mendelian segregation were estimated using the expected 1:1 ratio. The calculation of probability was based on χ^2 and 2 degrees of freedom. Based on the segregation distortion observed in recently published *L. angustifolius* linkage map versions [29,30,34], χ^2 *p*-value threshold of 1×10^{-7} was applied.

Segregation data for domestication traits and tightly linked SSR-derived markers were included in the study [13,21–23,25,52,53]. Moreover, to provide landmarks for chromosome map integration, recently published BAC-derived markers were incorporated [43–47,49,51]. Additionally, novel markers were developed using BAC-end sequences. PCR primers were designed using Primer3′lus [114]. Amplification was performed using DNA isolated from the parental lines of the *L. angustifolius* mapping population, 83A:476 and P27255. Amplicons were extracted directly from the post-reaction mixtures (QIAquick PCR Purification Kit; Qiagen) and sequenced using ABI PRISM 3130 Genetic Analyzer XL (Applied Biosystems, Hitachi, Tokyo, Japan). Allele-specific PCR (AS-PCR) polymorphisms were visualized by 1% agarose gel electrophoresis, whereas nucleotide substitution polymorphisms were revealed by the cleaved amplified polymorphic sequence (CAPS) and dCAPS approaches [115,116]. Restriction sites were identified using dCAPS Finder 2.0 and SNP2dCAPS [117,118]. Digestion products were separated by 2% agarose gel electrophoresis. Multipoint mapping (JoinMap 5, Kyazma, Wageningen, Netherlands) was performed after grouping under independence LOD of 9.5. Some inconsistency in segregation patterns were observed between previously published and newly developed marker sets. In such cases, marker segregation was tested using current DNA isolates (if possible) or questionable data was deleted from segregation file. Linkage group optimization was performed according to the procedure previously applied for white lupin by [54].

3.5. Expression Quantitative Trait Loci Mapping

Normalized gene expression values (continuous traits) obtained for RILs and mapping population parental lines were associated with *Ku*, *tardus*, *lentus*, *mollis*, *leucospermus*, *iucundus*, and *Lanr1* segregation data (binary trait) by t-Student test in two classes of polymorphism. Obtained p-values were false discovery rate (FDR) corrected [119]. Genes with corrected *p*-value ≤ 0.01 were subjected to composite interval mapping performed in Windows QTL Cartographer V2.5 (North Carolina State University, Raleigh, NC, USA) using 5 background control markers, window size 10 cM, walk speed 0.5 cM, and backward regression method. LOD threshold for QTL calling was established by permutation test ($N = 1000$) using the same parameters. Linkage groups and LOD graphs were drawn in MapChart [120]. Moreover, sets of genes with corrected *p*-value ≤ 0.01 were analyzed for gene ontology (GO) term enrichment by hypergeometric test with FDR correction in Bingo [121] using GO annotation of *L. angustifolius* genes obtained from Ensembl Plants Genes database (rel. 45, genome assembly LupAngTanjil v1.0). Whole-genome annotation was used as reference set. Results were provided as −log10 (corrected *p*-value).

4. Conclusions

1. The massive analysis of cDNA ends was revealed to be applicable for molecular marker development and linkage map construction, as well as for gene expression evaluation and expression quantitative trait loci mapping.

2. The analysis of vernalization independence *Ku* locus shed light on vernalization response via *FLOWERING LOCUS T* and *FD* regulon, providing transcriptomic evidence for contribution of several genes acting in C-repeat binding factor (*CBF*) cold responsiveness and in *UDP*-glycosyltransferases pathways. This information can be relevant to decipher vernalization pathway in legumes, because legume genomes do not contain a major vernalization-responsive gene *FLOWERING LOCUS C* (*FLC*) but other genes from this pathway, including activators and repressors of *FLC*, are present.

3. The study of low-alkaloid *iucundus* locus highlighted a high number of *cis*- and *trans*-regulated alkaloid biosynthesis genes with gene expression orchestrated by a regulatory agent localized at *iucundus* locus, supporting the concept that the *ETHYLENE RESPONSIVE TRANSCRIPTION FACTOR RAP2-7* gene may control low-alkaloid phenotype in narrow-leafed lupin.

4. Research on reduced pod shattering *lentus* locus selected a *DUF1218* domain homolog as a candidate gene controlling the orientation of the sclerified endocarp and a *DETOXIFICATION14* homolog for purplish hue of young pods.

5. An *ABCG* transporter gene was identified as a hypothetical contributor to sclerenchyma fortification underlying reduced pod shattering *tardus* locus.

Supplementary Materials: Supplementary materials can be found at http://www.mdpi.com/1422-0067/20/22/5670/s1.

Author Contributions: Conceptualization, M.K. and B.W.; methodology, M.K, B.W., and P.P.; software, P.P. and M.K.; validation, M.K. and P.P.; formal analysis, M.K and P.P.; investigation, S.R.-B., E.R., P.P., and M.K.; resources, S.R.-B., E.R., M.K., and P.P.; data curation, M.K and P.P.; writing—original draft preparation, M.K.; visualization, M.K.; supervision, B.W.; project administration, B.W. and M.K.; funding acquisition, B.W.

Funding: This research was funded by the Polish National Centre for Research and Development (Narodowe Centrum Badań i Rozwoju) grant number PBS3/A8/28/2015, SEGENMAS. The APC was funded by the Institute of Plant Genetics, Polish Academy of Sciences.

Acknowledgments: Authors thank Krzysztof Pudełko from the Wielkopolska Center of Advanced Technologies in Poznań, Poland for assistance in plant experiment. Moreover, authors acknowledge Magdalena Tomaszewska from the Department of Genomics of the Institute of Plant Genetics, Polish Academy of Sciences for help in plant sowing and leaf sampling, and Paweł Krajewski from the Department of Biometry and Bioinformatics, Institute of Plant Genetics, Polish Academy of Sciences, for help with MACE data processing and GO analyses. Authors would also like to thank Björn Rotter and Peter Winter from GenXPro GmBH, Frankfurt Main, Germany for executing MACE protocol, MACE marker identification, and production of annotated sequence read count and haplotype datasets. Part of the computations were performed at the Poznań Supercomputing and Networking Center (www.man.poznan.pl).

Abbreviations

2OG	DMR6-like OXYGENASE 2
AAP	Amino acid permease
ABCG5	G-family ATP-binding ABC transporter 5
AGL8	AGAMOUS-LIKE 8
AK1	Aspartate kinase 1
AP1	APETALA1
ASDH	Aspartate-semialdehyde dehydrogenase
BAC	Bacterial artificial chromosome
BAN	Anthocyanidin reductase
BolA2	BolA-like family protein 2
CBF	C-repeat binding factor
CCR1	Cinnamoyl-CoA reductase 1

CRLK1	CALCIUM/CALMODULIN-REGULATED RECEPTOR-LIKE KINASE 1
CXE1	Carboxylesterase 1
DAPDC1	Diaminopimelate decarboxylase 1
DapL	LL-diaminopimelate aminotransferase
DArT	Diversity Arrays Technology
DFR	Dihydroflavonol 4-reductase
DHDPS	4-Hydroxy-tetrahydrodipicolinate synthase
DTX14	MATE efflux family protein DETOXIFICATION14
DUF1218	Fiber protein Fb34, domain of unknown function 1218
eQTL	Expression quantitative trait locus
FKF1	FLAVIN-BINDING KELCH REPEAT, F-BOX 1
FLC	FLOWERING LOCUS C
FSD2	Fe superoxide dismutase 2
FT	FLOWERING LOCUS T
FUL	FRUITFULL
GAUT10	Galacturonosyltransferase 10-like
GLT1	Glutamate synthase 1
GO	GENE Gene ontology
GTF2H2	General transcription factor IIH subunit 2
GWAS	Genome-wide association study
ICE1	INDUCER OF CBF EXPRESSION 1
KAB1	Voltage-gated potassium channel subunit beta 1
KNAT1	Homeobox protein knotted-1-like
LaAT	HXXXD-type ACYL-TRANSFERASE
LanFTc1	*L. angustifolius* FLOWERING LOCUS T c1
LDC	Lysine/ornithine decarboxylase
LFY	LEAFY
MACE	Massive analysis of cDNA ends
MATE	Multidrug and toxic compound extrusion
MFLP	Molecular fragment length polymorphism
MHK10.21	Copper amine oxidase
MLP31	MAJOR LATEX PROTEIN 31
MLP423	MAJOR LATEX PROTEIN 423
MWL-1	MODIFYING WALL LIGNIN-1
MWL-2	MODIFYING WALL LIGNIN-2
MYB60	MYB transcription factor 60
NACA2	Nascent polypeptide-associated complex subunit alpha 2
NPF3.1	Protein NRT1/ PTR FAMILY 3.1-like
PGPS2	CDP-diacylglycerol–glycerol-3-phosphate 3-phosphatidyltransferase 2
PUP	Purine permease transporter
RAP2-7	ETHYLENE RESPONSIVE TRANSCRIPTION FACTOR RAP2-7
RIL	Recombinant inbred line
RPL40A	Ubiquitin-60S ribosomal protein L40A
SDR	Short-chain dehydrogenase reductase
SNP	Single nucleotide polymorphism
SOC1	SUPPRESSOR OF OVEREXPRESSION OF CONSTANS 1
UGT85A2	UDP-Glycosyltransferase 85A2
UGT87A2	UDP-Glycosyltransferase 87A2

References

1. Święcicki, W.; Święcicki, W.K. Domestication and breeding improvement of narrow-leafed lupin (*L. angustifolius* L.). *J. Appl. Genet.* **1995**, *36*, 155–167.
2. Landers, K.F. Vernalization responses in narrow-leafed lupin (*Lupinus angustifolius*) genotypes. *Aust. J. Agric. Res.* **1995**, *46*, 1011–1025. [CrossRef]

3. Adhikari, K.N.; Buirchell, B.J.; Sweetingham, M.W. Length of vernalization period affects flowering time in three lupin species. *Plant Breed.* **2012**, *131*, 631–636. [CrossRef]

4. Gladstones, J.; Hill, G. Selection for economic characters in *Lupinus angustifolius* and *L. digitatus*. 2. Time of flowering. *Aust. J. Exp. Agric.* **1969**, *9*, 213–220. [CrossRef]

5. Wink, M.; Meißner, C.; Witte, L. Patterns of quinolizidine alkaloids in 56 species of the genus *Lupinus*. *Phytochemistry* **1995**, *38*, 139–153. [CrossRef]

6. Bassoli, A.; Borgonovo, G.; Busnelli, G. Alkaloids and the bitter taste. In *Modern Alkaloids, Fattorusso, E., Taglialatela-Scafati, O.*; Wiley-VCH: Weinheim, Germany, 2007; pp. 53–72.

7. Matsuura, H.N.; Fett-Neto, A.G. Plant alkaloids: Main features, toxicity, and mechanisms of action. In *Plant Toxins*; Gopalakrishnakone, P., Carlini, C.R., Ligabue-Braun, R., Eds.; Springer: Dordrecht, The Netherlands, 2015; pp. 1–15.

8. Gladstones, J.S. Lupins as crop plants. *Field Crop Abstracts* **1970**, *23*, 26.

9. Von Sengbusch, R. Süßlupinen und Öllupinen. Die Entstehungsgeschichte einiger neuer Kulturpflanzen. *Landwirtsch. Jahrbücher* **1942**, *91*, 719–880.

10. Kamel, K.A.; Święcicki, W.; Kaczmarek, Z.; Barzyk, P. Quantitative and qualitative content of alkaloids in seeds of a narrow-leafed lupin (*Lupinus angustifolius* L.) collection. *Genet. Resour. Crop Evol.* **2016**, *63*, 711–719. [CrossRef]

11. Frick, K.M.; Kamphuis, L.G.; Siddique, K.H.; Singh, K.B.; Foley, R.C. Quinolizidine alkaloid biosynthesis in lupins and prospects for grain quality improvement. *Front. Plant Sci.* **2017**, *8*, 87. [CrossRef]

12. Gladstones, J. Selection for economic characters in *Lupinus angustifolius* and *L. digitatus*. *Aust. J. Exp. Agric.* **1967**, *7*, 360–366. [CrossRef]

13. Boersma, J.G.; Buirchell, B.J.; Sivasithamparam, K.; Yang, H. Development of two sequence-specific PCR markers linked to the *le* gene that reduces pod shattering in narrow-leafed lupin (*Lupinus angustifolius* L.). *Genet. Mol. Biol.* **2007**, *30*, 623–629. [CrossRef]

14. Mikolajczyk, J. Genetic studies in *Lupinus angustifolius*. 2. Inheritance of some morphological characters in blue lupine. *Genet. Pol.* **1966**, *7*, 153–180.

15. Li, X.; Buirchell, B.; Yan, G.; Yang, H. A molecular marker linked to the *mollis* gene conferring soft-seediness for marker-assisted selection applicable to a wide range of crosses in lupin (*Lupinus angustifolius* L.) breeding. *Mol. Breed.* **2012**, *29*, 361–370. [CrossRef]

16. Nirenberg, H.I.; Feiler, U.; Hagedorn, G. Description of *Colletotrichum lupini* comb. nov. in modern terms. *Mycologia* **2002**, *94*, 307–320. [CrossRef] [PubMed]

17. Yang, H.; Boersma, J.G.; You, M.; Buirchell, B.J.; Sweetingham, M.W. Development and implementation of a sequence-specific PCR marker linked to a gene conferring resistance to anthracnose disease in narrow-leafed lupin (*Lupinus angustifolius* L.). *Mol. Breed.* **2004**, *14*, 145–151. [CrossRef]

18. Yang, H.; Renshaw, D.; Thomas, G.; Buirchell, B.; Sweetingham, M. A strategy to develop molecular markers applicable to a wide range of crosses for marker assisted selection in plant breeding: A case study on anthracnose disease resistance in lupin (*Lupinus angustifolius* L.). *Mol. Breed.* **2008**, *21*, 473–483. [CrossRef]

19. Fischer, K.; Dieterich, R.; Nelson, M.N.; Kamphuis, L.G.; Singh, K.B.; Rotter, B.; Krezdorn, N.; Winter, P.; Wehling, P.; Ruge-Wehling, B. Characterization and mapping of *LanrBo*: A locus conferring anthracnose resistance in narrow-leafed lupin (*Lupinus angustifolius* L.). *Theor. Appl. Genet.* **2015**, *128*, 2121–2130. [CrossRef]

20. Yang, H.; Sweetingham, M.W.; Cowling, W.A.; Smith, P.M.C. DNA fingerprinting based on microsatellite-anchored fragment length polymorphisms, and isolation of sequence-specific PCR markers in lupin (*Lupinus angustifolius* L.). *Mol. Breed.* **2001**, *7*, 203–209. [CrossRef]

21. Boersma, J.G.; Pallotta, M.; Li, C.; Buirchell, B.J.; Sivasithamparam, K.; Yang, H. Construction of a genetic linkage map using MFLP and identification of molecular markers linked to domestication genes in narrow-leafed lupin (*Lupinus angustifolius* L.). *Cell. Mol. Biol. Lett.* **2005**, *10*, 331–344.

22. Li, X.; Yang, H.; Buirchell, B.; Yan, G. Development of a DNA marker tightly linked to low-alkaloid gene *iucundus* in narrow-leafed lupin (*Lupinus angustifolius* L.) for marker-assisted selection. *Crop Pasture Sci.* **2011**, *62*, 218–224. [CrossRef]

23. Boersma, J.G.; Buirchell, B.J.; Sivasithamparam, K.; Yang, H. Development of a sequence-specific PCR marker linked to the *Ku* gene which removes the vernalization requirement in narrow-leafed lupin. *Plant Breed.* **2007**, *126*, 306–309. [CrossRef]

24. Li, X.; Yang, H.; Yan, G. Development of a co-dominant DNA marker linked to the gene *lentus* conferring reduced pod shattering for marker-assisted selection in narrow-leafed lupin (*Lupinus angustifolius*) breeding. *Plant Breed.* **2012**, *131*, 540–544. [CrossRef]

25. Boersma, J.G.; Nelson, M.N.; Sivasithamparam, K.; Yang, H.A. Development of sequence-specific PCR markers linked to the *Tardus* gene that reduces pod shattering in narrow-leafed lupin (*Lupinus angustifolius* L.). *Mol. Breed.* **2009**, *23*, 259–267. [CrossRef]

26. Li, X.; Renshaw, D.; Yang, H.; Yan, G. Development of a co-dominant DNA marker tightly linked to gene *tardus* conferring reduced pod shattering in narrow-leafed lupin (*Lupinus angustifolius* L.). *Euphytica* **2010**, *176*, 49–58. [CrossRef]

27. Nelson, M.N.; Phan, H.T.T.; Ellwood, S.R.; Moolhuijzen, P.M.; Hane, J.; Williams, A.; O'Lone, C.E.; Fosu-Nyarko, J.; Scobie, M.; Cakir, M.; et al. The first gene-based map of *Lupinus angustifolius* L.-location of domestication genes and conserved synteny with *Medicago truncatula*. *Theor. Appl. Genet.* **2006**, *113*, 225–238. [CrossRef] [PubMed]

28. Nelson, M.N.; Moolhuijzen, P.M.; Boersma, J.G.; Chudy, M.; Lesniewska, K.; Bellgard, M.; Oliver, R.P.; Swiecicki, W.; Wolko, B.; Cowling, W.A.; et al. Aligning a new reference genetic map of *Lupinus angustifolius* with the genome sequence of the model legume, *Lotus japonicus*. *DNA Res.* **2010**, *17*, 73–83. [CrossRef] [PubMed]

29. Kamphuis, L.G.; Hane, J.K.; Nelson, M.N.; Gao, L.; Atkins, C.A.; Singh, K.B. Transcriptome sequencing of different narrow-leafed lupin tissue types provides a comprehensive uni-gene assembly and extensive gene-based molecular markers. *Plant Biotechnol. J.* **2015**, *13*, 14–25. [CrossRef]

30. Zhou, G.; Jian, J.; Wang, P.; Li, C.; Tao, Y.; Li, X.; Renshaw, D.; Clements, J.; Sweetingham, M.; Yang, H. Construction of an ultra-high density consensus genetic map, and enhancement of the physical map from genome sequencing in *Lupinus angustifolius*. *Theor. Appl. Genet.* **2018**, *131*, 209–223. [CrossRef]

31. Kasprzak, A.; Safár, J.; Janda, J.; Dolezel, J.; Wolko, B.; Naganowska, B. The bacterial artificial chromosome (BAC) library of the narrow-leafed lupin (*Lupinus angustifolius* L.). *Cell. Mol. Biol. Lett.* **2006**, *11*, 396–407. [CrossRef]

32. Gao, L.-L.; Hane, J.K.; Kamphuis, L.G.; Foley, R.; Shi, B.-J.; Atkins, C.A.; Singh, K.B. Development of genomic resources for the narrow-leafed lupin (*Lupinus angustifolius*): Construction of a bacterial artificial chromosome (BAC) library and BAC-end sequencing. *BMC Genom.* **2011**, *12*, 521. [CrossRef]

33. Yang, H.; Tao, Y.; Zheng, Z.; Zhang, Q.; Zhou, G.; Sweetingham, M.W.; Howieson, J.G.; Li, C. Draft genome sequence, and a sequence-defined genetic linkage map of the legume crop species *Lupinus angustifolius* L. *PLoS ONE* **2013**, *8*, e64799. [CrossRef] [PubMed]

34. Hane, J.K.; Ming, Y.; Kamphuis, L.G.; Nelson, M.N.; Garg, G.; Atkins, C.A.; Bayer, P.E.; Bravo, A.; Bringans, S.; Cannon, S.; et al. A comprehensive draft genome sequence for lupin (*Lupinus angustifolius*), an emerging health food: Insights into plant-microbe interactions and legume evolution. *Plant Biotechnol. J.* **2017**, *15*, 318–330. [CrossRef] [PubMed]

35. Zawada, A.M.; Rogacev, K.S.; Muller, S.; Rotter, B.; Winter, P.; Fliser, D.; Heine, G.H. Massive analysis of cDNA Ends (MACE) and miRNA expression profiling identifies proatherogenic pathways in chronic kidney disease. *Epigenetics* **2014**, *9*, 161–172. [CrossRef] [PubMed]

36. Bojahr, J.; Nhengiwa, O.; Krezdorn, N.; Rotter, B.; Saal, B.; Ruge-Wehling, B.; Struck, C.; Winter, P. Massive analysis of cDNA ends (MACE) reveals a co-segregating candidate gene for *LpPg1* stem rust resistance in perennial ryegrass (*Lolium perenne*). *Theor. Appl. Genet.* **2016**, *129*, 1915–1932. [CrossRef]

37. Zajac, B.K.; Amendt, J.; Horres, R.; Verhoff, M.A.; Zehner, R. De novo transcriptome analysis and highly sensitive digital gene expression profiling of *Calliphora vicina* (Diptera: Calliphoridae) pupae using MACE (Massive Analysis of cDNA Ends). *Forensic Science International: Genetics* **2015**, *15*, 137–146. [CrossRef]

38. Zhernakov, A.; Rotter, B.; Winter, P.; Borisov, A.; Tikhonovich, I.; Zhukov, V. Massive Analysis of cDNA Ends (MACE) for transcript-based marker design in pea (*Pisum sativum* L.). *Genom. Data* **2017**, *11*, 75–76. [CrossRef]

39. Keller, M.; Consortium, S.-I.; Simm, S. The coupling of transcriptome and proteome adaptation during development and heat stress response of tomato pollen. *BMC Genom.* **2018**, *19*, 447. [CrossRef]

40. Parreira, J.R.; Balestrazzi, A.; Fevereiro, P.; Araujo, S.S. Maintaining genome integrity during seed development in *Phaseolus vulgaris* L.: Evidence from a transcriptomic profiling study. *Genes* **2018**, *9*, 463. [CrossRef]

41. Oyiga, B.C.; Sharma, R.C.; Baum, M.; Ogbonnaya, F.C.; Léon, J.; Ballvora, A. Allelic variations and differential expressions detected at quantitative trait loci for salt stress tolerance in wheat. *Plant Cell Environ.* **2018**, *41*, 919–935. [CrossRef]

42. Kaczmarek, A.; Naganowska, B.; Wolko, B. Karyotyping of the narrow-leafed lupin (*Lupinus angustifolius* L.) by using FISH, PRINS and computer measurements of chromosomes. *J. Appl. Genet.* **2009**, *50*, 77–82. [CrossRef]

43. Wyrwa, K.; Książkiewicz, M.; Szczepaniak, A.; Susek, K.; Podkowiński, J.; Naganowska, B. Integration of *Lupinus angustifolius* L. (narrow-leafed lupin) genome maps and comparative mapping within legumes. *Chromosome Res.* **2016**, *24*, 355–378. [CrossRef] [PubMed]

44. Przysiecka, Ł.; Książkiewicz, M.; Wolko, B.; Naganowska, B. Structure, expression profile and phylogenetic inference of chalcone isomerase-like genes from the narrow-leafed lupin (*Lupinus angustifolius* L.) genome. *Front. Plant Sci.* **2015**, *6*, 268. [CrossRef] [PubMed]

45. Książkiewicz, M.; Zielezinski, A.; Wyrwa, K.; Szczepaniak, A.; Rychel, S.; Karlowski, W.; Wolko, B.; Naganowska, B. Remnants of the legume ancestral genome preserved in gene-rich regions: Insights from *Lupinus angustifolius* physical, genetic, and comparative mapping. *Plant Mol. Biol. Rep.* **2015**, *33*, 84–101. [CrossRef] [PubMed]

46. Książkiewicz, M.; Wyrwa, K.; Szczepaniak, A.; Rychel, S.; Majcherkiewicz, K.; Przysiecka, Ł.; Karlowski, W.; Wolko, B.; Naganowska, B. Comparative genomics of *Lupinus angustifolius* gene-rich regions: BAC library exploration, genetic mapping and cytogenetics. *BMC Genom.* **2013**, *14*, 79. [CrossRef] [PubMed]

47. Leśniewska, K.; Książkiewicz, M.; Nelson, M.N.; Mahé, F.; Aïnouche, A.; Wolko, B.; Naganowska, B. Assignment of 3 genetic linkage groups to 3 chromosomes of narrow-leafed lupin. *J. Hered.* **2011**, *102*, 228–236. [CrossRef]

48. Szczepaniak, A.; Książkiewicz, M.; Podkowiński, J.; Czyż, K.B.; Figlerowicz, M.; Naganowska, B. Legume cytosolic and plastid acetyl-coenzyme-A carboxylase genes differ by evolutionary patterns and selection pressure schemes acting before and after whole-genome duplications. *Genes* **2018**, *9*, 563. [CrossRef]

49. Nelson, M.N.; Książkiewicz, M.; Rychel, S.; Besharat, N.; Taylor, C.M.; Wyrwa, K.; Jost, R.; Erskine, W.; Cowling, W.A.; Berger, J.D.; et al. The loss of vernalization requirement in narrow-leafed lupin is associated with a deletion in the promoter and de-repressed expression of a *Flowering Locus T* (*FT*) homologue. *New Phytol.* **2017**, *213*, 220–232. [CrossRef]

50. Narożna, D.; Książkiewicz, M.; Przysiecka, Ł.; Króliczak, J.; Wolko, B.; Naganowska, B.; Mądrzak, C.J. Legume isoflavone synthase genes have evolved by whole-genome and local duplications yielding transcriptionally active paralogs. *Plant Sci.* **2017**, *264*, 149–167. [CrossRef]

51. Książkiewicz, M.; Rychel, S.; Nelson, M.N.; Wyrwa, K.; Naganowska, B.; Wolko, B. Expansion of the phosphatidylethanolamine binding protein family in legumes: A case study of *Lupinus angustifolius* L. *FLOWERING LOCUS T* homologs, *LanFTc1* and *LanFTc2*. *BMC Genom.* **2016**, *17*, 820.

52. Boersma, J.G.; Buirchell, B.J.; Sivasithamparam, K.; Yang, H. Development of a PCR marker tightly linked to *mollis*, the gene that controls seed dormancy in *Lupinus angustifolius* L. *Plant Breed.* **2007**, *126*, 612–616. [CrossRef]

53. You, M.; Boersma, J.G.; Buirchell, B.J.; Sweetingham, M.W.; Siddique, K.H.M.; Yang, H. A PCR-based molecular marker applicable for marker-assisted selection for anthracnose disease resistance in lupin breeding. *Cell. Mol. Biol. Lett.* **2005**, *10*, 123–134. [PubMed]

54. Książkiewicz, M.; Nazzicari, N.; Yang, H.A.; Nelson, M.N.; Renshaw, D.; Rychel, S.; Ferrari, B.; Carelli, M.; Tomaszewska, M.; Stawiński, S.; et al. A high-density consensus linkage map of white lupin highlights synteny with narrow-leafed lupin and provides markers tagging key agronomic traits. *Sci. Rep.* **2017**, *7*, 15335. [CrossRef] [PubMed]

55. Iqbal, M.M.; Huynh, M.; Udall, J.A.; Kilian, A.; Adhikari, K.N.; Berger, J.D.; Erskine, W.; Nelson, M.N. The first genetic map for yellow lupin enables genetic dissection of adaptation traits in an orphan grain legume crop. *BMC Genet.* **2019**, *20*, 68. [CrossRef] [PubMed]

56. Rychel, S.; Książkiewicz, M. Development of gene-based molecular markers tagging low alkaloid *pauper* locus in white lupin (*Lupinus albus* L.). *J. Appl. Genet.* **2019**, *60*, 269–281. [CrossRef] [PubMed]

57. Bunsupa, S.; Katayama, K.; Ikeura, E.; Oikawa, A.; Toyooka, K.; Saito, K.; Yamazaki, M. Lysine decarboxylase catalyzes the first step of quinolizidine alkaloid biosynthesis and coevolved with alkaloid production in Leguminosae. *Plant Cell* **2012**, *24*, 1202–1216. [CrossRef] [PubMed]

58. Frick, K.M.; Foley, R.C.; Kamphuis, L.G.; Siddique, K.H.M.; Garg, G.; Singh, K.B. Characterization of the genetic factors affecting quinolizidine alkaloid biosynthesis and its response to abiotic stress in narrow-leafed lupin (*Lupinus angustifolius* L.). *Plant Cell Environ.* **2018**, *41*, 2155–2168. [CrossRef]

59. Kroc, M.; Koczyk, G.; Kamel, K.A.; Czepiel, K.; Fedorowicz-Strońska, O.; Krajewski, P.; Kosińska, J.; Podkowiński, J.; Wilczura, P.; Święcicki, W. Transcriptome-derived investigation of biosynthesis of quinolizidine alkaloids in narrow-leafed lupin (*Lupinus angustifolius* L.) highlights candidate genes linked to *iucundus* locus. *Sci. Rep.* **2019**, *9*, 2231. [CrossRef]

60. Kroc, M.; Czepiel, K.; Wilczura, P.; Mokrzycka, M.; Swiecicki, W. Development and validation of a gene-targeted dCAPS marker for marker-assisted selection of low-alkaloid content in seeds of narrow-leafed lupin (*Lupinus angustifolius* L.). *Genes* **2019**, *10*, 428. [CrossRef]

61. Kato, K.; Shoji, T.; Hashimoto, T. Tobacco nicotine uptake permease regulates the expression of a key transcription factor gene in the nicotine biosynthesis pathway. *Plant Physiol.* **2014**, *166*, 2195–2204. [CrossRef]

62. Hildreth, S.B.; Gehman, E.A.; Yang, H.; Lu, R.-H.; Ritesh, K.C.; Harich, K.C.; Yu, S.; Lin, J.; Sandoe, J.L.; Okumoto, S.; et al. Tobacco nicotine uptake permease (NUP1) affects alkaloid metabolism. *Proc. Natl. Acad. Sci. USA* **2011**, *108*, 18179–18184. [CrossRef]

63. Yang, T.; Nagy, I.; Mancinotti, D.; Otterbach, S.L.; Andersen, T.B.; Motawia, M.S.; Asp, T.; Geu-Flores, F. Transcript profiling of a bitter variety of narrow-leafed lupin to discover alkaloid biosynthetic genes. *J. Exp. Bot.* **2017**, *68*, 5527–5537. [CrossRef] [PubMed]

64. Ha, J.; Kang, Y.-G.; Lee, T.; Kim, M.; Yoon, M.Y.; Lee, E.; Yang, X.; Kim, D.; Kim, Y.-J.; Lee, T.R.; et al. Comprehensive RNA sequencing and co-expression network analysis to complete the biosynthetic pathway of coumestrol, a phytoestrogen. *Sci. Rep.* **2019**, *9*, 1934. [CrossRef] [PubMed]

65. Goldsbrough, A.; Belzile, F.; Yoder, J.I. Complementation of the tomato anthocyanin without (aw) mutant using the dihydroflavonol 4-reductase gene. *Plant Physiol.* **1994**, *105*, 491–496. [CrossRef] [PubMed]

66. Bunsupa, S.; Okada, T.; Saito, K.; Yamazaki, M. An acyltransferase-like gene obtained by differential gene expression profiles of quinolizidine alkaloid-producing and nonproducing cultivars of *Lupinus angustifolius*. *Plant Biotechnol.* **2011**, *28*, 89–94. [CrossRef]

67. Wink, M. N-Methylation of quinolizidine alkaloids: An S-adenosyl-L-methionine: Cytisine N-methyltransferase from *Laburnum anagyroides* plants and cell cultures of *L. alpinum* and *Cytisus canariensis*. *Planta* **1984**, *161*, 339–344. [CrossRef]

68. Liscombe, D.K.; Usera, A.R.; O'Connor, S.E. Homolog of tocopherol C methyltransferases catalyzes N methylation in anticancer alkaloid biosynthesis. *Proc. Natl. Acad. Sci. USA* **2010**, *107*, 18793–18798. [CrossRef]

69. Li, Y.; Smolke, C.D. Engineering biosynthesis of the anticancer alkaloid noscapine in yeast. *Nat. Commun.* **2016**, *7*, 12137. [CrossRef]

70. Barker, S.J.; Si, P.; Hodgson, L.; Ferguson-Hunt, M.; Khentry, Y.; Krishnamurthy, P.; Averis, S.; Mebus, K.; O'Lone, C.; Dalugoda, D.; et al. Regeneration selection improves transformation efficiency in narrow-leaf lupin. *Plant Cell Tissue Organ Cult.* **2016**, *126*, 219–228. [CrossRef]

71. Yang, T.; Chaudhuri, S.; Yang, L.; Du, L.; Poovaiah, B.W. A calcium/calmodulin-regulated member of the receptor-like kinase family confers cold tolerance in plants. *J. Biol. Chem.* **2010**, *285*, 7119–7126. [CrossRef]

72. Kim, H.-J.; Hyun, Y.; Park, J.-Y.; Park, M.-J.; Park, M.-K.; Kim, M.D.; Kim, H.-J.; Lee, M.H.; Moon, J.; Lee, I.; et al. A genetic link between cold responses and flowering time through *FVE* in *Arabidopsis thaliana*. *Nat. Genet.* **2004**, *36*, 167–171. [CrossRef]

73. Lee, J.-H.; Jung, J.-H.; Park, C.-M. *INDUCER OF CBF EXPRESSION 1* integrates cold signals into *FLOWERING LOCUS* C-mediated flowering pathways in *Arabidopsis*. *Plant J.* **2015**, *84*, 29–40. [CrossRef] [PubMed]

74. Seo, E.; Lee, H.; Jeon, J.; Park, H.; Kim, J.; Noh, Y.S.; Lee, I. Crosstalk between cold response and flowering in *Arabidopsis* is mediated through the flowering-time gene *SOC1* and its upstream negative regulator *FLC*. *Plant Cell* **2009**, *21*, 3185–3197. [CrossRef] [PubMed]

75. Park, S.; Gilmour, S.J.; Grumet, R.; Thomashow, M.F. *CBF*-dependent and *CBF*-independent regulatory pathways contribute to the differences in freezing tolerance and cold-regulated gene expression of two *Arabidopsis* ecotypes locally adapted to sites in Sweden and Italy. *PLoS ONE* **2018**, *13*, e0207723. [CrossRef] [PubMed]

76. Hua, W.; Zhang, L.; Liang, S.; Jones, R.L.; Lu, Y.T. A tobacco calcium/calmodulin-binding protein kinase functions as a negative regulator of flowering. *J. Biol. Chem.* **2004**, *279*, 31483–31494. [CrossRef]

77. Hecht, V.; Foucher, F.; Ferrandiz, C.; Macknight, R.; Navarro, C.; Morin, J.; Vardy, M.E.; Ellis, N.; Beltran, J.P.; Rameau, C.; et al. Conservation of *Arabidopsis* flowering genes in model legumes. *Plant Physiol.* **2005**, *137*, 1420–1434. [CrossRef]

78. Akiyama, K.; Kurotani, A.; Iida, K.; Kuromori, T.; Shinozaki, K.; Sakurai, T. RARGE II: An integrated phenotype database of *Arabidopsis* mutant traits using a controlled vocabulary. *Plant Cell Physiol.* **2014**, *55*, e4. [CrossRef]

79. Yuan, N.; Balasubramanian, V.K.; Chopra, R.; Mendu, V. The photoperiodic flowering time regulator *FKF1* negatively regulates cellulose biosynthesis. *Plant Physiol.* **2019**, *180*, 2240–2253. [CrossRef]

80. Wang, B.; Jin, S.-H.; Hu, H.-Q.; Sun, Y.-G.; Wang, Y.-W.; Han, P.; Hou, B.-K. *UGT87A2*, an *Arabidopsis* glycosyltransferase, regulates flowering time via *FLOWERING LOCUS C*. *New Phytol.* **2012**, *194*, 666–675. [CrossRef]

81. Taylor, C.M.; Kamphuis, L.G.; Zhang, W.; Garg, G.; Berger, J.D.; Mousavi-Derazmahalleh, M.; Bayer, P.E.; Edwards, D.; Singh, K.B.; Cowling, W.A.; et al. INDEL variation in the regulatory region of the major flowering time gene *LanFTc1* is associated with vernalization response and flowering time in narrow-leafed lupin (*Lupinus angustifolius* L.). *Plant Cell Environ.* **2019**, *42*, 174–187. [CrossRef]

82. Abe, M.; Kobayashi, Y.; Yamamoto, S.; Daimon, Y.; Yamaguchi, A.; Ikeda, Y.; Ichinoki, H.; Notaguchi, M.; Goto, K.; Araki, T. *FD*, a bZIP protein mediating signals from the floral pathway integrator *FT* at the shoot apex. *Science* **2005**, *309*, 1052–1056. [CrossRef]

83. Andrés, F.; Romera-Branchat, M.; Martínez-Gallegos, R.; Patel, V.; Schneeberger, K.; Jang, S.; Altmüller, J.; Nürnberg, P.; Coupland, G. Floral induction in *Arabidopsis* by *FLOWERING LOCUS T* requires direct repression of *BLADE-ON-PETIOLE* genes by the homeodomain protein *PENNYWISE*. *Plant Physiol.* **2015**, *169*, 2187–2199. [CrossRef]

84. Ambawat, S.; Sharma, P.; Yadav, N.R.; Yadav, R.C. MYB transcription factor genes as regulators for plant responses: An overview. *Physiol. Mol. Biol. Plants* **2013**, *19*, 307–321. [CrossRef] [PubMed]

85. Wilson-Sánchez, D.; Martínez-López, S.; Navarro-Cartagena, S.; Jover-Gil, S.; Micol, J.L. Members of the *DEAL* subfamily of the *DUF1218* gene family are required for bilateral symmetry but not for dorsoventrality in *Arabidopsis* leaves. *New Phytol.* **2018**, *217*, 1307–1321. [CrossRef] [PubMed]

86. Persson, S.; Wei, H.; Milne, J.; Page, G.P.; Somerville, C.R. Identification of genes required for cellulose synthesis by regression analysis of public microarray data sets. *Proc. Natl. Acad. Sci. USA* **2005**, *102*, 8633–8638. [CrossRef] [PubMed]

87. Ubeda-Tomas, S.; Edvardsson, E.; Eland, C.; Singh, S.K.; Zadik, D.; Aspeborg, H.; Gorzsàs, A.; Teeri, T.T.; Sundberg, B.; Persson, P.; et al. Genomic-assisted identification of genes involved in secondary growth in *Arabidopsis* utilising transcript profiling of poplar wood-forming tissues. *Physiol. Plant.* **2007**, *129*, 415–428. [CrossRef]

88. Mewalal, R.; Mizrachi, E.; Coetzee, B.; Mansfield, S.D.; Myburg, A.A. The *Arabidopsis* Domain of Unknown Function 1218 (*DUF1218*) containing proteins, *MODIFYING WALL LIGNIN-1* and 2 (*At1g31720/MWL-1* and *At4g19370/MWL-2*) function redundantly to alter secondary cell wall lignin content. *PLoS ONE* **2016**, *11*, e0150254. [CrossRef]

89. Miyauchi, H.; Moriyama, S.; Kusakizako, T.; Kumazaki, K.; Nakane, T.; Yamashita, K.; Hirata, K.; Dohmae, N.; Nishizawa, T.; Ito, K.; et al. Structural basis for xenobiotic extrusion by eukaryotic MATE transporter. *Nat. Commun.* **2017**, *8*, 1633. [CrossRef]

90. Upadhyay, N.; Kar, D.; Deepak Mahajan, B.; Nanda, S.; Rahiman, R.; Panchakshari, N.A.; Bhagavatula, L.; Datta, S. The multitasking abilities of MATE transporters in plants. *J. Exp. Bot.* **2019**, *70*, 4643–4656. [CrossRef]

91. Yang, X.; Xia, X.; Zhang, Z.; Nong, B.; Zeng, Y.; Wu, Y.; Xiong, F.; Zhang, Y.; Liang, H.; Pan, Y.; et al. Identification of anthocyanin biosynthesis genes in rice pericarp using PCAMP. *Plant Biotechnol. J.* **2019**, *17*, 1700–1702. [CrossRef]

92. Chen, L.; Liu, Y.; Liu, H.; Kang, L.; Geng, J.; Gai, Y.; Ding, Y.; Sun, H.; Li, Y. Identification and expression analysis of MATE genes involved in flavonoid transport in blueberry plants. *PLoS ONE* **2015**, *10*, e0118578. [CrossRef]

93. Ellis, T.J.; Field, D.L. Repeated gains in yellow and anthocyanin pigmentation in flower colour transitions in the Antirrhineae. *Ann. Bot.* **2016**, *117*, 1133–1140. [CrossRef]

94. Shiono, K.; Ando, M.; Nishiuchi, S.; Takahashi, H.; Watanabe, K.; Nakamura, M.; Matsuo, Y.; Yasuno, N.; Yamanouchi, U.; Fujimoto, M.; et al. *RCN1/OsABCG5*, an ATP-binding cassette (ABC) transporter, is required for hypodermal suberization of roots in rice (*Oryza sativa*). *Plant J.* **2014**, *80*, 40–51. [CrossRef] [PubMed]

95. Landgraf, R.; Smolka, U.; Altmann, S.; Eschen-Lippold, L.; Senning, M.; Sonnewald, S.; Weigel, B.; Frolova, N.; Strehmel, N.; Hause, G.; et al. The ABC transporter ABCG1 is required for suberin formation in potato tuber periderm. *Plant cell* **2014**, *26*, 3403–3415. [CrossRef] [PubMed]

96. Alejandro, S.; Lee, Y.; Tohge, T.; Sudre, D.; Osorio, S.; Park, J.; Bovet, L.; Lee, Y.; Geldner, N.; Fernie, A.R.; et al. AtABCG29 is a monolignol transporter involved in lignin biosynthesis. *Curr. Biol.* **2012**, *22*, 1207–1212. [CrossRef]

97. Sibout, R.; Höfte, H. Plant Cell Biology: The ABC of monolignol transport. *Curr. Biol.* **2012**, *22*, R533–R535. [CrossRef]

98. Hinrichs, M.; Fleck, A.T.; Biedermann, E.; Ngo, N.S.; Schreiber, L.; Schenk, M.K. An ABC transporter is involved in the silicon-induced formation of Casparian Bands in the exodermis of rice. *Front. Plant Sci.* **2017**, *8*, 671. [CrossRef]

99. Forbes, I.; Wells, H.D. Hard and soft seededness in blue lupine, *Lupinus angustifolius* L.: Inheritance and phenotype classification. *Crop Sci.* **1968**, *8*, 195–197. [CrossRef]

100. Miao, Z.H.; Fortune, J.A.; Gallagher, J. Anatomical structure and nutritive value of lupin seed coats. *Aust. J. Agric. Res.* **2001**, *52*, 985–993. [CrossRef]

101. Smýkal, P.; Vernoud, V.; Blair, M.W.; Soukup, A.; Thompson, R.D. The role of the testa during development and in establishment of dormancy of the legume seed. *Front. Plant Sci.* **2014**, *5*, 351.

102. Silvestro, D.; Andersen, T.G.; Schaller, H.; Jensen, P.E. Plant sterol metabolism. Δ(7)-Sterol-C5-desaturase (*STE1/DWARF7*), Δ(5,7)-sterol-Δ(7)-reductase (*DWARF5*) and Δ(24)-sterol-Δ(24)-reductase (*DIMINUTO/DWARF1*) show multiple subcellular localizations in *Arabidopsis thaliana* (Heynh) L. *PLoS ONE* **2013**, *8*, e56429. [CrossRef] [PubMed]

103. DeBolt, S.; Scheible, W.-R.; Schrick, K.; Auer, M.; Beisson, F.; Bischoff, V.; Bouvier-Navé, P.; Carroll, A.; Hematy, K.; Li, Y.; et al. Mutations in UDP-Glucose:sterol glucosyltransferase in *Arabidopsis* cause transparent testa phenotype and suberization defect in seeds. *Plant Physiol.* **2009**, *151*, 78–87. [CrossRef] [PubMed]

104. Loubéry, S.; De Giorgi, J.; Utz-Pugin, A.; Demonsais, L.; Lopez-Molina, L. A maternally deposited endosperm cuticle contributes to the physiological defects of *transparent testa* seeds. *Plant Physiol.* **2018**, *177*, 1218–1233. [CrossRef] [PubMed]

105. Hellens, R.P.; Moreau, C.; Lin-Wang, K.; Schwinn, K.E.; Thomson, S.J.; Fiers, M.W.E.J.; Frew, T.J.; Murray, S.R.; Hofer, J.M.I.; Jacobs, J.M.E.; et al. Identification of Mendel's white flower character. *PLoS ONE* **2010**, *5*, e13230. [CrossRef] [PubMed]

106. Xu, X.; Zhang, X.-B.; Shi, Y.-F.; Wang, H.-M.; Feng, B.-H.; Li, X.-H.; Huang, Q.-N.; Song, L.-X.; Guo, D.; He, Y.; et al. A point mutation in an F-Box domain-containing protein is responsible for brown hull phenotype in rice. *Rice Sci.* **2016**, *23*, 1–8. [CrossRef]

107. Zhernakov, A.I.; Shtark, O.Y.; Kulaeva, O.A.; Fedorina, J.V.; Afonin, A.M.; Kitaeva, A.B.; Tsyganov, V.E.; Afonso-Grunz, F.; Hoffmeier, K.; Rotter, B.; et al. Mapping-by-sequencing using NGS-based 3'-MACE-Seq reveals a new mutant allele of the essential nodulation gene *Sym33* (IPD3) in pea (*Pisum sativum* L.). *Peer J.* **2019**, *7*, e6662. [CrossRef]

108. Berger, J.D.; Buirchell, B.J.; Luckett, D.J.; Nelson, M.N. Domestication bottlenecks limit genetic diversity and constrain adaptation in narrow-leafed lupin (*Lupinus angustifolius* L.). *Theor. Appl. Genet.* **2012**, *124*, 637–652. [CrossRef]

109. Cowling, W.A. Pedigrees and characteristics of narrow-leafed lupin cultivars released in Australia from 1967 to 1998. *Bull. Agric. West. Aust.* **1999**, *4365*, 4–11.

110. Cowling, W.A.; Buirchell, B.J.; Falk, D.E. A model for incorporating novel alleles from the primary gene pool into elite crop breeding programs while reselecting major genes for domestication or adaptation. *Crop Pasture Sci.* **2009**, *60*, 1009–1015. [CrossRef]

111. Mousavi-Derazmahalleh, M.; Bayer, P.E.; Nevado, B.; Hurgobin, B.; Filatov, D.; Kilian, A.; Kamphuis, L.G.; Singh, K.B.; Berger, J.D.; Hane, J.K.; et al. Exploring the genetic and adaptive diversity of a pan-Mediterranean crop wild relative: Narrow-leafed lupin. *Theor. Appl. Genet.* **2018**, *131*, 887–901. [CrossRef]

112. Mousavi-Derazmahalleh, M.; Nevado, B.; Bayer, P.E.; Filatov, D.A.; Hane, J.K.; Edwards, D.; Erskine, W.; Nelson, M.N. The western Mediterranean region provided the founder population of domesticated narrow-leafed lupin. *Theor. Appl. Genet.* **2018**, *131*, 2543–2554. [CrossRef]

113. Anders, S.; Huber, W. Differential expression analysis for sequence count data. *Genome Biol.* **2010**, *11*, R106. [CrossRef] [PubMed]

114. Untergasser, A.; Nijveen, H.; Rao, X.; Bisseling, T.; Geurts, R.; Leunissen, J.A.M. Primer3Plus, an enhanced web interface to Primer3. *Nucleic Acids Res.* **2007**, *35*, W71–W74. [CrossRef] [PubMed]

115. Konieczny, A.; Ausubel, F.M. A procedure for mapping *Arabidopsis* mutations using co-dominant ecotype-specific PCR-based markers. *Plant J.* **1993**, *4*, 403–410. [CrossRef] [PubMed]

116. Neff, M.M.; Neff, J.D.; Chory, J.; Pepper, A.E. dCAPS, a simple technique for the genetic analysis of single nucleotide polymorphisms: Experimental applications in *Arabidopsis thaliana* genetics. *Plant J.* **1998**, *14*, 387–392. [CrossRef]

117. Neff, M.M.; Turk, E.; Kalishman, M. Web-based primer design for single nucleotide polymorphism analysis. *Trends Genet.* **2002**, *18*, 613–615. [CrossRef]

118. Thiel, T.; Kota, R.; Grosse, I.; Stein, N.; Graner, A. SNP2CAPS: A SNP and INDEL analysis tool for CAPS marker development. *Nucleic Acids Res.* **2004**, *32*, e5. [CrossRef]

119. Benjamini, Y.; Hochberg, Y. Controlling the False Discovery Rate: A Practical and Powerful Approach to Multiple Testing. *J. R. Stat. Soc. Series B (Methodol.)* **1995**, *57*, 289–300. [CrossRef]

120. Voorrips, R.E. MapChart: Software for the graphical presentation of linkage maps and QTLs. *J. Hered.* **2002**, *93*, 77–78. [CrossRef]

121. Maere, S.; Heymans, K.; Kuiper, M. BiNGO: A Cytoscape plugin to assess overrepresentation of gene ontology categories in biological networks. *Bioinformatics* **2005**, *21*, 3448–3449. [CrossRef]

 International Journal of
Molecular Sciences

Article

Irrigation-Induced Changes in Chemical Composition and Quality of Seeds of Yellow Lupine (*Lupinus luteus* L.)

Justyna T. Polit [1,*], Iwona Ciereszko [2], Alina T. Dubis [3], Joanna Leśniewska [2], Anna Basa [4], Konrad Winnicki [1], Aneta Żabka [1], Marharyta Audzei [1], Łukasz Sobiech [5], Agnieszka Faligowska [5], Grzegorz Skrzypczak [5] and Janusz Maszewski [1]

[1] Department of Cytophysiology, Institute of Experimental Biology, Faculty of Biology and Environmental Protection, University of Lodz, Pomorska 141/143, 90-236 Lodz, Poland; konrad.winnicki@biol.uni.lodz.pl (K.W.); aneta.zabka@biol.uni.lodz.pl (A.Ż.); avdejka.margo@gmail.com (M.A.); janusz.maszewski@biol.uni.lodz.pl (J.M.)

[2] Department of Plant Biology and Ecology, Faculty of Biology, University of Bialystok, Ciołkowskiego1J, 15-245 Bialystok, Poland; icier@uwb.edu.pl (I.C.); joanles@uwb.edu.pl (J.L.)

[3] Department of Organic Chemistry, Faculty of Chemistry, University of Bialystok, Ciołkowskiego 1K, 15-245 Bialystok, Poland; alina@uwb.edu.pl

[4] Department of Physical Chemistry, Faculty of Chemistry, University of Bialystok, Ciołkowskiego 1K, 15-245 Białystok, Poland; abasa@uwb.edu.pl

[5] Agronomy Department, Poznań University of Life Sciences, Dojazd 11, 60-632 Poznań, Poland; lukasz.sobiech@up.poznan.pl (Ł.S.); agnieszka.faligowska@up.poznan.pl (A.F.); grzegorz.skrzypczak@up.poznan.pl (G.S.)

* Correspondence: justyna.polit@biol.uni.lodz.pl; Tel.: +48-42-635-45-14

Received: 30 September 2019; Accepted: 4 November 2019; Published: 6 November 2019

Abstract: The quality and amount of yellow lupine yield depend on water availability. Water scarcity negatively affects germination, flowering, and pod formation, and thus introduction of an artificial irrigation system is needed. The aim of this study was to evaluate the influence of irrigation on the quality of yellow lupine seeds. Raining was applied with a semi-solid device with sprinklers during periods of greatest water demand. It was shown that watered plants produced seeds of lesser quality, having smaller size and weight. To find out why seeds of irrigated plants were of poor quality, interdisciplinary research at the cellular level was carried out. DNA cytophotometry evidenced the presence of nuclei with lower polyploidy in the apical zone of mature seeds. This may lead to formation of smaller cells and reduce depositing of storage materials. The electrophoretic and Fourier transform infrared spectroscopic analyses revealed differences in protein and cuticular wax profiles, while scanning electron microscopy and energy dispersive spectroscopy revealed, among various chemical elements, decreased calcium content in one of seed zones (near plumule). Seeds from irrigated plants showed slightly higher germination dynamics but growth rate of seedlings was slightly lower. The studies showed that irrigation of lupine affected seed features and their chemical composition, an ability to germination and seedlings growth.

Keywords: endoreplication; FTIR; germination; mitotic activity; SEM-EDS; storage proteins

---

1. Introduction

Yellow lupine (*Lupinus luteus* L.) is a legume crop which has tremendous economic potential and is of great importance both in sustainable agriculture, particularly in reclamation of marginal lands, and as a natural source of nitrogen thus it could be one of the main species cultivated for green fertilizer, seeds, food and feed. As a rotation crop, it reconstructs the soil after cereals, thus it plays a phytosanitary role.

Perfectly developed pile root systems of lupine meliorate the soil, making its aeration and watering easier. In this way, it improves the water-air ratio and makes the damaged soil structure (resulting e.g., from cereal monoculture) more crumbly. As a result of symbiosis with papillary bacteria, lupine has the ability to bind indirectly free atmospheric nitrogen and, thus, to improve soil fertility. Its long roots take up ions of calcium, magnesium, potassium and phosphorus from deep layers of soil, inaccessible for other plant species. Thus, it increase the yield of follow-up crop, because the compounds stored in the tissues (macro- and micronutrients) return to the soil from crop residues (such as roots or straw), remaining after the harvest or when lupine plants are intended for plowing as a green fertilizer [1–5]. Yellow lupine contains a large amount of high quality proteins in its seeds and negligible amounts of harmful, bitter alkaloids. These proteins, due to the favorable amino acid composition, are of much higher quality than those derived from cereal grains. Therefore, the seeds are used as a protein source in the diets of livestock, and also as a component of food, especially functional food for people [1,4–9]. The lupine proteins have pharmaceutical qualities. They influence lipid and glucose metabolism as well as blood pressure. They may also affect inflammatory processes and changes in gut microbiome. This has a significant influence on the metabolism, nutrient absorption, and immune functions [4,10–12]. In addition, yellow lupine seeds are abundant in the Fe-rich ferritin and may be a safe way to increase dietary iron intake replacing traditional iron supplementation methods [13].

Yellow lupine is a species with the lowest soil requirements among other lupine plants, however, is characterized by a relatively long growing season until the seeds are produced. The atmospheric conditions prevailing at this time may both favor or impair the course of cultivation. During vegetation its yielding is unstable under unfavorable weather conditions, such as drought [5,9]. Due to this, it is not willingly grown despite its beneficial properties. Therefore, it is difficult to achieve such amounts of this high-protein crop seeds that could be competitive with soya, which now satisfies great part of nutritional needs, especially in Europe [4].

Thanks to the well-developed root system, plants of yellow lupine can cope with periodic water shortages, taking it from deeper soil layers, inaccessible for other herbaceous plants. However, long-lasting drought causes changes at the physiological and molecular level. Lack of water or its insufficient amount prolongs the flowering period. Water shortage activates stress responses and decreases the numbers of both flowers and developing pods, thus limiting lupine yield [5,9,14–16]. Drought inhibits the development of symbiotic bacteria from the *Rhizobium* group, and consequently decreases the total plant mass. In addition, water-deficit conditions can influence seed chemical composition, e.g., increasing alkaloid content in some sweet lupine varieties making them less attractive for farm animals [2,14–18]. Generally, shortage of water increases the production of reactive oxygen species in cells (which may cause damages in photosystems, especially PSII and in membranes of thylakoids), and decreases the rate of photosynthesis, due to low CO_2 uptake, a lowered activity of photosynthetic enzymes and reduced chlorophyll content [17,19–24]. All changes in plant metabolism which ensure survival of unfavorable conditions limit crop yield [18,21,25]. Every reduction of agricultural productivity causes economic losses among farmers and increases food prices [26,27].

Complexity of plant response to water deficit makes genetic research that could lead to obtaining drought-resistant crops difficult and time-consuming [4,20,21,24]. Thus, different methods of irrigation are still the most common approach to reduce adverse effects of drought in agriculture. It was found that they increase crop productivity and seeds quality, [18,25,28–33]. However, contradictory results concerning the lupine are also known, and they indicate that irrigation might reduce seed vigor, germination capacity and germination energy, but increase a share of mold, rotting, and dead seeds [34,35]. It is therefore necessary to analyze whether irrigation is beneficial in all circumstances, even when plants are exposed only to mild water stress. If the seeds are to be used for consumption purposes, a specially high level of quality is desired, but if they are treated as a planting material, it is possible that mild drought will increase plant resistance to stress and the memory of stress will help to tolerate unfavorable conditions by the next generation of plants.

The aim of the current research was to investigate cytological, chemical and biochemical traits which may be responsible for quality of seeds from irrigated plants of yellow lupine. The obtained results indicate that the seeds harvested from the non-irrigated and irrigated plants differ in size and weight, endopolyploidy level of cotyledon cells, content of storage proteins, protein composition and in cuticular wax profiles, as well as they differ in the germination capacity and growth rate of embryonic roots.

2. Results

2.1. Seed Yield

Irrigation did not improve the seed yield of the yellow lupine. The amount of yield collected from the main stem as well as from branches of irrigated plants was comparable to that obtained from the control (non-irrigated) plants (Table 1). However, the seed yield was visually of inferior quality. For clarity, in the following sections of the work the seeds collected from the plants growing under natural conditions (without additional irrigation) are referred to as "control seeds" whereas the seeds from the plants subjected to irrigation are referred to as "irrigated seeds".

Table 1. Seed yield of yellow lupine [g/plant] harvested from the main stems, branches and whole control (not irrigated) and irrigated plants.

	Main Stem	Branches	Plant
Control	3.68	0.38	4.06
Irrigation	4.14	0.94	5.08

All differences between presented pairs of mean values of seed weights are not statistically significant (the Student's *t*-test, $p < 0.01$).

2.2. Seed Morphology

In both control and irrigated plants there were seeds of different morphology and quality. Therefore the control and the irrigated seeds were divided into normal (correct) and abnormal (incorrect) groups due to differences in their morphological state (Figure 1a–d).

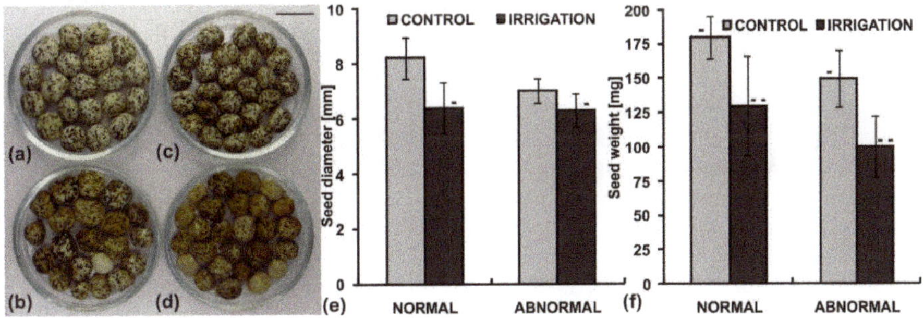

Figure 1. Seeds of yellow lupine collected from the control (not irrigated) and irrigated plants and sorted according to morphological features: (**a**) control—normal seeds, (**b**) control—abnormal seeds, (**c**) irrigated plants—normal seeds, (**d**) irrigated plants—abnormal seeds. Scale bar 10 mm, (**e**) seed size (diameter) measured along the long axis, (**f**) seed weight. Statistical significance between mean values of seed diameters and seed weights was assessed with the Mann–Whitney U test ($p < 0.01$) and Student's t test ($p < 0.01$), respectively. Error bars represent standard deviation (SD). Minuses and double minuses indicate pairs of statistically insignificant results.

The evaluation was based on seeds color, shape, estimated size and weight. Seeds from the first group were relatively large in size and weight, smooth, oval-shaped and slightly flattened, covered with a white seed coat with a specific regular marble pattern. The seeds from the second group were significantly smaller or lighter, with distorted oval shape, often stained brownish or without the clear marble pattern. The above-mentioned poor morphological features, occurred separately or accumulated in one seed.

The lupine seeds in each group were measured and weighed. It turned out that the yield of the irrigated plants was of inferior quality (Figure 1e,f). Their seed size and weight were smaller by about 25% and about 30%, respectively. Because the percentage of abnormal seeds in both groups of plants was similar, i.e., 32% and 30% for the control and irrigated plants, respectively, as well as about 35–40% of the abnormal ones did not germinate, in subsequent studies only normal seeds were taken into account.

2.3. DNA Content

Mature seeds of lupine have large cotyledons; their cells, depending on the area in which they are located, may be of different ploidy level and thus may occur at different sizes. Cytophotometric measurements of DNA content in the cell nuclei of two extremely situated cotyledon zones (basal and apical; Figure 2) in the control seeds did not reveal differences (Figure 2b,c). In both zones, besides the 2C and 4C DNA cells (nearly 70%), polyploid ones were also observed (about 30%). More than 20% of them passed the first round of endoreplication and contained 8C DNA, while about 7% passed two rounds of endoreplication reaching 16C DNA. A few (about 2%) contained 32C DNA.

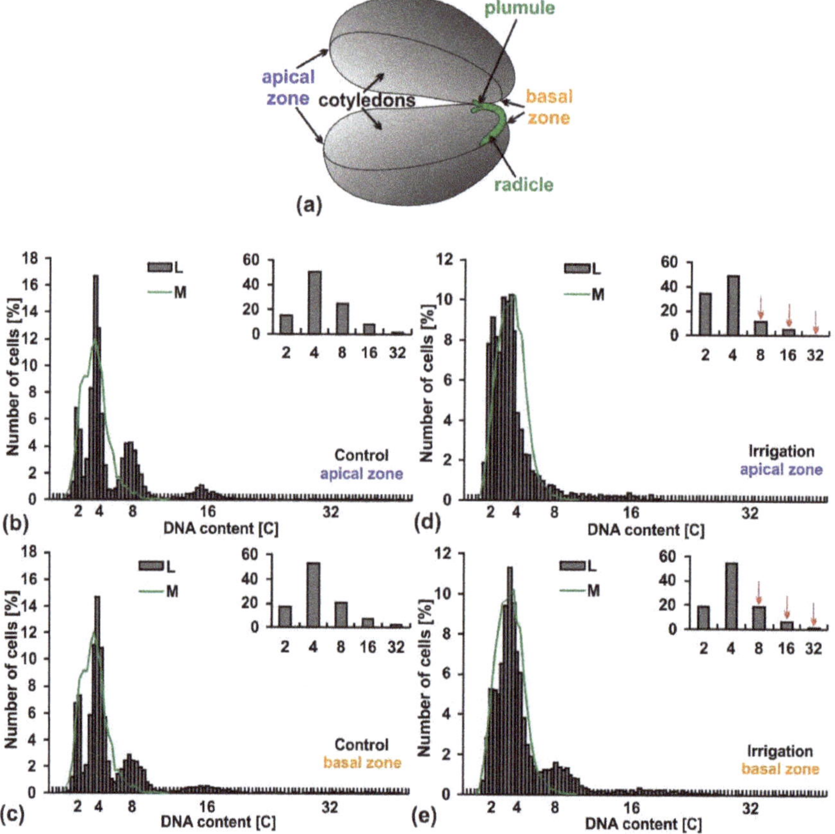

Figure 2. DNA content in the indicated zones of yellow lupine seeds. (**a**) Structure of lupine seed. (**b–e**) Frequency distribution [%] of nuclear DNA content in the selected zones: cotyledon zones (L) and root meristems (M) of yellow lupine; (**b**) Apical zone of control seeds from not irrigated plants. (**c**) Basal zone of control seeds from non-irrigated lupine plants. (**d**) Apical zone of seeds from irrigated plants. (**e**) Basal zone of seeds from irrigated plants. Inserted bar graphs show percentages of cells after successive rounds of endoreplication. Red arrows show a decrease in the number of polyploid cells in the seeds from irrigated plants.

In the irrigated seeds there was a weakly pronounced difference in the number of polyploid cells in the basal zone of cotyledons (less by only 3%), while a significantly lower number of them (less by 17%) was observed in apical zone (Figure 2d,e). In this zone, cells with 2–4C DNA content characteristic of the regular cell cycle accounted for 84% and polyploid ones for only 16% (Figure 2d). A decrease in the number of polyploid nuclei was mainly related to a significant quantitative reduction of the cells in the first round of endoreplication (Figure 2e).

2.4. Protein Profiles

Electrophoretic distribution of the proteins in the polyacrylamide gel allowed us to assess protein composition of the control and irrigated seeds (Figure 3a,b). The same number of distinguishable bands in both channels indicated the presence of similar protein composition in the tested seeds. The digital analysis of the intensity of their staining pointed to some differences in the amount of proteins present in them (Figure 3b). Even a small difference in the height of the bars (staining intensity) in each pair, e.g., band pairs 6 or 14 (Figure 3b) is clearly visible in the polyacrylamide gel (containing proteins of about 62 or 17 kDa, respectively; Figure 3a).

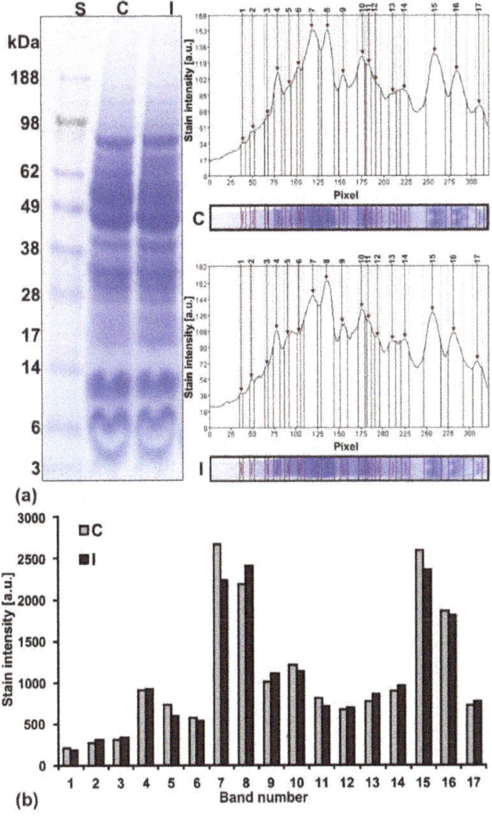

Figure 3. Protein profile in yellow lupine cotyledons from the seeds collected from not irrigated—(control C) and irrigated—(I) plants. (**a**) Electrophoretic separation of proteins in polyacrylamide gel (stained with Coomassie Blue) and computer analysis of staining intensity of the detected bands. Channel 1 shows protein mass standard (S), two and three show seed proteins from non-irrigated (control C) or irrigated (I) plants, respectively. (**b**) Comparison of protein contents in 17 detected bands.

The most stained bands (4, 7, 8, 15, 16, 17; Figure 3b) contained subunits of storage proteins (the most abundant in cotyledones of lupine): albumins, i.e., δ-conglutin, globulins, e.g., β-conglutin (vicilin-like), and α-conglutin (legumin-like), as well as probably a non-storage protein, γ-conglutin. In comparison with control plants, the decrease in storage protein content in some bands (seven cases), and the increase in others (eight cases) indicated modifications of their proportions due to irrigation.

2.5. Fourier Transform–Infrared Spectroscopy (FT-IR) Analysis of Lupine Seeds

The diffuse reflectance infrared spectroscopy (DRIFTS) FT-IR spectra of dry peeled lupine seeds from control and irrigated plants are shown in Figure 4a–c. The spectrum of the every dry lupine seed exhibit two prominent absorption bands at 3314 and 1674 cm^{-1} which could be assigned to N-H and C=O stretching bands (Figure 4a). There are also three prominent bands which appear in the 2955–2855 cm^{-1} range that originate from the hydrocarbon tails. For methyl (CH$_3$) and methylene (CH$_2$) groups, asymmetric and symmetric C-H stretching occur at 2955, 2925 and 2855 cm^{-1}, respectively. Triglyceride ester group show carbonyl C=O band at 1745 cm^{-1}. The major infrared modes due to protein give rises to amide carbonyl modes in the 1700–1620 cm^{-1} range. This region consists of some overlapping carbonyl bands which may be separated using the Fourier self-deconvolution mathematical method (FSD) [36]. One of the examples of improvement of information content by using FSD method is the estimation of protein secondary structure and conformations by the analysis of the resolution-enhanced amide I profile by FSD [37].

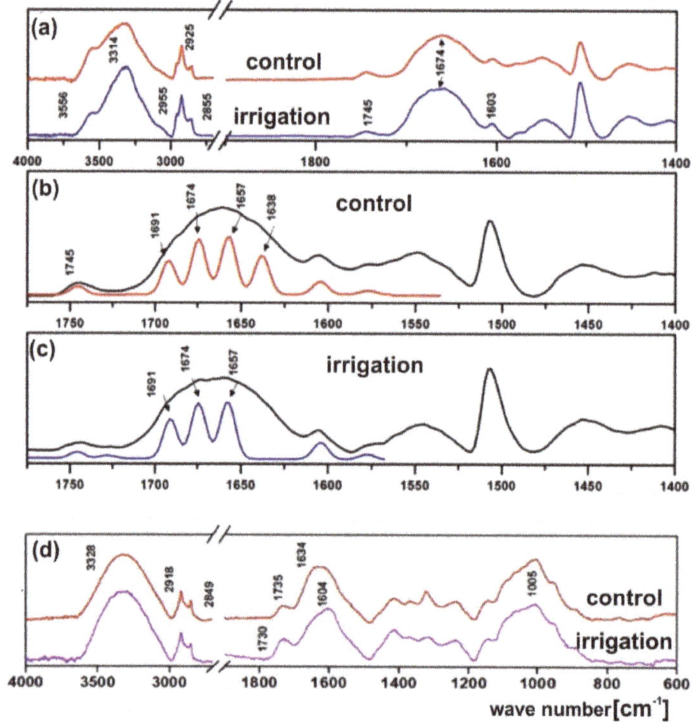

Figure 4. Fourier transform–infrared spectroscopy (FT-IR) spectrum of the yellow lupine seeds: (**a**) peeled seeds collected from the control (not irrigated) and irrigated lupine plants, (**b**) separation of the overlapping bands in the spectrum of the control seed—four Gaussian lines (red line) were found at 1691, 1674, 1657 and 1638 cm^{-1}, (**c**) separation of the overlapping bands in the spectrum of the irrigated seed—three Gaussian lines (navy line) were found at 1691, 1674, and 1657cm^{-1}. Spectra were recorded

at room temperature using the diffuse reflectance infrared spectroscopy (DRIFTS) module. (**d**) attenuated total reflection (ATR)/FT-IR spectrum of the lupine seed coat cutine from the control and irrigated material. Spectra were recorded at room temperature using ATR module.

The spectra of proteins exhibit absorption bands associated with amide groups. The exact wave numbers of C=O vibrations depend on the nature of hydrogen bonding interaction involving C=O and N-H groups. The characteristic bands of the amide groups of protein chains are similar to the absorption bands exhibited by secondary amides, and are labelled as amide I bands. It occurs between 1700 and 1600 cm^{-1}.As a consequence of inter- and intramolecular interactions, the amide I bands consist of a number of overlapping component bands. The FSD-IR was used to extract individual components from a complex composite band of C=O groups. Using the deconvolution method, the $\nu_{C=O}$ characteristic stretching bands at 1691, 1674, 1657, 1638 cm^{-1} were estimated in the control material (Figure 4b). It seems most likely that changes in the composition of the seed storage proteins are due to the irrigating process (Figure 4a, navy line). There was no absorption band at 1638 cm^{-1} (Figure 4C, navy line), as compared with the spectrum of the control lupine seeds (Figure 4b, red line). We believe that the difference of wave number reflects the structural nonequivalence of carbonyl groups. It means that various protein types are present in the lupine seeds.

Figure 4d shows attenuated total reflection (ATR)/FT-IR spectra of lupine seed coats. There are four main absorption bands. The broad and intense band at 3328 cm^{-1} was assigned to the O-H stretching modes of alcohols and fatty acids. The bands in the region of 2918–2849 cm^{-1} were assigned to the stretching of aliphatic CH$_2$ groups. The band at 1735 cm^{-1} was assigned to the C=O mode of carbonyl ester group. The broad band centered at 1634 cm^{-1} are due to proteins. The intense band at 1005 cm^{-1} is assigned to C–O vibration of cellulose [38].

The spectroscopic analysis demonstrated that the cuticular wax profiles of the irrigated seeds was different from the control ones. The absorption band assigned to the C=O mode of ester group was more intensive for the former ones. The broad band characteristic of proteins centered at 1634 cm^{-1} from non-irrigated seeds and at 1604 cm^{-1} from irrigated ones, as well as the broad band (characteristic of cellulose) centered at 1005 cm^{-1} also showed some differences.

2.6. Analysis of Chemical Elements by the SEM/Energy Dispersive Spectroscopy (EDS) Technique

In the seeds of yellow lupine three chemical elements, C, O and N, were dominant. Their contents in each analyzed zone were similar in the control and irrigated seeds (Figure 5). Other elements (Mg, Al, P, S, K, Ca), whose contents (expressed in weight %) oscillated on average around 0.5% showed no statistically significant changes after irrigation treatments (Figure 5c). Among these various studied elements only the calcium content decreased statistically significantly in one seed zone (in the cotyledon near plumule), probably as a result of plant irrigation.

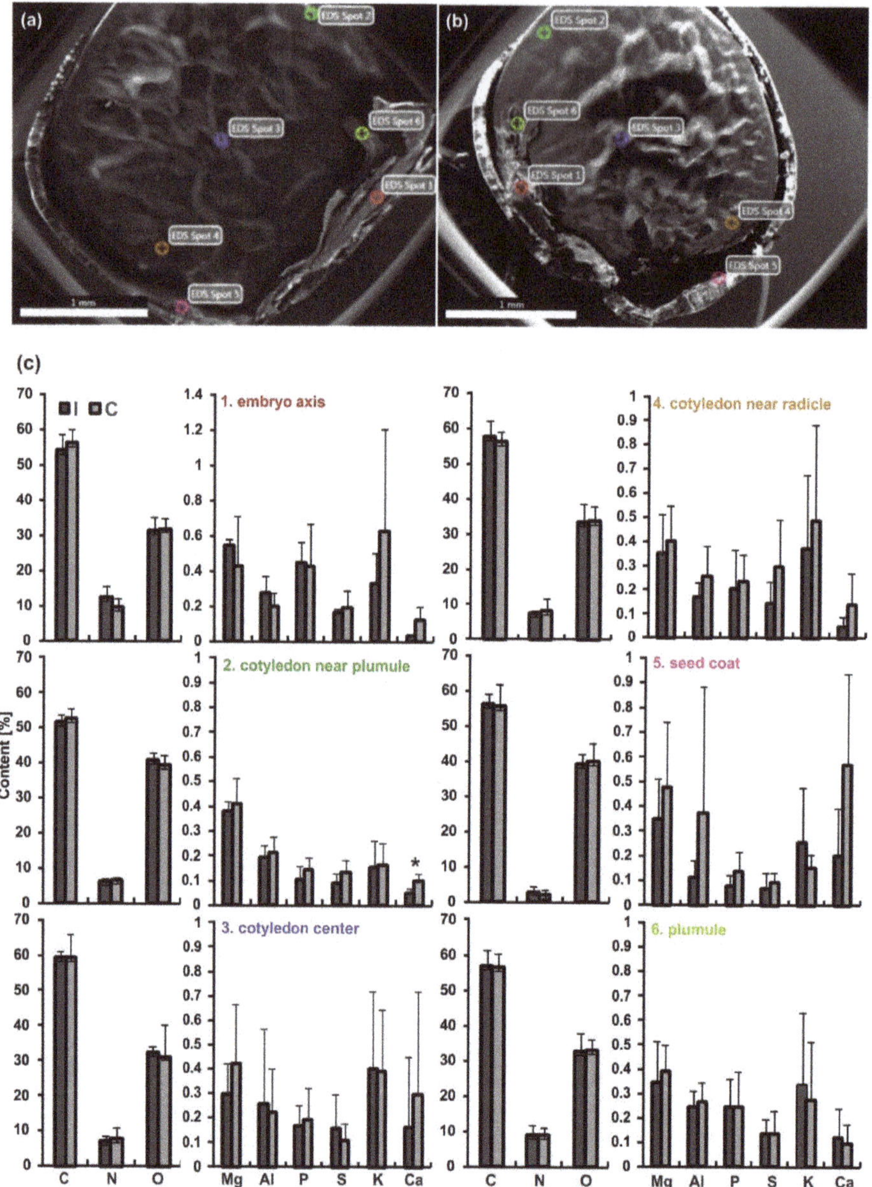

Figure 5. Scanning electron microscope (SEM) micrographs of yellow lupine half seeds: (**a**) Seed of a control plant. (**b**) Seed of an irrigated plant. The spots: 1-embryo axis, 2-cotyledon near plumule, 3-cotyledon center, 4-cotyledon near radicle, five-seed coat, six-plumule. (**c**) Corresponding content (weight %) of chemical elements (C, N, O, Mg, Al, P, S, K, Ca) in the indicated zones of seeds (C-control, I-irrigated plants, respectively). Statistical significance between mean values was assessed with the Student's t-test ($p = 0.008$). Error bars represent standard deviation (SD). An asterisk indicates statistically significant results.

2.7. Germination and Seedlings' Growth

Substances stored in a storage tissues (for example in the cotyledons) are used during germination and growth of young seedlings. The dynamics of germination and growth are the parameters that allow to evaluate seed quality. The seeds collected from control and irrigated plants (only correct, as described in the Section 2.1., Figure 1) were germinated for three days (Figure 6). The seeds that did not sprout after three days did not sprout at all; they constituted 13% and 20% in the control and irrigated seed lots, respectively. After the first day of germination as much as 23% of the irrigated seeds sprouted out, while in the control there was almost half as much, only 13% (Figure 6a). In both groups of seeds the vast majority sprouted after two days. However, the overall percentage of germinated seeds was higher in the control (72%) than those in the irrigated material (49%).

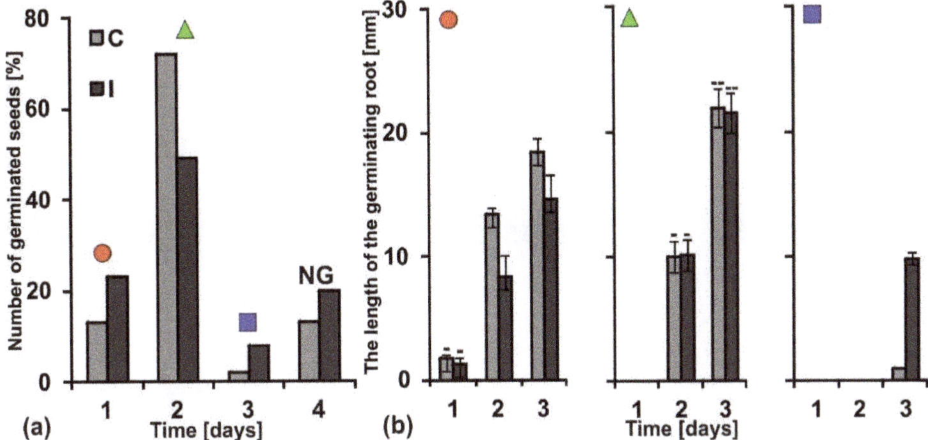

Figure 6. The dynamics of yellow lupine seed germination and growth of embryo roots. (**a**) Percentage of germinated seeds during three days. The seeds collected from not irrigated plants (control C). The seeds collected from irrigated plants (I). The seeds which remained non-germinated (NG) after four days. Black figures (circle, triangle, square) above bars indicate populations of germinated seeds whose root length is presented on the graphs marked with an adequate figure in part (B). (**b**) Dynamics of embryo roots growth during following days of germination. Statistical significance between mean values in diagram marked with black circle and triangle was assessed with the Mann–Whitney U test ($p < 0.01$) and the two-way ANOVA with the post-hoc unequal N HSD (honest significant difference) Tukey test ($p < 0.01$), respectively. Error bars represent standard deviation (SD). Minuses and double minuses indicate pairs of statistically insignificant results.

Among the seeds that germinated after the first day, the control seedlings grew faster, whereas among those that germinated later (after two days) both control and irrigated seedlings grew similarly and finally reached larger sizes than the first ones (Figure 6b).

2.8. Mitotic Activity

Cell proliferation in meristems is one of the main causes of seed germination and growth of the embryonic roots. Both after the first and the second day of germination, the differences between the mitotic indexes evaluated for root meristems in seeds of the control and irrigated plants were statistically not significant (Figure 7a). However, after the first day of germination in roots growing from the control seeds, cell divisions started with greater synchronization, as evidenced by the high prophase index (more than 45%; Figure 7b).

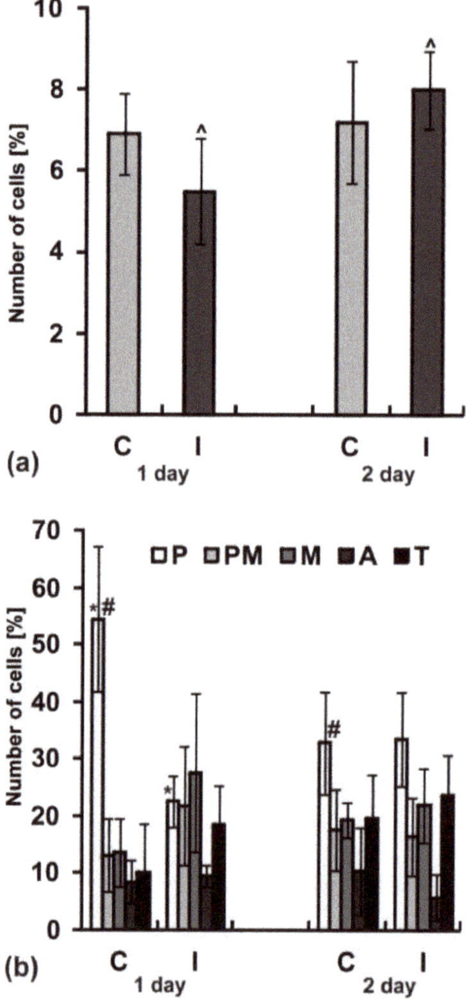

Figure 7. Mitotic activity in yellow lupine root meristems after one and two days of seed germination. The seeds collected from not irrigated (control—C) and irrigated—I plants. (**a**) Mitotic index, (**b**) phase index, P—prophase, PM—prometaphase, M—metaphase, A—anaphase, T—telophase. Statistical significance between mean values was assessed with the two-way ANOVA and post-hoc unequal N HSD Tukey test ($p < 0.01$). Error bars represent standard deviation (SD). Pairs of symbols ($\wedge$, $*$, #) over the bars indicate pairs of statistically significant results.

2.9. Detection of Hydrogen Peroxide

During germination of seeds, reactive oxygen species (e.g., hydrogen peroxide—H_2O_2) are produced in the embryo roots of young seedlings. Their appropriate level is necessary to promote changes in the structure of a cell wall and to facilitate elongation of cells. On the other hand, H_2O_2 is also a dangerous compound which adversely affects the cells. Too high a level of H_2O_2 causes double-strand DNA breaks and destroys the structure of chromosomes. Analyses of the H_2O_2 content (based on the 3,3-diaminobenzidine (DAB) polymerization method) revealed that its level was similar in the roots grown from the control and irrigated seeds (Figure 8).

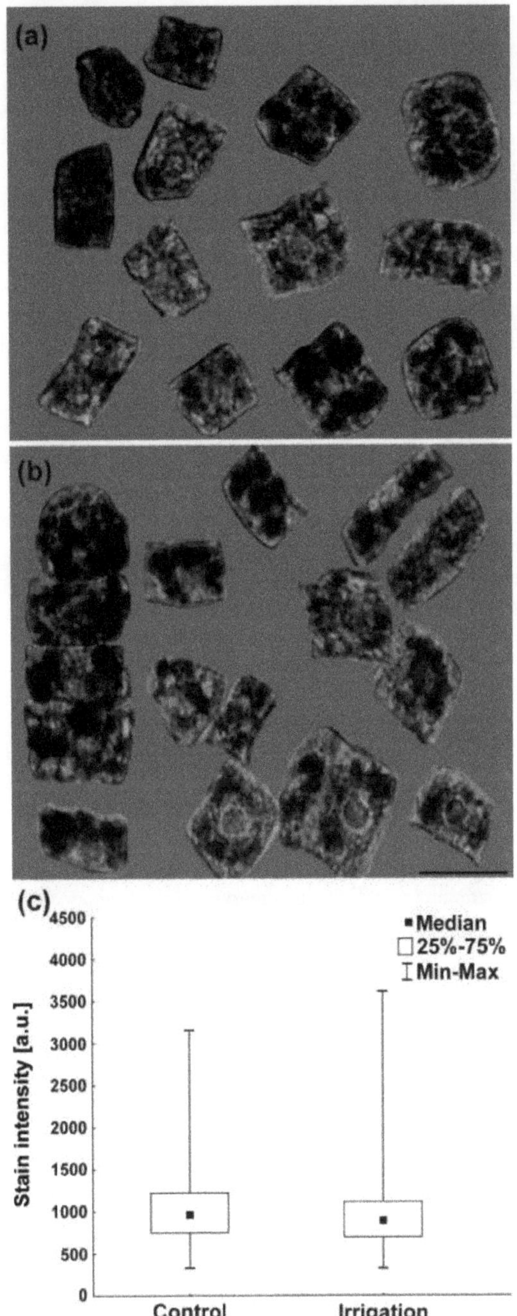

Figure 8. Identification of H_2O_2 in the form of dark 3,3-diaminobenzidine (DAB) polymers and the level of H_2O_2 in the cells of embryonic roots deriving from lupine seeds. (**a**) Not irrigated—control plants. (**b**) Irrigated plants. Scale bar 20 μm; (**c**) stain intensity (arbitrary units) in these cells. Statistical significance between median values was assessed with the Mann–Whitney U test ($p < 0.01$). Median values are statistically insignificant.

3. Discussion

Due to the advantages of yellow lupine cultivation, treatments aimed at counteracting adverse natural conditions as well as research studies that monitor plant reactions to the prevailing and modified growing conditions are justified. In the future they will allow to increase the yield or keep it at a predictable level for years characterized by changing weather, and thus will encourage farmers to grow this species. One of the commonly used agrotechnical methods that could prevent water shortages, ensure the correct rhythm of plant development and intensify yields, is crop irrigation [5,39]. It causes an increase in the yield of cereals (up to 27%) and also improves the amount and quality of some legume seeds like chickpea beans [31,39]. However, this common agronomic operation has not been optimized for lupine seed harvest yet. Generally, lupine plants are sensitive to water deficit and intolerant of waterlogging but stress response depend on the species and plant condition. It was found that plants belonging to one genus—*Lupinus* react differently to new conditions caused by unfavorable weather occurring over the growing season [5,18,40–42]. Therefore, it is extremely important to optimize the growing conditions for an individual species. Surprisingly, current studies showed that irrigation of yellow lupine (*L. luteus* L. cultivar Mister) did not significantly increase yields, while it weakened the quality of seeds. Seed size was smaller by about 25% and the weight—by about 30%. However, the yield of narrow-leaved lupine cultivated at the same time and under the same irrigation conditions was even 2.5 times higher than of the non-irrigated plants, although the quality of irrigated seeds was also worse in terms of size and weight [43]. Thus, yellow lupine seems to grow better under water shortage than under the conditions limiting this stress. In turn narrow-leaved lupine showed greater tolerance of the new created cultivation conditions, as the decreases in size and weight of seeds were smaller than in the case of yellow lupine. This could be the result of the anatomical structure of narrow-leaved lupine leaves, which allowed more efficient drying of the lupine field, and thus created better conditions for seed maturation.

Seed size and weight are important physical indicators of seed quality that affects vegetative growth of the next generation of plants (e.g., seedlings' vigour); both parameters are frequently related to the size of yield, market grade factors and harvest efficiency [44]. Generally, large seeds (e.g., of wheat, rice, oat, safflower, chickpea, sugar beet and many others species of plants) have better field performance than small seeds [44]. However, some researchers showed that cultivars of pea with lower seed mass displayed better germination than those with larger seeds [45]. Furthermore, small seeds of soybean had better germination and storage reserves utilization, as well as seedlings uniformity, which grown much faster than those from larger seeds [46]. Additionally, small seeds of safflower germinated faster and plants thereof grew higher under saline conditions [47].

To find out why yellow lupine seeds were of poor quality (mostly regarding their size and weight), the interdisciplinary research at the cellular level was carried out. To the best of our knowledge research involving microscopic, cytological, biochemical, and chemical analyses of seeds collected from the irrigated and non-irrigated yellow lupine plants has never been conducted so far.

Cytophotometric analyses of nuclei from cotyledon cells of seeds collected from the irrigated yellow lupine plants revealed lower ploidy level than those from the control plants. As demonstrated in numerous studies, polyploidization plays a key role during plant tissue and organ growth and development, both in favorable conditions and during environmental stress. A positive correlation between ploidy level and cell size, was observed in many plants, and was defined as the karyoplasmic ratio theory, which suggests that an increase in nuclear DNA content can be a driving force for cell expansion [48–50]. This mechanism seems to be advantageous especially when energy is limited, when rapid growth is necessary, or when terminal differentiation of some cells and their specialized functions are needed [51]. Cotyledons of lupine seeds are a reservoir of storage materials (mainly proteins) for developing embryos and growing young seedlings and should grow quickly during seed development to create space for the synthesized substances [51–53]. Endoreduplication associated with the production of storage materials is very common, although in some studies the correlation between endoreduplication and accumulation of storage proteins was not observed [54]. Different environmental factors can also have strong impact on the genome size [50,55–58]. Our research revealed that plant irrigation may be an inhibitory factor against switching of the classical cell cycle to the endocycles. Hence low level of ploidy in the cotyledon cell nuclei may be responsible for small seed sizes. The mechanism of this process is unknown, however it was suggested that only the ccs52 protein and protein inhibitors of cyclin-dependent kinases are of crucial importance in this case, as they inhibit cell entry into mitosis and promote endocycles [59,60]. The ploidy reduction in seeds of irrigated plants is not accidental because it was observed not only in yellow lupine but also in narrow-leaved lupine. Moreover, limitation of endoreplication, mainly in the apical zone of the seeds of both plant species, is of particular interest. This indicates the existence of a characteristic response mechanism which may be associated with the sequence of deposition of the storage compounds in specific seed areas.

Yellow lupine seeds are characterized by high protein content (44%), even higher than that in soybean (35%), white lupine (40%), and narrow-leaved lupine (34%), and thus, may be considered as a source of high quality storage proteins because of their nutritional, functional and chemical properties. Therefore preservation of the proper composition of proteins in lupine seeds during agro-technical treatments is of great importance. The lupine storage proteins are mainly globulins, which include α-, β- and γ-conglutin and their composition may differ in individual species of lupine [1,61–64]. Changes in DNA content in cotyledon cells caused by plant irrigation also encouraged us to make comparative analysis of protein profiles, because environmental stress factors may change gene expression, protein composition and their chemical structure [65–67]. The electrophoretic and spectroscopic (FTIR) analyses demonstrated that the seeds (cotyledons) of the non-irrigated and irrigated yellow lupine plants significantly differed with respect to their chemical composition. We believe that various protein types are present in the control and irrigated lupine seeds. However, at this stage of research, it is difficult to determine which of the observed changes are favorable or unfavorable for subsequent germination and seedling development, as well as for nutraceutical and taste properties of the seeds. This is an extremely interesting and important problem to be addressed in subsequent studies, all the more so, because the differences in the chemical composition of the seeds of irrigated and non-irrigated plants are species-specific [43].

Plant seeds are covered by seed coat and impregnated by cuticle and epicuticular waxes which protect them from environmental conditions, pathogens and insect attack [38,68,69]. This layer is also of great importance during the first stage of seed germination (imbibition). Our research revealed that irrigation of lupine plants during their cultivation affected the chemical composition of developing seeds coat. This modification influenced the subsequent germination of seeds. Similarly, as it was shown in the case of narrow-leaved lupine [43], the seeds produced by the irrigated yellow lupine plants also began to germinate faster. Due to the chemically changed coat of the seeds developed in the irrigated plants (revealed by spectroscopic analyses), the process of water absorption and seed imbibition may speed up, leading to quicker seed coat cracking and germination, similarly as it was observed in other seeds [69–71].

Imbibition of water causes the resumption of metabolic activity in the rehydrated seeds. During the next steps of germination catabolic enzymes (amylases, proteases) cause the breakdown of the stored substances (starch and proteins). After translocation of the hydrolyzed nutrients to the embryo proper and their subsequent assimilation, the cells of the embryo in the growing regions become metabolically very active, grow in size, begin proliferative activity and expansion to form the embryonic root and then young seedlings [72]. In order to mobilize storage substances and to make them available to the embryo axis, efficient functioning of a signaling network and activation of many genes associated with germination are necessary [73,74]. Different compounds are involved in the plant signaling network, among them sugars, hormones, nitric oxide, calcium ions (Ca^{2+}), hydrogen peroxide (H_2O_2), and others [75–78].

Our studies demonstrated that in seeds produced by the irrigated yellow lupine plants, which began to germinate faster (like narrow-leaved lupine seeds, just after the first day), the growth of embryonic root was weaker. Probably, these seeds were not fully ready for the next phases (catabolic and/or anabolic) of germination yet, which requires adequate resources of enzymes, regulatory and signal molecules. This may also be concluded from different contents of chemical elements in the seeds, i.e., nitrogen (in the embryo axis of narrow-leaved lupine) and calcium (in the cotyledon near plumule of the yellow lupine). The appropriate level of nitrogen and suitable carbon/nitrogen balance is crucial for the gene expression during germination and young seedling growth [74], while calcium signaling is, for example, involved in the regulation of cell cycle progression and gene expression in response to abiotic stresses [75]. Since calcium content was limited in the cotyledon near plumule of the irrigated yellow lupine seeds, their embryonic roots may have grown more slowly. However, analysis of mitotic activity in the meristems of the yellow lupine embryonic roots did not show statistically significant differences between the seeds of irrigated and non-irrigated plants (which were pronounced in narrow-leafed lupine), while changes between them were observed mainly in the proportions at the first stage of mitosis. Also, no statistically significant changes in H_2O_2 content (clearly visible in narrow-leafed lupine) were observed. H_2O_2 as one of the constitutive attributes of plant root physiology together with peroxidases (Clas III, E.C.1.11.1.7.), which catalyze the reduction of H_2O_2 or its formation (in the peroxidative or hydroxylic cycle, respectively). These processes are connected with cell wall loosening and root elongation during seed germination [79–82]. Such a result may suggest that irrigation during the growth and development of both species of plants (in an attempt to reduce drought stress) caused modifications in slightly different branches of signaling or metabolic networks and were reflected in different responses at the cell and tissue level.

4. Materials and Methods

4.1. Plant Cultivation

The research consisted in a field experiment carried out for three consecutive years at the Złotniki Research Station (*52°29′ N, 16°49′ E,*Poznań University of Life Sciences, Poland). The study was conducted as a stationary experiment (in a randomized complete block design with 4 replications) on grey-brown podzolic soil (pH = 4.8 measured in 1 M KCL; 1.3% organic matter: 50–110 mg kg^{-1} P, 115–195 mg kg^{-1} K) in 4-crop rotation. The yellow lupine (*L. luteus* L., cultivar Mister, certified seeds from PHR breeder, Poznań, Poland) was sown (150 kg ha^{-1}) in early April. Sowing depth was 4 cm and the row distance was 18 cm. The main plot treatments were natural rainfall (non-irrigated), and natural rainfall plus irrigations (irrigated). There was a gap of 6m in width between non-irrigated and irrigated parts of plots. Irrigations were applied during flowering, pod and seed ripening (May, June, July) when consumption of 30% of the readily available soil moisture (measured by the gravimetric method) was observed in the 0.30 m root zone. The irrigation water (of good quality, containing 114 Ca^{2+}, 7.4 Mg^{2+}, 0 Na^+, 0 K^+, <1 Fe^{3+}, 356 $CaCO_3$ mg·L^{-1}; pH 7.3) was taken from a small reservoir near the experimental site. Irrigation was performed using a water pump with aluminium outlet pipes (110 mm in diameter) and a rotary sprinkler. The diameters of the nozzles were 7 mm (NAAN 233/91) and the discharge rate was 5 L·h^{-1} (with the operating pressure of 0.35–0.4 MPa). The main pipes with the rotary sprinkler were placed in the middle of irrigated parts of plots. The mean dose of water and time of irrigation during vegetation period were 30–35 mm and 6–7 h, respectively, while the mean daily air temperatures and total precipitation in the vegetation periods in May, June and July were 15.3, 18.4, 17.5 °C and 17.5, 62.4, 214.8 mm, respectively (data from the Agrometeorological Observatory in Złotniki).

4.2. Yield Assessment

Ten plants of yellow lupine were collected randomly two days before harvest and were used to measure seed yield (expressed as g per plant).

4.3. Seed Germination for Cytological Study

Seeds of lupine were sown on wet filter paper in Petri dishes (10 seeds/∅ 15 cm) and germinated at room temperature for maximum 4 days in the dark.

4.4. Cytophotometry

Apical fragments of embryo roots and cotyledons were fixed in cold Carnoy's mixture (glacial acetic acid and absolute ethanol; 1:3; *v/v*) for 1 h. Following rehydration (70% ethanol, 30% ethanol, distilled water), the roots were hydrolyzed in 4 M HCl for 1 h and stained with Schiff's reagent (pararosaniline; Sigma-Aldrich, St. Louis, MO, USA) according to the standard methods [83]. After rinsing in SO_2-water and then in distilled water, fragments of cotyledons from the selected zones and 1.5-mm-long apical segments of the roots were cut off and squashed onto Super-Frost (Menzel-Gläser, Braunschweig, Germany) microscope slides. Following freezing with dry ice, cover slips were removed, and the dehydrated dry slides were embedded in Canada balsam. Nuclear DNA content was evaluated by means of microdensitometry using a Jenamed 2 microscope (Carl Zeiss, Jena, Germany) with the computer-aided Cytophotometer v1.2 (Forel, Lodz, Poland). The Feulgen-stained cell nuclei were measured at 565 nm. Microscopic slides were used also to analyze the mitotic and phase indexes.

4.5. Electrophoretic Separation of Proteins

P-PER Plant Protein Extraction Kit (Pierce) supplemented with Protease Inhibitor Cocktail (Sigma) was used for total protein extraction. The Lowry procedure was used to determine the total level of proteins in the solution [84]. Whole-cell protein extracts were fractionated on NuPAGE®® Novex®® 4–12% Bis-Tris gel, in NuPAGE®-MES SDS (50 mM MES, 50 mMTris, 0.1% SDS, 1 mM EDTA) buffer (pH 7.3; 200 V; 110–125 mA). Analysis of staining intensity (Coomassie™) of the bands obtained by the electrophoretic separation of proteins was carried out using the Gel Analyzer 2010a (http://www.gelanalyzer.com).

4.6. FTIR Analysis of Lupine Seeds

The Fourier transform infrared spectroscopy technique (FT-IR;, an analytical technique offering a possibility of chemical identification of samples) is based on the fact that chemical substances show selective absorption in infrared regions. The molecules vibrate, after absorption of IR radiations, giving rise to the spectrum of absorption [85]. The FTIR spectra were recorded in the range between 4000 and 500 cm^{-1} with a Nicolet™ 6700 spectrometer (Thermo Scientific, Waltham, MA, USA); a spectral resolution was 4 cm^{-1}. The spectra were obtained using ATR and DRIFTS techniques. Room temperature reflectance spectra were recorded using a Spectra-Tech DRIFTS and ATR accessory (Spectra-Tech Inc., Hanover Park, IL, USA). Eachsample was analyzed directly on the sample cup after roughing it with silicon carbide (SiC) paper. A small disc of SiC paper was used to rub off a small amount of sample. Pieces of clean SiC paper was used as the background. For the FT-IR/horizontal attenuated total reflectance (HATR) technique, a diamond crystal was used. HATR technique provides a simple means of direct handling of plant material. The lupine samples were placed in a HATR crystal and a beam of infrared radiation is directed onto a diamond crystal. The wave of radiation extends beyond the surface of the crystal and comes into the sample. The resultant radiation was measured and plotted as a function of the wave number.

4.7. SEM/EDS Microanalysis

Scanning electron microscope (SEM) which produces images of samples by scanning them with a focused beam of electrons (various characteristics of the sample e.g., size and shape) was used for morphological analysis of seed samples. The EDS technique (Energy Dispersive Spectroscopy), was used to identify different chemical elements present in lupine seeds as described by He and coworkers [86] and Psaras and Manetas [87], with modifications. Mature, dry seeds of yellow lupine from control and irrigated plants (five seeds of each kind) were cut on half and without sputter coating with gold were observed with a SEM, model FEI INSPECT S50 (FEI, Hillsboro, OR, USA). X-ray microanalyses were made with the EDS system (Ametek, Weiterstadt, Germany) connected to the SEM, in six selected points of each seed (embryo axis, cotyledon near plumule, cotyledon center, cotyledon near radicle, seed coat and plumule, Figure 7a,b). In all cases the voltage was 20 kV (for micrograhs 10 kV), the pressure 60 Pa, spod size 3 and live time 30s. EDS spectra were analyzed and elements whose presence was recorded in the form of peaks summarized in tables (eZAF Smart Quant Results). The content of chemical elements (weight %) were estimated statistically.

4.8. Histochemical Localization of H_2O_2

The generation of H_2O_2 was observed using peroxidase-catalyzed 3,3-diaminobenzidine (DAB; Sigma) polymerization test, according to Thordal–Christensen and cowerkers [88] with some modifications [89]. Seedlings of lupine were incubated for 12 h in a solution containing 1 mg·mL^{-1} DAB dissolved in Tris buffer (100 mM Tris, 10 mM EDTA-2Na, 100mM NaCl, pH 7.6). Additional "negative control" series comprised of lupine seedlings incubated with 1 mM ascorbic acid (AA; Sigma). Then the roots were fixed in PBS-buffered 3.7% paraformaldehyde solution for 40 min (4 °C), washed with PBS (three times) and incubated in a citric acid buffered digestion solution (pH 5.0) containing 2.5% pectinase, 2.5% cellulose and 2.5% pectolyase, at 37 °C for 30 min. Afterwards the roots were washed with PBS and squashed onto microscope glass slides in a mixture of glycerol and PBS (9:1; v/v). H_2O_2 was visualized under the SMZ-2T microscope (equipped with DXM 1200 CCD camera Nikon, Tokyo, Japan) as a reddish-brown coloration.

4.9. Statistical Analysis

The differences between values obtained in the particular experiments were assessed with the analysis of variance (ANOVA) and following post-hoc Tukey's test, the Student's *t*-test or the Mann–Whitney U test. The choice of the test to the individual experiment was indicated in the description of the graphs.

5. Conclusions

In conclusion, our research clearly indicates that irrigation of crops in drought conditions may prevent them from drying out, but due to the lack of appropriate parameters of this agrotechnical practice, it does not always lead to higher yields. Irrigation can affect seed formation, changes the level of ploidy of cotyledon cells. Furthermore, it may interfere with the quality of storage substances and influence seed germination. In connection with the above, we believe that the studies on the modifications of stressful environmental conditions on the arable crops are necessary and justified and that the agrotechnical procedure of plant irrigation (a subject of our current work) must be carefully selected and developed for the individual plant species.

Author Contributions: G.S. conceived and designed a field research. I.C. and J.T.P. conceived and designed a laboratory research. A.F. and Ł.S. conducted a field experiment and provided seeds. J.L., A.T.D., A.B., A.Ż., M.A., K.W. and J.P. conducted laboratory experiments. K.W. conducted statistical analysis. I.C. and J.P. analyzed the data. J.P. wrote the manuscript. J.M. conducted a language correction. All authors read and approved the manuscript.

Funding: Equipment for this research was partially financed by EU funds via the projects with contract numbers: POPW.01.03.00-20-034/09-00 and POPW.01.03.00-20-004/11-00. The paper has received financial support from: 1) Ministry of Science and Higher Education, programme "Regional Initiative of Excellence" in years 2019–2022, Project No. 005/RID/2018/19; 2) University of Łódź, statutory subsidy No. 506/1141, Project No. B1811000000053.01; 3) Polish Ministry of Science and Higher Education under subsidy for maintaining the research potential of the Faculty of Biology, University of Bialystok.

Acknowledgments: We thank M. Fronczak for help in preparing this manuscript in English.

Conflicts of Interest: The authors declare no conflict of interest.

Abbreviations

FTIR	Fourier transform infrared spectroscopy
SEM	Scanning electron microscope
EDS	Energy dispersive spectroscopy

References

1. Sujak, A.; Kotlarz, A.; Strobel, W. Compositional and nutritional evaluation of several lupin seeds. *Food Chem.* **2006**, *98*, 711–719. [CrossRef]

2. Fernández-Pascual, M.; PueyoMaría, J.J.; Lucas, M.M. Singular features of the Bradyrhizobium-Lupinus symbiosis. *Dyn. Soil Dyn. Plant* **2007**, *1*, 1–16.

3. Coba de la Peña, Y.; Pueyo, J.J. Legumes in the reclamation of marginal soils, from cultivar and inoculants selection to transgenic approaches. *Agron. Sustain. Dev.* **2012**, *32*, 65–91. [CrossRef]

4. Lucas, M.M.; Stoddard, F.L.; Annicchiarico, P.; Frías, J.; Martínez-Villaluenga, C.; Sussmann, D.; Duranti, M.; Seger, A.; Zander, P.M.; Pueyo, J.J. The future of lupin as a protein crop in Europe. *Front. Plant Sci.* **2015**, *6*, 705. [CrossRef] [PubMed]

5. Wilmowicz, E.; Kućko, A.; Burchardt, S.; Przywieczerski, T. Molecular and hormonal aspects of drought-triggered flower shedding in yellow lupine. *Int. J Mol. Sci.* **2019**, *20*, 3731. [CrossRef]

6. Graham, P.H.; Vance, C.P. Legumes: Importance and constraints to greater use. *Plant Physiol.* **2003**, *131*, 872–877. [CrossRef]

7. Straková, E.; Suchý, P.; Večerek, V.; Šerman, V.; Mas, N.; Jůzl, M. Nutritional composition of seeds of genus Lupinus. *Acta Vet. Brno* **2006**, *75*, 489–493. [CrossRef]

8. Ogura, T.; Ogihara, J.; Sunairi, M.; Takeishi, H.; Aizawa, T.; Olivos-Trujillo, M.R.; Maureira-Butler, I.J.; Salvo-Garrido, H.E. Proteomic characterization of seeds from yellow lupin (*Lupinus luteus* L.). *Proteomics* **2014**, *14*, 1543–1546. [CrossRef]

9. Juzoń, K.; Czyczyło-Mysza, I.; Marcińska, I.; Dziurka, M.; Waligórski, P.; Skrzypek, E. Polyamines in yellow lupin (*Lupinus luteus* L.) tolerance to soil drought. *Acta Physiol. Plant.* **2017**, *39*, 202. [CrossRef]

10. Bettzieche, A.; Brandsch, C.; Eder, K.; Stangl, G.I. Lupin protein acts hypocholesterolemic and increases milk fat content in lactating rats by influencing the expression of genes involved in cholesterol homeostasis and triglyceride synthesis. *Mol. Nutr. Food Res.* **2009**, *53*, 1134–1142. [CrossRef]

11. Duranti, M.; Morazzoni, P. Nutraceutical properties of lupin seed proteins. A great potential still waiting for full exploitation. *Agro Food Ind. Hi Technol.* **2011**, *22*, 20–23.

12. Walsh, C.J.; Guinane, C.M.; O'Toole, P.W.; Cotter, P.D. Beneficial modulation of the gut microbiota. *FEBS Lett.* **2014**, *588*, 4120–4130. [CrossRef] [PubMed]

13. Strozycki, P.M.; Szczurek, A.; Lotocka, B.; Figlerowicz, M.; Legocki, A.B. Ferritins and nodulation in *Lupinus luteus*: Iron management in indeterminate type nodules. *J. Exp. Bot.* **2007**, *58*, 3145–3153. [CrossRef] [PubMed]

14. Bieniaszewski, T.; Szwejkowski, Z.; Fordoński, G. Impact of temperature and rainfall distribution over 1989-1996 on the biometric and structural characteristics as well as on the 'Juno' yellow lupin yielding. *EJPAU* **2000**, *3*. Available online: http://www.ejpau.media.pl/volume3/issue2/agronomy/art-02.html (accessed on 5 November 2019).

15. Faligowska, A.; Szukała, J. Influence of irrigation and soil tillage systems on vigour and sowing value of yellow lupine seeds. *Sci. Nat. Technol.* **2012**, *2*, 26.

16. Podleśny, J.; Podleśna, A. Wpływ wysokiej temperatury w okresie kwitnienia na wzrost, rozwój i plonowanie łubinu żółtego. *Acta Agrophyscia* **2012**, *19*, 825–834.

17. Pszczółkowska, A.; Olszewski, J.; Płodzień, K.; Kulik, T.; Fordoński, G.; Żuk-Gołaszwska, K. Effect of the water stress on the productivity of selected genotypes of pea (*Pisum sativum* L.) and yellow lupin (*Lupinus luteus* L.). *Electron. J. Polish Agric. Univ, Agron.* **2003**, *6*, 1.

18. Gresta, F.; Wink, M.; Prins, U.; Abberton, M.; Capraro, J.; Scarafoni, A.; Hill, G. Lupins in european cropping system. In *Legumes in Cropping System*; Murphy-Bokern, D., Stoddard, F.L., Watson, C.A., Eds.; CABI Publishing: Wallingford, UK, 2017; pp. 88–108.

19. Okamoto, K.; Sagata, N. Mechanism for inactivation of the mitotic inhibitory kinase Wee1 at M phase. *Proc. Nat. Acad. Sci. USA* **2007**, *104*, 3753–3758. [CrossRef]

20. Farooq, M.; Wahid, A.; Kobayashi, N.; Basra, A.S.M. Plant drought stress: Effects, mechanisms and management. *Agron. Sustain. Dev.* **2009**, *29*, 185–212. [CrossRef]

21. Lisar, S.Y.S.; Motafakkerazad, R.; Hossain, M.M.; Rahman, I.M.M. Water Stress in Plants: Causes, Effects and Responses. In *Water Stress*; Rahman, I.M.M., Ed.; InTech: London, UK, 2012; pp. 1–16.

22. Osakabe, Y.; Osakabe, K.; Shinozaki, K.; Tran, L.S.P. Response of plants to water stress. *Front. Plant. Sci.* **2014**, *5*, 1–8. [CrossRef]

23. Shanker, A.K.; Maheswari, M.; Yadav, S.K.; Desai, S.; Bhanu, D.; Attal, N.B.; Venkateswarlu, B. Drought stress responses in crops. *Funct. Integr. Genomics* **2014**, *14*, 11–22. [CrossRef] [PubMed]

24. Yadav, S.; Sharma, K.D. Molecular and Morphophysiological Analysis of Drought Stress in Plants Summy. In *Plant Growth*; Rigobello, E., Ed.; InTech: London, UK, 2016; pp. 149–173.

25. Kang, Y.; Khan, S.; Ma, X. Climate change impacts on crop yield, crop water productivity and food security - A review. *Prog. Nat. Sci.* **2009**, *19*, 1665–1674. [CrossRef]

26. Morison, J.I.; Baker, N.; Mullineaux, P.; Davies, W. Improving water use in crop production. *Philos. Trans. R. Soc. Lond. B Biol. Sci.* **2008**, *363*, 639–658. [CrossRef] [PubMed]

27. Almer, C.; Laurent-Lucchetti, J.; Oechslin, M. Water scarcity and rioting: Disaggregated evidence from Sub-Saharan Africa. *J. Environ. Econ. Manage.* **2016**, *86*, 193–209. [CrossRef]

28. Cox, F.R.; Sullivan, G.A.; Martin, C.K. Effect of calcium and irrigation treatments on peanut yield, grade and seed quality 1. *Peanut Sci.* **1976**, *3*, 81–85. [CrossRef]

29. Holden, N.M.; Brereton, A.J. Adaptation of water and nitrogen management of spring barley and potato as a response to possible climate change in Ireland. *Agric. Water Manag.* **2006**, *82*, 297–317. [CrossRef]

30. Zhao, H.L.; Cui, J.Y.; Zhou, R.L.; Zhang, T.H.; Zhao, X.Y.; Drake, S. Soil properties, crop productivity and irrigation effects on five croplands of Inner Mongolia. *Soil Tillage Res.* **2007**, *93*, 346–355. [CrossRef]

31. Kassab, O.M.; Abo Ellil, A.A.; Abdallah, E.F.; Ibrahim, M.M. Performance of some chickpea cultivars under sprinkler irrigation treatments in sandy soil. *Aust. J. Basic App. Sci.* **2012**, *6*, 618–625.

32. Levidow, L.; Zaccaria, D.; Maia, R.; Vivas, E.; Todorovic, M.; Scardigno, A. Improving water-efficient irrigation: Prospects and difficulties of innovative practices. *Agric. Water Manag.* **2014**, *146*, 84–94. [CrossRef]

33. Breen, A.N.; Richards, J.H. Irrigation and fertilization effects on seed number, size, germination and seedling growth: Implications for desert shrub establishment. *Oecologia* **2008**, *157*, 13–19. [CrossRef]

34. Faligowska, A.; Panasiewicz, K.; Szymańska, G.; Bartos-Spychała, M. The seeds quality of yellow lupine depending on selected agrotechnical factors. *Prog. Plant Prot.* **2013**, *53*, 293–296.

35. Faligowska, A.; Panasiewicz, K.; Szukała, J.; Koziara, W. Germination and vigour of narrow-leaved lupin seeds as the effect of irrigation of parent plants and cultivation in different soil tillage systems. *Polish J. Agron.* **2016**, *24*, 3–8.

36. Tooke, P.B. Fourier self-deconvolution in IR spectroscopy. *Trends Analyt. Chem.* **1988**, *7*, 130–136. [CrossRef]

37. Lamba, O.P.; Borchman, D.; Sinha, S.K.; Shah, J.; Renugopalakrishnan, V.; Yappert, M.C. Estimation of the secondary structure and conformation of bovine lens crystallins by infrared spectroscopy: Quantitative analysis and resolution by Fourier self-deconvolution and curve fit. *Biochim. Biophys. Acta* **1993**, *1163*, 113–123. [CrossRef]

38. Ribeiro da Luz, B. Attenuated total reflectance spectroscopy of plant leaves: A tool for ecological and botanical studies. *New Phytol.* **2006**, *172*, 305–318. [CrossRef]

39. Żarski, J.; Dudek, S.; Kuśmierek-Tomaszewska, R.; Rolbiecki, R.; Rolbiecki, S. Prognozowanie efektów nawadniania roślin na podstawie wybranych wskaźników suszy meteorologicznej i rolniczej. *Rocznik Ochr. Środ.* **2013**, *15*, 2185–2203.

40. Davies, C.L.; Turner, D.W.; Dracup, M. Yellow lupin (*Lupinus luteus*) tolerates waterlogging better than narrow-leaved lupin (*L. angustifolius*). I. Schoot and root growth in a controlled environment. *Aust. J. Agric. Res.* **2000**, *51*, 701–709. [CrossRef]

41. Dracup, M.; Turner, N.C.; Tang, C.; Reader, M.; Palta, J. Responses to abiotic stresses. Chapter 8. In *Lupins as Crop Plants: Biology, Production and Utilization*; Gladstones, J.S., Atkins, C.A., Hamblin, J., Eds.; CABI: Wallingford, UK, 1998; pp. 227–262.

42. Płażek, A.; Dubert, F.; Kopeć, P.; Dziurka, M.; Kalandyk, A.; Pastuszak, J.; Wolko, B. Seed hydropriming and smoke water significantly improve low-temperature germination of *Lupinus angustifolius* L. *Int. J. Mol. Sci.* **2018**, *19*, 992. [CrossRef]

43. Winnicki, K.; Ciereszko, I.; Leśniewska, J.; Dubis, A.T.; Basa, A.; Żabka, A.; Hołota, M.; Sobiech, Ł.; Faligowska, A.; Skrzypczak, G.; et al. Irrigation affects characteristics of Barrow-leaved lupin (*Lupinus angustifolius* L.). *Planta* **2019**, *249*, 1731–1746. [CrossRef]

44. Ambika, S.; Manonmani, V.; Somasundaram, G. Review on effect of seed size on seedling vigour and seed yield. *Res. J. Seed Sci.* **2014**, *7*, 31–38. [CrossRef]

45. Peksen, E.; Peksen, A.; Bozoglu, H.; Gulumser, A. Some seed traits and their relationships to seed germination and field emergence in pea (*Pisum sativum* L.). *J. Agron.* **2004**, *3*, 243–246.

46. Rastegar, Z.; Kandi, M.A.S. The effect of salinity and seed size on seed reserve utilization and seedling growth of soybean (*Glycin max*). *Int. J. Agron. Plant Prod.* **2011**, *2*, 1–4.

47. Farhoudi, R.; Motamedi, M. Effect of salt stress and seed size on germination and early seedling growth of safflower (*Carthamus tinctorius* L.). *Seed Sci. Technol.* **2010**, *38*, 73–78. [CrossRef]

48. Bourdon, M.; Pirrello, J.; Cheniclet, C.; Coriton, O.; Bourge, M.; Brown, S.; Moïse, A.; Peypelut, M.; Rouyère, V.; Renaudin, J.P.; et al. Evidence for karyoplasmic homeostasis during endoreduplication and a ploidy-dependent increase in gene transcription during tomato fruit growth. *Development* **2012**, *139*, 3817–3826. [CrossRef]

49. Takahashi, N.; Umeda, M. Cytokinins promote onset of endoreplication by controlling cell cycle machinery. *Plant Signal Behav.* **2014**, *9*, e29396. [CrossRef]

50. Scholes, D.R.; Paige, K.N. Plasticity in ploidy: A generalized response to stress. *Trends Plant Sci.* **2015**, *20*, 165–175. [CrossRef]

51. Lee, H.O.; Davidson, J.M.; Duronio, R.J. Endoreplication: Polyploidy with purpose. *Genes Develop.* **2009**, *23*, 2461–2477. [CrossRef]

52. Knake-Sobkowicz, S.; Marciniak, K. Cellular accumulation of protein bodies and changes in DNA ploidy level during seed development of *Lathyrus tuberosus* L. *Acta Biol. Crac. Series Bot.* **2005**, *47*, 147–157.

53. Dante, R.A.; Larkins, B.A.; Sabelli, P.A. Cell cycle control and seed development. *Front. Plant. Sci.* **2014**, *5*, 1–14. [CrossRef]

54. Leiva-Neto, J.T.; Grafi, G.; Sabelli, P.A.; Dante, R.A.; Woo, Y.; Maddock, S.; Gordon-Kamm, W.J.; Larkins, B.A. A dominant negative mutant of cyclin-dependent kinase A reduces endoreduplication but not cell size or gene expression in maize endosperm. *Plant Cell* **2004**, *16*, 1854–1869. [CrossRef]

55. Joubès, J.; Chevalier, C. Endoredupication in higher plants. *Plant Mol. Biol.* **2000**, *43*, 735–745. [CrossRef] [PubMed]

56. González-Sama, A.; Coba de la Peña, T.; Kevei, Z.; Mergaert, P.; Lucas, M.M.; de Felipe, M.R.; Kondorosi, E.; Pueyo, J.J. Nuclear DNA endoreduplication and expression of the mitotic inhibitor Ccs52 associated to determinate and lupinoid nodule organogenesis. *Mol. Plant-Microbe Inter. J.* **2006**, *19*, 176–180. [CrossRef] [PubMed]

57. Park, S.; Yeung, E.C.; Paek, K. Endoredupication in *Phalaenopsis* is affected by light quality from light-emitting diodes during somatic embryogenesis. *Plant Biotechnol. Rep.* **2010**, *4*, 303–309. [CrossRef]

58. Chevalier, C.; Nafati, M.; Mathieu-Rivet, E.; Bourdon, M.; Frangne, N.; Cheniclet, C.; Renauldin, J.-P.; Gévaudant, F.; Hernould, M. Elucidating the functional role of endoreduplication in tomato fruit development. *Ann. Bot.* **2011**, *107*, 1159–1169. [CrossRef]

59. Kondorosi, E.; Roudier, F.; Gendreau, E. Plant cell size control: Growing by ploidy? *Curr. Opin. Plant Biol.* **2000**, *3*, 488–492. [CrossRef]

60. Breuer, C.; Braidwood, L.; Sugimoto, K. Endocycling in the path of plant development. *Curr. Opin. Plant Biol.* **2014**, *17*, 78–85. [CrossRef]

61. Lqari, H.; Pedroche, J.; Girón-Calle, J.; Vioque, J.; Millán, F. Purification and partial characterization of storage proteins in *Lupinus angustifolius* seeds. *Grasas y Aceites* **2004**, *55*, 364–369.

62. Duranti, M.; Consonni, A.; Magni, C.; Sessa, F.; Scarafoni, A. The major proteins of lupin seed: Characterisation and molecular properties for use as functional and nutraceutical ingredients. *Trends Food Sci. Technol.* **2008**, *19*, 624–633. [CrossRef]

63. Foley, R.C.; Jimenez-Lopez, C.; Kamphuis, L.G.; Hane, J.K.; Melser, S.; Singh, K.B. Analysis of conglutin seed storage proteins across lupin species using transcriptomic, protein and comparative genomic approaches. *BMC Plant Biol.* **2015**, *15*, 106. [CrossRef]

64. Jimenez-Lopez, J.C.; Melser, S.; DeBoer, K.; Thatcher, L.F.; Kamphuis, L.G.; Foley, R.C.; Singh, K.B. Narrow-leaved lupin (*Lupinus angustifolius*) β1- and β6-conglutin proteins exhibit antifungal activity, protecting plants against necrotrophic pathogen induced damage from *Sclerotinia sclerotiorum* and *Phytophthora nicotianae*. *Front. Plant. Sci.* **2016**, *7*, 1856. [CrossRef]

65. Barciszewska-Pacak, M.; Milanowska, K.; Knop, K.; Bielewicz, D.; Nuc, P.; Plewka, P.; Pacak, A.M.; Vazquez, F.; Karłowski, W.; Jarmołowski, A.; et al. Arabidopsis microRNA expression regulation in a wide range of abiotic stress responses. *Front Plant Sci.* **2015**, *6*, 410. [CrossRef] [PubMed]

66. Battaglia, M.; Covarrubis, A.A. Late Embryogenesis Abundant (LEA) proteins in legumes. *Front. Plant Sci.* **2013**, *4*, 190. [CrossRef] [PubMed]

67. Shrivastava, P.; Kumar, R. Soil salinity: A serious environmental issue and plant growth promoting bacteria as one of the tools for its alleviation. *Saudi. J. Biol. Sci.* **2015**, *22*, 123–131. [CrossRef] [PubMed]

68. Heredia, A. Biophysical and biochemical characteristics of cutin, a plant barrier biopolymer. *Bioch. Bioph. Acta* **2003**, *1620*, 1–7. [CrossRef]

69. Shao, S.; Meyer, C.J.; Ma, F.; Peterson, C.A.; Bernards, M.A. The outermost cuticle of soybean seeds: Chemical composition and function during imbibition. *J. Exp. Bot.* **2007**, *58*, 1071–1082. [CrossRef]

70. Clua, A.A.; Gimenez, D.O. Environmental factors during seed development of narrow-leaved bird's-foot-trefoil (*Lotus tenuis*) influences subsequent dormancy and germination. *Grass Forage Sci.* **2003**, *58*, 333–338. [CrossRef]

71. Clua, A.; Fernandez, G.; Ferro, L.; Dietrich, M. Drought stress conditions during seed development of narrowleaf birdsfoot trefoil (*Lotus glaber*) influences seed production and subsequent dormancy and germination. *Lotus Newsl.* **2006**, *36*, 58–63.

72. Ranal, M.A.; Santana, D.G. How and why to measure the germination process? *Rev. Bras. Bot.* **2006**, *29*, 1–11. [CrossRef]

73. Gallardo, K.; Job, C.; Groot, S.P.C.; Puype, M.; Demol, H.; Vandekerckhove, J.; Job, D. Proteomic analysis of *Arabidopsis* seed germination and priming. *Plant Physiol.* **2001**, *126*, 835–848. [CrossRef]

74. Osuna, D.; Prieto, P.; Aguilar, M. Control of seed germination and plant development by carbon and nitrogen availability. *Front. Plant Sci.* **2015**, *6*, 1023. [CrossRef]

75. Tuteja, N.; Mahajan, S. Calcium signaling network in plants. *Plant Signal Behav.* **2007**, *2*, 79–85. [CrossRef] [PubMed]

76. Polit, J.T.; Ciereszko, I. In situ activities of hexokinase and fructokinase in relation tophosphorylation status of root meristem cells of *Vicia faba* during reactivation from sugar starvation. *Physiol. Plant.* **2009**, *135*, 342–350. [CrossRef] [PubMed]

77. Polit, J.T.; Ciereszko, I. Sucrose synthase activity and carbohydrates content in relation to phosphorylation status of *Vicia faba* root meristems during reactivation from sugar depletion. *J. Plant Physiol.* **2012**, *169*, 1597–1606. [CrossRef] [PubMed]

78. Kurusu, T.; Kimura, S.; Tada, Y.; Kaya, H.; Kuchitsu, K. Plant signaling networks involving reactive oxygen species and Ca^{2+}. In *Handbook on reactive oxygen species (ROS): Formation mechanisms, physiological roles and common harmful effects*; Suzuki, M., Yamamoto, S., Eds.; Nova Science Publishers: Hauppauge, NY, USA, 2013; pp. 315–324.

79. Passardi, F.; Penel, C.; Dunand, C. Performing the paradoxical: How plant peroxidases modify the cell wall. *Trends Plant Sci.* **2004**, *9*, 534–540. [CrossRef] [PubMed]

80. Dunand, C.; Crèvecoeur, M.; Penel, C. Distribution of superoxide and hydrogen peroxide in *Arabidopsis* root and their influence on root development: Possible interaction with peroxidases. *New Physiol.* **2007**, *174*, 332–341. [CrossRef] [PubMed]

81. Liu, X.; Xing, D.; Li, L.; Zhang, L. Rapid determination of seed vigor based on the level of superoxide generation during early imbibitions. *Photochem. Photobiol. Sci.* **2007**, *6*, 767–774. [CrossRef]

82. Szopińska, D. Effects of hydrogen peroxide treatment on the germination, vigour and health of *Zinnia elegans* seeds. *Folia Hort.* **2014**, *26*, 19–29. [CrossRef]

83. Maszewski, J.; Kaźmierczak, A.; Polit, J. Cell cycle agrest in antheridial extract-treated Root meristems of *Allium cepa* and *Melandrium noctiflorum*. *Folia Histochem. Cytobiol.* **1998**, *36*, 35–43.

84. Lowry, O.H.; Rosenbrough, N.J.; Farr, A.L.; Randall, R.J. Protein measurement with the Folin Phenol Reagen. *J. Biol. Chem.* **1951**, *193*, 265–275.

85. Dubis, E.N.; Dubis, A.T.; Popławski, J. Determination of the aromatic compounds in plant cuticular waxes using FT-IR spectroscopy. *J. Mol. Struct.* **2001**, *596*, 83–88. [CrossRef]

86. He, H.; Veneklaas, E.J.; Kuo, J.; Lambers, H. Physiological and ecological significance of biomineralization in plants. *Trends Plant Sci.* **2014**, *19*, 166–174. [CrossRef] [PubMed]

87. Psaras, G.K.; Manetas, Y. Nickel localization in seeds of the metal hyperaccumulator *Thlaspi pindicum* Hausskn. *Ann. Bot.* **2001**, *88*, 513–516. [CrossRef]

88. Thordal-Christensen, H.; Zhang, Z.; Wei, Y.; Collinge, D.B. Subcellular localization of H2O2 in plants: H2O2 accumulation in papillae and hypersensitive response during the barley-powdery mildew interaction. *Plant J.* **1997**, *11*, 1187–1194. [CrossRef]

89. Żabka, A.; Polit, J.T.; Maszewski, J. DNA replication stress induces deregulation of the cell cycle events in root meristems of *Allium cepa*. *Ann. Bot.* **2012**, *110*, 1581–1591. [CrossRef]

International Journal of
Molecular Sciences

Article

Allelic Variants for Candidate Nitrogen Fixation Genes Revealed by Sequencing in Red Clover (*Trifolium pratense* L.)

Oldřich Trněný [1,*], David Vlk [2], Eliška Macková [2], Michaela Matoušková [1], Jana Řepková [2], Jan Nedělník [1], Jan Hofbauer [1], Karel Vejražka [1], Hana Jakešová [3], Jan Jansa [4], Lubomír Piálek [5] and Daniela Knotová [6]

[1] Agricultural Research, Ltd., Zahradní 1, 664 41 Troubsko, Czech Republic; matouskova@vupt.cz (M.M.); nedelnik@vupt.cz (J.N.); hofbauer@vupt.cz (J.H.); vejrazka@vupt.cz (K.V.)
[2] Department of Experimental Biology, Masaryk University, 625 00 Brno, Czech Republic; Vlk.DavidR@email.cz (D.V.); mackova.e.94@gmail.com (E.M.); repkova@sci.muni.cz (J.Ř.)
[3] Red Clover and Grass Breeding, 724 47 Hladké Životice, Czech Republic; hana.jakesova@tiscali.cz
[4] Institute of Microbiology of the Academy of Sciences of the Czech Republic, 142 20 Prague, Czech Republic; jansa@biomed.cas.cz
[5] Department of Zoology, Faculty of Science, University of South Bohemia, 370 05 České Budějovice, Czech Republic; lpialek@yahoo.com
[6] Research Institute for Fodder Crops, Ltd., 664 41 Troubsko, Czech Republic; knotova@vupt.cz
* Correspondence: trneny.oldrich@gmail.com

Received: 26 September 2019; Accepted: 30 October 2019; Published: 2 November 2019

Abstract: Plant–rhizobia symbiosis can activate key genes involved in regulating nodulation associated with biological nitrogen fixation (BNF). Although the general molecular basis of the BNF process is frequently studied, little is known about its intraspecific variability and the characteristics of its allelic variants. This study's main goals were to describe phenotypic and genotypic variation in the context of nitrogen fixation in red clover (*Trifolium pratense* L.) and identify variants in BNF candidate genes associated with BNF efficiency. Acetylene reduction assay validation was the criterion for selecting individual plants with particular BNF rates. Sequences in 86 key candidate genes were obtained by hybridization-based sequence capture target enrichment of plants with alternative phenotypes for nitrogen fixation. Two genes associated with BNF were identified: ethylene response factor required for nodule differentiation (*EFD*) and molybdate transporter 1 (*MOT1*). In addition, whole-genome population genotyping by double-digest restriction-site-associated sequencing (ddRADseq) was performed, and BNF was evaluated by the natural ^{15}N abundance method. Polymorphisms associated with BNF and reflecting phenotype variability were identified. The genetic structure of plant accessions was not linked to BNF rate of measured plants. Knowledge of the genetic variation within BNF candidate genes and the characteristics of genetic variants will be beneficial in molecular diagnostics and breeding of red clover.

Keywords: associated genes; associated polymorphisms; genome-wide association; biological nitrogen fixation; red clover

1. Introduction

The family Fabaceae, consisting of more than 750 genera and 19,000 species, is the third largest family of flowering plants and, in terms of agricultural importance, the second most important family, after Poaceae. Several species from this family serve as genetic model organisms (e.g., *Medicago truncatula* Gaertn., *Pisum sativum* L., and *Lotus corniculatus* L.). One of the largest genera of the Fabaceae family is the clover genus, *Trifolium* L., with more than 250 species [1,2]. This herbaceous genus, which acquired its name as a reference to the characteristic form of the leaf usually consisting of three leaflets (trifoliolate), includes both annual and perennial species and occurs natively in temperate and subtropical regions of the northern and southern hemispheres [3]. The importance of the genus *Trifolium* lies in its agricultural utilization. In addition to several species being cultivated extensively as fodder plants (such as *T. pratense* L., *T. repens* L., *T. hybridum* L., and *T. resupinatum* L.), fast-growing clovers are sown as green manure crops or mixed intercrops to enhance soil fertility and sustainability [4]. As typical for the majority of leguminous plants, *Trifolium* species can establish a mutualistic relationship with the root-nodulating bacteria *Rhizobium leguminosarum* bv. *trifolii*. This initiates a complex process of biological (atmospheric) nitrogen fixation (BNF). In this relationship, the plant provides the bacteria a source of carbon and energy, in addition to phosphorus and other mineral nutrients and also anoxic shelter, and the bacteria supply the plant with nitrogen acquired from the atmosphere, converted into organic compounds utilizable in plant metabolism [5]. BNF in legumes constitutes an irreplaceable nitrogen source for both ecosystems and circulation in nature. Soil N enrichment due to effective BNF is environmentally more sustainable than application of synthetic N fertilizers depending on utilization of nonrenewable sources of energy.

Interactions between nodulating bacteria and root plant systems are highly specific. In many cases, some variant of bacterial strain, or biovar, is able to create functional nodules with only one or several plant species [6] and, to a considerable extent, this determines the efficiency of nodulation and nitrogen fixation [7]. In addition to complexities due to this high specificity, plant breeding directed to the enhancement of nitrogen fixing ability is further complicated by the complexity of the phenotypic trait, as it involves an estimated several hundred genes in nodulation and nitrogen fixation [8]. Red clover (*T. pratense*), with a reported BNF level in aboveground plant tissues as great as 373 kg N·ha^{-1}·year^{-1} [9], and other *Trifolium* species having high rates of BNF heritability [10] are promising for purposes of plant breeding directed to enhancing nitrogen fixing rates.

Cloned and characterized genes responsible for symbioses are involved in recognition of rhizobial nodulation signals, early symbiotic signaling cascades, infection and nodulation processes, and regulation of nitrogen fixation [8]. Plant–bacteria interaction is initiated by phenolic compounds exuded by the plant rhizosphere and which attract rhizobacteria present in soil. Moreover, these phenolic compounds bind to bacterial transcriptional regulator nodD and induce activation of *nod* genes [11]. Products of *nod* genes, termed nodulation (Nod) factors, are lipochitooligosaccharide signaling molecules. Nod factors are specifically bound to receptors on the root surface inducing morphological alteration and activation of root-specific cascades that enable initiation of nodulation, whereby specific kinases and transcription factors participate [12–14]. The systemic signals enable plants to control the number of nodules they form depending upon the number of existing nodules and availability of soil nitrogen [15,16].

Morphological alteration of the root surfaces includes both induction of cell division and curling of root hair to enable bacterial infection of the plant. Infection continues with the creation of an infection thread, which enables bacterial invasion into the cells of the inner cortex. The invasion is followed by activation of gene expression in inner cortex cells, promoting nodule formation and development [17]. The process of nitrogen fixation is performed by a nitrogenase enzyme complex encoded by bacterial *nif* genes [5]. Because reduction of atmospheric nitrogen to ammonia is associated with high energy consumption, it is limited by the availability in soil of phosphorus, a critical component of adenosine triphosphate (ATP). Nitrogenase is extremely sensitive to oxygen exposure, with even low concentrations resulting in irreversible denaturation. To supply bacteria with oxygen for the respiration process while at the same time keeping nitrogenase protected from denaturation, a hemoprotein called leghemoglobin carries oxygen to the peribacteroid membrane while allowing a nearly anaerobic environment to be maintained inside bacteroids [18,19].

Using both forward (mostly chemical mutagenesis) and reverse genetics approaches (such as insertional mutagenesis or gene silencing), some of these genes were already identified. Mostly, this was in model organisms *M. truncatula* and *L. japonicus* [8,20–25], thus enabling scientists to search for orthologous genes in other legumes, including *T. pratense*. Using DNA markers and genetic mapping, candidate legume genes likely participating in different signaling pathways were gradually identified by Cregan et al. [26], Santos et al. [27,28], and Nicolás et al. [29] in the model crop soybean. Recently, key regulating genes conferring nodulation and nitrogen fixation were revealed using comparative genomic and transcriptomic analyses, mainly in *Pisum sativum* L., *Glycine max* (L.) Merr., and *Phaseolus vulgaris* L. [30–32].

Searching for genes involved in symbiotic nitrogen fixation in red clover is facilitated by its small genome size (estimated 418 Mb [33]). Genomic data are available for two varieties, tetraploid Tatra (~314.6 Mbp [34]) and diploid Milvus (~309 Mbp [35]). Both were recently de novo sequenced using next-generation sequencing (NGS). The acquired genome sequences were subsequently annotated, resulting in annotation of 47,398 protein-coding genes from 64,761 predicted genes in variety Tatra [34]. Moreover, several gene families characteristic for red clover were revealed, including 11 leghemoglobin genes and 542 nodule-specific cysteine-rich peptides [34]. For the variety Milvus, 22,042 of a total 40,868 annotated genes were located on seven pseudomolecules (chromosomes) and, using *M. truncatula* as a reference sequence [36], a physical map was constructed.

NGS methods enable genome-wide mining of DNA polymorphisms associated with the traits analyzed. Genome-wide association studies with high-throughput genotyping by sequencing to identify loci associated with nitrogen fixation efficiency were applied in legumes such as *M. truncatula* [37,38] and soybean [39].

Reduced-representation NGS-based genotyping methods, such as double-digest restriction-site-associated sequencing (ddRADseq) [40], were also proven to be beneficial for detecting genome-wide allele frequency fingerprints [41] of populations. Allelic variants such as single-nucleotide polymorphisms (SNPs) and insertion/deletion variations (InDels) make it possible to reveal genetic structure, identify population-specific variants, and find genotype–phenotype associations. Not only do NGS methods allow for genome-wide study, they also look into sequences within every particular candidate gene using bulk target sequencing approaches, such as hybridization-based sequence capture (SeqCap) target enrichment [42].

Here, we describe within- and between-population variability in nitrogen fixation capacity in red clover and demonstrate the utility of several NGS methods for the purpose of key genes and population genotyping. Our present study was based on two sequencing methods, SeqCap using a hybridization-based strategy and ddRADseq. Our goals were to (i) characterize variability that appears in nitrogen fixation candidate genes in red clover populations, (ii) assess this variability in the context of nitrogen fixation efficiency in various red clover accessions, (iii) analyze how level of variance in host candidate genes explains efficiency of biological nitrogen fixation, and (iv) identify allelic variants present in red clover populations and associated with nitrogen fixation level.

2. Results

2.1. Nitrogen Fixation Assays

In total, 1426 individual plants of 12 diploid and 16 tetraploid accessions were measured in three sets using an acetylene reduction assay (ARA). The characteristics of the intrapopulation distribution of nitrogen fixation level depended on the genotypes of the population (Figure 1).

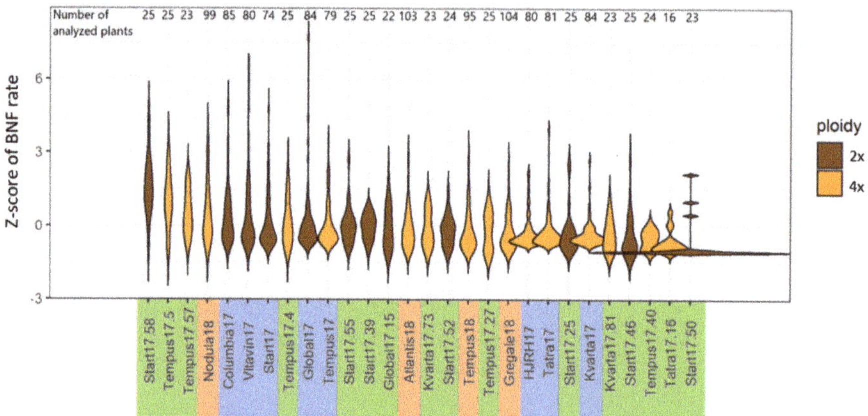

Figure 1. Distribution of Z-score for nitrogen fixation rate evaluated in red clover plants using acetylene reduction assay. On the *x*-axis, genotypes are ordered by mean values of nitrogen fixation. Diploid (brown) and tetraploid (yellow) red clover plants were measured in three sets: Set 1 (blue labels with suffix 17), Set 2 (orange labels, suffix 18), and Set 3 (green labels, suffix 17.xx—progeny of selected contrasting plants from Set 1).

There were significant differences in BNF rates among accessions within all three plant sets (Table 1). In Set 1, approximately 80 plants per accession were without extreme values of fixation. The Columbia17 accession with the highest nitrogen activity differed significantly ($p < 0.01$) from accessions HJRH17 and Kvarta17 (Table 1). In Set 2, approximately 100 plants per accession were evaluated. From Sets 1 and 2 together, accession Nodula18 was the best fixator according to the mean value of BNF rate, which was among the four highest mean values across all accessions (Figure 1). Remaining accessions only showed nitrogen fixation values near the mean. Progeny of eight high- and eight low-BNF plants from Set 1 were retested in Set 3 (Figure 1; Suffix 17.xx). There were significant differences in BNF level among the offspring both of high and low fixators. Examining more closely the progeny of high fixators, multiple comparison revealed significant differences ($p < 0.01$) between Start17.58 and nine accessions and between Tempus17.5 and four accessions. Among progeny of low fixators, significant differences were confirmed between Start17.50 and six accessions (Table 1). As visible in the Figure 1 violin plot, there exist individual plants in most populations that are highly effective BNF rate outliers with several times greater fixation efficiency relative to others.

Table 1. Statistically significant differences of nitrogen fixation capacity within three evaluated sets of red clover plants using acetylene reduction assay.

Plant Set	*p*-Value [1]	Different Pairs of Accessions [2]
1	3.413×10^{-6}	Columbia17-HJRH17, Columbia17-Kvarta17
2	1.151×10^{-6}	Nodula18-Gregale18, Nodula18-Tempus18
3	2.2×10^{-16}	Kvarta17.73-Start17.58, Kvarta17.81-Start17.58, Start17.25-Start17.58, Start17.39-Start17.50, Start17.46-Start17.58, Start17.46-Tempus17.5, Start17.50-Start17.55, Start17.50-Start17.58, Start17.50-Tempus17.4, Start17.50-Tempus17.5, Start17.50-Tempus17.57, Start17.52-Start17.58, Start17.58-Tatra17.16, Start17.58-Tempus17.27, Start17.58-Tempus17.40, Tatra17.16-Tempus17.5, Tempus17.40-Tempus17.5

[1] Kruskal–Wallis test; [2] statistical significance for $p < 0.01$.

For ARA validation, we included measuring of an ethylene standard and measuring the same accession in two consecutive years. The regularly measured ethylene control varied little. Coefficients of variation of standardized ethylene control (97.5 ppm) measurement were 12.5%, 5.6%, and 5.2% in sets 1, 2, and 3, respectively. Tempus plants were planted in both Set 1 (79 plants) and Set 2 (95 plants) as a control variety for nitrogen fixation measurement. In both years of analysis, the results for Tempus accession were similar; comparison of the two plant collections showed no statistically significant differences (Table 1), and both mean values of nitrogen fixation (Tempus17 and Tempus18) were in the middle part of the distribution plot (Figure 1).

2.2. Candidate Gene Target Sequencing

Two panels of selected BNF candidate genes were compiled, and DNAs from plants with contrasting BNF level were sequenced. Panel 1 contained 17 genes with key roles in BNF studied on a model organism (Supplementary Table S1). In this panel, 24 high-BNF and 24 low-BNF plants (Supplementary Table S2) were selected according to ARA and then sequenced. The number of polymorphisms per candidate gene varied between 220 and 887. Polymorphisms were associated with BNF phenotypes while correcting for genetic structure and plant kinship.

The gene ethylene response factor required for nodule differentiation (*EFD*) from the ethylene response factor (ERF) family that was found in targeted sequence Tp_3333 had the most closely associated polymorphisms (Supplementary Table S3) with BNF phenotypes in Panel 1 (Figure 2).

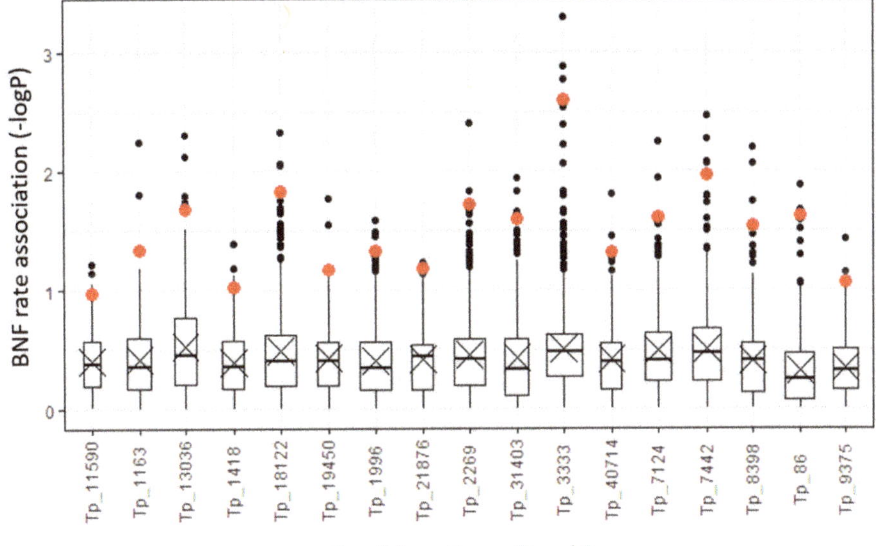

Figure 2. Boxplot of −logP association values for Panel 1 of nitrogen fixation candidate gene polymorphisms based on mixed linear model incorporating both population structure and relationships among accessions. Red dots show mean values for the 10 highest −logP values of each associated polymorphism. The highest mean value was that of Tp_3333, which is the sequence with the ethylene response factor required for nodule differentiation (*EFD*) gene [35].

Panel 2 consisted of 69 candidate genes, which were predominantly selected according to literature specifications with prevalent expression in *M. truncatula* nodules [43]. DNA samples from 25 high-BNF and 25 low-BNF tetraploid plants were sequenced (Supplementary Table S2). Coverage along capture sequences varied among samples (Supplementary Figure S1). Gene polymorphisms were called with high quality and homogenously along the sequences due to the sufficient coverage.

The number of polymorphisms ranged from 18 to 696 per candidate gene sequence. The gene coding molybdate transporter type 1 (*MOT1*) on targeted sequence Tp_34389 was evaluated as having strong effect on the BNF phenotype. This was proven by the highest mean *p*-value among 10 polymorphisms (Supplementary Table S3) with the highest association levels (Figure 3). *MOT1* [35] plays a key role in the BNF process, and its main function is to provide molybdenum for synthesis of the iron–molybdenum cofactor of nitrogenase [44].

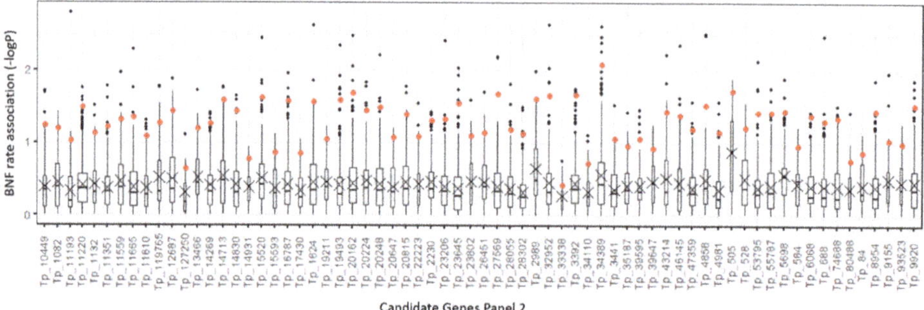

Figure 3. Boxplot of −logP association values for Panel 2 of nitrogen fixation candidate gene polymorphisms based upon a mixed linear model incorporating both population structure and relationship between accessions. Red dots show mean values for the 10 highest −logP values of each associated polymorphism. The highest mean value was for Tp_34389, which is the sequence carrying the molybdate transporter type 1 (*MOT1*) gene [35].

Expected heterozygosity (Hs) was used as a criterion for assessing diversity levels of candidate genes alleles. In the candidate genes of Panel 1, the sequences with the three highest mean Hs values were Tp_2269 with the gene nod factor perception (*NFP*), Tp_21876 with the gene partner of NOB1-like (*PNO1-like*), and Tp_1418 with the gene cytokinin response 1 (*CRE*) cytokinin receptor kinase/nodule organogenesis (Figure 4). In any of the candidate genes of Panel 1, there was no obvious difference between the expected and observed heterozygosity found. In the candidate genes of Panel 2, the two targeted sequences with the highest level of diversity (Hs = 0.23) were Tp_16787, which encodes the gene for nuclear transcription factor Y subunit C2 (NF-YC2), and Tp_20162, encoding flotillin (FLOT) protein. The means of both genes were shown to be close to their medians, indicating symmetrical distribution of their Hs values (Figure 5). In comparison with other genes of Panel 2, there was an obvious difference found between the expected and observed heterozygosity in two of the sequences with candidate genes (Tp_33338 and Tp_84). We found significantly higher values of the expected heterozygosity than values of the observed heterozygosity ($p < 0.01$) in both of the genes using a Mann–Whitney U test. The gene coding *MOT1* on targeted sequence Tp_34389, which manifested the strongest association with the BNF rate phenotype, had a modest diversity level (Hs = 0.164). Among targeted sequences with small diversity were Tp_127250 with the gene non-symbiotic hemoglobin 2 and Tp_2989 with the gene rac-like GTP-binding protein (*ARAC10*). These targeted sequences had low numbers of polymorphisms with low mean Hs, thus implying conserved region and importance of the genes (Figure 5). From seven targeted sequences for leghemoglobins, we could distinguish three groups. Sequences Tp_119765 and Tp_127250, with leghemoglobins genes, were in the first group having low polymorphism counts with low diversity. Sequence Tp_93523, with a leghemoglobin gene, had a low polymorphism count but the highest diversity level among leghemoglobin genes. Leghemoglobin sequences from the third group (Tp_1132, Tp_13466, Tp_14713, Tp_3441) had high polymorphism counts per sequence with medium genetic diversity levels.

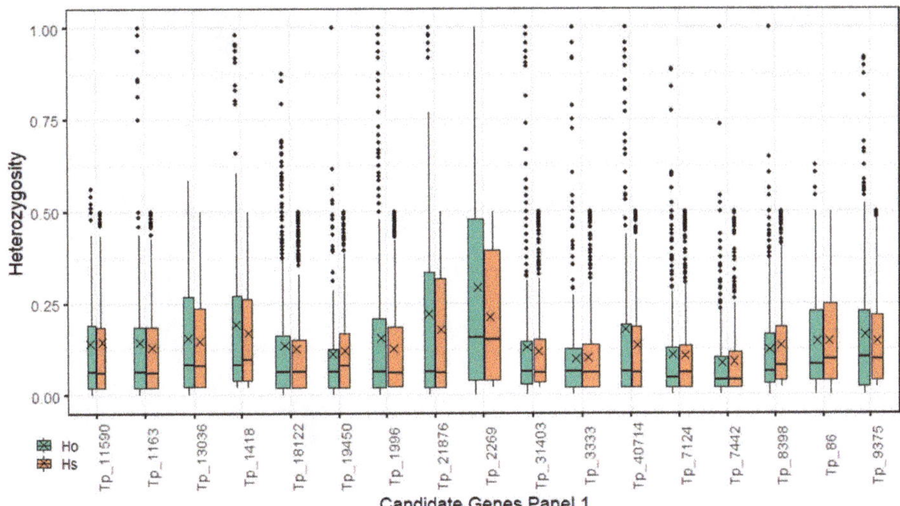

Figure 4. Boxplot of observed (Ho; blue boxes) and expected (Hs; orange boxes) heterozygosity of Panel 1 nitrogen fixation candidate genes. Hs expresses the level of genetic variability. Crosses indicate mean values. Horizontal lines in boxes indicate medians. Bottoms and tops of boxes indicate the first and third quartiles of the dataset. Whiskers indicate range of data but the maximum length of each is 1.5 times greater than the height of its box. Remaining points are outliers. The boxes are drawn with widths proportional to the square roots of the numbers of polymorphisms in targeted sequences.

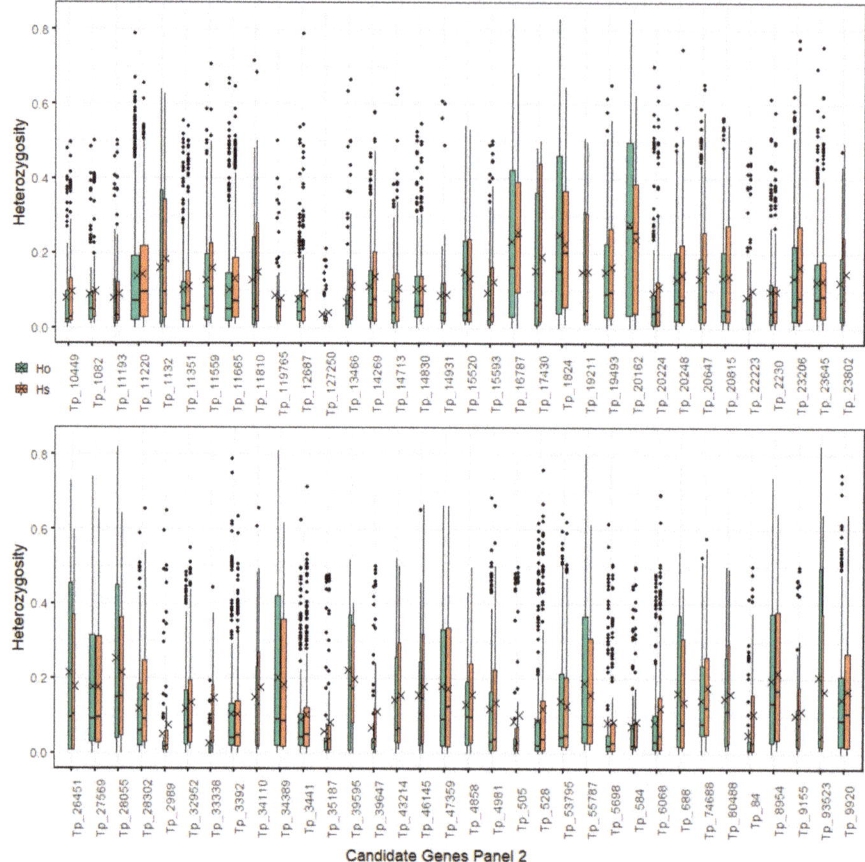

Figure 5. Boxplot of observed (Ho; blue boxes) and expected (Hs; orange boxes) heterozygosity of Panel 2 nitrogen fixation candidate genes. Hs expresses the level of genetic variability. Crosses indicate mean values. Horizontal lines in boxes indicate medians. Bottoms and tops of boxes indicate first and third quartiles of the dataset. Whiskers indicate range of data but the maximum length of each is 1.5 times greater than the height of its box. Remaining points are outliers. The boxes are drawn with widths proportional to the square roots of the numbers of polymorphisms in targeted sequences.

2.3. ddRADseq and N Isotopic Composition

In addition to the targeted sequencing approach that assesses variability of BNF key genes, we harnessed the power of high-throughput sequencing to assess complex whole-genome genotype. Ninety-one *T. pratense* diploid accessions were genotyped at population level using the ddRADseq approach and were phenotypically analyzed for N isotopic composition (indicative of BNF) using the natural ^{15}N abundance method, using isotope ratio mass spectrometry (Figure 6). The first three accessions with the highest BNF level were the variety Start and two wild accessions, TROU 33/96 and CZETROU 15/93. N concentration was measured together with isotope composition. No obvious correlation between isotope composition and N concentration in the leaves was found.

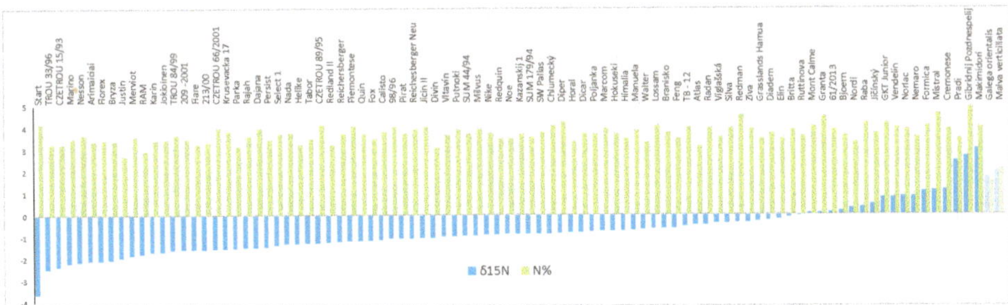

Figure 6. Interpopulation diversity of biological nitrogen fixation as revealed by natural ^{15}N abundance measurement of red clover leaves (δ^{15}N values are shown in blue). Alongside, N concentrations in the leaves are displayed in green (weight %). The control non-nitrogen symbiotic plant (*Malva verticillata*) and leguminous plant *Galega orientalis* uninoculated by symbiotic partner are located on the right side.

Altogether, 91,589 polymorphisms (Supplementary Table S4 were identified with a maximum of 50% missing information, and the minor allele occurred for more than 5% of samples. Sixty-one percent of polymorphisms were mapped to seven linkage groups on the red clover reference genome and 39% of them were mapped to the remaining contigs. The mean coverage of polymorphism was 39.7 × per accession. The mean Hs of polymorphisms was 0.23, which points to a high level of diversity in *T. pratense* populations and corresponds to red clover's cross-pollination system.

In order to assess genetic diversity and its comparison to BNF level, principal component analysis (PCA) was performed. The first two principal components (PCs) of the PCA (Figure 7) explained just 5.6% (3.0% and 2.6% for PC1 and PC2, respectively) of genotypic variance. Despite the weak determination of variance by the first and second PCs, they did distinguish a basic pattern of genetic diversity among the accessions. While the first PC separated in particular wild-type accessions, the second separated varieties. The rest of the accessions formed the main group. Evidently, BNF level did not correspond with this main diversity pattern in the first two PCs, although accession TROU 33/96, which had the second highest BNF rate, was genetically the most different from the others according to the first PC. Moreover, correlation analysis of other PCs up to PC30 revealed no strong correlation level between any genetic structure pattern and phenotype (Supplementary Figure S3), although some PC correlations did show closer relationships with phenotype in comparison with those of other PCs.

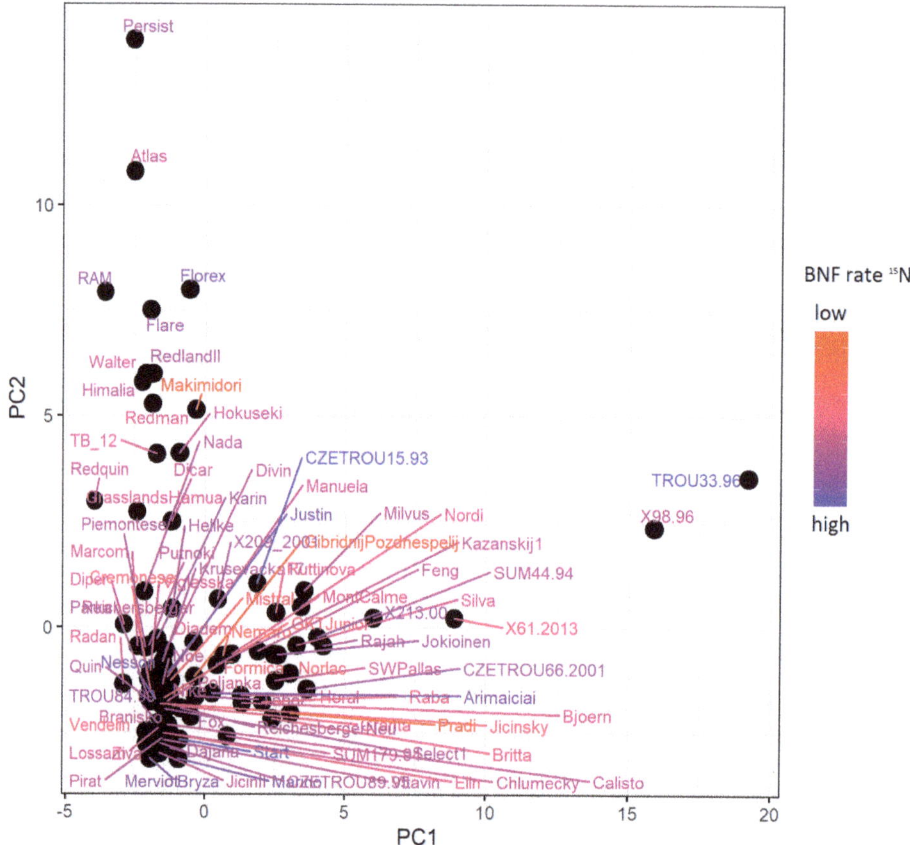

Figure 7. Principal component analysis (PCA) plot of genetic structure of genotype data using 91 samples from red clover populations. PC1 and PC2 indicate principal components. Color scale shows delta ^{15}N value that corresponds to biological nitrogen fixation (BNF) level (red color indicates low BNF level, blue indicates high BNF level).

In order to find associations between genotype and phenotypes, an association study was conducted using the FarmCPU algorithm [45]. We identified three SNPs and one InDel variant that were significantly associated with BNF phenotype (Figure 8) (false discovery rate-adjusted *p*-value < 0.05). Two SNPs lay on linkage group 4 (LG4), one InDel lay on LG1, and one SNP lay on an unmapped contig. Some of the variants were located near genes with functions in the BNF process (Supplementary Table S3). The first mapped significant associated SNP was identified in LG4 position 6,307,333 bp within an intergenic region between genes annotated as mitochondrial rho GTPase 1-like protein and auxin response factor and near the gene for sulfotransferase. The second associated mapped SNP, in LG4 position 12,136,158 bp, lay in an exon of an uncharacterized protein in the neighborhood of two ethylene-responsive transcription factor 3-like genes. The InDel positioned on LG1 at 6,268,253 bp was located in an intron of the gene for lipid phosphate phosphatase 2-like protein and near to several genes for amino-acid permease BAT1-like protein. The third associated SNP had an association level very close to the threshold of association and lay on unmapped contig FKJA01001578.1 at 124 bp, near to the gene for transcription factor DIVARICATA-like protein.

In order to assess the proportion of the total variance explained by the genetic variance, we estimated marker-based narrow-sense heritability from genotype polymorphisms and ^{15}N BNF rate phenotype data. The ^{15}N BNF rate-estimated heritability was 84.7%.

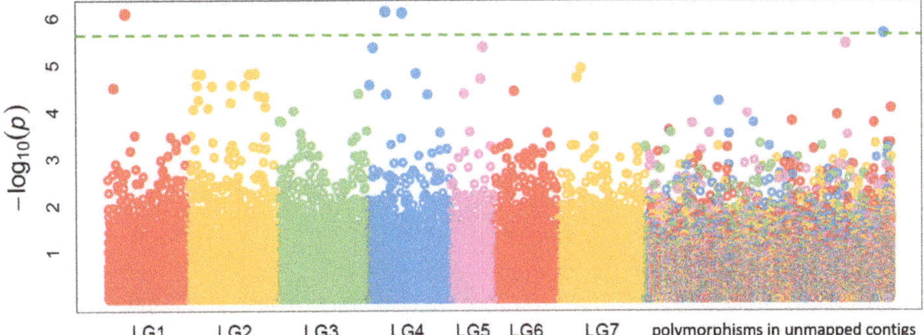

Figure 8. Manhattan plot of genome-wide association of allele frequency and ^{15}N nitrogen fixation phenotypes using the FarmCPU algorithm. The different colors (LG1–LG7) indicate different linkage groups [35]. The segment to the far right shows the polymorphisms unmapped to the linkage groups. The green line indicates the false discovery rate-adjusted *p*-value of 0.05 using the Benjamini–Hochberg correction [46].

2.4. Polymorphism Annotations

Annotation of variants in candidate genes for BNF (Figure 9a,b) and whole-genome population genotyping (Figure 9c,d) were obtained. SNPs were revealed as the most frequent variants. Other variants resulted from length differences (deletion and insertion), and the rest of the variants were based on sequence alterations (Figure 9a,c). In target sequencing of Panels 1 and 2, we found a greater part of sequence alterations than in ddRADseq population genotyping. From the perspective of consequences, half of the variants from targeted sequencing belonged to genic regions (Figure 9b), while ddRADseq population genic variants (Figure 9d) formed only one-quarter of the total variants. Variants of Panel 1 were 29% from genic regions in comparison with variants from Panel 2 that constituted 60% of genic variants. Consequently, Panel 1 was focused on the sequencing of 17 candidate genes and their broad surroundings, but Panel 2 was focused on a higher number of genes and their near-adjacent sequences. Missense variants formed a similar part of variants, as did synonymous variants under both genotyping approaches. For targeted sequencing and ddRADseq population genotyping, we identified a minority of genic variants, such as frameshifts (2% and 1%, i.e., 491 and 431, respectively; Figure 9b,d), stop gained (122 and 160), stop lost (27 and 17), and start lost variants (13 and 16), with severe impact on gene expression.

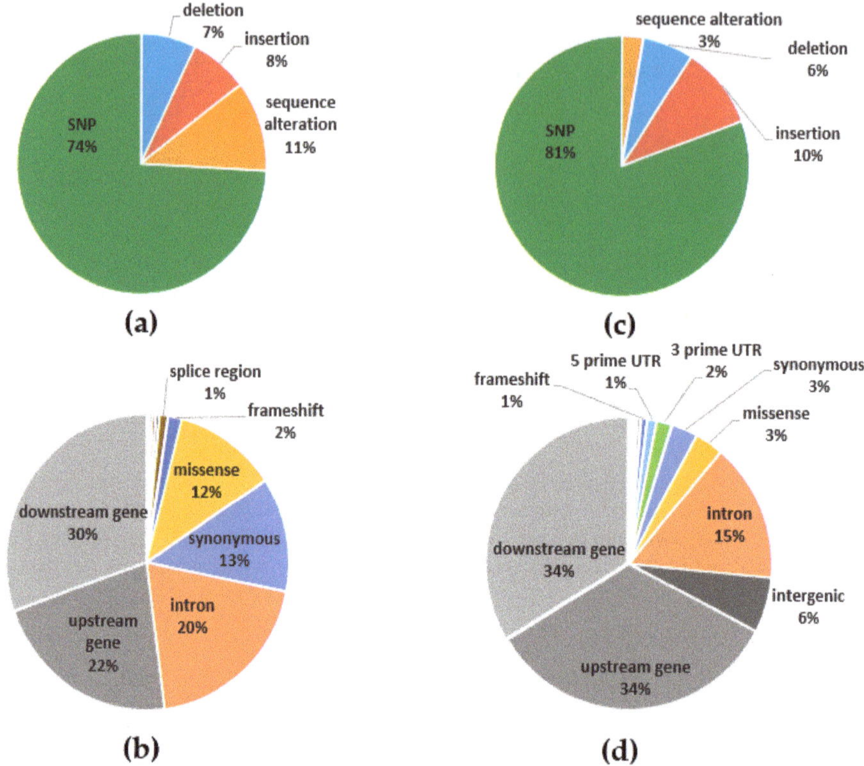

Figure 9. Polymorphism annotation of candidate genes for nitrogen fixation in red clover evaluated by hybridization-based sequence capture (SeqCap) (**a,b**) and whole-genome population genotyping double-digest restriction-site-associated sequencing (ddRADseq) (**c,d**). Distribution of polymorphism classes (**a,c**) and polymorphism consequences (**b,d**).

2.5. Validation of Selected InDel Polymorphisms

Length of validated InDels obtained by targeted sequencing ranged from 9 to 289 bp. From 10 designed primer pairs (Supplementary Table S5), nine gave specific products. The primer for InDel in position 5325 within the *NSP2* gene (targeted sequence Tp_7442) generated no product, but its existence was demonstrated by another primer pair. Analysis of the targeted sequence Tp_19450 with the defective in nitrogen fixation (*DNF2)* gene confirmed the existence of an InDel, but its length was about 200 bp longer than the expected length. Analysis of the remaining InDels confirmed their existence and validated the sequencing data. The lengths of the amplified products were in line with those of the expected products (Supplementary Figure S2).

3. Discussion

BNF is a complex process wherein many genes participate along with the context of environmental conditions [47]. The potential amounts of nitrogen that can be fixed are several times greater than the amounts of nitrogen usually fixed in the fields. The amount of nitrogen fixed by legume–rhizobia symbioses may be increased by as much as 300% through plant breeding and crop management [48]. The potential that plant selection for symbiotic activity may be highly effective is also supported by the data on high heritability. In a relatively stable field environment, the broad-sense heritability of nodulation traits in soybean may exceed 0.8 [47,49], suggesting that nodulation traits are mainly controlled by genetic loci and are useful for breeding varieties with high BNF capacity. In *Trifolium incarnatum* inbred lines, broad-sense heritability was estimated to be similarly high (up to 0.91) [50].

Although the genic nature of BNF efficiency is undeniable, it is the complexity and difficulty of phenotyping that prevented the breeding of red clover for BNF efficiency from being accomplished successfully [51]. With the availability of high-throughput target and genome-wide genotyping approaches, however, new ways were opened for dealing with complex polygenic traits. Recent omics studies revealed deep complexity of the nitrogen fixation process.

Various legume species perform differently in fixing nitrogen, and interspecies variability is well known [10,52]. A study comparing fixation efficiency between model plant *M. truncatula* and fodder crop *M. sativa* showed several-fold lower efficiency in *M. truncatula* than in *M. sativa* [53]. Significant intraspecific variability in BNF efficiency in red clover was frequently observed and evaluated, and phenotypic variability does not appear to be related to ploidy level [54,55]. Here, we evaluated intraspecific variability in symbiotic activity and BNF capacity in red clover, and two methods were applied, indirect (acetylene reduction; ARA) and (isotopic; ^{15}N) estimation of nitrogenase activity [56,57]. ARA was an effective criterion for red clover populations and selection of individual plants with high rates of fixation. Based on ARA of nearly 1500 red clover plants, we observed differences among varieties and among individuals within a variety. The distribution of actual fixation level had a specific characteristic. The largest proportion of plants had low fixation efficiency up to the mean level, while a smaller proportion of plants had higher efficiency, but nearly all of the plants were outperformed by a couple of plants having fixation efficiency several times greater than the mean value of the measured population. This was seen mainly in default populations from Sets 1 and 2. Populations from Set 3 were influenced by selection and, therefore, interpopulation variability in Set 3 was also the highest.

Our research highlights that the breeding value of a plant should be based on progeny performance, and especially so in self-sterile species such as red clover when breeding for a trait as complex as BNF efficiency. Three populations—Start17.58, Tempus17.5, and Tempus17.57 (Figure 1)—were evaluated as being the best fixators among progeny of the selected best BNF plants from Set 1. Even though Start17.58 and Tempus17.5 were the offspring from high-BNF rate plants from Set 1, population Tempus 17.57 was the offspring from low-BNF rate selected plants from Set 1. This confirms the need to select plants based on progeny performance, which is feasible due to the perennial character of red clover. All other red clover population studies showed mostly plants with low fixation rate and rarely plants with high fixation rate. Outlier plants that outperformed the others contributed greatly to the population mean BNF level, but it is probably not achievable to select a population consisting solely of superior plants on the highest performance level. Superior plants occurred in most populations, and, in addition to the additive effect of many genes, their superiority can be derived also from non-additive effects such as a heterotic effect [51]. A high nitrogen fixation rate was confirmed for the Nodula accession, a variety bred for high BNF efficiency. In accord with our previous experience with the Columbia accession, it stood among the most BNF-efficient genotypes. This study's results in Columbia17 confirmed this disposition. The idea of breeding for nitrogen fixation efficiency is limited by the cost of BNF for plants. Leguminous plants have an effective mechanism for holding BNF at the right level [58].

Different types of BNF evaluation methods with many variations were designed [52], but the ARA and natural ^{15}N-abundance methods are commonly used. Each method has its own advantages and difficulties that must be considered. ARA was the subject of much discussion because many factors influence BNF rate, such as temperature [59] and light [60], but uniform measurement conditions allow the relative assessment of BNF rate [61]. ARA is focused on instantaneous measurement, and it is suitable for the comparison of actual BNF levels in specific time. The natural ^{15}N-abundance method is time-integrated, and it inherently assesses the total amount of N fixed for the sample growth period. The natural ^{15}N-abundance method is, therefore, appropriate when we assess interpopulation BNF rate, because we can filter out the influence of actual environment–genotype interaction (such as phenological stages). It assesses a total growth period, whereas ARA is suitable for high-throughput selection in populations where genetic differences among plants are smaller and we can perform selection based on relative comparison. The difference between these two methods is one of the factors why the results of genome-wide associations do not correspond with the results of association of candidate genes. Another reason for non-corresponding results could be that target sequencing analysis was focused on individual samples within the population, while the genome-wide association sample set consisted of bulked population samples where the genotype was expressed as the allelic frequency in a population.

The natural ^{15}N-abundance method was applied for a collection of diverse populations. The populations with the second and third highest fixation rate (Figure 6) were wild accessions. This was in agreement with Provorov and Tikhonovich [10], who concluded that symbiotic potential in the wild-growing (local) varieties is usually greater than that in commercial varieties. Not all of the wild accessions belong to the best nitrogen fixators, however, because there is BNF rate variation within wild accessions, as well as within cultivated accessions. The same conclusion arose from assessment of the ^{15}N BNF rate and genetic structure. The ^{15}N BNF rates were not clearly distributed according to the first two PCs of the PCA that corresponded to the main genetic pattern of collection (Figure 7); however, the unequal correlation level between other PCs and the BNF rate (Supplementary Figure S3) suggests low but possible influence of genetic structure on BNF rate. The best nitrogen fixator was the variety Start, which was also a progenitor of the highest BNF level populations in Set 3. Both natural ^{15}N-abundance and ARA methods confirmed the Start variety to be appropriate default material for BNF rate selection.

The complexity of genetic control over BNF corresponds to the complexity of the symbiotic BNF process. The contact among plants and bacteria precedes the establishment of a successful symbiosis. The host plant must discern the right partner within the soil biome. It must distinguish and select the rhizobia partner from pathogens and also from among distinct rhizobia species and inappropriate strains. Successful infection is followed by nodule organogenesis. Both processes are driven and regulated by orchestration of gene expression. More than 4000 differentially expressed transcripts were identified in nodules and roots, and more than 500 transcripts were exclusively detected in nodules of the model organism *M. truncatula* [43]. Red clover, a non-model organism, is a significant fodder crop whose breeding for high nitrogen fixation capacity would be valuable, without molecular approaches, albeit difficult and slow. BNF seems to be a polygenic trait [51] that is based on a couple of essential genes [58] that are themselves modulated by many genes with a potential effect on BNF rate [43]. We took the first steps to identifying red clover key genes playing central roles in the formation of root nodules and nitrogen fixation variability. We used an association study based on hybridization-based sequence capture target enrichment and a genome-wide approach, focused on finding variants and genome locations where genetic variance meets phenotype variance and they influence one another.

One of the genes having strong polymorphism association with BNF that arose from the analysis of candidate genes Panel 1 was ethylene response factor required for nodule differentiation (*EFD*). This gene belongs to the ethylene response factor (ERF) family that is a part of the AP2/ERF superfamily (containing the APETALA2 DNA binding domain) [62,63]. The ERF gene family includes plant-specific transcription factors that play roles in response to biotic and abiotic stress, control of organ development, and cell division and differentiation [62,64]. *EFD* is located in the nucleus. It is most expressed in nodule primordia and at the border of infection zones I and II. *EFD* activity is probably not induced by ethylene. The *EFD* role in nodule development and differentiation is dual. *EFD* negatively regulates the nodulation process, affecting the number of infections, but *EFD* also positively influences bacterial and plant cell differentiation in the late stages of nodule development. It was detected in mutant *efd-1* plants, for example, where it causes a later onset of nodule senescence. *EFD* also plays a role in regulation of the pathway of cytokines that influence nodule meristem activity [64].

The analysis of candidate genes in Panel 2 revealed another gene strongly associated with BNF, the molybdate transporter 1 [35]. Molybdenum is an essential plant micronutrient involved in nitrogen fixation and in some other plant enzymatic processes like nitrate assimilation, phytohormone biosynthesis, purine metabolism, sulfite detoxification, and amidoxime reduction [65]. Molybdenum is present in soil in the form of oxyanion molybdate, and the intake of this nutrient is managed by molybdate transporters. The molybdate transporter type 1 family is involved in molybdate transport to the cytoplasm of nodule cells. These transporters are located in the plasma membrane of infected and uninfected cells within the interzone and early fixation zone of the nodule. From the cell cytoplasm, molybdate must be transported across the symbiosome membrane. This transport is presumed to be performed by the symbiotic sulfate transporter SST1 [66], after which ATP-binding cassette transporter (ModABC) transfers molybdate into the bacteroid [67,68]. The molybdenum in a plant cell is a component of the iron–molybdenum cofactor (FeMoco) of nitrogenase. In knockout *M. truncatula* line mot1.3-1, lower nitrogenase activity and reduced plant growth as a result of a lack of nitrogen were observed. Under non-symbiotic conditions, *M. truncatula* plants showed no physiological or phenotypical difference from a control group, and this result was consistent with a hypothesis that the MOT1 transporter is evolutionarily specialized to provide molybdenum for symbiotic nitrogen fixation [44].

A part of the analysis of candidate genes in Panel 2 was an analysis of leghemoglobin genes. Leghemoglobin proteins play an important role in the activity of the oxygen labile enzyme nitrogenase [69]. Leghemoglobins maintain the low free oxygen level in the nodule-infected zone [70], and they also transport oxygen to sites of respiration, thus enabling ATP production in a low-oxygen environment [71]. In *M. truncatula*, genes for leghemoglobin are among the most strongly expressed genes in nodule tissue [72]. Ištvánek et al. [34] identified in red clover a similar number of leghemoglobin genes as found in *M. truncatula*. The number of nine leghemoglobin genes in red clover coincides with the number in *M. sativa*. The family of non-symbiotic hemoglobin genes shows only limited amino-acid sequence similarity to the symbiotic hemoglobins. Genes encoding this type of hemoglobin were cloned from the nitrogen-fixing species [73] and from plants that do not fix nitrogen, including monocots [74] and *Arabidopsis thaliana* [75]. These non-symbiotic hemoglobins are typically expressed at low levels in roots and leaves [76,77]. Functions of non-symbiotic hemoglobins are not yet clearly understood [78], although they may play a role in plant survival by increasing the energy status of the cells under hypoxic conditions [79,80]. Seven genes for leghemoglobins were analyzed as a part of candidate gene Panel 2. They can be distinguished into three groups according to the levels of their genetic diversity.

Target sequencing of BNF candidate genes of plants with alternative phenotypes for nitrogen fixation and whole-genome population genotyping using ddRADseq demonstrated two complementary methods for using knowledge about known key genes from related model organisms and simultaneously assessing whole-genome genotype information to exploit complex genetic information from species of interest. Polymorphism annotation (Figure 9) and diversity assessment (Figure 4; Figure 5) revealed that the allelic diversity in genic regions of BNF key genes and potential BNF key genes in populations of red clover is sufficient, satisfying that prerequisite for high phenotype variability and, ultimately, BNF selection. For the candidate genes in Panel 1 and 2, expected and observed heterozygosity was calculated. In the candidate genes in Panel 1, no obvious differences between expected and observed heterozygosity were found. We assume that the analyzed plants do not deviate from Hardy–Weinberg equilibrium in the studied genes in Panel 1. In Panel 2, the difference between the expected and observed heterozygosity was found in two of the candidate genes sequences (Tp_33338, Tp_84). We can conclude that these genes do not meet the assumptions of Hardy–Weinberg equilibrium, especially the assumption that the genes are not under selection. These two genes may be subject to selection; however, this selection does not correspond to BNF rate because variants in these genes are not associated with BNF rate. Nevertheless, the specific function of this genes should be checked by gene function analysis. In addition, these genes have a low level of diversity and a low number of polymorphisms.

The discovery-driven approach of the genome-wide association study complemented the results gained by the hypothesis-driven approach of target sequencing of candidate genes. This exploratory analysis of tens of the populations using genome-wide association studies was not robust enough to clearly identify causal genes, but the results could be valuable for a breeding purpose. Although our dataset was not capable of comparing the genome-wide association studies with hundreds of samples, it was sufficient to reveal potentially associated alleles with a large effect on complex traits. Rather than finding new genes in the BNF process, our study focused on highlighting loci in the red clover genome that are potentially beneficial for BNF, and which should be selected as fixed in starting plant material for breeding new high-BNF rate varieties. On the other hand, the associated alleles of the candidate genes should be used for fine-tuning of the BNF rate red clover phenotype. Moreover, the relevance of an association signal is supported by the location of some variants in the vicinity of a gene that potentially has a role in the BNF process. In our case, we detected two significantly associated SNPs and one InDel mapped on linkage groups (Supplementary Table S3). The first associated SNP on LG4 is linked with the gene for auxin response factor and sulfotransferase. Auxin response factors are among the regulators of auxin response genes, and they play roles in various processes of plant growth and development [81]. According to Breakspear et al. [82], auxin is involved through its regulation of cell-wall remodeling in the initiation of rhizobial infection and growth of infection thread. The role of sulfotransferases is potentially connected to nitrogen fixation. Sulfotransferases enable the transfer of a sulfuryl group from a donor to an acceptor. The nitrogenase consists of two proteins, dinitrogenase reductase (Fe protein) and dinitrogenase (MoFe protein), whose structures are rich in sulfur, thus indicating that this element could be limiting in rhizobial symbiosis. Sulfur is also a part of the amino acids cysteine and methionine, and nodules contain a cysteine-rich protein, ferredoxin, which operates as an electron transporter and donates electrons to nitrogenase. Sulfur deficiency in nodulated legumes negatively affects nodulation, causing reduction in nodule number and in nodule mass per unit root length. This directly inhibits N fixation and alters the nodule metabolism. A sufficient sulfur supply contributes to increased nodulation and symbiotic nitrogen fixation [67,83]. Sulfate intake is provided by symbiotic sulfate transporters (SST), and the sulfate is reduced to organic sulfide. The symbiotic function of sulfur in the bacteroid is the sulfation of Nod factors and of cell-surface polysaccharides. The process is catalyzed by the sulfotransferase activity of NodH [83,84].

The second associated SNP on LG4 is placed near genes for ethylene-responsive transcription factor 3 (*ERF3*). The *ERF3* gene belongs to the AP2/ERF superfamily of transcription factors [62,63], and it plays a key role in crown root development and elongation. Through its interaction with cytokinin-responsive gene *RR2* from type-A RR genes, ERF3 acts as a repressor of cytokinin signaling that results in crown root initiation. In the crown root meristem, a WUSCHEL-related homeobox gene (*WOX11*) is expressed and it binds to the complex RR2/ERF3. This process leads to inhibition of ERF3 and RR2 and results in increased cytokinin signaling and crown root elongation [85].

The associated InDel on LG1 is near several genes for bidirectional amino-acid transporter 1 (BAT1). BAT1 serves as a transmembrane protein that transports amino acids in both directions through the plasma membrane. This process is necessary for amino-acid transport between xylem and phloem [86]. In the process of BNF, the nitrogen is reduced to ammonia and, using glutamate synthetase, it is incorporated into glutamate [87]. According to Dündar and Bush [86], glutamate, together with amino acids such as alanine, arginine, and lysine, is transported by BAT1.

In order to estimate the strength of the connection between genetic polymorphism variance and ^{15}N BNF rate phenotype variance, we estimated marker-based narrow-sense heritability. We estimated that 84.7% of phenotypic variance is due to additive genetic effects expressed in genotypic polymorphism data. The high level of BNF rate heritability corresponds to the high levels of heritability mentioned in earlier results [10,50], and it predetermines associated polymorphisms to be good genetic markers for the prospective genomic selection of a new variety with high BNF rate that is based on the assessed collection of populations.

In conclusion, knowledge of genotype–phenotype associations led to a deeper understanding of how genotype leads to phenotype, and DNA markers could be developed based on characterized gene polymorphisms. Due to the statistical approach of association studies, functional validation of candidate polymorphisms will be essential for their implementation. SNP microarrays and InDel-specific markers will be designed for genotyping and co-segregation studies in red clover. Both provide an important resource in the form of beneficial alleles for efficient marker-assisted selection and application in red clover breeding for improved nitrogen fixation capacity. To link theory with practice, the results of this study will be used as input molecular markers for a high-throughput genotyping platform using a DNA microarray. The DNA microarray platform will be used as a tool in BNF rate breeding program of red clover. In particular, the associated polymorphisms from the population genome-wide association study could be used as markers for the pre-selection of appropriate input red clover populations for breeding on BNF efficiency. On the other hand, the associated variants from the candidate genes panels will be used to fix the beneficial alleles of BNF candidate genes in breeding populations. Finally, the association level of selected polymorphisms will have to be validated in practice using the first generation of the mentioned DNA microarray before it can be implemented in a real red clover BNF breeding program.

4. Materials and Methods

4.1. Plant Materials

Three plant sets and one plant population for BNF rate evaluation were prepared, the former for ARA and the latter for the natural ^{15}N-abundance method. Sets 1, 2, and 3 of plants were grown in 2017, 2018, and 2019, respectively: in 2017, 647 plants of four diploid (Start, Vltavín, Columbia, Global) and four tetraploid (Tatra, Tempus, Kvarta, HJRH) accessions; in 2018, 401 plants of four tetraploid accessions (Nodula, Gregale, Atlantis, Tempus); and, in 2019, 378 plants as offspring of 16 parents selected in Set 1. In total, 1426 plants were grown and the number of plants per accession varied between years. In Sets 1 and 2, higher numbers of plants per accession were grown to assess intrapopulation BNF diversity and to find high- and low-BNF rate plants among broad input populations. In Set 3, we used a smaller number of plants per accession to assess how real selection works.

In order to the BNF evaluation by natural ^{15}N-abundance approach, population samples consisted of 91 diploid accessions and originated from the Czech core collection of *T. pratense* within the Czech national seed bank, which is maintained by the Crop Research Institute (Prague, Czech Republic). The list contained varieties and wild accessions. *Galega orientalis* Lam. uninoculated by *Neorhizobium galegae* and non-nitrogen symbiotic plants *Malva verticillata* L. were used as controls. Red clover accessions and their characteristics are summarized in Supplementary Table S2.

4.2. Growth Conditions and Evaluation of Nitrogen Fixation by Acetylene Reduction Assay

The red clover seeds were scarified and germinated on wet perlite. Sprouted seeds were planted in individual pots filled with perlite and inoculated with rhizobia by adding 1 mL of *Rhizobium leguminosarum* bv. *trifolii* inoculum, which was provided by the Crop Research Institute (Prague, Czech Republic). Different rhizobia strains were applied for diploid and tetraploid varieties as recommended by the collection's curator. Plants were grown hydroponically in a greenhouse within individual pots filled with perlite. They were watered with a nutrient solution containing 870 mg/L K_2HPO_4, 135 mg/L $FeCl_3 \cdot 6H_2O$, 735 mg/L $CaCl_2 \cdot 2H_2O$, 246 mg/L $MgSO_4 \cdot 7H_2O$, 0.123 mg/L $Na_2MoO_4 \cdot H_2O$, 0.486 mg/L H_3BO_3, 0.055 mg/L $CuSO_4 \cdot 5H_2O$, 0.25 mg/L $MnCl_2 \cdot 4H_2O$, and 0.06 mg/L $ZnSO_4 \cdot 7H_2O$. No nitrogen was supplied exogenously, and the pH was 6.5–6.8. The solution was replenished as necessary and exchanged once a week. ARA was used for evaluating the efficiency of nitrogen fixation in individual plants through analyzing nitrogenase activity [56]. ARA was carried out approximately 100 days after sowing. The results were expressed as concentration of ethylene C_E (μmol/mL) in a jar after 0.5 h of incubation.

4.3. Evaluation of Nitrogen Fixation by Natural ^{15}N-Abundance Method

The 15 bulked plants per accession were grown in pots with soil from local field with red clover. The plants were sampled at the beginning of flowering of early accessions. The nitrogen (N) and carbon (C) concentrations and their isotopic compositions in red clover shoots (ground to a fine powder using a Retsch MM200 ball mill, sample weights 3–4 mg, packed in tin capsules) were measured using a Flash EA 2000 elemental analyzer coupled with a Delta V Advantage isotope ratio mass spectrometer (both Thermo Scientific, Waltham MA, USA). Elemental composition was calibrated using certified standards from Elemental Microanalysis (Okehampton, UK). Isotopic composition was assessed by comparison with certified standards from the International Atomic Energy Agency (Vienna, Austria).

4.4. Selection of Candidate Genes and Procedure of Targeted Sequencing

Selection of candidate genes was carried out based on the annotated genome of the model legume *M. truncatula*. The genes essential for the nodulation process and nitrogen fixation were chosen for sequencing. Overall, 17 and 69 chosen candidate genes from Panels 1 and 2, respectively, included genes for transcription factors, receptor-like kinases (*RLK*), leghemoglobins, and cytokinin receptors (Supplementary Table S1). Many of these genes were functionally characterized for their roles in the nitrogen fixation process. Sequences of these genes extracted from the GeneBank database (https://www.ncbi.nlm.nih.gov) were aligned to the genome sequence of *T. pratense* variety Tatra [34] using BLAST+ (ver. 2.8.1, [88]). Sequences with highest similarity (>90%) were chosen for further analysis. The Panel 1 span was 95,000 bp and that of Panel 2 was 98,464 bp of the red clover genome.

Forty-eight and 50 plants from Sets 1 and 2, respectively, with the most contrasting BNF values were used for SeqCap. One hundred milligrams of fresh leaves were collected, and flash-frozen in liquid nitrogen. DNA was isolated using a DNeasy Plant Mini Kit (Qiagen, Germany) according to the manufacturer's protocol and following the cetyl trimethylammonium bromide (CTAB) method [89]. DNA quality was checked on a 3% agarose gel, and DNA concentration was quantified by NanoDrop 2000c spectrophotometer (Thermo Fisher Scientific, USA) and by a Qubit fluorometer (Invitrogen/Thermo Fisher Scientific, USA).

Probe design was performed with Roche NimbleGen's custom probe design pipeline (Roche Diagnostic, USA; http://www.nimblegen.com/products/seqcap/ez/designs/). Two gene panels were designed (Supplementary Table S1). Gene Panel 1 spanned 95 kbp of the selected genomic sequences, including the 17 candidate genes. Gene Panel 2 spanned 99.5 kbp and the 69 genes. Forty-eight and 50 DNA samples were sequenced for Panels 1 and 2, respectively. Libraries of both panels were prepared using the SeqCap EZ HyperCap procedure (Roche Diagnostic, USA) while following the manufacturer's instructions, and the libraries were sequenced for 150-bp reads with paired-end sequencing on a NextSeq 500 sequencer (Illumina, San Diego, CA, USA). Library preparation and sequencing were performed at Core Facility Genomics CEITEC MU (Brno, Czech Republic).

4.5. ddRADseq Library Preparation and Sequence Processing of T. pratense Population Set

Ninety-six batch samples of 15 plants per sample (Supplementary Table S2) were processed together into one final ddRADseq library. Library preparation followed a slightly modified protocol by Peterson et al. [40]. Three hundred nanograms of genomic DNA from each population was digested with two restriction enzymes, *Sph*I and *Mlu*CI, in one 30-µL reaction. P1 and P2 "flex" adapters were ligated in a 40-µL reaction with 100 ng of the digestion product. The total volume of 48 ligation products differing in adapter barcode were pooled together into a "sublibrary", and two sublibraries in total were prepared. The order of samples was randomized between and within sublibraries. Automated size selection of a fraction of 220–320 bp separately from each sublibrary was performed on the Pippin Prep laboratory platform using a Pippin Prep 2010 kit (Sage Science, Beverly, MA, USA). PCR amplification with primers bearing the multiplexing indices and Illumina flow cell annealing regions was done in several 50-µL reactions (separately for each sublibrary). PCR products were purified on

AMPure XP beads and combined in equimolar ratios to compose the final library. Sequencing was performed using 125-bp paired-end reads on a HiSeq 2500 (Illumina) at the EMBL Genomic Core Facility, Heidelberg, Germany.

4.6. Bioinformatic Analysis

Basic characteristics of the reads obtained were reviewed in FastQC v0.10.1 [90]. Barcode sorting was performed in process_radtags, a pipeline component of Stacks v2.3 [91]. A reference-based strategy was used for assembling the targeted sequences and ddRADseq sequences obtained. Reads were firstly qualitatively filtered and trimmed using Trimmomatic v0.38 [92], and then aligned onto the genomes of *T. pratense* [34,35] reference genomes with Milvus and Tatra varieties using the BWA-MEM algorithm from BWA v0.7.17 assembler [93]. Sequence data from target sequencing were randomly downsampled to 150× coverage. GATK (Genome Analysis Toolkit) v4.1.0.0 [94] was used for base quality score recalibration and performing SNP and InDel variant calling across samples of target sequencing and ddRADseq population genotyping as well. Variants were filtered using standard hard filtering parameters according to GATK Best Practices recommendations [95,96].

In order to express genotypes information of bulked samples in ddRADseq population genotyping, continuous numerical genotypes were computed as frequencies of allelic depth counted from allelic depths and read depth in variant positions. The polymorphisms that were identified with a maximum of 50% missing information and polymorphisms that were polymorphic in more than 5% of called population numeric genotypes were used for the analysis. Missing population genotypes were imputed before association analysis as means of continuous numerical genotypes of the variants.

For target sequencing, Panel 1 and 2 genotypes were called in diploid and tetraploid states. All variants from candidate gene panels sequencing and also from ddRADseq genotyping were annotated using Variant Effect Predictor (VEP) [97]. Called final variants of Panel 1, Panel 2, and the population ddRADseq genotype are stored and presented in Supplementary Table S4.

4.7. Statistical Analysis

Results of ARA were expressed as ethylene molar concentration (C_E) values that were computed from ethylene peak area in accordance with Unkovich et al. [52]. The C_E value was standardized to Z-score within measuring sets in order to compare BNF rate among different sets. Differences in nitrogen fixation rate measured using ARA among different populations were tested with the nonparametric Kruskal–Wallis test and subsequent nonparametric post hoc comparisons.

Polymorphism diversity level was expressed as expected heterozygosity (Hs). This was computed as if the species were diploid, because it is also appropriate for diversity comparison for polyploid cases [98]. In order to assess if the genes meet the assumptions of Hardy–Weinberg equilibrium, observed heterozygosity for candidate genes was calculated as well. To test differences between expected heterozygosity and observed heterozygosity, we used a Mann–Whitney U test in R. Genetic diversity pattern was assessed by principal component analysis using the pcaMethods R package [99].

The association analyses of variants from candidate genes in Panels 1 and 2 were conducted using the mixed linear model algorithm [100] in GAPIT in R.

The genome-wide association study for variants from population genotyping were conducted using the statistical method FarmCPU [45], and estimation of marker-based heritability was performed in GAPIT in R [101]. The significance threshold was set to the false discovery rate-adjusted *p*-value of 0.05 using the Benjamini–Hochberg correction [46].

4.8. Validation of Selected InDel Polymorphisms

For validation, 10 InDels for six different candidate genes (Supplementary Table S6) from Panel 1 were chosen. Genotypes used for validation are given in Supplementary Table S6. Validation was performed by means of allele-specific PCR and 3% agarose gel electrophoresis; surrounding primers were designed for InDels longer than 50 bp, and the products were clearly distinguished according to the length of the PCR products. For InDels shorter than 50 bp, one of the primers hybridized to the sequence of the InDels and the other one matched the sequence adjacent to the InDel. In this case, PCR products were only visible if the genotypes contained the desired InDels. Specificity of the designed primers was verified using BLAST+ (ver. 2.8.1, [88]) with *T. pratense* var. Tatra [34] as a database.

5. Conclusions

Red clover plants with high BNF rate contribute more to the accumulation of biogenic nitrogen in the soil to improve sustainability in agriculture. We performed genome-wide and targeted association studies and described phenotypic and genotypic variation of BNF in red clover, which allowed finding key candidate genes responsible for this complex polygenic trait. We identified polymorphisms in key genes strongly associated with BNF rate: *EFD*, which negatively regulates the nodulation process and positively influences cell differentiation in the late stages of nodulation, and *MOT1*, which is responsible for molybdate intake of nodule cells. Our population genotyping data confirmed polymorphisms strongly associated with BNF and located near the genes for auxin response factor, which regulates the cell-wall remodeling, and sulfotransferase involved in the process of sulfur metabolism, and also near *ERF3* regulating the crown root development and *BAT1* ensuring bidirectional transport of amino acids between xylem and phloem.

In comparison with conventional breeding of red clover, breeding based on genomic data can be effective in dealing with complex polygenic traits like BNF. It can help to identify and select additive genes or beneficial recessive alleles even at tetraploid varieties of cross-pollinating species. Because of the statistical approach of association studies, functional validation of those candidate polymorphisms found will be essential for confirming the biological importance of the alleles identified to be beneficial for efficient red clover selection and breeding for improved nitrogen fixation capacity. The practical outcome of this study will provide input molecular markers for the high-throughput DNA microarray genotyping platform that will be used for breeding of new red clover varieties with higher BNF rate.

Supplementary Materials: Supplementary materials can be found at http://www.mdpi.com/1422-0067/20/21/5470/s1. Supplementary Figure S1. targeted sequences coverage. Supplementary Figure S2. InDel validation. Supplementary Figure S3. correlation of phenotype and diversity pattern. Supplementary Table S1. Targeted sequence. Supplementary Table S2. Plant material. Supplementary Table S3. Associated polymorphisms. Supplementary Table S4. Genotypic tables. Supplementary Table S5. Polymorphisms indel validation.

Author Contributions: J.Ř. and J.N. designed the study and supervised analyses; O.T. performed most of the experiments and processed sequencing data; J.J. measured the natural ^{15}N-abundance method; L.P. prepared the ddRADseq library; J.H., K.V., and M.M. evaluated ARA nitrogen fixation; E.M. and D.V. validated data; H.J. and D.K. prepared plant material; O.T., M.M., and J.Ř. wrote the manuscript. All authors approved the final manuscript.

Funding: This research was funded by the Technology Agency of the Czech Republic (grant no. TH02010351), the Ministry of Education, Youth, and Sports (grant no. MUNI/A/0958/2018), and the Ministry of Agriculture of the Czech Republic (MZE-RO1719) and National program of conservation and utilization of plant genetic resources and agro-biodiversity, Ref. 51834/2017-MZE17253/6.2.2.

Acknowledgments: The authors are thankful for and greatly appreciate access to computing and storage facilities owned by parties and projects contributing to the National Grid Infrastructure MetaCentrum that was provided under the program "Projects of Large Research, Development, and Innovations Infrastructures" (CESNET LM2015042). In particular, we greatly appreciate access to the CERIT-SC computing and storage facilities provided by the CERIT-SC Center, provided under the program "Projects of Large Research, Development, and Innovations Infrastructures" (Cerit Scientific Cloud LM2015085). Seeds were procured from the GeneBank of Crop Research Institute Ltd., Prague-Ruzyně, Czech Republic. The authors are grateful to the EMBL Genomic Core Facility in Heidelberg (Germany) for its Illumina sequencing of the ddRADseq library.

Conflicts of Interest: Some authors are employed by the company Agricultural Research, Ltd., Troubsko, Czech Republic (O. Trněný, J. Nedělník, J. Hofbauer, K. Vejražka, M. Matoušková), and author D. Knotová is employed by the company Research Institute for Fodder Crops, Ltd., Troubsko, Czech Republic. The remaining authors declare no competing interests.

Abbreviations

AP2	APETALA2
ARA	Acetylene reduction assay
ATP	Adenosine triphosphate
BNF	Biological fixation of atmospheric nitrogen
C_E	Ethylene molar concentration
CRE	Cytokinin response 1
CTAB	Cetyl trimethylammonium bromide
ddRADseq	Double-digest RAD sequencing
DNF2	Defective in nitrogen fixation 2
EFD	Ethylene response factor required for nodule differentiation
ERF	Ethylene response factor
ERF3	Ethylene-responsive transcription factor 3
FLOT	Flotillin
GATK	Genome Analysis Toolkit
Hs	Expected heterozygosity
InDel	Insertion/deletions
LG	Linkage group
MOT1	Molybdate transporter 1
ModABC	ATP-binding cassette transporter involved in molybdate transport
Mt	*Medicago truncatula*
NFP	Nod factor perception
NF-YC2	Nuclear transcription factor Y subunit C2
NGS	Next-generation sequencing
Nod factors	Nodulation factors
PC	Principal component
PCA	Principal component analysis
PNO1	partner of NOB1-like
RLK	Receptor-like kinases
RR2	Cytokinin responsive gene
SeqCap	Hybridization-based sequence capture
SNP	Single-nucleotide polymorphism
SST	Symbiotic sulfate transporter
Tp	*Trifolium pratense*
VEP	Variant effect predictor
WOX11	WUSCHEL-related homeobox gene

References

1. Zohary, M.; Heller, D. *The genus Trifolium*; Israel Academy of Sciences and Humanities: Jerusalem, Israel, 1984.
2. Gillett, J.M.; Taylor, N.L. *The World of Clovers*; Iowa State University Press: Ames, Iowa, IA, USA, 2001.
3. Ellison, N.W.; Liston, A.; Steiner, J.J.; Williams, W.M.; Taylor, N.L. Molecular phylogenetics of the clover genus (Trifolium-Leguminosae). *Mol. Phylogenet. Evol.* **2006**, *39*, 688–705. [CrossRef] [PubMed]
4. Kintl, A.; Elbl, J.; Lošák, T.; Vaverková, M.D.; Nedělník, J. Mixed intercropping of wheat and white clover to enhance the sustainability of the conventional cropping system: Effects on biomass production and leaching of mineral nitrogen. *Sustain* **2018**, *10*, 3367. [CrossRef]
5. Hauer, R.F.; Lamberti, G.A. *Methods in Stream Ecology: Volume 1: Ecosystem Structure*; Academic Press: Cambridge, MA, USA, 2017; ISBN 9780124165786.
6. Lerouge, P. Symbiotic host specificity between leguminous plants and rhizobia is determined by substituted and acylated glucosamine oligosaccharide signals. *Glycobiology* **1994**, *4*, 127–134. [CrossRef] [PubMed]
7. Luna, R.; Planchon, C. Genotype x Bradyrhizobium japonicum strain interactions in dinitrogen fixation and agronomic traits of soybean (Glycine max L. Merr.). *Euphytica* **1995**, *86*, 127–134. [CrossRef]

8. Kouchi, H.; Imaizumi-Anraku, H.; Hayashi, M.; Hakoyama, T.; Nakagawa, T.; Umehara, Y.; Suganuma, N.; Kawaguchi, M. How many peas in a pod? Legume genes responsible for mutualistic symbioses underground. *Plant Cell Physiol.* **2010**, *51*, 1381–1397. [CrossRef]

9. Carlsson, G.; Huss-Danell, K. Nitrogen fixation in perennial forage legumes in the field. *Plant Soil* **2003**, *253*, 353–372. [CrossRef]

10. Provorov, N.A.; Tikhonovich, I.A. Genetic resources for improving nitrogen fixation in legume-rhizobia symbiosis. *Genet. Resour. Crop. Evol.* **2003**, *50*, 89–99. [CrossRef]

11. Freiberg, C.; Fellay, R.; Bairoch, A.; Broughton, W.J.; Rosenthal, A.; Perret, X. Molecular basis of symbiosis between Rhizobium and legumes. *Nature* **1997**, *387*, 394–401. [CrossRef]

12. Catoira, R.; Galera, C.; De Billy, F.; Penmetsa, R.V.; Journet, E.P.; Maillet, F.; Rosenberg, C.; Cook, D.; Gough, C.; Denarie, J. Four genes of Medicago truncatula controlling components of a Nod factor transduction pathway. *Plant Cell* **2000**, *12*, 1647–1665. [CrossRef]

13. Oldroyd, G.E.D.; Long, S.R. Identification and characterization of nodulation-signaling pathway 2, a gene of Medicago truncatula involved in nod factor signaling. *Plant Physiol.* **2003**, *131*, 1027–1032. [CrossRef]

14. Gleason, C.; Yang, T. Nodulation independent of rhizobia induced by a calcium-activated kinase lacking autoinhibition Article W342F Mutation in CCaMK Enhances Its Affinity to Calmodulin But Compromises Its Role in Supporting Root Nodule Symbiosis in Medicago truncatula View pr. *Nature* **2006**, *441*, 1149–1152. [CrossRef] [PubMed]

15. Delves, A.C.; Mathews2, A.; Day, D.A.; Carter, A.S.; Carroll, B.J.; Gresshoff, P.M. Regulation of the Soybean-Rhizobium Nodule Symbiosis by Shoot and Root Factors'. *Plant Physiol.* **1986**, *82*, 588–590. [CrossRef] [PubMed]

16. Caetano-Anolles, G.; Gresshoff, P.M. Plant Genetic Control of Nodulation. *Annu. Rev. Microbiol.* **1991**, *45*, 345–382. [CrossRef] [PubMed]

17. Gage, D.J. 2004 Infection and Invasion of Roots by Symbiotic, Nitrogen-Fixing Rhizobia during Nodulation of Temperate Legumes. *Microbiol. Mol. Biol. Rev.* **2017**, *68*, 203.

18. Marchal, K.; Vanderleyden, J. The "oxygen paradox" of dinitrogen-fixing bacteria. *Biol. Fertil. Soils* **2000**, *30*, 363–373. [CrossRef]

19. Kundu, S.; Trent, J.T.; Hargrove, M.S. Plants, humans and hemoglobins. *Trends Plant Sci.* **2003**, *8*, 387–393. [CrossRef]

20. Penmetsa, R.V.; Cook, D.R. A legume ethylene-insensitive mutant hyperinfected by its rhizobial symbiont. *Science* **1997**, *275*, 527–530. [CrossRef]

21. Mitra, R.M.; Gleason, C.A.; Edwards, A.; Hadfield, J.; Downie, J.A.; Oldroyd, G.E.D.; Long, S.R. A Ca^{2+}/calmodulin-dependent protein kinase required for symbiotic nodule development: Gene identification by transcript-based cloning. *Proc. Natl. Acad. Sci. USA* **2004**, *101*, 4701–4705. [CrossRef]

22. Domonkos, A.; Horvath, B.; Marsh, J.F.; Halasz, G.; Ayaydin, F.; Oldroyd, G.E.D.; Kalo, P. The identification of novel loci required for appropriate nodule development in Medicago truncatula. *BMC Plant Biol.* **2013**, *13*, 157. [CrossRef]

23. Kang, Y.; Li, M.; Sinharoy, S.; Verdier, J. A snapshot of functional genetic studies in Medicago truncatula. *Front. Plant Sci.* **2016**, *7*, 1175. [CrossRef]

24. Veerappan, V.; Jani, M.; Kadel, K.; Troiani, T.; Gale, R.; Mayes, T.; Shulaev, E.; Wen, J.; Mysore, K.S.; Azad, R.K.; et al. Rapid identification of causative insertions underlying Medicago truncatula Tnt1 mutants defective in symbiotic nitrogen fixation from a forward genetic screen by whole genome sequencing. *BMC Genom.* **2016**, *17*, 141. [CrossRef] [PubMed]

25. Yano, K.; Aoki, S.; Liu, M.; Umehara, Y.; Suganuma, N.; Iwasaki, W.; Sato, S.; Soyano, T.; Kouchi, H.; Kawaguchi, M. Function and evolution of a Lotus japonicus AP2/ERF family transcription factor that is required for development of infection threads. *DNA Res.* **2017**, *24*, 193–203. [PubMed]

26. Cregan, P.B.; Jarvik, T.; Bush, A.L.; Shoemaker, R.C.; Lark, K.G.; Kahler, A.L.; Kaya, N.; VanToai, T.T.; Lohnes, D.G.; Chung, J. An integrated genetic linkage map of the soybean genome. *Crop Sci.* **1999**, *39*, 1464–1490. [CrossRef]

27. Santos, M.A.D.; Nicolás, M.F.; Hungria, M. Identificação de QTL associados à simbiose entre Bradyrhizobium japonicum, B. elkanii e soja. *Pesqui. Agropecuária Bras.* **2006**, *41*, 67–75. [CrossRef]

28. Santos, M.A.; Geraldi, I.O.; Garcia, A.A.F.; Bortolatto, N.; Schiavon, A.; Hungria, M. Mapping of QTLs associated with biological nitrogen fixation traits in soybean. *Hereditas* **2013**, *150*, 17–25. [CrossRef]

29. Nicolás, M.F.; Hungria, M.; Arias, C.A.A. Identification of quantitative trait loci controlling nodulation and shoot mass in progenies from two Brazilian soybean cultivars. *Field Crop. Res.* **2006**, *95*, 355–366. [CrossRef]

30. Kim, D.H.; Parupalli, S.; Azam, S.; Lee, S.H.; Varshney, R.K. Comparative sequence analysis of nitrogen fixation-related genes in six legumes. *Front. Plant Sci.* **2013**, *4*, 300. [CrossRef]

31. Alves-Carvalho, S.; Aubert, G.; Carrère, S.; Cruaud, C.; Brochot, A.L.; Jacquin, F.; Klein, A.; Martin, C.; Boucherot, K.; Kreplak, J.; et al. Full-length de novo assembly of RNA-seq data in pea (Pisum sativum L.) provides a gene expression atlas and gives insights into root nodulation in this species. *Plant J.* **2015**, *84*, 1–19. [CrossRef]

32. Qiao, Z.; Pingault, L.; Nourbakhsh-Rey, M.; Libault, M. Comprehensive comparative genomic and transcriptomic analyses of the legume genes controlling the nodulation process. *Front. Plant Sci.* **2016**, *7*, 34. [CrossRef]

33. Vižintin, L.; Javornik, B.; Bohanec, B. Genetic characterization of selected Trifolium species as revealed by nuclear DNA content and ITS rDNA region analysis. *Plant Sci.* **2006**, *170*, 859–866. [CrossRef]

34. Ištvánek, J.; Jaroš, M.; Krenek, A.; Řepková, J. Genome assembly and annotation for red clover (Trifolium pratense; Fabaceae). *Am. J. Bot.* **2014**, *101*, 327–337. [CrossRef] [PubMed]

35. De Vega, J.J.; Ayling, S.; Hegarty, M.; Kudrna, D.; Goicoechea, J.L.; Ergon, Å.; Rognli, O.A.; Jones, C.; Swain, M.; Geurts, R.; et al. Red clover (Trifolium pratense L.) draft genome provides a platform for trait improvement. *Sci. Rep.* **2015**, *5*, 1–10. [CrossRef] [PubMed]

36. Young, N.D.; Debellé, F.; Oldroyd, G.E.D.; Geurts, R.; Cannon, S.B.; Udvardi, M.K.; Benedito, V.A.; Mayer, K.F.X.; Gouzy, J.; Schoof, H.; et al. The Medicago genome provides insight into the evolution of rhizobial symbioses. *Nature* **2011**, *480*, 520–524. [CrossRef] [PubMed]

37. Stanton-Geddes, J.; Paape, T.; Epstein, B.; Briskine, R.; Yoder, J.; Mudge, J.; Bharti, A.K.; Farmer, A.D.; Zhou, P.; Denny, R.; et al. Candidate Genes and Genetic Architecture of Symbiotic and Agronomic Traits Revealed by Whole-Genome, Sequence-Based Association Genetics in Medicago truncatula. *PLoS ONE* **2013**, *8*, e65688. [CrossRef] [PubMed]

38. Curtin, S.J.; Tiffin, P.; Guhlin, J.; Trujillo, D.; Burghart, L.; Atkins, P.; Baltes, N.J.; Denny, R.; Voytas, D.F.; Stupar, R.M.; et al. Validating genome-wide association candidates controlling quantitative variation in nodulation. *Plant Physiol.* **2017**, *173*, 921–931. [CrossRef] [PubMed]

39. Grunvald, A.K.; Torres, A.R.; Luiz de Lima Passianotto, A.; Santos, M.A.; Jean, M.; Belzile, F.; Hungria, M. Identification of QTLs associated with biological nitrogen fixation traits in soybean using a genotyping-by-sequencing approach. *Crop. Sci.* **2018**, *58*, 2523–2532. [CrossRef]

40. Peterson, B.K.; Weber, J.N.; Kay, E.H.; Fisher, H.S.; Hoekstra, H.E. Double digest RADseq: An inexpensive method for de novo SNP discovery and genotyping in model and non-model species. *PLoS ONE* **2012**, *7*, e37135. [CrossRef]

41. Byrne, S.; Czaban, A.; Studer, B.; Panitz, F.; Bendixen, C.; Asp, T. Genome Wide Allele Frequency Fingerprints (GWAFFs) of Populations via Genotyping by Sequencing. *PLoS ONE* **2013**, *8*, e57438. [CrossRef]

42. Kozarewa, I.; Armisen, J.; Gardner, A.F.; Slatko, B.E.; Hendrickson, C.L. Overview of target enrichment strategies. *Curr. Protoc. Mol. Biol.* **2015**, *112*, 7–21.

43. Roux, B.; Rodde, N.; ßoise Jardinaud, M.-F.; Timmers, T.; Sauviac, L.; Cottret, L.; Ebastien Carr Ere, S.; Sallet, E.; Courcelle, E.; Moreau, S.; et al. An integrated analysis of plant and bacterial gene expression in symbiotic root nodules using laser-capture microdissection coupled to RNA sequencing. *Plant J.* **2014**, *77*, 817–837. [CrossRef]

44. Tejada-Jiménez, M.; Gil-Díez, P.; León-Mediavilla, J.; Wen, J.; Mysore, K.S.; Imperial, J.; González-Guerrero, M. Medicago truncatula Molybdate Transporter type 1 (MtMOT1.3) is a plasma membrane molybdenum transporter required for nitrogenase activity in root nodules under molybdenum deficiency. *New Phytol.* **2017**, *216*, 1223–1235. [CrossRef] [PubMed]

45. Liu, X.; Huang, M.; Fan, B.; Buckler, E.S.; Zhang, Z. Iterative Usage of Fixed and Random Effect Models for Powerful and Efficient Genome-Wide Association Studies. *PLoS Genet.* **2016**, *12*, e1005767. [CrossRef] [PubMed]

46. Benjamini, Y.; Hochberg, Y. Controlling the False Discovery Rate: A Practical and Powerful Approach to Multiple Testing. *J. R. Stat. Soc. Ser. B* **1995**, *57*, 289–300. [CrossRef]

47. Yang, Y.; Zhao, Q.; Li, X.; Ai, W.; Liu, D.; Qi, W.; Zhang, M.; Yang, C.; Liao, H. Characterization of genetic basis on synergistic interactions between root architecture and biological nitrogen fixation in soybean. *Front. Plant Sci.* **2017**, *8*, 1466. [CrossRef] [PubMed]

48. Vance, C.P. Legume Symbiotic Nitrogen Fixation: Agronomic Aspects. In *The Rhizobiaceae*; Springer Netherlands: Heidelberg, Germany, 1998; pp. 509–530.

49. Yang, Q.; Yang, Y.; Xu, R.; Lv, H.; Liao, H. Genetic analysis and mapping of QTLs for soybean biological nitrogen fixation traits under varied field conditions. *Front. Plant Sci.* **2019**, *10*, 75. [CrossRef] [PubMed]

50. Smith, G.R.; Knight, W.E.; Peterson, H.H. Variation among Inbred Lines of Crimson Clover for N2 Fixation (C2H2) Efficiency. *Crop Sci.* **1982**, *22*, 716–719. [CrossRef]

51. Nutman, P.S. Improving nitrogen fixation in legumes by plant breeding; the relevance of host selection experiments in red clover (Trifolium pratense L.) and subterranean clover (T. subterraneum L.). *Plant Soil* **1984**, *82*, 285–301. [CrossRef]

52. Unkovich, M.; Herridge, D.; Peoples, M.; Cadisch, G.; Boddey, B.; Giller, K.; Alves, B.; Chalk, P. Measuring plant-associated nitrogen fixation in agricultural systems. *Aust. Cent. Int. Agric. Res.* **2008**, *136*, 132–188.

53. Sulieman, S.; Schulze, J. The efficiency of nitrogen fixation of the model legume Medicago truncatula (Jemalong A17) is low compared to Medicago sativa. *J. Plant Physiol.* **2010**, *167*, 683–692. [CrossRef]

54. Thilakarathna, M.S.; Papadopoulos, Y.A.; Rodd, A.V.; Grimmett, M.; Fillmore, S.A.E.; Crouse, M.; Prithiviraj, B. Nitrogen fixation and transfer of red clover genotypes under legume–grass forage based production systems. *Nutr. Cycl. Agroecosystems* **2016**, *106*, 233–247. [CrossRef]

55. Thilakarathna, M.S.; Papadopoulos, Y.A.; Grimmett, M.; Fillmore, S.A.E.; Crouse, M.; Prithiviraj, B. Red Clover Varieties with Nitrogen Fixing Advantage during the Early Stages of Seedling Development. *Can. J. Plant Sci.* **2018**, *98*, 517–526. [CrossRef]

56. Hardy, R.W.F.; Burns, R.C.; Holsten, R.D. Applications of the acetylene-ethylene assay for measurement of nitrogen fixation. *Soil Biol. Biochem.* **1973**, *5*, 47–81. [CrossRef]

57. Mckenna, P.; Cannon, N.; Dooley, J.; Conway, J. The use of red clover (Trifolium pratense) in soil fertility-building: A Review. *Field Crop. Res.* **2018**, *221*, 38–49. [CrossRef]

58. Ferguson, B.J.; Mens, C.; Hastwell, A.H.; Zhang, M.; Su, H.; Jones, C.H.; Chu, X.; Gresshoff, P.M. Legume nodulation: The host controls the party. *Plant Cell Environ.* **2018**, *42*, 41–51. [CrossRef]

59. Roughley, R.J.; Dart, P.J. Reduction of acetylene by nodules of Trifolium subterraneum as affected by root temperature, Rhizobium strain and host cultivar. *Arch. Mikrobiol.* **1969**, *69*, 171–179. [CrossRef]

60. Bergersen, F.J. The Quantitative Relationship Between Nitrogen Fixation And The Acetylene-Reduction assay. *Aust. J. Biol. Sci.* **1970**, *23*, 1015–1026. [CrossRef]

61. Vessey, J.K. Measurement of nitrogenase activity in legume root nodules: In defense of the acetylene reduction assay. *Plant Soil* **1994**, *158*, 151–162. [CrossRef]

62. Nakano, T.; Suzuki, K.; Fujimura, T.; Shinshi, H. Genome-wide analysis of the ERF gene family in arabidopsis and rice. *Plant Physiol.* **2006**, *140*, 411–432. [CrossRef]

63. Riechmann, J.L.; Heard, J.; Martin, G.; Reuber, L.; Jiang, C.Z.; Keddie, J.; Adam, L.; Pineda, O.; Ratcliffe, O.J.; Samaha, R.R.; et al. Arabidopsis transcription factors: Genome-wide comparative analysis among eukaryotes. *Science* **2000**, *290*, 2105–2110. [CrossRef]

64. Vernié, T.; Moreau, S.; De Billy, F.; Plet, J.; Combier, J.P.; Rogers, C.; Oldroyd, G.; Frugier, F.; Niebel, A.; Gamas, P. EFD is an ERF transcription factor involved in the control of nodule number and differentiation in Medicago truncatula. *Plant Cell* **2008**, *20*, 2696–2713. [CrossRef]

65. Tejada-Jiménez, M.; Chamizo-Ampudia, A.; Galván, A.; Fernández, E.; Llamas, Á. Molybdenum metabolism in plants. *Metallomics* **2013**, *5*, 1191–1203. [CrossRef] [PubMed]

66. Krusell, L.; Krause, K.; Ott, T.; Desbrosses, G.; Krämer, U.; Sato, S.; Nakamura, Y.; Tabata, S.; James, E.K.; Sandal, N.; et al. The Sulfate Transporter SST1 Is Crucial for Symbiotic Nitrogen Fixation in Lotus japonicus Root Nodules. *Plant Cell* **2005**, *17*, 1625–1636. [CrossRef] [PubMed]

67. Cheng, G.; Karunakaran, R.; East, A.K.; Poole, P.S. Multiplicity of Sulfate and Molybdate Transporters and Their Role in Nitrogen Fixation in Rhizobium leguminosarum bv. viciae Rlv3841. *Mol. Plant Microbe. Interact.* **2016**, *29*, 143–152. [CrossRef] [PubMed]

68. Delgado, M.J.; Tresierra-Ayala, A.; Talbi, C.; Bedmar, E.J. Functional characterization of the Bradyrhizobium japonicum modA and modB genes involved in molybdenum transport. *Microbiology* **2006**, *152*, 199–207. [CrossRef]

69. Avenhaus, U.; Cabeza, R.A.; Liese, R.; Lingner, A.; Dittert, K.; Salinas-Riester, G.; Pommerenke, C.; Schulze, J. Short-term molecular acclimation processes of legume nodules to increased external oxygen concentration. *Front. Plant Sci.* **2016**, *6*. [CrossRef]

70. Minchin, F.R. Regulation of oxygen diffusion in legume nodules. *Soil Biol. Biochem.* **1997**, *29*, 881–888. [CrossRef]

71. Mylona, P.; Pawlowski, K.; Bisseling, T. Symbiotic Nitrogen Fixation. *Plant Cell* **1995**, *7*, 869–885. [CrossRef]

72. Cabeza, R.; Koester, B.; Liese, R.; Lingner, A.; Baumgarten, V.; Dirks, J.; Salinas-Riester, G.; Pommerenke, C.; Dittert, K.; Schulze, J. An RNA Sequencing Transcriptome Analysis Reveals Novel Insights into Molecular Aspects of the Nitrate Impact on the Nodule Activity of Medicago truncatula. *Plant Physiol.* **2014**, *164*, 400–411. [CrossRef]

73. Bogusz, D.; Appleby, C.A.; Landsmann, J.; Dennis, E.S.; Trinick, M.J.; Peacock, W.J. Functioning haemoglobin genes in non-nodulating plants. *Nature* **1988**, *331*, 178–180. [CrossRef]

74. Arredondo-Peter, R.; Hargrove, S.; Sarath, C.; Moran, J.F.; Lohrman, J.; Olson, J.S.; Klucas, R. V Rice Hemoglobins. *Plant Physiol.* **1997**, *115*, 1259–1266. [CrossRef]

75. Trevaskis, B.; Watts, R.A.; Andersson, C.R.; Llewellyn, D.J.; Hargrove, M.S.; Olson, J.S.; Dennis, E.S.; Peacock, W.J. Two hemoglobin genes in Arabidopsis thaliana: The evolutionary origins of leghemoglobins. *Proc. Natl. Acad. Sci. USA* **1997**, *94*, 12230–12234. [CrossRef] [PubMed]

76. Andersson, C.R.; Ostergaard Jensen, E.; Llewellyn, D.J.; Dennis, E.S.; Peacock, W.J. A new hemoglobin gene from soybean: A role for hemoglobin in all plants (nonsymbiotic/leghemoglobin/evolution). *Plant Biol.* **1996**, *93*, 5682–5687.

77. Bustos-Sanmamed, P.; Tovar-Méndez, A.; Crespi, M.; Sato, S.; Tabata, S.; Becana, M. Regulation of nonsymbiotic and truncated hemoglobin genes of Lotus japonicus in plant organs and in response to nitric oxide and hormones. *New Phytol.* **2011**, *189*, 765–776. [CrossRef] [PubMed]

78. Calvo-Begueria, L.; Cuypers, B.; Van Doorslaer, S.; Abbruzzetti, S.; Bruno, S.; Berghmans, H.; Dewilde, S.; Ramos, J.; Viappiani, C.O.; Becana, M. Characterization of the heme pocket structure and ligand binding kinetics of non-symbiotic hemoglobins from the model legume lotus japonicus. *Front. Plant Sci.* **2017**, *8*, 1–14. [CrossRef]

79. Igamberdiev, A.U. Nitrate, NO and haemoglobin in plant adaptation to hypoxia: An alternative to classic fermentation pathways. *J. Exp. Bot.* **2004**, *55*, 2473–2482. [CrossRef]

80. Gupta, K.J.; Mur, L.A.J.; Wany, A.; Kumari, A.; Fernie, A.R.; Ratcliffe, R.G. The role of nitrite and nitric oxide under low oxygen conditions in plants. *New Phytol.* **2019**. [CrossRef]

81. Guilfoyle, T.J.; Hagen, G. Auxin response factors. *Curr. Opin. Plant Biol.* **2007**, *10*, 453–460. [CrossRef]

82. Breakspear, A.; Liu, C.; Roy, S.; Stacey, N.; Rogers, C.; Trick, M.; Morieri, G.; Mysore, K.S.; Wen, J.; Oldroyd, G.E.D.; et al. The root hair "infectome" of medicago truncatula uncovers changes in cell cycle genes and reveals a requirement for auxin signaling in rhizobial infectionw. *Plant Cell* **2014**, *26*, 4680–4701. [CrossRef]

83. Becana, M.; Wienkoop, S.; Matamoros, M.A. Sulfur Transport and Metabolism in Legume Root Nodules. *Front. Plant Sci.* **2018**, *9*, 1–10. [CrossRef]

84. Ehrhardt, D.W.; Atkinson, M.E.; Faull, K.F.; Freedberg, D.I.; Sutherlin, D.P.; Armstrong, R.; Long, S.R. In vitro sulfotransferase activity of NodH, a nodulation protein of Rhizobium meliloti required for host-specific nodulation. *J. Bacteriol.* **1995**, *177*, 6237–6245. [CrossRef]

85. Zhao, Y.; Cheng, S.; Song, Y.; Huang, Y.; Zhou, S.; Liu, X.; Zhou, D.-X. The Interaction between Rice ERF3 and WOX11 Promotes Crown Root Development by Regulating Gene Expression Involved in Cytokinin Signaling. *Plant Cell* **2015**, *27*, 2469–2483. [CrossRef] [PubMed]

86. Dündar, E.; Bush, D.R. BAT1, a bidirectional amino acid transporter in Arabidopsis. *Planta* **2009**, *229*, 1047–1056. [CrossRef] [PubMed]

87. Miflin, B.J.; Habash, D.Z. The role of glutamine synthetase and glutamate dehydrogenase in nitrogen assimilation and possibilities for improvement in the nitrogen utilization of crops. *J. Exp. Bot.* **2002**, *53*, 979–987. [CrossRef] [PubMed]

88. Camacho, C.; Coulouris, G.; Avagyan, V.; Ma, N.; Papadopoulos, J.; Bealer, K.; Madden, T.L. BLAST+: Architecture and applications. *BMC Bioinform.* **2009**, *10*, 421. [CrossRef] [PubMed]

89. Rogers, S.O.; Bendich, A.J. Extraction of DNA from plant tissues. In *Plant Molecular Biology Manual*; Gelvin, S.B., Schilperoort, R.A., Verma, D.P.S., Eds.; Springer Netherlands: Heidelberg, Germany, 1989; pp. 73–83. ISBN 978-94-009-0951-9.

90. Andrews, S. Babraham Bioinformatics-FastQC A Quality Control tool for High Throughput Sequence Data. Available online: http://www.bioinformatics.babraham.ac.uk/projects/fastqc/ (accessed on 9 September 2019).

91. Catchen, J.M.; Amores, A.; Hohenlohe, P.; Cresko, W.; Postlethwait, J.H. Stacks: Building and Genotyping Loci De Novo From Short-Read Sequences. *G3 Genes Genomes Genet.* **2011**, *1*, 171–182. [CrossRef]

92. Bolger, A.M.; Lohse, M.; Usadel, B. Trimmomatic: A flexible trimmer for Illumina sequence data. *Bioinformatics* **2014**, *30*, 2114–2120. [CrossRef]

93. Li, H.; Durbin, R. Fast and accurate long-read alignment with Burrows-Wheeler transform. *Bioinformatics* **2010**, *26*, 589–595. [CrossRef]

94. McKenna, A.; Hanna, M.; Banks, E.; Sivachenko, A.; Cibulskis, K.; Kernytsky, A.; Garimella, K.; Altshuler, D.; Gabriel, S.; Daly, M.; et al. The genome analysis toolkit: A MapReduce framework for analyzing next-generation DNA sequencing data. *Genome Res.* **2010**, *20*, 1297–1303. [CrossRef]

95. Depristo, M.A.; Banks, E.; Poplin, R.; Garimella, K.V.; Maguire, J.R.; Hartl, C.; Philippakis, A.A.; Del Angel, G.; Rivas, M.A.; Hanna, M.; et al. A framework for variation discovery and genotyping using next-generation DNA sequencing data. *Nat. Genet.* **2011**, *43*, 491–501. [CrossRef]

96. Van der Auwera, G.A.; Carneiro, M.O.; Hartl, C.; Poplin, R.; del Angel, G.; Levy-Moonshine, A.; Jordan, T.; Shakir, K.; Roazen, D.; Thibault, J.; et al. From fastQ data to high-confidence variant calls: The genome analysis toolkit best practices pipeline. *Curr. Protoc. Bioinform.* **2013**, *43*, 11.10.1–11.10.33.

97. McLaren, W.; Gil, L.; Hunt, S.E.; Riat, H.S.; Ritchie, G.R.S.; Thormann, A.; Flicek, P.; Cunningham, F. The Ensembl Variant Effect Predictor. *Genome Biol.* **2016**, *17*, 1–14. [CrossRef] [PubMed]

98. Meirmans, P.G.; Liu, S.; Van Tienderen, P.H. The Analysis of Polyploid Genetic Data. *J. Hered.* **2018**, *109*, 283–296. [CrossRef] [PubMed]

99. Stacklies, W.; Redestig, H.; Scholz, M.; Walther, D.; Selbig, J. pcaMethods-A bioconductor package providing PCA methods for incomplete data. *Bioinformatics* **2007**, *23*, 1164–1167. [CrossRef] [PubMed]

100. Yu, J.; Pressoir, G.; Briggs, W.H.; Vroh Bi, I.; Yamasaki, M.; Doebley, J.F.; McMullen, M.D.; Gaut, B.S.; Nielsen, D.M.; Holland, J.B.; et al. A unified mixed-model method for association mapping that accounts for multiple levels of relatedness. *Nat. Genet.* **2006**, *38*, 203–208. [CrossRef] [PubMed]

101. Lipka, A.E.; Tian, F.; Wang, Q.; Peiffer, J.; Li, M.; Bradbury, P.J.; Gore, M.A.; Buckler, E.S.; Zhang, Z. GAPIT: Genome association and prediction integrated tool. *Bioinformatics* **2012**, *28*, 2397–2399. [CrossRef]

International Journal of
Molecular Sciences

Integrated Analysis of Small RNA, Transcriptome and Degradome Sequencing Provides New Insights into Floral Development and Abscission in Yellow Lupine (*Lupinus luteus L.*)

Paulina Glazińska [1,2,*], Milena Kulasek [1,2], Wojciech Glinkowski [1,2], Waldemar Wojciechowski [1,2] and Jan Kosiński [3]

[1]   Department of Plant Physiology and Biotechnology, Faculty of Biological and Veterinary Sciences, Nicolaus Copernicus University in Torun, 87-100 Torun, Poland; milena.kulasek@gmail.com (M.K.); w_glinkowski@o2.pl (W.G.); wwojc@umk.pl (W.W.)
[2]   Centre for Modern Interdisciplinary Technologies, Nicolaus Copernicus University in Torun, 87-100 Torun, Poland
[3]   Department of Computational Biology, Institute of Molecular Biology and Biotechnology, Faculty of Biology, Adam Mickiewicz University, 61-712 Poznan, Poland; kosinski@ideas4biology.com
*   Correspondence: paulina.glazinska@umk.pl

Received: 29 August 2019; Accepted: 14 October 2019; Published: 16 October 2019

Abstract: The floral development in an important legume crop yellow lupine (*Lupinus luteus* L., Taper cv.) is often affected by the abscission of flowers leading to significant economic losses. Small non-coding RNAs (sncRNAs), which have a proven effect on almost all developmental processes in other plants, might be of key players in a complex net of molecular interactions regulating flower development and abscission. This study represents the first comprehensive sncRNA identification and analysis of small RNA, transcriptome and degradome sequencing data in lupine flowers to elucidate their role in the regulation of lupine generative development. As shedding in lupine primarily concerns flowers formed at the upper part of the inflorescence, we analyzed samples from extreme parts of raceme separately and conducted an additional analysis of pedicels from abscising and non-abscising flowers where abscission zone forms. A total of 394 known and 28 novel miRNAs and 316 phased siRNAs were identified. In flowers at different stages of development 59 miRNAs displayed differential expression (DE) and 46 DE miRNAs were found while comparing the upper and lower flowers. Identified tasiR-ARFs were DE in developing flowers and were strongly expressed in flower pedicels. The DEmiR-targeted genes were preferentially enriched in the functional categories related to carbohydrate metabolism and plant hormone transduction pathways. This study not only contributes to the current understanding of how lupine flowers develop or undergo abscission but also holds potential for research aimed at crop improvement.

Keywords: yellow lupine;   miRNA;   phased   siRNA;   RNA-seq;   degradome;   flower development; abscission

---

1. Introduction

Yellow lupine is a crop plant with remarkable economic potential. Because of the symbiotic bond with nitrogen-fixing *Rhizobium* bacteria it does not need fertilizers, and its protein-rich seeds may be an excellent source of protein for both human consumption and animal feed [1–3]. *Lupinus luteus* flowers are stacked in whorls along the common stem forming a raceme. Pods are formed at the lowest whorls, while the flowers above them fall off [4]. The estimated percentage of dropped flowers is 60% at the 1st (and lowest) whorl, 90% at the 2nd whorl, and ~100% at the whorls above them. Thus, the problem of flower abscission generates large economic losses in agriculture [1]. Precise control of flower emergence and development is crucial for plant's reproductive cycle. This is especially true for crop plants, as it is directly tied to potential yield. Molecular basis for flower formation has been extensively studied for many years across different plant species, and described collectively by ABCDE model (reviewed in [5]), with slight modifications depending on either species or flower shape [6]. Mutations that occur in genes governing flower formation cause various morphogenetic aberrations, including changes in the identity, number, and positioning of floral organs [7]. Proper development of already established flower elements is equally important. Numerous factors are involved in flower development, such as plant hormones (for example GA, IAA, JA [8]), numerous genes [9] and microRNAs, [10]. All of these components create a complex regulatory network, malfunction of which can cause a variety of abnormalities with the loss of fertility being the most detrimental [11,12].

Plant organ abscission is an element of the developmental strategy related to reproduction, defense mechanisms or disposal of unused organs [13,14]. In most species, the key components involved in the activation of the abscission zone (AZ) are plant hormones, in particular, auxin (IAA) and ethylene (ET) [15,16].

Our previous transcriptome-wide study [17] proved that the abscission of yellow lupine flowers and pods is associated, *inter alia*, with intensive changing of auxin catabolism and signaling. Genes encoding auxin response factors *ARF4* and *ARF2* were objectively more expressed in generative organs that were maintained on the plant, in contrast to the mRNA encoding auxin receptor *TIR1* (*TRANSPORT INHIBITOR RESPONSE 1*), which is accumulated in larger quantities in shed organs [17]. Since (i) some micro RNAs (miRNAs) and small interfering RNAs (siRNAs) restrict the activity of certain *ARFs* [18,19] and members of the *TAAR (TIR1/AFB AUXIN RECEPTOR)* family encoding auxin receptors [20], and since (ii) we proved that the precursor of miR169 is accumulated in increased quantities in yellow lupine's generative organs undergoing abscission [17], we predict that sRNAs play significant roles in orchestrating organ abscission in *L. luteus*.

MiRNAs are 21-22-nt-long regulatory RNAs formed as a result of the activity of *MIR* genes in certain tissues and at certain developmental stages [21–23] and also in response to environmental stimuli [24–26]. *MIR* genes encode two consecutively formed precursor RNAs, first pri-miRNAs and then pre-miRNAs, which are subsequently processed by DCL1 (Dicer-like) into mature miRNAs [27,28]. *MIR* genes are often divided into small families encoding nearly or completely identical mature miRNAs [29]. miRNA sequences of 19–21 nucleotides are long enough to enable binding particular mRNAs by complementary base pairing, and allow either for cutting within a recognized sequence or for translational repression [30]. Plant miRNAs are involved in, for instance, regulating leaf morphogenesis, the establishment of flower identity, and stress response [10,24–26,31,32]. Some of them also form a negative feedback loop by influencing their own biogenesis, as well as the biogenesis of some 21-nt-long siRNAs called trans-acting siRNAs (ta-siRNAs). Ta-siRNAs are processed from non-coding *TAS* mRNAs, which contain a sequence complementary to specific miRNAs [33,34]. There is also a large group of plant sRNAs that are referred to as phased siRNA, which are formed from long, perfectly double-stranded transcripts of various origins, mainly processed by DCL4 [35,36].

Studies on sRNA in legumes (e.g., *Glycine max, G. soja* [37], *Medicago truncatula* [38], *M. sativa* [39], *Arachis hypogaea* [40], *Lotus japonicus* [41] and *Phaseolus vulgaris* [42]) have primarily focused on stress response or nodulation. Only three studies on miRNAs have been conducted so far using only two species of *Lupinus* genus: *Lupinus albus* (white lupine) and *Lupinus angustifolius* (narrow-leafed lupine). These studies were focused on small RNA sequences isolated from phloem exudate [43], global expression of miRNAs during phosphate deficiency [44], and gene regulatory networks during seed development [45]. Unfortunately, the knowledge on the roles of mi- and siRNAs function during flower development in leguminous plants is still incomplete [43]. Moreover, the involvement of regulatory sRNAs in mechanisms responsible for the maintenance/abscission of generative organs in the *Fabaceae* family has never been explored before.

Our observations of *L. luteus* generative development suggest that the fate of flowers (pod set or shedding) is determined on a molecular level during flower development. This study aims to characterize and investigate the role of these important molecules and their target genes during flower development and abscission.

In order to achieve this goal, an integrated analysis of small non-coding RNAs (sncRNA), transcriptome and degradome sequencing data was performed. We identified both known and presumably new miRNAs and siRNAs from flowers at different developmental stages, specifically the lower flowers (usually maintained and developed into pods) and the upper flowers (usually shed before fruit setting). Moreover, in our comparisons of libraries from the upper and lower flowers, differentially expressed miRNAs were found. In order to identify the miRNAs involved exclusively in flower abscission, we compared sRNA libraries from the pedicels of flowers that were maintained on the plant and those that were shed. A transcriptome- and a degradome-wide analysis was carried out to identify the target genes for the conserved or new *L. luteus* sRNAs. The targeted transcripts were then functionally annotated to outline the putative regulatory network in which these sRNAs might have a role to play. Our results of next-generation sequencing (NGS) analysis indicate that the identified miRNA-targeted modules may be vital in regulating yellow lupine flower development, both generally and depending on the flower location on the inflorescence. Furthermore, these scnRNA also display differential accumulation during flower abscission in this plant.

2. Results

2.1. Sequencing and Annotation of Yellow Lupine sRNAs from Flowers and Flower Pedicels

Flowers collected from the top and bottom parts of the inflorescence were separated into four categories based on the progression of their development, and thus: Stage 1—closed green buds, parts of which were still elongating. Stage 2—closed yellow buds, around the time of anther opening. Stage 3—flowers in full anthesis. Stage 4—flowers with enlarged gynoecia from the lower parts of the inflorescence, or aging flowers from the upper parts of the inflorescence. Based on their position on the inflorescence, flowers in each of the stages were additionally tagged as either upper (UF) or lower flowers (LF), resulting in eight different variants: UF1, UF2, UF3, UF4, LF1, LF2, LF3 and LF4 (Figure 1, Table S1). Flower pedicels from flowers undergoing abscission (FPAB) or maintained on the plant (FPNAB) were also collected, as they had been in our previous study [17]. This division resulted in ten variants of small RNA libraries, which were subjected to single-end deep sequencing performed on the Illumina HiSeq4000 platform (Illumina, Great Abington Cambridge, United Kingdom). After removing low-quality reads, a total of 303,267,263 reads (from 14,186,278 to 15,504,860 reads per library) and 128,060,403 unique reads (from 5,677,701 to 6,990,061 per library) were obtained (Table S2).The length distribution of the small RNAs (15–30 nt) revealed that a length of 24 nt was the most frequent and that of 21 nt was the second most abundant class of the clean and redundant reads (Figure 2), which was compliant with many other RNA-Seq experiments [46–48] and correlated with the abundance of siRNAs and miRNAs, respectively.

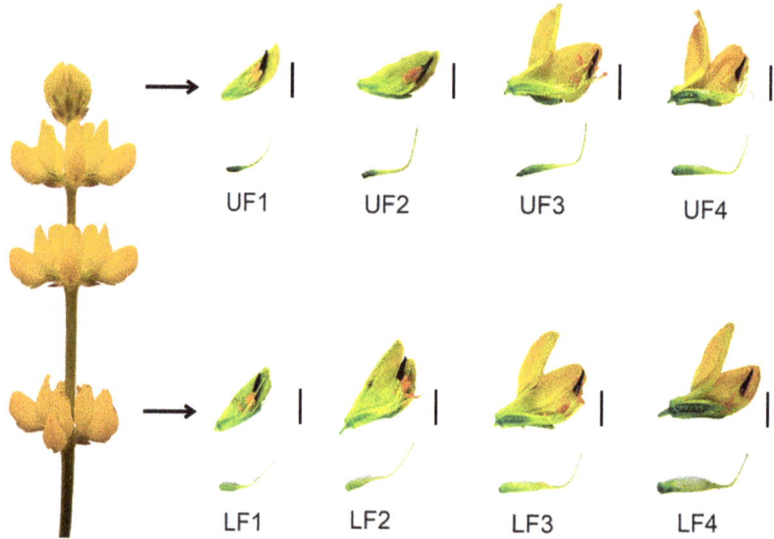

Figure 1. Development of *Lupinus luteus* flowers from the upper and lower part of the raceme. An isolated pistil from a given developmental stage is shown under each flower. LF—lower flower, UF—upper flower. Bar 5 mm.

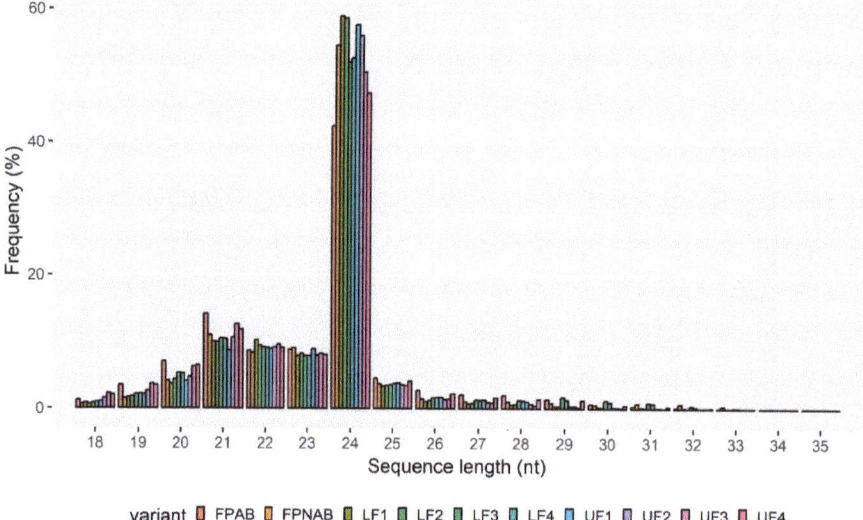

Figure 2. Nucleotide length distribution of small RNAs from all ten libraries: *Y*-axis represents the percentage frequency of the sRNA sequences identified in this study, the X-axis represents sRNA length.

The unique reads were annotated against Rfam [49,50] and miRBase [51] databases, and from the latter both mature (named in tables as 'miRBase') and precursor sequences (named as 'Hairpin') were taken into account. However, many of them remained unassigned (Table 1).

Table 1. Summary of reads and general annotation of small RNA-seq data.

	FPAB	FPNAB	LF1	LF2	LF3	LF4	UF1	UF2	UF3	UF4
					All reads					
unique	5,915,879.5	6,387,744.5	6,755,903	6,725,750	6,401,622.5	6,434,224.5	6,633,250	6,525,844.5	6,061,243.5	6,188,739.5
redundant	15,042,451	14,794,948	15,357,830	15,397,345	15,132,153	15,067,593	15,368,697	15,249,713	14,996,762	15,226,142
					Annotation					
					Unique					
miRBase	424	388	449	399	346	412	467	336	360	410
hairpin	2001	1832	1713	1610	1738	1801	1815	1699	1750	1995
Rfam	45,858	31,044	25,877	29,300	34,998	33,577	31,265	33,221	36,206	43,875
unknown	5,867,598	6,354,480.5	6,727,865	6,694,441.5	6,364,541.5	6,398,435	6,599,704	6,490,589.5	6,022,928	6,142,461
					All					
miRBase	580,674	562,932	364,583	351,641	394,114	410,044	368,739	448,377	571,963	471,398
hairpin	298,173	319,855	208,266	234,335	286,889	274,200	192,098	236,605	298,894	301,915
Rfam	731,119	493,959	299,502	386,808	581,709	528,515	522,096	499,417	555,178	727,483
unknown	13,432,486	13,418,203	14,485,480	14,424,562	13,869,443	13,854,835	14,285,765	14,065,314	13,570,728	13,725,347

The unique sequences were annotated into different RNA classes against the Rfam database using BLAST [52] such as known miRNAs, rRNA, tRNA, sn/snoRNA and others (Table 2). A total of 690,436 sRNAs were annotated into all libraries, with the highest number observed in the upper flowers and abscising pedicles. Between these libraries, the most abundant classes were rRNAs and tRNAs, with average values of 26,390 and 3,726 sequences, respectively, followed by snoRNAs and different subtypes of snRNAs with average values ranging from 863 to 876 sequences (Table 2).

Table 2. Rfam annotation summary.

	FPAB	FPNAB	LF1	LF2	LF3	LF4	UF1	UF2	UF3	UF4
tRNA	4742	3467	2537	3245	3771	4209	2914	3617	4115	4645
rRNA	33,810	23,320	19,921	22,331	27,302	25,453	24,092	25,477	27,557	34,641
snoRNA	2164	893	496	561	540	591	1125	683	827	748
Intro	1480	1238	1094	1201	1403	1340	1166	1339	1407	1546
Retro	829	800	681	707	792	742	803	766	751	852
U1	415	100	64	103	83	66	62	84	124	124
U2	620	323	263	261	286	294	275	308	312	346
U3	433	244	150	172	169	163	189	184	245	215
U4	248	61	51	63	54	54	58	67	91	82
U5	69	10	12	15	16	16	13	9	13	21
U6	349	81	52	81	64	58	102	76	90	108
Total	45,858	31,044	25,877	29,300	34,998	33,577	31,265	33,221	36,206	43,875

2.2. Identification of Known, Conserved miRNAs

After analyzing the results of the alignment against miRBase [51], 394 unique miRNAs containing 366 conserved miRNAs were identified (Table S2). The number of identified miRNAs in each library is shown in Table S1. The identified miRNAs belonged to more than 67 families (Table S2), while most of them belonged to the MIR156, MIR159, and MIR166 families, with more than 35 members in each (Figure 3a). Each discovered miRNA received an identification number in the following format: Ll-miR(number). In case of miRNAs displaying identity to sequences from miRBase, annotation Ll-miR(number)/miRBase annotation is used, for example, Ll-miR224/miR393.

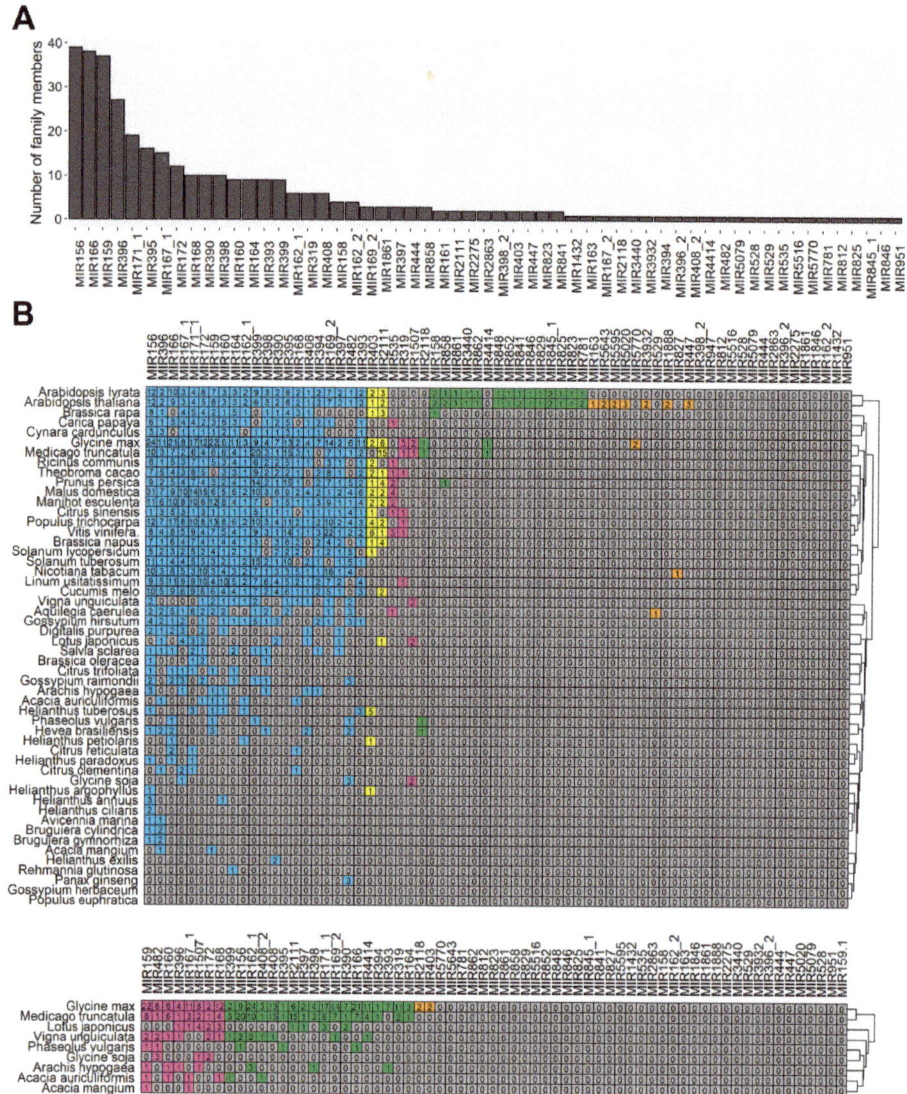

Figure 3. Identification and evolutionary conservation of known miRNA families in *Lupinus luteus*. (**A**) The distribution of known miRNA family sizes in *L. luteus*. (**B**) Comparison of known miRNA families in *L. luteus* and their 52 homologs in *Eudicotyledons* species present in miRBase (upper panel) and 9 *Fabaceae* species (lower panel). Known miRNA families of *L. luteus* identified from small RNA-seq are listed in the top row. The colors represent relative miRNA families classified into different groups with similar conservation. Blue, yellow, magenta, green and orange represent relative miRNA families with homologs across more than 20, 10–19, 5–9, 2–4 species and in 1 species, respectively.

2.3. Evolutionary Conservation of microRNAs Identified in Lupinus luteus

Since this study is the first wide-scale analysis of yellow lupine miRNAs, we decided to explore the evolutionary characteristics of these sequences when compared to the data of almost all [52] *Eudicotyledons* species present in miRBase [51]. The same analysis was performed exclusively against nine *Fabaceae* species. As shown in Figure 3b, the 67 known miRNA families exhibited different numbers of homologous sequences in both of the comparisons. Twenty of them were the most conserved ones, i.e., had homologues in over 20 species (Figure 3b, shaded in blue). Our comparison across legumes revealed that 8 miRNA families were highly conserved in this taxon, i.e., had homologues in 5–9 species out of 9 (Figure 3b, lower panel, shaded in magenta), 18 had homologues in 2–4 legumes (Figure 3b, shaded in green), and 2 had homologues only in one plant, *Glycine max* (Figure 3b, shaded in orange).

A surprisingly high number (39) of miRNA families identified in yellow lupine flowers were not conserved across *Fabaceae*, probably due to a still incomplete list of miRNAs in these taxa.

2.4. Identification of Novel miRNAs

With the use of the ShortStack software (https://github.com/MikeAxtell/ShortStack/) [53], 28 candidates for novel miRNAs were identified (Table 3). This tool identifies miRNAs based on their mapping against a reference genome. Since there was no genome available for the studied species, we used a transcriptome instead (statistical data on de novo assembly is shown in Table S3). The results obtained were filtered against mature miRNAs from miRBase, and unique sequences received names in the following format: "Ll-miRn(number)", (for example, Ll-miRn1). All of these 28 sequences were 21–24 nt in length, with 68% of them being 21 nt long (Table 3).

The expression of novel miRNAs was also highly diversified across all the libraries. Ll-miRn26 was present only in the LF1 sample, while Ll-miRn21 was present in all the sRNA libraries and had an expression ranging from 3,982.82 to 11,421.55 RPM (Table S4).

Table 3. Novel miRNAs identified in *Lupinus luteus*.

miRNA ID	Sequence (5′–3′)	Size (nt)	Precursor (RNA-seq ID)	LP (nt)	MFE (kcal/mol)	Target Description (degradome/psRNAtarget)
Ll-miRn1	TTGCCAATTCCACCCATGCCTA	22	TRINITY_DN58100_c0_g3_i1	125	−59.90	**SUPPRESSOR OF GENE SILENCING 3**
Ll-miRn2	TCACTCCAACTTTGACCTTCT	21	TRINITY_DN50576_c0_g2_i1	215	−84.70	**65-kDa microtubule-associated protein 7**
Ll-miRn3	TGAAGAGGGAGGGAGACTGATG	22	TRINITY_DN77107_c0_g1	185	−86.50	**SRC2 homolog**
Ll-miRn4	GTAGCACCATCAAGATTCACA	21	TRINITY_DN43941_c0_g1_i1	151	−60.30	RCC1 and BTB domain-containing protein 2
Ll-miRn5	TGGAATAGTGAATGAGACATC	21	TRINITY_DN52736_c3_g2_i2	102	−38.70	Probable cinnamyl alcohol dehydrogenase 9
Ll-miRn6	TGCTATCCATCCTGAGTTTCA	21	TRINITY_DN54182_c6_g1_i1	133	−47.90	Probable amino acid permease 7
Ll-miRn7	AGAGGTGTATGGCACAAGAGA	21	TRINITY_DN53175_c1_g6_i1	85	−36.60	Probable protein phosphatase 2C
Ll-miRn8	TGAAGTGTTTGGGGGAACTCC	21	TRINITY_DN44441_c0_g1_i1	102	−37.40	**ATP sulfurylase 1**
Ll-miRn9	TCGGACCAGGCTTTATTCCTT	21	TRINITY_DN50586_c0_g1_i3	167	−65.60	**Homeobox-leucine zipper protein REVOLUTA**
Ll-miRn10	ATGTTGTGATGGGAATCAATG	21	TRINITY_DN67022_c0_g1_i1	84	−43.50	**CBL-interacting serine/threonine-protein kinase 6**
Ll-miRn11	TAAAGACCTCATTCTCTCATG	21	TRINITY_DN31556_c0_g1_i1	130	−62.80	**Vacuolar protein sorting-associated protein 62**
Ll-miRn12	AGGTCATCTTGCAGCTTCAAT	21	TRINITY_DN52990_c2_g1_i5	71	−36.84	DNA-directed RNA polymerase I subunit 1
Ll-miRn15	TTCGGCTTTCTACTTCTCATG	21	TRINITY_DN54101_c8_g2_i10	56	−66.20	**Transcription termination factor MTERF8**
Ll-miRn16	AGTTCTTTAGATGGGCTGGACGCC	24	TRINITY_DN52523_c6_g2_i1	83	−36.50	Amino acid transporter AVT6A
Ll-miRn17	TGTCTCATTCTCTATCTCAAG	21	TRINITY_DN51068_c0_g1_i2	142	−64.30	IST1-like protein
Ll-miRn18	AATAGGGCACATCTCTCATGG	22	TRINITY_DN46596_c0_g1_i1	112	−49.00	E3 ubiquitin-protein ligase HOS1
Ll-miRn19	TCCAAAGGGATCGCATTGATTT	22	TRINITY_DN53637_c4_g2_i4	110	−48.10	**AUXIN SIGNALING F-BOX 3**
Ll-miRn21	TGAGCATGAGGATAAGGACGG	21	TRINITY_DN50271_c0_g3_i1	246	−144.90	**Tetratricopeptide repeat protein 1**
Ll-miRn22	TATCATTCCATACATCGTCTCG	21	TRINITY_DN50592_c0_g4_i2	80	−33.60	Putative disease resistance RPP13-like protein 1
Ll-miRn24	ATTGTCACTGTATCATTCACCATT	24	TRINITY_DN52987_c0_g1_i1	104	−32.30	Zinc finger CCCH domain-containing protein 55
Ll-miRn25	TGGTACAAAAAGTGGGGCAAC	21	TRINITY_DN48871_c3_g1_i9	151	−43.90	Nuclear transcription factor Y subunit A-9
Ll-miRn26	TGTTGTTTTCTGGTAAAAATA	21	TRINITY_DN58488_c1_g2_i4	99	−33.80	Auxin-responsive protein IAA27
Ll-miRn27	ATTAGATCATGTGGCAGTTTCACC	24	TRINITY_DN51506_c3_g2_i5	77	−36.60	U-box domain-containing protein 33
Ll-miRn28	TACGGGTGTCCTCACCTCTGAT	22	TRINITY_DN70730_c0_g1_i1	98	−36.90	**ISWI chromatin-remodeling complex ATPase**
Ll-miRn29	TGGGATAGAGAGTGAGATACC	21	TRINITY_DN51068_c0_g1_i2	125	−67.80	Ethylene-responsive transcription factor ERF017
Ll-miRn30	TTCGTTTGTGTGCAGACTCTGT	22	TRINITY_DN57730_c1_g9_i2	105	−42.70	**Endoribonuclease Dicer homolog 2**
Ll-miRn31	GCGTACCAGGAGCCATGCATG	21	TRINITY_DN58934_c0_g4_i1	149	−60.20	Calcium-transporting ATPase 4
Ll-miRn32	AAGGGTTGTTTACAGAGTTTA	21	TRINITY_DN51330_c0_g1_i1	128	−55.40	26S proteasome regulatory subunit 7

2.5. Analysis of the Expression Abundance of Known miRNA Families

Since miRNA expression across all libraries displayed high variation, we put the data into five categories based on the maximum value (Figure 4). Two miRNAs, namely Ll-miR341/miR319 and Ll-miRn21, showed expression maxima of over 10,000 RPM. The maximum expression of another two, Ll-miR260/miR166 and Ll-miR384/miR396, ranged from 2000 to 10,000 RPM. Thirteen miRNAs showed expression maxima ranging from 100 to 2000 RPM. The most numerous category, with 33 elements was the one for miRNAs with expression maxima ranging from 10 to 100 RPM. Another 24 miRNAs were expressed with the maximal RPM values between 1 and 10. The expression value of the five least abundant miRNAs did not exceed 1 RPM (Figure 4, Table S5).

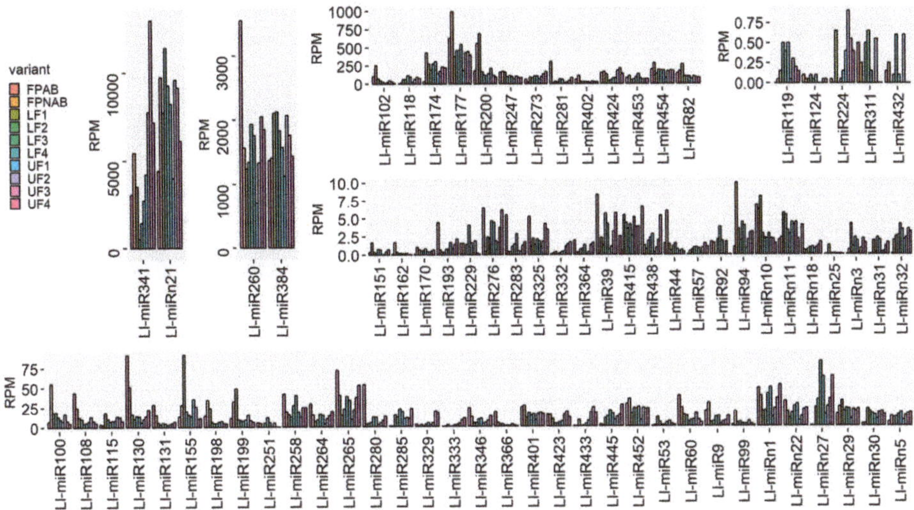

Figure 4. Diversity of miRNA expression (reads per million, RPM) in yellow lupine flowers. Complete data concerning differential miRNA expression in the experiment described herein, divided into six groups, depending on their expression maxima listed in order of appearance from left to right, and top to bottom: over 10,000 RPM, 2000–10,000 RPM, 100–2000 RPM, up to 1 RPM, 1–10 RPM, 10–100 RPM.2.6. Identification of phased siRNA in Yellow Lupine.

Numerous reports and studies indicate the importance of phased siRNA not only in stress response mechanisms but also in growth regulation [54]. Therefore, we decided to investigate the role of siRNAs during yellow lupine inflorescence development. To achieve this, ShortStack (https://github.com/MikeAxtell/ShortStack/) [53] was used to identify small RNAs that were being cut in phase from longer precursors. We identified 316 siRNA ranging from 21 to 25 nt in length, of which 71% were 24 nt long (Table S6, Figure S1). The identified siRNAs received names in the following format: "Ll-siR(number)", (for example Ll-siR1) and displayed a highly differential expression pattern (Table S7). Some of the sequences showed organ-specific expression, for example, Ll-siR4, -13, -173 were present only in the pedicels of abscising flowers (FPAB), while Ll-siR308 showed an elevated expression in the pedicels (FPAB and FPNAB). On the other hand, Ll-siR246, -291 and -56 were present almost exclusively in the youngest flowers in the lower part of inflorescence (LF1) (Table S7).

2.6. Analysis of the Expression Profile of the Identified sRNAs During Yellow Lupine Flower Development

To gain better insight into the dynamics expression of the identified sRNAs during floral development in yellow lupine, we established a wide scope comparison of the following growth stages of flowers from the upper (UF2 vs UF1, UF3 vs UF2 and UF4 vs UF3) and lower (LF2 vs LF1, LF3 vs LF2 and LF4 vs LF3) parts of the inflorescence (Table 4, Figure 5).

The analyses resulted in the identification of 30 differentially expressed miRNAs (DEmiRs) in the lower and 29 in the upper flowers (Table 4), as well as 14 and 7 DE siRNAs, respectively (Table S8). Between UF2 and UF1, there was a change in the expression of 8 miRNAs, 2 sequences belonging to MIR359 and MIR166 families each, as well as one representative of each of the MIR159, MIR167, MIR396 families and novel Ll-miRn10. Ten DE miRNAs were identified in a comparison of UF3 vs UF2, of which only one Ll-miR258/miR166 was up-regulated. The remaining miRNAs were downregulated and consisted of 3 sequences belonging to the MIR390 and MIR396 families each, and single miRNAs from the MIR168, MIR408, and MIR396 families. A comparison of the UF4 vs UF3 libraries revealed 12 DEmiRs. The most numerous group were members of the MIR390 family, followed by 2 members of MIR167 and MIR319, and singular representatives of MIR398, MIR164, and MIR858, with one novel Ll-miRn11 (Table 4).

During the development of flowers from the lower part of the inflorescence, the miRNAs accumulation dynamics were different. The highest number of the identified DEmiRs was found comparing the youngest flowers (LF2 vs LF1), while, interestingly, a complete lack of DE miRNAs was found when comparing the oldest flowers: LF4 vs LF3 (Table 4). In our comparison of LF2 vs LF1, among the 18 DEmiRs, the most numerous group were novel miRNAs, followed by members of the MIR396 family. Between the LF2 and LF3 stages we confirmed that there was a change in the accumulation of 11 miRNAs, and this pertained to two members of the MIR166 and MIR399 families each, Ll-miRn1 and Ll-miRn22, which were followed by single representatives of the MIR390, MIR395, MIR858, MIR398, MIR408 families (Table 4).

In order to identify miRNAs the presence of which is either common or unique depending on the developmental stage of the upper and lower flowers in lupine, Venn diagrams were constructed (Figure 6a and Table S9) using Venny 2.1 (https://bioinfogp.cnb.csic.es/tools/venny/) [55]. The results of these analyses revealed that approximately 70% of the identified miRNAs were common in all developmental stages of both the upper (Figure 6a) and lower flowers (Figure 5b). However, miRNAs unique to certain developmental stages were also found (Figure 6 and Table S9).

In regard to siRNAs during flower development in yellow lupine, almost every differentially expressed siRNA was up-regulated. In the lower part of the inflorescence, similarly to miRNAs, there were no differences between the LF4 and LF3 stages. During the upper flower development, most DEsiRs were identified in a comparison of UF2 vs UF1, and the least (only one) when comparing UF3 vs UF2. One noteworthy observation was the presence of the same siRNAs in the comparisons of UF2 vs UF1 and LF2 vs LF1, namely Ll-siR281, -308. and -249, which suggests that an increase in their accumulation is important during phase 1 to phase 2 transition in the development of yellow lupine flowers, regardless of their position on the inflorescence. The complete dataset can be found in (Table S8).

Table 4. Expressed miRNAs identified in comparisons of flower development stages between lower and upper parts of the raceme with padj < 0.05.

			Flower Development			
ID	miRNA Sequence	miRBase Annotation	log_2FC	*p*-value	padj	Target Description (psRNAtarget/degradome)
			Lower flowers development LF2 vs LF1			
Ll-miRn22	TATCATTCCATACATCGTCTCG	new	0.71	0.0000	0.0004	Putative disease resistance RPP13-like protein 1
Ll-miR401	TTGACAGAAGAGAGTGAGCAC	gma-miR156k	0.61	0.0003	0.0032	**Squamosa promoter-binding-like protein 2**
Ll-miR265	TCGGACCAGGCTTCATTCCTT	ata-miR166c-3p	0.61	0.0001	0.0007	**Homeobox-leucine zipper protein ATHB-15**
Ll-miRn27	ATTAGATCATGTGGCAGTTTCACC	new	0.51	0.0028	0.0241	U-box domain-containing protein 33
Ll-miR454	TTTGGATTGAAGGGAGCTCTT	aly-miR159b-3p	0.47	0.0000	0.0000	Transcription factor GAMYB
Ll-miR124	TGACAGAAGAGAGTGAGCAC	ama-miR156	0.43	0.0021	0.0193	**Squamosa promoter-binding-like protein 13B**
Ll-miR247	TCGCTTGGTGCAGGTCGGGAA	aly-miR168a-5p	−0.68	0.0000	0.0000	AGO1
Ll-miR102	CGCTGTCCATCCTGAGTTTCA	bra-miR390-3p	−0.69	0.0000	0.0003	TAS3
Ll-miR115	CTCGTTGTCTGTTCGACCTTG	ppe-miR858	−0.76	0.0000	0.0001	**Transcription repressor MYB5**
Ll-miR53	ATCATGCTATCCCTTTGGATT	gma-miR393c-3p	−1.02	0.0034	0.0278	**Midasin-like**
Ll-miR82	CCCGCCTTGCATCAACTGAAT	aly-miR168a-3p	−1.31	0.0000	0.0000	FAD-linked sulfhydryl oxidase ERV1
Ll-miRn10	ATGTTGTGATGGGAATCAATG	new	−1.37	0.0000	0.0000	**CBL-interacting Ser/Thr -protein kinase 6**
Ll-miRn3	TGAAGAGGGAGGGAGACTGATG	new	−1.43	0.0006	0.0060	**SRC2 homolog**
Ll-miR198	GTTCAATAAAGCTGTGGGAA	osa-miR396a-3p	−1.75	0.0000	0.0000	ECERIFERUM 1
Ll-miRn31	GCGTACCAGGAGCCATGCATG	new	−1.99	0.0001	0.0007	Calcium-transporting ATPase 4
Ll-miR200	GTTCAATAAAGCTGTGGGAAG	aly-miR396a-3p	−2.00	0.0000	0.0000	ECERIFERUM 1
Ll-miR155	GCTCAAGAAAGCTGTGGGAGA	gma-miR396b-3p	−2.04	0.0000	0.0000	**Lysine-specific demethylase JMJ25**
Ll-miR199	GTTCAATAAAGCTGTGGGAAA	ata-miR396e-3p	−2.31	0.0000	0.0000	ECERIFERUM 1
			LF3 vs LF2			
Ll-miR258	TCGGACCAGGCTTCATTCCCG	cpa-miR166d	0.91	0.0000	0.0001	Homeobox-leucine zipper protein ATHB-15
Ll-miRn1	TTGCCAATTCCACCCATGCCTA	new	0.87	0.0000	0.0008	**SUPPRESSOR OF GENE SILENCING 3**
Ll-miR9	AAGCTCAGGAGGGATAGCGCC	aly-miR390a-5p	0.78	0.0005	0.0100	**TAS3**
Ll-miR118	CTGAAGTGTTTGGGGGAACTC	aly-miR395d-3p	0.70	0.0000	0.0007	**ATP sulfurylase 1, chloroplastic**
Ll-miR264	TCGGACCAGGCTTCATTCCTC	aqc-miR166a	0.66	0.0037	0.0401	**Homeobox-leucine zipper protein ATHB-15**
Ll-miRn22	TATCATTCCATACATCGTCTCG	new	0.46	0.0032	0.0401	Putative disease resistance RPP13-like protein 1
Ll-miR384	TTCCACAGCTTTCTTGAACTT	aly-miR396b-5p	−0.35	0.0033	0.0401	MPE-cyclase
Ll-miR115	CTCGTTGTCTGTTCGACCTTG	ppe-miR858	−0.59	0.0048	0.0426	**Transcription repressor MYB5**
Ll-miR200	GTTCAATAAAGCTGTGGGAAG	aly-miR396a-3p	−0.62	0.0000	0.0000	ECERIFERUM 1
Ll-miR108	CGTGTTCTCAGGTCGCCCCTG	ppe-miR398b	−1.08	0.0044	0.0426	Plastocyanin
Ll-miR60	ATGCACTGCCTCTTCCCTGGC	ahy-miR408-3p	−1.09	0.0024	0.0385	**Basic blue protein**
			Lower flowers development LF4 vs LF3			
ND						
			Upper flowers development UF2 vs UF1			
Ll-miR119	TGAAGTGTTTGGGGGAACTCC	sly-miR395a	1.22	0.0000	0.0019	**ATP sulfurylase 1, chloroplastic**
Ll-miR452	TTTGGATTGAAGGGAGCTCTC	lus-miR159b	0.74	0.0001	0.0019	**Gamma-glutamyl peptidase 5**

Table 4. *Cont.*

			Flower Development			
ID	miRNA Sequence	miRBase Annotation	$\log_2$FC	*p*-value	padj	Target Description (psRNAtarget/degradome)
			Lower flowers development LF4 vs LF3			
ND						
			Upper flowers development UF2 vs UF1			
Ll-miR118	CTGAAGTGTTTGGGGGAACTC	aly-miR395d-3p	0.64	0.0001	0.0019	ATP sulfurylase 1, chloroplastic
Ll-miR177	GGAATGTTGTCTGGCTCGAGG	aly-miR166a-5p	0.50	0.0002	0.0033	Transcription factor RADIALIS
Ll-miR174	GGAATGTTGGCTGGCTCGAGG	mtr-miR166e-5p	−0.45	0.0002	0.0033	Nucleolar GTP-binding protein 1
Ll-miR285	TGAAGCTGCCAGCATGATCTTA	mdm-miR167h	−0.56	0.0023	0.0286	Auxin response factor 6
Ll-miRn10	ATGTTGTGATGGGAATCAATG	new	−0.96	0.0008	0.0117	CBL-interacting Ser/Thr-protein kinase 6
Ll-miR155	GCTCAAGAAAGCTGTGGGAGA	gma-miR396b-3p	−1.17	0.0000	0.0000	Lysine-specific demethylase JMJ25
			UF3 vs UF2			
Ll-miR258	TCGGACCAGGCTTCATTCCCG	cpa-miR166d	0.70	0.0022	0.0323	Homeobox-leucine zipper protein ATHB-15
Ll-miR247	TCGCTTGGTGCAGGTCGGGAA	aly-miR168a-5p	−0.40	0.0005	0.0113	AGO1
Ll-miR102	CGCTGTCCATCCTGAGTTTCA	bra-miR390-3p	−0.96	0.0000	0.0004	TAS3
Ll-miR199	GTTCAATAAAGCTGTGGGAAA	ata-miR396e-3p	−0.97	0.0000	0.0001	ECERIFERUM 1
Ll-miR60	ATGCACTGCCTCTTCCCTGGC	ahy-miR408-3p	−1.01	0.0008	0.0145	Basic blue protein
Ll-miR200	GTTCAATAAAGCTGTGGGAAG	aly-miR396a-3p	−1.02	0.0000	0.0000	ECERIFERUM 1
Ll-miR100	CGCTATCCATCCTGAGTTTCA	aly-miR390a-3p	−1.10	0.0001	0.0021	TAS3
Ll-miR99	CGCTATCCATCCTGAGTTTC	gma-miR390a-3p	−1.16	0.0014	0.0228	TAS3
Ll-miR53	ATCATGCTATCCCTTTGGATT	gma-miR393c-3p	−1.19	0.0001	0.0017	Midasin-like
Ll-miR155	GCTCAAGAAAGCTGTGGGAGA	gma-miR396b-3p	−1.45	0.0000	0.0000	Lysine-specific demethylase JMJ25
			UF4 vs UF3			
Ll-miR438	TTGTGTTCTCAGGTCACCCCT	stu-miR398b-3p	1.09	0.0003	0.0098	Probable nucleoredoxin 1
Ll-miR281	TGAAGCTGCCAGCATGATCTGA	ata-miR167b-5p	0.92	0.0001	0.0048	Auxin response factor 6 and ARF8
Ll-miR445	TTTGGACTGAAGGGAGCTCCT	atr-miR319b	0.84	0.0000	0.0009	Transcription factor TCP4
Ll-miR9	AAGCTCAGGAGGGATAGCGCC	aly-miR390a-5p	0.78	0.0004	0.0098	TAS3
Ll-miR346	TGGAGAAGCAGGGCACGTGCA	aly-miR164a-5p	0.73	0.0029	0.0359	CUP-SHAPED COTYLEDON 2
Ll-miR280	TGAAGCTGCCAGCATGATCTG	atr-miR167	0.66	0.0010	0.0207	Auxin response factor 6 and ARF8
Ll-miR130	CTTGGACTGAAGGGAGCTCCC	ppt-miR319c	0.65	0.0001	0.0034	Transcription factor MYB33
Ll-miR115	CTCGTTGTCTGTTCGACCTTG	ppe-miR858	−0.67	0.0019	0.0333	Transcription repressor MYB5
Ll-miR100	CGCTATCCATCCTGAGTTTCA	aly-miR390a-3p	−0.70	0.0025	0.0337	TAS3
Ll-miRn11	TAAAGACCTCATTCTCTCATG	new	−0.83	0.0037	0.0386	Vacuolar protein sorting-associated protein 62
Ll-miR99	CGCTATCCATCCTGAGTTTC	gma-miR390a-3p	−0.86	0.0038	0.0386	TAS3
Ll-miR102	CGCTGTCCATCCTGAGTTTCA	bra-miR390-3p	−0.91	0.0025	0.0337	TAS3

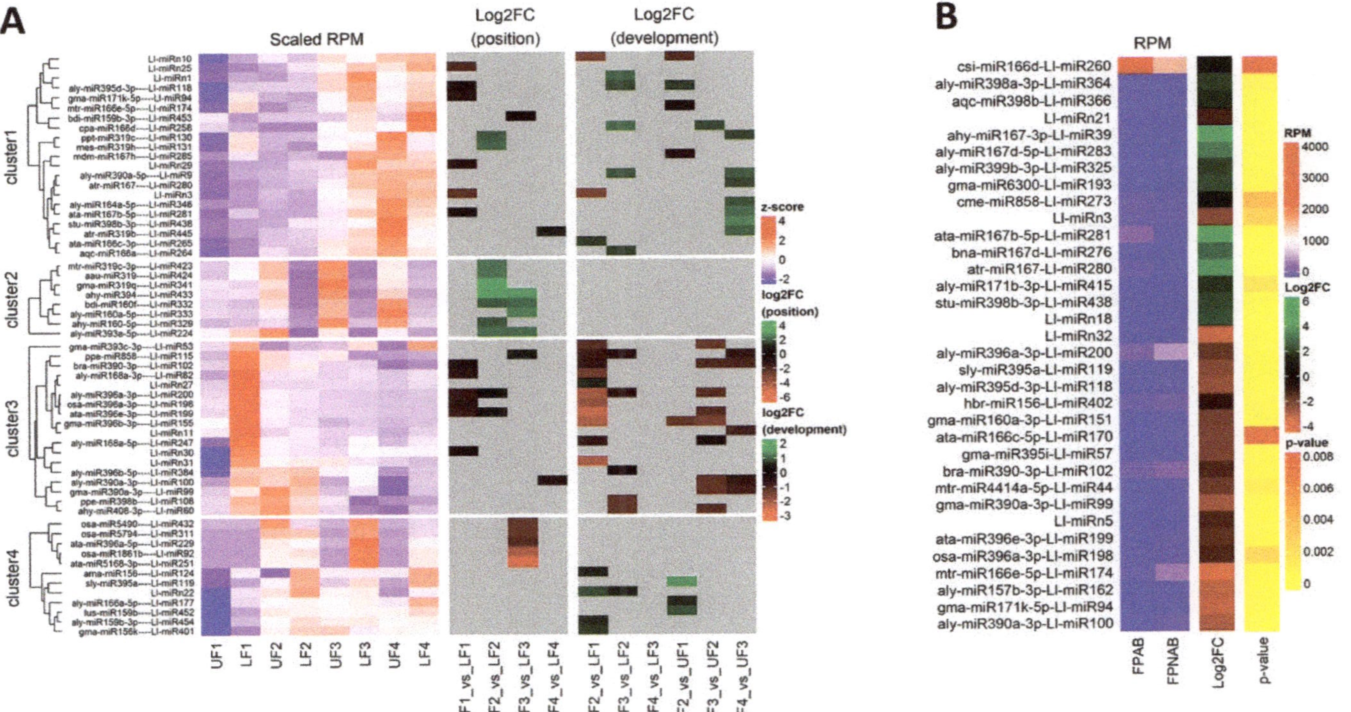

Figure 5. Differential miRNA expression in lupine flowers and flower pedicels. (**A**) Heatmaps of z-scaled miRNA expression (scaled RPM) and $\log_2$ fold changes for either position of the flower on the raceme (Log_2FC position) or identified between consecutive stages of flower development (Log_2FC development). Grey indicates insignificant changes. (**B**) Heatmaps of miRNA expression, $\log_2$ fold changes (Log_2FC) and p-values for flower pedicels with an active or inactive abscission zone. The miRNA names are shown on the right vertical axis. Red and green represent the up-regulated and down-regulated miRNAs, respectively.

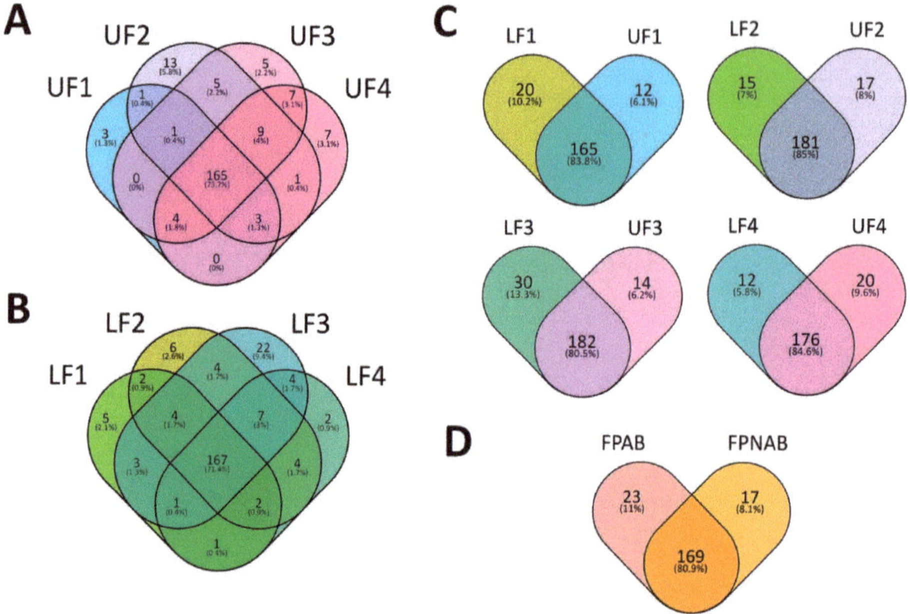

Figure 6. Diagrams showing distribution of yellow lupine miRNAs in (**A**) upper flowers, (**B**) lower flowers, (**C**) both upper and lower flowers at particular stages of their development, (**D**) pedicels of abscising flowers or flowers maintained on the plant.

2.7. Comparison of Differentially Expressed sRNAs Between Developing Flowers From the Lower and Upper Whorls of the Raceme

In order to determine the differences in sRNA expression in developing yellow lupine flowers, comparative analyses of both the upper and lower flowers were performed for each developmental stage of the inflorescence (LF1 vs UF1, LF2 vs UF2, LF3 vs UF3 and LF4 vs UF4) (Table 5, Figure 5). In general, 46 DEmiRs were identified (Table 5). In the first stage of development, the most numerous group of DEmiRs was that of the novel sequences (Ll-miRn3, -25, -29 and -30), followed by sequences annotated as miR396 (3 miRNAs). In the second stage of flower development, miRNAs belonging to the MIR319 family were identified as the largest group (5 sequences), followed by two DE miRNAs annotated as miR160 (Ll-miR329/miR160-5p, Ll-miR332/miR160f) and miR396 (Ll-miR199/miR396e-3p, Ll-miR200/miR396a-3p), respectively. The third stage turned out to be the most diverse, with 2 representatives of the MIR160 (Ll-miR333/miR160a-5p and Ll-miR332/bdi-miR160f) family, followed by single sequences annotated as Ll-miR433/miRr394, Ll-miR224/miR393a-5p, Ll-miR115/miR858, Ll-miR453/miR19b-3p, Ll-miR229/miR396a-5p, Ll-miR432/miR490 and Ll-miR92/miR5168-3p.

Regarding the phased siRNAs, only 4 of them displayed differential expression, namely Ll-siR119 at stage 1 and Ll-siR224, -100 and -146 at stage 4. These results might suggest that, firstly, miRNAs display differential expression in each and every stage of flower development, regardless of flower position on the inflorescence, and secondly, that miRNAs seem to be much more impactful in comparison with phased siRNA in regards to yellow lupine flower differentiation.

Table 5. Differentially expressed miRNAs identified in comparisons between flowers from lower and upper parts of the raceme with padj < 0.05.

				Flowers From Upper and Lower Parts of Receme		
ID	miRNA sequence	miRBase annotation	log$_2$FC	p-value	padj	Target description (psRNAtarget/degradome)
			UF1 vs LF1			
Ll-miR281	TGAAGCTGCCAGCATGATCTGA	ata-miR167b-5p	−0.47	0.0002	0.0054	Auxin response factor 6 and ARF8
Ll-miRn30	TTCGTTTGTGTGCAGACTCTGT	new	−0.49	0.0009	0.0221	**Endoribonuclease Dicer homolog 2**
Ll-miR118	CTGAAGTGTTTGGGGGAACTC	aly-miR395d-3p	−0.58	0.0000	0.0000	**ATP sulfurylase 1, chloroplastic**
Ll-miR102	CGCTGTCCATCCTGAGTTTCA	bra-miR390-3p	−0.71	0.0000	0.0000	TAS3
Ll-miRn29	TGGGATAGAGAGTGAGATACC	new	−1.00	0.0000	0.0000	Ethylene-responsive transcription factor ERF017
Ll-miR94	CGATGTTGGTGAGGTTCAATC	gma-miR171k-5p	−1.09	0.0012	0.0253	Transcription factor MYB4
Ll-miR198	GTTCAATAAAGCTGTGGGAA	osa-miR396a-3p	−1.10	0.0000	0.0001	ECERIFERUM 1
Ll-miR82	CCCGCCTTGCATCAACTGAAT	aly-miR168a-3p	−1.25	0.0000	0.0000	FAD-linked sulfhydryl oxidase ERV1
Ll-miR200	GTTCAATAAAGCTGTGGGAAG	aly-miR396a-3p	−1.44	0.0000	0.0000	ECERIFERUM 1
Ll-miR199	GTTCAATAAAGCTGTGGGAAA	ata-miR396e-3p	−1.47	0.0000	0.0000	ECERIFERUM 1
Ll-miRn25	TGGTACAAAAAGTGGGGCAAC	new	−1.82	0.0001	0.0030	Nuclear transcription factor Y subunit A-9
Ll-miRn3	TGAAGAGGGAGGGAGACTGATG	new	−2.11	0.0000	0.0000	**SRC2 homolog**
			UF2 vs LF2			
Ll-miR433	TTGGCATTCTGTCCACCTCC	ahy-miR394	3.11	0.0000	0.0000	**F-box only protein 6**
Ll-miR341	TGGACTGAAGGGAGCTCCTTC	gma-miR319q	3.07	0.0000	0.0000	**Transcription factor TCP2**
Ll-miR424	TTGGACTGAAGGGAGCTCCCT	aau-miR319	1.98	0.0000	0.0000	Transcription factor TCP4
Ll-miR423	TTGGACTGAAGGGAGCTCCCA	mtr-miR319c-3p	1.91	0.0000	0.0000	Transcription factor TCP4
Ll-miR224	TCCAAAGGGATCGCATTGATCC	aly-miR393a-5p	1.88	0.0002	0.0052	**TRANSPORT INHIBITOR RESPONSE 1**
Ll-miR131	TTGGACTGAAGGGAGCTCCT	mes-miR319h	1.78	0.0000	0.0000	Transcription factor TCP2
Ll-miR332	TGCCTGGCTCCCTGTATGCC	bdi-miR160f	1.63	0.0015	0.0262	**Auxin response factor 18**
Ll-miR130	TTGGACTGAAGGGAGCTCCC	ppt-miR319c	1.60	0.0000	0.0000	**Transcription factor MYB33**
Ll-miR329	TGCCTGGCTCCCTGAATGCCA	ahy-miR160-5p	1.56	0.0024	0.0381	**Auxin response factor 16**
Ll-miR199	GTTCAATAAAGCTGTGGGAAA	ata-miR396e-3p	0.66	0.0005	0.0095	ECERIFERUM 1
Ll-miR200	GTTCAATAAAGCTGTGGGAAG	aly-miR396a-3p	0.31	0.0001	0.0030	ECERIFERUM 1
			UF3 vs LF3			
Ll-miR433	TTGGCATTCTGTCCACCTCC	ahy-miR394	2.59	0.0000	0.0009	**F-box only protein 6**
Ll-miR333	TGCCTGGCTCCCTGTATGCCA	aly-miR160a-5p	2.18	0.0002	0.0072	**Auxin response factor 18**
Ll-miR332	TGCCTGGCTCCCTGTATGCC	bdi-miR160f	1.88	0.0006	0.0210	**Auxin response factor 18**
Ll-miR224	TCCAAAGGGATCGCATTGATCC	aly-miR393a-5p	1.87	0.0018	0.0498	**TRANSPORT INHIBITORRESPONSE 1**
Ll-miR115	CTCGTTGTCTGTTCGACCTTG	ppe-miR858	0.78	0.0001	0.0072	**Transcription repressor MYB5**
Ll-miR453	TTTGGATTGAAGGGAGCTCTG	bdi-miR159b-3p	−0.57	0.0002	0.0072	**RING-type zinc-finger**
Ll-miR229	TCCACAGGCTTTCTTGAACTG	ata-miR396a-5p	−1.94	0.0013	0.0404	**Growth-regulating factor 5**
Ll-miR432	TTGGATTTTTATTTAGGACGG	osa-miR5490	−2.29	0.0001	0.0072	**Acid phosphatase 1**
Ll-miR311	TGAGGAATCACTAGTAGTCGT	osa-miR5794	−2.32	0.0001	0.0072	**Uncharacterized WD repeat-containing protein**
Ll-miR92	CGATCTTGAGGCAGGAACTGAG	osa-miR1861b	−4.11	0.0000	0.0000	Clathrin interactor EPSIN 2
Ll-miR251	TCGGACCAGGCTTCAATCCCT	ata-miR5168-3p	−5.13	0.0000	0.0000	Homeobox-leucine zipper protein ATHB-15
			UF4 vs LF4			
Ll-miR445	TTTGGACTGAAGGGAGCTCCT	atr-miR319b	0.75	0.0001	0.0078	Transcription factor TCP4
Ll-miR100	CGCTATCCATCCTGAGTTTCA	aly-miR390a-3p	−0.97	0.0006	0.0452	TAS3

Analyses of the Venn diagrams we created (Figure 6c), displaying the presence profiles for the library miRNAs, revealed that in each comparison between the upper and lower flowers (UF1 vs LF1, etc.) around 80% of the identified sequences were common for both the upper and lower flowers (Figure 6c). However, in each comparison, we were able to identify miRNAs unique to each stage of the development and each flower position. For example, 20 miRNAs were exclusively present in LF1, while 12 miRNAs were unique to UF1. The detailed information on these comparisons can be found in Table S9.

Based on the data received, we suggest that differences in miRNA expression between lower and upper flowers may be related to the fate of these organs (pod formation/flower abscission). To further confirm this function, we performed an experiment in which flowers were removed from the lower whorls, leaving only flower buds from the last, top whorl (Figure S2). Removing the lower flowers causes maintenance of flowers on the plant and their development into pods, unlike flowers from this whorl in control plants. Thus, their fate seems to be associated with the location in the inflorescence changes. Then, the expression of selected lupine DEmiRs and their target genes were compared during the development of upper flowers after removal of the lower flowers (UFR) in the development stages of S1-S4, with control upper (UF S1–S4) and lower (LF S1–S4) flowers, respectively (Figure S3). The obtained results show that the removal of lower flowers caused a change in the levels of chosen sRNAs in upper flowers and it similar in this respect to flowers from the lower part of raceme. This indicates a link between these genes and the fate of the flowers.

2.8. Comparison of Differentially Expressed sRNAs Between Flower Pedicels with Active And Inactive Abscission Zones

To identify sRNAs possibly involved in yellow lupine flower abscission, mi- and siRNA expression patterns for flower pedicels with an active abscission zone (AZ) (FPAB) and inactive AZ (FPNAB) were compared. As a result, 34 DE miRNAs (including 5 novel ones) (Table 6) and 20 DE phased siRNAs (Table S8) were identified. 14 miRNAs and 9 siRNAs were up-regulated, while the rest remained down-regulated in FPNAB. Among the up-regulated miRNAs, the most numerous family was MIR167 (5 members), followed by MIR398 (3 members). Among the down-regulated miRNAs, the most abundant were MIR390, MIR396 and MIR395 families with 3 members each (Table 6, Figure 5b). With regard to siRNAs, the most up-regulated in FPANB were Ll-siR173, -4 and -13, and the most down-regulated was Ll-siR208 (Table S8).

Table 6. Differentially expressed miRNAs identified in comparisons between pedicels collected from abscised or non-abscised flowers (FPAB vs FPNAB) with padj < 0.05.

			Flower Pedicels			
ID	miRNA sequence	miRBase annotation	log$_2$FC	*p*-value	padj	Target description (psRNAtarget/degradome)
			FPAB vs FPNAB			
Ll-miR281	TGAAGCTGCCAGCATGATCTGA	ata-miR167b-5p	4.77	0.0000	0.0000	Auxin response factor 6 and ARF8
Ll-miR39	AGATCATGTGGCAGTTTCACC	ahy-miR167-3p	4.65	0.0000	0.0000	Transcription repressor OFP14
Ll-miR280	TGAAGCTGCCAGCATGATCTG	atr-miR167	3.73	0.0000	0.0000	Auxin response factor 6 and ARF8
Ll-miR283	TGAAGCTGCCAGCATGATCTGG	aly-miR167d-5p	3.19	0.0000	0.0000	Auxin response factor 6 and ARF8
Ll-miR276	TGAAGCTGCCAGCATGATCT	bna-miR167d	2.74	0.0000	0.0005	Auxin response factor 6 and ARF8
Ll-miRn18	AATAGGGCACATCTCTCATGG	new	2.26	0.0001	0.0006	E3 ubiquitin-protein ligase HOS1
Ll-miR193	GTCGTTGTAGTATAGTGG	gma-miR6300	2.18	0.0001	0.0006	-
Ll-miR438	TTGTGTTCTCAGGTCACCCCT	stu-miR398b-3p	2.09	0.0000	0.0000	**Probable nucleoredoxin 1**
Ll-miR325	TGCCAAAGGAGAGTTGCCCTG	aly-miR399b-3p	1.95	0.0000	0.0000	Inorganic phosphate transporter 1–4
Ll-miR364	TGTGTTCTCAGGTCACCCCTT	aly-miR398a-3p	1.94	0.0010	0.0071	**Superoxide dismutase [Cu-Zn]**
Ll-miR415	TTGAGCCGTGCCAATATCACG	aly-miR171b-3p	1.80	0.0017	0.0117	**Scarecrow-like protein 6**
Ll-miR366	TGTGTTCTCAGGTCGCCCCTG	aqc-miR398b	1.66	0.0005	0.0042	Superoxide dismutase [Cu-Zn]
Ll-miR260	TCGGACCAGGCTTCATTCCCT	csi-miR166d	1.07	0.0078	0.0450	**Homeobox-leucine zipper protein ATHB-14**
Ll-miR273	TCTCGTTGTCTGTTCGACCTT	cme-miR858	1.02	0.0037	0.0223	**Transcription factor MYB78**
Ll-miR402	TTGACAGAAGATAGAGAGC	hbr-miR156	−0.89	0.0008	0.0060	Squamosa promoter-binding protein 1
Ll-miRn21	TGAGCATGAGGATAAGGACGG	new	−1.18	0.0000	0.0000	**Tetratricopeptide repeat protein 1**
Ll-miR198	GTTCAATAAAGCTGTGGGAA	osa-miR396a-3p	−1.24	0.0024	0.0159	ECERIFERUM 1
Ll-miRn5	TGGAATAGTGAATGAGACATC	new	−1.25	0.0003	0.0024	Probable cinnamyl alcohol dehydrogenase 9
Ll-miR102	CGCTGTCCATCCTGAGTTTCA	bra-miR390-3p	−1.27	0.0003	0.0027	TAS3
Ll-miR199	GTTCAATAAAGCTGTGGGAAA	ata-miR396e-3p	−1.49	0.0005	0.0039	ECERIFERUM 1
Ll-miR200	GTTCAATAAAGCTGTGGGAAG	aly-miR396a-3p	−1.61	0.0001	0.0009	ECERIFERUM 1
Ll-miR44	AGCTGCTGACTCGTTGGTTCA	mtr-miR4414a-5p	−1.76	0.0012	0.0084	Non-specific phospholipase C1
Ll-miR151	GCGTATGAGGAGCCAAGCATA	gma-miR160a-3p	−1.90	0.0007	0.0055	E3 ubiquitin-protein ligase RFWD3
Ll-miR57	TGAAGTGTTTGGGGGAACTC	gma-miR395i	−2.01	0.0000	0.0000	**ATP sulfurylase 1, chloroplastic**
Ll-miR170	GGAACGTTGGCTGGCTCGAGG	ata-miR166c-5p	−2.02	0.0072	0.0425	Probable methyltransferase PMT21
Ll-miR119	TGAAGTGTTTGGGGGAACTCC	sly-miR395a	−2.23	0.0000	0.0000	**ATP sulfurylase 1, chloroplastic**
Ll-miRn3	TGAAGAGGGAGGGAGACTGATG	new	−2.26	0.0027	0.0171	**SRC2 homolog**
Ll-miR118	CTGAAGTGTTTGGGGGAACTC	aly-miR395d-3p	−2.47	0.0000	0.0000	**ATP sulfurylase 1, chloroplastic**
Ll-miR99	CGCTATCCATCCTGAGTTTC	gma-miR390a-3p	−2.65	0.0001	0.0006	TAS3
Ll-miR100	CGCTATCCATCCTGAGTTTCA	aly-miR390a-3p	−2.75	0.0000	0.0000	TAS3
Ll-miR162	GCTCTCTAAGCTTCTGTCATCA	aly-miR157b-3p	−2.90	0.0000	0.0003	Dr1 homolog
Ll-miR94	CGATGTTGGTGAGGTTCAATC	gma-miR171k-5p	−3.01	0.0000	0.0000	Transcription factor MYB4
Ll-miRn32	AAGGGTTGTTTACAGAGTTTA	new	−3.18	0.0000	0.0000	26S proteasome regulatory subunit 7
Ll-miR174	GGAATGTTGGCTGGCTCGAGG	mtr-miR166e-5p	−3.46	0.0000	0.0000	**Nucleolar GTP-binding protein 1**

An analysis of the Venn diagrams based on the presence of the identified miRNAs revealed that approx. 80% of the miRNAs were present in both abscising and non-abscising flower pedicles (Figure 6d). However, 23 miRNAs remained unique to FPAB and 17 to FPNAB (Figure 6d, Table S9).

2.9. Validation of the Identified sRNAs in RNA-seq

Stem-loop RT-qPCR technique [56,57] was employed in order to validate the data generated using deep sequencing technology and to confirm the expression patterns of the identified sRNA. Eight identified sRNAs (six conserved miRNAs, one novel miRNA, and one siRNA) were used for this task (Table S10). The qPCR results were similar to sRNA-seq data (Figure 7). For example, in the RT-qPCR analysis, the Ll-siR254 expression increased as the flower developed, showing a positive correlation with the deep sequencing results. Ll-siR249 was preferentially accumulated in yellow lupine pedicels, both in qPCR and RNA-seq. The results of the expression analysis of these sRNAs supported the validity of our sRNA-Seq.

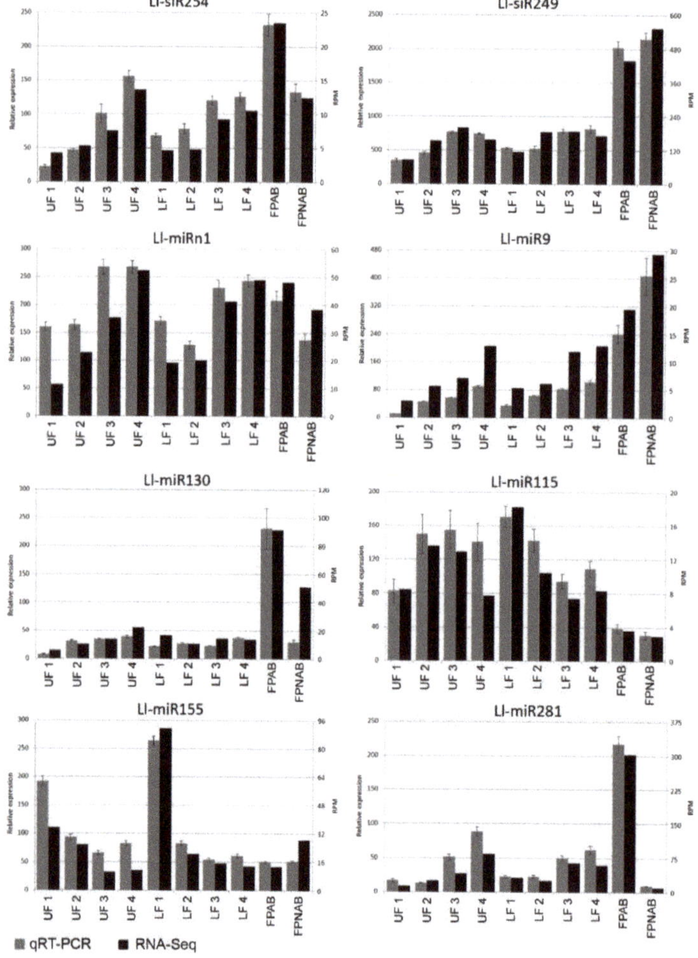

Figure 7. SL RT-qPCR validation of selected sRNAs in *L. lupinus*. Grey indicates the miRNA expression levels determined by qPCR. Black indicates the miRNA expression levels determined by deep sequencing. Vertical bars indicate standard errors.

2.10. Identification of sRNA Target Genes using Degradome and psRNATarget Analysis

In order to estimate accurately the biological function and impact of certain miRNAs, their target genes needed to be identified. To achieve this, we constructed degradome libraries from pooled samples of stage 3 upper and lower parts of the inflorescence. Through total degradome library sequencing, 19,353,278 raw reads were obtained (Table S11). After quality filtering, the degradome data were aligned to the reference transcriptome with CleaveLand 4 [58] to find sliced miRNA and siRNA targets. After processing and analysis, a total of 14,077 targets were identified, and after filtering with a *p*-value < 0.05, 538 targets emerged (501 targets for 178 known miRNAs and 37 targets for 13 novel miRNAs) (Table S12). For the phased siRNAs, 3,340 targets were initially identified, and after similar filtering, their number dropped to 89 targets for 46 siRNAs (Table S13). Exemplary target t-plots and sequences of the miRNAs and target mRNAs are shown in Figure 8.

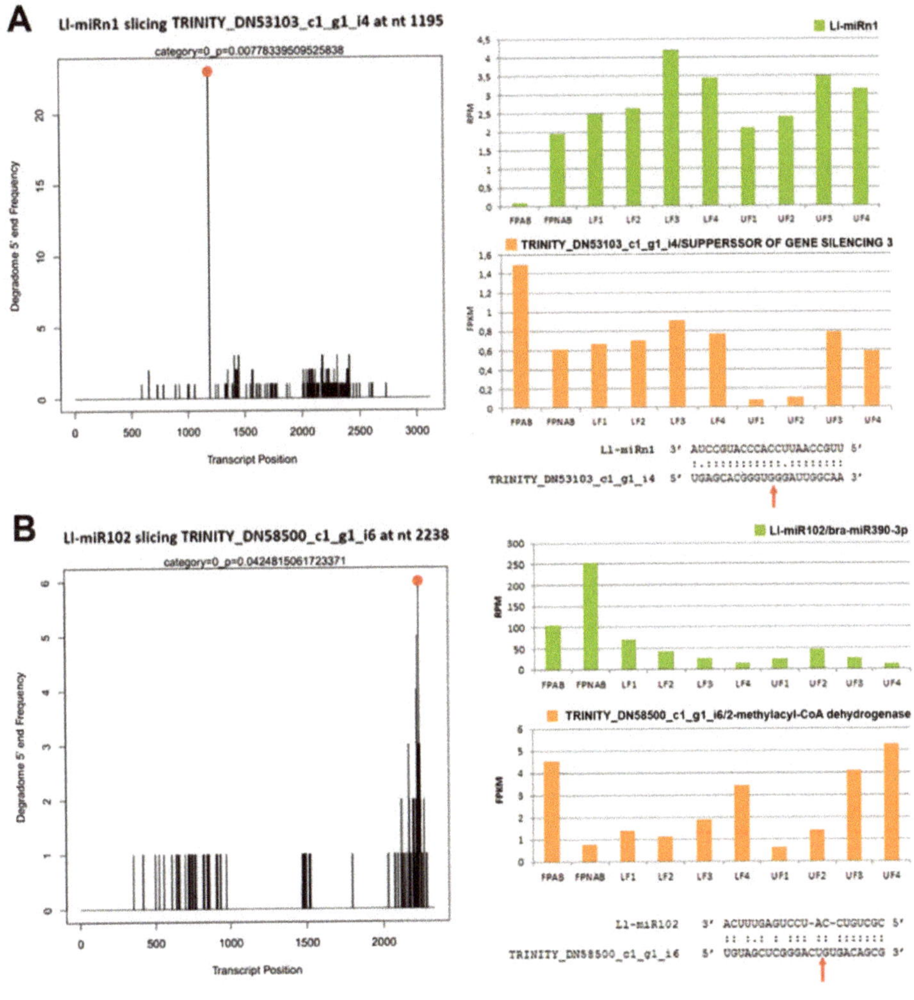

Figure 8. *Cont.*

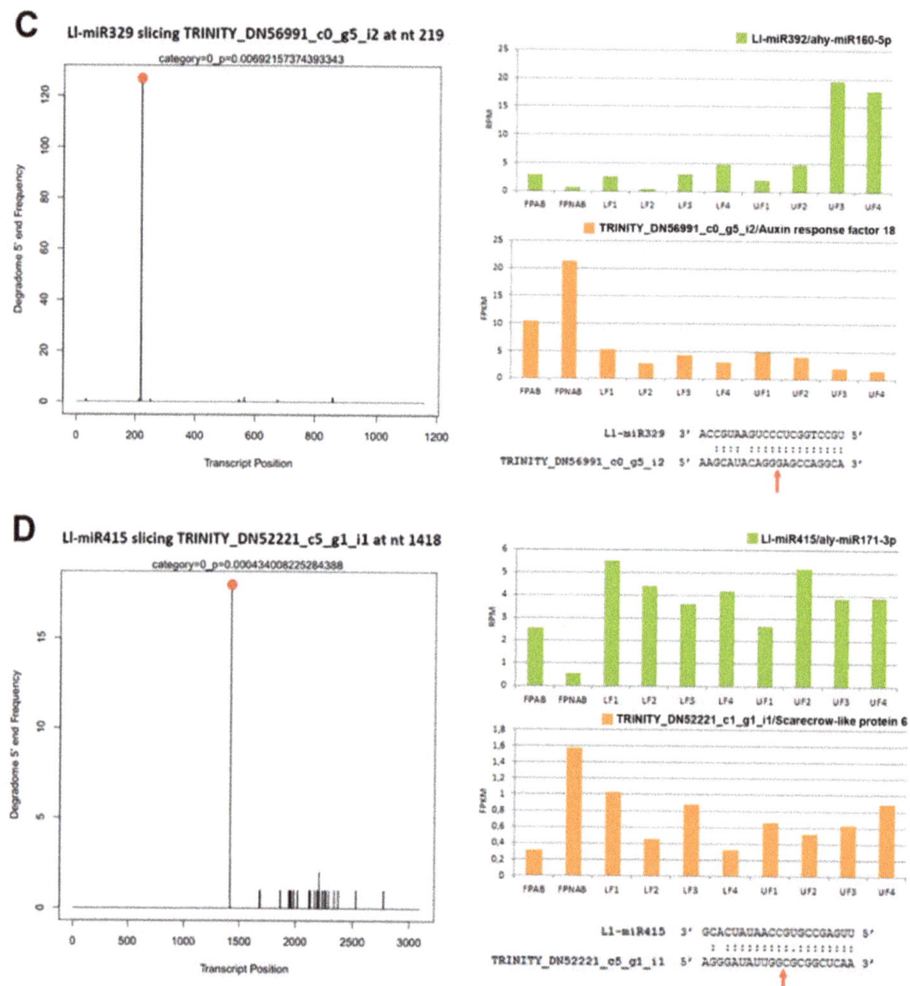

Figure 8. Examples of post-transcriptional regulation of miRNA targets in yellow lupine. (**A**) Ll-miRn1 and *SGS3* mRNA, (**B**) Ll-miR102 and *2-methylacyl-CoA dehydrogenese* mRNA, (**C**) Ll-miR392 and *ARF18* mRNA, (**D**) Ll-miR415 and *SCL6* mRNA. The T-plots show the distribution of the degradome tags along the full length of the target gene sequence. The cleavage site of each transcript is indicated by a red dot. Comparison of the expression levels of miRNAs and their targets in flowers from upper and lower whorls of yellow lupine racemes, and flower pedicels, as determined by deep sequencing. In miRNA-mRNA alignments, the red arrows indicate the cleavage site of the target gene transcript.

As expected, many of the targets for evolutionarily conserved miRNAs were compliant with literature data. For example, Ll-miR329/miR160-5p targeted *ARF16* and *ARF18*, the Ll-miR415/miR171b targeted *SCL6*, Ll-miR341/miR319q targeted *TCP2*, Ll-miR224/miR393a-5p targeted *TIR1*, etc. (Table S12).

A comparison of the expression of four exemplary miRNAs and their target genes (Figure 8) confirmed the reverse-correlation in the accumulation of miRNAs and an abundance of mRNA target genes, especially in flower pedicels. In the flowers, this correlation was not so obvious, presumably because of the organ's more complex nature (with its various elements, such as the stamen and the pistil), where regulation could be tissue specific.

In the case of some of the identified mi- and siRNAs, we were unable to determine the targets with a degradome analysis, which might have been caused by the lack of a sufficient amount of cleavage products ensuing from using only stage 3 flowers to construct the library. In order to find the putative missing target genes, the psRNATarget tool [59] was employed, which rendered plausible target genes through a comparison of the sRNAs with the reference transcriptome containing data from all of the samples. Using this method, we managed to establish putative target genes for most of the mi- and siRNAs, obtaining 66,102 miRNA and 32,725 siRNA targeted transcripts. A full list of the targets identified using the psRNATarget or degradome analysis for DE miRNAs, siRNAs, and novel miRNAs is contained in Table 3, Table 4, Table 5, Table 6 and Table S13. Targets for all of them are shown in Tables S14, S15, S16, S17 and S18.

2.11. Function of the miRNAs Potential Targets

Gene Ontology (GO) analysis was performed in order to investigate the functions of the miRNAs targets identified in yellow lupine flowers. Among the 27,547 targets of known and novel miRNAs identified with psRNATarget 26,230 targets exhibited GO terms (Table S17). 23,092 genes were categorized into 'Cellular component', 23,501 into 'Molecular function', and 22,939 into 'Biological process'. Figure 9 shows target gene percentages for each GO category. The largest number of targets classified as 'Cellular component' was attributable to 'cell', 'cell part' and 'organelle'. The majority of targets of the 'Molecular function' category were classified as 'binding' and 'catalytic activity'. Within the 'Biological process', most of the targets were categorized as 'cellular' and 'metabolic process' (Figure 9, Table S17). Within the 'Flower development' category, the targets of 37 miRNAs fit within GO terms related to phytohormones (Figure S4a), and the targets of 69 miRNAs were placed into the category of GO terms related to the development of flower parts (Figure S4b).

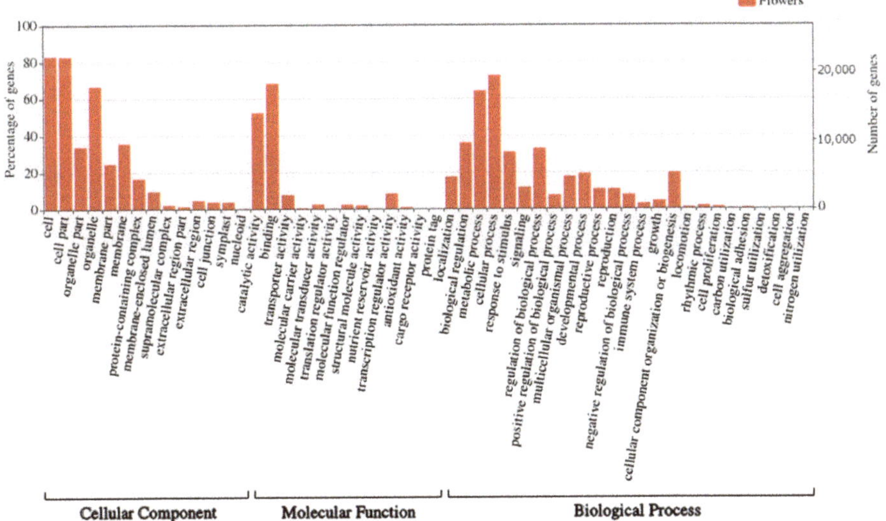

Figure 9. Visualization of GO categories annotated to predicted target genes of known and novel miRNAs in yellow lupine.

Our miRNAs targets analysis against the Kyoto Encyclopedia of Genes and Genomes (KEGG) revealed that most of the sequences in the main KEGG categories belonged to Metabolism (15,856), followed by Genetic information processing (5,267), Environmental information processing (1,517), Cellular processes (1,326) and Organismal Systems (800) (Figure S5). A full list of KEGG pathways and numbers of assigned sequences is shown in (Figures S6 and S7). One of the most represented sub-categories was Signal transduction (1,484), with over 700 putative targets in the Plant Hormone Signal Transduction pathway, where almost every sequence was frequently targeted with multiple miRNAs (Figure 10). The second most notable pathway was mitogen-activated protein kinase (MAPK) signaling, which was associated with different abiotic and biotic stress factors, with 350 putative targets distributed across every described stress response (Figure S8). A complete dataset on the KEGG analysis can be seen in (Table S16).

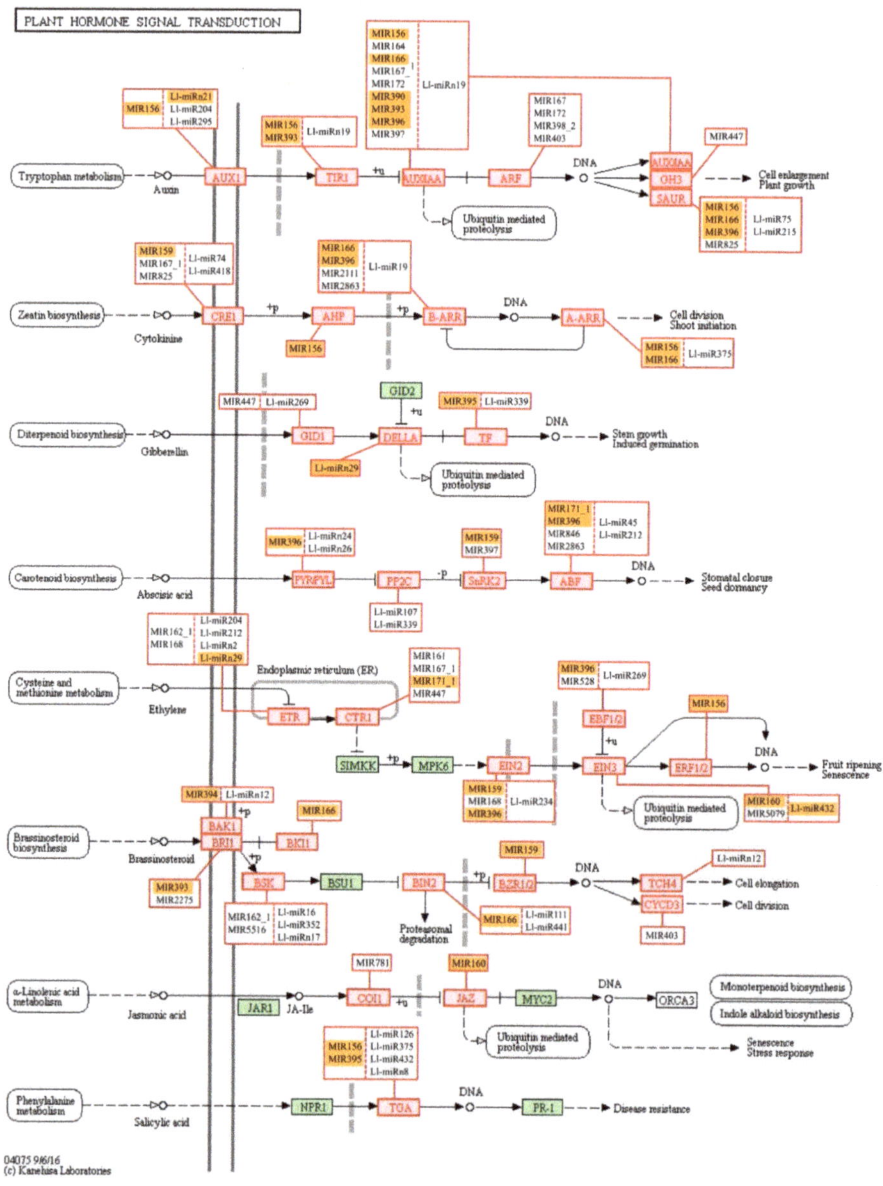

Figure 10. KEGG pathways related to plant hormone signal transduction targeted by known and novel miRNAs. Orange indicates DE miRNAs.

3. Discussion

Yellow lupine has great potential to become one of the leading legumes in Europe in both animal and human nutrition. Reduction the economic drawbacks resulting from excessive flower abscission would be the most convincing argument for lupine cultivation. However, this can only be achieved if we gain a deeper understanding of the plant's biology and insight into the molecular basis for the development and maintenance of lupine flowers. Therefore, we believe that the pathways controlling these processes deserve intensive research focus. Our previous analyses of yellow lupine transcriptomes resulted in the identification of transcripts of many genes involved in flower and pod abscission and suggested sRNA involvement in this process [17]. Notably, our observations of *L. lupinus* floral development indicate that their fate (abscission or pod formation) is determined prior to AZ activation. Therefore, we decided to perform comparative analyses between sRNAs from flowers developing on the upper and lower parts of the raceme. Identifying the miRNAs and their target genes involved in the above-mentioned processes will further our knowledge of the biology of not only lupines but plants in general since the role played by sRNA in organ abscission is still obscure.

Our sRNA-seq analyses shed more light on the molecular mechanisms that control flower development of *L. luteus* and confirmed the involvement of known miRNAs, such as miR159, miR167 or miR172, in this process [60], but we have also explored the roles of sRNAs in flower abscission and identified species-specific miRNAs.

3.1. Known sRNAs and Their Target Genes Are Involved in Regulating Flower Development in Yellow Lupine

Among the known and conserved miRNAs a number of miRNAs commonly associated with flower morphogenesis and development, belonging to, *inter alia*, the MIR156/157, MIR159, MIR165/166, MIR167 and MIR172 families [10] were spotted.

Studies have shown that miR156 is necessary for maintaining anther fertility in *Arabidopsis*, by orchestrating the development of primary tapetum cells and primary sporogenous cells [61]. In *A. thaliana*, *SPL13B* expression is strictly limited by miR156 to anther tapetum in young buds, while *SPL2* is weakly expressed in parietal and sporogenous cells and the surrounding cell layers in young flowers [61], where it is targeted by miR156 to regulate pollen maturation [62]. MiR159 was shown to target the conserved *GAMYB-like* genes that are a part of the GA signaling pathway [63,64]. In *A. thaliana* miR159 regulates the morphogenesis of the stamen, and male fertility [65]. Two transcription factors involved in pistil and stamen development in various plant species, *ARF6* and *ARF8*, contain the target site for miR167 [66–68]. For *Arabidopsis*, it has been proven that both these genes are involved in stamen filament elongation, anther dehiscence, stamen maturation and anthesis [69]. In tomato, a reduction in the accumulation of the miR167-targeted *ARF6* and *ARF8* leads to the lack of trichomes on the style surface, failed pollen germination and, consequently, sterility [11]. Recent research into multiple plant species has shown that miR172 targets genes belonging to the *APETALA2* (*AP2*, *TOE1*, *TOE2*, *TOE3*) family. MiR172 is part of the photoperiodic flower induction pathway and is associated with the functioning of the ABCDE model of floral development [70]. Overexpression of *MIR172* causes formation of a phenotype characterized by the absence of perianth, transformation of sepals into pistils and early flowering [70].

Our study showed the presence of at least one member of all these families in flowers (Figure 3, Table S5), which indicated that in lupine how crucial the families were for generative development in lupine, as well. MIR156 and MIR159 are the most numerous families in *L. luteus*, which suggests they play fundamental roles in its flower development processes.

The differentially expressed miRNAs identified in yellow lupine flowers were clustered by the dynamics of their expression (Figure 5). The first cluster comprised miRNAs, the accumulation of which increased as the flowers developed, and contained miRNAs belonging to the MIR166, MIR167, MIR319, MIR390, and MIR395 families. The first of these families include Ll-miR177, which guides the cleavage of *RADIALIS*, a transcription factor from the MYB family that controls the asymmetric flower shape in *Antirrhinum majus* [71,72], as well as Ll-miR258 and Ll-miR265, which probably target the Homeobox-leucine zipper protein ATHB-15. In *Arabidopsis*, both miR165 and miR166 target the same *HD-ZIP III* genes: *ATHB15, ATHB8, REVOLUTA (REV), PHABULOSA (PHB)*, and *PHAVOLUTA (PHV)* to regulate gynoecium and microspore development [28,73]. In lupine the MIR167 family members that accumulate in larger quantities during flower development are Ll-miR280, Ll-miR281, and Ll-miR285, which probably target *ARF6* and *ARF8*. Ll-miR445 and Ll-miR130 are members of the MIR319 family, while their putative target genes are *TCP4* and *MYB33*, respectively. In *Arabidopsis*, the miR319a/TCP4 regulatory module is necessary for petal growth and development. Moreover, the overexpression of *MIR319* reduces male fertility, and this defect is hypothesized to be caused by the cross-regulation of *MYB33* and *MYB65* by miR319 and miR159. As the miR319 target site within the *MYB33* and *MYB65* transcripts exhibit a lower match with miRNA than the miR159 target site, the latter is more efficient at regulating these genes and miR319 is their secondary regulator [74]. This regulatory network is even more complex. In *A. thaliana*, cooperation of three miRNAs and their target genes, namely miR159/*MYB*, miR167/*ARF6/ARF8*, and miR319/*TCP4*, is a prerequisite for proper sepal, petal and anther development, and maturation. miR159 and miR319 influence the expression of *MIR167* genes, which in turn affect each other. These miRNAs orchestrate plant development by regulating the activity of the phytohormones GA, JA, and auxin [75]. Increased accumulation of miR167 and miR319 in the late stages of yellow lupine flower development could also be associated with regulating the growth and development of petals and anthers. Another miRNA showing a similar expression profile is Ll-miR9/miR390-5p. In lupine, it targets the *TAS3* transcript, which in turn is a source of tasiR-ARF, a negative regulator of *ARF2*, *ARF3* and *ARF4* activity. This regulatory cascade plays a vivid role in development of many plant species [76]. The expression level of miR390 derived from *MIR390b* reflects auxin concentration in organs, while the repression of *ARF2*, *ARF3*, and *ARF4* by tasiR-ARF are important for lateral organ development [18,77], and flower formation [78]. Ll-miR118 and Ll-miR119, which target ATP sulfurylase (*ATPS*) according to our degradome data, belong to the MIR395 family. In *Arabidopsis*, miR395 targets two gene families, ATP sulfurylases and sulfate transporter 2:1 (*SULTR2:1*), which are elements of the sulfate metabolism pathway [79]. ATPS regulates glutathione synthesis and is an essential enzyme in the sulfur-assimilatory pathway [80]. In cotton, the miR395-APS1 module is engaged in drought and salt stress response [81]. Sulfate is the main source of sulfur and is taken up by roots, transported throughout the plant and used for assimilation. Sulfate limitation forces a significant up-regulation of miR395 expression [82]. Presumably, during yellow lupine flower development, the demand for sulfur increases and the plant activates mechanisms for its efficient uptake.

Within the cluster of miRNAs, the expression of which decreased as the flowers developed, there were homologues of miR390-3p, miR858, miR396-3p, miR168, miR408-3p and miR398 (Figure 5). Ll-miR99, Ll-miR100, and Ll-miR102 are identical to miR390-3p (the so-called passenger strand, former star strand). However, their expression showed an opposite trend to that of miR390-5p. The differential expression and functioning of passenger miRNAs have already been described. The research carried out by Xie and Zhang in 2015 on cotton showed that the formation of some miRNA*s, such as miR172* and miR390*, was associated with the phases of the plant's growth [83]. Therefore, miRNA*s can be specifically expressed in various tissues to maintain the steady state of the organism. Our degradome analysis for yellow lupine showed that Ll-miR9/miR390-5p was able to guide the cleavage of the *TAS3* transcript. There is no certainty as to the status of its passenger strand, which suggests its locally limited activity or its involvement in regulation of other targets and further research is required to identify its accumulation and function in the organs concerned. Another miRNA from the cluster is Ll-miR155/miR396-3p (passenger strand), which guides cleavage of JMJ25 demetyhylase mRNA (confirmed in degradomes), involved in preserving the active chromatin state [84]. *ECERIFERUM1 (CER1)*, the target gene in lupine for another two homologues of miR396-3p, Ll-miR199 and Ll-miR200, is a homologue encoding an enzyme involved in alkane biosynthesis, and in cucumber is engaged both in wax synthesis and ensuring pollen viability [85]. This cluster also included a miRNA that negatively regulates elements involved in miRNA and ta-siRNA functioning, namely Ll-miR247/miR168 targeting *AGO1* mRNA [86]. Another miRNA clustered here was the highly conserved Ll-miR60/miR408-3p, which guides the processing of the mRNA of the copper-binding Basic Blue protein homologue (plantacyanin, PC). In *Arabidopsis*, PC plays a role in fertility, exhibiting the highest expression in the inflorescence, especially in the transmitting tract. [87]. Transgenic *Arabidopsis* plants over-expressing *MIR408* displayed altered morphology, including significantly enlarged organs, resulting in enhanced biomass and seed yield. Plant enlargement was shown to be primarily caused by cell expansion rather than cell proliferation, and in transgenic plants it was correlated with stronger accumulation of the myosin-encoding transcript and gibberellic acid [88]. It seems that high expression levels of miRNAs grouped in the cluster are correlated with intensive growth and differentiation of young floral tissues.

Among the miRNAs identified in yellow lupine several that seemed to be crucial in particular stages of the plant's development were spotted (Figure 4, Table 4, Table S7). For example, the largest quantities of miR159 (Ll-miR452 and Ll-miR454) were accumulated in stages 2 and 3 of the plant's development. According to degradome data they targeted *GGP-5* (*GAMMA-GLUTAMYL PEPTIDASE 5*) of an undefined function in plants, and an evolutionarily conserved target for *GAMYB*, respectively. As already mentioned, this could be associated with miRNA family cooperating with miR167 and miR319 in regulating *L. luteus* anther maturation. The accumulation of Ll-miR251/miR5168-3p, Ll-miR92/miR1861b, Ll-miR229/miR369-5p, and Ll-miR311/miR5794 increased in stage 2 upper and lower flowers, while – interestingly – in the later stages these miRNAs were only present in lower flowers. According to degradome analysis, Ll-miR251/miR5168 guides cleavage of the mRNAs of the genes encoding the Homeobox-leucine zipper protein ATHB-14 and the chaperone protein dnaJ 13. The miR5168 sequence displays a great similarity to that of miR166, thanks to which they may perhaps share the same target gene *ATHB-14*, the putative transcription factor engaged in the adaxial-abaxial polarity determination in the ovule primordium in *A. thaliana* [89]. As confirmed by yellow lupine degradome sequencing, Ll-miR229/miR396-5p targets *GROWTH-REGULATING FACTOR 5* (*GRF5*) and *GRF4* transcripts. In *Arabidopsis*, *GRF5* is expressed in anthers at early stages of flower development and in gynoecia throughout the whole flower development, and transcripts of *GRF4* accumulate later in sepals, tapetum, and endocarpic tissues of ovary valves [90]. Transgenic rice with Os-miR396 overexpression and *GRF6* knock-down suffers from open husks and sterile seeds [91]. *GRF6* cooperates with *GRF10* to transactivate the *JMJC* gene *706* (*OsJMJ706*) and *CRINKLY4 RECEPTOR-LIKE KINASE* (*OsCR4*) responding to GA, which is a prerequisite for the flower to successfully develop into a normal seed [91]. An increased share of miRNAs involved in cell division, namely miR396, miR319, and miR164, in NGS analyses was also observed in early grain development in wheat [92].The presence of these miRNAs in yellow lupine flowers suggests that their regulation of cell proliferation also plays an important role in development of generative organs.

3.2. Involvement of New miRNAs in L. luteus Flower Development

Using ShortStack [53] software we predicted 28 candidates for new miRNAs (Table 3). Interestingly, many of these novel miRNAs showed similarity to precursor miRNAs from miRBase, which leads to the conclusion that they might be new members of the already known families, for example MIR167 (Ll-miRn12 and Ll-miRn27), MIR172 (Ll-miRn4), MIR393 (Ll-miRn19) or MIR169 (Ll-miRn3, Ll-miRn11, and Ll-miRn15) (Table S6).The other 13 had no homologues among known miRNAs and were recognized as lupine-specific miRNAs. Some of the new miRNAs displayed differential expression during *L. luteus* flower development. Ll-miRn3, which shows similarity to pre-miR169, displayed differential expression in UF1 vs LF1 and LF2 vs LF1 library comparisons, wherein it is the most accumulated in LF1, and in flower pedicels (up-regulated in FPNAB). According to degradome data, this miRNA targets *SCARECROW2* (*SCR2*) homologue, a putative activator of the calcium-dependent activation of *RBOHF* that enhances reactive oxygen species (ROS) production and may be involved in cold stress response [93]. In rice *SCR2* expression is relatively high in flower buds and flowers, and after flowering rises in the leaves and roots [94]. In yellow lupine, this gene may be involved in intense cell divisions during early flower development and is down-regulated in the pedicels with an active AZ to stop its growth. Another frequently encountered novel DEmiR was Ll-miRn22, which shows sequence similarity to pre-miR1507, is up-regulated in LF3 vs LF2 and LF2 vs LF1 library comparisons, and its expression escalates with flower development in the bottom whorl. The MiR1507 family is annotated as legume-specific [95]. Through analyses of our degradome data we have not found its target gene, and the psRNATarget hit was the putative disease resistance RPP13-like protein 1. Unfortunately, this protein has been poorly described, therefore it is difficult to determine its function in yellow lupine flowers. Noteworthily, the target genes of Ll-miRn1 and Ll-miRn30 identified through degradome sequencing are *SGS3* and *DCL2*, respectively, and the miRNAs are up-regulated in LF3 vs LF2 comparisons and down-regulated in UF1 vs LF1 comparisons, respectively. *SGS3*- and *DCL2*-encoded proteins are involved in sRNA biogenesis [96]. Importantly, novel miRNA identified in soybean Soy_25 displays high sequence similarity to Ll-miRn1 and also targets *SGS3*, which indicates that this regulatory feedback loop for sRNA biogenesis is common for *Fabaceae* [97]. These results indicate that *L. luteus* miRNAs play a regulatory role in siRNA biogenesis in early flower development.

3.3. miRNA Accumulation Varies in Lower and Upper Flowers in Different Stages of Development

One of our goals was to identify the sRNAs engaged in yellow lupine flower development, with a particular emphasis on the differences between flowers from lower and upper parts of the inflorescence, in order to gain an insight into how early the flower fate is determined.

In our study, we spotted differences in miRNA accumulation patterns as early as the first stage of flower development.

Flowers collected from the lower whorls displayed higher accumulations of sequences corresponding to miR5490, miR5794, miR1861, miR396-5p, miR395, miR166, and miR159-3p (Table 5). miR1861 and miR396 were recognized as positive cell proliferation and development regulators [98–100]. In rice, for example, miR1861 exhibited differential expression during grain filling [101], and its expression was higher in superior grains in comparison to inferior ones [102]. This is consistent with our hypothesis, that a higher occurrence of miR1861 and miR396 in lower flowers may be an indication of the plant investing more supplies in this part of the inflorescence.

From the second stage until the end of their development, upper flowers accumulated more miRNAs corresponding to miR319, miR394, miR160, and miR393 (Figure 4, Table 5). MiR393 regulates the accumulation of transcripts encoding auxin receptors belonging to the TAAR family. Changes in receptor abundance affect the sensitivity of the given tissue to auxin and this is how this molecule influences plant development [102]. In *A. thaliana*, miR160 directly controls three *ARF* genes, namely: *ARF10, ARF16* and *ARF17* [103]. In tomato, sly-miR160 is abundant in ovaries, and changes in its expression affect plant fertility [12]. Down-regulation of sly-miR160 caused improper ovary patterning and thinning of the placenta already prior to anthesis [12]. In view of these facts, higher expression of miR160 in lupine upper flowers in their development means that a slightly different organization of the gynoecia may be one of the crucial determinants of flower fate. Additionally, the elevated expression levels of miR160 and miR393 in upper flowers of lupine suggest a reduction in the abundance of the transcripts of their target genes encoding auxin receptors and auxin response factors. This, in turn, may have led to a reduction in auxin sensitivity. Decreasing the number of transcription factors belonging to the TCP family (targeted by miR319), probably caused different cell proliferation profiles in flowers collected from the upper whorls.

Additional expression studies of selected miRNA (Ll-miR281/miR167, Ll-miR224/miR393, Ll-miR333/miR160, Ll-miR329/miR160) carried out in the upper flowers of yellow lupin developing after removal of the lower ones (UFR) (Figure S2), and consequently with a changed, when compared to the original, fate, provide additional confirmation of the results obtained from RNA-seq analysis (Figure S3).

3.4. sRNAs Are Involved in Flower Abscission in L. luteus

Little is known about sRNA engagement in flower abscission. Research on the involvement of miRNAs in this process has been already carried out in cotton [104], tomato [12,105], and sugarcane [106].

For a genome-wide investigation of miRNAs involved in the formation of the abscission layer in cotton, two sRNA libraries were constructed using the abscission zones (AZ) of cotton pedicels treated with ethephon or water. Among the 460 identified miRNAs, only gra-MIR530b and seven novels showed differential expression in abscission tissues [104], and these miRNAs have no homologues in our dataset.

Besides ovary patterning in tomato, sly–miR160 regulates other two auxin-mediated developmental processes: floral organ abscission and lateral organ lamina outgrowth [12]. In that study, down-regulation of sly-miR160 and the resulting higher expression of its target genes, transcriptional repressors of auxin response *ARF10* and *ARF17*, also resulted in the narrowing of leaves, sepals and petals and an impeded shedding of the perianth after successful pollination [12]. This was consistent with the higher accumulation of Ll-miR329/miR160-5p, Ll-miR332/miR160-5p, and Ll-miR333/miR160-5p in upper flowers designated to fall off in yellow lupine. As these miRNAs showed no differential expression in flower pedicels, it probably does not play a role in the executory module of abscission itself but is rather a part of a mechanism that determines flower fate.

Another research on tomato using sRNA and degradome sequencing libraries explored the roles of sRNAs in AZ formation in the early and late stages of the process additionally accelerated or not by ethylene or control treatment [107]. The study showed that in tomato pedicels, the accumulation levels of, *inter alia*, miR156, miR166, miR167, miR169, miR171, and miR172 rose in late stages of abscission, while the abundance of miR160, miR396 and miR477 dropped [107]. Although it is difficult to compare ethylene-treated tomato pedicel results to our data, it is worth noting that in the corresponding FPAB vs. FPNAB comparison in our study, the accumulation of some miRNAs was similar: miR396 level was lower, and the levels of miRNAs annotated as miR167 and miR166 were higher in FPAB (Table 6).

It has been proven for sugarcane that among others both mature (5p) and passenger (3p) miRNAs from MIR167 family were up-regulated in 'leaf abscission sugarcane plants' comparing to 'leaf packaging sugarcane plants' (which corresponds to the FPAB vs. FPNAB comparison in our study) [106]. In our study, both mature and passenger members of the MIR167 family were leaders among DEmiRs, too, (Table 6) pointing to their crucial role in both vegetative and generative organ abscission. Significantly, this applies to evolutionarily distant taxa: both monocots and dicots.

In our paper, among the up-regulated miRNAs, the most numerous family besides already mentioned MIR167 was MIR398 with 3 members being among top-regulated ones. Among the down-regulated miRNAs, the members of MIR390, MIR396 and MIR395 families were most abundant. It was shown for other plant species, that these miRNAs are engaged in the regulation of auxin signal transduction pathway (miR167 and miR390 [108]), regulation of cell division (miR396 [100]) and stress response (miR395 [81,82]).

It is worth noting, that in comparisons of *Lupinus* pedicel libraries there are novel miRNAs: three are down-regulated in FPAB and one is up-regulated. Furthermore, Ll-miRn3 is up-regulated in both, young flowers designated to be maintained on the plant (LF1) and pedicels with inactive AZ (Table 6), which may indicate its role in preventing flower abscission. In the future, it is worth examining the role of its target gene, which encodes a protein that does not resemble any known protein.

With regard to siRNAs, the most up-regulated ones in FPNAB were: Ll-siR173, Ll-siR4 and Ll-siR13, and the most down-regulated one were Ll-siR208. Unfortunately, the lack of literature data on their targets makes it impossible for the specifics of their function in the studied process. However, it is worth mentioning, that in pedicels high levels of accumulation are displayed by siR249/tasiR-ARF and siR308/tasiR-ARF, which target transcripts encoding ARF2, ARF3 (confirmed in degradomes). These results strongly suggest the involvement of siRNAs in the functioning of lupine pedicels.

3.5. Possible miRNA-dependent Regulatory Pathways That Participate in Development and Abscission of Yellow Lupine Flowers

Recent studies have shown that sRNA activity is associated with the hormonal regulation of plant development through influencing the spatio-temporal localization of the hormone response pathway [109].

The auxin signal transduction pathway mainly consists of three elements. Auxin is perceived by members of the TAAR family. There are AUX/IAA repressor proteins and ARF transcription factors downstream of these receptors [110–112]. The expression of *TAAR* receptors is regulated by miR393 and secondary ta-siRNA derived from their own transcripts [20]. miR167 and miR160 affect the *ARF6, ARF8* [67] *ARF10, ARF16* and *ARF17* [113] transcript accumulation, respectively. It has been proven that the expression of *ARF2*, together with *ARF3* and *ARF4*, is regulated by the ta-siRNA/miR390 module [114]. In the two-hit model, ta-siRNA-containing the *TAS* transcript is recognized by two miR390 molecules, one of which guides its cleavage, and the other, in a complex with AGO7, serves as a primer for complementary strand synthesis, with its subsequent processing ultimately resulting in ARF-targeting siRNA biogenesis [115].

In our study, among the differentially expressed sRNAa in flowers and flower pedicels, there were members of the MIR167, MIR160, MIR393 and MIR390 families, as well as phased siRNAs targeting *ARF2, ARF3,* and *ARF4*. This fact suggests a vivid role of auxin-related sRNAs in flower development and abscission in *L. luteus* and confirms our previously published results of transcriptome-wide analyses, where we observed differences in expression levels of genes encoding several elements of the auxin signal transduction pathway [17]. The relatively high number of members of the MIR167 family showing differential expression in the studied variants indicates that miR167 is one of the key regulators of flower development and abscission in yellow lupine.

*Lupinus Ll*ARF2, *Ll*ARF3, and *Ll*ARF4 transcripts are possibly down-regulated in the processing that is guided by Ll-siR249 and Ll-siR308 (Table 4), which are identical to tasiR-ARFs in many plant species according to the tasiRNAdb database [116]. These tasiR-ARFs probably originate from *TAS3* transcript (TRINITY_DN55534_c4_g1) containing two binding sites for miR390 (Figure S9a). Ll-miR9/miR390and surprisingly also Ll-siR240, guide the cleavage of another *TAS3* mRNA (TRINITY_DN54998_c6_g5_i2) (Figure S10) which contains only one target site for miR390 (Figure S9b). This is the first report on *TAS3* processing regulated by siRNA. The target site for Ll-siR240 is shifted by 10 nucleotides relative to the target site for Ll-miR9/miR390 (Figure S10). The expression of Ll-siR249, Ll-siR308, and Ll-miR9 showed a similar profile, as it rose during flower development and was the highest in the pedicels (Figure 7). Ll-siR240 accumulated proportionally to *TAS3* with only one target site for miR390, which means that it was least expressed in the pedicels, while in flowers its expression increased with time (Table S18). The identified target transcripts belonging to the *ARF2, ARF3,* and *ARF4* gene families showed differential expression but with no clear trend (Table S18). This may indicate that these siRNAs act locally, repressing only a pool of transcripts expressed in a given tissue, while in other flower parts activity of these genes is regulated in other ways. The presence of all the elements of the miR390/TAS3/tasiR-ARF module among the DE sRNAs in yellow lupine suggests that alterations in its functioning have a great impact on *L. luteus* flower development. The additional element in the form of siRNA that processes *TAS3* mRNA seems to be a new species-specific adjuster of this regulation module.

We have also performed GO enrichment analysis of the target genes for sRNAs identified in flowers of yellow lupine (Figure 9, Figure S4a,b, Table S10). What is most interesting is that quite a considerable number of target genes fell within the 'response to stimulus' and 'signaling' categories, which means that miRNAs modulated the way the plant adapted to environmental stimuli (Figure 9). An in-depth analysis of GO terms concerning plant hormones (Figure S4a) showed that most of the miRNAs identified in yellow lupine modulated more than one hormone signaling pathway. For example, Ll-miR181 belonging to the MIR166 family modulated processes associated with four hormones, namely auxin, gibberellin, jasmonic acid, and salicylic acid, by targeting not only transcription factor

AS1, a central cell division regulator [117], but also Cullin-3A, an element of the ubiquitination complex [118]. Another two members of this family, Ll-miR173 and Ll-miR177, targeted the same gene, *26S PROTEASOME NON-ATPASE REGULATORY SUBUNIT 8 HOMOLOG A (RPN12A)*, involved in the ATP-dependent degradation of ubiquitinated proteins during auxin and cytokinin response [119]. Our GO analysis for yellow lupine flowers additionally showed that miRNAs were responsible for guiding the processing of genes simultaneously involved in multiple processes associated with flower development (Figure S4b). For example, in many plants *AP2* is involved in the specification of floral organ identity [120], as well as ovule [121] and seed development [122,123], and in our study, it was targeted by ten lupine miRNAs. On the other hand, seven of these miRNAs additionally targeted a homologue of negative flower development regulator, *LIKE HETEROCHROMATIN PROTEIN 1 (LHP1)* [124]. This highly degenerated and ambiguous model of gene regulation by lupine miRNAs shows that in this plant the adjustment of key biological processes related to fertility is a complex network of interconnected factors.

We have also conducted KEGG functional analysis of the putative targets identified for miRNAs in lupine which indicated their engagement in regulating a number of metabolic pathways—especially 'carbohydrate metabolism' and 'nucleotide metabolism' (Figure S5). 'Carbohydrate metabolism' was also one of the most enriched KEGG pathways in our previous *L. luteus* transcriptome analysis [17], and its activation may be an indication of cell walls being rebuilt or changes in nutrient supply. The next most numerous group of miRNA targets was categorized into the 'Genetic information processing' KEGG pathways, namely, 'spliceosome', 'RNA transport', and 'ubiquitin proteolysis'. This suggests that in yellow lupine flowers most miRNAs regulate processes related to post-transcriptional events and protein degradation. Three KEGG categories within the 'Environmental information processing' category is extremely important in terms of plant development, and they are 'Signal transduction pathways' comprising the MAPK cascade, 'phosphatidylinositol' and 'plant hormone' signaling pathways (Figure 10, Figure S6, S7, S8). The MAPK pathway is involved in regulating several processes, such as biotic and abiotic stress response (reviewed in [125,126]), and associated with the functioning of hormones such as ethylene [127] and abscisic acid, engaged in organ abscission and other processes (reviewed in [128,129]). The MAPK cascade is also an element of the positive feedback loop amplifying the abscission signal [130]. Auxin seems to be major target of sRNAs in yellow lupine. However, KEGG enrichment analyses of the identified target genes for lupine miRNAs indicated that the signal transduction pathways of gibberellin, cytokinin, the already mentioned ethylene, and ABA were potentially modulated by miRNAs in *L. luteus*, as well, but in less extent (Figure 10).

Interestingly, like in the case of GO analysis, KEGG analysis for the MIR166 family showed that it was involved in the auxin, cytokinin, and brassinosteroid signal transduction pathways (Figure 10). These data show again how the fine-tuning of expression of phytohormone-related genes by sRNAs is important for growth and development regulation.

4. Materials and Methods

4.1. Plant Material

Commercially available seeds of yellow lupine cv. Taper were obtained from the Breeding Station Wiatrowo (Poznań Plant Breeders LTD. Tulce, Poland). Seeds were treated with 3,5ml/kg Vitavax 200FS solution (Chemtura AgroSolutions, Middlebury, United States) to prevent fungal infections and inoculated with cultures of *Bradyrhizobium lupine* contained in Nitragina (BIOFOOD s.c., Walcz, Poland) according to seed producer's recommendations [131]. All the research material used for RNA isolation was collected from field-grown plants cultivated in the Nicolaus Copernicus University's experimental field (in the area of the Centre for Astronomy, Piwnice near Torun, Poland, 53°05′42.0″N 18°33′24.6″E) according to producer's agricultural recommendations [131] until the time of flowering. Flowers were collected 50 to 54 days after germination from the top and bottom parts of the inflorescence and were separated into four categories based on the progression of their development. Flower pedicels from flowers undergoing abscission or maintained on the plant were also collected, as in our previous study [17].

Plants with the same number of flower whorls were selected for the flower removal experiment and control. All plants were grown as described above up to the flowering stage. When plants reached the stage at which the top-most flowers were in the developmental stage S1, other flowers were removed from the inflorescence (UFR samples). The samples were collected for the gene expression analysis in the stages S1–S4. As a control, flowers from stages S1–S4 from upper (UF) and lower (LF) part of the inflorescence were collected.

4.2. RNA isolation, Library Construction, and RNA Sequencing

Total RNA from at least 5 plants (25 flowers or pedicels) for each variant was performed using the miRNeasy Mini Kit (Qiagen, Venlo, the Netherlands) including on-column DNA digestion with the RNase-Free DNase Set (Qiagen, Venlo, the Netherlands). The total RNA quality and quantity were evaluated with agarose gel electrophoresis and Nanodrop ND-1000 spectrophotometer (Thermo Scientific Waltham, MA, USA). Both the RNA Integrity Number (RIN), and RNA concentration were measured with the 2100 Bioanalyzer (Agilent Santa Clara, CA, USA) using the Small RNA Kit (Agilent Santa Clara, CA, USA). All the samples had adequate concentrations of RNA and RIN ranging from 8.9 to 10.0 and were sent for library construction to Genomed S.A (Warszawa, Poland) and sequencing BGI (Shenzhen, China).

Small RNA libraries were prepared from the total RNA using the NEBNext Multiplex Small RNA Library Prep kit for Illumina (New England Biolabs, Ipswich, MA, USA) and subsequently sequenced on the HiSeq4000 platform (Illumina, San Diego, CA, USA) in the 50 single-end mode. All libraries were constructed in two biological replications resulting in a total number of 20 sRNA libraries.

The total RNA extracted from pooled material derived from three biological replicates was used to prepare ten transcript libraries using the NEBNext Ultra Directional RNA Library Prep Kit for Illumina (New England Biolabs, Ipswich, MA, USA) and sequenced on the HiSeq4000 platform in the 100 paired-end mode.

For degradome sequencing, total RNA from three biological samples of UF3 and LF3 was pooled to maximize the amount of required material. The protocol for degradome library preparation comprised the following steps: (i) mRNA isolation, where poly (A)-containing mRNA molecules are purified from total RNA using poly(dT)oligo-attached magnetic beads, (ii) synthesis and subjugation of cDNA to ligate 5′ adaptors, and purification of the resulting products with TAE-agarose gel electrophoresis, (iii) PCR amplification to enrich the final products, (iv) library-quality validation on the Agilent Technologies 2100 Bio-analyzer and using the ABI StepOnePlus Real-Time PCR System (Applied Biosystems, Foster City, CA, USA), and (v) sequencing of the prepared library on the HiSeq4000 platform in the 50 single-end mode.

4.3. De Novo Transcriptome Assembly and Gene Expression Analysis

The transcriptome was assembled *de novo* from RNA-Seq data using Trinity v 2.4.0 (https://github.com/trinityrnaseq/trinityrnaseq/releases) with default settings as in our previous study [17]. The expression level was estimated at both the unigene and isoform levels and described by FPKM (Fragments Per Kilobase Of Exon Per Million Fragments Mapped): the number of reads per unigene normalized to the library size and transcript length using RSEM [132] as previously described [17].

4.4. Identification of Known and Potentially Novel miRNAs and Phased siRNA

Adapter-free sRNA reads were subjected to quality filtering with fastq_quality_filter from the FASTX-Toolkit package (http://hannonlab.cshl.edu/fastx_toolkit/) using -p 95 and -q 20 parameters (http://hannonlab.cshl.edu/fastx_toolkit/commandline.html#fastq_quality_filter_usage). Then, redundant and counting read occurrences (i.e., raw expression values) were identified with the fastx_collapser from the same package.

Short reads were compared against noncoding RNAs from Rfam [49,50] and both mature miRNAs and their precursors from miRBase [51]. The comparison was performed with Bowtie [133] allowing for no mismatches.

To identify phylogenetically conserved mature miRNAs with sequences and lengths identical to known plant miRNAs we searched miRBase for similarity at the mature miRNA level.

To predict potential novel miRNAs we applied ShortStack [53] with default settings. This tool identifies miRNAs based on their mapping against a reference genome. Since no genome was available for the studied species, we used de novo approach for transcriptome assembly instead. The latter method allowed for identification of miRNAs that showed no similarity to miRNAs annotated in miRBase and these miRNAs were assigned as new.

ShortStack [53] was used to identify small RNAs that were being cut in phase from longer precursors (phased siRNAs) with transcriptomes used as references. The top 200 candidates were selected from each sample, based on the phased score value provided by ShortStack. Finally, lists of such sRNAs from all samples were merged into a single dataset of non-redundant phased siRNAs (Table S6). The expression values were calculated as in the case of miRNAs.

4.5. Small RNA Expression Quantifications and Analysis of Differentially Expressed si- and miRNAs

MiRNA counts within each sample were first normalized to RPM values (reads per million values) and then a differential expression analysis was performed with the DESeq2 R package [134]. The files containing raw read counts for miRNAs/siRNAs from treatment and control replicates were used as input, and only candidates with an adjusted p-value (q-value) below 0.05 were retained for further analysis.

4.6. Identification of sRNA Targets

For target prediction using degradome analyses after sequencing, the reads were filtered using fastq_quality_filter from the Fastx-Toolkit package (http://hannonlab.cshl.edu/fastx_toolkit/) with at least 95% of nucleotides in each read demonstrating quality >= 20 (Phred Quality Score) with -p 95 and -q 20. The filtered Degradome-seq data, sequences of mature miRNA/siRNA and the assembled transcriptomes were processed with the CleaveLand4 package [58] to determine the cleavage sites for sRNA using default program settings. The final results were filtered based on the p-value < 0.05.

To predict targets for known or novel miRNAs, and phased siRNAs, we used also the psRNATarget tool [59] querying the assembled transcriptomes with the default Schema V2 (2017 release) and an expectation score of up to 4.

4.7. Evolutionary Conservation of miRNAs

L. luteus miRNAs were assigned to miRNA families based on miRBase classification, and the same was done for the sequences of all *Eudicotyledons* species present in miRBase, with the exclusion of *Gossypium arboretum* (which has only one sequence deposited in the database that cannot be classified as belonging to any known miRNA family). miRNAs from 52 species were compared against *L. luteus* miRNAs in order to count the numbers of miRNA family members shared amongst the species. The same analysis was performed with data narrowing to nine *Fabaceae* species.

4.8. Expression analysis with RT-qPCR

MiRNAs and siRNAs expression analysis was performed using the Stem Loop RT-qPCR technique, according to [56] with some modifications. An RT primer specific for each sRNA was used for the reverse transcription using total RNA of each sample and the SuperScript III Reverse Transcriptase (Thermo Fisher Scientific, Waltham, MA, USA) in a 20 μL reaction volume. To increase the accuracy and efficiency of the reaction, the pulse RT approach [57] was applied to the reverse transcription which consisted of two steps: 30 min of pre-incubation at 16 °C, followed by 60 cycles at 30 °C for 30 s, 42 °C for 30s and 50 °C for 1 s. qPCR was subsequently performed using specific primers designed according to [57], modified so that the UPL9 hydrolysis probe (Roche, Basel, Switzerland) was used for maximization of accuracy and background reduction. This reaction was performed using the SensiFAST Probe No-ROX kit (Bioline meridian bioscience Cincinnati, OH, USA) and the LightCycler 480 (Roche, Basel, Switzerland). Each 20 μL reaction contained: 1 μL cDNA template (transcribed from ~100 ng of total RNA for less expressed miRNAs and 25 ng of total RNA for more expressed miRNAs), 1 μL of 10 μM qPCR specific forward primer, 1 μL of 10 μM Universal-qPCR primer, 10 μL of 2× SensiFAST Probe No-ROX Mix, 0.2 μL of 10 μM UPL9 probe and 6.8 μL ddH$_2$O. qPCR was executed by pre-incubation at 95 °C for 10 min, followed by 45 cycles of 95 °C for 10 s, 59 °C for 30 s, and 72 °C for 1 s. Target gene expression was performed as in [17]. Each experiment consisted of three biological and technical replicates. The relative expression levels were calculated using the $2^{-\Delta\Delta Ct}$ method, and the data were normalized to the CT values for the *LlActin* reference gene (according to [17]). All primer sequences are given in Table S10.

4.9. Gene Ontology (GO) and KEGG Analysis of Target Genes

In order to estimate the potential roles of *L. luteus* sRNAs in biological processes, GO annotations of their target genes were downloaded from the Gene Ontology using NCBI or UniProt identifiers The Bioconductor GOseq package [135] was used for GO enrichment analysis. KEGG annotation and enrichment analysis were performed to determine the metabolic pathways. The GO terms and KEGG pathways were considered to be significantly enriched with the corrected *p*-value of 0.05, which was calculated using a hypergeometric test [136].

4.10. Data submission to Sequence Read Archive NCBI

The RNA-Seq and small RNA-Seq data have been uploaded to the SRA database and are available under BioProject ID PRJNA419564 and Submission ID SUB3230840.

5. Conclusions

In this paper, we present the first case of identification and integrated analysis of small ncRNA, transcriptome, and degradome sequencing data, which allowed us to identify known and novel miRNAs, siRNAs and their target genes probably involved in regulating yellow lupine flower development and abscission. These miRNAs and siRNAs, by controlling the expression of their target genes, may have an impact on the development and fate of flowers growing in particular parts of the inflorescence (Figure 11). There appear to be microRNAs controlling auxin signal transduction elements and proliferation regulators in n the central node of the regulatory network controlling flower development or abscission. In addition to the purely cognitive aspects of describing the evolutionary conservation and the species specificity of important mechanisms regulating plant development, this work may contribute to the optimization of field crops and to monitoring the impact of various factors on flowering in yellow lupine. The use of the NGS technique allows for a detailed analysis of the regulatory networks which include sRNAs and their target genes. However, the results of sRNA-seq also contain a large number of uncharacterized sRNAs, the function of which may also have significance for the studied processes. More experimental and bioinformatic research is needed to fully describe the complex mechanisms of plant development regulation by low-molecular-weight regulatory RNAs.

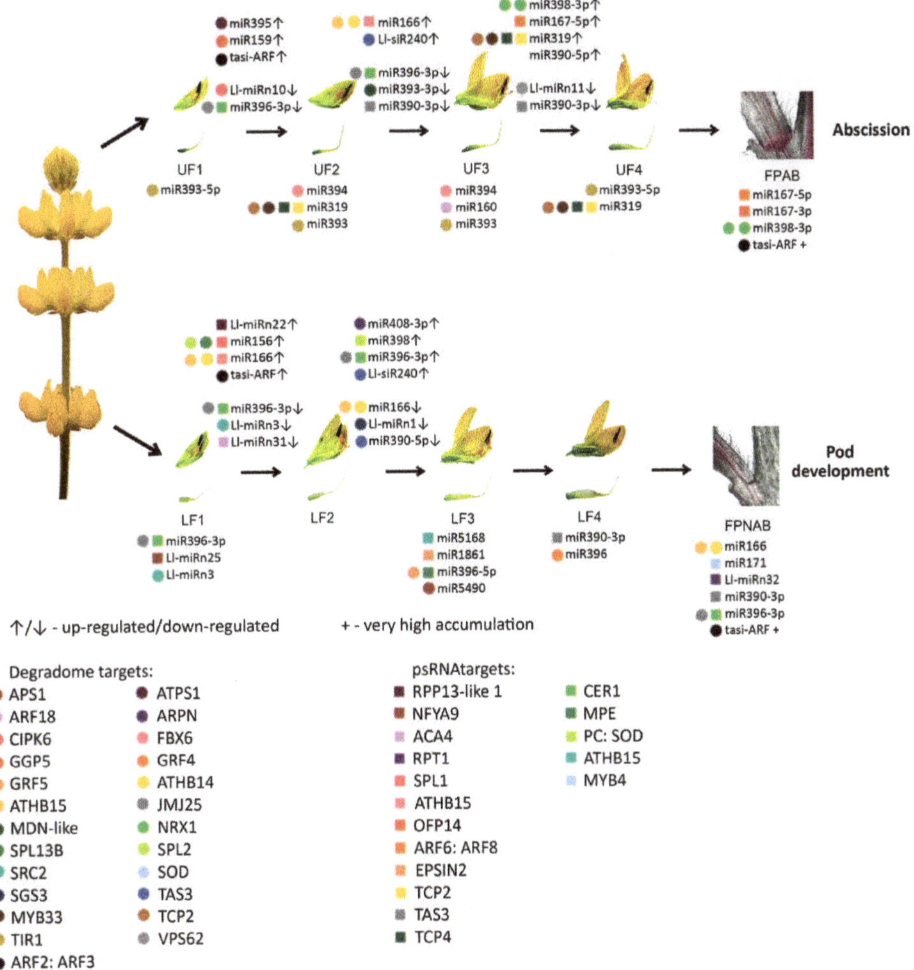

Figure 11. MiRNAs and siRNAs participating in yellow lupine flower development and abscission. Scheme based on the results of the current analysis. Arrows pointing upwards and downwards represent sRNAs that are up or downregulated in the transition between two developmental stages, respectively. The plus sign marks significantly expressed sRNAs. Colored circles represent targets found in the degradome, colored squares represent targets found using psRNATarget, as listed below. Multiple markers indicate that the sequence has multiple targets. Abbreviated gene names were acquired from UniProt database, where full target names can be found. Pictures from left to right are as follows: 4-whorl inflorescence of yellow lupine, flower cross-sections and isolated pistils for each stage of development, cross-sections of abscissing and non-abscissing flower pedicels stained with phloroglucinol-HCL solution.

Supplementary Materials: Supplementary materials can be found at http://www.mdpi.com/1422-0067/20/20/5122/s1.

Author Contributions: Conceptualization, P.G.; Data curation, P.G. and J.K.; Funding acquisition, P.G.; Investigation, P.G., M.K., W.G. and W.W.; Methodology, P.G.; Software, J.K.; Supervision, P.G.; Visualization, P.G., M.K. and W.G.; Writing—original draft, P.G., M.K. and W.G.; Writing—review & editing, P.G.

Int. J. Mol. Sci. **2019**, *20*, 5122

Funding: This research was funded by The National Science Centre SONATA grant No. 2015/19/D/NZ9/03601 and by the Program Supported by Resolution of the Council of Ministers (RM-111-222-15) in association with the Institute of Plant Genetics, Polish Academy of Sciences.

Acknowledgments: We would like to thank Michal Szczesniak and coworkers from Ideas4biology Sp. z o. o. for performing the RNA sequencing data analysis and consulting.

Conflicts of Interest: The authors declare no conflict of interest. The funders had no role in the design of the study; in the collection, analyses, or interpretation of data; in the writing of the manuscript, or in the decision to publish the results.

Abbreviations

AZ	abscission zone
DE	differentially expressed/differential expression
FPAB	flower pedicel abscissed
FPKM	fragments per kilobase of transcript per million mapped reads
FPNAB	flower pedicel non abscissed
GO	gene ontology
KEGG	kyoto encyclopedia of genes and genomes
LF	lower flower
miRNAs	microRNAs
ncRNA	non coding RNA
NGS	next generation sequencing
Pajd	adjusted p-value
RPM	reads per million
siRNA	small interfering RNA
sRNA	small RNA
UF	upper flower
UFR	upper flower after removing lower flowers

References

1. Prusiński, J. Degree of success of legume cultivars registered by the center for cultivar testing over the period of market economy. *Acta Sci. Pol.* **2007**, *6*, 3–16.
2. Lucas, M.M.; Stoddard, F.L.; Annicchiarico, P.; Frías, J.; Martínez-Villaluenga, C.; Sussmann, D.; Duranti, M.; Seger, A.; Zander, P.M.; Pueyo, J.J. The future of lupin as a protein crop in Europe. *Front. Plant Sci.* **2015**, *6*, 705. [CrossRef] [PubMed]
3. Ogura, T.; Ogihara, J.; Sunairi, M.; Takeishi, H.; Aizawa, T.; Olivos-Trujillo, M.R.; Maureira-Butler, I.J.; Salvo-Garrido, H.E. Proteomic characterization of seeds from yellow lupin (*Lupinus luteus* L.). *Proteomics* **2014**, *14*, 1543–1546. [CrossRef] [PubMed]
4. Van Steveninck, R.F. Abscission-accelerators in lupins (*Lupinus luteus* L.). *Nature* **1959**, *183*, 1246–1248. [CrossRef] [PubMed]
5. Ali, Z.; Raza, Q.; Atif, R.M.; Aslam, U.; Ajmal, M.; Chung, G. Genetic and molecular control of floral organ identity in cereals. *Int. J. Mol. Sci.* **2019**, *20*, 2743. [CrossRef] [PubMed]
6. Irish, V. The ABC model of floral development. *Curr. Biol.* **2017**, *27*, R887–R890. [CrossRef]
7. Robles, P.; Pelaz, S. Flower and fruit development in *Arabidopsis thaliana*. *Int. J. Dev. Biol.* **2005**, *49*, 633–643. [CrossRef]
8. Chandler, J.W. The hormonal regulation of flower development. *J. Plant Growth Regul.* **2011**, *30*, 242–254. [CrossRef]
9. Krishnamurthy, K.V.; Bahadur, B. Genetics of flower development. In *Plant Biology and Biotechnology*; Bhadur, B., Rajam, M.V., Sahijram, L., Krishnamurthy, K.V., Eds.; Springer India: New Delhi, India, 2015; pp. 385–407. ISBN 978-81-322-2286-6.
10. Luo, Y.; Guo, Z.; Li, L. Evolutionary conservation of microRNA regulatory programs in plant flower development. *Dev. Biol.* **2013**, *380*, 133–144. [CrossRef]

11. Liu, N.; Wu, S.; Van Houten, J.; Wang, Y.; Ding, B.; Fei, Z.; Clarke, T.H.; Reed, J.W.; van der Knaap, E. Down-regulation of AUXIN RESPONSE FACTORS 6 and 8 by microRNA 167 leads to floral development defects and female sterility in tomato. *J. Exp. Bot.* **2014**, *65*, 2507–2520. [CrossRef]

12. Damodharan, S.; Zhao, D.; Arazi, T. A common miRNA160-based mechanism regulates ovary patterning, floral organ abscission and lamina outgrowth in tomato. *Plant J.* **2016**, *86*, 458–471. [CrossRef] [PubMed]

13. Ascough, G.D.; Nogemane, N.; Mtshali, N.P.; van Staden, J.; Bornman, C.H. Flower abscission: Environmental control, internal regulation and physiological responses of plants. *South Afr. J. Bot.* **2005**, *71*, 287–301. [CrossRef]

14. Estornell, L.H.; Agustí, J.; Merelo, P.; Talón, M.; Tadeo, F.R. Elucidating mechanisms underlying organ abscission. *Plant Sci.* **2013**, *199–200*, 48–60. [CrossRef]

15. Basu, M.M.; González-Carranza, Z.H.; Azam-Ali, S.; Tang, S.; Shahid, A.A.; Roberts, J.A. The manipulation of auxin in the abscission zone cells of *Arabidopsis* flowers reveals that indole acetic acid signaling is a prerequisite for organ shedding. *Plant Physiol.* **2013**, *162*, 96–106. [CrossRef] [PubMed]

16. Patterson, S.E.; Bleecker, A.B. Ethylene-dependent and -independent processes associated with floral organ abscission in *Arabidopsis*. *Plant Physiol.* **2004**, *134*, 194–203. [CrossRef] [PubMed]

17. Glazinska, P.; Wojciechowski, W.; Kulasek, M.; Glinkowski, W.; Marciniak, K.; Klajn, N.; Kesy, J.; Kopcewicz, J. *De novo* transcriptome profiling of flowers, flower pedicels and pods of *Lupinus luteus* (yellow lupine) reveals complex expression changes during organ abscission. *Front. Plant Sci.* **2017**, *8*, 641. [CrossRef] [PubMed]

18. Marin, E.; Jouannet, V.; Herz, A.; Lokerse, A.S.; Weijers, D.; Vaucheret, H.; Nussaume, L.; Crespi, M.D.; Maizel, A. miR390, *Arabidopsis* TAS3 tasiRNAs, and their AUXIN RESPONSE FACTOR targets define an autoregulatory network quantitatively regulating lateral root growth. *Plant Cell* **2010**, *22*, 1104–1117. [CrossRef] [PubMed]

19. Williams, L.; Carles, C.C.; Osmont, K.S.; Fletcher, J.C. A database analysis method identifies an endogenous trans-acting short-interfering RNA that targets the *Arabidopsis ARF2, ARF3*, and *ARF4* genes. *Proc. Natl. Acad. Sci. USA* **2005**, *102*, 9703–9708. [CrossRef]

20. Si-Ammour, A.; Windels, D.; Arn-Bouldoires, E.; Kutter, C.; Ailhas, J.; Meins, F.; Vazquez, F. miR393 and secondary siRNAs regulate expression of the TIR1/AFB2 auxin receptor clade and auxin-related development of *Arabidopsis* leaves. *Plant Physiol.* **2011**, *157*, 683–691. [CrossRef]

21. D'Ario, M.; Griffiths-Jones, S.; Kim, M. Small RNAs: Big Impact on Plant Development. *Trends Plant Sci.* **2017**, *22*, 1056–1068. [CrossRef]

22. Liu, H.; Yu, H.; Tang, G.; Huang, T. Small but powerful: Function of microRNAs in plant development. *Plant Cell Rep.* **2018**, *37*, 515–528. [CrossRef] [PubMed]

23. Bhogale, S.; Mahajan, A.S.; Natarajan, B.; Rajabhoj, M.; Thulasiram, H.V.; Banerjee, A.K. MicroRNA156: A potential graft-transmissible microRNA that modulates plant architecture and tuberization in *Solanum tuberosum* ssp. andigena. *Plant Physiol.* **2014**, *164*, 1011–1027. [CrossRef] [PubMed]

24. Sun, X.; Fan, G.; Su, L.; Wang, W.; Liang, Z.; Li, S.; Xin, H. Identification of cold-inducible microRNAs in grapevine. *Front. Plant Sci.* **2015**, *6*, 595. [CrossRef] [PubMed]

25. Koroban, N.V.; Kudryavtseva, A.V.; Krasnov, G.S.; Sadritdinova, A.F.; Fedorova, M.S.; Snezhkina, A.V.; Bolsheva, N.L.; Muravenko, O.V.; Dmitriev, A.A.; Melnikova, N.V. The role of microRNA in abiotic stress response in plants. *Mol. Biol.* **2016**, *50*, 337–343. [CrossRef]

26. Jin, D.; Wang, Y.; Zhao, Y.; Chen, M. MicroRNAs and their cross-talks in plant development. *J. Genet. Genom.* **2013**, *40*, 161–170. [CrossRef]

27. Kurihara, Y.; Watanabe, Y. *Arabidopsis* micro-RNA biogenesis through Dicer-like 1 protein functions. *Proc. Natl. Acad. Sci. USA* **2004**, *101*, 12753–12758. [CrossRef]

28. Reinhart, B.J.; Weinstein, E.G.; Rhoades, M.W.; Bartel, B.; Bartel, D.P. MicroRNAs in plants. *Genes Dev.* **2002**, *16*, 1616–1626. [CrossRef]

29. Voinnet, O. Origin, biogenesis, and activity of plant microRNAs. *Cell* **2009**, *136*, 669–687. [CrossRef]

30. Yu, Y.; Jia, T.; Chen, X. The "how" and "where" of plant microRNAs. *New Phytol.* **2017**, *216*, 1002–1017. [CrossRef]

31. Teotia, S.; Tang, G. To bloom or not to bloom: Role of microRNAs in plant flowering. *Mol. Plant* **2015**, *8*, 359–377. [CrossRef]

32. Yang, T.; Wang, Y.; Teotia, S.; Zhang, Z.; Tang, G. The making of leaves: How small RNA networks modulate leaf development. *Front. Plant Sci.* **2018**, *9*, 824. [CrossRef] [PubMed]

33. Vazquez, F.; Hohn, T. Biogenesis and biological activity of secondary siRNAs in plants. *Scientifica* **2013**, *2013*, 783253. [CrossRef] [PubMed]

34. Singh, A.; Gautam, V.; Singh, S.; Sarkar Das, S.; Verma, S.; Mishra, V.; Mukherjee, S.; Sarkar, A.K. Plant small RNAs: Advancement in the understanding of biogenesis and role in plant development. *Planta* **2018**, *248*, 545–558. [CrossRef] [PubMed]

35. Deng, P.; Muhammad, S.; Cao, M.; Wu, L. Biogenesis and regulatory hierarchy of phased small interfering RNAs in plants. *Plant Biotechnol. J.* **2018**, *16*, 965–975. [CrossRef]

36. Lee, C.H.; Carroll, B.J. Evolution and diversification of small RNA pathways in flowering plants. *Plant Cell Physiol.* **2018**, *59*, 2169–2187. [CrossRef]

37. Sun, Y.; Mui, Z.; Liu, X.; Yim, A.K.-Y.; Qin, H.; Wong, F.-L.; Chan, T.-F.; Yiu, S.-M.; Lam, H.-M.; Lim, B.L. Comparison of small RNA profiles of *Glycine max* and *Glycine soja* at early developmental stages. *Int. J. Mol. Sci.* **2016**, *17*, 2043. [CrossRef]

38. Zhou, Z.S.; Huang, S.Q.; Yang, Z.M. Bioinformatic identification and expression analysis of new microRNAs from *Medicago truncatula*. *Biochem. Biophys. Res. Commun.* **2008**, *374*, 538–542. [CrossRef]

39. Pokoo, R.; Ren, S.; Wang, Q.; Motes, C.M.; Hernandez, T.D.; Ahmadi, S.; Monteros, M.J.; Zheng, Y.; Sunkar, R. Genotype- and tissue-specific miRNA profiles and their targets in three alfalfa (*Medicago sativa* L) genotypes. *BMC Genomics* **2018**, *19*, 913. [CrossRef]

40. Fletcher, S.J.; Shrestha, A.; Peters, J.R.; Carroll, B.J.; Srinivasan, R.; Pappu, H.R.; Mitter, N. The tomato spotted wilt virus genome is processed differentially in its plant host *Arachis hypogaea* and its thrips vector *Frankliniella fusca*. *Front. Plant Sci.* **2016**, *7*, 1349. [CrossRef]

41. Tsikou, D.; Yan, Z.; Holt, D.B.; Abel, N.B.; Reid, D.E.; Madsen, L.H.; Bhasin, H.; Sexauer, M.; Stougaard, J.; Markmann, K. Systemic control of legume susceptibility to rhizobial infection by a mobile microRNA. *Science* **2018**, *362*, 233–236. [CrossRef]

42. Wu, J.; Wang, L.; Wang, S. MicroRNAs associated with drought response in the pulse crop common bean (*Phaseolus vulgaris* L.). *Gene* **2017**, *628*, 78–86. [CrossRef] [PubMed]

43. Rodriguez-Medina, C.; Atkins, C.A.; Mann, A.J.; Jordan, M.E.; Smith, P.M. Macromolecular composition of phloem exudate from white lupin (*Lupinus albus* L.). *Bmc Plant Biol.* **2011**, *11*, 36. [CrossRef] [PubMed]

44. Zhu, Y.Y.; Zeng, H.Q.; Dong, C.X.; Yin, X.M.; Shen, Q.R.; Yang, Z.M. microRNA expression profiles associated with phosphorus deficiency in white lupin (*Lupinus albus* L.). *Plant Sci.* **2010**, *178*, 23–29. [CrossRef]

45. DeBoer, K.; Melser, S.; Sperschneider, J.; Kamphuis, L.G.; Garg, G.; Gao, L.-L.; Frick, K.; Singh, K.B. Identification and profiling of narrow-leafed lupin (*Lupinus angustifolius*) microRNAs during seed development. *BMC Genomics* **2019**, *20*, 135. [CrossRef] [PubMed]

46. Tang, C.-Y.; Yang, M.-K.; Wu, F.-Y.; Zhao, H.; Pang, Y.-J.; Yang, R.-W.; Lu, G.-H.; Yang, Y.-H. Identification of miRNAs and their targets in transgenic *Brassica napus* and its acceptor (Westar) by high-throughput sequencing and degradome analysis. *Rsc Adv.* **2015**, *5*, 85383–85394. [CrossRef]

47. Zhou, R.; Wang, Q.; Jiang, F.; Cao, X.; Sun, M.; Liu, M.; Wu, Z. Identification of miRNAs and their targets in wild tomato at moderately and acutely elevated temperatures by high-throughput sequencing and degradome analysis. *Sci. Rep.* **2016**, *6*, 33777. [CrossRef]

48. Fang, Y.-N.; Zheng, B.-B.; Wang, L.; Yang, W.; Wu, X.-M.; Xu, Q.; Guo, W.-W. High-throughput sequencing and degradome analysis reveal altered expression of miRNAs and their targets in a male-sterile cybrid pummelo (*Citrus grandis*). *BMC Genomics* **2016**, *17*, 591. [CrossRef]

49. Kalvari, I.; Nawrocki, E.P.; Argasinska, J.; Quinones-Olvera, N.; Finn, R.D.; Bateman, A.; Petrov, A.I. Non-coding RNA analysis using the Rfam database. *Curr. Protoc. Bioinforma.* **2018**, *62*, e51. [CrossRef]

50. Kalvari, I.; Argasinska, J.; Quinones-Olvera, N.; Nawrocki, E.P.; Rivas, E.; Eddy, S.R.; Bateman, A.; Finn, R.D.; Petrov, A.I. Rfam 13.0: Shifting to a genome-centric resource for non-coding RNA families. *Nucleic Acids Res.* **2018**, *46*, D335–D342. [CrossRef]

51. Griffiths-Jones, S.; Saini, H.K.; van Dongen, S.; Enright, A.J. miRBase: Tools for microRNA genomics. *Nucleic Acids Res.* **2008**, *36*, D154–D158. [CrossRef]

52. Altschul, S.F.; Gish, W.; Miller, W.; Myers, E.W.; Lipman, D.J. Basic local alignment search tool. *J. Mol. Biol.* **1990**, *215*, 403–410. [CrossRef]

53. Axtell, M.J. ShortStack: Comprehensive annotation and quantification of small RNA genes. *RNA* **2013**, *19*, 740–751. [CrossRef] [PubMed]

54. Chen, C.; Zeng, Z.; Liu, Z.; Xia, R. Small RNAs, emerging regulators critical for the development of horticultural traits. *Hortic. Res.* **2018**, *5*, 63. [CrossRef] [PubMed]

55. Oliveros, J. VENNY. An interactive tool for comparing lists with Venn diagrams. Available online: http://bioinfogp.cnb.csic.es/tools/venny/index.html (accessed on 6 March 2019).

56. Kramer, M.F. Stem-Loop RT-qPCR for miRNAs. *Curr. Protoc. Mol. Biol.* **2011**, *95*, 15.10.1–15.10.15. [CrossRef] [PubMed]

57. Varkonyi-Gasic, E.; Hellens, R.P. Quantitative Stem-Loop RT-PCR for detection of microRNAs. In *RNAi and plant gene function analysis. Methods in Molecular Biology (Methods and Protocols)*; Kodama, H., Komamine, A., Eds.; Humana Press: Totowa, NJ, USA, 2011; Volume 744, pp. 145–157. ISBN 978-1-61779-123-9.

58. Addo-Quaye, C.; Miller, W.; Axtell, M.J. CleaveLand: A pipeline for using degradome data to find cleaved small RNA targets. *Bioinformatics* **2009**, *25*, 130–131. [CrossRef]

59. Dai, X.; Zhuang, Z.; Zhao, P.X. psRNATarget: A plant small RNA target analysis server (2017 release). *Nucleic Acids Res.* **2018**, *46*, W49–W54. [CrossRef]

60. Nag, A.; Jack, T. Sculpting the flower; the role of microRNAs in flower development. In *Plant Development*; Timmermans, M.C.P., Ed.; Elsevier: Amsterdam, The Netherlands, 2010; Volume 91, pp. 349–378. ISBN 978-0-12-380910-0.

61. Xing, S.; Salinas, M.; Höhmann, S.; Berndtgen, R.; Huijser, P. miR156-targeted and nontargeted SBP-box transcription factors act in concert to secure male fertility in *Arabidopsis*. *Plant Cell* **2010**, *22*, 3935–3950. [CrossRef]

62. Wang, Z.; Wang, Y.; Kohalmi, S.E.; Amyot, L.; Hannoufa, A. SQUAMOSA PROMOTER BINDING PROTEIN-LIKE 2 controls floral organ development and plant fertility by activating ASYMMETRIC LEAVES 2 in *Arabidopsis thaliana*. *Plant Mol. Biol.* **2016**, *92*, 661–674. [CrossRef]

63. Murray, F.; Kalla, R.; Jacobsen, J.; Gubler, F. A role for HvGAMYB in anther development. *Plant J.* **2003**, *33*, 481–491. [CrossRef]

64. Gubler, F.; Chandler, P.M.; White, R.G.; Llewellyn, D.J.; Jacobsen, J.V. Gibberellin signaling in barley aleurone cells. Control of *SLN1* and *GAMYB* expression. *Plant Physiol.* **2002**, *129*, 191–200. [CrossRef]

65. Achard, P.; Herr, A.; Baulcombe, D.C.; Harberd, N.P. Modulation of floral development by a gibberellin-regulated microRNA. *Development* **2004**, *131*, 3357–3365. [CrossRef] [PubMed]

66. Ru, P.; Xu, L.; Ma, H.; Huang, H. Plant fertility defects induced by the enhanced expression of microRNA167. *Cell Res.* **2006**, *16*, 457–465. [CrossRef] [PubMed]

67. Wu, M.-F.; Tian, Q.; Reed, J.W. *Arabidopsis* microRNA167 controls patterns of *ARF6* and *ARF8* expression, and regulates both female and male reproduction. *Development* **2006**, *133*, 4211–4218. [CrossRef] [PubMed]

68. Barik, S.; Kumar, A.; Sarkar Das, S.; Yadav, S.; Gautam, V.; Singh, A.; Singh, S.; Sarkar, A.K. Coevolution pattern and functional conservation or divergence of miR167s and their targets across diverse plant species. *Sci. Rep.* **2015**, *5*, 14611. [CrossRef]

69. Cecchetti, V.; Altamura, M.M.; Falasca, G.; Costantino, P.; Cardarelli, M. Auxin regulates *Arabidopsis* anther dehiscence, pollen maturation, and filament elongation. *Plant Cell* **2008**, *20*, 1760–1774. [CrossRef]

70. Chen, X. A microRNA as a translational repressor of APETALA2 in *Arabidopsis* flower development. *Science* **2004**, *303*, 2022–2025. [CrossRef]

71. Almeida, J.; Rocheta, M.; Galego, L. Genetic control of flower shape in *Antirrhinum majus*. *Development* **1997**, *124*, 1387–1392.

72. Galego, L.; Almeida, J. Role of DIVARICATA in the control of dorsoventral asymmetry in *Antirrhinum flowers*. *Genes Dev.* **2002**, *16*, 880–891. [CrossRef]

73. Du, Q.; Wang, H. The role of HD-ZIP III transcription factors and miR165/166 in vascular development and secondary cell wall formation. *Plant Signal. Behav.* **2015**, *10*, e1078955. [CrossRef]

74. Palatnik, J.F.; Wollmann, H.; Schommer, C.; Schwab, R.; Boisbouvier, J.; Rodriguez, R.; Warthmann, N.; Allen, E.; Dezulian, T.; Huson, D.; et al. Sequence and expression differences underlie functional specialization of *Arabidopsis* microRNAs miR159 and miR319. *Dev. Cell* **2007**, *13*, 115–125. [CrossRef]

75. Rubio-Somoza, I.; Weigel, D. Coordination of flower maturation by a regulatory circuit of three microRNAs. *Plos Genet.* **2013**, *9*, e1003374. [CrossRef] [PubMed]

76. Xia, R.; Xu, J.; Meyers, B.C. The emergence, evolution, and diversification of the miR390-TAS3-ARF pathway in land plants. *Plant Cell* **2017**, *29*, 1232–1247. [CrossRef] [PubMed]

77. Chitwood, D.H.; Guo, M.; Nogueira, F.T.S.; Timmermans, M.C.P. Establishing leaf polarity: The role of small RNAs and positional signals in the shoot apex. *Development* **2007**, *134*, 813–823. [CrossRef] [PubMed]

78. Matsui, A.; Mizunashi, K.; Tanaka, M.; Kaminuma, E.; Nguyen, A.H.; Nakajima, M.; Kim, J.-M.; Nguyen, D.V.; Toyoda, T.; Seki, M. tasiRNA-ARF pathway moderates floral architecture in *Arabidopsis* plants subjected to drought stress. *Biomed Res. Int.* **2014**, *2014*, 303451. [CrossRef]

79. Liang, G.; Yu, D. Reciprocal regulation among miR395, APS and SULTR2;1 in *Arabidopsis thaliana*. *Plant Signal. Behav.* **2010**, *5*, 1257–1259. [CrossRef]

80. Herrmann, J.; Ravilious, G.E.; McKinney, S.E.; Westfall, C.S.; Lee, S.G.; Baraniecka, P.; Giovannetti, M.; Kopriva, S.; Krishnan, H.B.; Jez, J.M. Structure and mechanism of soybean ATP sulfurylase and the committed step in plant sulfur assimilation. *J. Biol. Chem.* **2014**, *289*, 10919–10929. [CrossRef]

81. Wang, M.; Wang, Q.; Zhang, B. Response of miRNAs and their targets to salt and drought stresses in cotton (*Gossypium hirsutum* L.). *Gene* **2013**, *530*, 26–32. [CrossRef]

82. Kawashima, C.G.; Yoshimoto, N.; Maruyama-Nakashita, A.; Tsuchiya, Y.N.; Saito, K.; Takahashi, H.; Dalmay, T. Sulphur starvation induces the expression of microRNA-395 and one of its target genes but in different cell types. *Plant J.* **2009**, *57*, 313–321. [CrossRef]

83. Xie, F.; Zhang, B. microRNA evolution and expression analysis in polyploidized cotton genome. *Plant Biotechnol. J.* **2015**, *13*, 421–434. [CrossRef]

84. Inagaki, S.; Miura-Kamio, A.; Nakamura, Y.; Lu, F.; Cui, X.; Cao, X.; Kimura, H.; Saze, H.; Kakutani, T. Autocatalytic differentiation of epigenetic modifications within the *Arabidopsis* genome. *EMBO J.* **2010**, *29*, 3496–3506. [CrossRef]

85. Bourdenx, B.; Bernard, A.; Domergue, F.; Pascal, S.; Léger, A.; Roby, D.; Pervent, M.; Vile, D.; Haslam, R.P.; Napier, J.A.; et al. Overexpression of *ArabidopsisECERIFERUM1* promotes wax very-long-chain alkane biosynthesis and influences plant response to biotic and abiotic stresses. *Plant Physiol.* **2011**, *156*, 29–45. [CrossRef] [PubMed]

86. Vaucheret, H.; Mallory, A.C.; Bartel, D.P. AGO1 homeostasis entails coexpression of MIR168 and AGO1 and preferential stabilization of miR168 by AGO1. *Mol. Cell* **2006**, *22*, 129–136. [CrossRef] [PubMed]

87. Dong, J.; Kim, S.T.; Lord, E.M. Plantacyanin plays a role in reproduction in *Arabidopsis*. *Plant Physiol.* **2005**, *138*, 778–789. [CrossRef] [PubMed]

88. Song, Z.; Zhang, L.; Wang, Y.; Li, H.; Li, S.; Zhao, H.; Zhang, H. Constitutive expression of miR408 improves biomass and seed yield in *Arabidopsis*. *Front. Plant Sci.* **2017**, *8*, 2114. [CrossRef] [PubMed]

89. Sieber, P.; Gheyselinck, J.; Gross-Hardt, R.; Laux, T.; Grossniklaus, U.; Schneitz, K. Pattern formation during early ovule development in *Arabidopsis thaliana*. *Dev. Biol.* **2004**, *273*, 321–334. [CrossRef] [PubMed]

90. Lee, S.-J.; Lee, B.H.; Jung, J.-H.; Park, S.K.; Song, J.T.; Kim, J.H. GROWTH-REGULATING FACTOR and GRF-INTERACTING FACTOR specify meristematic cells of gynoecia and anthers. *Plant Physiol.* **2018**, *176*, 717–729. [CrossRef] [PubMed]

91. Liu, H.; Guo, S.; Xu, Y.; Li, C.; Zhang, Z.; Zhang, D.; Xu, S.; Zhang, C.; Chong, K. OsmiR396d-regulated OsGRFs function in floral organogenesis in rice through binding to their targets OsJMJ706 and OsCR4. *Plant Physiol.* **2014**, *165*, 160–174. [CrossRef]

92. Li, T.; Ma, L.; Geng, Y.; Hao, C.; Chen, X.; Zhang, X. Small RNA and degradome sequencing reveal complex roles of miRNAs and their targets in developing wheat grains. *Plos ONE* **2015**, *10*, e0139658. [CrossRef]

93. Kawarazaki, T.; Kimura, S.; Iizuka, A.; Hanamata, S.; Nibori, H.; Michikawa, M.; Imai, A.; Abe, M.; Kaya, H.; Kuchitsu, K. A low temperature-inducible protein AtSRC2 enhances the ROS-producing activity of NADPH oxidase AtRbohF. *Biochim. Biophys. Acta* **2013**, *1833*, 2775–2780. [CrossRef]

94. Wang, H.; Niu, Q.-W.; Wu, H.-W.; Liu, J.; Ye, J.; Yu, N.; Chua, N.-H. Analysis of non-coding transcriptome in rice and maize uncovers roles of conserved lncRNAs associated with agriculture traits. *Plant J.* **2015**, *84*, 404–416. [CrossRef]

95. Jagadeeswaran, G.; Zheng, Y.; Li, Y.-F.; Shukla, L.I.; Matts, J.; Hoyt, P.; Macmil, S.L.; Wiley, G.B.; Roe, B.A.; Zhang, W.; et al. Cloning and characterization of small RNAs from *Medicago truncatula* reveals four novel legume-specific microRNA families. *New Phytol.* **2009**, *184*, 85–98. [CrossRef] [PubMed]

96. Xie, Z.; Johansen, L.K.; Gustafson, A.M.; Kasschau, K.D.; Lellis, A.D.; Zilberman, D.; Jacobsen, S.E.; Carrington, J.C. Genetic and functional diversification of small RNA pathways in plants. *Plos Biol.* **2004**, *2*, E104. [CrossRef] [PubMed]

97. Song, Q.-X.; Liu, Y.-F.; Hu, X.-Y.; Zhang, W.-K.; Ma, B.; Chen, S.-Y.; Zhang, J.-S. Identification of miRNAs and their target genes in developing soybean seeds by deep sequencing. *BMC Plant Biol.* **2011**, *11*, 5. [CrossRef] [PubMed]

98. Peng, T.; Sun, H.; Qiao, M.; Zhao, Y.; Du, Y.; Zhang, J.; Li, J.; Tang, G.; Zhao, Q. Differentially expressed microRNA cohorts in seed development may contribute to poor grain filling of inferior spikelets in rice. *BMC Plant Biol.* **2014**, *14*, 196. [CrossRef] [PubMed]

99. Zhu, Q.-H.; Spriggs, A.; Matthew, L.; Fan, L.; Kennedy, G.; Gubler, F.; Helliwell, C. A diverse set of microRNAs and microRNA-like small RNAs in developing rice grains. *Genome Res.* **2008**, *18*, 1456–1465. [CrossRef]

100. Rodriguez, R.E.; Mecchia, M.A.; Debernardi, J.M.; Schommer, C.; Weigel, D.; Palatnik, J.F. Control of cell proliferation in *Arabidopsis thaliana* by microRNA miR396. *Development* **2010**, *137*, 103–112. [CrossRef]

101. Yi, R.; Zhu, Z.; Hu, J.; Qian, Q.; Dai, J.; Ding, Y. Identification and expression analysis of microRNAs at the grain filling stage in rice (*Oryza sativa* L.) via deep sequencing. *Plos ONE* **2013**, *8*, e57863. [CrossRef]

102. Chen, Z.-H.; Bao, M.-L.; Sun, Y.-Z.; Yang, Y.-J.; Xu, X.-H.; Wang, J.-H.; Han, N.; Bian, H.-W.; Zhu, M.-Y. Regulation of auxin response by miR393-targeted Transport Inhibitor Response protein 1 is involved in normal development in *Arabidopsis*. *Plant Mol. Biol.* **2011**, *77*, 619–629. [CrossRef]

103. Bonnet, E.; van de Peer, Y.; Rouzé, P. The small RNA world of plants. *New Phytol.* **2006**, *171*, 451–468. [CrossRef]

104. Guo, N.; Zhang, Y.; Sun, X.; Fan, H.; Gao, J.; Chao, Y.; Liu, A.; Yu, X.; Cai, Y.; Lin, Y. Genome-wide identification of differentially expressed miRNAs induced by ethephon treatment in abscission layer cells of cotton (*Gossypium hirsutum*). *Gene* **2018**, *676*, 263–268. [CrossRef]

105. Hu, G.; Fan, J.; Xian, Z.; Huang, W.; Lin, D.; Li, Z. Overexpression of *SlREV* alters the development of the flower pedicel abscission zone and fruit formation in tomato. *Plant Sci.* **2014**, *229*, 86–95. [CrossRef] [PubMed]

106. Li, M.; Liang, Z.; He, S.; Zeng, Y.; Jing, Y.; Fang, W.; Wu, K.; Wang, G.; Ning, X.; Wang, L.; et al. Genome-wide identification of leaf abscission associated microRNAs in sugarcane (*Saccharum officinarum* L.). *BMC Genomics* **2017**, *18*, 754. [CrossRef] [PubMed]

107. Xu, T.; Wang, Y.; Liu, X.; Lv, S.; Feng, C.; Qi, M.; Li, T. Small RNA and degradome sequencing reveals microRNAs and their targets involved in tomato pedicel abscission. *Planta* **2015**, *242*, 963–984. [CrossRef] [PubMed]

108. Teotia, P.S.; Mukherjee, S.K.; Mishra, N.S. Fine tuning of auxin signaling by miRNAs. *Physiol. Mol. Biol. Plants* **2008**, *14*, 81–90. [CrossRef] [PubMed]

109. Curaba, J.; Singh, M.B.; Bhalla, P.L. miRNAs in the crosstalk between phytohormone signalling pathways. *J. Exp. Bot.* **2014**, *65*, 1425–1438. [CrossRef]

110. Guilfoyle, T.J.; Hagen, G. Auxin response factors. *Curr. Opin. Plant Biol.* **2007**, *10*, 453–460. [CrossRef]

111. Lau, S.; Jurgens, G.; de Smet, I. The evolving complexity of the auxin pathway. *Plant Cell* **2008**, *20*, 1738–1746. [CrossRef]

112. Quint, M.; Gray, W.M. Auxin signaling. *Curr. Opin. Plant Biol.* **2006**, *9*, 448–453. [CrossRef]

113. Rhoades, M.W.; Reinhart, B.J.; Lim, L.P.; Burge, C.B.; Bartel, B.; Bartel, D.P. Prediction of plant microRNA targets. *Cell* **2002**, *110*, 513–520. [CrossRef]

114. Allen, E.; Xie, Z.; Gustafson, A.M.; Carrington, J. CmicroRNA-directed phasing during trans-acting siRNA biogenesis in plants. *Cell* **2005**, *121*, 207–221. [CrossRef]

115. Axtell, M.J.; Jan, C.; Rajagopalan, R.; Bartel, D.P. A two-hit trigger for siRNA biogenesis in plants. *Cell* **2006**, *127*, 565–577. [CrossRef] [PubMed]

116. Zhang, C.; Li, G.; Zhu, S.; Zhang, S.; Fang, J. tasiRNAdb: A database of ta-siRNA regulatory pathways. *Bioinformatics* **2014**, *30*, 1045–1046. [CrossRef] [PubMed]

117. Sun, Y.; Zhou, Q.; Zhang, W.; Fu, Y.; Huang, H. *ASYMMETRIC LEAVES1*, an *Arabidopsis* gene that is involved in the control of cell differentiation in leaves. *Planta* **2002**, *214*, 694–702. [CrossRef] [PubMed]

118. Dieterle, M.; Thomann, A.; Renou, J.-P.; Parmentier, Y.; Cognat, V.; Lemonnier, G.; Müller, R.; Shen, W.-H.; Kretsch, T.; Genschik, P. Molecular and functional characterization of *Arabidopsis* Cullin 3A. *Plant J.* **2004**, *41*, 386–399. [CrossRef] [PubMed]

119. Smalle, J.; Kurepa, J.; Yang, P.; Babiychuk, E.; Kushnir, S.; Durski, A.; Vierstra, R.D. Cytokinin growth responses in *Arabidopsis* involve the 26S proteasome subunit RPN12. *Plant Cell* **2002**, *14*, 17–32. [CrossRef]

120. Krogan, N.T.; Hogan, K.; Long, J.A. APETALA2 negatively regulates multiple floral organ identity genes in *Arabidopsis* by recruiting the co-repressor TOPLESS and the histone deacetylase HDA19. *Development* **2012**, *139*, 4180–4190. [CrossRef]

121. Elliott, R.C.; Betzner, A.S.; Huttner, E.; Oakes, M.P.; Tucker, W.Q.; Gerentes, D.; Perez, P.; Smyth, D.R. *AINTEGUMENTA*, an *APETALA2*-like gene of *Arabidopsis* with pleiotropic roles in ovule development and floral organ growth. *Plant Cell* **1996**, *8*, 155–168.

122. Ohto, M.; Floyd, S.K.; Fischer, R.L.; Goldberg, R.B.; Harada, J.J. Effects of APETALA2 on embryo, endosperm, and seed coat development determine seed size in *Arabidopsis*. *Sex. Plant Reprod.* **2009**, *22*, 277–289. [CrossRef]

123. Ohto, M.A.; Fischer, R.L.; Goldberg, R.B.; Nakamura, K.; Harada, J.J. Control of seed mass by APETALA2. *Proc. Natl. Acad. Sci. USA* **2005**, *102*, 3123–3128. [CrossRef]

124. Gaudin, V.; Libault, M.; Pouteau, S.; Juul, T.; Zhao, G.; Lefebvre, D.; Grandjean, O. Mutations in *LIKE HETEROCHROMATIN PROTEIN 1* affect flowering time and plant architecture in *Arabidopsis*. *Development* **2001**, *128*, 4847–4858.

125. Sinha, A.K.; Jaggi, M.; Raghuram, B.; Tuteja, N. Mitogen-activated protein kinase signaling in plants under abiotic stress. *Plant Signal. Behav.* **2011**, *6*, 196–203. [CrossRef] [PubMed]

126. Jalmi, S.K.; Sinha, A.K. ROS mediated MAPK signaling in abiotic and biotic stress- striking similarities and differences. *Front. Plant Sci.* **2015**, *6*, 769. [CrossRef] [PubMed]

127. Ouaked, F.; Rozhon, W.; Lecourieux, D.; Hirt, H. A MAPK pathway mediates ethylene signaling in plants. *Embo J.* **2003**, *22*, 1282–1288. [CrossRef] [PubMed]

128. Finkelstein, R. Abscisic acid synthesis and response. *Arab. B.* **2013**, *11*, e0166. [CrossRef] [PubMed]

129. Sawicki, M.; Aït Barka, E.; Clément, C.; Vaillant-Gaveau, N.; Jacquard, C. Cross-talk between environmental stresses and plant metabolism during reproductive organ abscission. *J. Exp. Bot.* **2015**, *66*, 1707–1719. [CrossRef]

130. Patharkar, O.R.; Walker, J.C. Floral organ abscission is regulated by a positive feedback loop. *Proc. Natl. Acad. Sci. USA* **2015**, *112*, 2906–2911. [CrossRef]

131. Poznan Plant Breeding Sp. z o.o. Information on yellow lupine cv. Taper. Available online: http://phr.pl/wp-content/uploads/2017/07/Taper.pdf (accessed on 5 August 2019).

132. Haas, B.J.; Papanicolaou, A.; Yassour, M.; Grabherr, M.; Blood, P.D.; Bowden, J.; Couger, M.B.; Eccles, D.; Li, B.; Lieber, M.; et al. *De novo* transcript sequence reconstruction from RNA-seq using the Trinity platform for reference generation and analysis. *Nat. Protoc.* **2013**, *8*, 1494–1512. [CrossRef]

133. Langmead, B. *Aligning Short Sequencing Reads with Bowtie. Current Protocols in Bioinformatics*; John Wiley & Sons, Inc.: Hoboken, NJ, USA, 2010; Chapter 11; Volume 32.

134. Love, M.I.; Huber, W.; Anders, S. Moderated estimation of fold change and dispersion for RNA-seq data with DESeq2. *Genome Biol.* **2014**, *15*, 550. [CrossRef]

135. Young, M.D.; Wakefield, M.J.; Smyth, G.K.; Oshlack, A. Gene ontology analysis for RNA-seq: Accounting for selection bias. *Genome Biol.* **2010**, *11*, R14. [CrossRef]

136. Huo, Z.; Tang, S.; Park, Y.; Tseng, G. P-value evaluation, variability index and biomarker categorization for adaptively weighted Fisher's meta-analysis method in omics applications. *Bioinformatics* **2019**, btz589. [CrossRef]

International Journal of
Molecular Sciences

Article

Two Novel *er1* Alleles Conferring Powdery Mildew (*Erysiphe pisi*) Resistance Identified in a Worldwide Collection of Pea (*Pisum sativum* L.) Germplasms

Suli Sun [1], Dong Deng [1], Canxing Duan [1], Xuxiao Zong [1], Dongxu Xu [2], Yuhua He [3] and Zhendong Zhu [1,*]

[1] National Key Facility for Crop Gene Resources and Genetic Improvement, Institute of Crop Sciences, Chinese Academy of Agricultural Sciences, Beijing 100081, China; sulisun@caas.cn (S.S.); q2499363366@163.com (D.D.); duancanxing@caas.cn (C.D.); zongxuxiao@caas.cn (X.Z.)
[2] Zhangjiakou Academy of Agricultural Sciences, Zhangjiakou 075000, China; xudongxu1972@163.com
[3] Yunnan Academy of Agricultural Sciences, Kunming 650205, China; erbio@163.com
*   Correspondence: zhuzhendong@caas.cn; Tel.: +86-10-82109609; Fax: +86-10-82109608

Received: 22 August 2019; Accepted: 30 September 2019; Published: 12 October 2019

Abstract: Powdery mildew caused by *Erysiphe pisi* DC. severely affects pea crops worldwide. The use of resistant cultivars containing the *er1* gene is the most effective way to control this disease. The objectives of this study were to reveal *er1* alleles contained in 55 *E. pisi*-resistant pea germplasms and to develop the functional markers of novel alleles. Sequences of 10 homologous *PsMLO1* cDNA clones from each germplasm accession were used to determine their *er1* alleles. The frame shift mutations and various alternative splicing patterns were observed during transcription of the *er1* gene. Two novel *er1* alleles, *er1-8* and *er1-9*, were discovered in the germplasm accessions G0004839 and G0004400, respectively, and four known *er1* alleles were identified in 53 other accessions. One mutation in G0004839 was characterized by a 3-bp (GTG) deletion of the wild-type *PsMLO1* cDNA, resulting in a missing valine at position 447 of the PsMLO1 protein sequence. Another mutation in G0004400 was caused by a 1-bp (T) deletion of the wild-type *PsMLO1* cDNA sequence, resulting in a serine to leucine change of the PsMLO1 protein sequence. The *er1-8* and *er1-9* alleles were verified using resistance inheritance analysis and genetic mapping with respectively derived F_2 and $F_{2:3}$ populations. Finally, co-dominant functional markers specific to *er1-8* and *er1-9* were developed and validated in populations and pea germplasms. These results improve our understanding of *E. pisi* resistance in pea germplasms worldwide and provide powerful tools for marker-assisted selection in pea breeding.

Keywords: *Erysiphe pisi*; *er1-8*; *er1-9*; KASPar marker; pea

1. Introduction

Pea (*Pisum sativum* L.) is a widely distributed legume crop, which frequently suffers from various stresses, including abiotic and biotic factors in the season of growth [1,2]. Powdery mildew, induced by *Erysiphe pisi* DC., severely reduces the yield and quality of pea crops worldwide [3–5]. Severe *E. pisi* infections of peas can lead to yield losses of up to 80% in regions which are suitable for disease development [5,6]. The use of resistant cultivars carrying the *E. pisi*-resistant gene *er1* has been considered to be the most effective and environmentally friendly way to prevent this disease to date [6,7].

Formerly, *E. pisi* infection was the only known cause of pea powdery mildew. However, since 2005, two other *Erysiphe* species, *Erysiphe trifolii* and *Erysiphe baeumleri*, have been reported to also infect peas and induce the same powdery mildew symptoms as *E. pisi* in some regions [8–10]. Previous studies of pea powdery mildew have primarily focused on breeding peas resistant to *E. pisi*. Their results have indicated that resistance to *E. pisi* is controlled by two single recessive genes (*er1* and *er2*) and one dominant gene (*Er3*) [11–14]. The *er1*, *er2*, and *Er3* genes have been mapped using linked markers [15–27]. The genes *er1* and *er2* map to pea linkage groups (LGs) VI and III, respectively [17,28]. *Er3*, which was isolated from wild pea (*Pisum fulvum*), was initially mapped on an uncertain pea LG, but it was more recently assigned to pea LG IV [29].

As *er1* confers high resistance or complete immunity to *E. pisi* in most pea germplasms, it is currently the most widely used gene in pea production [30]. In contrast, *er2* is only found in a few pea germplasms resistant to *E. pisi* [30]. Moreover, the efficacy of *E. pisi* resistance conferred by *er2* varies with leaf development stage and plant location [12,30–32]. *Er3* was known from wild pea (*P. fulvum*), and there have not been extensive studies conducted to date [13,33].

Gene *er1* confers stable, durable, and broadly effective resistance to *E. pisi*. This gene inhibits the incursion of *E. pisi* into pea epidermal cells [32]. Recent studies have shown that the *er1*-resistant phenotype is caused by loss-of-function mutations in the pea MLO (Mildew Resistance Locus O) homolog (*PsMLO1*). The MLO gene family has been identified in both dicots (e.g., *Arabidopsis thaliana* and tomato: *Solanum lycopersicum*) and monocots (e.g., barley: *Hordeum vulgare*) [14,34–39].

To date, nine *er1* alleles resistant to *E. pisi* have been identified in *E. pisi*-resistant pea germplasms: *er1*-1 (also known as *er1mut1*) [14,21,25,40,41], *er1*-2 [14,24,25], *er1*-3 [14], *er1*-4 [14], *er1*-5 [38], *er1*-6 [27], *er1*-7 [26], *er1*-10 (also known as *er1mut2*) [21,40,42], and *er1*-11 [42,43]. Each *er1* allele corresponds to a different *PsMLO1* mutation site and pattern. Among the nine *er1* alleles identified, only *er1*-1 and *er1*-2 are commonly applied in pea breeding programs [14,38]. Several studies have attempted to design functional markers of *er1* alleles to allow for the rapid selection of pea germplasms resistant to *E. pisi* [24,26,27,38,42–44].

The yield and quality of the Chinese pea crop are severely damaged by powdery mildew [2], with the disease affecting up to 100% of pea plants in some regions of China [4]. Several studies have focused on the identification of Chinese pea germplasms resistant to *E. pisi* [41,44–49]. In the Chinese pea cultivars X9002 and Xucai 1, *E. pisi* resistance is conferred by the *er1*-2 allele [24,25,47], while in some Chinese pea landraces from Yunnan Province, *E. pisi* resistance is conferred by the *er1*-6 allele [27,48]. *E. pisi* resistance in the Indian pea cultivar DDR11 is conferred by the *er1*-7 allele [26]. Thus, natural resistance to *E. pisi* conferred by the *er1* gene has been observed in pea germplasms worldwide, providing a rich source of genetic material that can be used to improve the *E. pisi* resistance of Chinese pea cultivars [41,46,48,50]. Allelic diversity of this locus in the cultivated pea has been well characterized; however, relatively few studies have investigated and characterized *E. pisi*-resistant pea germplasms in an international collection. Thus, this study aimed to identify and characterize the *E. pisi*-resistant alleles at the *er1* locus in a worldwide collection of pea germplasms resistant to *E. pisi*. Additionally, any novel *er1* alleles were genetically mapped, and functional markers specific to these novel *er1* alleles were developed to improve marker-assisted selection in pea breeding programs.

2. Results

2.1. Phenotypic Evaluation

Fifty-five *E. pisi*-immune or -resistant pea germplasm accessions from 13 countries were re-evaluated for their resistance to the *E. pisi* isolate EPYN. At 10 days post-inoculation, the *E. pisi* disease severity of all susceptible controls (Bawan 6 and Longwan 1) were rated as score 4. In contrast, the 55 *E. pisi*-resistant germplasm accessions appeared to be either immune (symptom-free; disease severity 0) or resistant (slight infection; disease severity 1–2) to *E. pisi* isolate EPYN. Of the 55 resistant germplasm accessions, 46 were classified as immune and nine as resistant to *E. pisi* (Table 1). To provide comprehensive information for the resistance of a worldwide collection of 86 pea germplasms to *E. pisi*, the phenotypes of 31 resistant pea germplasms carrying known *er1* alleles are also shown in Table 1.

Table 1. Information about phenotype and the resistance gene at the *er1* locus of the 86 *Erysiphe pisi*-resistant and the two *E. pisi*-susceptible controls (two controls are bolded).

No.	Accession No./Germplasm Name	Origin	Phenotype	*er1* Allele	Reference
1	G0004389	Afghanistan	I	er1-8	This study
2	G0004382	Australia	I	er1-1	This study
3	G0004400	Australia	I	er1-9	This study
4	G0004417	Australia	I	er1-2	This study
5	G0004434	Australia	I	er1-2	This study
6	G0004448	Australia	I	er1-2	This study
7	G0004450	Australia	I	er1-2	This study
8	G0002102	Canada	I	er1-6	This study
9	G0006514	Canada	R	er1-2	This study
10	G0006515	Canada	R	er1-2	This study
11	G0006516	Canada	I	er1-2	This study
12	G0006519	Canada	I	er1-2	This study
13	G0003925	Canada	I	er1-1	[41]
14	Cooper	Canada	I	er1-1	[41]
15	G0005576	China, Chongqing	I	er1-2	[27]
16	G0006273	China, Gansu	I	er1-2	[24]
17	20012	China, Gansu	I	er1-1	This study
18	Jia2	China, Gansu	I	er1-2	This study
19	Texuan11	China, Gansu	I	er1-2	This study
20	Hehuan66	China, Gansu	R	er1-1	This study
21	**Longwan 1**	**China, Gansu**	**S**	**Er1**	[51]
22	PI391630	China, Guangdong	I	er1-4	[14]
23	Xucai1	China, Hebei	I	er1-2	[25]
24	G0003694	China, Hebei	R	er1-6	[27]
25	**Bawan 6**	**China, Hebei**	**S**	**Er1**	[24]
26	L0314	China, Yunnan	I	er1-1	[51]
27	L1332	China, Yunnan	I	er1-2	[51]
28	L1335	China, Yunnan	I	er1-2	[51]
29	G0001747	China, Yunnan	R	er1-6	This study
30	G0001752	China, Yunnan	I	er1-6	[27]
31	G0001763	China, Yunnan	I	er1-6	[27]
32	G0001764	China, Yunnan	I	er1-6	[27]
33	G0001767	China, Yunnan	I	er1-6	[27]
34	G0001768	China, Yunnan	I	er1-6	[27]
35	G0001773	China, Yunnan	I	er1-6	This study
36	G0001777	China, Yunnan	I	er1-6	[27]
37	G0001778	China, Yunnan	I	er1-6	[27]
38	G0001780	China, Yunnan	I	er1-6	[27]
39	G0003824	China, Yunnan	R	er1-6	[27]
40	G0003825	China, Yunnan	I	er1-6	[27]
41	G0003826	China, Yunnan	I	er1-6	[27]
42	G0003831	China, Yunnan	R	er1-6	[27]
43	G0003834	China, Yunnan	R	er1-6	[27]
44	G0003836	China, Yunnan	R	er1-6	[27]
45	G0003839	China, Yunnan	R	er1-6	This study
46	G0005117	China, Yunnan	I	er1-6	This study
47	G0003974	China, Yunnan	I	er1-7	This study
48	G0003975	China, Yunnan	I	er1-7	This study
49	Yunwan4	China, Yunnan	R	er1-1	This study
50	Yunwan18	China, Yunnan	R	er1-2	This study
51	Yunwan35	China, Yunnan	I	er1-2	This study
52	Yunwan37	China, Yunnan	I	er1-6	This study
53	L2157	China, Yunnan	I	er1-2	This study
54	G0002848	Denmark	I	er1-2	This study
55	G0002971	England	I	er1-2	This study
56	G0002859	Germany	I	er1-2	This study
57	G0002860	Germany	I	er1-2	This study
58	G0002883	Germany	I	er1-2	This study
59	G0003895	ICRISAT	I	er1-7	[26]
60	G0003897	ICRISAT	I	er1-2	This study
61	G0003899	ICRISAT	I	er1-7	[26]
62	G0003907	ICRISAT	I	er1-2	This study
63	G0003911	ICRISAT	I	er1-2	This study
64	G0003961	India	I	er1-2	This study
65	G0003967	India	I	er1-7	[26]
66	G0003958	India	I	er1-7	[26]
67	G0006285	Japan	R	er1-2	This study
68	G0004332	Mexico	R	er1-1	This study
69	G0004394	Nepal	R	er1-7	[26]
70	G0002980	Unknown country	I	er1-2	This study

Table 1. *Cont.*

No.	Accession No./Germplasm Name	Origin	Phenotype	*er1* Allele	Reference
71	G0003931	Unknown country	I	*er1-7*	[26]
72	G0003935	Unknown country	I	*er1-2*	This study
73	G0003936	Unknown country	I	*er1-7*	[26]
74	G0003942	Unknown country	I	*er1-1*	This study
75	G0003943	Unknown country	I	*er1-1*	This study
76	G0002128	USA	I	*er1-2*	This study
77	G0002129	USA	I	*er1-2*	This study
78	G0002131	USA	I	*er1-2*	This study
79	G0002132	USA	I	*er1-2*	This study
80	G0002134	USA	I	*er1-2*	This study
81	G0002137	USA	I	*er1-2*	This study
82	G0002183	USA	I	*er1-2*	This study
83	G0002235	USA	I	*er1-6*	This study
84	G0002250	USA	I	*er1-2*	This study
85	G0002602	USA	I	*er1-2*	This study
86	G0002608	USA	I	*er1-2*	This study
87	G0002847	USA	I	*er1-2*	This study
88	G0002960	USA	I	*er1-2*	This study

"R", "I", and "S" stand for resistant, immune, and susceptible, respectively.

2.2. PsMLO1 Sequence Analysis

The *PsMLO1* cDNA sequence of Bawan 6 and Longwan 1, the susceptible controls, was consistent with that of the wild-type *PsMLO1* cDNA (Table 1). Among the 55 resistant pea germplasms with previously unknown *er1* alleles, *er1-1* was identified in seven germplasm accessions, *er1-2* in 37, *er1-6* in seven, and *er1-7* in two (Tables 1 and 2).

Table 2. The distribution and numbers of pea germplasm accessions carrying *er1* alleles.

Country	No. of Pea Germplasm Accessions Contained *er1* Alleles									
	er1-1	*er1-2*	*er1-3*	*er1-4*	*er1-5*	*er1-6*	*er1-7*	*er1-8*	*er1-9*	Total
USA	-	12	-	-	-	1	-	-	-	13
Canada	-	4	-	-	-	1	-	-	-	5
Germany	-	3	-	-	-	-	-	-	-	3
ICRISAT	-	3	-	-	-	-	-	-	-	3
India	-	1	-	-	-	-	-	-	-	1
Australia	1	4	-	-	-	-	-	-	1	6
England	-	1	-	-	-	-	-	-	-	1
Denmark	-	1	-	-	-	-	-	-	-	1
Nepal	-	-	-	-	-	-	-	-	-	0
Japan	-	1	-	-	-	-	-	-	-	1
Afghanistan	-	-	-	-	-	-	-	1	-	1
Mexico	1	-	-	-	-	-	-	-	-	1
China	3	5	-	-	-	5	2	-	-	15
Unknown country	2	2	-	-	-	-	-	-	-	4
Total	7	37	-	-	-	7	2	1	1	55

"-" indicates there was no pea germplasm containing this *er1* allele.

Two novel *er1* alleles were discovered in the two remaining germplasms: G0004389 (from Afghanistan) and G0004400 (from Australia). A novel mutation pattern was found in the G0004389 cDNA fragment homologous to *PsMLO1*: a 3-bp deletion (GTG) corresponding to positions 1339–1341 in exon 15 (the final exon) of the *PsMLO1* cDNA sequence. This deletion caused the loss of the amino acid valine at position 447 of the PsMLO1 protein sequence, probably resulting in a functional change (Figure 1A). This mutation differed from all known *er1* alleles, indicating that the *E. pisi* resistance of G0004389 was controlled by a novel allele of *er1*. This novel allele was designated *er1-8*, following the accepted nomenclature [14,26,27,42,44,51]. In pea germplasm G0004400, a 1-bp deletion (T) was identified in a previously unreported position homologous to position 928 in exon 10 of the *PsMLO1* cDNA sequence. This deletion caused a substitution of the amino acid serine with leucine at position

310 of the PsMLO1 protein sequence (Figure 1B). This change caused the early termination of protein translation, probably also resulting in a functional change of PsMLO1 (Figure 1B). Thus, *E. pisi* resistance in G0004400 was also controlled by a novel *er1* allele, herein designated *er1-9*.

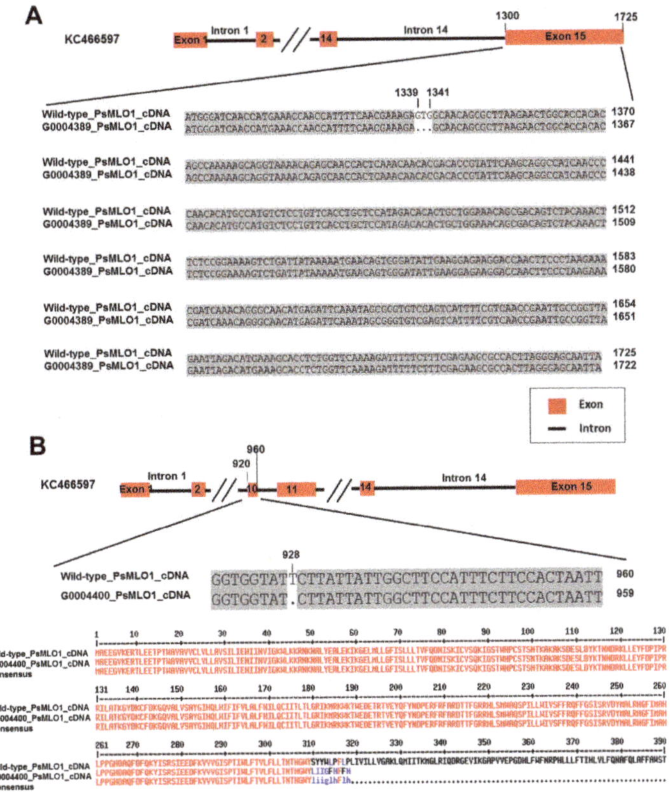

Figure 1. *PsMLO1* cDNA sequences from the powdery mildew-resistant pea germplasms G0004389 and the wild-type pea cultivar Sprinter (GenBank accession number: FJ463618.1), and *PsMLO1* cDNA sequences from G0004400 and amino acid sequence difference caused by mutation. (**A**) There is a 3-bp deletion (GTG) in the *PsMLO1* cDNA of G0004389 at positions 1339–1341 of exon 15. (**B**), there is a single base deletion (T) in the *PsMLO1* cDNA sequence of G0004400 at position 928 in exon 10, the lower figure shows the difference of amino acid sequence from G0004389 and the wild-type pea cultivar Sprinter. The two mutation sites are indicated in the respective cDNA sequences.

Interestingly, frame shift mutations, where small fragments are deleted or inserted, were identified in the cloned sequences of several pea germplasms. The fragments homologous to the wild-type *PsMLO1* cDNA in seven pea germplasms (G0002602, G0006515, G0002883, G0004448, G0002848, G0003935, and G0005117) had 5-bp deletions (GTTAG) at positions 700–704 of wild-type *PsMLO1* cDNA, while three pea germplasm accessions (G0002883, G0002971, and L0368) had another 5-bp deletion (TAGGG) at positions 1235–1239 of the wild-type *PsMLO1* cDNA. In accession G0006514, there was a 4-bp deletion (GGAG) at positions 181–184 of the wild-type *PsMLO1* cDNA. In four pea accessions (G0002847, G0004434, G0003974, and Texuan 11) and two pea accessions (G0002235 and G0002848), there were a 16-bp deletion (CTCATCTTCCTCCAGG) at positions 776–791 and a 16-bp insertion (AATTTTTCTGTTTCAG) at position 1171 of the wild-type *PsMLO1* cDNA, respectively. In germplasm accession Jia 2, there was a 7-bp insertion (TAATAAG) at position 921 of the wild-type

PsMLO1 cDNA. It was probable that these indels resulted from aberrant splicing events during transcription. Each frame shift mutation was observed in only one or two of ten cloned *PsMLO1* cDNA sequences per germplasm accession.

Various alternative splicing patterns, including intron retention and exon skipping, were also observed in multiple PsMLO1 sequences cloned from the 55 resistant pea germplasm accessions. The eight introns retained were 1, 2, 4, 6, 7, 9, 12, and 13, and the three exons skipped were 4, 10, and 11 of the wild-type PsMLO1. Each intron retention and exon skipping event were discovered in only one or two of ten cloned PsMLO1 cDNA sequences.

2.3. Genetic Analysis and Mapping of er1-8 and er1-9

As expected, the two resistant pea parents, G0004389 and G0004400, were immune to *E. pisi* infection (disease severity 0), while the two susceptible parents (Bawan 6 and WSU 28) were heavily infected (disease severity 4) (Figure 2). The segregation patterns of *E. pisi* resistance in the F_1, F_2, and $F_{2:3}$ populations derived from the crosses WSU 28 × G0004389 and Bawan 6 × G0004400 are presented in Table S1.

Figure 2. Phenotypic evaluation of the *Erysiphe pisi*-resistant pea germplasms G0004389 and G0004400, as well as the *E. pisi*-susceptible cultivars WSU 28 and Bawan 6, after inoculation with *E. pisi* isolate EPYN. (**A**) G0004389 and *E. pisi*-susceptible cultivar WSU 28. (**B**) G0004400 and *E. pisi*-susceptible cultivar Bawan 6.

Six F_1 plants produced from the cross WSU 28 × G0004389 were susceptible to *E. pisi* (Table S1). One of the six plants generated 120 F_2 and $F_{2:3}$ offspring through self-pollination. Of these 120 F_2 plants, 30 were resistant (R) to *E. pisi*, and 90 were susceptible (S) to *E. pisi*-. This indicates that the segregation ratio (resistance:susceptibility) in the F_2 population was exactly 1:3 ($x^2 = 0.01$; $P = 0.92$), indicating recessive heredity of a single gene. Moreover, a segregation ratio of 30 (homozygous resistant): 63 (segregating): 27 (homozygous susceptible) in the $F_{2:3}$ population fitted well with the genetic model of 1:2:1 ratio ($x^2 = 0.48$, $P = 0.79$) (Table S1), confirming that the *E. pisi* resistance in G0004389 was controlled by a single recessive gene.

The cross of Bawan 6 × G0004400 generated five F_1 plants which showed *E. pisi*-susceptibility (Table S1). One of five F_1 plants generated 119 F_2 offspring. 32 of 119 were resistant, and 87 of 119 were susceptible to *E. pisi*. The segregation ratio in the F_2 population of resistance to susceptibility fitted a genetic model ratio of 1:3 ($x^2 = 0.14$; $P = 0.71$), also indicating recessive heredity of a single gene. Moreover, a segregation ratio of 32 (homozygous resistant): 64 (segregating): 23 (homozygous susceptible) in the $F_{2:3}$ population (119 families) fitted well with the genetic model of 1:2:1 ratio (χ^2 = 2.51; P = 0.29), indicating that *E. pisi* resistance in G0004400 was also controlled by a single recessive gene (Table S1).

Of the 20 markers tested, five (c5DNAmet, AD160, AA200, AA224, and PSMPSA5) were polymorphic between parents WSU 28 and G0004389, and seven (AC74, AD160, PSMPSAD51, ScOPD10-650, ScOPX04-880, ScOPE16-1600, and AD59) were polymorphic between Bawan 6 and G0004400, indicating that these markers were likely linked to the *E. pisi* resistance gene. Thus, the five and the seven parental polymorphic markers were used to confirm the genotypes of each F_2 plant derived from WSU 28 × G0004389 and Bawan 6 × G0004400, respectively. This genetic linkage analysis suggested that three markers (c5DNAmet, AA200, and AA224) and six markers (AD160, PSMPSAD51, ScOPD10-650, ScOPX04-880, ScOPE16-1600, and AD59) were linked to the resistance gene *er1* in G0004389 and G0004400, respectively (Figure 3). Our results also indicated that the resistance genes in both germplasm accessions were located in the *er1* region. In G0004389, the linkage map indicated that the markers (c5DNAmet and AA200) were mapped on both sides of the target gene with 9.6 cM and 3.5 cM genetic distances, respectively (Figure 3A). In G0004400, two other markers (PSMPSAD51 and ScOPX04-880) were located on both sides of the target gene with 12.2 cM and 4.2 cM genetic distances, respectively (Figure 3B). Our linkage and genetic map analyses confirmed that *er1-8* and *er1-9* controlled *E. pisi* resistance in G0004389 and G0004400, respectively (Figure 3).

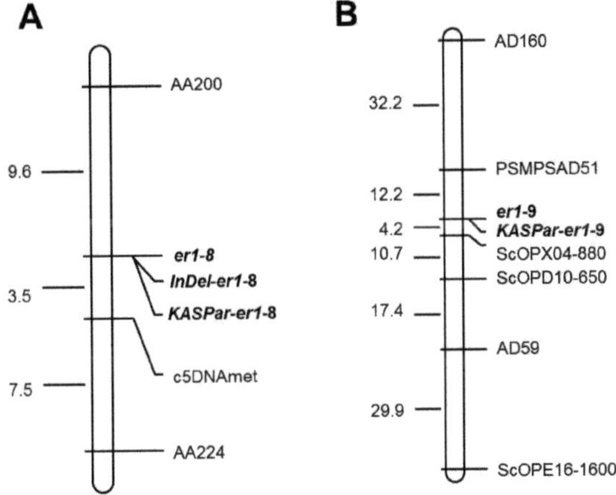

Figure 3. Genetic linkage maps constructed using the *er1*-linked markers and the functional markers for *er1-8* and *er1-9*, based on the F_2 populations derived from (**A**) WSU 28 × G0004389 and (**B**) Bawan 6 × G0004400. Map distances and loci order were determined with MAPMAKER v3.0 (Lander et al. 1993). Estimated genetic distances between loci are shown to the left of the maps in centiMorgans (cM).

2.4. Development of Functional Markers for er1-8 and er1-9

The indel marker, InDel-*er1-8* flanking the 3-bp deletion in *er1-8*, amplified 231-bp and 228-bp fragments in the parents WSU 28 and G0004389, respectively. The amplicons were clearly polymorphic between the contrasting parents, as visualized on an 8% polyacrylamide gel (Figure S1A). InDel-*er1-8* was then used to identify the genotypes of the 120 F_2 plants derived from WSU 28 × G0004389. Three distinct electrophoretic bands corresponding to the homozygous resistant (R), homozygous susceptible (S), and heterozygous (H) genotypes were observed (Figure S1A). Each F_2 genotype corresponded to a phenotype of the 120 $F_{2:3}$ families. A chi-squared (χ^2) test showed that the segregation ratio of InDel-*er1-8* in the $F_{2:3}$ population derived from WSU 28 × G0004389 fit a 1:2:1 (χ^2 = 0.48; P = 0.79). All results suggested that the marker InDel-*er1-8* co-segregated with gene *er1-8*, indicating a co-dominant marker.

In the Kompetitive allele-specific PCR (KASPar) assay, KASPar-*er1-8* and KASPar-*er1-9* successfully distinguished the contrasting parents (WSU 28 and G0004389, Bawan 6 and G0004400) into two different clusters corresponding to the FAM-labeled and HEX-labeled groups, respectively (Figure S2). When KASPar-*er1-8* and KASPar-*er1-9* were used to analyze the 120 and 119 F_2 progeny derived from WSU 28 × G0004389 and Bawan 6 × G0004400, the KASPar markers clearly separated the F_2 progeny into three clusters corresponding to three genotypes: homozygous resistant, homozygous susceptible, and heterozygous (Figure S2). In the F_2 population derived from WSU 28 × G0004389, 30 plants were identified as homozygous resistant, 63 were heterozygous, and 27 were homozygous susceptible. In the F_2 population derived from Bawan 6 × G0004400, 32 plants were homozygous resistant, 64 were heterozygous, and 23 were homozygous susceptible. These results were completely consistent with the phenotypes of both $F_{2:3}$ populations, suggesting that KASPar-*er1-8* and KASPar-*er1-9* co-segregated with *er1-8* and *er1-9*, respectively. A chi-squared (χ^2) test showed that both segregation ratios of KASPar-*er1-8* and KASPar-*er1-9* in respective F_2 populations fit 1:2:1 (KASPar-*er1-8*: $\chi^2 = 0.48$, $P = 0.79$; KASPar-*er1-9*: $\chi^2 = 2.51$; $P = 0.29$), indicating co-dominant markers.

2.5. Validation and Application of Functional Markers

Of the 169 germplasm accessions selected and tested for their phenotypic resistance to *E. pisi* isolate EPYN (Table S2), 19 were phenotypically immune to *E. pisi*, 22 were resistant, and 128 were susceptible (Table S2).

Among the 169 germplasms genotyped with InDel-*er1-8*, the 228-bp fragment corresponding to *er1-8* was only amplified in G0004839 (Figure S1B). In all of the other tested germplasm accessions, a 231-bp fragment was consistently amplified by InDel-*er1-8*, indicating that no accessions besides G0004839 carried *er1-8* (Figure S1B; Table S1).

When the 169 germplasm accessions were genotyped with KASPar-*er1-8*, two distinct clusters were recovered, with one gene (*er1-8*) corresponding to G0004389 and the other (non-*er1-8*) to the other germplasms, respectively. Similarly, when the germplasms were genotyped with KASPar-*er1-9*, two distinct clusters were recovered, corresponding to G0004400 and all of the other germplasms, respectively (Figure S2; Table S1). Thus, markers KASPar-*er1-8* and KASPar-*er1-9* effectively identified pea germplasms carrying the *er1-8* and *er1-9* alleles, respectively. Our results also showed that none of the other 169 pea germplasm accessions carried the *er1-8* or *er1-9* alleles.

3. Discussion

Powdery mildew induced by *E. pisi* DC. is a major disease on pea and causes considerable yield losses worldwide. The resistance gene *er1* is the most widely deployed gene controlling powdery mildew in pea cultivars worldwide. Furthermore, *er1* allelic diversity has been widely reported in pea [14,21,25–27,38,40–44,51].

To date, more than 40 *MLO* mutant alleles have been described in the monocotyledonous plant barley [52]. It is predicted that additional *er1* alleles resulting from natural mutations would be present among pea germplasms from around the world. As expected, we not only encountered the four known *er1* alleles (*er1-1*, *er1-2*, *er1-6*, and *er1-7*) across the 53 *E. pisi*-resistant pea germplasms, but we also discovered two novel *er1* alleles: *er1-8* in germplasm G0004389 from Afghanistan and *er1-9* in germplasm G0004400 from Australia (Table 1).

Among the nine known *er1* alleles, *er1-1* and *er1-2* are most commonly used in pea breeding programs because they confer stable resistance to *E. pisi* [14,25,38,51]. Our results indicated that these two alleles were common in the tested pea germplasm accessions resistance to *E. pisi*. The *er1-1* allele was found in seven accessions (12.73%), and *er1-2* was found in 37 accessions (67.27%) (Table 2). Among the 86 *E. pisi*-resistant pea accessions, *er1-1* and *er1-2* were identified in 10 (11.62%) and 42 (48.84%) accessions, respectively (Table 1). Previously, *er1-1* has been identified in four *E. pisi*-resistant pea cultivars (JI1559, Tara, and Cooper from Canada; and Yunwan 8 from China), while *er1-2* has been identified in seven *E. pisi*-resistant pea cultivars (Stratagem, Franklin, Dorian, Nadir, X9002, Xucai 1, and G0005576) [14,24,25,27,38]. Here, more *E. pisi*-resistant germplasm accessions carrying the *er1-1* and *er1-2* alleles were identified.

At the genomic level, seven alleles (*er1-1/er1mut1*, *er1-3*, *er1-4*, *er1-5*, *er1-6*, *er1-9*, and *er1-10/er1mut2*) are the result of point mutations in the exons of wild-type *PsMLO1*. Four alleles result from single base substitutions in wild-type *PsMLO1* cDNA: in *er1-1*, a C→G at position 680 (exon 6); in *er1-5*, a G→A at position 570 (exon 5); in *er1-6*, a T→C at position 1121 (exon 11); and in *er1-10*, a G→A at position 939 (exon 10) (Figure S3) [14,27,38,40]. Three alleles result from single base deletions in wild-type *PsMLO1* cDNA, including ΔG at position 862 (exon 8) in *er1-3*; ΔA at position 91 (exon 1) in *er1-4*; and ΔT at position 928 (exon 10) in *er1-9* identified in this study [14] Two alleles result from small fragment deletions in wild-type *PsMLO1* cDNA, including a 10-bp deletion of positions 111–120 (exon 1) in *er1-7* [26]; and a 3-bp deletion of positions 1339–1341 (exon 15) in *er1-8*. To date, only the *er1-11* mutation is known to have resulted from an intron mutation in *PsMLO1* (a 2-bp insertion in intron 14) [42,43], and only *er1-2* results from a large indel of unknown size in wild-type *PsMLO1* cDNA [14,24,27].

Previous studies have indicated that the *er1-2* allele produces three distinct *PsMLO1* transcripts [14,25,27,51]. Interestingly, this study observed that the *er1-2* carried by the pea germplasm accession G0002860 produced four distinct *PsMLO1* transcripts. One of these transcripts was characterized by a 129-bp deletion, corresponding to the deletion of exon 13 (68 bp) and exon 14 (61 bp) from wild-type *PsMLO1* cDNA, indicating alternative splicing of exon skipping. Previously, two transcripts of *er1-2* were observed to have large insertions (155-bp and 220-bp) based on comparisons with the transcripts of wild-type *PsMLO1* cDNA [14,24,25,27,51]. Here, we discovered that the 155-bp "insertion" in *er1-2* resulted from a 192-bp insertion at position 1263 and a 37-bp deletion of positions 1263–1299 in exon 14 of wild-type *PsMLO1*, while the 220-bp "insertion" resulted from a 257-bp insertion at position 1263 and a 37-bp deletion of positions 1263–1299 in exon 14 of wild-type *PsMLO1*. Another alternative transcript of *er1-2*, an 87-bp "insertion", was observed and resulted from a 192-bp insertion and a 37-bp deletion in exon 14 and a 68-bp deletion corresponding to exon 13 of wild-type *PsMLO1*. Our blast analysis indicated that the 192- and 257-bp insertions had 95% sequence identity with a five-part repetition in the pea genomic BAC sequence (GenBank accession number CU655882). These insertions were also highly similar (~85–87% identity) to a portion of the giant *Ogre* retrotransposons in the pea genome (GenBank accession numbers AY299395, AY299398, AY299397, and AY299394).

Based on 10 cloned sequences, several pea germplasms had frame shift mutations with small fragment indels (4-bp, 5-bp, or 16-bp) in one or two cloned *PsMLO1* cDNA sequences. Previously, a 5-bp (GTTAG) insertion was identified in G0001763 and G0003831; 11-bp (GTAGGAATAAG) and 13-bp (GTAATCTTATTAG) deletions were identified in G0003831 and G0001778; and a 16-bp (CTCATCTTCCTCCAGG) deletion was detected in G0001778 [27]. These small fragment indels in the *PsMLO1* cDNAs were assumed to have resulted from aberrant splicing events during transcription [27].

Alternative splicing in eukaryotes is a pervasive molecular mechanism that significantly increases transcriptome and proteome complexity [53]. Four main types of alternative splicing are known: exon skipping, alternative 5′ splice sites, alternative 3′ splice sites, and intron retention [54]. Exon skipping is common in humans, while intron retention is common in plants [55]. Alternative splicing is involved in many physiological processes, including response to biotic and abiotic stressors [56]. In the pea germplasms, three types of alternative splicing, intron retention, exon skipping, and alternative 5′ splice site selection, were observed in this study. Interestingly, pea germplasms carrying identical *er1* alleles varied in their resistance to *E. pisi*, from immune (disease severity of 0) to merely resistant (disease severity of 1–2) (Table 1). Alternative splicing in response to biotic stress may affect the expression of regulatory genes. Thus, it is speculated that the alternative splicing of *er1* alleles might affect the expression of the *E. pisi* resistance genes *er1*. In addition, the different levels of resistance to *E. pisi* might result from other related gene regulation. It is possible that multiple molecular processes and pathways contribute to *MLo*-based *E. pisi* resistance in peas.

Several functional markers specific to the previously recognized *er1* alleles have already been developed to facilitate marker-assisted breeding of pea cultivars resistant to *E. pisi* [14,24,26,27,40,42–44]. Pavan et al. [38] developed a functional cleaved amplified polymorphic sequence (CAPS) marker for *er1-5*, while Pavan et al. [44] developed functional markers for the five *er1* alleles, *er1-1* through *er1-5*. Santo et al. [40] developed functional markers for *er1mut1* and *er1mut2*, and Wang et al. [24] developed a dominant marker for *er1-2*. Sudheesh et al. [43] developed a functional marker for *er1-11*, while Sun et al. [26,27] developed co-dominant functional markers for *er1-6* and *er1-7*. More recently, Ma et al. [42] developed eight KASPar markers for eight known *er1* alleles, excluding *er1-2*.

This study discovered two novel *er1* alleles resulting from novel mutations of wild-type *PsMLO1* cDNA: *er1-8* was generated by a 3-bp deletion in exon 15, and *er1-9* was generated by a 1-bp deletion in exon 10. The co-dominant functional markers specific to *er1-8* (InDel-*er1-8* and KASPar-*er1-8*) and to *er1-9* (KASPar-*er1-9*) were developed. These markers were validated in genetic populations and in pea germplasms. Our results are vital for future studies of powdery mildew resistance and for the development of *E. pisi*-resistant pea cultivars. The novel *er1* alleles and the corresponding co-dominant functional markers developed herein could constitute efficient and powerful tools for the breeding of *E. pisi*-resistant peas.

4. Materials and Methods

4.1. Plant Material and E. pisi Isolate

Previously, 86 pea germplasms had been found to be *E. pisi*-resistant in screenings of over 1000 pea accessions in a worldwide collection [27,48,50]. And, 31 of 86 resistant pea germplasms had been previously identified the *E. pisi*-resistant *er1* allele [24–27,48,51]. In this study, the remaining 55 of the 86 *E. pisi*-resistant pea germplasms from the United States of America, Canada, Germany, India, Australia, Columbia, England, Denmark, Nepal, Japan, Afghanistan, and Mexico, as well as data from the International Crop Research Institute for Semi-arid Tropics (ICRISAT) and conserved in the China National Genebank (http://www.cgris.net/), were used as research materials to reveal their *E. pisi*-resistant genes at *er1* locus (Table 1). The Chinese pea cultivars Bawan 6 and Longwan 1, which carry the *E. pisi*-susceptible gene *Er1*, were used as susceptible controls [24,51]. The Chinese pea cultivars Xucai 1, carrying *er1-2*, and YI (JI1591), carrying *er1-4*, were used as *E. pisi*-resistant controls [14,25].

The *E. pisi* isolate EPYN from Yunnan Province of China was used as the inoculum [26,27,41,48,50,51]. The EPYN isolate was maintained through continuous re-inoculation of seedlings of the pea cultivar Longwan 1 under controlled conditions. The inoculated plants were incubated in a growth chamber to prevent contamination with other isolates [25].

4.2. Phenotypic Evaluation

Twenty seeds were planted from each of the 55 *E. pisi*-resistant pea germplasm accessions, from the susceptible controls Bawan 6 and Longwan 1, and from the resistant controls Xucai 1 and YI [27]. The seedlings were thinned to 15 per pot before the phenotypic evaluation. Three replications were planted. Seeded pots were placed in a greenhouse maintained at 18 to 26 °C. At the same time, the *E. pisi* inoculum was prepared by inoculating the 10-day-old seedlings of the susceptible pea cultivar Longwan 1, which were incubated in a growth chamber at 20 ± 1 °C with a 12-h photoperiod. Two weeks later, the 14-day-old seedlings of 55 germplasm accessions and controls were inoculated by gently shaking off conidia of the Longwan 1 plants. Inoculated plants were incubated in a growth chamber at 20 ± 1 °C with a 12-h photoperiod. Ten days later, disease severity was rated based on a scale (0–4 scale) [27]. Plants with a score of 0 were considered *E. pisi*-immune, while those with scores of 1 and 2, 3 and 4 were considered as *E. pisi*-resistant and *E. pisi*-susceptible, respectively. For those identified as immune or resistant to *E. pisi*, repeated identification was performed.

4.3. RNA Extraction and PsMLO1 Sequence Analysis

The extraction of total RNA and synthesis of cDNA from the 55 pea germplasms and controls were completed according to our previous studies [25–27].

To identify the resistance alleles at the *er1* loci, the full-length cDNAs of the *PsMLO1* homologs were amplified using the primers specific for *PsMLO1* [14]. The PCR cycling conditions were as follows: 95 °C for 5 min; then 35 cycles of denaturation at 94 °C for 30 s, annealing at 58 °C for 45 s, and extension at 72 °C for 1 min; and a final extension at 72 °C for 10 min. The purified amplicons were cloned with a pEasy-T5 vector (TransGen Biotech, Beijing, China). The sequencing reactions of 10 clones per germplasm (including controls) were performed by the Shanghai Shenggong Biological Engineering Co., Ltd. (Shanghai, China). The resulting sequences were aligned with wild-type *PsMLO1* of pea (NCBI accession number: FJ463618.1) using DNAMAN v6.0 (Lynnon Biosoft, Quebec, Canada).

4.4. Genetic Analysis of Pea Germplasms Carrying Novel Alleles

To confirm the resistance genes, *er1-8* and *er1-9*, G0004389 and G0004400 were crossed with the *E. pisi*-susceptible cultivars WSU 28 and Bawan 6, respectively, to generate genetic populations. The derived F_1, F_2, and $F_{2:3}$ populations from both crosses (WSU 28 × G0004389 and Bawan 6 × G0004400) were used to evaluate the *E. pisi* resistance and genetic analysis of G0004389 and G0004400. The four parents and the derived F_1 and F_2 populations were planted in a propagation greenhouse to generate F_2 and $F_{2:3}$ family seeds, respectively.

Plants of the F2 populations at the fourth or fifth leaf stage were inoculated with the *E. pisi* isolate EPYN using the detached leaf method [25–27,57]. After inoculation, the treated leaves were placed in a growth chamber at 20 °C with a 14-h photoperiod. The four parents (WSU 28, G0004389, Bawan 6, and G0004400) were also inoculated as controls. Ten days after inoculation, disease severity was rated based on a scale of 0–4 as described above. Plants with scores of 0–2 and 3–4 were classified as resistant and susceptible, respectively [25–27,31,58]. Those plants identified as *E. pisi*-resistant were tested again to confirm their resistance.

Twenty-five seeds were selected randomly from each of the 120 $F_{2:3}$ families derived from WSU 28 × G0004389, and from each of the 119 $F_{2:3}$ families derived from Bawan 6 × G0004400. These seeds were planted and cultivated together with their parents, following previously published protocols [25–27]. Disease severity was scored 10 days after inoculation using the 0–4 scale, as described above for the phenotypic identification of the pea germplasms. The $F_{2:3}$ families with scores of 0–2 and 3–4 were classified as homozygous resistant and homozygous susceptible, respectively. Families with scores of 0–2 and 3–4 were considered segregated to *E. pisi* resistance [27,31,58]. The families identified as homozygous resistant or resistance segregated were subjected to repeated testing.

A chi-squared (χ^2) analysis was used to evaluate the goodness-of-fit to Mendelian segregation ratio of the F_2 and $F_{2:3}$ phenotypes derived from WSU 28 × G0004389 and Bawan 6 × G0004400.

4.5. Genetic Mapping of the Resistance Alleles er1-8 and er1-9

The Genomic DNA was isolated from the leaves of the F_2 populations and of their parents using the cetyltrimethylammonium bromide (CTAB) extraction method [59]. The DNA solution was diluted and stored at −20 °C until use.

To map the novel *er1* alleles *er1-8* and *er1-9*, the 10 known *er1*-linked markers on the pea LG VI, including four sequence-characterized amplified region (SCAR) markers [ScOPD10-650 [17], ScOPE16-1600 [18], ScOPO18-1200 [18], and ScOPX04-880 [23]; five simple sequence repeat (SSR) markers (PSMPSAD51, PSMPSA5, PSMPSAD60, i.e., AD60, PSMPSAA374e, and PSMPSAA369); a gene marker [Cytosine-5, DNA-methyltransferase (c5DNAmet)] [20,24–27,48,60]; and 10 additional molecular markers on the pea LG VI (AD160, AC74, AC10_1, AA224, AA200, AD159, AD59, AB71, AA335, and AB86), were used to screen for polymorphisms between the crossed parents (i.e., WSU 28 and G0004389; Bawan 6 and G0004400) [61]. The parental polymorphic markers were then used for genetic linkage analysis based on the genotype of each F_2 plant. PCR amplification of each marker was conducted in a total volume of 20 µL according to the previous descriptions [25–27]. PCR reactions were performed in a thermal cycler (Biometra, Göttingen, Germany) [25–27]. The PCR products were separated on 6% polyacrylamide gels.

The segregation data of the polymorphic markers in the F_2 populations were evaluated for goodness-of-fit to Mendelian segregation patterns with a chi-squared (χ^2) test. Genetic linkage analyses were completed using MAPMAKER/EXP version 3.0b. A logarithm of odds (LOD) score > 3.0 and a distance < 50 cM were used as the thresholds to determine the linkage groups [62]. Genetic distances were determined using the Kosambi mapping function [63]. The genetic linkage map was constructed using the Microsoft Excel macro MapDraw [64].

4.6. Development of Functional Markers for er1-8 and er1-9

Primers flanking the mutation site (GTG/—) were designed based on the *PsMLO1* gene sequence (GenBank accession number KC466597), using Primer Premier v5.0, to develop an insertion/deletion (indel) functional marker specific to allele *er1-8*, InDel-*er1-8* (Table 3). The marker InDel-*er1-8* was used to determine the genotypes of the 120 F_2 offspring derived from WSU 28 × G0004389. PCR amplification was performed as described above on a thermal cycler with the following cycling program: 95 °C for 5 min; 35 cycles of 94 °C for 30 s, 55 °C for 30 s, and 72 °C for 30 s; and 72 °C for 7 min. PCR products were separated on 8% polyacrylamide gels.

Table 3. Sequence information for the indel and Kompetitive allele-specific PCR (KASPar) markers specific to *er1-8*, and for the KASPar marker specific to *er1-9*.

Markers	Primers	Sequence Information (5′-3′)	Annealing Tm
InDel-*er1-8*	Forward	GTTTTGACTGATATGACAGATGGGA	55 °C
	Reverse	GTTTGTAGACTGTCGCTGTTTCC	
KASPar-*er1-8*	Forward-TGG	TGGCAACAGCGCTTAAGAACTGG	65–57 °C touchdown
	Forward	GAGCAACAGCGCTTAAGAACTGG	
	Common reverse	TGGTTGGTTTCATGGTTGATCCCATC	
KASPar-*er1-9*	Forward-T	TTTTGTTATATGGGCAGGGTGGTATT	65–57 °C touchdown
	Forward	TGTTATATGGGCAGGGTGGTATC	
	Common reverse	CAAAATGTAGATTATGCTTACAATTAGTGGA	

Based on allele *er1-8* indels (a 3-bp deletion) and *er1-9* SNPs (1-bp deletion) in *PsMLO1*, the forward primers and the common reverse primers specific to *er1-8* (KASPar-*er1-8*) and *er1-9* (KASPar-*er1-9*) were designed for Kompetitive allele-specific PCR (KASPar) markers by LGC KBioscience (KBioscience, Hoddesdon, UK), respectively. In brief, two KASPar markers (KASPar-*er1-8* and KASPar-*er1-9*) were used to detect parental polymorphisms (WSU 28 × G0004389, and Bawan 6 × G0004400), and then used to analyze the genotypes of the F_2 offspring (WSU 28 × G0004389: 120 F_2 individuals; Bawan 6 × G0004400: 119 F_2 individuals).

KASPar markers were amplified with a Douglas Scientific Array Tape Platform (China Golden Marker, Beijing, Biotech Co., Ltd.) in a 0.8 µL Array Tape reaction volume with 10 ng dry DNA, 0.8 µL 2 × KASP master mix, and 0.011 µL primer mix (KBioscience, Hoddesdon, UK). A Nexar Liquid handling instrument was used to add the PCR solution to the Array Tape (Douglas Scientific). PCRs were performed on a Soellex PCR Thermal Cycler with the following conditions: initial denaturation at 94 °C for 15 min; followed by 10 cycles of denaturation at 94 °C for 20 s, and 65 °C for 60 s at an annealing temperature that decreased by 0.8 °C per cycle; and then 26 cycles of denaturation at 94 °C for 20 s and 57 °C for 60 s; and a final cooling to 4 °C. A fluorescent end-point reading was completed with the Araya fluorescence detection system (part of the Douglas Scientific Array Tape Platform). Genotypes and clusters were visualized with Kraken (http://ccb.jhu.edu/software/kraken/MANUAL.html).

4.7. Validation and Application of Functional Markers

To test the efficacy of the novel functional markers specific to *er1-8* (InDel-*er1-8* and KASPar-*er1-8*) and *er1-9* (KASPar-*er1-9*), 169 pea germplasm accessions were tested for (a) their phenotypic resistance to *E. pisi* isolate EPYN and (b) whether they carried the *er1* alleles *er1-8* or *er1-9* (Table S2). The four parent cultivars (WSU 28, G0004389, Bawan 6, and G0004400) were used as contrasting controls, and seven cultivars, including Tara (*er1-1*) [41], Xucai 1 (*er1-2*) [25], JI210 (*er1-3*) [14], YI (*er1-4*) [14], G0001778 (*er1-6*) [27], DDR11 (*er1-7*) [26], and GI2480 (*er2*) [28], were used as positive controls (Table S2).

DNA was extracted from the 169 selected pea germplasm accessions and the 11 controls (four parents and seven resistant cultivars with known *er1* alleles) using the CTAB method (Shure et al. 1983). PCR amplifications of the indel and KASPar markers were performed as described above (in the section "Development of functional *er1-8* and *er1-9* markers").

Supplementary Materials: Supplementary materials can be found at http://www.mdpi.com/1422-0067/20/20/5071/s1.

Author Contributions: Z.Z. conceived and designed the experiments. S.S., D.D., C.D., X.Z., D.X., and Y.H. performed the experiments. S.S. analyzed the data and wrote the manuscript. Z.Z. revised the manuscript. All authors read and approved the manuscript.

Funding: This study was supported by the Modern Agro-industry Technology Research System (CARS-09) and the Crop Germplasm Conservation and Utilization Program from the Ministry of Agriculture of China (2019NWB030-12), the National Infrastructure for Crop Germplasm Resources (NICGR2019-008), and the Scientific Innovation Program of the Chinese Academy of Agricultural Sciences from the Institute of Crop Sciences, Chinese Academy of Agricultural Sciences.

Acknowledgments: We sincerely thank Clarice J. Coyne (Plant Germplasm Introduction and Testing, USDA ARS, Pullman, WA 99164, USA) and Rebecca J. McGee (Grain Legume Genetics and Physiology Research Unit, USDA ARS, Pullman, WA 99164, USA) for providing the pea germplasm WSU 28.

Conflicts of Interest: The authors declare no conflict of interest.

Abbreviations

SSR	Simple sequence repeat
SNP	Single nucleotide polymorphism
InDel	Insertion/deletion
KASPar	Kompetitive allele-specific PCR

References

1. Ali, S.M.; Sharma, B.; Ambrose, M.J. Current status and future strategy in breeding pea to improve resistance to biotic and abiotic stresses. *Euphytica* **1993**, *1*, 115–126.
2. Wang, X.; Zhu, Z.; Duan, C.; Zong, X. *Identification and Control Technology of Disease and Pest on Faba Bean and Pea*; Chinese Agricultural Science and Technology Press: Beijing, China, 2007.
3. Gritton, E.T.; Ebert, R.D. Interaction of planting date and powdery mildew on pea plant performance. *Am. Soc. Horti. Sci.* **1975**, *100*, 137–142.
4. Peng, H.X.; Yao, G.; Jia, R.L.; Liang, H.Y. Identification of pea germplasm resistance to powdery mildew. *J. Southwest Agric. Univ.* **1991**, *13*, 384–386. (In Chinese)
5. Smith, P.H.; Foster, E.M.; Boyd, L.A.; Brown, J.K.M. The early development of *Erysiphe pisi* on *Pisum sativum* L. *Plant. Pathol.* **1996**, *45*, 302–309. [CrossRef]
6. Ghafoor, A.; McPhee, K. Marker assisted selection (MAS) for developing powdery mildew resistant pea cultivars. *Euphytica* **2012**, *186*, 593–607. [CrossRef]
7. Fondevilla, S.; Rubiales, D. Powdery mildew control in pea, a review. *Agron. Sustain. Dev.* **2012**, *32*, 401–409. [CrossRef]
8. Ondřej, M.; Dostálová, R.; Odstrčilová, L. Response of *Pisum sativum* germplasm resistant to *Erysiphe pisi* to inoculation with *Erysiphe baeumleri*, a new pathogen of peas. *Plant. Prot. Sci.* **2005**, *41*, 95–103. [CrossRef]
9. Attanayakea, R.N.; Glaweab, D.A.; McPheec, K.E.; Dugand, F.M.; Chend, W. *Erysiphe trifolii*—A newly recognized powdery mildew pathogen of pea. *Plant. Pathol.* **2010**, *59*, 712–720. [CrossRef]

10. Fondevilla, S.; Chattopadhyay, C.; Khare, N.; Rubiales, D. *Erysiphe trifolii* is able to overcome *er1* and *Er3*, but not *er2*, resistance genes in pea. *Eur. J. Plant. Pathol.* **2013**, *136*, 557–563. [CrossRef]

11. Harland, S.C. Inheritance of immunity to mildew in Peruvian forms of *Pisum sativum*. *Heredity* **1948**, *2*, 263–269. [CrossRef]

12. Heringa, R.J.; Van Norel, A.; Tazelaar, M.F. Resistance to powdery mildew (*Erysiphe polygoni* D.C.) in peas (*Pisum sativum* L.). *Euphytica* **1969**, *18*, 163–169. [CrossRef]

13. Fondevilla, S.; Torres, A.M.; Moreno, M.T.; Rubiales, D. Identification of a new gene for resistance to powdery mildew in *Pisum fulvum*, a wild relative of pea. *Breed. Sci.* **2007**, *57*, 181–184. [CrossRef]

14. Humphry, M.; Reinstädler, A.; Ivanov, S.; Bisseling, T.; Panstruga, R. Durable broad-spectrum powdery mildew resistance in pea *er1* plants is conferred by natural loss-of-function mutations in *PsMLO1*. *Mol. Plant. Pathol.* **2011**, *12*, 866–878. [CrossRef] [PubMed]

15. Sarala, K. Linkage Studies in Pea (*Pisum sativum* L.) with Reference to Er Gene for Powdery Mildew Resistance and Other Genes. Ph.D. Thesis, Indian Agricultural Research Institute, New Delhi, India, 1993.

16. Dirlewanger, E.; Isaac, P.G.; Ranade, S.; Belajouza, M.; Cousin, R.; Vienne, D. Restriction fragment length polymorphism analysis of loci associated with disease resistance genes and developmental traits in *Pisum sativum* L. *Theor. Appl. Genet.* **1994**, *88*, 17–27. [CrossRef]

17. Timmerman, G.M.; Frew, T.J.; Weeden, N.F. Linkage analysis of *er1*, a recessive *Pisum sativum* gene for resistance to powdery mildew fungus (*Erysiphe pisi* DC). *Theor. Appl. Genet.* **1994**, *88*, 1050–1055. [CrossRef]

18. Tiwari, K.R.; Penner, G.A.; Warkentin, T.D. Identification of coupling and repulsion phase RAPD markers for powdery mildew resistance gene *er1* in pea. *Genome* **1998**, *41*, 440–444. [CrossRef]

19. Janila, P.; Sharma, B. RAPD and SCAR markers for powdery mildew resistance gene *er* in pea. *Plant. Breed.* **2004**, *123*, 271–274. [CrossRef]

20. Ek, M.; Eklund, M.; von Post, R.; Dayteg, C.; Henriksson, T.; Weibull, P.; Ceplitis, A.; Isaac, P.; Tuvesson, S. Microsatellite markers for powdery mildew resistance in pea (*Pisum sativum* L.). *Hereditas* **2005**, *142*, 86–91. [CrossRef]

21. Pereira, G.; Marques, C.; Ribeiro, R.; Formiga, S.; Dâmaso, M.; Sousa, T.; Farinhó, M.; Leitão, J.M. Identification of DNA markers linked to an induced mutated gene conferring resistance to powdery mildew in pea (*Pisum sativum* L.). *Euphytica* **2010**, *171*, 327–335. [CrossRef]

22. Tonguc, M.; Weeden, N.F. Identification and mapping of molecular markers linked to *er1* gene in pea. *J. Plant. Mol. Biol. Biotech.* **2010**, *1*, 1–5.

23. Srivastava, R.K.; Mishra, S.K.; Singh, K.; Mohapatra, T. Development of a coupling-phase SCAR marker linked to the powdery mildew resistance gene *er1* in pea (*Pisum sativum* L.). *Euphytica* **2012**, *86*, 855–866. [CrossRef]

24. Wang, Z.; Fu, H.; Sun, S.; Duan, C.; Wu, X.; Yang, X.; Zhu, Z. Identification of powdery mildew resistance gene in pea line X9002. *Acta Agron. Sin.* **2015**, *41*, 515–523, (In Chinese with English abstract). [CrossRef]

25. Sun, S.; Wang, Z.; Fu, H.; Duan, C.; Wang, X.; Zhu, Z. Resistance to powdery mildew in the pea cultivar Xucai 1 is conferred by the gene *er1*. *Crop. J.* **2015**, *3*, 489–499. [CrossRef]

26. Sun, S.; Deng, D.; Wang, Z.; Duan, C.; Wu, X.; Wang, X.; Zong, X.; Zhu, Z. A novel *er1* allele and the development and validation of its functional marker for breeding pea (*Pisum sativum* L.) resistance to powdery mildew. *Appl. Genet.* **2016**, *129*, 909–919.

27. Sun, S.; Fu, H.; Wang, Z.; Duan, C.; Zong, X.; Zhu, Z. Discovery of a novel *er1* allele conferring powdery mildew resistance in Chinese pea (*Pisum sativum* L.) landraces. *PLoS ONE* **2016**, *11*, e0147624. [CrossRef]

28. Katoch, V.; Sharma, S.; Pathania, S.; Banayal, D.K.; Sharma, S.K.; Rathour, R. Molecular mapping of pea powdery mildew resistance gene *er2* to pea linkage group III. *Mol. Breed.* **2010**, *25*, 229–237. [CrossRef]

29. Cobos, M.J.; Rubiales, D.; Fondevilla, S. Er3 gene conferring resistance to *Erysiphe pisi* is located in pea LGIV. In Proceedings of the Second International Legume Society Conference, Troia, Portugal, 11–14 October 2016.

30. Tiwari, K.R.; Penner, G.A.; Warkentin, T.D. Inheritance of powdery mildew resistance in pea. *Can. J. Plant. Sci.* **1997**, *77*, 307–310. [CrossRef]

31. Vaid, A.; Tyagi, P.D. Genetics of powdery mildew resistance in pea. *Euphytica* **1997**, *96*, 203–206. [CrossRef]

32. Fondevilla, S.; Carver, T.L.W.; Moreno, M.T.; Rubiales, D. Macroscopic and histological characterisation of genes *er1* and *er2* for powdery mildew resistance in pea. *Eur. J. Plant. Pathol.* **2006**, *115*, 309–321. [CrossRef]

33. Fondevilla, S.; Cubero, J.I.; Rubiales, D. Confirmation that the *Er3* gene, conferring resistance to *Erysiphe pisi* in pea, is a different gene from *er1* and *er2* genes. *Plant. Breed.* **2011**, *130*, 281–282. [CrossRef]

34. Bai, Y.; Pavan, S.; Zheng, Z.; Zappel, N.F.; Reinstadler, A.; Lotti, C.; DeGiovanni, C.; Ricciardi, L.; Lindhout, P.; Visser, R.; et al. Naturally occurring broad-spectrum powdery mildew resistance in a central American tomato accession is caused by loss of MLO1 function. *Mol. Plant. Microbe Interact.* **2008**, *21*, 30–39. [CrossRef] [PubMed]

35. Büschges, R.; Hollricher, K.; Panstruga, R.; Simons, G.; Wolter, M.; Frijters, A.; van Daelen, R.; van der Lee, T.; Diergaarde, P.; Groenendijk, J. The barley MLO gene, a novel control element of plant pathogen resistance. *Cell* **1997**, *88*, 695–705. [CrossRef]

36. Consonni, C.; Humphry, M.E.; Hartmann, H.A.; Livaja, M.; Durner, J.; Westphal, L.; Vogel, J.; Lipka, V.; Kemmerling, B.; Schulze-Lefert, P.; et al. Conserved requirement for a plant host cell protein in powdery mildew pathogenesis. *Nat. Genet.* **2006**, *38*, 716–720. [CrossRef] [PubMed]

37. Devoto, A.; Hartmann, H.A.; Piffanelli, P.; Elliott, C.; Simmons, C.; Taramino, G.; Goh, C.S.; Cohen, F.E.; Emerson, B.C.; Schulze-Lefert, P.; et al. Molecular phylogeny and evolution of the plant-specific seven-transmembrane MLO family. *J. Mol. Evol.* **2003**, *56*, 77–88. [CrossRef] [PubMed]

38. Pavan, S.; Schiavulli, A.; Appiano, M.; Marcotrigiano, A.R.; Cillo, F.; Visser, R.G.F.; Bai, Y.; Lotti, C.; Luigi Ricciardi, L. Pea powdery mildew *er1* resistance is associated to loss-of-function mutations at a MLO homologous locus. *Appl. Genet.* **2011**, *123*, 1425–1431. [CrossRef] [PubMed]

39. Rispail, N.; Rubiales, D. Genome-wide identification and comparison of legume *MLO* gene family. *Sci. Rep.* **2016**, *6*, 32673. [CrossRef]

40. Santo, T.; Rashkova, M.; Alabaca, C.; Leitao, J. The ENU–induced powdery mildew resistant mutant pea (*Pisum sativum* L.) lines S (*er1mut1*) and F (*er1mut2*) harbour early stop codons in the *PsMLO1* gene. *Mol. Breed.* **2013**, *32*, 723–727. [CrossRef]

41. Fu, H.; Sun, S.; Zhu, Z.; Duan, C.; Yang, X. Phenotypic and genotypic identification of powdery mildew resistance in pea cultivars or lines from Canada. *J. Plant. Genet. Resour.* **2014**, *15*, 1028–1033. (In Chinese with English abstract).

42. Ma, Y.; Coyne, C.J.; Main, D.; Pavan, S.; Sun, S.; Zhu, Z.; Zong, X.; Leitão, J.; McGee, R.J. Development and validation of breeder-friendly KASPar markers for *er1*, a powdery mildew resistance gene in pea (*Pisum sativum* L.). *Mol. Breed.* **2017**, *37*, 151. [CrossRef]

43. Sudheesh, S.; Lombardi, M.; Leonforte, A.; Cogan, N.O.I.; Materne, M.; Forster, J.W.; Kaur, S. Consensus Genetic Map Construction for Field Pea (*Pisum sativum* L.), Trait dissection of biotic and abiotic stress tolerance and development of a diagnostic marker for the *er1* powdery mildew resistance gene. *Plant. Mol. Biol. Rep.* **2015**, *33*, 1391–1403. [CrossRef]

44. Pavan, S.; Schiavulli, A.; Appiano, M.; Miacola, C.; Visser, R.G.F.; Bai, Y.L.; Lotti, C.; Ricciardi, L. Identification of a complete set of functional markers for the selection of *er1* powdery mildew resistance in *Pisum sativum* L. *Mol. Breed.* **2013**, *31*, 247–253. [CrossRef]

45. Peng, H.X.; Yao, G. On resistance to powdery mildew of pea varieties Chinese. *Acta Phytopathol Sin.* **1993**, *23*, 62. (In Chinese)

46. Liu, A.A. Identification method of resistance of pea to powdery mildew using detached leaves. *Acta Phytophylacica Sin.* **2002**, *29*, 19–123. (In Chinese with English abstract).

47. Zeng, L.; Li, M.Q.; Yang, X.M. Identification of resistance of peas resources to powdery mildew. *Grassl. Turf.* **2012**, *32*, 35–38. (In Chinese with English abstract).

48. Wang, Z.; Bao, S.; Duan, C.; Zong, X.; Zhu, Z. Screening and molecular identification of resistance to powdery mildew in pea germplasm. *Acta Agron. Sin.* **2013**, *39*, 1030–1038. (In Chinese with English abstract). [CrossRef]

49. Lu, J.; Yang, X.; Wang, C.; Yang, F.; Zhang, L. Screening for pea resources resistant to pea powdery mildew in field. *Gansu Agr. Sci. Technol.* **2015**, *41*, 154–158, (In Chinese with English abatract).

50. Fu, H. Phenotyping and Genotyping Powdery Mildew Resistance in Pea. Ph.D. Thesis, Gansu Agricultural University, Gansu, China, 2014. (In Chinese with English abstract).

51. Sun, S.; He, Y.; Dai, C.; Duan, C.; Zhu, Z. Two major *er1* alleles confer powdery mildew resistance in three pea cultivars bred in Yunnan Province, China. *The Crop. J.* **2016**, *4*, 353–359. [CrossRef]

52. Kusch, S.; Panstruga, R. Mlo–based resistance, an apparently universal "Weapon" to defeat powdery mildew disease. *MPMI* **2017**, *30*, 179–189. [CrossRef]

53. Kim, E.; Magen, A.; Ast, G. Different levels of alternative splicing among eukaryotes. *Nucl. Acids Res.* **2007**, *35*, 125–131. [CrossRef]

54. Reddy, A.S.N. Alternative splicing of pre-messenger RNAs in plants in the genomic era. *Annu. Rev. Plant. Biol.* **2007**, *58*, 267–294. [CrossRef]

55. Ner–Gaon, H.; Halachmi, R.; Savaldi–Goldstein, S.; Rubin, E.; Ophir, R.; Fluhr, R. Intron retention is a major phenomenon in alternative splicing in Arabidopsis. *Plant. J.* **2004**, *39*, 877–885. [CrossRef] [PubMed]

56. Barbazuk, W.B.; Fu, Y.; McGinnis, K.M. Genome-wide analyses of alternative splicing in plants, opportunities and challenges. *Genome Res.* **2008**, *18*, 1381–1392. [CrossRef] [PubMed]

57. Rubiales, D.; Brown, J.K.M.; Martín, A. Hordeum chilense resistance to powdery mildew and its potential use in cereal breeding. *Euphytica* **1993**, *67*, 215–220. [CrossRef]

58. Rana, J.C.; Banyal, D.K.; Sharma, K.D.; Sharma, M.K.; Gupta, S.K.; Yadav, S.K. Screening of pea germplasm for resistance to powdery mildew. *Euphytica* **2013**, *189*, 271–282. [CrossRef]

59. Shure, M.; Wessler, S.; Fedoroff, N. Molecular-identification and isolation of the waxy locus in maize. *Cell* **1983**, *35*, 225–233. [CrossRef]

60. Bordat, A.; Savois, V.; Nicolas, M.; Salse, J.; Chauveau, A.; Bourgeois, M.; Potier, J.; Houtin, H.; Rond, C.; Murat, F.; et al. Translational genomics in legumes allowed placing in silico 5460 unigenes on the pea functional map and identified candidate genes in *Pisum sativum* L. *Genes Genome Genet.* **2011**, *1*, 93–103. [CrossRef]

61. Loridon, K.; McPhee, K.; Morin, J.; Dubreuil, P.; Pilet-Nayel, M.L.; Aubert, G.; Rameau, C.; Baranger, A.; Coyne, C.; Lejeune-Hénaut, I.; et al. Microsatellite marker polymorphism and mapping in pea (*Pisum sativum* L.). *Theor. Appl. Genet.* **2005**, *111*, 1022–1031. [CrossRef]

62. Lander, E.S.; Daly, M.J.; Lincoln, S.E. Constructing genetic linkage maps with MAPMAKER/EXP Version 3.0, a tutorial and reference manual. In *Institute for Biomedical Research Technical Report*, 3rd ed.; Whitehead, A., Ed.; Whitehead Institute for Biomedical Research: Cambridge, MA, USA, 1993.

63. Kosambi, D.D. The estimation of map distances from recombination values. *Ann. Eugen.* **1944**, *12*, 172–175. [CrossRef]

64. Liu, R.H.; Meng, J.L. MapDraw, A Microsoft excel macro for drawing genetic linkage maps based on given genetic linkage data. *Hereditas* **2003**, *25*, 317–321. (In Chinese with English abstract).

International Journal of
Molecular Sciences

Article

Characterization and Rapid Gene-Mapping of Leaf Lesion Mimic Phenotype of *spl-1* Mutant in Soybean (*Glycine max* (L.) Merr.)

G M Al Amin [1,2,†], Keke Kong [1,†], Ripa Akter Sharmin [1], Jiejie Kong [1], Javaid Akhter Bhat [1,*] and Tuanjie Zhao [1,*]

1   National Center for Soybean Improvement, Key Laboratory of Biology and Genetics and Breeding for Soybean, Ministry of Agriculture, State Key Laboratory of Crop Genetics and Germplasm Enhancement, Nanjing Agricultural University, Nanjing 210095, China; alamin25@gmail.com (G.M.A.A.); 2017201030@njau.edu.cn (K.K.); ripa.sharmin@gmail.com (R.A.S.); 2012094@njau.edu.cn (J.K.)

2   Department of Botany, Jagannath University, Dhaka 1100, Bangladesh

*   Correspondence: javid.akhter69@gmail.com (J.A.B.); tjzhao@njau.edu.cn (T.Z.)

†   These authors contributed equally to this work.

Received: 19 March 2019; Accepted: 30 April 2019; Published: 3 May 2019

Abstract: In plants, lesion mimic mutants (LMMs) reveal spontaneous disease-like lesions in the absence of pathogen that constitutes powerful genetic material to unravel genes underlying programmed cell death (PCD), particularly the hypersensitive response (HR). However, only a few LMMs are reported in soybean, and no related gene has been cloned until now. In the present study, we isolated a new LMM named spotted leaf-1 (*spl-1*) from NN1138-2 cultivar through ethyl methanesulfonate (EMS) treatment. The present study revealed that lesion formation might result from PCD and excessive reactive oxygen species (ROS) accumulation. The chlorophyll content was significantly reduced but antioxidant activities, viz., superoxide dismutase (SOD), peroxidase (POD) and catalase (CAT), as well as the malondialdehyde (MDA) contents, were detected higher in *spl-1* than in the wild-type. According to segregation analysis of mutant phenotype in two genetic populations, viz., W82×*spl-1* and PI378692×*spl-1*, the spotted leaf phenotype of *spl-1* is controlled by a single recessive gene named *lm1*. The *lm1* locus governing mutant phenotype of *spl-1* was first identified in 3.15 Mb genomic region on chromosome 04 through MutMap analysis, which was further verified and fine mapped by simple sequence repeat (SSR) marker-based genetic mapping. Genetic linkage analysis narrowed the genomic region (*lm1* locus) for mutant phenotype to a physical distance of ~76.23 kb. By searching against the Phytozome database, eight annotated candidate genes were found within the *lm1* region. qRT-PCR expression analysis revealed that, among these eight genes, only *Glyma.04g242300* showed highly significant expression levels in wild-type relative to the *spl-1* mutant. However, sequencing data of the CDS region showed no nucleotide difference between *spl-1* and its wild type within the coding regions of these genes but might be in the non-coding regions such as 5′ or 3′ UTR. Hence, the data of the present study are in favor of *Glyma.04g242300* being the possible candidate genes regulating the mutant phenotype of *spl-1*. However, further validation is needed to prove this function of the gene as well as its role in PCD, which in turn would be helpful to understand the mechanism and pathways involved in HR disease resistance of soybean.

Keywords: soybean; spotted leaf mutant; physio-chemical performance; MutMap mapping; candidate gene

1. Introduction

Plants have evolved complicated signaling pathways and defense system for protecting themselves against pathogen attack. Among them, hypersensitive response (HR) is the most efficient and prominent response characterized by the rapid death of plants cells that come in direct contact or are close to a pathogen. Extensive efforts have been made to identify the signaling pathway as well as to identify candidate genes involved in the control and execution of the hypersensitive cell death [1–4]. Isolation and identification of mutants in which cell death is misregulated are one of the approaches used for this study. These mutants are named as lesion mimic mutants (LMMs) showing either unregulated or constitutive cell death formation that mimic the pathogen-inducible, HR cell death [5]. Previously, LMMs have been characterized and extensively analyzed in many plant species for their responses to different plant hormones as well as modes of inheritances including groundnut [6], maize [7], *Arabidopsis thaliana* [8], rice [9,10] and barley [11,12]. However, the different pathways engaged for initiation and developments of the lesion or molecular mechanisms involved in lesion mimic, as well as basic function of the wild-type allele at a mutant locus are not well defined. Initiation and propagation of lesion on leaves of LMM plants are regulated with the age of plants, i.e., developmentally regulated [13]. Generally, the lesions first appear in the older leaves and then progress to young upper leaves [14]. Hence, LMMs are very promising genetic materials for exploring the regulatory mechanisms of plant growth and defense response.

The genes related to lesion mimic are reported to have diverse functions including a transcription factor regulating membrane receptors, superoxide dismutase, salicylate and sphingolipid signaling [15]. Genes underlying lesion mimic phenotypes appear to play direct roles in the maintenance of cellular homeostasis. Some of the lesion mimic mutant genes that have been cloned plays important role in programmed cell death (PCD) such as Ca^{2+} ion influx (*dnd1*, *dnd2/hlm1*, *cpr227* and *cpn1/bon1*), sphingolipid metabolism (*acd5* and *acd11*), ROS formation/sensing (*lsd1*), and porphyrin/chlorophyll biosynthesis and catabolism (*acd1*, *acd2*, *lin2*, *les22* and *flu1*) [16]. However, few LMMs, e.g., *lsd1*, accelerate the PCD with the HR-inducing bacteria, and some show normal growth, e.g., *acd5* and *cpr22* [16]. Hence, LMMs are an important tool for identifying and characterizing genes that are directly or indirectly associated with the regulation and execution of PCD in crop plants.

In soybean, T363 was the first LMM mutant to be characterized and named as *dlm* (disease-lesion mimic) [14]. The mutation in *dlm* results in the formation of small necrotic spots surrounded by chlorotic halos on leaves and is controlled by single recessive gene according to segregation ratios of the LMM trait in genetic populations [14]. Subsequently, the *dlm* phenotype was found to be light-dependent and associated with chloroplast function [15]. Inheritance of *dlm* mutant phenotype and some leaf morphological traits were carried out, and it was reported the *dlm* allele inherited independently with that of *P1*, *y9*, *f*, *lf2* and *ti* alleles controlling glabrous, chlorophyll-deficient leaf, stem trait, seven leaflet and trypsin inhibitors traits in soybean, respectively [17,18]. Wang et al. [19] found a new LMM in soybean with rugose leaf phenotype controlled by two duplicated genes, *rf1*, and *rf2*, which were mapped on chromosome 18 and 08, respectively. Although few LMMs have been characterized in soybean, the genes underlying the mutant phenotype have not been cloned yet. In addition, very little information is available about the phenomic characteristics, molecular mechanism of LMMs in soybean as well as how genes regulate PCD in soybean. In this regard, efforts are required to identify genes governing the mutant phenotype of LMMs in soybean, and to understand the molecular mechanism regulating the HR and PCD in soybean.

Map-based cloning method has been successfully implemented to identify agronomically important QTLs/candidate genes in various crops, such as wheat [20] and rice [21,22]. However, this approach has limitations being laborious, low-throughput and time-consuming in specific crops such as soybean [23]. In this regard, BSA-based whole genome re-sequencing (WGRS) approaches such as MutMap and QTL-seq methods permits rapid isolation of the genes or genomic locus responsible for the causative mutation of the phenotypes, and have been confirmed to be promising gene mapping approaches in crop plants [24,25]. Using this approach candidate genes has been successfully detected for important phenotypic characters in different crops, such as chickpea [26], barley [27], maize [28], soybean [23,29], tomato [30] and cucumber [31]. Hence, this application of WGRS in detecting the causative genes underlying mutant phenotype of crop traits will be of considerable significance in this challenging time of global hunger and the alarming global population increase [25].

In the present study, we isolated an LMM named spotted leaf-1 (*spl-1*) from the progeny of an elite soybean cultivar NN1138-2 treated with ethyl methanesulfonate (EMS), and investigated its detailed morphological and physiological characters, and then used the combined strategy of MutMap and map-based cloning mapping approaches to identify the candidate genes underlying the lesion mimic mutant phenotype of *spl-1*. The *lm1* locus controlling the *spl-1* phenotype was mapped in a 76.23 kb genomic regions on chromosome 04 harboring eight candidate genes, of which *Glyma.04g242300* was screened out as the possible target genes for *spl-1* mutant phenotype based on the qRT-PCR analysis. Hence, our findings provide new gene resources, and functional analysis of these genes will help to understand the pathway mechanism of lesion mimic as well as how plants can develop an innate immune response named hypersensitive response (HR) and programmed cell death (PCD) defense in the whole life of plants mainly from biotic and abiotic constrained.

2. Results

2.1. Phenotypic Characterization of spl-1 Mutant

Under natural field condition and environment, the typical tiny brown spot lesions first appeared on the lower leaves (older-leaf) of *spl-1* mutant plants after 2–3 weeks of sowing, i.e., trifoliate stage, and then progressively lesions formed on every leaf up of the plant body when the plant began to flower. The characteristic phenotype of the mutant at maturity stage were the older leaves revealed highly brown necrotic lesions of chlorotic leaves in the absence of pathogens, and early senescence was apparent, whole mutant leaves became tan, and, eventually, some died at a mature stage, unlike the wild-type. The number and size of the spots continued to increase as the leaf grew older and covered the entire leaf surface (Figure 1A–D). These results suggest that lesions on the *spl-1* mutant were developmentally regulated. The mutant phenotype was also observed to be environmentally-sensitive; the appearance of mutant phenotype was more prominent under summer-planting condition compared to spring-planting condition. Similar to our findings, Kim et al. [15] also reported that phenotype of the soybean disease-lesion mimic (*dlm1*) mutant is dependent on the light intensity, temperature, relative humidity and day length for affecting the development of cell death phenotype.

Figure 1. Morphological characteristics of wild-type and *spl-1* mutant soybean genotypes: (**A,C**) leaf phenotype characteristic of wild-type; and (**B,D**) leaf phenotype characteristic of *spl-1* mutant plants. Scale bars: (**A,B**) 1.0 cm; and (**C,D**) 1.5 cm.

2.2. Leaf Pigment Content and Histological Analysis

Chlorophyll degradation is an integral part and indicator for the degree of leaf senescence. In this study, contents of chlorophyll a, chlorophyll b, carotenoids, and total chlorophylls showed no significant difference between *spl-1* mutant and wild-type plants at the seedling stage (Figure 2A). In contrast, at maturity stage *spl-1* mutant plants showed a considerable reduction in the contents of photosynthetic pigments, with chlorophyll a, chlorophyll b and total chlorophyll decreased by 52%, 56%, and 54.6%, respectively (Figure 2B). However, no significant differences for carotenoids were found between *spl-1* mutant and wild-type at maturity stage. This result suggests that pigment accumulation in leaves was largely influenced by lesion formation in *spl-1* soybean mutant.

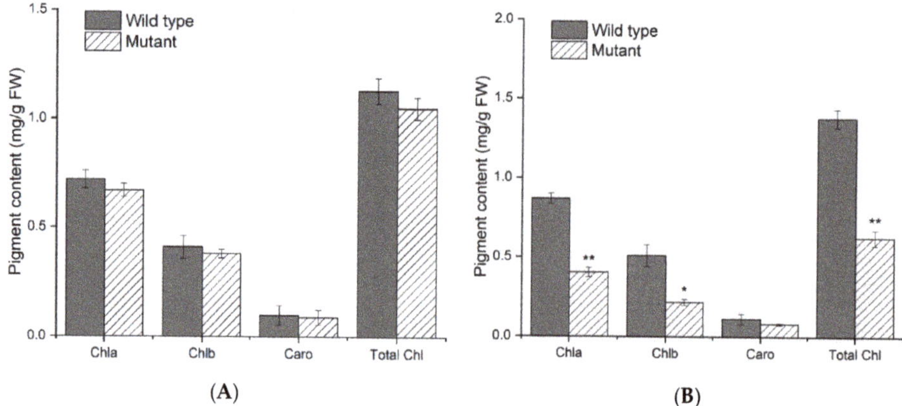

Figure 2. Comparison of leaf photosynthetic pigment contents in wild-type and *spl-1* mutant plants at: seedling stage (**A**); and mature stage (**B**). Chla, chlorophyll a; Chlb, chlorophyll b; Caro, carotenoids; Total Chl, total chlorophyll; FW, Fresh weight. The error bars indicate the mean ± SE ($n = 3$). SPSS software was used for statistical analysis. * significantly different at $p < 0.05$; ** significantly different at $p < 0.01$.

To elucidate the leaf anatomical differences between mutant and wild-type, transverse sections of leaves from both soybean genotypes were used for histological observation (Figure 3A–D). Leaf photosynthetic mesophyll cells contain chloroplasts and are usually arranged in palisade and spongy parenchyma. In wild-type soybean, the arrangement of mesophyll cells was normal and uniform, they were well separated from each other, and both the spongy and palisade parenchyma were loosely expanded (Figure 3A). In contrast, the arrangement of the palisade and spongy parenchyma cells was highly disoriented and compact in mutant soybean (Figure 3B). In addition, our results reveal a poorly developed vascular bundle in a mutant plant, compared to normal plant that was discordant with the mesophyll expansion of mutant (Figure 3C,D). Stomata play a vital role in the gaseous exchange between leaf and outside atmosphere due to the presence of air space among the mesophyll parenchyma, which is essential for normal leaf photosynthesis [32]. Hence, compactly arranged leaf mesophyll restricted the gaseous exchanges, which in turn reduced the leaf photosynthesis resulting in chlorophyll degradation and early leaf senescence. Kura-Hotta et al. [33] reported that inactivation of photosynthesis is closely related to loss of reaction center complexes during leaf senescence of rice seedlings because the leaf hydraulic conductance (K_{leaf}) is strictly determined by leaf venation/vascular bundle that has a strong influence on the degree to which the stomata may remain open for photosynthesis without desiccating the leaf [34]. Within and across species, K_{leaf} correlates strongly with stomatal pore area per leaf area, stomatal conductance and light-saturated photosynthetic rate per leaf area [34–37]. It suggests that distorted leaf anatomy might lead to reduced photosynthesis, chlorophyll degradation, lesion mimic phenotype and early leaf senescence of soybean *spl-1* mutant.

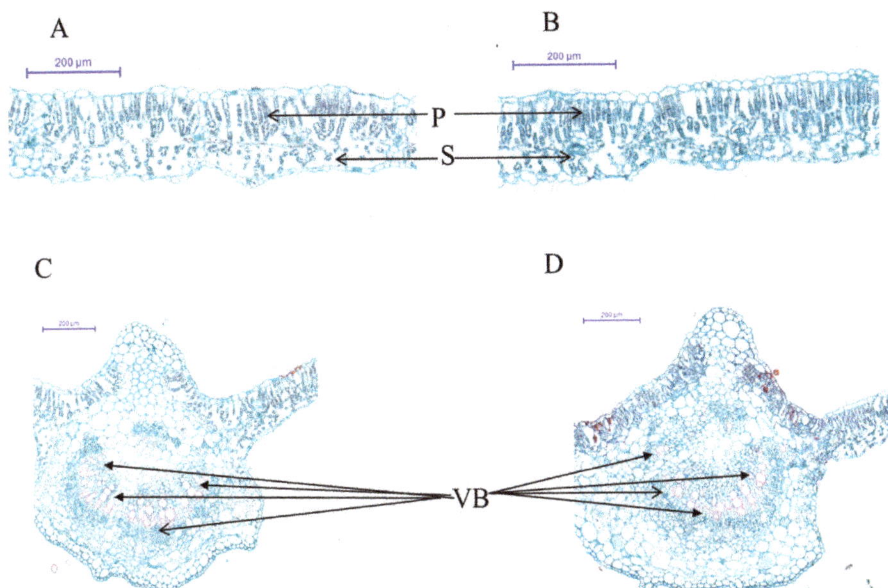

Figure 3. Leaf anatomical structure of wild-type and *spl-1* mutant soybean genotypes: (**A,C**) leaf anatomical structure of wild-type plant; and (**B,D**) leaf anatomical structure of *spl-1* mutant plants. *p*, palisade parenchyma; S, spongy parenchyma; VB, vascular bundle. Arrows show linear arrangement of VB in wild-type (**C**) and non-linear/distorted arrangement of VB in *spl-1* mutant (**D**).

2.3. Physiochemical Analysis for PCD, H_2O_2 Accumulation and Antioxidants

Necrotic lesion formation usually results from PCD and ROS accumulation [38]. To determine cell death and ROS accumulation, we performed traditional methods of Trypan blue and Diaminobenzidine (DAB) staining assays, respectively [5]. After staining with Trypan blue, the leaves of *spl-1* mutant showed deep blue spots at the site of lesions, whereas the surrounding normal cells of *spl-1* mutant, as well as the whole leaf of wild-type plant, exhibited negative staining. Trypan blue staining (Figure 4A) suggested that PCD occurred during lesion formation in the *spl-1* mutant. To confirm that ROS accumulation was accompanied by PCD, we performed a DAB staining assay to assess H_2O_2 accumulation. After DAB staining, the leaves of *spl-1* mutant exhibited many reddish-brown spots only at necrotic sites, and dark brownish staining appeared with increasing severity of necrosis (Figure 4B), indicating a high level of H_2O_2 accumulation in the *spl-1* mutant. This result indicates that ROS accumulation in cells might be responsible for cell death and lesion formation, and the staining assay confirmed that the *spl-1* mutant suffered from a hypersensitive reaction and exhibited PCD with a visible phenotype at necrotic sites.

For further insights, we also examined some physiological changes in wild-type and mutant genotype, and evaluated the activities of some key enzymatic antioxidants, viz., SOD, POD and CAT, and also estimated the lipid peroxide content (Malondialdehyde (MDA)) at different level of lesion appearances (high lesion mimic (HLM) and low lesion mimic (LLM)) in mutant plants (Figure 5A–D). Our results show that activities of SOD, POD, and CAT were significantly higher in the *spl-1* mutant than in wild-type plants for both HLM and LLM, except for CAT that exhibited significantly lower activity at LLM in the *spl-1* mutant (Figure 5C). Furthermore, MDA content was significantly higher in *spl-1* mutant than in wild-type for HLM but revealed no significant difference at LLM. These results reveal the increased accumulation of ROS and lipid peroxides as well as the activities of antioxidants, and suggest that mutation in *spl-1* plants resulted in oxidative stress that in turn led to PCD and brown necrotic lesions on the leaf surface [39].

Figure 4. Histochemical staining analysis for leaves of wild-type and *spl-1* mutant soybean genotypes: (**A**) Trypan blue staining for cell death. The spots indicated the ROS accumulated area in the *spl-1* mutant. (**B**) DAB staining for H_2O_2 accumulation. Scale bars: (**A**,**B**) 1 cm.

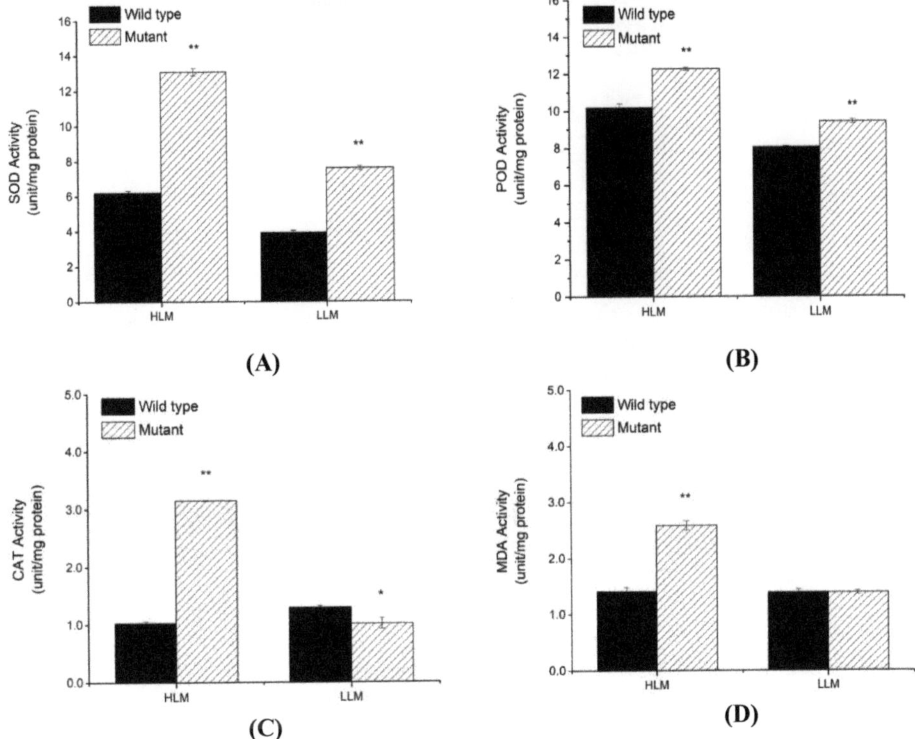

Figure 5. Graph showing different physiological characteristics/parameters determined for both the wild-type and *spl-1* mutant plants: (**A**) activity of superoxide dismutase (SOD); (**B**) activity of peroxidase (POD); (**C**) activity of Catalase (CAT); and (**D**) content of the malondialdehyde (MDA). The upper second leaves (lower lesion mimic (LLM)) and upper third leaves (higher lesion mimic (HLM)) of plants were used the estimation of these parameters at six weeks after sowing in pots. The data represent the means ± SE of three replicates. * significantly different at $p < 0.05$; ** significantly different at $p < 0.01$.

2.4. Inheritance for Spotted Leaf Trait of spl-1 Mutant

The inheritance of mutant phenotype was determined by evaluating the presence and absence of brown necrotic lesions on the leaves of F_2 and $F_{2:3}$ populations that were derived from the two different crosses, viz., W82×*spl-1* and PI 378692×*spl-1* (Table 1). Genetic analysis of the segregated populations revealed that F_2 populations of W82×*spl-1* (310 plants showed wild-type phenotype and 90 plants exhibited the *spl-1* phenotype, $\chi^2 = 1.20 < \chi^2_{0.05} = 3.84$, $p = 0.27$) and PI378692×*spl-1* (609 wild-type and 184 *spl-1* mutant, $\chi^2 = 1.27 < \chi^2_{0.05} = 3.84$, $p = 0.26$) crosses fitted an expected 3:1 segregation ratio of wild-type to mutant. In the $F_{2:3}$ populations of both crosses, viz., W82×*spl-1* and PI378692×*spl-1*, wild-type non-segregating and segregating lines fit 1:2 ratio (Table 1), suggesting that mutant phenotype is controlled by a single nuclear recessive gene, which is designated as *lm1*.

Table 1. Chi-square test for segregation ratio of normal and mutant plants in the F_2 and $F_{2:3}$ lines in two crosses viz., W82×*spl-1* and PI378692×*spl-1*.

Cross.	Generation	No. of Plants/Lines				Expected Ratio	χ^2	p
		Total	Wild Type	Segregation	Mutant			
W82×*spl-1*	F_2	400	310	-	90	3:1	1.20	0.27
	$F_{2:3}$ line	20	6	14	0	1:2	0.01	0.94
PI378692×*spl-1*	F_2	793	609	-	184	3:1	1.27	0.26
	$F_{2:3}$ line	13	4	9	0	1:2	0.01	0.92

2.5. MutMap Analysis for Identification of lm1 Locus

For accelerating the mapping and identification of the genomic region for target traits, the combined strategy of WGRS and traditional map-based cloning approach with BSA have been performed. Based on mutant phenotypic data evaluation of F_2 population derived from W82×*spl-1* cross, 20 F_2 individuals each for wild-type and mutant were selected, and their DNA was bulked to constitute DNA Pool A (wild-type) and Pool B (mutant) for the WGRS/sequencing. After filtering, 19.36 Gb of clean data were obtained with average Q20 of 99.59% and Q30 of 98.0%, indicating the high quality of the sequencing data (Table S1). A total of 94,700,266 and 93,193,018 short reads (150 bp in length) were obtained for Pool A (96.87% coverage) and Pool B (96.96% coverage), respectively. These short reads of both pools were aligned with references genome of Williams 82, and the match rates were 95.96 and 95.84% respectively (Table S1).

To identify candidate genomic region associated with the mutant phenotype, SNP-index of each SNP locus in Pool A and Pool B was calculated using high-quality SNPs, those with quality score $\geq$ 100 and read depth $\geq$ 10. The average SNP-index in Pool A and Pool B and Δ (SNP-index) between Pool A and Pool B across a 2-Mb genomic interval were measured using a 50-kb sliding window and plotted for all 20 chromosomes of the soybean genome (Figure S1). Test of significance (Fisher's exact test) was also conducted at each SNP locus for Pool A and Pool B, and the average p-values were calculated for SNPs located in each sliding window. In the SNP-index plotting of Pool A and Pool B, many peaks were identified. The SNP index plotting for 20 chromosomes of both the wild-type and mutant pools are provided in Figure S1, but statistical significant (p-value > 0.05) of only one major peak was identified in Δ (SNP-index) association analysis and were assigned as the candidate region of the gene controlling leaf lesion mimic mutant phenotype in *spl-1* mutant (Figure 6A–C and Table S2). This candidate region covers the genomic physical distance of 3.15 Mb on chromosome 04 between 45.84 and 48.95 Mb (version *Glycine max*, Wm82. a1.v1), and has Δ (SNP index) value significantly different from 0. These results indicate that a major genomic region governing lesion mimic mutant phenotype of *spl-1* was at the 3.15 Mb region of chromosome 04.

2.6. Validation and Fine Mapping of the lm1 Locus

To validate and narrow the candidate genomic region, i.e., *lm1* locus identified by MutMap method, we initially conducted preliminary mapping analysis by using 90 mutant plants from the F_2 population of W82×*spl-1* cross. Out of the total 90 SSR markers in the target genomic region detected by MutMap, 18 SSR markers exhibited polymorphism between W82 and *spl-1*. Linkage analysis of segregation data by MapMaker 3.0 software revealed that the *lm1* gene was primarily located between the markers BARCSOYSSR_04_1385 and BARCSOYSSR_04_1435 in a physical distance of 655.8 kb region on chromosome 04 lying correctly in the same candidate region as identified through MutMap, and hence confirmed the results of MutMap.

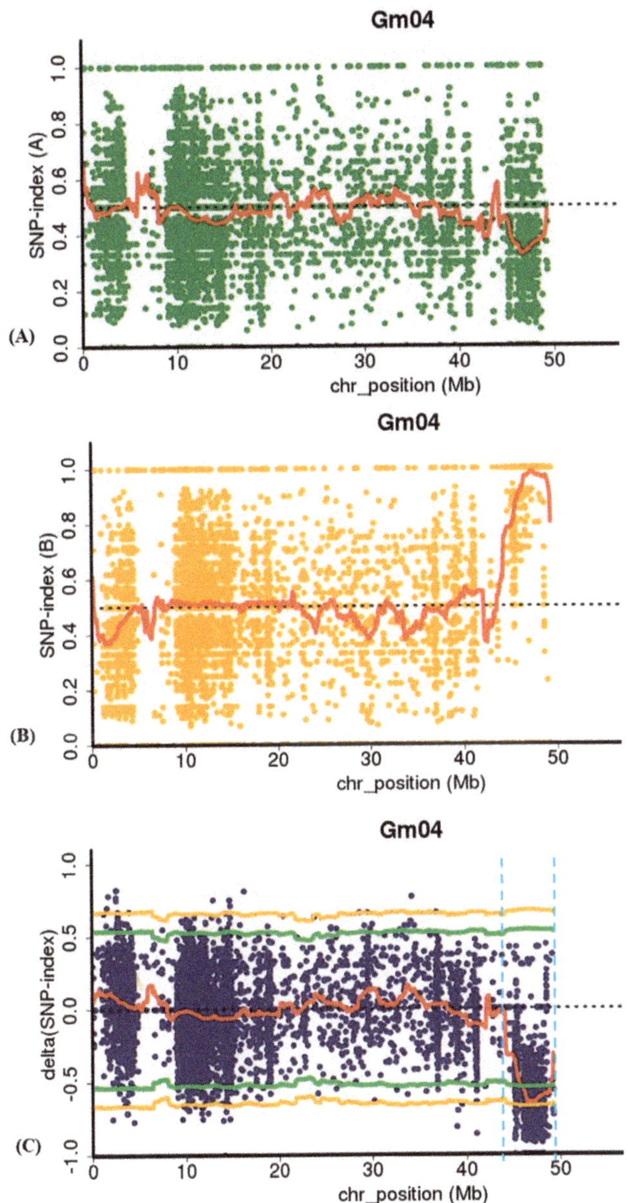

Figure 6. Identification of candidate genomic region (*lm1* locus) through MutMap analysis at a genomic interval of 45.80–48.95 Mb (Version Glyma v1.a1) on chromosome 04 of soybean: (**A,B**) the SNP-index of wild-type (A-Pool) and *spl-1* mutant (B-Pool) pools, respectively, for chromosome 04; and (**C**) the Δ (SNP-index) plot for chromosome 04. x-axis indicates the physical position of chromosome and y-axis indicates the average SNP-index in a 2-Mb interval with a 50-kb sliding window. The Δ (SNP-index) graph was plotted with statistical confidence intervals under the null hypothesis of no QTL ($p < 0.05$). The candidate region (*lm1* locus) identified for *spl-1* mutant phenotype is marked by two red dash border lines in Δ (SNP-index) plot.

The above genomic region detected through preliminary mapping was further fine-mapped by using 197 F_2 and $F_{2:3}$ *spl-1* mutant lines from the cross of PI378692×*spl-1*. By selecting randomly 40 pairs of SSR markers within the chromosome region identified through preliminary mapping, seven were polymorphic between PI378692 and *spl-1* and were used for further analysis (Table 2). Using genetic linkage analysis, the *lm1* gene was positioned between BARCSOYSSR_04_1429 and BARCSOYSSR_04_1435 markers, covering the physical distance of ~76.23 kb (Figure 7). By considering the reference genome sequence of Williams 82 [40] (Version Glyma 2.0), eight candidate genes were present in the genomic region of the *lm1* locus, which was narrowed to a 76.23 kb interval by fine mapping (Table 3). These genes include *Glyma.04g242300*, *Glyma.04g242400*, *Glyma.04g242500*, *Glyma.04g242600*, *Glyma.04g242700*, *Glyma.04g242800*, *Glyma.04g242900*, and *Glyma.04g243000* (Table 3). Among these eight genes, the functional annotation of six genes are known, whereas it is not available for two genes, viz., *Glyma.04g242400* and *Glyma.04g242600*, in public database (Table 3). Based on the function, *Glyma.04g242300* was considered as the probable candidate being a member of plantacyanin gene family, which belongs to sub-family of blue copper proteins, which functions in the electron transport chain during photosynthesis [41]. To further clarify it, we subject all eight candidate genes to qRT-PCR expression analysis, as discussed below.

Table 2. Polymorphic simple sequence repeat (SSR) markers used to narrow down the *lm1* locus.

Marker	Chromosome	Start *	End		Primer (F/R) Sequences
BARCSOYSSR_04_1390	Gm04	50549768	50549787	F	CCCGGTACAGTTGAGATGGA
		50550014	50549995	R	TTGCACTTCAGTAGGCCCTC
BARCSOYSSR_04_1391	Gm04	50588500	50588519	F	AGATGGTGGTGTTCTCAGGG
		50588766	50588747	R	ACCATCACCAACATGCAGAT
BARCSOYSSR_04_1418	Gm04	50942037	50942061	F	TTTTTCTTCAGAAACTTGAAACATT
		50942254	50942234	R	TGCATTTCTGAAACAAGGCAT
BARCSOYSSR_04_1420	Gm04	50953149	50953174	F	AAGTGATCAATGTTATCGATGAAGTA
		50953433	50953409	R	TTTGTCTCAATTAGTGTGGAATTTG
BARCSOYSSR_04_1426	Gm04	51011052	51011071	F	ATCAGAGGTCTGCCACCAAT
		51011271	51011252	R	CGCTGACAGACACCAAGAGA
BARCSOYSSR_04_1429	Gm04	51035485	51035504	F	TTTGCTACAGTGCTATCGGC
		51035766	51035747	R	TGCCAGCCGCTTATCTATCT
BARCSOYSSR_04_1435	Gm04	51111716	51111735	F	GTCCGTGCCAGTTTTTCATT
		51111960	51111941	R	TGCTGCACTTTCTCCTGATG

* Location has taken from Glyma2.0; F (forward primer), R (reverse primer).

Table 3. Functional annotation of eight candidate genes located within the *lm1* locus/region identified through fine mapping.

Gene ID	Position (bp)	Direction	Function Annotation
Glyma.04g242300	51036742-51037897	Forward	Plantacyanin
Glyma.04g242400	51047485-51048880	Forward	Unknown
Glyma.04g242500	51053346-51056055	Reverse	Flavin-binding monooxygenase family protein
Glyma.04g242600	51061920-51064096	Forward	Unknown
Glyma.04g242700	51064744-51067380	Reverse	F-box/RNI-like superfamily protein
Glyma.04g242800	51082744-51092913	Reverse	ACT domain repeat 3
Glyma.04g242900	51103163-51108385	Reverse	Protein kinase superfamily protein
Glyma.04g243000	51109534-51115501	Reverse	Thiamin diphosphate-binding fold (THDP-binding) superfamily protein

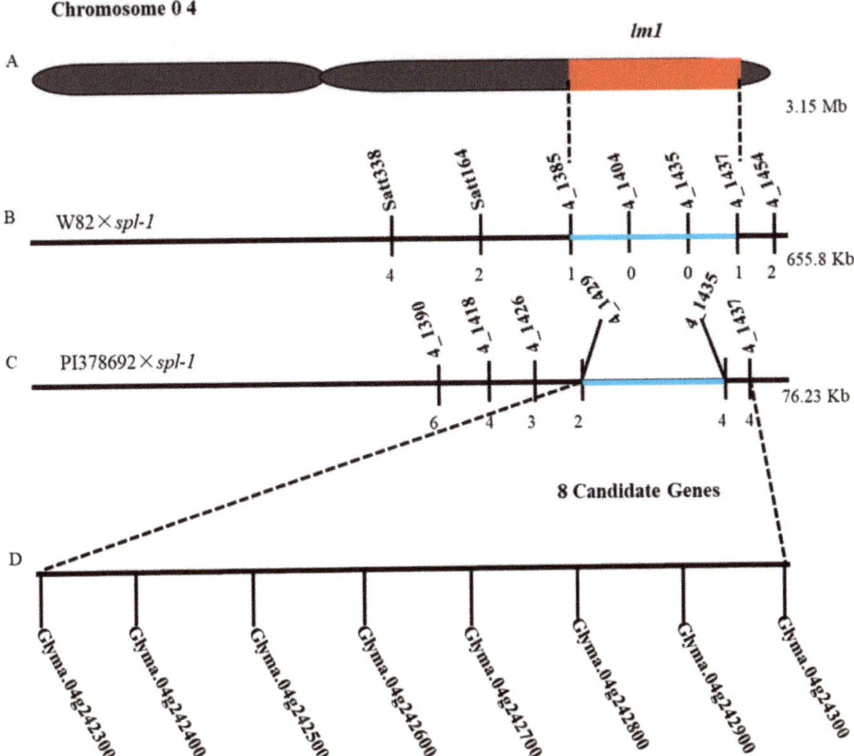

Figure 7. Mapping and fine mapping of *lm1* locus. (**A**) Location of *lm1* locus identified by MutMap-based BSA method on chromosome 04. (**B**) Dashes line indicated rough mapping of *spl-1* locus from cross of W82×*spl-1*. Vertical lines indicate polymorphic markers. Names of markers are shown above the line and the recombinants between *lm1* and each marker are shown below the line. (**C**) Fine mapping of *lm1* with genotyping data from newly developed polymorphic markers in the cross of PI378692×*spl-1*. (**D**) Eight candidate genes in the fine-mapped region.

2.7. qRT-PCR Expression and Sequences Analysis of Candidate Genes

To identify the candidate gene underlying the *lm1* locus of *spl-1* mutant, the expression patterns of all the eight candidate genes were tested in leaf tissues of wild-type and mutant parents at three different growth developmental stages, viz., V1, V3 and R1 [42], using qRT-PCR analysis. The oligo-nucleotides primers used for the qRT-PCR analysis are listed in Table S3. Out of these eight candidate genes, *Glyma.04g242300* showed significantly higher gene expression in wild-type relative to mutant genotype, and the expression was considerably lower in the *spl-1* mutant at all studied growth stages (Figure 8). The remaining seven candidate genes within the *lm1* locus revealed non-significant gene expression differences between wild-type and *spl-1* genotype at all stages. Therefore, the highly significant differential expression of *Glyma.04g242300* between wild-type and mutant genotype provided evidence for being the possible candidate genes responsible for leaf lesion mimic mutant phenotype of *spl-1* in soybean. To further clarify the sequence differences of the eight candidate genes, we sequenced the cDNA sequences of these genes. However, we did not find any nucleotide differences between the CDS sequences of *spl-1* and wild-type parents within the exons, and hence the difference might be in the 5′ or 3′ UTR region or non-coding region of this gene (the portion of gene not sequenced). Finally, depending on the above results, *Glyma.04g242300* is most likely to be involved in the lesion mimic phenotype of the *spl-1* mutant. However, it needs further functional validation to prove this function of *Glyma.04g242300*.

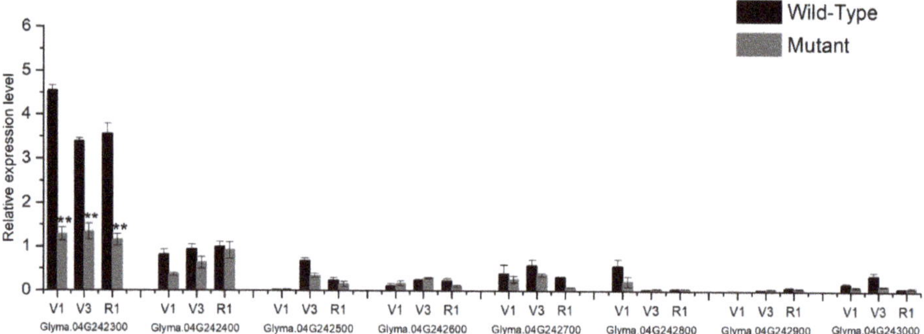

Figure 8. Relative gene expression of eight candidate genes in the leaves of wild-type and mutant (*spl-1*) plants at three developmental growth stages V1, V3 and R1 using qRT-PCR. Mean values of expression data of wild-type and *spl-1* mutant plants were analyzed for statistical significance at $p < 0.01$ (**) level, as indicated by asterisks on top of bars.

3. Discussion

3.1. spl-1 Is a New Soybean Leaf Lesion Mutant with Special Characteristics

To protect themselves against pathogen attack, plants have developed complicated defense mechanisms and signaling pathways. The HR results due to different pathways, and is a component of an effective defense system against biotrophic and hemibiotrophic pathogens. However, the molecular mechanisms and genes controlling the HR remain largely unknown [43]. In this regard, LMMs represent a broad group of phenotypes showing spontaneous cell death in the absence of disease pathogen, are interesting genetic materials for elucidating the molecular mechanism, pathways, and genes underlying HR and disease resistance. The appearance of lesions in different LMMs differs in induction conditions, timing, the extent of lesion spreading, color and size [44]. Thus far, some genes related to lesion mimic phenotype have been identified and cloned in different crop species, and their functions were also found diverse [2]. The results of these studies have indicated that lesion mimic phenotypes are regulated by different biological processes, thus hinting at the complexity of molecular mechanisms and signaling networks involved in HR and disease resistance [8].

Although a few LMMs have been characterized in soybean, the genetic mechanisms and pathways have not been well understood, and the genes regulating the mutant phenotype have not been cloned [14]. In this regard, the present study used a combined strategy of MutMap and traditional map-based cloning mapping to identify the candidate genomic region and genes underlying the LMM phenotype of *spl-1*. Chlorophyll a, chlorophyll b and total chlorophyll were significantly decreased in *spl-1* mutant relative to wild-type. Tetrapyrrole biosynthesis pathway leads to the production of chlorophyll a/b [45]. Hence, disruption of tetrapyrroles biosynthesis pathway at different stages leads to abnormal accumulation of photo-reactive molecules, which in turn leads to lesion-mimic phenotypes. For example, in the mutant *rugosa1* (*rug1*) tetrapyrroles biosynthesis pathway is affected at porphobilinogen deaminase (PGBD) that results in the accumulation of porphobilinogen [46]. Similarly, accumulation of protochlorophyllide (Pchlide) in *flu* and *oep16* mutants [47,48], uroporphyrinogen III in lesion 22 (*les22*) mutant [49] and coproporphyrinogen III in *lesion initiation 2/rice lesion initiation 1(lin2/rlin1)* mutants [50,51] leads to cell death phenotypes. Interestingly, cell death also results due to defects in chlorophyll catabolism. Indeed, disruption of two enzymes involved in the degradation of chlorophyll generates spontaneous lesions in the *accelerated cell death 1/lethal-leaf spot 1(acd1/lls1)* and *accelerated cell death 2 (acd2)* LMM [52–54]. Because of the known role of ROS during HR and in cells undergoing PCD, we further investigated the production of H_2O_2 as well as activities of antioxidants, viz., SOD, POD and CAT, and the content of lipid peroxidation (MDA) at different leaf position following the appearance of lesions in the lesion-mimic plants. Our study revealed that activities of all antioxidants, viz., SOD, POD, and CAT, were significantly increased in *spl-1* mutant in case of both HLM and LLM leaves except for CAT whose activity is reduced in LLM leaf of *spl-1* mutant compared to wild-type. This can be explained because lesions are present throughout the HLM leaf from bottom to top and are at final stage of development, whereas in the case of LLM leaf lesion mimic mutant phenotype is in its initial stage, i.e., yet to be developing, and lesions are very less in number usually at the bottom of leaf as well as small in size. Thus, it can be suggested that the CAT activity in leaves of *spl-1* depends upon the intensity and degree of lesion development.

Hence, the present study reported substantial accumulation of ROS, antioxidants and lipid peroxide in the leaves of LMM compared to wild-type, which is similar to the findings reported by Anand et al. [55], and thus suggested that mutation in *spl-1* mutant results in oxidative stress leading to PCD and brown necrotic lesions on the leaf surface [39].

3.2. Deploying MutMap and Traditional Mapping Methods to Accelerate Gene Identification

Conventional mapping is an important and effective strategy for identifying and isolating candidate genomic regions and genes for many crops. However, the general strategy for conventional mapping is time-consuming and laborious [23]. For example, *D53* gene encoding a protein that acted as a repressor of strigolactones in rice was identified by using 12,000 F_2 plants [56]; Similarly, for the fine mapping of the recessive dialytic gene, *dl*, in tomatoes, 2248 F_2 individuals were used [57]. In soybean, E1 a maturity locus gene involved in flowering time was delimited by using a very large number of individuals from $F_{2:3}$ to $F_{2:5}$ generations of soybean [58]. However, for the case of soybean, which is a larger crop plant requiring a huge area for sowing, it is impractical for growing so many progenies in the field. In this context, the last few decades have witnessed many reverse-genetic approaches that have become increasingly popular in some species, but map-based cloning is still an important approach for identifying candidate regions; however, BSA-seq methods facilitate and accelerate the gene identification process.

Therefore, in this study, we used an improved BSA-seq (MutMap) method that integrates the traditional BSA method with WGRS to rapidly identify specific genomic regions for the *spl-1* mutant phenotype of soybean. Moreover, the combination of MutMap and map-based cloning could effectively detect and fine map the QTL of interest. In the present study, major candidate genomic region underlying *spl-1* mutant phenotype was identified and mapped into a 76.23 kb genomic region on chromosome 04 by using an F_2 and $F_{2:3}$ mapping population via combined approaches of whole-genome NGS-based high-throughput MutMap and traditional mapping. MutMap analysis detected candidate genomic region, i.e., *lm1* locus based on Δ (SNP-index) that were further validated and fine mapped by SSR traditional map-based cloning, and were mapped between the SSR markers BARCSOYSSR_04_1429 and BARCSOYSSR_04_1435, which suggests the validity and robustness of MutMap-seq as a strategy for quick and efficient scanning of major genomic region for mutant phenotype on a genome-wide scale in soybean. The merits of BSA-seq method relative to other traditional mapping approaches for identifying the major genomic regions governing plant height, seed weight, seedling vigor and flowering time in soybean, chickpea and rice have been recently reported [26,59,60]. MutMap takes advantage of the high-throughput WGRS and BSA. In addition, the use of an SNP-index provides accurate quantitative evaluation for the parental alleles' frequencies, and also the genomic contribution from the two parents to F_2 individuals. Hence, the above characteristics of MutMap make it a very efficient and faster approach for identifying genomic regions underlying mutant phenotype in soybean.

3.3. Candidate Genes for spl-1 Phenotype

In the present study, the major genomic region (*lm1* locus) governing mutant phenotype were delimited in a 76.23 kb physical interval on chromosome 04 by using the combined strategy of MutMap and traditional map-based cloning analysis. Eight genes were predicated within this region, and the functional annotations of six genes are known, whereas the annotation of the remaining two genes was not available (Table 3). Based on the functional annotation and qRT-PCR expression analysis *Glyma.04g242300* was suggested to be a possible candidate gene for governing lesion mimic phenotype of *spl-1* mutant. The *Glyma.04g242300* is a family protein gene in *Arabidopsis* (At2G02850) that belongs to plantacyanin (PLC), which is a plant-specific phytocyanin (PC) sub-family of blue copper proteins functioning in the electron transport chain of photosynthesis [41,61]. It serves as an electron transfer agent in the cytochrome complex which follows Photosystem II and the entry point to Photosystem I of the non-cyclic electron transfer process. Defects in photosynthetic electron transport will affect photosynthesis process as well as chlorophyll catabolism and cell death [62]. Recently, it has been revealed that *OsUCL8* (*Oryza sativa* Uclacyanin like protein 8), a rice plantacyanin gene could regulate grain yield and photosynthesis [63]. *OsUCL8* cleaved by *miR408* affects copper homeostasis in the plant cell, which in turn affects the abundance of plastocyanin proteins and photosynthesis [63]. In the present study, soybean *spl-1* mutant revealed a significant reduction of chlorophyll content compared to wild-type, which indicates that the degradation of chlorophyll generates spontaneous chlorotic leaves. Previous studies have indicated that PCs are involved in various plant activities, including cell differentiation and reorganization [64], pollen tube germinating and anther pollination [41,65]. Hence, these studies indicate that the PC gene family can have multiple functions during plant development. Several researchers have indicated that salt and drought stresses can induce the expression of PC genes, suggesting the potential response to abiotic stresses [66,67]. Microarray data also suggest that plantacyanins may be stress-related proteins and be involved in plant defense responses [68,69]. It is assumed that plantacyanin is one of the targets of microRNAs that regulates transcription factors involved in different aspects of plant development [70,71], and miR408 regulates photosynthesis via plantacyanin [63,72]. Therefore, it is reasonable to postulate that *Glyma.04g242300* is the candidate gene for lesion mimic mutant phenotype of *spl-1* in soybean. However, further evidence is needed to functionally validate this hypothesis.

Hence, the results of the present study provide new gene resources in soybean that might regulate the leaf LMM phenotype of *spl-1* mutant. This increases our current knowledge of genes involved in PCD and HR in soybean. By analyzing the function, these genes will help to elucidate the mechanism as well as pathways involved in the development of lesion mimic phenotype in soybean. This in turn will provide explanation how plants regulate PCD as well as develop HR for resistance against biotic stresses, and hence will greatly help to develop disease-resistant soybean varieties to overcome the losses that occur due to disease constraint.

4. Materials and Methods

4.1. Plant Materials and Phenotypic Evaluation

Plant material of the present study included soybean accessions, viz., NN 1138-2, Williams 82 (W82) and PI378692, which were obtained from National Center for Soybean Improvement, Nanjing Agricultural University, Nanjing, Jiangsu Province, China. The leaf lesion mimic mutant (LMM) called spotted leaf-1 (*spl-1*) was identified from the EMS-induced mutational library of the cultivar "NN 1138-2" that was treated with 0.5% (*w/v*) EMS for 12 h. The M_1 seeds were harvested and pooled together. Subsequently, M_2 plants were individually harvested. Furthermore, $M_{2:3}$ lines showed segregation for mutant and normal phenotype at V1 and V2 stages [42]. Seeds from 10 individual plants that had normal leaves were harvested and sown for M_4 generation. Progeny obtained from the normal heterozygous $M_{2:3}$ lines also exhibited segregation of normal and lesion mimic phenotypes. The same selection and planting procedures were conducted through M_4 to M_{10} generations. In each generation, some lines showed segregation of the normal and disease-like leaf phenotypes, with a 3:1 ratio, indicating that a single recessive allele might control the disease-like leaf trait. Through consecutive selfing and selection, we achieved $M_{9:10}$ lines, and the mutant plants that have 99.8%

homogeneity to the normal $M_{9:10}$ plants were bulked together. These mutant lines were named as *spl-1* (spotted leaf-1) and were obtained from heterozygous individuals.

Seeds of mutant *spl-1*, W82, and PI378692 were planted at Jiangpu Agricultural Experiment Station, Nanjing Agricultural University in 2015, and two crosses were made at flowering time, viz., W82×*spl-1* and PI378692×*spl-1*. Mutant parent (*spl-1*) was used as a male parent in both crosses. The F_1 seeds obtained from each cross were planted at Jiangpu Station in next year cropping season, i.e., 2016, and no F_1 plants showed mutant phenotype, indicating recessive nature of the mutant trait. F_2 seeds derived from F_1 plant were harvested separately from both crosses. The F_2 population and $F_{2:3}$ families of each cross along with their parents were grown in the cropping seasons of 2017 and 2018, respectively, at the Jiangpu Station. Phenotypic data (normal and lesion mimic) of parents, F_1, F_2, and $F_{2:3}$ plants were recorded at V1–V5 and R1 growth stages of soybean under normal field conditions [42]. These F_2 and $F_{2:3}$ populations derived from W82×*spl-1* and PI378692×*spl-1* were used for mapping of the mutant locus. Chi-square analysis was applied to test the goodness-of-fit of observed to the expected ratio for independent assortment or linkage in all populations.

4.2. Leaf Pigment Quantification and Histological Analyses

Leaf photosynthetic pigments were extracted from leaves at seedling and mature stages collected from the same position/rank of both wild-type and mutant plants [42]. Fresh leaf sample of 0.1 g was taken and cut into small pieces, then steeped in 80% acetone at room temperature for 24 h. The quantification of pigments was performed using a Tecan Infinite Pro Microplate Reader (Tecan Austria GmbH, Grodig, Austria) following the method reported previously [73]. Pigments measurement was conducted for three independent experiment repeats. All these experiment operations were carried out in the dark to avoid degradation of photosynthetic pigments.

For histological analysis, third top leaves of both mutant and the wild-type plants were collected from the 35-day-old plants when lesions mimic phenotypes fully appeared. The 10 µm leaf sections of both wild-type and mutant were obtained using ultra-microtome (Leica EM UC7, Leica Microsystems Inc., Buffalo Grove, IL, USA) with three replicates. Leaf sections were prepared for histological analysis following the method of Carland and McHale [74] with some modification, and further leaf sections were stained with 0.1% safranine. Images were observed with an optical microscope under different magnification (Zeiss Axioplan, Jena, Germany), and were captured by a digital camera connected with the microscope. Parameters, viz., central meta-xylem shape, number of xylem and phloem vessels, spongy and palisade mesophyll parenchyma, were observed and recorded for comparative analysis.

4.3. Leaf Histochemical and Physiological Analyses

4.3.1. Trypan Blue Staining of Cell Death

Leaf samples of both the mutant and wild-type plants were collected at the same position/rank for Trypan blue staining when mutant phenotype appeared on the leaves of mutant plants at mature developmental stage. Leaves were stained with Trypan blue (Sigma-Aldrich, St. Louis, MO, USA) according to the method previously described by Chen et al. [75], with some modifications. Briefly, plant tissues were submerged in a 70 °C Trypan blue solution (2.5 mg of Trypan blue per mL, 25% (wt/vol) lactic acid, 23% water-saturated phenol, 25% glycerol, and H_2O) for 10 min, and then heated over boiling water for 2 min and left to stain overnight. Stained leaves were washed several times with absolute ethanol to remove Trypan blue solution until the leaves become colorless. Finally, ethanol was discarded, and the leaf samples were covered with 70% glycerol for visualization of cell death under microscopic analysis. The staining procedure was done in triplicate (three times).

4.3.2. H_2O_2 Detection by DAB

Hydrogen peroxide (H_2O_2) was detected by submerging the leaf samples of both wild-type and *spl-1* mutant in a 3,3′-diaminobenzidine (DAB) solution according to Rahman et al. [76], with some modifications. Briefly, leaves from wild-type and mutant plants were taken for DAB staining after lesions appeared (30 days and 60 days after sowing), and incubated in a 0.1% (*w/v*) DAB (10 mM MES, pH 7.0) solution at 25 °C temperature in the dark with gentle shaking for 12 h or more depending upon visibility of spots. Leaves were thoroughly washed in ddH$_2$O several times until DAB solutions were completely removed. Then, the chlorophyll was cleared by treating with 95% (*v/v*) ethanol boiling for 10 min. The transparent leaves were observed and photographed in 70% glycerol.

4.3.3. Antioxidants Activities and Lipid Peroxidation Determination

Activities of antioxidants, viz., Superoxide dismutase (SOD), Peroxidase (POD) and Catalase (CAT), as well as the content of lipid peroxidation (MDA), were determined for both the wild-type and mutant plants. High-lesion mimic (HLM) and low-lesion mimic (LLM) leaves were collected at the third and second position from the top of the same mutant plant, respectively, and were compared with the corresponding leaves of wild-type plants from the same position. In the case of HLM leaves, the lesion mimic mutant phenotype was well developed, and lesions were present throughout the leaf surface, whereas for LLM leaves, lesion were yet to be developed, were much fewer in number, and were not present throughout the leaf, usually on the bottom of leaf. The activities of SOD, POD and CAT as well as MDA content of the leaves were determined using SOD Assay Kit (T-SOD, A001-1), POD Assay Kit (A084-3), CAT Assay Kit (A007-1) and MDA Assay Kit (A003) by following the manufacturer's protocol (Nanjing Jiancheng Bioengineering Institute, Nanjing, China) and Li et al. [77]. Briefly, fresh leaf samples (1.0 g) of six-week-old plants were sliced and homogenized in mortar and pestle with 9 mL ice-cool 10× phosphate buffered saline (PBS) (pH 7.2–7.4) (Beijing Solarbio Science & Technology Co., Ltd., Beijing, China). The homogenates were further centrifuged at 3500 rpm for 10 min at 4 °C, and the supernatants were collected and used as crude extracts for above-cited assays by using a UV-1800 Shimadzu, spectrophotometer (SHIMADZU Corporation, Kyoto, Japan). Three independent samples were assayed, and standard errors (SE) among them were calculated.

4.4. MutMap Analysis

4.4.1. Construction of MutMap Libraries and Illumina Sequencing

Genomic DNA was isolated from young leaves of soybean using DNAquick Plant System (TIANGEN Biotech, Beijing, China) according to the manufacturer's protocol. DNA samples were quantified using Qubit® 2.0 Fluorometer (Thermo Scientific, Waltham, MA, USA). Two DNA bulks/pools, viz., wild-type pool (Pool A), and mutant pool (Pool B), were generated for Illumina libraries by pooling equal amounts of DNA from 20 wild-type and 20 mutant F$_2$ genotypes of W82×*spl-1* cross. About 5–10 µg of DNA from two pools were used to construct paired-end sequencing libraries, which were sequenced with an IlluminaHiSeq® 2500 (Illumina Inc., San Diego, CA, USA) NGS platform. FASTQ raw sequence reads with a minimum phred Q-score of 30 across >95% of nucleotide sequences were considered as high quality. The quality of these sequences was further checked by using FASTQC v0.10.1 (Babraham Institute, Cambridge, UK). High-quality FASTQ filtered sequences obtained from two DNA pools were aligned and mapped to the *Glycine max* Wm82.a1.v1 reference genome from Phytozome [40] using Burrows–Wheeler alignment tool (BWA v0.7.10, Cambridge, UK) with default parameters [78]. High-quality SNPs (minimum sequence read depth: 7 with SNP base quality ≥20) were discovered using SAM tools (Cambridge, UK) [79] by following the detailed procedure of Takagi et al. [80] and Lu et al. [31].

4.4.2. SNP Index Analysis

SNP-index was calculated at each SNP position for both Pool A and Pool B, which represented the parental alleles of the population [81,82]. A Δ (SNP-index) was calculated by subtraction of the Pool A SNP index from the Pool B SNP index [26,80,83,84]. Hence, the SNP locus with high Δ (SNP-index) value is an indicator an allele was highly common in Pool A and depleted in Pool B. If there is no major candidate region/locus of the target gene in a genomic region, the Δ (SNP-index) value should not be significantly different from 0. Using a null hypothesis of no QTLs, 95% confidence intervals of the Δ (SNP-index) for all the SNP positions were calculated with given read depths and plotted these against the Δ (SNP-index) [80].

4.4.3. Sliding-Window Analysis

In a given genomic interval, the average distributions of the SNP-index and Δ (SNP-index) were estimated by using sliding window approach with a 2-Mb window size and 50-kb sliding step, and these data were used to plot SNP-index plots for all soybean chromosomes. Genomic regions that showed average Δ (SNP-index) significantly higher than surrounding region and windows revealed an average p-value < 0.05 were considered candidate genomic regions harboring a locus associated with the mutant phenotype of *spl-1* soybean mutant [80].

4.5. Fine Mapping of lm1 Locus

To verify the accuracy of the *lm1* genomic region identified by MutMap analysis and establish mapping reliability of this approach, a traditional map-based cloning genetic linkage method was performed to find out the linkage of molecular markers and phenotypic loci of *spl-1*. The two F_2 populations, viz., W82×*spl-1* and PI378692×*spl-1*, were used for this purpose, and a total of 130 SSR markers in the predicted region on chromosome 04 were selected to survey the polymorphism between the wild-type and *spl-1* mutant lines [85]. Polymorphic markers that may be linked with the mutated genes were screened using the method of bulked-segregant analysis (BSA), as proposed by Michelmore et al. [86]. Both wild-type and mutant groups contained 10 randomly selected F_2 individuals, and the protocol was according to Wang et al. [19]. Within each group, the DNA from all individuals was pooled using an equal amount of DNA from each plant. The mapping steps involve as: the polymorphic SSR markers between two parents of cross populations were identified; then, the pools of wild-type and mutant plants were screened with those SSR markers that were polymorphic between the parents to identify markers for screening the F_2 populations. Mapmaker 3.0 software (Whitehead Institute for Biomedical Research, Cambridge, MA, USA) was used to identify the linkage between SSR markers and target genes [87]. A total of 307 recessive mutant plants from the F_2 and $F_{2:3}$ populations of W82×*spl-1* and PI378692×*spl-1* crosses were used for preliminary and fine mapping (Table 1), and the protocol was according to Wang et al. [19].

For the PCR amplification of marker genotyping, 10 ng DNA was used in 10 µL system under the instruction of the Taq Master Mix (Novoprotein, Shanghai, China). PCR thermal cycler was programmed as follows: initial denaturation at 95 °C for 5 min; followed by 32 cycles of denaturation at 95 °C for 30 s, annealing at 55 °C for 40 s and extension at 72 °C for 50 s; with final incubation at 72 °C for 10 min before hold to 4 °C. The amplification product was separated on 8% non-denaturing polyacrylamide gels that were stained with 1 g·L^{-1} AgNO$_3$ for 15 min before visualizing with 16 g·L^{-1} NaOH plus 11 mL·L^{-1} CH$_3$OH for 10 min.

4.6. Expression and Sequence Analysis of Candidate Genes

To analyze the candidate genes underlying the *lm1* locus of *spl-1* mutant, we investigated the expression pattern of all the eight genes present within *lm1* locus using real-time quantitative PCR (qRT-PCR). Young leaf samples at three different growth stages, viz., V1, V3, and R1, were collected from the wild-type and mutant parents. Total RNAs from the leaves were isolated by using the RNA prepare plant kit (TIANGEN, Beijing, China). First-strand cDNA was synthesized using two-step PrimerScript^{TM} RT reagent Kit gDNA Eraser (TaKaRa, Kusatsu, Shiga, Japan) according to the manufacturer's instructions. Real-time qRT-PCR was performed using a ChamQ SYBR qPCR Master Mix (Vazyme, Jiangsu, Nanjing, China) on a Bio-Rad system. Primers were designed by Beacon Designer 7.9 software (Premier Biosoft International, Palo Alto, CA, USA). *GmActin11* was used as an internal control for the qRT-PCR analysis, and three biological replicates were used for each reaction.

Average relative expression levels for wild-type and mutant parent were calculated. One-way ANOVA tests were performed by IBM-SPSS software to test the significance of differences in expression levels among different samples.

For further verification, we sequenced the coding sequence (CDS) of eight candidate genes, viz., *Glyma.04g242300*, *Glyma.04g242400*, *Glyma.04g242500*, *Glyma.04g242600*, *Glyma.04g242700*, *Glyma.04g242800*, *Glyma.04g242900* and *Glyma.04g243000*, in the wild-type and mutant parents for the identification of nucleotide differences and possible candidate gene responsible for lesion mimic mutant phenotype of *spl-1*. The homologous localized region and sequences of candidate genes were obtained from the database of Phytozome (https://phytozome.jgi.doe.gov), and SoyBase (http://soybase.org/). RNA was isolated according to the above-mentioned protocol. Transcript two-step gDNA Removal was used for reverse transcription, and Prime Script™ RT Reagent Kit (TaKaRa, Kusatsu, Shiga, Japan) was used for cDNA synthesis. Primers for qRT-PCR were designed by Primer Premier 5.0 software (Premier Biosoft International). The target gene was subjected to PCR by using Phanta® Max SuperFidelity DNA Polymerase from Vazyme, and sent to GeneScript®, China for sequencing. The alignments of the nucleotide sequences were performed using BioXM software.

5. Conclusions

In this study, we isolated a new soybean leaf lesion mimic (*spl-1*) mutant, in which necrotic lesions started to first visualize on the aged/older leaves, and, finally, the whole leaf became chlorotic yellow. The *lm1* locus controlling mutant phenotype of *spl-1* was fine-mapped in a 76.23 kb genomic region harboring eight candidate genes, and among them, *Glyma.04g242300* was considered to be the possible candidate gene for the mutant phenotype of *spl-1*. We speculate that mutation in this gene affected chlorophyll degradation, resulted in oxidative stress and increased antioxidant activities, which in turn led to necrotic lesions and PCD, and we also suggest this gene may be related for resistance to disease and stress. However, further studies are required for detailed investigation of the actual molecular mechanism and signaling pathways involved in the PCD. The results obtained in this study provide a foundation for the cloning and validating the *lm1* gene of *spl-1* mutant.

Supplementary Materials: Supplementary Materials can be found at http://www.mdpi.com/1422-0067/20/9/2193/s1.

Author Contributions: T.Z. and J.A.B. conceived and designed the experiments. G.M.A.A., K.K., and R.A.S. performed the experiments. G.M.A.A., K.K., and J.K. analyzed the data. G.M.A.A. and J.A.B. drafted the manuscript. T.Z. and J.A.B. revised the paper.

Funding: This work was supported by the National Key Research and Development Program (2018YFD0201006), the National Natural Science Foundation of China (Grant No. 31871646 and 31571691), the MOE Program for Changjiang Scholars and Innovative Research Team in University (PCSIRT_17R55), the Fundamental Research Funds for the Central Universities (KYT201801), and the Jiangsu Collaborative Innovation Center for Modern Crop Production (JCIC-MCP) Program.

Conflicts of Interest: The authors declare no conflict of interest.

Abbreviations

LMMs	lesion mimic mutants
spl-1	spotted Leaf-1
lm1	lesion mimic 1
PCD	programmed cell death
HR	hypersensitive response
ROS	reactive oxygen species
SOD	superoxide dismutase
POD	peroxidase
CAT	catalase
MDA	malondialdehyde
SSR	simple sequence repeat

PLC	plantacyanin
PC	phytocyanin
qRT-PCR	quantitative real-time PCR
NGS	next-generation sequencing
WGRS	whole genome re-sequencing
DAB	3,3′-diaminobenzidine
SNP	Single Nucleotide Polymorphisms

References

1. Li, R.; Chen, S.; Liu, G.; Han, R.; Jiang, J. Characterization and identification of a woody lesion mimic mutant *lmd*, showing defence response and resistance to *Alternaria alternate* in birch. *Sci. Rep.* **2017**, *7*, 11308. [CrossRef]

2. Zhou, Q.; Zhang, Z.; Liu, T.; Gao, B.; Xiong, X. Identification and map-based cloning of the *Light-Induced Lesion Mimic Mutant 1 (LIL1)* gene in rice. *Front. Plant Sci.* **2017**, *8*, 2122. [CrossRef] [PubMed]

3. Chen, P.; Hu, H.; Zhang, Y.; Wang, Z.; Dong, G.; Cui, Y.; Qian, Q.; Ren, D.; Guo, L. Genetic analysis and fine-mapping of a new rice mutant, white and lesion mimic leaf1. *Plant Growth Regul.* **2018**, *85*, 425–435. [CrossRef]

4. Wang, L.P.; Wen, R.; Wang, J.H.; Xiang, D.Q.; Wang, Q.; Zang, Y.P.; Wang, Z.; Huang, S.; Li, X.; Datla, R.; et al. Arabidopsis UBC13 differentially regulates two programmed cell death pathways in responses to pathogen and low-temperature stress. *New Phytol.* **2019**, *221*, 919–934. [CrossRef]

5. Dietrich, R.A.; Delaney, T.P.; Uknes, S.J.; Ward, E.R.; Ryals, J.A.; Dangl, J.L. Arabidopsis mutants simulating disease resistance response. *Cell* **1994**, *77*, 565–577. [CrossRef]

6. Badigannavar, A.M.; Kale, D.M.; Eapen, S.; Murty, G.S.S. Inheritance of disease lesion mimic leaf trait in groundnut. *J. Hered.* **2002**, *93*, 50–52. [CrossRef]

7. Walbot, V. Maize Mutants for the 21st Century. *Plant Cell* **1991**, *3*, 851–856. [CrossRef]

8. Lorrain, S.; Vailleau, F.; Balaque, C.; Roby, D. Lesion mimic mutants: Keys for deciphering cell death and defense pathways in plants? *Trends Plant Sci.* **2003**, *8*, 263–271. [CrossRef]

9. Mizobuchi, R.; Hirabayashi, H.; Kaji, R.; Nishizawa, Y.; Satoh, H.; Ogawa, T.; Okamoto, M. Differential expression of disease resistance in rice lesion-mimic mutants. *Plant Cell Rep.* **2002**, *2*, 390–396. [CrossRef]

10. Wu, C.J.; Bordeos, A.; Madamba, M.R.S.; Baraoidan, M.; Ramos, M.; Wang, G.L.; Leach, J.E.; Leung, H. Rice lesion mimic mutants with enhanced resistance to diseases. *Mol. Genet. Genom.* **2008**, *279*, 605–619. [CrossRef] [PubMed]

11. Buschges, R.; Hollricher, K.; Panstruga, R.; Simons, G.; Wolter, M.; Frijters, A.; van Daelen, R.; van der Lee, T.; Diergaarde, P.; Groenendijk, J.; et al. The barley *mlo* gene: A novel control element of plant pathogen resistance. *Cell* **1997**, *88*, 695–705. [CrossRef]

12. Rostoks, N.; Schmierer, D.; Mudie, S.; Drader, T.; Brueggeman, R.; Caldwell, D.G.; Waugh, R.; Kleinhofs, A. Barley necrotic locus *nec1* encodes the cyclic nucleotide-gated ion channel 4 homologous to the Arabidopsis *HLM1*. *Mol. Genet. Genom.* **2006**, *275*, 159–168. [CrossRef] [PubMed]

13. Johal, G.S.; Hulbert, S.H.; Briggs, S.P. Disease lesion mimics of maize: A model for cell death in plants. *Bioessays* **1995**, *17*, 685–692. [CrossRef]

14. Chung, J.; Staswick, P.E.; Graef, G.L.; Wysong, D.S.; Specht, J.E. Inheritance of a disease lesion mimic mutant in soybean. *J. Hered.* **1998**, *89*, 363–365. [CrossRef]

15. Kim, H.K.; Kim, J.K.; Paek, K.B.; Kim, Y.J.; Chung, J. The phenotype of the soybean disease-lesion imic *(dlm)* mutant is light-dependentand associated with chloroplast function. *Plant Path. J.* **2005**, *21*, 395–401. [CrossRef]

16. Moeder, W.; Yoshioka, K. Lesion mimic mutants A classical, yet still fundamental approach to study programmed cell death. *Plant Sig. Behav.* **2008**, *3*, 764–767. [CrossRef]

17. Jeong, W.H.; Lee, K.J.; Park, M.S.; Nam, K.C.; Kim, M.S.; Chung, J.I. Independent Inheritance of *dlm* Allele with *lf2* and *P1* alleles in soybean (*Glycine max* L.). *Korean J. Breed. Sci.* **2007**, *39*, 232–235.

18. Sung, M.K.; Kim, M.H.; Seo, H.J.; Chung, J.I. Inheritance of *dlm* and *ti* genes in soybean. *Plant Breed. Biotechnol.* **2013**, *1*, 9–13. [CrossRef]

19. Wang, Y.; Chen, W.; Zhang, Y.; Liu, M.; Kong, J.; Yu, Z.; Jaffer, A.M.; Gai, J.; Zhao, T. Identification of two duplicated loci controlling a disease-like rugose leaf phenotype in soybean. *Crop Sci.* **2016**, *56*, 1611–1618. [CrossRef]

20. Lu, P.; Qin, J.; Wang, G.; Wang, L.; Wang, Z.; Wu, Q.; Xie, J.; Liang, Y.; Wang, Y.; Zhang, D.; et al. Comparative fine mapping of the *Wax 1* (*W1*) locus in hexaploid wheat. *Theor. Appl. Genet.* **2015**, *128*, 1595–1603. [CrossRef]

21. Xu, J.; Wang, B.; Wu, Y.; Du, P.; Wang, J.; Wang, M.; Yi, C.; Gu, M.; Liang, G. Fine mapping and candidate gene analysis of *ptgms2-1*, the photoperiod-thermo-sensitive genic male sterile gene in rice (*Oryza sativa* L.). *Theor. Appl. Genet.* **2011**, *122*, 365–372. [CrossRef] [PubMed]

22. Fang, Y.; Hu, J.; Xu, J.; Yu, H.; Shi, Z.; Xiong, G.; Zhu, L.; Zeng, D.; Zhang, G.; Gao, Z.; et al. Identification and characterization of *Mini1*, a gene regulating rice shoot development. *J. Integr. Plant Biol.* **2015**, *57*, 151–161. [CrossRef] [PubMed]

23. Song, J.; Li, Z.; Liu, Z.; Guo, Y.; Qiu, L. Next-generation sequencing from Bulked-Segregant Analysis accelerates the simultaneous identification of two qualitative genes in soybean. *Front. Plant Sci.* **2017**, *8*, 919. [CrossRef] [PubMed]

24. Schneeberger, K.; Weigel, D. Fast-forward genetics enabled by new sequencing technologies. *Trends Plant Sci.* **2011**, *16*, 282–288. [CrossRef] [PubMed]

25. Zegeye, W.A.; Zhang, Y.; Cao, L.; Cheng, S. Whole Genome Resequencing from bulked populations as a rapid QTL and gene identification method in rice. *Inter. J. Mol. Sci.* **2018**, *19*, 4000. [CrossRef] [PubMed]

26. Das, S.; Upadhyaya, H.D.; Bajaj, D.; Kujur, A.; Badoni, S.; Laxmi; Kumar, V.; Tripathi, S.; Gowda, C.L.L.; Sharma, S.; et al. Deploying QTL-seq for rapid delineation of a potential candidate gene underlying major trait-associated QTL in chickpea. *DNA Res.* **2015**, *22*, 193–203. [CrossRef] [PubMed]

27. Mascher, M.; Jost, M.; Kuon, J.E.; Himmelbach, A.; Aßfalg, A.; Beier, S.; Scholz, U.; Graner, A.; Stein, N. Mapping-by-sequencing accelerates forward genetics in barley. *Genome Biol.* **2014**, *15*, R78. [CrossRef]

28. Liu, S.; Yeh, C.T.; Tang, H.M.; Nettleton, D.; Schnable, P.S. Gene mapping via Bulked Segregant RNA-Seq (BSR-Seq). *PLoS ONE* **2012**, *7*, e36406. [CrossRef]

29. Zhong, C.; Sun, S.; Li, Y.; Duan, C.; Zhu, Z. Next-generation sequencing to identify candidate genes and develop diagnostic markers for a novel *Phytophthora* resistance gene, *RpsHC18*, in soybean. *Theor. Appl. Genet.* **2018**, *131*, 525–538. [CrossRef]

30. Illa-Berenguer, E.; Houten, J.V.; Huang, Z.; van der Knaap, E. Rapid and reliable identification of tomato fruit weight and locule number loci by QTL-seq. *Theor. Appl. Genet.* **2015**, *128*, 1329–1342. [CrossRef]

31. Lu, H.; Lin, T.; Klein, J.; Wang, S.; Qi, J.; Zhou, Q.; Sun, J.; Zhang, Z.; Weng, Y.; Huang, S. QTL-seq identifies an early flowering QTL located near flowering locus *T* in cucumber. *Theor. Appl. Genet.* **2014**, *127*, 1491–1499. [CrossRef] [PubMed]

32. Pruyn, M.L.; Spicer, R. Parenchyma. *e LS* **2001**. [CrossRef]

33. Kura-Hotta, M.; Satoh, K.; Katoh, S. Relationship between photosynthesis and chlorophyll content during leaf senescence of rice seedlings. *Plant Cell Physiol.* **1987**, *28*, 1321–1329.

34. Sack, L.; Cowan, P.D.; Jaikumar, N.; Holbrook, N.M. The 'hydrology' of leaves: Co-ordination of structure and function in temperate woody species. *Plant Cell Environ.* **2003**, *26*, 1343–1356. [CrossRef]

35. Sack, L.; Holbrook, N.M. Leaf hydraulics. *Annu. Rev. Plant Biol.* **2006**, *57*, 361–381. [CrossRef]

36. Brodribb, T.J.; Feild, T.S.; Sack, L. Viewing leaf structure and evolution from a hydraulic perspective. *Funct. Plant Biol.* **2010**, *37*, 488–498. [CrossRef]

37. Brodribb, T.J.; Holbrook, N.M. Stomatal closure during leaf dehydration, correlation with other leaf physiological traits. *Plant Physiol.* **2003**, *132*, 2166–2173. [CrossRef]

38. Wang, J.; Ye, B.; Yin, J.; Yuan, C.; Zhou, X.; Li, W.; He, M.; Wang, J.; Chen, W.; Qin, P.; et al. Characterization and fine mapping of a light-dependent leaf lesion mimic mutant 1 in rice. *Plant Physiol. Biochem.* **2015**, *97*, 44–51. [CrossRef]

39. Van Breusegem, F.; Dat, J.F. Reactive Oxygen Species in Plant Cell Death. *Plant Physiol.* **2006**, *141*, 384–390. [CrossRef]

40. Schmutz, J.; Cannon, S.B.; Schlueter, J.; Ma, J.; Mitros, T.; Nelson, W.; Hyten, D.L.; Song, Q.; Thelen, J.J.; Cheng, J.; et al. Genome sequence of the palaeopolyploid soybean. *Nature* **2010**, *463*, 178–183. [CrossRef]

41. Dong, J.; Kim, S.T.; Lord, E.M. Plantacyanin plays a role in reproduction in Arabidopsis. *Plant Physiol.* **2005**, *138*, 778–789. [CrossRef]

42. Fehr, W.R.; Caviness, C.E.; Burmood, D.T.; Pennington, J.S. Stage of development descriptions for soybeans, *Glycine max* (L.) Merrill. *Crop Sci.* **1971**, *11*, 929–931. [CrossRef]

43. Hunt, M.D.; Neuenschwander, U.H.; Delaney, T.P.; Weymann, K.B.; Friedrich, L.B.; Lawton, K.A.; Steiner, H.Y.; Ryals, J.A. Recent advances in systemic acquired resistance research—A review. *Gene* **1996**, *179*, 89–95. [CrossRef]

44. Yin, Z.C.; Chen, J.; Zeng, L.R.; Goh, M.L.; Leung, H.; Khush, G.S.; Wang, G.L. Characterizing rice lesion mimic mutants and identifying a mutant with broad-spectrum resistance to rice blast and bacterial blight. *Mol. Plant-Microbe Interact.* **2000**, *13*, 869. [CrossRef] [PubMed]

45. Mochizuki, N.; Tanaka, R.; Grimm, B.; Masuda, T.; Moulin, M.; Smith, A.G.; Tanaka, A.; Terry, M.J. The cell biology of tetrapyrroles: A life and death struggle. *Trends Plant Sci.* **2010**, *15*, 488–498. [CrossRef] [PubMed]

46. Quesada, V.; Sarmiento-Manus, R.; Gonzalez-Bayon, R.; Hricova, A.; Ponce, M.R.; Micol, J.L. PORPHOBILINOGEN DEAMINASE deficiency alters vegetative and reproductive development and causes lesions in arabidopsis. *PLoS ONE* **2013**, *8*, e53378. [CrossRef] [PubMed]

47. Meskauskiene, R.; Nater, M.; Goslings, D.; Kessler, F.; den Camp, R.O.; Apel, K. FLU: A negative regulator of chlorophyll biosynthesis in *Arabidopsis thaliana*. *Proc. Natl. Acad. Sci. USA* **2001**, *98*, 12826–12831. [CrossRef]

48. Samol, I.; Buhr, F.; Springer, A.; Pollmann, S.; Lahroussi, A.; Rossig, C.; von Wettstein, D.; Reinbothe, C.; Reinbothe, S. Implication of the *oep16-1* Mutation in a flu-Independent, singlet oxygen-regulated cell death pathway in *Arabidopsis thaliana*. *Plant Cell Physiol.* **2011**, *52*, 84–95. [CrossRef]

49. Hu, G.S.; Yalpani, N.; Briggs, S.P.; Johal, G.S. A porphyrin pathway impairment is responsible for the phenotype of a dominant disease lesion mimic mutant of maize. *Plant Cell* **1998**, *10*, 1095–1105. [CrossRef]

50. Ishikawa, A.; Okamoto, H.; Iwasaki, Y.; Asahi, T. A deficiency of coproporphyrinogen III oxidase causes lesion formation in Arabidopsis. *Plant J.* **2001**, *27*, 89–99. [CrossRef]

51. Sun, C.; Liu, L.; Tang, J.; Lin, A.; Zhang, F.; Fang, J.; Zhang, G.; Chu, C. *RLIN1*, encoding a putative coproporphyrinogen III oxidase, is involved in lesion initiation in rice. *J. Genet. Genom.* **2011**, *38*, 29–37. [CrossRef]

52. Greenberg, J.T.; Ausubel, F.M. Arabidopsis mutants compromised for the control of cellular damage during pathogenesis and aging. *Plant J.* **1993**, *4*, 327–341. [CrossRef] [PubMed]

53. Mach, J.M.; Castillo, A.R.; Hoogstraten, R.; Greenberg, J.T. The Arabidopsis accelerated cell death gene *ACD2* encodes red chlorophyll catabolite reductase and suppresses the spread of disease symptoms. *Proc. Natl. Acad. Sci. USA* **2001**, *98*, 771–776. [CrossRef]

54. Tanaka, R.; Hirashima, M.; Satoh, S.; Tanaka, A. The Arabidopsis-accelerated cell death gene *ACD1* is involved in oxygenation of pheophorbide a: Inhibition of the pheophorbide a oxygenase activity does not lead to the "stay-green" phenotype in Arabidopsis. *Plant Cell Physiol.* **2003**, *44*, 1266–1274. [CrossRef] [PubMed]

55. Anand, A.; Schmelz, E.A.; Muthukrishnan, S. Development of a lesion-mimic phenotype in a transgenic wheat line overexpressing genes for pathogenesis-related (PR) proteins is dependent on salicylic acid concentration. *Mol. Plant-Microbe Interact.* **2003**, *16*, 916–925. [CrossRef] [PubMed]

56. Zhou, F.; Lin, Q.; Zhu, L.; Ren, Y.; Zhou, K.; Shabek, N.; Wu, F.; Mao, H.; Dong, W.; Gan, L.; et al. D14-SCFD3-dependent degradation of D53 regulates strigolactone signalling. *Nature* **2013**, *504*, 406. [CrossRef] [PubMed]

57. Chang, J.; Yu, T.; Gao, S.; Xiong, C.; Xie, Q.; Li, H.; Ye, Z.; Yang, C. Fine mapping of the dialytic gene that controls multicellular trichome formation and stamen development in tomato. *Theor. Appl. Genet.* **2016**, *129*, 1531–1539. [CrossRef] [PubMed]

58. Xia, Z.J.; Watanabe, S.; Yamada, T.; Tsubokura, Y.; Nakashima, H.; Zhai, H.; Anai, T.; Sato, S.; Yamazaki, T.; Lu, S.X.; et al. Positional cloning and characterization reveal the molecular basis for soybean maturity locus E1 that regulates photoperiodic flowering. *Proc. Natl. Acad. Sci. USA* **2012**, *109*, E2155–E2164. [CrossRef] [PubMed]

59. Zhang, X.; Wang, W.; Guo, N.; Zhang, Y.; Bu, Y.; Zhao, J.; Xing, H. Combining QTL-seq and linkage mapping to fine map a wild soybean allele characteristic of greater plant height. *BMC Genom.* **2018**, *19*, 226. [CrossRef] [PubMed]

60. Kadambari, G.; Vemireddy, L.R.; Srividhya, A.; Nagireddy, R.; Jena, S.S.; Gandikota, M.; Patil, S.; Veeraghattapu, R.; Deborah, D.A.K.; Reddy, G.E.; et al. QTL-Seq-based genetic analysis identifies a major genomic region governing dwarfness in rice (*Oryza sativa* L.). *Plant Cell Rep.* **2018**, *37*, 677–689. [CrossRef]

61. Ryden, L.G.; Hunt, L.T. Evolution of protein complexity - the blue copper-containing oxidases and related proteins. *J. Mol. Evol.* **1993**, *36*, 41–66. [CrossRef]

62. Roach, T.; Krieger-Liszkay, A. Regulation of photosynthetic electron transport and photoinhibition. *Curr. Protein Pept. Sci.* **2014**, *15*, 351–362. [CrossRef] [PubMed]

63. Zhang, J.P.; Yu, Y.; Feng, Y.Z.; Zhou, Y.F.; Zhang, F.; Yang, Y.W.; Lei, M.Q.; Zhang, Y.C.; Chen, Y. MiR408 pregulates grain yield and photosynthesis via a phytocyanin rotein. *Plant Physiol.* **2017**, *175*, 1175–1185. [CrossRef]

64. Fedorova, M.; van de Mortel, J.; Matsumoto, P.A.; Cho, J.; Town, C.D.; VandenBosch, K.A.; Gantt, J.S.; Vance, C.P. Genome-wide identification of nodule-specific transcripts in the model legume Medicago truncatula. *Plant Physiol.* **2002**, *130*, 519–537. [CrossRef]

65. Kim, S.; Mollet, J.C.; Dong, J.; Zhang, K.L.; Park, S.Y.; Lord, E.M. Chemocyanin, a small basic protein from the lily stigma, induces pollen tube chemotropism. *Proc. Natl. Acad. Sci. USA* **2003**, *100*, 16125–16130. [CrossRef]

66. Ma, H.; Zhao, H.; Liu, Z.; Zhao, J. The Phytocyanin Gene Family in Rice (*Oryza sativa* L.): Genome-wide identification, classification and transcriptional analysis. *PLoS ONE* **2011**, *6*, e25184. [CrossRef] [PubMed]

67. Ozturk, Z.N.; Talame, V.; Deyholos, M.; Michalowski, C.B.; Galbraith, D.W.; Gozukirmizi, N.; Tuberosa, R.; Bohnert, H.J. Monitoring large-scale changes in transcript abundance in drought- and salt-stressed barley. *Plant Mol. Biol.* **2002**, *48*, 551–573. [CrossRef]

68. Kreps, J.A.; Wu, Y.J.; Chang, H.S.; Zhu, T.; Wang, X.; Harper, J.F. Transcriptome changes for Arabidopsis in response to salt, osmotic, and cold stress. *Plant Physiol.* **2002**, *130*, 2129–2141. [CrossRef] [PubMed]

69. Provart, N.J.; Gil, P.; Chen, W.Q.; Han, B.; Chang, H.S.; Wang, X.; Zhu, T. Gene expression phenotypes of Arabidopsis associated with sensitivity to low temperatures. *Plant Physiol.* **2003**, *132*, 893–906. [CrossRef]

70. Chen, X.M. A microRNA as a translational repressor of *APETALA2* in Arabidopsis flower development. *Science* **2004**, *303*, 2022–2025. [CrossRef]

71. Llave, C.; Xie, Z.X.; Kasschau, K.D.; Carrington, J.C. Cleavage of Scarecrow-like mRNA targets directed by a class of Arabidopsis miRNA. *Science* **2002**, *297*, 2053–2056. [CrossRef] [PubMed]

72. Sunkar, R.; Zhu, J.K. Novel and stress-regulated microRNAs and other small RNAs from Arabidopsis. *Plant Cell* **2004**, *16*, 2001–2019. [CrossRef] [PubMed]

73. Lichtenthaler, H.K. Chlorophylls and carotenoids-pigments of photosynthetic biomembranes. *Methods Enzymol.* **1987**, *148*, 350–382.

74. Carland, F.M.; McHale, N.A. LOP1: A gene involved in auxin transport and vascular patterning in Arabidopsis. *Development* **1996**, *122*, 1811–1819.

75. Chen, F.; Wang, F.; Wu, F.; Mao, W.; Zhang, G.; Zhou, M. Modulation of exogenous glutathione in antioxidant defense system against Cd stress in the two barley genotypes differing in Cd tolerance. *Plant Physiol. Biochem.* **2010**, *48*, 663–672. [CrossRef] [PubMed]

76. Rahman, A.; Nahar, K.; Hasanuzzaman, M.; Fujita, M. Calcium supplementation improves Na+/K+ ratio, antioxidant defense and glyoxalase systems in salt-stressed rice seedlings. *Front. Plant Sci.* **2016**, *7*, 609. [CrossRef]

77. Li, Z.; Zhang, Y.; Liu, L.; Liu, Q.; Bi, Z.; Yu, N.; Cheng, S.; Cao, L. Fine mapping of the lesion mimic and early senescence 1 (*lmes1*) in rice (*Oryza sativa*). *Plant Physiol. Biochem.* **2014**, *80*, 300–307. [CrossRef] [PubMed]

78. Langmead, B.; Salzberg, S.L. Fast gapped-read alignment with Bowtie 2. *Nat. Methods* **2012**, *9*, 357–359. [CrossRef]

79. Li, H.; Durbin, R. Fast and accurate short read alignment with Burrows-Wheeler transform. *Bioinformatics* **2009**, *25*, 1754–1760. [CrossRef]

80. Takagi, H.; Abe, A.; Yoshida, K.; Kosugi, S.; Natsume, S.; Mitsuoka, C.; Uemura, A.; Utsushi, H.; Tamiru, M.; Takuno, S.; et al. QTL-seq: Rapid mapping of quantitative trait loci in rice by whole genome resequencing of DNA from two bulked populations. *Plant J.* **2013**, *74*, 174–183. [CrossRef]

81. Takagi, H.; Uemura, A.; Yaegashi, H.; Tamiru, M.; Abe, A.; Mitsuoka, C.; Utsushi, H.; Natsume, S.; Kanzaki, H.; Matsumura, H.; et al. MutMap-Gap: Whole-genome resequencing of mutant F_2 progeny bulk combined with de novo assembly of gap regions identifies the rice blast resistance gene *Pii*. *New Phytol.* **2013**, *200*, 276–283. [CrossRef]

82. Abe, A.; Kosugi, S.; Yoshida, K.; Natsume, S.; Takagi, H.; Kanzaki, H.; Matsumura, H.; Yoshida, K.; Mitsuoka, C.; Tamiru, M.; et al. Genome sequencing reveals agronomically important loci in rice using MutMap. *Nat. Biotechnol.* **2012**, *30*, 174–178. [CrossRef] [PubMed]

83. Fekih, R.; Takagi, H.; Tamiru, M.; Abe, A.; Natsume, S.; Yaegashi, H.; Sharma, S.; Sharma, S.; Kanzaki, H.; Matsumura, H.; et al. MutMap plus: Genetic mapping and mutant identification without crossing in rice. *PLoS ONE* **2013**, *8*, e68529. [CrossRef] [PubMed]

84. Singh, V.K.; Khan, A.W.; Saxena, R.K.; Kumar, V.; Kale, S.M.; Sinha, P.; Chitikineni, A.; Pazhamala, L.T.; Garg, V.; Sharma, M.; et al. Next-generation sequencing for identification of candidate genes for Fusarium wilt and sterility mosaic disease in pigeonpea (*Cajanus cajan*). *Plant Biotechnol. J.* **2016**, *14*, 1183–1194. [CrossRef] [PubMed]

85. Song, Q.J.; Jia, G.F.; Zhu, Y.L.; Grant, D.; Nelson, R.T.; Hwang, E.Y.; Hyten, D.L.; Cregan, P.B. Abundance of SSR motifs and development of candidate polymorphic SSR markers (BARCSOYSSR_1.0) in soybean. *Crop Sci.* **2010**, *50*, 1950–1960. [CrossRef]

86. Michelmore, R.W.; Paran, I.; Kesseli, R.V. Identification of markers linked to disease-resistance genes by bulked segregant analysis: A rapid method to detect markers in specific genomic regions by using segregating populations. *Proc. Natl. Acad. Sci. USA* **1991**, *88*, 9828–9832. [CrossRef]

87. Lander, E.S.; Green, P.; Abrahamson, J.; Barlow, A.; Daly, M.J.; Lincoln, S.E.; Newberg, L.A. MAPMAKER: An interactive computer package for constructing primary genetic linkage maps of experimental and natural populations. *Genomics* **1987**, *1*, 1174–1181. [CrossRef]

International Journal of
Molecular Sciences

Article

Allelic Diversity of Acetyl Coenzyme A Carboxylase *accD/bccp* Genes Implicated in Nuclear-Cytoplasmic Conflict in the Wild and Domesticated Pea (*Pisum* sp.)

Eliška Nováková [1,†], Lenka Zablatzká [1,†], Jan Brus [2], Viktorie Nesrstová [3], Pavel Hanáček [4], Ruslan Kalendar [5,6], Fatima Cvrčková [7], Ľuboš Majeský [1] and Petr Smýkal [1,*]

[1] Department of Botany, Faculty of Sciences, Palacký University, 78371 Olomouc, Czech Republic; eli.novakova@seznam.cz (E.N.); zablatzka.lenka@seznam.cz (L.Z.); lubos.majesky@upol.cz (Ľ.M.)

[2] Department of Geoinformatics, Faculty of Sciences, Palacký University, 78371 Olomouc, Czech Republic; jan.brus@upol.cz

[3] Department of Mathematical Analysis and Applications of Mathematics, Palacký University, 78371 Olomouc, Czech Republic; viktorie.nesrstova@gmail.com

[4] Department of Plant Biology, Faculty of Agronomy, Mendel University, 61300 Brno, Czech Republic; hanacek@mendelu.cz

[5] National Center for Biotechnology, Astana 010000, Kazakhstan; ruslan.kalendar@mail.ru

[6] Department of Agricultural Sciences, Viikki Plant Science Centre and Helsinki Sustainability Centre, University of Helsinki, FI-00014 Helsinki, Finland

[7] Department of Experimental Plant Biology, Faculty of Sciences, Charles University, 12844 Prague, Czech Republic; fatima.cvrckova@natur.cuni.cz

* Correspondence: petr.smykal@upol.cz

† These authors contributed equally to this work.

Received: 22 March 2019; Accepted: 8 April 2019; Published: 10 April 2019

Abstract: Reproductive isolation is an important component of species differentiation. The plastid *accD* gene coding for the acetyl-CoA carboxylase subunit and the nuclear *bccp* gene coding for the biotin carboxyl carrier protein were identified as candidate genes governing nuclear-cytoplasmic incompatibility in peas. We examined the allelic diversity in a set of 195 geographically diverse samples of both cultivated (*Pisum sativum, P. abyssinicum*) and wild (*P. fulvum* and *P. elatius*) peas. Based on deduced protein sequences, we identified 34 *accD* and 31 *bccp* alleles that are partially geographically and genetically structured. The *accD* is highly variable due to insertions of tandem repeats. *P. fulvum* and *P. abyssinicum* have unique alleles and combinations of both genes. On the other hand, partial overlap was observed between *P. sativum* and *P. elatius*. Mapping of protein sequence polymorphisms to 3D structures revealed that most of the repeat and indel polymorphisms map to sequence regions that could not be modeled, consistent with this part of the protein being less constrained by requirements for precise folding than the enzymatically active domains. The results of this study are important not only from an evolutionary point of view but are also relevant for pea breeding when using more distant wild relatives.

Keywords: acetyl-CoA carboxylase; hybrid incompatibility; hybrid necrosis; nuclear-cytoplasmic conflict; pea; reproductive isolation; speciation

1. Introduction

Reproductive isolation is an important component of species differentiation. Mechanisms that create reproductive barriers between once-conspecific organisms have long been a focus of evolutionary biology [1]. Although geographical separation plays a vital role in speciation [2], ecological factors also contribute [3]. Ecological selection favoring a particular cytoplasm has been described from various taxa [4,5]. Hybrid incompatibility due to the genetic divergence between the hybridizing parents has been theorized already by Bateson [6], Dobzhansky [7], and Muller [8]. Hybrid incompatibilities are proposed to be among the first genetic barriers to arise during speciation [9]. Although interspecific hybridization seems to be relatively frequent in plants, comparatively less is known about the reproductive barriers within species [10]. The most classical definition of the species relies on reproductive isolation, namely the inability to produce a viable offspring from inter-species hybridization [11,12]. Reproductive barriers might be broadly classified into prezygotic (pre-pollination) and postzygotic (post-pollination) ones [13]. Pre-pollination isolation mechanisms, such as habitat divergence, temporal isolation, pollinator isolation, and mating system divergence, are usually more effective than post-pollination isolation [2].

Interactions among nuclear-encoded genes can lead to diverse forms of hybrid incompatibility via multiple gametophytic and sporophytic mechanisms [9]. The identification of so-called 'speciation genes' is of interest because their knowledge would offer clues to the ecological settings, evolutionary forces, and molecular mechanisms that drive the divergence of populations and species [12,14]. A speciation gene can be strictly defined as a gene that contributes to the splitting of two lineages by reducing the amount of gene flow between them [12].

Until recently, characterization of genetic incompatibility has largely focused on the differences between species and on nuclear incompatibilities [2,12,15–17]. As a result, the importance of cytonuclear incompatibility (i.e., incompatibility between the nuclear and organelle genomes) in driving the early stages of speciation received less attention [10]. There has been long co-evolution between the nuclei and organelles. Molecular data indicate a large degree of interdependence between the cellular sub-genomes [18]. The subdivided eukaryotic genome has resulted from a massive restructuring and intermixing of the genomes of the initially free-living symbiotic partner cells with loss, intracellular transfer, and gain of genetic information, with resulting high interdependence and mutual "fine tuning" of both genomes that can easily become disrupted upon intraspecific hybridization. Cytonuclear incompatibilities are predisposed to be substantial contributors to reproductive isolation and speciation [19,20]. Empirical studies have shown that intrinsic postzygotic barriers to reproduction—hybrid inviability and hybrid sterility—evolve through mechanisms consistent with the classic Bateson–Dobzhansky–Muller model [9]. As adaptive or nearly neutral substitutions accumulate in diverging lineages, these may in a particular lineage become fixed in a state incompatible with that in the other lineage. As a result, the hybrid dysfunction occurs when such incompatible alleles are brought together. The genetic basis for hybrid sterility has been studied in several plants, such as *Solanum* [21], *Oryza* [22], *Mimulus* [23], *Oenothera* [24], *Arabidopsis lyrata* [25], and *A. thaliana* [26,27]. There are two classes of cytonuclear hybrid incompatibility: cytoplasmic male sterility (CMS), due to mitochondrial-nuclear mismatch, and cytonuclear chlorosis, caused by plastome–nuclear incompatibilities. Organelle genomes have a reduced population size and lack sexual recombination [28]. These characteristics both increase genetic drift, and lead to potential accumulation of deleterious mutations and selection for compensatory evolution in interacting nuclear genes. Due to these factors, cytonuclear incompatibilities have been proposed to be among the first genetic incompatibilities to arise, influencing the earliest stages of speciation [10,19,20,24,29].

Most known plant sterility loci have been found in the mitochondrial genome, causing CMS characterized by the absence of viable pollen. The genetics of hybrid CMS are remarkably conserved across flowering plants. Molecular genetic studies indicate that CMS typically results from rearrangements in the mitochondrial genome [30,31]. As mitochondria are usually maternally inherited, CMS is typically transmitted through the ovules. In contrast, nuclear genes are transmitted through both ovules and pollen. This difference in inheritance patterns creates a genetic conflict between nuclear and cytoplasmic genes. Hybrid nucleo-organelle dysfunction can result in post-zygotic hybridization barriers that usually manifest as differences in the offspring of reciprocal crosses owing to non-Mendelian inheritance of organelles. Asymmetry in reproductive isolation appears to be common and taxonomically widespread among plant species. Plastids can also contribute to nucleo-cytoplasmic incompatibility. Although cytonuclear chlorosis or albinism of hybrids is not as common as CMS, these have been widely observed, and their implications for speciation were recognized early on [32–35]. The role of plastids in speciation processes is known from species with a biparental mode of plastid inheritance, e.g., *Geranium*, *Pelargonium* and *Medicago* [36], and mainly from genus *Oenothera*, which became one of the models for studying plant evolution [24]. Various incompatible phenotypes have also been reported from *Rhododendron*, *Hypericum*, *Trifolium*, *Zantedeschia*, and *Pisum* [24]. Cyto-nuclear co-adaptation has been described in *Arabidopsis thaliana* [18] and demonstrated to affect its adaptive traits [37]. Interestingly, crop domestication may also increase the likelihood that genes causing incompatibility become fixed in the population through genetic hitchhiking [38].

The plastid *accD* gene coding for the acetyl-CoA carboxylase beta subunit and the nuclear gene *bccp* coding for the biotin carboxyl carrier protein of acetyl-CoA carboxylase were nominated as candidate genes responsible for nuclear-cytoplasmic incompatibility in peas based on data from crosses between wild and domesticated pea forms [39]. Incompatible hybrids exhibit chlorophyll deficiency, reduction of leaf size low pollen fertility, low seed set, and poorly developed roots [40]. The acetyl-CoA carboxylase (ACCase) complex is involved in the biosynthesis of fatty acids, which takes place in the plastids [40]. ACCase belongs to a group of biotin dependent carboxylases, catalyzing acetyl-coenzyme A carboxylation to malonyl coenzyme A and providing the only entry point for all carbon atoms in the fatty acid synthesis pathway [41]. Uniquely in Eukaryota, plants have two distinct ACCases: one eukaryotic-like homomeric multidomain ACCase in the cytosol and a bacterial-like heteromeric ACCase within the plastids [41]. The heteromeric form of ACCase is found in prokaryotes and the plastids of Viridiplantae. Presumably, all genes encoding ACCase subunits initially resided in the plastid genome after the original endosymbiotic event in algae and underwent sequential transfer to the nuclear genome [42]. Plastid ACCase participates in fatty acid synthesis, whereas the cytosolic enzyme is engaged in the synthesis of very long chain fatty acids, phytoalexins, flavonoids, and anthocyanins. Plastid-localized ACCD enzyme is responsible for catalyzing the initial tightly-regulated and rate-limiting step in fatty acid biosynthesis. Nuclear encoded Biotin Carboxyl Carrier Protein (BCCP) is a part of the enzyme Acetyl-CoA carboxylase complex and serves as a carrier protein for biotin and carboxybiotin throughout the ATP-dependent carboxylation of acetyl-CoA to form malonyl-CoA. The resulting Acetyl-CoA carboxylase is a heterohexamer composed of the biotin carboxyl carrier protein, biotin carboxylase, and two subunits each of the ACCase subunit alpha and the ACCase plastid-coded subunit beta [40].

The plastid ACCase of legumes (Papilionoideae) consists of four subunits, each coded by a separate gene: biotin carboxylase (*accC*), biotin carboxyl carrier protein (*accB=bccp*), alpha-carboxyltransferase (*accA*), and beta-carboxyltransferase (*accD*). The genes coding *accC*, *accB*, and *accA* are localized in the nuclear genome, whereas the *accD* gene is localized in the plastid genome [42]. Multiple independent lineages have experienced accelerated rates of substitution in similar subsets of non-photosynthetic genes, including *accD* (in legumes [43–45] and in Oleaceae [46]). In *Silene* (Caryophyllaceae) species with accelerated plastid genome evolution, the nuclear-encoded subunits of the ACCase complexes are also evolving rapidly, indicating a strong positive selection [47]. Such patterns of molecular evolution in these plastid–nuclear complexes are unusual for ancient conserved enzymes but resemble cases of antagonistic coevolution between pathogens and host immune genes. Genetic characterization of hybrid necrosis in crosses between tomato species [48] and between *Arabidopsis* ecotypes [49,50] has revealed that incompatibilities among complementary disease resistance genes might play such a role in the evolution of hybrid inviability [51].

In this work, we explored the allelic diversity of *accD*/*bccp* in the geographically diverse set of wild pea (*Pisum* sp.). The *accD*/*bccp* are recently identified genes underlying nuclear-cytoplasmic incompatibility in *Pisum* sp. [39]. We sought to map the allelic combinations of *accD*/*bccp* occurring in nature to determine geographic patterns in their distribution, and to identify possible relationships to pea genetic diversity.

2. Results

2.1. Structure and Variation of accD Gene

The *accD* gene is located between positions 70,882 and 72,654 in the *P. sativum* cv. Feltham First (HM029370) reference chloroplast genome, resulting in a 1772 bp DNA encoding a protein of 432 amino acid residues. The primers used in our study were designed to match the most conserved region and were located close to the ends of the *accD* coding sequence. Consequently, we did not capture the very 5′ and 3′ end of the coding sequence due to quality trimming. The beginning and end of the *accD* sequence, comprising 48 nt from the start codon and 58 nt from the stop codon, consequently missing the first 16 and last 19 codons, were thus excluded from the subsequent analysis.

The length of the *accD* gene within our studied material ranged from 1403 bp to 1859 bp at DNA level and from 467 to 619 amino acid residues, respectively (GenBank accession numbers MK619486—MK619678). In the studied set of 195 accessions, there was extraordinary variation in the gene length, due to the occurrence of 13 indels whose length varied between 3 and 167 nucleotides. This variation is due to insertions consisting of tandem repeats of 10-150 bp units present in 1 to 37 nearly identical copies, all in the same (i.e., direct) orientation relative to each other (Figure 1). The repetitive sequences can be divided into 6 categories. In the shortest 1403 bp allele (JI1010, *P. fulvum*) there are four, three, and one repeats of 9 to 12 bp long. These expand in the longest 1859 bp allele (JI267, *P. elatius*), which has 37 repeats of 10 to 33 bp, 1 repeat of 57 bp, 1 repeat of 102 bp, and 1 repeat of 149 bp. We identified the main five longest tandem repeats blocks, which consist of two or three individual blocks of different lengths and degrees of identity. These blocks are not identical and contain many nucleotide changes and triplet duplications. Such repeats were identified by the presence of small, almost identical blocks, that are part of larger tandem repeats. The first tandem repeat block is the most complex and most degenerate, consisting of three sequential blocks (highlighted in yellow in Figure 1, Figure S1). These blocks are of different lengths and are degenerate to varying degrees from each other. The most similar are first two blocks (89%), which differ by 3 amino acids and by the insertion L-I-L-I for a total of 64 amino acid residues. Characteristic for this tandem repeat is the presence of multiple duplications of three amino acids D-T-N alone or together with D-I-S. The complex, degenerate, and mixed tandem repeat is also the penultimate (3 and 4 grey blocks). This tandem repeat has multiple duplications of five amino acid stretch of S-E-E-E-K. The remaining repeats consist of two blocks separated from each other by 7 or 9 amino acids (Figure 1).

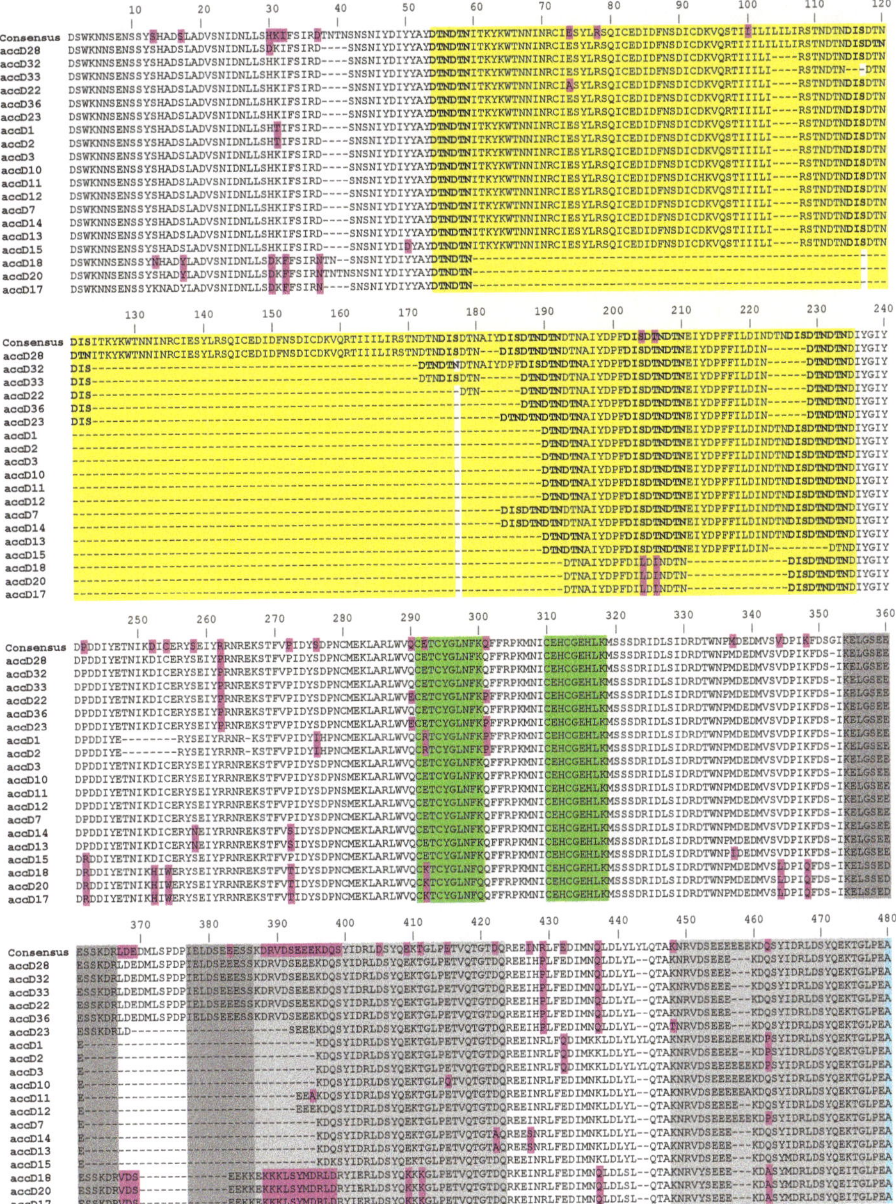

Figure 1. The alignment of amino acid sequences of all identified *accD* alleles. The figure only shows the region from 1 to 480 amino acid residues. The colored regions show the 5 translated repeats, polymorphic amino acid exchanges (in magenta), Zn-finger (boxed), acetyl-CoA binding (light blue), coA carboxylation catalytic (dark blue), and carboxybiotin binding (in green) sites. Residues in purple are point mutations in at least one haplotype. There are no indels after position 480 (for full see Figure S1).

2.2. Variation in Nuclear bccp Gene

The predicted ORF of the *bccp* gene encoding the biotin carboxyl carrier protein of *P. sativum* cv. Cameor from the pea RNA atlas is 873 bp long and encodes a protein of 290 amino acids. In the pea RNA atlas, this is represented by the ubiquitously expressed PsCam051640 transcript, which corresponds to Tayeh et al. (2015) map PsCam051640 at LGIII. The genomic DNA extracted from the shotgun genome sequence is 5906 bp, with 9 exons interspersed by 8 introns (Exon 1 is 234 bp, exon 2 is 206 bp, exon 3 is 76 bp, exon 4 is 54 bp, exon 5 is 262 bp, exon 6 is 62 bp, exon 7 is 69 bp, exon 8 is 46, and exon 9 is 265 bp). The respective introns are 1170, 541, 263, 874, 111, 856, 84, and 733 bp. The following analysis was conducted on cDNA, avoiding introns. The detected polymorphism, thus, only concerns the coding sequence, and is correspondingly lower than that expected for the complete locus. Notably, to obtain sufficient PCR product we had to perform out two consecutive nested PCR amplifications. This likely reflects the relatively low expression level of the gene in young leaf tissue. There were altogether 39 variable positions and no indels in a total of 195 studied accessions (NCBI accession numbers MK644626—MK644819). These identified 31 protein *bccp* variants (Table S1). Sixteen analyzed *P. fulvum* accessions had three *bccp* alleles (*bccp*1/2/3) separated by 4 to 10 amino acid changes from the nearest *P. elatius* alleles. From domesticated *P. sativum* landraces (60 acc.), 16 had the *bccp_22*, and six had the *bccp_18* allele. From the independently domesticated Ethiopian pea *P. abyssinicum* (24 acc.), 19 had the specific *bccp_26* allele, shared with two *P. elatius* accessions (PI343978, PI343979 from Turkey), four had the *bccp_20* allele, separated by one or two amino acid exchanges from nearest *P. elatius*. Ninety-five analyzed *P. elatius* accessions had the largest diversity (all together 28 distinct *bccp* alleles, Table S1).

2.3. Network and Maximum Parsimony Analyses

Various approaches in the visualization of the data through networks and maximum parsimony (MP) analysis produced a very similar view, with only minimal differences. For further interpretation of clustering of identified alleles into larger groups, the consensus maximum parsimony tree method was used. This produced a very similar clustering of alleles as inspected networks (Median network, NeighborNet, SplitDecomposition networks; not shown). The MP analysis found 18 equally parsimonious trees for the *accD* gene (length 73 steps) and 19 for the *bccp* (42 steps) (Figure 2, Figure 3). The resulting trees contained several polytomies. This is because of a large part of the total sequence variability being due to indels in the case of *accD*, and this information was not included into the MP analysis. In addition, a number of homoplasious mutations were also excluded, with the resulting trees contained several polytomies. However, as we were not interested in the assessment of the gene phylogeny, we did not try to interpret these polytomies. Produced clades (with a rather high bootstrap support) were very similar to groups inferred from the network analyses. Based on the similarity in the grouping of alleles between inspected networks and the MP analysis, the groups of alleles were inferred from the consensus MP tree for both investigated genes. For the *accD* gene 10 groups (A–J) were inferred; 15 groups were inferred for the *bccp* gene (A–O) were inferred (Figure 2, Figure 3, Table S1). The *accD* gene group D (comprising alleles *accD_13* and *accD_14*) was specific for *P. abyssinicum*, except for one sample of *P. sativum* from Montenegro (accession n° PI357292), which also possessed the *accD_14* allele. Group F (comprising alleles accD_17/18/19/20/21) was specific for *P. fulvum*. Accessions of *P. elatius* and landraces of *P. sativum* were represented by multiple alleles belonging to different groups.

In the case of the *bccp* gene, *P. abyssinicum* was represented by groups J (allele *bccp_26*) and G (alleles *bccp_20/22*). However, in contrast to the *accD* gene, inferred alleles were not specific for *P. abyssinicum*, but were also found within *P. elatius* and samples of *P. sativum* (Figure 3, Table S1). The three identified alleles observed for *P. fulvum* (*bccp_1/2/3*) clustered together and represented group A. Two of these alleles were specific (*bccp_1/2*) for *P. fulvum* and one (*bccp_3*) was shared with two samples of *P. sativum* from Greece (JI1525 and JI2573). The identified alleles for the investigated accessions of *P. elatius* fall within 12 groups and for *P. sativum* within six groups, which were shared between these two species (Table S1).

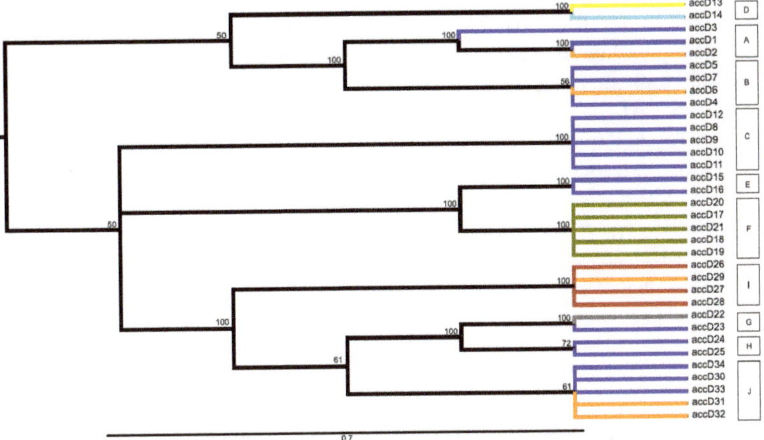

Figure 2. Midpoint-rooted consensus tree for the *accD* gene presenting the most parsimonious relationships among the identified 34 alleles within the studied world-wide pea collection. The consensus tree is build up from the 18 equally parsimonious trees (length 73, consistency index 0.900; retention index 0.972; composite index 0.892). Branch coloring follows the species presence of particular alleles: olive green = alleles observed only within *P. fulvum*; grey = alleles shared among *P. fulvum* and *P. elatius*; orange = alleles shared among *P. sativum* and *P. elatius*; red = alleles observed only for *P. sativum*; turquoise = alleles shared among *P. abyssinicum* and *P. sativum*; yellow = alleles observed only within *P. abyssinicum*; blue = alleles observed only within *P. elatius*. Bootstrap support ≥ 50 is shown above branches.

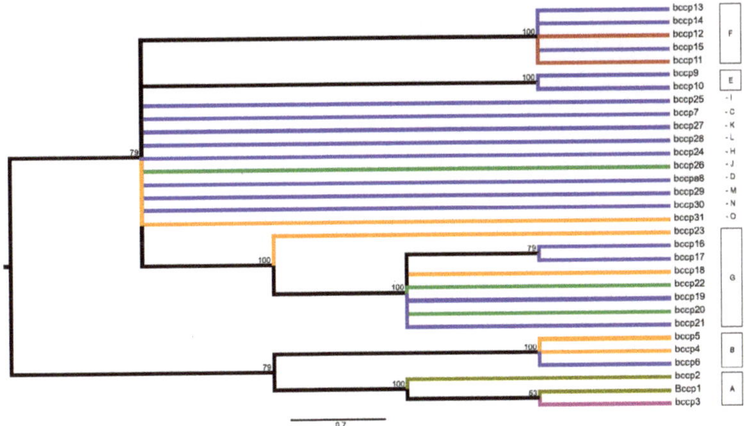

Figure 3. Midpoint-rooted consensus tree for the *bccp* gene presenting parsimonious relationships among the identified 31 alleles within the studied world-wide pea collection. The consensus tree built from the 19 equally parsimonious trees (length 42, consistency index 0.762; retention index 0.900; composite index 0.793). Branch coloring follows species presence of particular alleles: olive green = alleles observed only within *P. fulvum*; magenta = alleles shared among *P. fulvum* and *P. sativum*; orange = alleles shared among *P. sativum* and *P. elatius*; red = alleles observed only for *P. sativum*; green = alleles shared among *P. abyssinicum*, *P. sativum* and *P. elatius*; blue = alleles observed only within *P. elatius*. Bootstrap support ≥ 50 is shown above branches.

2.4. Frequency of Amino Acid Substitutions and Their Distribution

Analysis of the nuclear encoded *bccp* gene in a panel of 179 samples of 809 sites resulting in 269 analyzed codons revealed 196 synonymous sites (Pi(s): 0,00616 Pi(s), Jukes & Cantor: 0,00620) and 610 non-synonymous sites (Pi(a): 0,00553, Pi(a), Jukes & Cantor: 0,00556). This resulted in a Ka/Ks ratio of 0.895. Despite the presence of frequent insertions and deletions, the *accD* sequence could be translated into protein. The analysis covered 1306 sites (e.g., 425 codons). Nucleotide diversity analysis of *accD* showed 278 synonymous sites (Pi(s): 0,00390, Pi(s), Jukes & Cantor: 0,00391) and 997 non-synonymous sites (Pi(a): 0,01048, Pi(a), Jukes & Cantor: 0,01064). This resulted in a high Ka/Ks ratio of 2.726, which indicates positive selections and accelerated evolutions.

Analysis of protein sequence revealed that the ACCD protein has a ClpP protease/crotonase domain (IPRO 29045; region of 251 to 296 and 384 to 584 amino acids), coiled coil domain (region of 380 to 407 amino acids), an acetyl-CoA-carboxyltransferase N terminal domain (IPRO 11762; in region of 226 to 590 amino acids), and a zinc finger (230–252 amino acids) domain (Figure 1). The BCCP protein has a biotin/lipoyl attachment (IPRO 000089) domain (region of 207 to 280 amino acids) and a carboxytransferase (CT) interaction site (239G-284F-249G-250A-257D), where 249G is a conserved biotinylation site.

We next attempted to investigate the location of the individual amino acid substitutions, and the conspicuous indels found in *accD*. This was performed with respect to the 3D folding of both ACCD and BCCP proteins, to the extent that we were able to predict their spatial structure by threading on experimentally characterized related templates. We could produce only partial models for both proteins (File S1, S2 For ACCD, the model covered approximately 43% of the sequence, corresponding to the C-terminal portion of the protein). The N-terminal region and an additional loop within the modelled segment were disordered in the prediction. For the BCCP protein, approximately 45% of the sequence was covered by the best templates but only two short separate fragments from this domain could be reliably modeled; the rest of the molecule was disordered in the prediction (Figure 4, Table 1).

Table 1. Distribution of protein sequence polymorphisms in structurally modelled versus non-modelled parts of the ACCD and BCCP protein sequences.

Protein	Substitutions/Alignment Length		Indels/Alignment Length	
	Modelled	Not Modelled	Modelled	Not Modelled
accD	36/299	50/256*	2/299	17/256**
bccp	17/134	19/138	0/134	1/138

Asterisks denote significant differences in the frequency of the given category of mutations in non-modelled (disordered) parts of the protein compared to the modelled ones (*—$p < 0.05$, **—$p < 0.01$).

Remarkably, mapping of the identified protein sequence polymorphisms revealed that most of the above-described repeat and indel polymorphisms in the ACCD sequence map to sequence regions could not be modelled due to the lack of suitable templates and intrinsic disorder. This is consistent with this part of the protein being less constrained by requirements for precise folding than the enzymatically active domain. Point mutations were also somewhat enriched in the part of the ACCD protein that was not modeled. However, no such bias was detected for BCCP (Figure 4, Table 1).

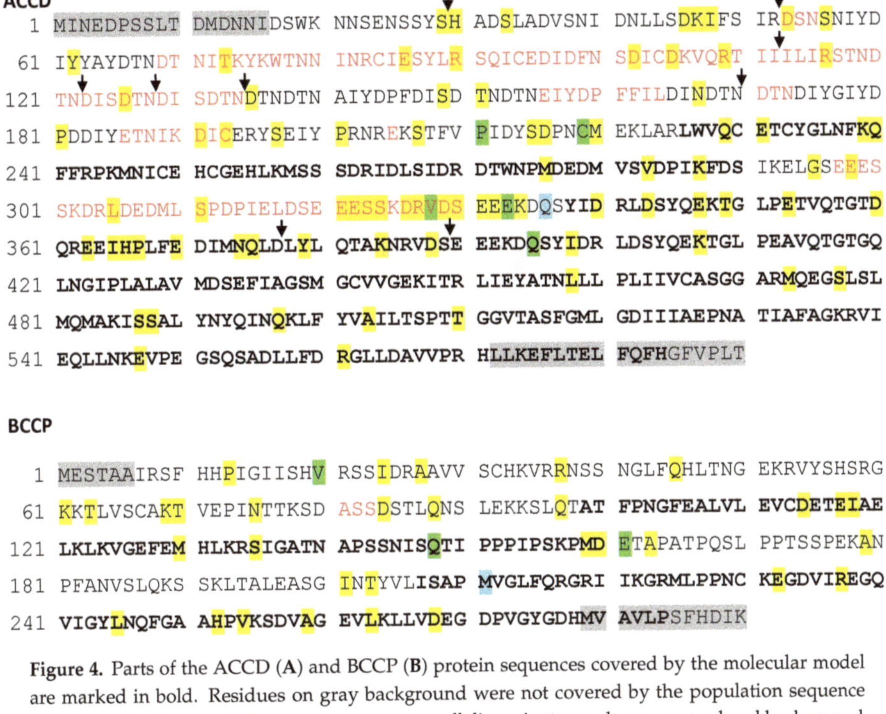

Figure 4. Parts of the ACCD (**A**) and BCCP (**B**) protein sequences covered by the molecular model are marked in bold. Residues on gray background were not covered by the population sequence alignment. Residues exhibiting one, two, or more allelic variants are shown on a colored background. Residues shown in red are deleted only in some alleles. Black arrows indicate the location of insertions in some alleles.

2.5. Allelic accD/bccp Combinations

We found 34 *accD* and 31 *bccp* alleles yielding altogether 1054 possible combinations. Within the wild pea (*P. elatius*) we detected 61 combinations (Table S2). Most of these combinations (45) were found only once. Cultivated *P. sativum* landraces had 20 combinations; the most frequent were *accD_29/bccp_22* (30), followed by *accD_29/bccp_18* (8). *P. abyssinicum* accessions had 4 distinct combinations, with *accD_14/bccp_26* being predominant (17). *P. fulvum* had 9 combinations, *accD_21/bccp_1* (4), *accD_20/bccp_1* (3), and *accD_17/bccp_3* (2). The only exception in our *P. fulvum* set was JI2539 from Israel, which had *accD_22* (*accD_G* lineage) shared with *P. elatius*. There were two *bccp* alleles (*bccp_22* and *bccp_31*) that formed the highest number of combinations with 18 and 10 *accD* alleles, respectively. Conversely, two *accD* alleles, *accD_29*, *accD_25*, and *bccp_22*, *bccp_31* formed 8, 9, and 19, 10 combinations, respectively. Notably, the most frequent combination found in *P. sativum* landraces *accD_29/bccp_22* was found in these high occurrence alleles (Figure 5).

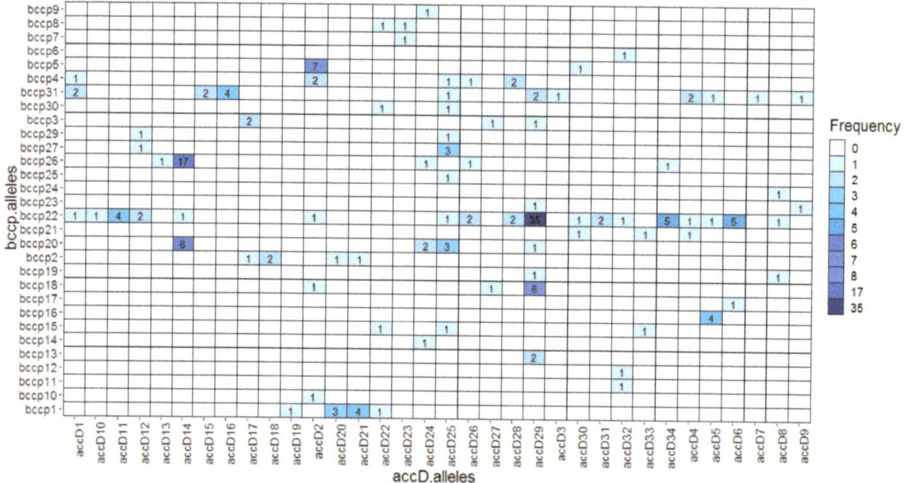

Figure 5. Heatmap of identified pairwise ACCD/BCCP allelic combinations.

2.6. Relationship to Pisum Genetic Diversity

Having previously analyzed genetic diversity based on genome-wide sampled polymorphism [52,53], we examined the distribution of both *accD* and *bccp* alleles within respective genetic groups. Cultivated *Pisum sativum* accessions can be divided into two (nr. 3 and 6) equally abundant (24 and, 27 accessions, respectively) groups. The independently domesticated Ethiopian pea (*P. abyssinicum*) forms a separate (nr. 7) group (Table S1). With respect to *accD/bccp* alleles, *accD_29* and *bccp_22* alleles predominate in 60 analyzed *P. sativum* accessions (41, 38 accessions respectively) (Supplementary Table S1), while all 24 *P. abyssinicum* accessions had single unique *accD_14* and *bccp_26* (17 acc.), *bccp_20* (5 acc.) and *bccp_22* (JI1974) alleles corresponding to its separate domestication history and associated bottleneck. *P. fulvum* as a separate species forms a separate genetic group (nr. 2) and has also distinct and the most distant *accD* (*accD_17-21*) and *bccp* (*bccp_1-3*) alleles, separated by 39 to 40, and 7 to 8 amino acids, respectively, from the closest *P. elatius* alleles. On the contrary, wild *P. elatius* is genetically the most diverse and has seven genetic groups (Trněný et al. 2018), one of which (nr. 3) overlaps with *P. sativum*. This diversity is also reflected with 22 different *accD* and 25 *bccp* alleles, respectively. The most abundant are *accD_25* (13 acc.), *accD_29* (9 acc.), *accD_2* (11 acc.), and *bccp_22* (28 acc.), and *bccp_31* (16 acc.) (Table S1). There is only a partial relationship between the genome wide DARTseq and *accD/bccp* based diversity. Genetic group nr. 10 of *P. elatius* accessions from the Caucasus region has the most distinct *accD_30, 31, 34*, but not *bccp* alleles. Similarly, genetic groups nr. 4 and 5 have a high proportion of *accD_2* (8 acc.) and *accD_15/16* (5 acc.) alleles in samples from Israel or eastern Turkey and Georgia, respectively. No clear genetic group assignment was found for *bccp* alleles within *P. elatius* accessions.

2.7. Geographic Distribution of accD/bccp Alleles

Pisum fulvum (16 acc.) is geographically restricted to Israel (7 acc.), Syria (7 acc.), Jordan (1 acc.), and southeastern Turkey (1 acc.), and displays distinct *accD/bccp* alleles. Genetically and geographically the most diverse set is from *P. elatius* (96 acc.). Of these, there were 34 accessions from Turkey, which had the highest genetic diversity (Figure 6, Table S1). These accessions have various *accD/bccp* alleles, although the combination *accD_25* and *bccp_20* is the most frequent (10). The next large group is *P. elatius* from Israel, which had 25 accessions that belong to various genetic groups. These also have different *accD_2* and *bccp_5* (22 alleles occurring in 13 accessions). European samples cover a large region of Western (Spain, Portugal, France), Central (Italy), and Eastern (Greece, Hungary, Serbia) Europe (Table S1). The later samples are distinct by both by genome wide analyses and by *accD/bccp* alleles analysis. Finally, the most separate group of *P. elatius* is from Armenia, with unique *accD*_34 and *bccp*_21/22 alleles (Figure 6).

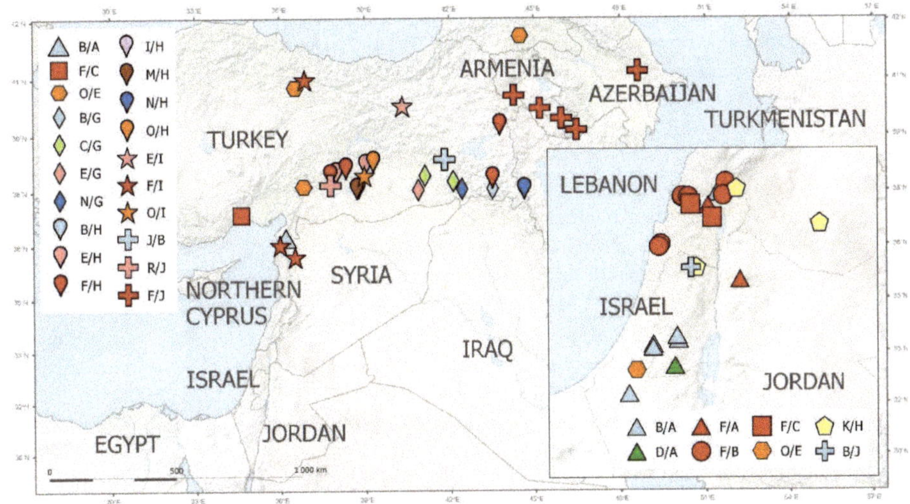

Figure 6. Geographic distribution of ACCD/BCCP allelic combinations assigned to large groups (for details see Table S1) within the Middle East.

The cultivated pea is geographically less precisely localized, except for *P. abyssinicum*, which is found only in Ethiopia and Yemen. All *P. abyssinicum* accessions have *accD*_14/*bccp*_20/26 alleles. Landraces of *P. sativum* originate from 24 countries and span a large geographical area from the Western Mediterranean to Central and Southern Asia. They predominantly have *accD*_29 (41 acc.) and *bccp*_22 (36 acc.) alleles typical for cultivated pea. There are few distinct accessions that have different alleles. Two were from Algeria (*accD*_32/*bccp*_11/12), and two accessions were from Greece (specific *bccp*_3 allele). Two accessions from China (ATC6925, ATC6937) have a *accD*_6 allele shared with *P. elatius*, while PI560969 from Nepal has distinct *accD*_2/*bccp*_5 alleles (Table S1).

3. Discussion

Here we report the allelic composition and geographical distribution of two genes involved in postzygotic reproductive isolation in the pea [39]. Taking advantage of the available germplasm resources [52,53], we analyzed the allelic composition of chloroplast localized *accD* and nuclear encoded *bccp* genes. Our results extend the experimental data of Bogdanova et al. [39]. We analyzed the allelic composition of accessions collected from the wild (including all recognized *Pisum* species) and domesticated peas of various geographical origins.

Postzygotic reproductive isolation, expressed as hybrid sterility or inviability, hybrid weakness or necrosis, and hybrid breakdown, is considered one of the two major fundamental processes leading to speciation [2,9]. The plastome–genome dysfunctions concern various kinds of albinism. Generally, incompatible hybrid materials suffer from reduced pigment content, lower rates of photosynthesis, and an impaired thylakoid structure. We detected the occurrence of albinotic plants in crosses of wild *Pisum fulvum* or *P. elatius* with the cultivated pea *P. sativum*, which upon identification of the respective genes [39] prompted this study.

3.1. Hypervariability of the Chloroplast accD Gene

The region of the chloroplast genome around the *accD* gene has been found to be prone to accumulation of repeats, resulting in high interspecific variability in numerous species (*Pisum* and

Lathyrus [45], *Capsicum* [54], *Glycine* [43], *Silene* [47], *Oenothera* [55,56], Cupressophytes [57]) but much less variability at the intraspecific level (*Medicago truncatula* [44], tea, *Camellia sinensis* [58], and pea, *Pisum sp.* [39,40]). Our present study substantially expands the previous reports [39,40] by analyzing 195 pea samples covering the entire geographical and species range [52,59]. Our results on the ratios of nonsynonymous to synonymous substitutions (*Ka/Ks*) in the pea *accD* gene agree with data from *Oenothera, Silene,* and Cupressophytes [47,55,57]. This indicates positive selection, since *Ka/Ks* values significantly above 1 are unlikely to occur without at least some of the mutations being advantageous. The large variation in plastid-encoded *accD* gene sequences, both between and within the *Pisum* species, is consistent with findings in *Silene*, where positive selection in the phylogenetic context has been detected [47]. In many cases of plastid genome evolution, mutations have disproportionately affected nonsynonymous sites, resulting in elevated ratios of nonsynonymous to synonymous substitution rates. Notably, plastid genome comparison between *Lathyrus sativus* and *Pisum sativum* resulted in identification of a region spanning the *accD* gene with increased mutation rate [45]. Analysis of publicly available *accD* sequences for *Lathyrus* and *Vicia* species supported these findings (unpublished).

Variation detected in the *Pisum* sp. *accD* sequence is mainly caused by the insertion of multiple tandem repeated sequences, as found in Cupressophytes [57] and *Medicago* [44]. In particular, the later study corresponds well to our pea *accD* data, since each of the 24 studied *Medicago truncatula* genotypes appears to have a different *accD* sequence, yet with maintained reading frames despite the high variability. Mapping of the insertion sites onto the predicted protein structure indicated their clustering within the N-terminal part of the ACCD protein that could not be reliably modelled due to intrinsic disorder. Such disordered protein regions are known to be extremely flexible and dynamic, alleviating some structural constraints [60], and were reported to be prone to insertions and deletions [61]. It has been suggested that regions surrounding tandem repeats evolve faster than other non-repeat-containing regions, which results in increased frequency of substitutions near the flanking sequences [62]. As shown in tobacco, a functional *accD* is essential for development [63]. Interestingly, the relationship to biparental inheritance of plastids was proposed to be related to the plastid competition [56]. Since about 20% of all angiosperms contain plastid DNA in the sperm cell, it is likely that this mechanism of cytonuclear conflict is also present in other systems [64–67].

3.2. Allelic accD/bccp Combinations Found in Wild and Domesticated Peas

One of our major aims was to detect allelic combinations of both genes occurring in wild peas, as well as in cultivated pea crop. Altogether we found 36 *accD* and 35 *bccp* alleles in the set of 195 accessions. Within the wild pea (*P. elatius*) these occurred in 60 out of 671 possible combinations, indicating a high diversity, while both domesticated *P. sativum* and *P. abyssinicum* had only a reduced subset. There was no overlap between *P. fulvum* and *P. elatius*, except for one *P. fulvum* JI2539 accession from Israel, which had *accD_22* (G lineage) allele shared with three *P. elatius* samples from Turkey. Notably, in our previous study [52], we have found in this accession a typical *P. elatius trnSG_E6* allele, suggesting some past hybridization event between *P. fulvum* and *P. elatius*. Interestingly, in another two *P. fulvum* accessions (JI2510, JI2521) that also have the *trnSG_E6* allele [52], the *accD* allele was canonical to *P. fulvum* (*accD_20, 21*, e.g., F lineage). *P. abyssinicum* had *accD* alleles and combinations distinct from *P. sativum*, supporting its independent domestication [53]. The *accD_14* allele of *P. abyssinicum* was not found in any of *P. elatius* or *P. sativum* samples. Notably, two of the most frequent alleles of each gene, *accD_29* and *bccp_22*, contributed to the most frequent combination of *accD_30/bccp_25* found in domesticated *P. sativum*.

It remains to be experimentally tested by crosses if the allelic combinations detected in the natural conditions create barriers against gene flow in natural pea populations. Some experimental crosses between cultivated pea and selected *P. fulvum* and *P. elatius* accessions were conducted by Bogdanova et al. [68]. These crosses revealed hybrid sterility, ultimately leading to identification of the respective genes [39]. In our work, we made reciprocal crosses between *P. elatius* L100 (*accD_2/bccp_5*) and *P. sativum* cv. Cameor (*accD_29/bccp_22*), which resulted in the appearance of albinotic plants (Smýkal, unpublished), while a cross between *P. elatius* JI64 (*accD_30/bccp_5*) and *P. sativum* JI92 (*accD_29/bccp_22*) was fully viable and fertile [69,70]. This corresponds to the findings of Bogdanova et al. (2015) [39] of a incompatible cross between *P. elatius* L100 (*accD_2/bccp_5*) and *P. sativum* WL12238 (*accD_29/bccp_22*); a cross between *P. elatius* JI1794 (*accD_25/bccp_27*), 721 (*accD_5/bccp_22*), and *P. abyssinicum* VIR 2759 (*accD_14/bccp_26*) were compatible with the cultivated pea *P. sativum* WL12238 (*accD_29/bccp_22*) [68]. Moreover, the existence of a second, unlinked, and yet unidentified nuclear *scs2* locus also involved in nuclear-cytoplasmic conflict has been proposed [39]. In this study, the authors proposed a model of determinants, based on seven substitutions and three deletions in ACCD and four amino acid substitutions in the biotinyl domain of BCCP protein. The results of our study add to this complexity, as there are far more possible combinations.

3.3. Domestication and Hybrid Incompatibility

In crops, artificial selection and hybridization accelerate the evolutionary process [71]. The majority of economically important crops were isolated from their progenitors through the existence of prezygotic or postzygotic reproductive barriers (or both), even though geographic isolation was absent during the domestication [38]. The reproductive barriers between wild crop progenitors and domesticated crops might be attributed to several mechanisms, including differences in karyotype or chromosomal rearrangements. Such karyotype differences are reported between *P. fulvum* and *P. elatius*, *P. sativum*, and between *P. sativum* and *P. abyssinicum* [72,73], and contribute to the partial fertility of the respective hybrids. Much less is known about the interactions between nuclear and cytoplasmic genomes. To date, only a few genes implicated in hybrid incompatibility have been isolated in crops. In maize, *Tcb1*, *Ga1*, and *Ga2* alleles influence interaction of pollen tubes with silk tissue and confer prezygotic barriers in crosses between cultivated *Zea mays* and the wild teosinte *Z. m. mexicana* [74]. About 50 loci controlling postzygotic reproductive barriers between rice subspecies have been identified and molecular products of some genes have been characterized [22]. For example, the *S5* locus, a determinant of *japonica-indica* sterility, is located in proximity to the domestication *OsC1* gene [75]. Similarly, the *Gn1a* gene involved in rice yield formation is linked with *S35*, which determines pollen sterility of *japonica-indica* hybrids [76]. Another example was shown in the tomato, where the *Cf-2* gene from wild *Lycopersicon pimpinellifolium* confers resistance to the fungus *Cladosporium fulvum* in an *Rcr3* dependent manner [48]; these two genes interact with each other to induce hybrid necrosis syndrome in the hybrids. Although the occurrence of albino plants in many interspecific crosses in crops is widely documented [77,78], its causes have not been studied in most cases. Notably, crosses between cultivated chickpea (*Cicer arietinum*) and its progenitor (*C. reticulatum*) yielded yellow and albino plants and a biparental plastid inheritance [77,78]. We speculate that this was caused by a similar mechanism as in the pea.

The results of this study might be relevant for breeding, particularly using more distant crop wild relatives, as well as hybrid crop breeding [79,80], but it remains to be tested by experimental crosses to identify causal effectors.

4. Material and Methods

4.1. Plant Material

We analyzed 195 previously described pea accessions (Smýkal et al. 2017, 2018, Trněný et al. 2018) [52,53,59], consisting of wild *P. elatius* (95) and *P. fulvum* (16) accessions (Table S1). Sixty domesticated *P. sativum* landraces and 24 domesticated *P. abyssinicum* accessions were selected to maximize the genetic diversity and to cover the entire range of the wild and landrace pea habitats. This span is approximately 5000 km in longitude from Morocco to Iran, and in latitude from Tunisia to Hungary; altitude ranged from sea level to about 2000 m. This material was previously morphologically described and assessed for its genetic diversity structure [52,53]. Plants were grown in 5 L pots with peat-sand (90:10) substrate mix (Florcom Profi, BB Com Ltd. Letohrad, Czech Republic), in glasshouse conditions (UP campus, Olomouc, Czech Republic).

4.2. DNA and RNA Analysis

Genomic DNA was isolated from a single plant per accession from approximately 100 mg of dry leaf material using the Invisorb Plant Genomic DNA Isolation kit (Invisorb, Berlin, Germany) and standard protocol [52,59]. Total RNA was isolated from young leaves using plant RNA kit (Macherey-Nagel, Düren, Germany). Isolated RNA was treated with DNaseI to remove genomic DNA. The *accD* gene was amplified directly from genomic DNA using primers (F1—GCATTAGTTTTCATTTTCAGTCC located 27 bp upstream of stop codon, R4—CTTTAATAGGGGTTTAGAATACA, located 94 bp upstream of ATG codon) [39]. We used cDNA as a template to avoid large intron sequences present in the *bccp3* gene. One microgram of a total RNA was reversely transcribed with Oligo(dT) primer and AMV reverse transcriptase (Promega, Madison, USA) according to manufacturer's protocol (Hradilová et al. 2017) [71]. Two step nested PCR amplification was used. After the first PCR (with primers F—CTAATGAAAGTGGCGGAAATC, R—CCTTATTACGCGTCTTAGTGAATG), the product was diluted (1:100) and the second PCR was performed (F33—CCATTCTCTGCACTCCCTTTCGCG, R1113—CAATTATTTCTCAATCTATTCAAA ACG), using the conditions as described in Hradilová et al. [71]. PCR products were verified on a 1.5% agarose gel, treated with Exonuclease-Alkaline Phosphatase (Thermo Scientific, Brno, Czech Republic) and sequenced at Macrogene.

4.3. Sequence Analysis

For initial analysis, Geneious 7.1.7 (Biomatters Ltd., Auckland, New Zealand) was used to edit and align sequences. Due to the presence of large gaps in the *accD* gene, sequences were translated into protein sequences, which reduced the overall length of the *accD* nucleotide alignment and partially helped to eliminate large gaps. This procedure reduced the complexity of the *accD* sequences. Sequences of the *bccp* gene were treated in the same manner, although these sequences were largely devoid of large indels. The translated protein sequences were aligned in Geneious using the MAFFT algorithm and the final alignment was manually adjusted. From the final alignment, different alleles and their frequencies were identified using the online tool FABOX [81].

To explore possible connections or relationships among the identified alleles, the reduced dataset (including each allele defined only once) was used for the network analysis. Several approaches of network construction were used (based on characters, Median network, Median-joining; based on distances, Neighbor network, Split decomposition) and implemented in SplitsTree [82]. The results were then compared. To compare the results of network analysis with a classically constructed bifurcating tree, a maximum parsimony (MP) tree was built using MEGA 6 with 1000 bootstrap replicates [83]. Because of the complex pattern of gaps within the *accD* gene, indels were treated as "partially deleted" (pairwise deletion, option implemented in MEGA) during the MP analysis. The final consensus tree was computed from all the equally parsimonious trees found during the analysis and was midpoint rooted. The tree topology was compared against the constructed networks. To simplify or reduce the number of identified alleles, groups of related alleles were inferred based on the constructed networks and the final consensus MP tree for both investigated genes. DnaSP v5.10 was used to determine nucleotide diversity and synonymous/non-synonymous sites ratios [84]. All studied *accD* and *bccp* sequences were deposited in the GenBank database under the accession numbers MK619486 to MK619678, and MK644626 to MK644819, respectively.

4.4. Tandem Repeat Analysis

Tandem repeats within DNA and protein sequences were identified in a combination of two algorithms (FastPCR [85] and RADAR [86]). The consensus DNA sequence of *accD* gene was first scanned by FastPCR at a repeat length $\geq$20 bp (k-mer = 12 with a tolerance for up to one mismatch within k-mer) with a similarity of above 70%. Potential tandem repeats for consensus protein sequence were further identified by RADAR software. Both methods complemented each other, since the boundaries of some degenerate and mixed tandem repeats were difficult to identify separately.

4.5. Protein Sequence Analysis and Structure Modelling

To identify the domains we used InterPro (www.ebi.ac.uk/interpro) and SMART databases (http://smart.embl-heidelberg.de). To generate molecular models of both proteins, standard sequences of the pea *accD* (GenBank YP_003587558.1) and *bccp* (GenBank DR89228.1) were used as queries to identify suitable templates and to perform molecular modelling by threading using Phyre2 in "normal" mode [87]. Only a partial model was generated for each protein, as portions of the sequence predicted to be disordered or lacking a suitable template (including some internal loops) could not be reliably modeled. In the case of ACCD, the structure of *Staphylococcus* acetyl-CoA carboxylase carboxyltransferase (PDB 2F9I) was identified as the best template. The second best template (PDB 2F9Y, also of bacterial origin) yielded a model of similar coverage and spatial organization. A similar model, also based on the PDB 2F9I template, was obtained for the same part of ACCD using another algorithm, RaptorX [88]. For BCCP, the best template identified by Phyre2 was the pyruvate carboxylase from *Methylobacillus flagellatus* (PDB 5KS8). The same template was also found by RaptorX as second best; namely, pyruvate carboxylase from *Listeria monocytogenes* (PDB 4QSH) yielded a spatially similar model. The Phyre2-generated models were subjected to additional refinement in the DeepView environment [89] to eliminate amino acid sidechain clashes. Subsequent evaluation of the resulting models using the WHAT_CHECK tools [90] revealed no critical errors, with scores for some parameters only slightly poorer than observed for the template for both proteins.

4.6. Mapping Protein Sequence Polymorphisms on Predicted Structure

Unique protein sequences encoded by alleles, each of the two loci were identified within aligned protein sequence sets using the ElimDupes tool at the Los Alamos HIV database website (https://www.hiv.lanl.gov/content/sequence/elimdupesv2/elimdupes.html). A map of polymorphisms was then generated manually from the resulting unique sequence alignments. A distribution of the polymorphisms between the modeled and non-modeled portions of the protein was statistically evaluated using the Chi-square test.

Supplementary Materials: The following are available online at http://www.mdpi.com/1422-0067/20/7/1773/s1. Table S1: List and description of analyzed material, Table S2: Table of *accD*/*bccp* combinations, Figure S1: The alignment of amino acid sequences of all identified *accD* alleles, File S1: Theoretical model of pea BCCP protein structure, (co-ordinates in standard PDB format), File S2: Model of BCCP protein structure. Theoretical model of pea ACCD protein structure, (co-ordinates in standard PDB format). File S3: Theoretical model of pea BCCP protein structure, (co-ordinates in standard PDB format).

Author Contributions: Conceptualization, P.S.; methodology, L.M., J.B., F.C., and P.S.; formal analysis, E.N., L.Z., J.B., R.K., V.N., F.C., L.M., and P.H.; resources, P.S.; data curation, P.S.; writing—original draft preparation, P.S., F.C., and L.M.; writing—review and editing, P.S., F.C., and L.M.; funding acquisition, P.S.

Funding: This work received funding from the European Community's Seventh Framework Programme (FP7/2007-2013) under the grant agreement n°FP7-613551, LEGATO project. R.K. was supported by the Science Committee of the Ministry of Education and Science of the Republic of Kazakhstan in the framework of program funding for research (BR05236334 and BR06349586). P.S., L.Z., and E.N. were supported from Grant Agency of Palacký University, IGA 001-2018, IGA 004-2019 projects, F.C. by the Ministry of Education, Youth and Sports of the Czech Republic NPUI LO1417 project.

Acknowledgments: We are thankful to Clarice Coyne, Mike Ambrose, and Bob Redden, genebank curators for stimulating discussion, and sharing their knowledge and selection of pea germplasm.

Conflicts of Interest: The authors declare no conflict of interest. The funders had no role in the design of the study; in the collection, analyses, or interpretation of data; in the writing of the manuscript, or in the decision to publish the results.

References

1. Coyne, J.A. Genetics and speciation. *Nature* **1992**, *355*, 511–515. [CrossRef]
2. Coyne, J.A.; Orr, H.A. *Speciation. Sinauer, Sunderland*; Oxford University Press: New York, NY, USA, 2004; pp. 1–545.
3. Givnish, T.J. Ecology of plant speciation. *Taxon* **2010**, *59*, 1326–1366. [CrossRef]
4. Case, A.L.; Finseth, F.R.; Barr, C.M.; Fishman, L. Selfish evolution of cytonuclear hybrid incompatibility in *Mimulus. Proc. Biol. Sci.* **2016**, *283*, 20161493. [CrossRef]
5. Sambatti, J.B.M.; Ortiz-Barrientos, D.; Baack, E.J.; Rieseberg, L.H. Ecological selection maintains cytonuclear incompatibilities in hybridizing sunflowers. *Ecol. Lett.* **2008**, *11*, 1082–1091. [CrossRef]
6. Bateson, W. *Mendel's Principles of Heredity*; Cambridge University Press: Cambridge, UK, 1909.
7. Dobzhansky, T. Genetics and the Origin of Species. In *Columbia Biological Series*; Columbia University Press: New York, NY, USA, 1937; Volume 9, pp. 1–364.
8. Muller, H.J. Isolating mechanisms, evolution, and temperature. *Biol. Symp.* **1942**, *6*, 71–125.
9. Fishman, L.; Sweigart, A.L. When Two Rights Make a Wrong: The Evolutionary Genetics of Plant Hybrid Incompatibilities. *Annu. Rev. Plant Biol.* **2018**, *69*, 707–731. [CrossRef]
10. Barnard-Kubow, K.B.; So, N.; Galloway, L.F. Cytonuclear incompatibility contributes to the early stages of speciation. *Evolution* **2016**, *70*, 2752–2766. [CrossRef] [PubMed]
11. Mayr, E. *Systematics and the Origin of Species*; Columbia University Press: New York, NY, USA, 1942.
12. Rieseberg, L.H.; Blackman, B.K. Speciation genes in plants. *Ann. Bot.* **2010**, *106*, 439–455. [CrossRef] [PubMed]
13. Chen, C.; Lin, H.-X. Evolution and Molecular Control of Hybrid Incompatibility in Plants. *Front. Plant Sci.* **2016**, *7*, 1208. [CrossRef]
14. Orr, H.A.; Masly, J.P.; Presgraves, D.C. Speciation genes. *Curr. Opin. Genet. Dev.* **2004**, *14*, 675–679. [CrossRef] [PubMed]
15. Bomblies, K. Doomed lovers: Mechanisms of isolation and incompatibility in plants. *Annu. Rev. Plant Biol.* **2010**, *61*, 109–124. [CrossRef]
16. Sweigart, A.L.; Willis, J.H. Molecular evolution and genetics of postzygotic reproductive isolation in plants. *F1000 Biol. Rep.* **2012**, *4*, 23. [CrossRef]
17. Baack, E.; Melo, M.C.; Rieseberg, L.H.; Ortiz-Barrientos, D. The origins of reproductive isolation in plants. *New Phytol.* **2015**, *207*, 968–984. [CrossRef]
18. Moison, M.; Roux, F.; Quadrado, M.; Duval, R.; Ekovich, M.; Lê, D.-H.; Verzaux, M.; Budar, F. Cytoplasmic phylogeny and evidence of cyto-nuclear co-adaptation in *Arabidopsis thaliana. Plant J.* **2010**, *63*, 728–738. [CrossRef]
19. Levin, D.A. The cytoplasmic factor in plant speciation. *Syst. Bot.* **2003**, *28*, 5–11.
20. Burton, R.S.; Pereira, R.J.; Barreto, F.S. Cytonuclear Genomic Interactions and Hybrid Breakdown. *Annu. Rev. Ecol. Evol. Syst.* **2013**, *44*, 281–302. [CrossRef]
21. Moyle, L.C.; Nakazato, T. Complex Epistasis for Dobzhansky–Muller Hybrid Incompatibility in *Solanum. Genetics* **2009**, *181*, 347–351. [CrossRef]
22. Ouyang, Y.; Liu, Y.-G.; Zhang, Q. Hybrid sterility in plant: Stories from rice. *Curr. Opin. Plant Biol.* **2010**, *13*, 186–192. [CrossRef]
23. Barr, C.M.; Fishman, L. The Nuclear Component of a Cytonuclear Hybrid Incompatibility in Mimulus Maps to a Cluster of Pentatricopeptide Repeat Genes. *Genetics* **2010**, *184*, 455–465. [CrossRef]
24. Greiner, S.; Rauwolf, U.; Meurer, J.; Herrmann, R.G. The role of plastids in plant speciation. *Mol. Ecol.* **2011**, *20*, 671–691. [CrossRef]
25. Leppälä, J.; Savolainen, O. Nuclear-Cytoplasmic Interactions Reduce Male Fertility in Hybrids of Arabidopsis Lyrata Subspecies. *Evolution* **2011**, *65*, 2959–2972. [CrossRef]
26. Törjék, O.; Witucka-Wall, H.; Meyer, R.C.; von Korff, M.; Kusterer, B.; Rautengarten, C.; Altmann, T. Segregation distortion in Arabidopsis C24/Col-0 and Col-0/C24 recombinant inbred line populations is due to reduced fertility caused by epistatic interaction of two loci. *Theor. Appl. Genet.* **2006**, *113*, 1551–1561. [CrossRef]
27. Durand, S.; Bouché, N.; Perez Strand, E.; Loudet, O.; Camilleri, C. Rapid Establishment of Genetic Incompatibility through Natural Epigenetic Variation. *Curr. Biol.* **2012**, *22*, 326–331. [CrossRef]

28. Birky, C.W. The Inheritance of Genes in Mitochondria and Chloroplasts: Laws, Mechanisms, and Models. *Annu. Rev. Genet.* **2001**, *35*, 125–148. [CrossRef]

29. Fishman, L.; Willis, J.H. A cytonuclear incompatibility causes anther sterility in *Mimulus* hybrids. *Evolution* **2006**, *60*, 1372–1381. [CrossRef]

30. Hanson, M.R.; Bentolila, S. Interactions of mitochondrial and nuclear genes that affect male gametophyte development. *Plant Cell* **2004**, *16* (Suppl. 1), S154–S169. [CrossRef]

31. Chen, L.; Liu, Y.-G. Male Sterility and Fertility Restoration in Crops. *Annu. Rev. Plant Biol.* **2014**, *65*, 579–606. [CrossRef]

32. Rhoades, M.M. The cytoplasmic inheritance of male sterility in *Zea mays*. *J. Genet.* **1931**, *27*, 71–93. [CrossRef]

33. Renner, O. Die pflanzlichen Plastiden als selbstandige Elemente der genetischen Konstitution. *Ber. Math. Phys. Kl. Sachs. Akad.* **1934**, *86*, 241–266.

34. Stebbins, G.L. *Variation and Evolution in Plants*; Columbia University Press: New York, NY, USA, 1950.

35. Stubbe, W. The role of the plastome in evolution of the genus *Oenothera*. *Genetica* **1964**, *35*, 28–33. [CrossRef]

36. Crosby, K.; Smith, D.R. Does the mode of plastid inheritance influence plastid genome architecture? *PLoS ONE* **2012**, *7*, e46260. [CrossRef]

37. Roux, F.; Mary-Huard, T.; Barillot, E.; Wenes, E.; Botran, L.; Durand, S.; Villoutreix, R.; Martin-Magniette, M.-L.; Camilleri, C.; Budar, F. Cytonuclear interactions affect adaptive traits of the annual plant *Arabidopsis thaliana* in the field. *Proc. Natl. Acad. Sci. USA* **2016**, *113*, 3687–3692. [CrossRef]

38. Dempewolf, H.; Hodgins, K.A.; Rummell, S.E.; Ellstrand, N.C.; Rieseberg, L.H. Reproductive isolation during domestication. *Plant Cell* **2012**, *24*, 2710–2717. [CrossRef]

39. Bogdanova, V.S.; Zaytseva, O.O.; Mglinets, A.V.; Shatskaya, N.V.; Kosterin, O.E.; Vasiliev, G.V. Nuclear-cytoplasmic conflict in pea (*Pisum sativum* L.) is associated with nuclear and plastidic candidate genes encoding acetyl-CoA carboxylase subunits. *PLoS ONE* **2015**, *10*, e0119835. [CrossRef]

40. Sasaki, Y.; Nagano, Y. Plant acetyl-CoA carboxylase: Structure, biosynthesis, regulation, and gene manipulation for plant breeding. *Biosci. Biotechnol. Biochem.* **2004**, *68*, 1175–1184. [CrossRef]

41. Nikolau, B.J.; Ohlrogge, J.B.; Wurtele, E.S. Plant biotin-containing carboxylases. *Arch. Biochem. Biophys.* **2003**, *414*, 211–222. [CrossRef]

42. Szczepaniak, A.; Książkiewicz, M.; Podkowiński, J.; Czyż, K.B.; Figlerowicz, M.; Naganowska, B. Legume Cytosolic and Plastid Acetyl-Coenzyme—A Carboxylase Genes Differ by Evolutionary Patterns and Selection Pressure Schemes Acting before and after Whole-Genome Duplications. *Genes* **2018**, *9*, 563. [CrossRef]

43. Asaf, S.; Khan, A.L.; Aaqil Khan, M.; Muhammad Imran, Q.; Kang, S.-M.; Al-Hosni, K.; Jeong, E.J.; Lee, K.E.; Lee, I.-J. Comparative analysis of complete plastid genomes from wild soybean (*Glycine soja*) and nine other *Glycine* species. *PLoS ONE* **2017**, *12*, e0182281. [CrossRef]

44. Gurdon, C.; Maliga, P. Two Distinct Plastid Genome Configurations and Unprecedented Intraspecies Length Variation in the *accD* Coding Region in *Medicago truncatula*. *DNA Res.* **2014**, *21*, 417–427. [CrossRef]

45. Magee, A.M.; Aspinall, S.; Rice, D.W.; Cusack, B.P.; Sémon, M.; Perry, A.S.; Stefanović, S.; Milbourne, D.; Barth, S.; Palmer, J.D.; et al. Localized hypermutation and associated gene losses in legume chloroplast genomes. *Genome Res.* **2010**, *20*, 1700–1710. [CrossRef]

46. Ha, Y.-H.; Kim, C.; Choi, K.; Kim, J.-H. Molecular Phylogeny and Dating of Forsythieae (Oleaceae) Provide Insight into the Miocene History of Eurasian Temperate Shrubs. *Front. Plant Sci.* **2018**, *9*, 99. [CrossRef]

47. Rockenbach, K.; Havird, J.C.; Monroe, J.G.; Triant, D.A.; Taylor, D.R.; Sloan, D.B. Positive Selection in Rapidly Evolving Plastid-Nuclear Enzyme Complexes. *Genetics* **2016**, *204*, 1507–1522. [CrossRef]

48. Krüger, J.; Thomas, C.M.; Golstein, C.; Dixon, M.S.; Smoker, M.; Tang, S.; Mulder, L.; Jones, J.D.G. A tomato cysteine protease required for Cf-2-dependent disease resistance and suppression of autonecrosis. *Science* **2002**, *296*, 744–747. [CrossRef]

49. Rooney, H.C.; Van't Klooster, J.W.; van der Hoorn, R.A.; Joosten, M.H.; Jones, J.D.; de Wit, P.J. *Cladosporium Avr2* inhibits tomato Rcr3 protease required for Cf-2-dependent disease resistance. *Science* **2005**, *308*, 1783–1786. [CrossRef]

50. Bomblies, K.; Lempe, J.; Epple, P.; Warthmann, N.; Lanz, C.; Dangl, J.L.; Weigel, D. Autoimmune response as a mechanism for a Dobzhansky-Muller-type incompatibility syndrome in plants. *PLoS Biol.* **2007**, *5*, e236. [CrossRef]

51. Bomblies, K.; Weigel, D. Hybrid necrosis: Autoimmunity as a potential gene-flow barrier in plant species. *Nat. Rev. Genet.* **2007**, *8*, 382–393. [CrossRef]

52. Smýkal, P.; Hradilová, I.; Trněný, O.; Brus, J.; Rathore, A.; Bariotakis, M.; Das, R.R.; Bhattacharyya, D.; Richards, C.; Coyne, C.J.; et al. Genomic diversity and macroecology of the crop wild relatives of domesticated pea. *Sci. Rep.* **2017**, *7*, 17384. [CrossRef]

53. Trněný, O.; Brus, J.; Hradilová, I.; Rathore, A.; Das, R.R.; Kopecký, P.; Coyne, C.J.; Reeves, P.; Richards, C.; Smýkal, P. Molecular Evidence for Two Domestication Events in the Pea Crop. *Genes (Basel)* **2018**, *9*, 535. [CrossRef]

54. D'Agostino, N.; Tamburino, R.; Cantarella, C.; De Carluccio, V.; Sannino, L.; Cozzolino, S.; Cardi, T.; Scotti, N. The Complete Plastome Sequences of Eleven Capsicum Genotypes: Insights into DNA Variation and Molecular Evolution. *Genes* **2018**, *9*, 503. [CrossRef]

55. Greiner, S.; Wang, X.; Herrmann, R.G.; Rauwolf, U.; Mayer, K.; Haberer, G.; Meurer, J. The complete nucleotide sequences of the 5 genetically distinct plastid genomes of *Oenothera*, subsection Oenothera: II. A microevolutionary view using bioinformatics and formal genetic data. *Mol. Biol. Evol.* **2008**, *25*, 2019–2030. [CrossRef]

56. Sobanski, J.; Giavalisco, P.; Fischer, A.; Kreiner, J.; Walther, D.; Schoettler, M.A.; Pellizzer, T.; Golczyk, H.; Obata, T.; Bock, R.; et al. Chloroplast competition is controlled by lipid biosynthesis in evening primroses. *Proc. Natl. Acad. Sci. USA* **2019**, *116*, 5665–5674. [CrossRef]

57. Li, J.; Su, Y.; Wang, T. The Repeat Sequences and Elevated Substitution Rates of the Chloroplast *accD* Gene in Cupressophytes. *Front. Plant Sci.* **2018**, *9*, 533. [CrossRef]

58. Ujihara, T.; Hayashi, N.; Kohata, K.; Matsushita, S.; Kitajima, S. Intraspecific Sequence Variation in the rbcL-*accD* Region of the Chloroplast Genome in Tea (*Camellia sinensis*). *Tea Res. J.* **2007**, *104*, 15–23. [CrossRef]

59. Smýkal, P.; Trněný, O.; Brus, J.; Hanáček, P.; Rathore, A.; Roma, R.D.; Pechanec, V.; Duchoslav, M.; Bhattacharyya, D.; Bariotakis, M.; et al. Genetic structure of wild pea (*Pisum sativum* subsp. *elatius*) populations in the northern part of the Fertile Crescent reflects moderate cross-pollination and strong effect of geographic but not environmental distance. *PLoS ONE* **2018**, *13*, e0194056.

60. Wright, P.E.; Dyson, H.J. Intrinsically disordered proteins in cellular signalling and regulation. *Nat. Rev. Mol. Cell Biol.* **2015**, *16*, 18–29. [CrossRef]

61. Light, S.; Sagit, R.; Sachenkova, O.; Ekman, D.; Elofsson, A. Protein expansion is primarily due to indels in intrinsically disordered regions. *Mol. Biol. Evol.* **2013**, *30*, 2645–2653. [CrossRef]

62. Simon, M.; Hancock, J.M. Tandem and cryptic amino acid repeats accumulate in disordered regions of proteins. *Genome Biol.* **2009**, *10*, R59. [CrossRef]

63. Kode, V.; Mudd, E.A.; Iamtham, S.; Day, A. The tobacco plastid *accD* gene is essential and is required for leaf development. *Plant J.* **2005**, *44*, 237–244. [CrossRef]

64. Corriveau, J.L.; Coleman, A.W. Rapid Screening Method to Detect Potential Biparental Inheritance of Plastid DNA and Results for Over 200 Angiosperm Species. *Am. J. Bot.* **1988**, *75*, 1443–1458. [CrossRef]

65. Zhang, Q.; Sodmergen. Why does biparental plastid inheritance revive in angiosperms? *J. Plant Res.* **2010**, *123*, 201–206. [CrossRef]

66. Greiner, S.; Sobanski, J.; Bock, R. Why are most organelle genomes transmitted maternally? *Bioessays* **2015**, *37*, 80–94. [CrossRef]

67. Christie, J.R.; Beekman, M. Uniparental Inheritance Promotes Adaptive Evolution in Cytoplasmic Genomes. *Mol. Biol. Evol.* **2017**, *34*, 677–691. [CrossRef]

68. Bogdanova, V.S.; Kosterin, O.E.; Yadrikhinskiy, A.K. Wild peas vary in their cross-compatibility with cultivated pea (*Pisum sativum* subsp. *sativum* L.) depending on alleles of a nuclear-cytoplasmic incompatibility locus. *Theor. Appl. Genet.* **2014**, *127*, 1163–1172. [CrossRef]

69. North, N.; Casey, R.; Domoney, C. Inheritance and mapping of seed lipoxygenase polypeptides in *Pisum*. *Theor. Appl. Genet.* **1998**, *77*, 805–808. [CrossRef]

70. Hradilová, I.; Trněný, O.; Válková, M.; Cechová, M.; Janská, A.; Prokešová, L.; Aamir, K.; Krezdorn, N.; Rotter, B.; Winter, P.; et al. A Combined Comparative Transcriptomic, Metabolomic, and Anatomical Analyses of Two Key Domestication Traits: Pod Dehiscence and Seed Dormancy in Pea (*Pisum* sp.). *Front. Plant Sci.* **2017**, *8*, 542. [CrossRef]

71. Meyer, R.S.; Purugganan, M.D. Evolution of crop species: Genetics of domestication and diversification. *Nat. Rev. Genet.* **2013**, *14*, 840–852. [CrossRef]

72. Ben-Ze'Ev, N.; Zohary, D. Species relationships in the genus *Pisum* L. *Isr. J. Bot.* **1973**, *22*, 73–91.

73. Errico, A.; Conicella, C.; Taliercio, U. Cytological and Morphological Characterization of *Pisum sativum* and *Pisum fulvum* Tetraploids. *Plant Breed.* **1991**, *106*, 141–148. [CrossRef]

74. Lu, Y.; Kermicle, J.L.; Evans, M.M.S. Genetic and cellular analysis of cross-incompatibility in *Zea mays*. *Plant Reprod.* **2014**, *27*, 19–29. [CrossRef]

75. Saitoh, K.; Onishi, K.; Mikami, I.; Thidar, K.; Sano, Y. Allelic diversification at the C (OsC1) locus of wild and cultivated rice: Nucleotide changes associated with phenotypes. *Genetics* **2004**, *168*, 997–1007. [CrossRef]

76. Kubo, T. Genetic mechanisms of postzygotic reproductive isolation: An epistatic network in rice. *Breed. Sci.* **2013**, *63*, 359–366. [CrossRef]

77. Kumari, M.; Clarke, H.J.; Small, I.; Siddique, K.H.M. Albinism in Plants: A Major Bottleneck in Wide Hybridization, Androgenesis and Doubled Haploid Culture. *Crit. Rev. Plant Sci.* **2009**, *28*, 393–409. [CrossRef]

78. Kumari, M.; Clarke, H.J.; des Francs-Small, C.C.; Small, I.; Khan, T.N.; Siddique, K.H.M. Albinism does not correlate with biparental inheritance of plastid DNA in interspecific hybrids in *Cicer* species. *Plant Sci.* **2011**, *180*, 628–633. [CrossRef]

79. Bohra, A.; Jha, U.C.; Adhimoolam, P. Cytoplasmic male sterility (CMS) in hybrid breeding in field crops. *Plant Cell Rep.* **2016**, *35*, 967–993. [CrossRef]

80. Kim, Y.-J.; Zhang, D. Molecular Control of Male Fertility for Crop Hybrid Breeding. *Trends Plant Sci.* **2018**, *23*, 53–65. [CrossRef]

81. Villesen, P. FaBox: An online toolbox for fasta sequences. *Mol. Ecol. Notes* **2007**, *7*, 965–968. [CrossRef]

82. Huson, D.H.; Bryant, D. Application of phylogenetic networks in evolutionary studies. *Mol. Biol. Evol.* **2006**, *23*, 254–267. [CrossRef]

83. Kumar, S.; Stecher, G.; Tamura, K. MEGA7: Molecular Evolutionary Genetics Analysis version 7.0 for bigger datasets. *Mol. Biol. Evol.* **2016**, *33*, 1870–1874. [CrossRef]

84. Librado, P.; Rozas, J. DnaSP v5: A software for comprehensive analysis of DNA polymorphism data. *Bioinformatics* **2009**, *25*, 1451–1452. [CrossRef]

85. Kalendar, R.; Khassenov, B.; Ramanculov, E.; Samuilova, O.; Ivanov, K.I. FastPCR: An in silico tool for fast primer and probe design and advanced sequence analysis. *Genomics* **2017**, *109*, 312–319. [CrossRef]

86. Heger, A.; Holm, L. Rapid automatic detection and alignment of repeats in protein sequences. *Proteins* **2000**, *41*, 224–237. [CrossRef]

87. Kelley, L.A.; Mezulis, S.; Yates, C.M.; Wass, M.N.; Sternberg, M.J. The Phyre2 web portal for protein modeling, prediction and analysis. *Nat. Protoc.* **2015**, *10*, 845–858. [CrossRef]

88. Källberg, M.; Wang, H.; Wang, S.; Peng, J.; Wang, Z.; Lu, H.; Xu, J. Template-based protein structure modeling using the RaptorX web server. *Nat. Protoc.* **2012**, *7*, 1511–1522. [CrossRef]

89. Johansson, M.U.; Zoete, V.; Michielin, O.; Guex, N. Defining and searching for structural motifs using DeepView/Swiss-PdbViewer. *BMC Bioinform.* **2012**, *13*, 173. [CrossRef]

90. Hooft, R.W.; Vriend, G.; Sander, C.; Abola, E.E. Errors in protein structures. *Nature* **1996**, *381*, 272. [CrossRef]

MDPI

St. Alban-Anlage 66

4052 Basel

Switzerland

Tel. +41 61 683 77 34

Fax +41 61 302 89 18

www.mdpi.com

International Journal of Molecular Sciences Editorial Office

E-mail: ijms@mdpi.com

www.mdpi.com/journal/ijms